Periodic Table of the Elements

Key

An element: 79 (Atomic number), Au (Symbol), Gold (Name), 196.9665 (Atomic mass)

- Metals
- Semimetals
- Nonmetals

Group number, U.S. system / IUPAC system; Period number

Period	1A (1)	2A (2)	3B (3)	4B (4)	5B (5)	6B (6)	7B (7)	8B (8)	8B (9)	8B (10)	1B (11)	2B (12)	3A (13)	4A (14)	5A (15)	6A (16)	7A (17)	8A (18)	Period
1	1 H Hydrogen 1.0079																	2 He Helium 4.0026	1
2	3 Li Lithium 6.941	4 Be Beryllium 9.0122											5 B Boron 10.811	6 C Carbon 12.011	7 N Nitrogen 14.0067	8 O Oxygen 15.9994	9 F Fluorine 18.9984	10 Ne Neon 20.1797	2
3	11 Na Sodium 22.9898	12 Mg Magnesium 24.3050											13 Al Aluminum 26.9815	14 Si Silicon 28.0855	15 P Phosphorus 30.9738	16 S Sulfur 32.066	17 Cl Chlorine 35.4527	18 Ar Argon 39.948	3
4	19 K Potassium 39.0983	20 Ca Calcium 40.078	21 Sc Scandium 44.9559	22 Ti Titanium 47.88	23 V Vanadium 50.9415	24 Cr Chromium 51.9961	25 Mn Manganese 54.9380	26 Fe Iron 55.847	27 Co Cobalt 58.9332	28 Ni Nickel 58.693	29 Cu Copper 63.546	30 Zn Zinc 65.39	31 Ga Gallium 69.723	32 Ge Germanium 72.61	33 As Arsenic 74.9216	34 Se Selenium 78.96	35 Br Bromine 79.904	36 Kr Krypton 83.80	4
5	37 Rb Rubidium 85.4678	38 Sr Strontium 87.62	39 Y Yttrium 88.9059	40 Zr Zirconium 91.224	41 Nb Niobium 92.9064	42 Mo Molybdenum 95.94	43 Tc Technetium (98)	44 Ru Ruthenium 101.07	45 Rh Rhodium 102.9055	46 Pd Palladium 106.42	47 Ag Silver 107.8682	48 Cd Cadmium 112.411	49 In Indium 114.82	50 Sn Tin 118.710	51 Sb Antimony 121.757	52 Te Tellurium 127.60	53 I Iodine 126.9045	54 Xe Xenon 131.29	5
6	55 Cs Cesium 132.9054	56 Ba Barium 137.327	57 La Lanthanum 138.9055	72 Hf Hafnium 178.49	73 Ta Tantalum 180.9479	74 W Tungsten 183.85	75 Re Rhenium 186.207	76 Os Osmium 190.2	77 Ir Iridium 192.22	78 Pt Platinum 195.08	79 Au Gold 196.9665	80 Hg Mercury 200.59	81 Tl Thallium 204.3833	82 Pb Lead 207.2	83 Bi Bismuth 208.9804	84 Po Polonium (209)	85 At Astatine (210)	86 Rn Radon (222)	6
7	87 Fr Francium (223)	88 Ra Radium 227.0278	89 Ac Actinium (227)	104 Rf Rutherfordium (261)	105 Db Dubnium (262)	106 Sg Seaborgium (263)	107 Bh Bohrium (262)	108 Hs Hassium (265)	109 Mt Meitnerium (266)	110 Ds Darmstadtium (269)	111 Rg Roentgenium (272)								7

	Period															Period
Lanthanides	6	58 Ce Cerium 140.115	59 Pr Praseodymium 140.9076	60 Nd Neodymium 144.24	61 Pm Promethium (145)	62 Sm Samarium 150.36	63 Eu Europium 151.965	64 Gd Gadolinium 157.25	65 Tb Terbium 158.9253	66 Dy Dysprosium 162.50	67 Ho Holmium 164.9303	68 Er Erbium 167.26	69 Tm Thulium 168.9342	70 Yb Ytterbium 173.04	71 Lu Lutetium 174.967	6
Actinides	7	90 Th Thorium 232.0381	91 Pa Protactinium 231.0359	92 U Uranium 238.00289	93 Np Neptunium (237)	94 Pu Plutonium (244)	95 Am Americium (243)	96 Cm Curium (247)	97 Bk Berkelium (247)	98 Cf Californium (251)	99 Es Einsteinium (252)	100 Fm Fermium (257)	101 Md Mendelevium (258)	102 No Nobelium (259)	103 Lr Lawrencium (260)	7

Numbers in parentheses are mass numbers of radioactive isotopes.

Organic KNOWLEDGE TOOLS

Succeed in your course with the help of this best-selling text and its powerful **Organic Knowledge Tools**. John McMurry's **Organic Chemistry** carefully and thoroughly integrates powerful multimedia learning components with the text, providing a seamless learning system that will improve your understanding of course concepts.

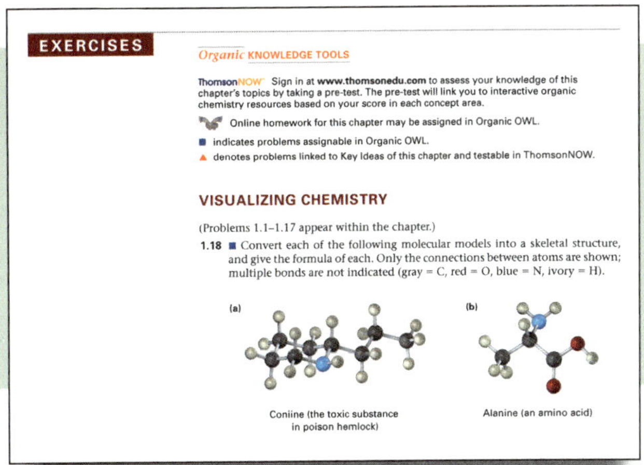
EXERCISES

Organic KNOWLEDGE TOOLS

ThomsonNOW™ Sign in at **www.thomsonedu.com** to assess your knowledge of this chapter's topics by taking a pre-test. The pre-test will link you to interactive organic chemistry resources based on your score in each concept area.

Online homework for this chapter may be assigned in Organic OWL.

■ indicates problems assignable in Organic OWL.

▲ denotes problems linked to Key Ideas of this chapter and testable in ThomsonNOW.

VISUALIZING CHEMISTRY

(Problems 1.1–1.17 appear within the chapter.)

1.18 ■ Convert each of the following molecular models into a skeletal structure, and give the formula of each. Only the connections between atoms are shown; multiple bonds are not indicated (gray = C, red = O, blue = N, ivory = H).

(a) Coniine (the toxic substance in poison hemlock)

(b) Alanine (an amino acid)

ThomsonNOW!™ for *Organic Chemistry*

Link to interactive tutorials based on your unique level of understanding! **ThomsonNOW** is an online, assessment-centered learning tool that gives you a *Personalized Study* plan based on a diagnostic pre-test. Your unique study plan can help you target your study needs and visualize, practice, and master the material in the text.

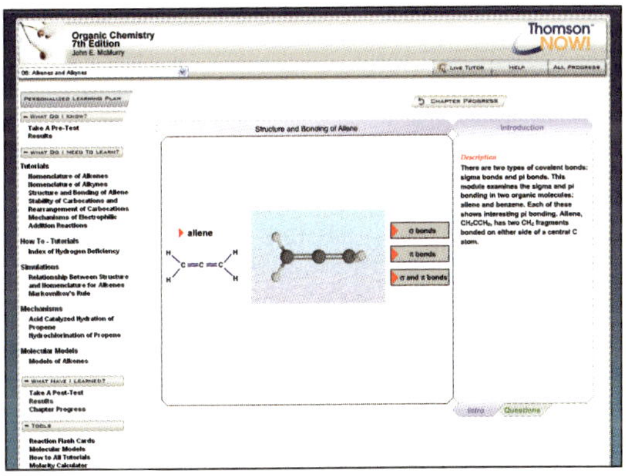

How does it work?

Just log on to **ThomsonNOW** by using the access code packaged with your text.* With a click of the mouse, you can build a complete *Personalized Study* plan for yourself by taking advantage of the program's three powerful components.

What I Know

First, a *Pre-Test* helps you identify what more you need to study—you receive your score automatically. Once you've completed the *Pre-Test*, a detailed *Personalized Study* plan (based on your test results) outlines the concepts you need to review and presents links to specific pages in your textbook that discuss that concept.

What I Need to Know

Working from your *Personalized Study* plan, you are guided to media-enhanced activities, such as problem solving with *Organic Interactive*, working through conceptual problems with *Building Block Questions*, interacting with text illustrations using *Active Figures*, and visualizing models with the *Molecular Models* tool.

What I've Learned

An optional *Post-Test* ensures that you've mastered the concepts in each chapter. As with the *Pre-Test*, your individual results may be e-mailed to your instructor to help you both assess your progress. If you need to improve your score, **ThomsonNOW** will work with you as you continue to build your knowledge.

Sign in at **www.thomsonedu.com** to view organic chemistry tutorials and simulations, develop problem-solving skills, and test your knowledge with these interactive self-study resources.

* If you purchased a new book, you may already have access to ThomsonNOW. Register by following the directions on the card that came with this text. If you did not purchase a new book, you can still obtain an access code by visiting www.thomsonedu.com/thomsonnow/buy.

Turn the page to learn about Organic OWL.

Organic KNOWLEDGE TOOLS

Organic OWL

Developed at the University of Massachusetts, Amherst, **Organic OWL** is a customizable, flexible, web-based homework system and assessment tool.

Instructors . . .

Now a more robust course solution, **OWL** offers fully class-tested content, unsurpassed ease of use, reliability, and dedicated training and service! With **Organic OWL** you can select from various types of assignments, including tutorials, simulations, and parameterized, short-answer questions. Instructors across the country are discovering **Organic OWL** and its ability to reduce faculty workload and facilitate instruction.

Online homework for this and other chapters may be assigned in Organic OWL.

Organic OWL offers thousands of questions as well as MarvinSketch, a Java™ applet for viewing and drawing chemical structures. **Organic OWL** also uses the MDL Chime application to assist students with viewing structures of organic compounds. The text's end-of-chapter exercises include problems that are assignable in **Organic OWL**; these are noted with a ■.

A new OWL option: A Complete e-Book!

The McMurry **e-Book in Organic OWL** includes the complete textbook as an assignable resource that is fully linked to **Organic OWL** homework content. This new **e-Book in Organic OWL** is an exclusive option that will be available to all your students if you choose it. This version can be bundled with the text and/or ordered as a text replacement. Please consult your Thomson representative for pricing details.

Students . . .

Organic OWL is the only system specifically designed to support mastery learning, where you work as long as you need using instantaneous feedback to master each chemical concept and skill. Easy to use, **Organic OWL** offers various types of assignments and tools, including tutors, simulations, and animations, created specifically for this text.

To learn more about **Organic OWL**, visit **http://owl.thomsonlearning.com** or contact your Thomson Brooks/Cole representative.

* A fee-based code is required for access to Organic OWL. If you purchased a new book, you may already have access. Register by following the directions on the card that came with this text. If you did not purchase a new book, you can still obtain an access code by visiting http://owl.thomsonlearning.com.

Organic Chemistry

Volume 1

John McMurry
Cornell University

THOMSON
BROOKS/COLE

Australia · Canada · Mexico · Singapore · Spain · United Kingdom · United States

THOMSON
BROOKS/COLE

Organic Chemistry
John McMurry

Executive Editors:
Michele Baird, Maureen Staudt &
Michael Stranz

Project Development Manager:
Linda deStefano

Sr. Marketing Coordinators:
Lindsay Annett and Sara Mercurio

Production/Manufacturing Manager:
Donna M. Brown

Production Editorial Manager:
Dan Plofchan

Pre-Media Services Supervisor:
Becki Walker

Rights and Permissions Specialist:
Kalina Ingham Hintz

Cover Image
Getty Images*

Printed in the
United States of America
1 2 3 4 5 6 7 10 09 08 07

For more information, please contact Thomson Custom Solutions, 5191 Natorp Boulevard, Mason, OH 45040. Or you can visit our Internet site at www.thomsoncustom.com

The Adaptable Courseware Program consists of products and additions to existing Brooks/Cole products that are produced from camera-ready copy. Peer review, class testing, and accuracy are primarily the responsibility of the author(s).

Organic Chemistry / John McMurry

ISBN-13: 978-0-495-45255-3
ISBN-10: 0-495-45255-6

International Divisions List

Asia (Including India):
Thomson Learning
(a division of Thomson Asia Pte Ltd)
5 Shenton Way #01-01
UIC Building
Singapore 068808
Tel: (65) 6410-1200
Fax: (65) 6410-1208

Australia/New Zealand:
Thomson Learning Australia
102 Dodds Street
Southbank, Victoria 3006
Australia

Latin America:
Thomson Learning
Seneca 53
Colonia Polano
11560 Mexico, D.F., Mexico
Tel (525) 281-2906
Fax (525) 281-2656

Canada:
Thomson Nelson
1120 Birchmount Road
Toronto, Ontario
Canada M1K 5G4
Tel (416) 752-9100
Fax (416) 752-8102

UK/Europe/Middle East/Africa:
Thomson Learning
High Holborn House
50-51 Bedford Row
London, WC1R 4LS
United Kingdom
Tel 44 (020) 7067-2500
Fax 44 (020) 7067-2600

Spain (Includes Portugal):
Thomson Paraninfo
Calle Magallanes 25
28015 Madrid
España
Tel 34 (0)91 446-3350
Fax 34 (0)91 445-6218

Table of Contents

Preface

I love to write. I get real pleasure from taking a complicated subject, turning it around until I see it clearly, and then explaining it in simple words. I write to explain chemistry to students today the way I wish it had been explained to me years ago.

The enthusiastic response to the six previous editions has been very gratifying and suggests that this book has served students well. I hope you will find that this seventh edition of *Organic Chemistry* builds on the strengths of the first six and serves students even better. I have made every effort to make this new edition as effective, clear, and readable as possible; to show the beauty and logic of organic chemistry; and to make organic chemistry enjoyable to learn.

Organization and Teaching Strategies This seventh edition, like its predecessors, blends the traditional functional-group approach with a mechanistic approach. The primary organization is by functional group, beginning with the simple (alkenes) and progressing to the more complex. Most faculty will agree that students new to the subject and not yet versed in the subtleties of mechanism do better this way. In other words, the *what* of chemistry is generally easier to grasp than the *why*. Within this primary organization, however, I place heavy emphasis on explaining the fundamental mechanistic similarities of reactions. This emphasis is particularly evident in the chapters on carbonyl-group chemistry (Chapters 19–23), where mechanistically related reactions like the aldol and Claisen condensations are covered together. By the time students reach this material, they have seen all the common mechanisms and the value of mechanisms as an organizing principle has become more evident.

The Lead-Off Reaction: Addition of HBr to Alkenes Students usually attach great importance to a text's lead-off reaction because it is the first reaction they see and is discussed in such detail. I use the addition of HBr to an alkene as the lead-off to illustrate general principles of organic chemistry for several reasons: the reaction is relatively straightforward; it involves a common but important functional group; no prior knowledge of stereochemistry or kinetics in needed to understand it; and, most important, it is a *polar* reaction. As such, I believe that electrophilic addition reactions represent a much more useful and realistic introduction to functional-group chemistry than a lead-off such as radical alkane chlorination.

Reaction Mechanisms In the first edition of this book, I introduced an innovative format for explaining reaction mechanisms in which the reaction steps are printed vertically, with the changes taking place in each step described next to the reaction arrow. This format allows a reader to see easily what is occurring at each step without having to flip back and forth between structures and text. Each successive edition has seen an increase in the number and quality of these vertical mechanisms, which are still as fresh and useful as ever.

Organic Synthesis Organic synthesis is treated in this text as a teaching device to help students organize and deal with a large body of factual information—the same skill so critical in medicine. Two sections, the first in Chapter 8 (Alkynes) and the second in Chapter 16 (Chemistry of Benzene), explain the thought processes involved in working synthesis problems and emphasize the value of starting from what is known and logically working backward. In addition, *Focus On* boxes, including The Art of Organic Synthesis, Combinatorial Chemistry, and Enantioselective Synthesis, further underscore the importance and timeliness of synthesis.

Modular Presentation Topics are arranged in a roughly modular way. Thus, certain chapters are grouped together: simple hydrocarbons (Chapters 3–8), spectroscopy (Chapters 12–14), carbonyl-group chemistry (Chapters 19–23), and biomolecules (Chapters 25–29). I believe that this organization brings to these subjects a cohesiveness not found in other texts and allows the instructor the flexibility to teach in an order different from that presented in the book.

Basic Learning Aids In writing and revising this text, I consistently aim for lucid explanations and smooth transitions between paragraphs and between topics. New concepts are introduced only when they are needed, not before, and they are immediately illustrated with concrete examples. Frequent cross-references to earlier material are given, and numerous summaries are provided to draw information together, both within and at the ends of chapters. In addition, the back of this book contains a wealth of material helpful for learning organic chemistry, including a large glossary, an explanation of how to name polyfunctional organic compounds, and answers to all in-text problems. For still further aid, an accompanying *Study Guide and Solutions Manual* gives summaries of name reactions, methods for preparing functional groups, functional-group reactions, and the uses of important reagents.

Changes and Additions for the Seventh Edition

The primary reason for preparing a new edition is to keep the book up to date, both in its scientific coverage and in its pedagogy. My overall aim is always to refine the features that made earlier editions so successful, while adding new ones.

- **The writing** has again been revised at the sentence level, streamlining the presentation, improving explanations, and updating a thousand small details. Several little-used reactions have been deleted (the alkali fusion of arenesulfonic acids to give phenols, for instance), and a few new ones have been added (the Sharpless enantioselective epoxidation of alkenes, for instance).

- Other notable **content changes** are:

  Chapter 2, *Polar Covalent Bonds; Acids and Bases*—A new Section 2.13 on noncovalent interactions has been added.

  Chapter 3, *Organic Compounds: Alkanes and Their Stereochemistry*—The chapter has been revised to focus exclusively on open-chain alkanes.

  Chapter 4, *Organic Compounds: Cycloalkanes and Their Stereochemistry*—The chapter has been revised to focus exclusively on cycloalkanes.

  Chapter 5, *An Overview of Organic Reactions*—A new Section 5.11 comparing biological reactions and laboratory reactions has been added.

Chapter 7, *Alkenes: Reactions and Synthesis*—Alkene epoxidation has been moved to Section 7.8, and Section 7.11 on the biological addition of radicals to alkenes has been substantially expanded.

Chapter 9, *Stereochemistry*—A discussion of chirality at phosphorus and sulfur has been added to Section 9.12, and a discussion of chiral environments has been added to Section 9.14.

Chapter 11, *Reactions of Alkyl Halides: Nucleophilic Substitutions and Eliminations*—A discussion of the E1cB reaction has been added to Section 11.10, and a new Section 11.11 discusses biological elimination reactions.

Chapter 12, *Structure Determination: Mass Spectrometry and Infrared Spectroscopy*—A new Section 12.4 discusses mass spectrometry of biological molecules, focusing on time-of-flight instruments and soft ionization methods such as MALDI.

Chapter 20, *Carboxylic Acids and Nitriles*—A new Section 20.3 discusses biological carboxylic acids and the Henderson–Hasselbalch equation.

Chapter 24, *Amines and Heterocycles*—This chapter now includes a discussion of heterocycles, and a new Section 24.5 on biological amines and the Henderson–Hasselbalch equation has been added.

Chapter 25, *Biomolecules: Carbohydrates*—A new Section 25.7 on the eight essential carbohydrates has been added, and numerous content revisions have been made.

Chapter 26, *Biomolecules: Amino Acids, Peptides, and Proteins*—The chapter has been updated, particularly in its coverage of solid-phase peptide synthesis.

Chapter 27, *Biomolecules: Lipids*—The chapter has been extensively revised, with increased detail on prostaglandins (Section 27.4), terpenoid biosynthesis (Section 27.5), and steroid biosynthesis, (Section 27.7).

Chapter 28, *Biomolecules: Nucleic Acids*—Coverage of heterocyclic chemistry has been moved to Chapter 24.

Chapter 29, *The Organic Chemistry of Metabolic Pathways*—The chapter has been reorganized and extensively revised, with substantially increased detail on important metabolic pathways.

Chapter 30, *Orbitals and Organic Chemistry: Pericyclic Reactions*—All the art in this chapter has been redone.

- **The order of topics** remains basically the same but has been changed to devote Chapter 3 entirely to alkanes and Chapter 4 to cycloalkanes. In addition, epoxides are now introduced in Chapter 7 on alkenes, and coverage of heterocyclic chemistry has been moved to Chapter 24.

- **The problems** within and at the end of each chapter have been reviewed, and approximately 100 new problems have been added, many of which focus on biological chemistry.

- ***Focus On* boxes** at the end of each chapter present interesting applications of organic chemistry relevant to the main chapter subject. Including topics from biology, industry, and day-to-day life, these applications enliven and reinforce the material presented within the chapter. The boxes have been updated, and new ones added, including Where Do Drugs Come From? (Chapter 5),

Green Chemistry (Chapter 11), X-Ray Crystallography (Chapter 22), and Green Chemistry II: Ionic Liquids (Chapter 24).

- **Biologically important molecules and mechanisms** have received particular attention in this edition. Many reactions now show biological counterparts to laboratory examples, many new problems illustrate reactions and mechanisms that occur in living organisms, and enhanced detail is given for major metabolic pathways.

More Features

NEW!

- Why do we have to learn this? I've been asked this question so many times by students that I thought that it would be appropriate to begin each chapter with the answer. The *Why This Chapter?* section is a short paragraph that appears at the end of the introduction to every chapter and tells students why the material about to be covered is important.

NEW!

- Thirteen Key Ideas are highlighted in the book. These include topics pivotal to students' development in organic chemistry, such as Curved Arrows in Reaction Mechanisms (Chapter 5) and Markovnikov's Rule (Chapter 6). These Key Ideas are further reinforced in end-of-chapter problems marked with a ▲ icon. A selection of these problems are also assignable in OWL, denoted by a ■.

- Worked Examples are now titled to give students a frame of reference. Each Worked Example includes a Strategy and a worked-out Solution, and then is followed by problems for students to try on their own. This book has more than 1800 in-text and end-of-chapter problems.

- An overview chapter, *A Preview of Carbonyl Chemistry,* follows Chapter 18 and highlights the author's belief that studying organic chemistry requires both summarizing and looking ahead.

NEW!
Organic KNOWLEDGE TOOLS

- Thorough media integration with Organic Knowledge Tools: ThomsonNOW for Organic Chemistry and Organic OWL are provided to help students practice and test their knowledge of the concepts. ThomsonNOW is an online assessment program for self-study with interactive tutorials. Organic OWL is an online homework learning system. Icons throughout the book direct students to ThomsonNOW at **www.thomsonedu.com**. A fee-based access code is required for Organic OWL.

NEW!

- About 15 to 20 end-of-chapter problems per chapter, denoted with a ■ icon, are assignable in the OWL online homework system. These questions are algorithmically generated, allowing students more practice.

- OWL (Online Web-based Learning) for Organic Chemistry, developed at the University of Massachusetts, Amherst; class-tested by thousands of students; and used by more than 50,000 students, provides fully class-tested questions and tutors in an easy-to-use format. OWL is also customizable and cross-platform. The OWL Online Web-based Learning system provides students with instant grading and feedback on homework problems, modeling questions, and animations to accompany this text. With parameterization, OWL for Organic Chemistry offers nearly 6000 different questions as well as MarvinSketch for viewing and drawing chemical structures.

- A number of the figures are animated in ThomsonNOW. These are designated as **Active Figures** in the figure legends.
- The Visualizing Chemistry Problems that begin the exercises at the end of each chapter offer students an opportunity to see chemistry in a different way by visualizing molecules rather than by simply interpreting structural formulas.
- Summaries and Key Word lists help students by outlining the key concepts of the chapter.
- Summaries of Reactions, at the ends of appropriate chapters, bring together the key reactions from the chapter in one complete list.

Companions to This Text

Supporting instructor materials are available to qualified adopters. Please consult your local Thomson Brooks/Cole representative for details.

Visit **www.thomsonedu.com** to:

Find your local representative
Download electronic files of text art and ancillaries
Request a desk copy

Ancillaries for Students

Study Guide and Solutions Manual, by Susan McMurry, provides answers and explanations to all in-text and end-of-chapter exercises. (0-495-11268-2)

ThomsonNOW

ThomsonNOW Website To further student understanding, the text features sensible media integration through **ThomsonNOW**, a powerful online learning companion that helps students determine their unique study needs and provides them with individualized resources. This dynamic learning companion combines with the text to provide students with a seamless, integrated learning system. The access code that is required for ThomsonNOW may be included with a new copy of the text or purchased separately. Visit **www.thomsonedu.com** to register for access (or to purchase access) to ThomsonNOW.

OWL for Organic Chemistry, authored by Steve Hixson and Peter Lillya of the University of Massachusetts, Amherst, and William Vining of the State University of New York at Oneonta. Class-tested by thousands of students and used by more than 50,000 students, OWL (Online Web-based Learning) provides fully class-tested content in an easy-to-use format. OWL is also customizable and cross-platform. The OWL Online Web-based Learning system provides students with instant analysis and feedback on homework problems, modeling questions, and animations to accompany this text. With parameterization, OWL for Organic Chemistry offers more than 6000 questions as well as MarvinSketch, a Java applet for viewing and drawing chemical structures.

This powerful system maximizes the students' learning experience and, at the same time, reduces faculty workload and helps facilitate instruction. OWL also uses the MDL Chime application to assist students with viewing structures of organic compounds. New to this edition are 15 to 20 end-of-chapter problems per chapter, denoted by a ■ icon, which are assignable in OWL. A fee-based access code is required for OWL.

Pushing Electrons: A Guide for Students of Organic Chemistry, third edition, by Daniel P. Weeks. A workbook designed to help students learn techniques of electron pushing, its programmed approach emphasizes repetition and active participation. (0-03-020693-6)

NEW! **Spartan Model Electronic Modeling Kit**, A set of easy-to-use builders allow for the construction and 3-D manipulation of molecules of any size or complexity—from a hydrogen atom to DNA and everything in between. This kit includes the SpartanModel software on CD-ROM, an extensive molecular database, 3-D glasses, and a *Tutorial and Users Guide* that includes a wealth of activities to help you get the most out of your course. (0-495-01793-0)

Ancillaries for Instructors

PowerLecture A dual-platform digital library and presentation tool that provides art and tables from the main text in a variety of electronic formats that are easily exported into other software packages. Also contains simulations, molecular models, and QuickTime movies to supplement lectures as well as electronic files of various print supplements. Slides use the full power of Microsoft PowerPoint® and incorporate videos, animations, and other media assets from ThomsonNOW. Instructors can customize their lecture presentations by adding their own slides or by deleting or changing existing slides (0-495-11265-8). PowerLecture also includes:

- **ExamView Testing** This easy-to-use software, containing questions and problems authored specifically for the text, allows professors to create, deliver, and customize tests in minutes.
- **JoinIn on Turning Point for Organic Chemistry** Book-specific JoinIn™ content for Response Systems tailored to *Organic Chemistry* allows you to transform your classroom and assess your students' progress with instant in-class quizzes and polls. Our exclusive agreement to offer TurningPoint software lets you pose book-specific questions and display students' answers seamlessly within the Microsoft PowerPoint slides of your own lecture, in conjunction with the "clicker" hardware of your choice. Enhance how your students interact with you, your lecture, and one another. Contact your local Thomson representative to learn more.

WebCT/NOW Integration Instructors and students enter **ThomsonNOW** through their familiar Blackboard or WebCT environment without the need for a separate user name or password and can access all of the **ThomsonNOW** assessments and content. Contact your local Thomson representative to learn more.

Transparency Acetates Approximately 200 full-color transparency acetates of key text illustrations, enlarged for use in the classroom and lecture halls. (0-495-11260-7)

Organic Chemistry Laboratory Manuals Brooks/Cole is pleased to offer a choice of organic chemistry laboratory manuals catered to fit individual needs. Visit **www.thomsonedu.com**. Customizable laboratory manuals also can be assembled—contact your Thomson representative to learn more.

Acknowledgments

I thank all the people who helped to shape this book and its message. At Brooks/Cole they include: David Harris, publisher; Sandra Kiselica, senior development editor; Amee Mosley executive marketing manager; Teresa Trego, project manager; Lisa Weber; technology project manager; and Sylvia Krick, assistant editor, along with Suzanne Kastner and Gwen Gilbert at Graphic World.

I am grateful to colleagues who reviewed the manuscript for this book and participated in a survey about its approach. They include:

Manuscript Reviewers

Arthur W. Bull, Oakland University
Robert Coleman, Ohio State University
Nicholas Drapela, Oregon State University
Christopher Hadad, Ohio State University
Eric J. Kantorowski, California Polytechnic State University
James J. Kiddle, Western Michigan University
Joseph B. Lambert, Northwestern University
Dominic McGrath, University of Arizona
Thomas A. Newton, University of Southern Maine
Michael Rathke, Michigan State University
Laren M. Tolbert, Georgia Institute of Technology

Reviewers of Previous Editions

Wayne Ayers, East Carolina University
Kevin Belfield, University of Central Florida-Orlando
Byron Bennett, University of Las Vegas
Robert A. Benkeser, Purdue University
Donald E. Bergstrom Purdue University
Christine Bilicki, Pasedena City College
Weston J. Borden, University of North Texas
Steven Branz, San Jose State University
Larry Bray, Miami-Dade Community College
James Canary, New York University
Ronald Caple, University of Minnesota-Duluth
John Cawley, Villanova University
George Clemans, Bowling Green State University
Bob Coleman, Ohio State University
Paul L. Cook, Albion College
Douglas Dyckes, University of Colorado-Denver
Kenneth S. Feldman, Pennsylvania State University
Martin Feldman, Howard University
Kent Gates, University of Missouri-Columbia
Warren Gierring, Boston University
Daniel Gregory, St. Cloud State University
David Hart, Ohio State University
David Harpp, McGill University
Norbert Hepfinger, Rensselaer Polytechnic Institute
Werner Herz, Florida State University
John Hogg, Texas A&M University
Paul Hopkins, University of Washington
John Huffman, Clemson University
Jack Kampmeier, University of Rochester
Thomas Katz, Columbia University
Glen Kauffman, Eastern Mennonite College
Andrew S. Kendle, University of North Carolina- Wilmington
Paul E. Klinedinst, Jr., California State University- Northridge
Joseph Lamber, Northwestern University
John T. Landrum, Florida International University

Peter Lillya, University of Massachusetts
Thomas Livinghouse, Montana State University
James Long, University of Oregon
Todd Lowary, University of Alberta
Luis Martinez, University of Texas, El Paso
Eugene A. Mash, University of Arizona
Fred Matthews, Austin Peay State University
Guy Matson, University of Central Florida
Keith Mead, Mississippi State University
Michael Montague-Smith, University of Maryland
Andrew Morehead, East Carolina University
Harry Morrison, Purdue University
Cary Morrow, University of New Mexico
Clarence Murphy, East Stroudsburg University
Roger Murray, St. Joseph's University
Oliver Muscio, Murray State University
Ed Neeland, University of British Columbia
Jacqueline Nikles, University of Alabama
Mike Oglioruso, Virginia Polytechnic Institute and State University
Wesley A. Pearson, St. Olaf College
Robert Phillips, University of Georgia
Carmelo Rizzo, Vanderbilt University
William E. Russey, Juniata College
Neil E. Schore, University of California-Davis
Gerald Selter, California State University- San Jose
Eric Simanek, Texas A&M University
Jan Simek, California Polytechnic State University
Ernest Simpson, California State Polytechnic University- Pomona
Peter W. Slade, University College of Fraser Valley
Gary Snyder, University of Massachusetts
Ronald Starkey, University of Wisconsin- Green Bay
J. William Suggs, Brown University
Michelle Sulikowski, Vanderbilt University
Douglas Taber, University of Delaware
Dennis Taylor, University of Adelaide
Marcus W. Thomsen, Franklin & Marshall College
Walter Trahanovsky, Iowa State University
Harry Ungar, Cabrillo College
Joseph J. Villafranca, Pennsylvania State University
Barbara J. Whitlock, University of Wisconsin-Madison
Vera Zalkow, Kennesaw College

1 Structure and Bonding

***Organic* KNOWLEDGE TOOLS**

ThomsonNOW™ Throughout this chapter, there are opportunities for online self-study, linking you to interactive tutorials based on your level of understanding. Sign in at **www.thomsonedu.com** to view organic chemistry tutorials and simulations, develop problem-solving skills, and test your knowledge with these interactive self-study resources.

Online homework for this and other chapters may be assigned in Organic OWL.

What is organic chemistry, and why should you study it? The answers to these questions are all around you. Every living organism is made of organic chemicals. The proteins that make up your hair, skin, and muscles; the DNA that controls your genetic heritage; the foods that nourish you; and the medicines that heal you are all organic chemicals. Anyone with a curiosity about life and living things, and anyone who wants to be a part of the many exciting developments now happening in medicine and the biological sciences, must first understand organic chemistry. Look at the following drawings for instance, which show the chemical structures of some molecules whose names might be familiar to you.

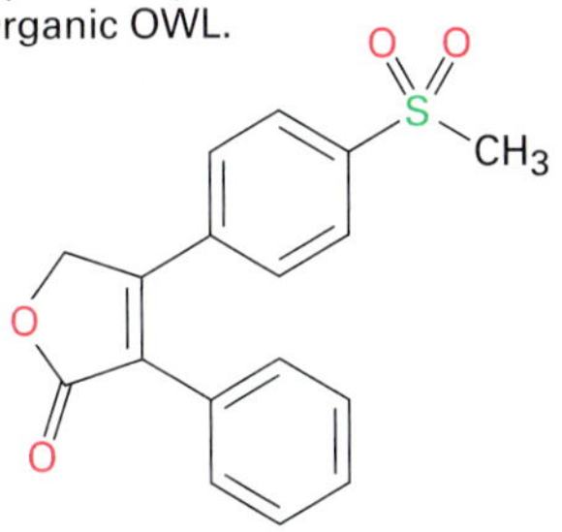

Rofecoxib (Vioxx)

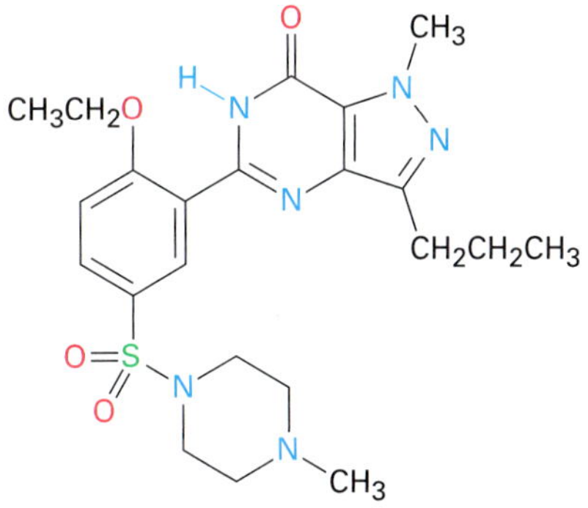

Sildenafil (Viagra)

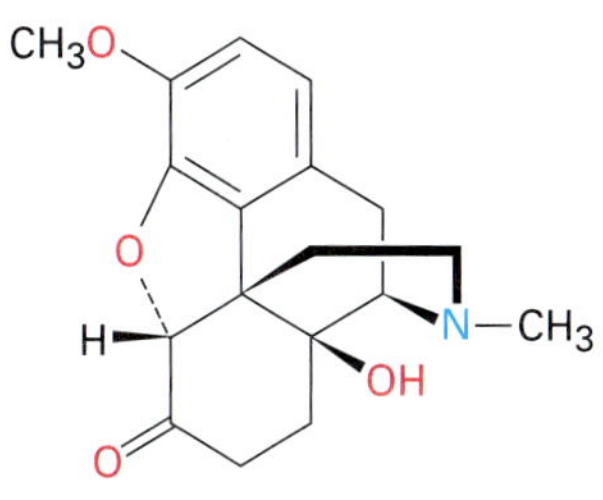

Oxycodone (OxyContin)

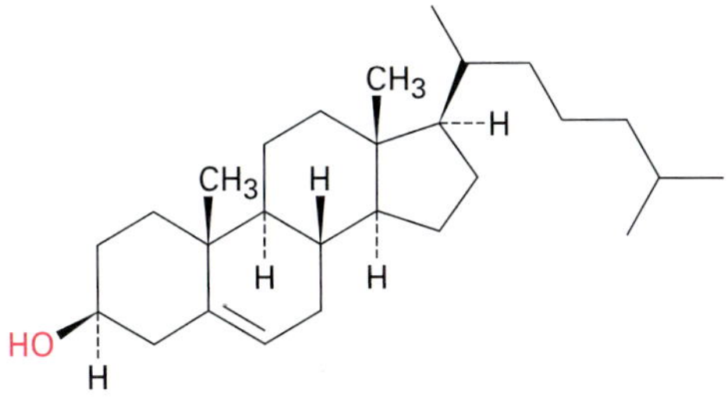

Cholesterol

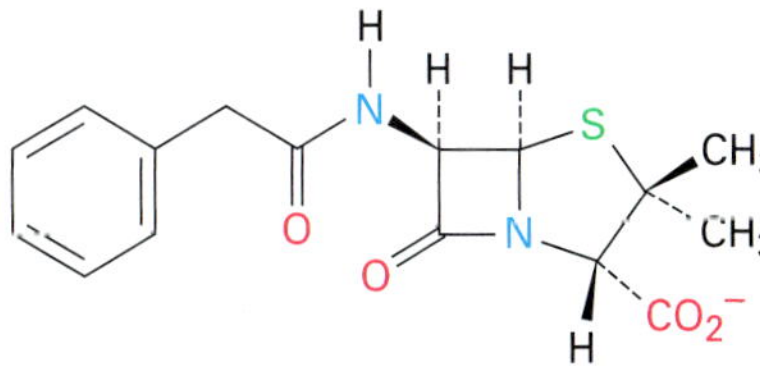

Benzylpenicillin

Although the drawings may appear unintelligible at this point, don't worry. Before long they'll make perfectly good sense and you'll be drawing similar structures for any substance you're interested in.

The foundations of organic chemistry date from the mid-1700s, when chemistry was evolving from an alchemist's art into a modern science. At that time, unexplainable differences were noted between substances obtained from living sources and those obtained from minerals. Compounds obtained from plants and animals were often difficult to isolate and purify. Even when pure, they were often difficult to work with, and they tended to decompose more easily than compounds obtained from minerals. The Swedish chemist Torbern Bergman in 1770 was the first to express this difference between "organic" and "inorganic" substances, and the term *organic chemistry* soon came to mean the chemistry of compounds found in living organisms.

Michel-Eugène Chevreul

Michel-Eugène Chevreul (1786–1889) was born in Angers, France. After studies at the Collège de France in Paris, he became professor of physics at the Lycée Charlemagne in 1813 and professor of chemistry in 1830. Chevreul's studies of soaps and waxes led him to patent a method for manufacturing candles. He also published work on the psychology of color perception and of aging. All France celebrated his 100th birthday in 1886.

To many chemists of the time, the only explanation for the differences in behavior between organic and inorganic compounds was that organic compounds must contain a peculiar "vital force" as a result of their origin in living sources. One consequence of this vital force, chemists believed, was that organic compounds could not be prepared and manipulated in the laboratory as could inorganic compounds. As early as 1816, however, this vitalistic theory received a heavy blow when Michel Chevreul found that soap, prepared by the reaction of alkali with animal fat, could be separated into several pure organic compounds, which he termed *fatty acids*. For the first time, one organic substance (fat) was converted into others (fatty acids plus glycerin) without the intervention of an outside vital force.

$$\text{Animal fat} \xrightarrow[H_2O]{NaOH} \text{Soap} + \text{Glycerin}$$

$$\text{Soap} \xrightarrow{H_3O^+} \text{"Fatty acids"}$$

Friedrich Wöhler

Friedrich Wöhler (1800–1882) was born in Eschersheim, Germany, and studied at Heidelberg under Leopold Gmelin. From 1836 to 1882, he was professor of chemistry at Göttingen. Wöhler developed the first industrial method for preparing aluminum metal, and he discovered several new elements. In addition, he wrote textbooks about both inorganic and organic chemistry.

Little more than a decade later, the vitalistic theory suffered still further when Friedrich Wöhler discovered in 1828 that it was possible to convert the "inorganic" salt ammonium cyanate into the "organic" substance urea, which had previously been found in human urine.

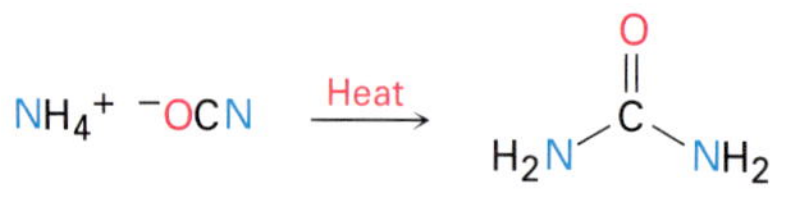

Ammonium cyanate **Urea**

William Thomas Brande

William Thomas Brande (1788–1866) was born in London, England. Trained as an apothecary, he became a lecturer in chemistry at the University of London in 1808 and was a professor at the Royal Institution from 1813 to 1852. His scientific achievements were modest, although he was the first person to discover naphthalene, now used in mothballs.

By the mid-1800s, the weight of evidence was clearly against the vitalistic theory. As William Brande wrote in 1848, "No definite line can be drawn between organic and inorganic chemistry. . . . Any distinctions . . . must for the present be merely considered as matters of practical convenience calculated to further the progress of students." Chemistry today is unified, and the same principles explain the behaviors of all substances, regardless of origin or complexity. The only distinguishing characteristic of organic chemicals is that *all contain the element carbon.*

Organic chemistry, then, is the study of carbon compounds. But why is carbon special? Why, of the more than 30 million presently known chemical compounds, do more than 99% of them contain carbon? The answers to these questions come from carbon's electronic structure and its consequent position in the periodic table (Figure 1.1). As a group 4A element, carbon can share four valence electrons and form four strong covalent bonds. Furthermore, carbon atoms can bond to one another, forming long chains and rings. Carbon, alone of all elements, is able to form an immense diversity of compounds, from the simple to the staggeringly complex—from methane, with one carbon atom, to DNA, which can have more than *100 hundred million* carbons.

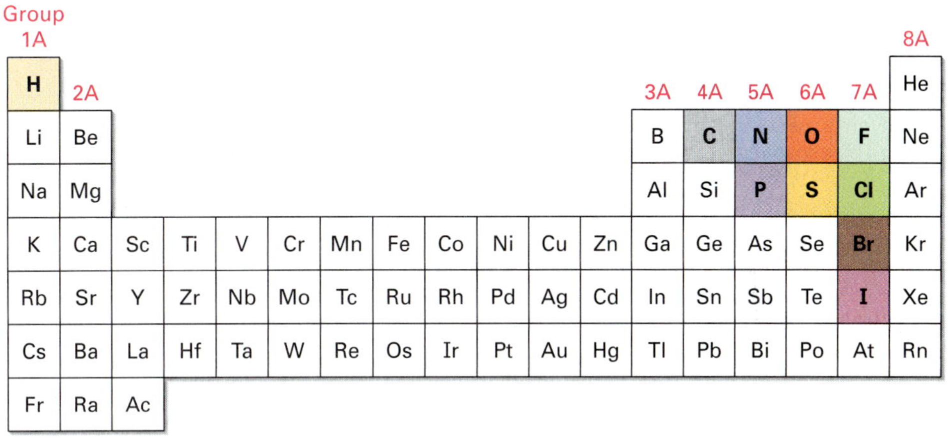

Figure 1.1 The position of carbon in the periodic table. Other elements commonly found in organic compounds are shown in the colors typically used to represent them.

Not all carbon compounds are derived from living organisms, of course, and chemists over the years have developed a remarkably sophisticated ability to design and synthesize new organic compounds. Medicines, dyes, polymers, food additives, pesticides, and a host of other substances are now prepared in the laboratory. Organic chemistry touches the lives of everyone. Its study is a fascinating undertaking.

WHY THIS CHAPTER?

We'll ease into the study of organic chemistry by first reviewing some ideas about atoms, bonds, and molecular geometry that you may recall from your general chemistry course. Much of the material in this chapter and the next is likely to be familiar to you, but it's nevertheless a good idea to make sure you understand it before going on.

1.1 Atomic Structure: The Nucleus

As you probably know, an atom consists of a dense, positively charged *nucleus* surrounded at a relatively large distance by negatively charged *electrons* (Figure 1.2). The nucleus consists of subatomic particles called *neutrons,* which are electrically neutral, and *protons,* which are positively charged. Because an atom is neutral overall, the number of positive protons in the nucleus and the number of negative electrons surrounding the nucleus are the same.

Although extremely small—about 10^{-14} to 10^{-15} meter (m) in diameter—the nucleus nevertheless contains essentially all the mass of the atom. Electrons have negligible mass and circulate around the nucleus at a distance of approximately 10^{-10} m. Thus, the diameter of a typical atom is about 2×10^{-10} m, or 200 *picometers* (pm), where 1 pm = 10^{-12} m. To give you an idea of how small this is, a thin pencil line is about 3 million carbon atoms wide. Many organic chemists and biochemists, particularly in the United States, still use the unit *angstrom* (Å) to express atomic distances, where 1 Å = 10^{-10} m = 100 pm, but we'll stay with the SI unit picometer in this book.

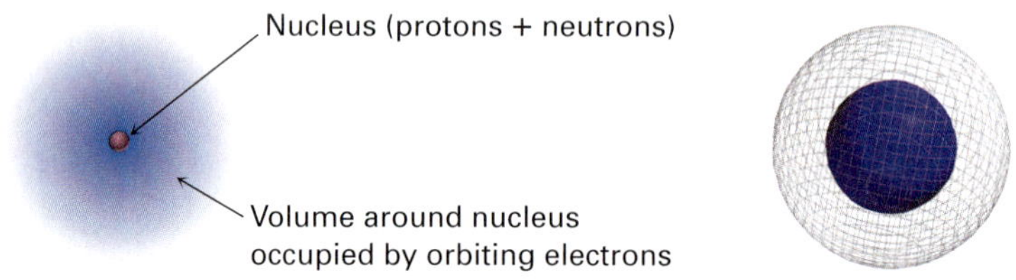

Figure 1.2 A schematic view of an atom. The dense, positively charged nucleus contains most of the atom's mass and is surrounded by negatively charged electrons. The three-dimensional view on the right shows calculated electron-density surfaces. Electron density increases steadily toward the nucleus and is 40 times greater at the blue solid surface than at the gray mesh surface.

A specific atom is described by its *atomic number* (*Z*), which gives the number of protons in the atom's nucleus, and its *mass number* (*A*), which gives the total of protons plus neutrons in its nucleus. All the atoms of a given element have the same atomic number—1 for hydrogen, 6 for carbon, 15 for phosphorus, and so on—but they can have different mass numbers, depending on how many neutrons they contain. Atoms with the same atomic number but different mass numbers are called **isotopes**. The weighted average mass in atomic mass units (amu) of an element's naturally occurring isotopes is called the element's *atomic mass* (or *atomic weight*)—1.008 amu for hydrogen, 12.011 amu for carbon, 30.974 amu for phosphorus, and so on.

1.2 Atomic Structure: Orbitals

How are the electrons distributed in an atom? You might recall from your general chemistry course that, according to the quantum mechanical model, the behavior of a specific electron in an atom can be described by a mathematical expression called a *wave equation*—the same sort of expression used to describe the motion of waves in a fluid. The solution to a wave equation is called a *wave function*, or **orbital**, and is denoted by the Greek letter psi, ψ.

By plotting the square of the wave function, ψ^2, in three-dimensional space, the orbital describes the volume of space around a nucleus that an electron is most likely to occupy. You might therefore think of an orbital as looking like a photograph of the electron taken at a slow shutter speed. The orbital would appear as a blurry cloud indicating the region of space around the nucleus where the electron has been. This electron cloud doesn't have a sharp boundary, but for practical purposes we can set the limits by saying that an orbital represents the space where an electron spends most (90%–95%) of its time.

What do orbitals look like? There are four different kinds of orbitals, denoted *s*, *p*, *d*, and *f*, each with a different shape. Of the four, we'll be concerned primarily with *s* and *p* orbitals because these are the most common in organic and biological chemistry. The *s* orbitals are spherical, with the nucleus at their center; *p* orbitals are dumbbell-shaped; and four of the five *d* orbitals are cloverleaf-shaped, as shown in Figure 1.3. The fifth *d* orbital is shaped like an elongated dumbbell with a doughnut around its middle.

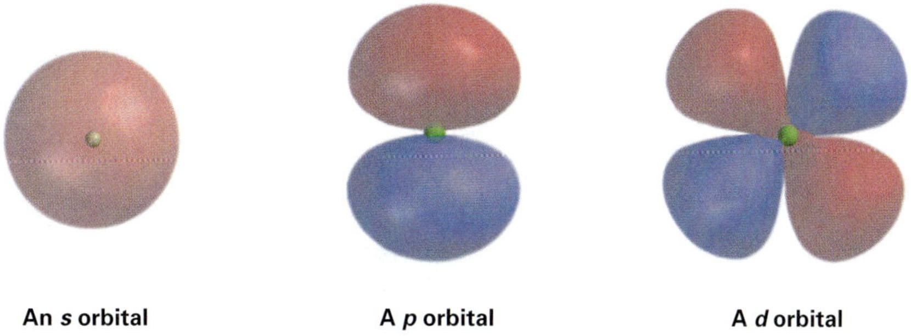

Figure 1.3 Representations of *s*, *p*, and *d* orbitals. The *s* orbitals are spherical, the *p* orbitals are dumbbell-shaped, and four of the five *d* orbitals are cloverleaf-shaped. Different lobes of *p* orbitals are often drawn for convenience as teardrops, but their true shape is more like that of a doorknob, as indicated.

The orbitals in an atom are organized into different layers, or **electron shells**, of successively larger size and energy. Different shells contain different numbers and kinds of orbitals, and each orbital within a shell can be occupied by two electrons. The first shell contains only a single *s* orbital, denoted 1*s*, and thus holds only 2 electrons. The second shell contains one 2*s* orbital and three 2*p* orbitals and thus holds a total of 8 electrons. The third shell contains a 3*s* orbital, three 3*p* orbitals, and five 3*d* orbitals, for a total capacity of 18 electrons. These orbital groupings and their energy levels are shown in Figure 1.4.

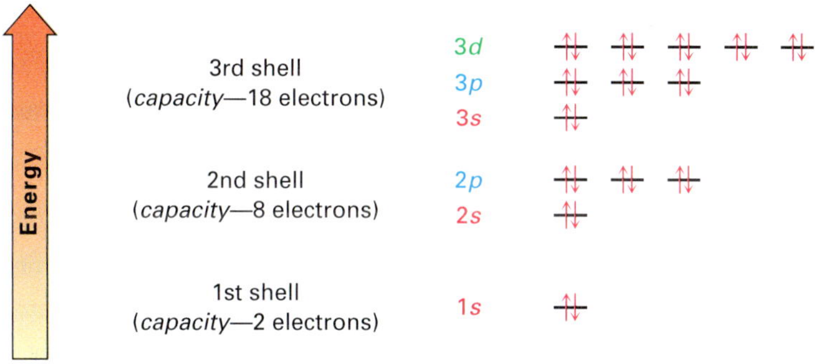

Figure 1.4 The energy levels of electrons in an atom. The first shell holds a maximum of 2 electrons in one 1*s* orbital; the second shell holds a maximum of 8 electrons in one 2*s* and three 2*p* orbitals; the third shell holds a maximum of 18 electrons in one 3*s*, three 3*p*, and five 3*d* orbitals; and so on. The two electrons in each orbital are represented by up and down arrows, ↑↓. Although not shown, the energy level of the 4*s* orbital falls between 3*p* and 3*d*.

The three different *p* orbitals within a given shell are oriented in space along mutually perpendicular directions, denoted p_x, p_y, and p_z. As shown in Figure 1.5, the two lobes of each *p* orbital are separated by a region of zero electron density called a **node**. Furthermore, the two orbital regions separated by the node have different algebraic signs, + and −, in the wave function. As we'll see in Section 1.11, the algebraic signs of the different orbital lobes have important consequences with respect to chemical bonding and chemical reactivity.

Figure 1.5 Shapes of the 2*p* orbitals. Each of the three mutually perpendicular, dumbbell-shaped orbitals has two lobes separated by a node. The two lobes have different algebraic signs in the corresponding wave function, as indicated by the different colors.

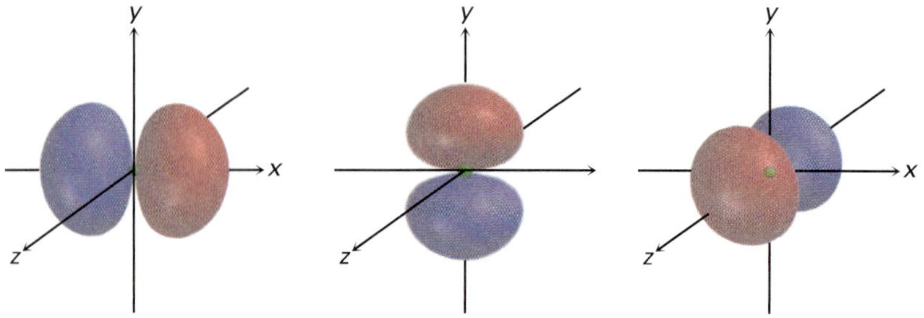

A $2p_x$ orbital **A $2p_y$ orbital** **A $2p_z$ orbital**

1.3 Atomic Structure: Electron Configurations

The lowest-energy arrangement, or **ground-state electron configuration**, of an atom is a listing of the orbitals occupied by its electrons. We can predict this arrangement by following three rules.

Rule 1 The lowest-energy orbitals fill up first, according to the order 1*s* → 2*s* → 2*p* → 3*s* → 3*p* → 4*s* → 3*d*, a statement called the *aufbau principle*. Note that the 4*s* orbital lies between the 3*p* and 3*d* orbitals in energy.

Rule 2 Electrons act as if they were spinning around an axis, in much the same way that the earth spins. This spin can have two orientations, denoted as up ↑ and down ↓. Only two electrons can occupy an orbital, and they must be of opposite spin, a statement called the *Pauli exclusion principle*.

Rule 3 If two or more empty orbitals of equal energy are available, one electron occupies each with spins parallel until all orbitals are half-full, a statement called *Hund's rule*.

Some examples of how these rules apply are shown in Table 1.1. Hydrogen, for instance, has only one electron, which must occupy the lowest-energy orbital. Thus, hydrogen has a 1*s* ground-state configuration. Carbon has six electrons and the ground-state configuration $1s^2\ 2s^2\ 2p_x^{\ 1}\ 2p_y^{\ 1}$, and so forth. Note that a superscript is used to represent the number of electrons in a particular orbital.

Table 1.1 Ground-State Electron Configurations of Some Elements

Element	Atomic number	Configuration		Element	Atomic number	Configuration	
Hydrogen	1	1*s*	↑	Phosphorus	15	3*p*	↑ ↑ ↑
						3*s*	↑↓
Carbon	6	2*p*	↑ ↑ —			2*p*	↑↓ ↑↓ ↑↓
		2*s*	↑↓			2*s*	↑↓
		1*s*	↑↓			1*s*	↑↓

Problem 1.1 Give the ground-state electron configuration for each of the following elements:
(a) Oxygen (b) Silicon (c) Sulfur

Problem 1.2 How many electrons does each of the following elements have in its outermost electron shell?

(a) Magnesium (b) Molybdenum (c) Selenium

1.4 Development of Chemical Bonding Theory

By the mid-1800s, the new science of chemistry was developing rapidly and chemists had begun to probe the forces holding compounds together. In 1858, August Kekulé and Archibald Couper independently proposed that, in all its compounds, carbon is *tetravalent*—it always forms four bonds when it joins other elements to form stable compounds. Furthermore, said Kekulé, carbon atoms can bond to one another to form extended chains of linked atoms.

Shortly after the tetravalent nature of carbon was proposed, extensions to the Kekulé–Couper theory were made when the possibility of *multiple* bonding between atoms was suggested. Emil Erlenmeyer proposed a carbon–carbon triple bond for acetylene, and Alexander Crum Brown proposed a carbon–carbon double bond for ethylene. In 1865, Kekulé provided another major advance when he suggested that carbon chains can double back on themselves to form *rings* of atoms.

Although Kekulé and Couper were correct in describing the tetravalent nature of carbon, chemistry was still viewed in a two-dimensional way until 1874. In that year, Jacobus van't Hoff and Joseph Le Bel added a third dimension to our ideas about organic compounds when they proposed that the four bonds of carbon are not oriented randomly but have specific spatial directions. Van't Hoff went even further and suggested that the four atoms to

Friedrich August Kekulé

Friedrich August Kekulé (1829–1896) was born in Darmstadt, Germany. He entered the University of Giessen in 1847 intending to become an architect but soon switched to chemistry. After receiving his doctorate under Liebig and doing further study in Paris, Kekulé became a lecturer at Heidelberg in 1855 and a professor of chemistry at Ghent (1858) and Bonn (1867). His realization that carbon can form rings of atoms is said to have come to him in a dream in which he saw a snake biting its tail.

Archibald Scott Couper

Archibald Scott Couper (1831–1892) was born in Kirkintilloch, Scotland, and studied at the universities of Glasgow, Edinburgh, and Paris. Although his scientific paper about the ability of carbon to form four bonds was submitted prior to a similar paper by Kekulé, Couper never received credit for his work. His health began to decline after the rejection of his achievements, and he suffered a nervous breakdown in 1858. He then retired from further scientific work and spent the last 30 years of his life in the care of his mother.

Richard A. C. E. Erlenmeyer

Richard A. C. E. Erlenmeyer (1825–1909) was born in Wehen, Germany. He studied in Giessen and in Heidelberg, intending originally to be a pharmacist, and was professor of chemistry at Munich Polytechnicum from 1868 to 1883. Much of his work was carried out with biological molecules, and he was the first to prepare the amino acid tyrosine.

Alexander Crum Brown

Alexander Crum Brown (1838–1922) was born in Edinburgh, the son of a Presbyterian minister. He studied at Edinburgh, Heidelberg, and Marburg and was professor of chemistry at Edinburgh from 1869 to 1908. Crum Brown's interests were many. He studied the physiology of the canals in the inner ear, he was proficient in Japanese, and he had a lifelong interest in knitting.

Jacobus Hendricus van't Hoff

Jacobus Hendricus van't Hoff (1852–1911) was born in Rotterdam, Netherlands, and studied at Delft, Leyden, Bonn, Paris, and Utrecht. Widely educated, he served as professor of chemistry, mineralogy, and geology at the University of Amsterdam from 1878 to 1896 and later became professor at Berlin. In 1901, he received the first Nobel Prize in chemistry for his work on chemical equilibrium and osmotic pressure.

which carbon is bonded sit at the corners of a regular tetrahedron, with carbon in the center.

A representation of a tetrahedral carbon atom is shown in Figure 1.6. Note the conventions used to show three-dimensionality: solid lines represent bonds in the plane of the page, the heavy wedged line represents a bond coming out of the page toward the viewer, and the dashed line represents a bond receding back behind the page, away from the viewer. These representations will be used throughout the text.

Joseph Achille Le Bel

Joseph Achille Le Bel (1847–1930) was born in Péchelbronn, France, and studied at the École Polytechnique and the Sorbonne in Paris. Freed by his family's wealth from the need to earn a living, he established his own private laboratory.

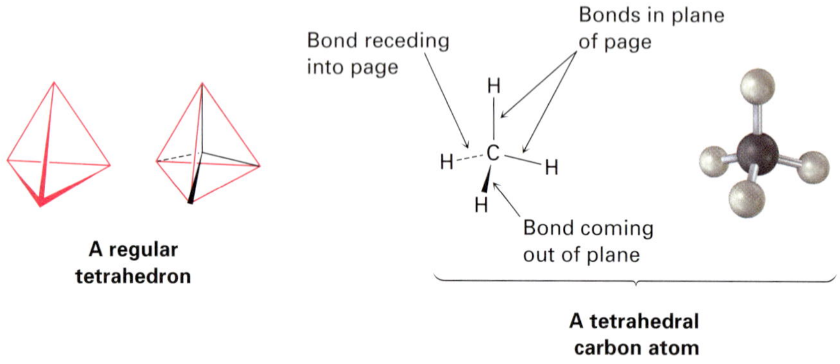

Figure 1.6 A representation of Van't Hoff's tetrahedral carbon atom. The solid lines are in the plane of the paper, the heavy wedged line comes out of the plane of the page, and the dashed line goes back behind the plane of the page.

Why, though, do atoms bond together, and how can bonds be described electronically? The *why* question is relatively easy to answer. Atoms bond together because the compound that results is lower in energy, and thus more stable, than the separate atoms. Energy (usually as heat) always flows out of the chemical system when a chemical bond forms. Conversely, energy must be put into the system to break a chemical bond. Making bonds always releases energy, and breaking bonds always absorbs energy. The *how* question is more difficult. To answer it, we need to know more about the electronic properties of atoms.

We know through observation that eight electrons (an electron *octet*) in an atom's outermost shell, or **valence shell**, impart special stability to the noble-gas elements in group 8A of the periodic table: Ne (2 + *8*); Ar (2 + 8 + *8*); Kr (2 + 8 + 18 +*8*). We also know that the chemistry of main-group elements is governed by their tendency to take on the electron configuration of the nearest noble gas. The alkali metals in group 1A, for example, achieve a noble-gas configuration by losing the single *s* electron from their valence shell to form a cation, while the halogens in group 7A achieve a noble-gas configuration by gaining a *p* electron to fill their valence shell, thereby forming an anion. The resultant ions are held together in compounds like $Na^+ Cl^-$ by an electrostatic attraction that we call an *ionic bond.*

Gilbert Newton Lewis

Gilbert Newton Lewis (1875–1946) was born in Weymouth, Massachusetts, and received his Ph.D. at Harvard in 1899. After a short time as professor of chemistry at the Massachusetts Institute of Technology (1905–1912), he spent the rest of his career at the University of California at Berkeley (1912–1946). In addition to his work on structural theory, Lewis was the first to prepare "heavy water," D_2O, in which the two hydrogens of water are the ^2H isotope, deuterium.

But how do elements closer to the middle of the periodic table form bonds? Look at methane, CH_4, the main constituent of natural gas, for example. The bonding in methane is not ionic because it would take too much energy for carbon ($1s^2\ 2s^2\ 2p^2$) either to gain or lose four electrons to achieve a noble-gas configuration. As a result, carbon bonds to other atoms, not by gaining or losing electrons, but by sharing them. Such a shared-electron bond, first proposed in 1916 by G. N. Lewis, is called a **covalent bond**. The neutral collection of atoms held together by covalent bonds is called a **molecule**.

A simple way of indicating the covalent bonds in molecules is to use what are called *Lewis structures,* or **electron-dot structures**, in which the valence electrons of an atom are represented as dots. Thus, hydrogen has one dot representing its 1*s* electron, carbon has four dots ($2s^2\ 2p^2$), oxygen has six dots ($2s^2\ 2p^4$), and so on. A stable molecule results whenever a noble-gas configuration is achieved for all the atoms—eight dots (an octet) for main-group atoms or two dots for hydrogen. Simpler still is the use of *Kekulé structures,* or **line-bond structures**, in which a two-electron covalent bond is indicated as a line drawn between atoms.

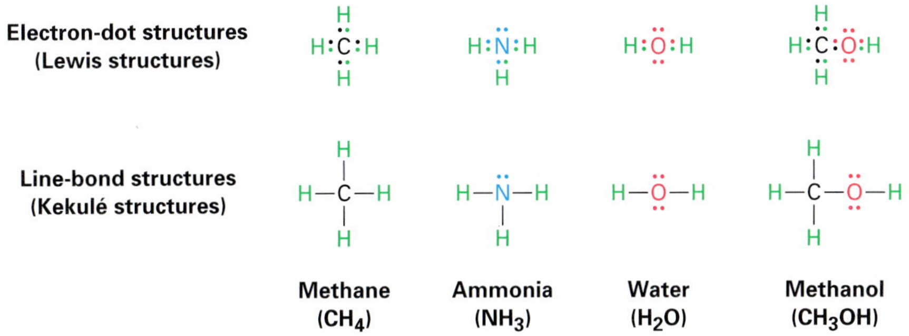

The number of covalent bonds an atom forms depends on how many additional valence electrons it needs to reach a noble-gas configuration. Hydrogen has one valence electron (1*s*) and needs one more to reach the helium configuration ($1s^2$), so it forms one bond. Carbon has four valence electrons ($2s^2\ 2p^2$) and needs four more to reach the neon configuration ($2s^2\ 2p^6$), so it forms four bonds. Nitrogen has five valence electrons ($2s^2\ 2p^3$), needs three more, and forms three bonds; oxygen has six valence electrons ($2s^2\ 2p^4$), needs two more, and forms two bonds; and the halogens have seven valence electrons, need one more, and form one bond.

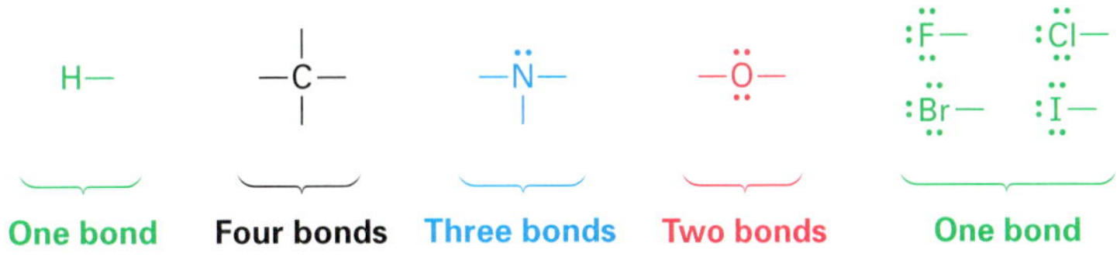

Valence electrons that are not used for bonding are called **lone-pair electrons**, or *nonbonding electrons.* The nitrogen atom in ammonia, for instance, shares six valence electrons in three covalent bonds and has its remaining two valence electrons in a nonbonding lone pair. As a time-saving shorthand, nonbonding electrons are often omitted when drawing line-bond structures, but you still have to keep them in mind since they're often crucial in chemical reactions.

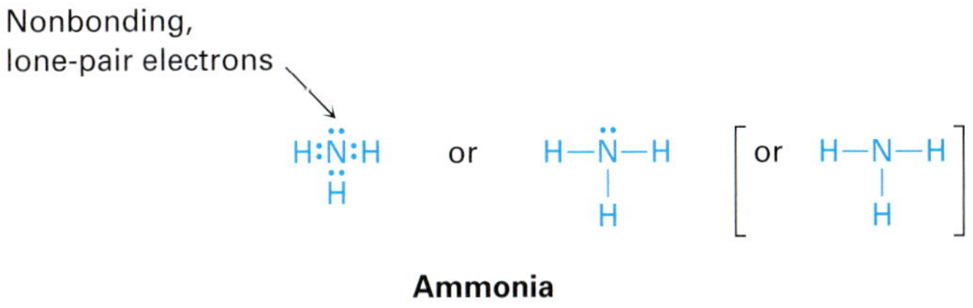

WORKED EXAMPLE 1.1 ***Predicting the Number of Bonds Formed by Atoms in a Molecule***

How many hydrogen atoms does phosphorus bond to in forming phosphine, $PH_{?}$?

Strategy Identify the periodic group of phosphorus, and tell from that how many electrons (bonds) are needed to make an octet.

Solution Phosphorus is in group 5A of the periodic table and has five valence electrons. It thus needs to share three more electrons to make an octet and therefore bonds to three hydrogen atoms, giving PH_3.

Problem 1.3 Draw a molecule of chloroform, $CHCl_3$, using solid, wedged, and dashed lines to show its tetrahedral geometry.

Problem 1.4 Convert the following representation of ethane, C_2H_6, into a conventional drawing that uses solid, wedged, and dashed lines to indicate tetrahedral geometry around each carbon (gray = C, ivory = H).

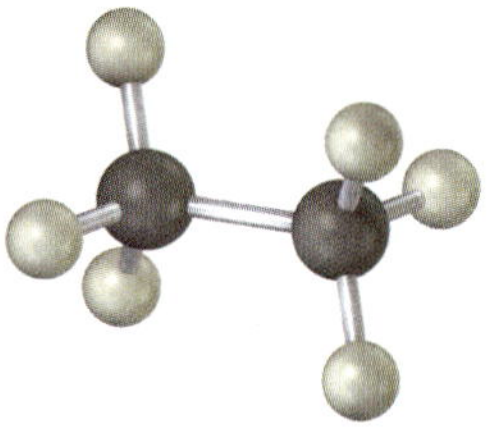

Ethane

Problem 1.5 What are likely formulas for the following substances?

(a) $GeCl_{?}$ **(b)** $AlH_{?}$ **(c)** $CH_{?}Cl_2$ **(d)** $SiF_{?}$ **(e)** $CH_3NH_{?}$

Problem 1.6 Write line-bond structures for the following substances, showing all nonbonding electrons:

(a) $CHCl_3$, chloroform **(b)** H_2S, hydrogen sulfide
(c) CH_3NH_2, methylamine **(d)** CH_3Li, methyllithium

Problem 1.7 Why can't an organic molecule have the formula C_2H_7?

1.5 The Nature of Chemical Bonds: Valence Bond Theory

How does electron sharing lead to bonding between atoms? Two models have been developed to describe covalent bonding: *valence bond theory* and *molecular orbital theory.* Each model has its strengths and weaknesses, and chemists tend to use them interchangeably depending on the circumstances. Valence bond theory is the more easily visualized of the two, so most of the descriptions we'll use in this book derive from that approach.

According to **valence bond theory**, a covalent bond forms when two atoms approach each other closely and a singly occupied orbital on one atom *overlaps* a singly occupied orbital on the other atom. The electrons are now paired in the overlapping orbitals and are attracted to the nuclei of both atoms, thus bonding the atoms together. In the H_2 molecule, for example, the H–H bond results from the overlap of two singly occupied hydrogen 1*s* orbitals.

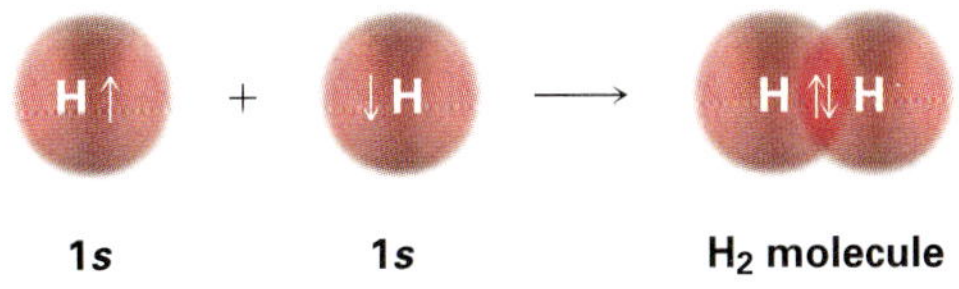

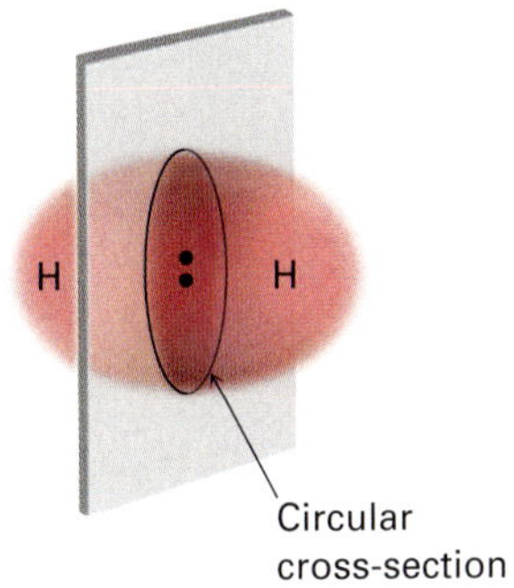

Figure 1.7 The cylindrical symmetry of the H–H σ bond in an H_2 molecule. The intersection of a plane cutting through the σ bond is a circle.

The overlapping orbitals in the H_2 molecule have the elongated egg shape we might get by pressing two spheres together. If a plane were to pass through the middle of the bond, the intersection of the plane and the overlapping orbitals would be a circle. In other words, the H–H bond is *cylindrically symmetrical,* as shown in Figure 1.7. Such bonds, which are formed by the head-on overlap of two atomic orbitals along a line drawn between the nuclei, are called **sigma (σ) bonds**.

During the bond-forming reaction 2 H· → H_2, 436 kJ/mol (104 kcal/mol) of energy is released. Because the product H_2 molecule has 436 kJ/mol less energy than the starting 2 H· atoms, we say that the product is more stable than the reactant and that the H–H bond has a **bond strength** of 436 kJ/mol. In other words, we would have to put 436 kJ/mol of energy *into* the H–H bond to break the H_2 molecule apart into H atoms (Figure 1.8.) [For convenience, we'll generally give energies in both kilocalories (kcal) and the SI unit kilojoules (kJ): 1 kJ = 0.2390 kcal; 1 kcal = 4.184 kJ.]

Figure 1.8 Relative energy levels of H atoms and the H_2 molecule. The H_2 molecule has 436 kJ/mol (104 kcal/mol) less energy than the two H atoms, so 436 kJ/mol of energy is released when the H–H bond forms. Conversely, 436 kJ/mol must be added to the H_2 molecule to break the H–H bond.

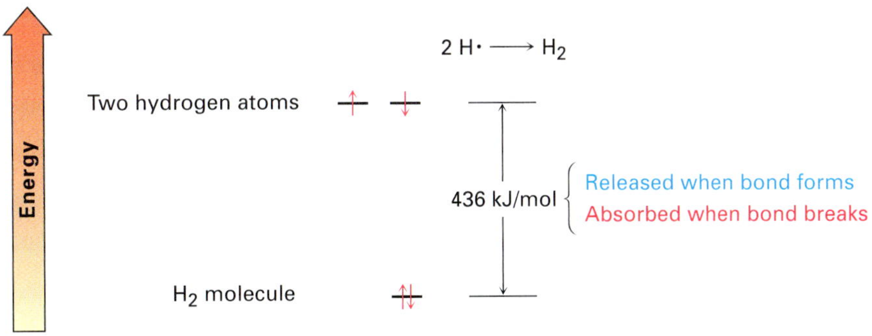

How close are the two nuclei in the H_2 molecule? If they are too close, they will repel each other because both are positively charged, yet if they're too far apart, they won't be able to share the bonding electrons. Thus, there is an optimum distance between nuclei that leads to maximum stability (Figure 1.9). Called the **bond length**, this distance is 74 pm in the H_2 molecule. Every covalent bond has both a characteristic bond strength and bond length.

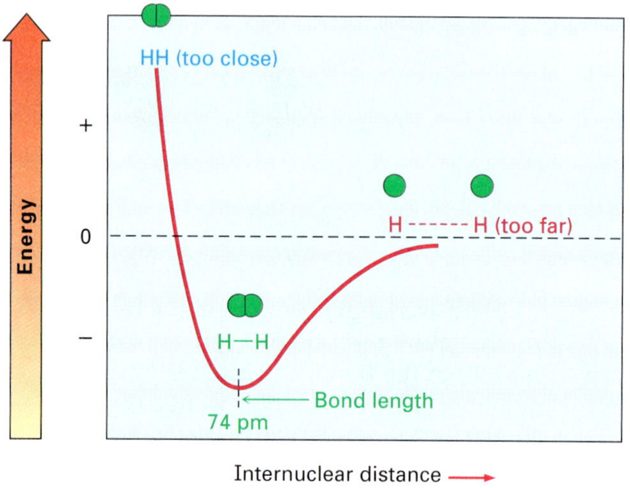

Figure 1.9 A plot of energy versus internuclear distance for two hydrogen atoms. The distance between nuclei at the minimum energy point is the bond length.

1.6 *sp*³ Hybrid Orbitals and the Structure of Methane

The bonding in the hydrogen molecule is fairly straightforward, but the situation is more complicated in organic molecules with tetravalent carbon atoms. Take methane, CH_4, for instance. As we've seen, carbon has four valence electrons ($2s^2\ 2p^2$) and forms four bonds. Because carbon uses two kinds of orbitals for bonding, 2*s* and 2*p*, we might expect methane to have two kinds of C–H bonds. In fact, though, all four C–H bonds in methane are identical and are spatially oriented toward the corners of a regular tetrahedron (Figure 1.6). How can we explain this?

An answer was provided in 1931 by Linus Pauling, who showed how an *s* orbital and three *p* orbitals on an atom can combine mathematically, or *hybridize*, to form four equivalent atomic orbitals with tetrahedral orientation. Shown in Figure 1.10, these tetrahedrally oriented orbitals are called ***sp*³ hybrids**. Note that the superscript 3 in the name sp^3 tells how many of each type of atomic orbital combine to form the hybrid, not how many electrons occupy it.

The concept of hybridization explains *how* carbon forms four equivalent tetrahedral bonds but not *why* it does so. The shape of the hybrid orbital suggests the answer. When an *s* orbital hybridizes with three *p* orbitals, the resultant sp^3 hybrid orbitals are unsymmetrical about the nucleus. One of the two

Linus Carl Pauling

Linus Carl Pauling (1901–1994) was born in Portland, Oregon, the son of a pharmacist. After obtaining a B.S. degree at Oregon State University, he received a Ph.D. from the California Institute of Technology in 1925. He was professor of chemistry from 1925 to 1967 at the California Institute of Technology and then from 1974 to 1994 at the University of California in San Diego and Stanford University.

Pauling was a scientific giant, who made fundamental discoveries in fields ranging from chemical bonding to molecular biology to medicine. A lifelong pacifist, Pauling is the only solo winner of two Nobel Prizes in different fields: the first in 1954 for chemistry and the second in 1963 for peace.

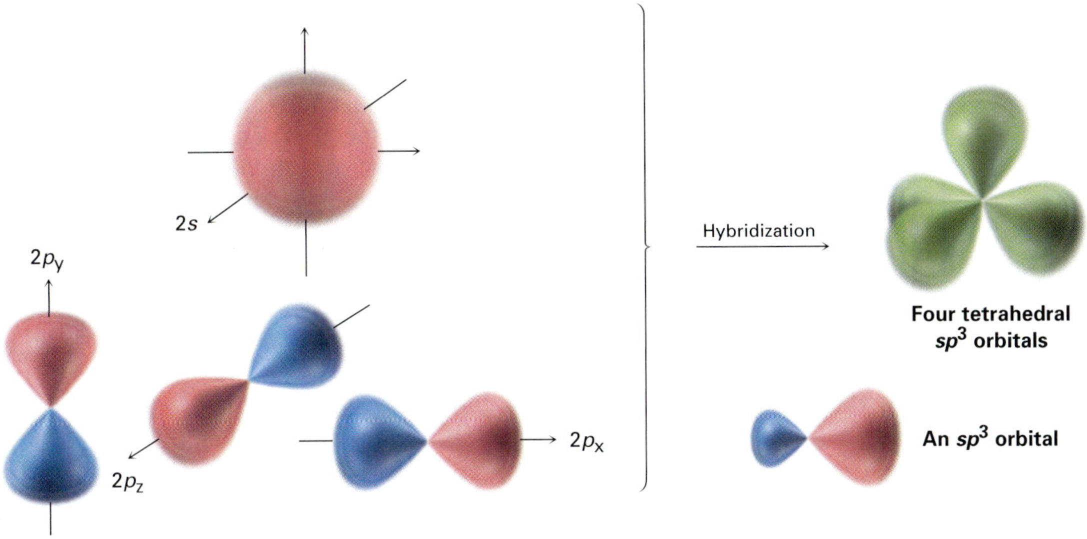

Active Figure 1.10 Four sp^3 hybrid orbitals (green), oriented to the corners of a regular tetrahedron, are formed by combination of an *s* orbital (red) and three *p* orbitals (red/blue). The sp^3 hybrids have two lobes and are unsymmetrical about the nucleus, giving them a directionality and allowing them to form strong bonds to other atoms. *Sign in at* **www.thomsonedu.com** *to see a simulation based on this figure and to take a short quiz.*

lobes is much larger than the other and can therefore overlap more effectively with an orbital from another atom when it forms a bond. As a result, sp^3 hybrid orbitals form stronger bonds than do unhybridized *s* or *p* orbitals.

The asymmetry of sp^3 orbitals arises because, as noted previously, the two lobes of a *p* orbital have different algebraic signs, + and −. Thus, when a *p* orbital hybridizes with an *s* orbital, the positive *p* lobe adds to the *s* orbital but the negative *p* lobe subtracts from the *s* orbital. The resultant hybrid orbital is therefore unsymmetrical about the nucleus and is strongly oriented in one direction.

When each of the four identical sp^3 hybrid orbitals of a carbon atom overlaps with the 1*s* orbital of a hydrogen atom, four identical C–H bonds are formed and methane results. Each C–H bond in methane has a strength of 436 kJ/mol (104 kcal/mol) and a length of 109 pm. Because the four bonds have a specific geometry, we also can define a property called the **bond angle**. The angle formed by each H–C–H is 109.5°, the so-called tetrahedral angle. Methane thus has the structure shown in Figure 1.11.

Active Figure 1.11 The structure of methane, showing its 109.5° bond angles. *Sign in at* **www.thomsonedu.com** *to see a simulation based on this figure and to take a short quiz.*

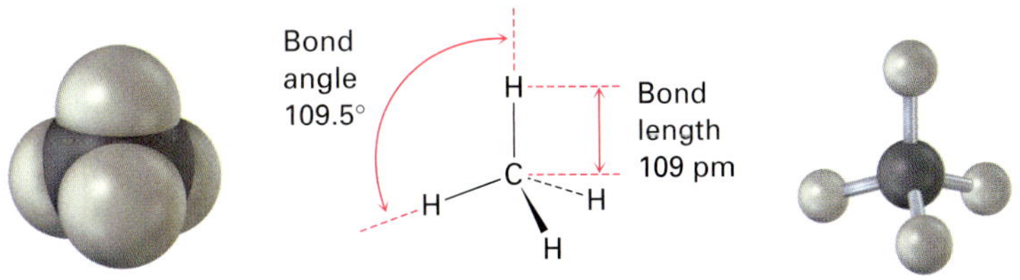

1.7 *sp*^{3} Hybrid Orbitals and the Structure of Ethane

The same kind of orbital hybridization that accounts for the methane structure also accounts for the bonding together of carbon atoms into chains and rings to make possible many millions of organic compounds. Ethane, C_2H_6, is the simplest molecule containing a carbon–carbon bond.

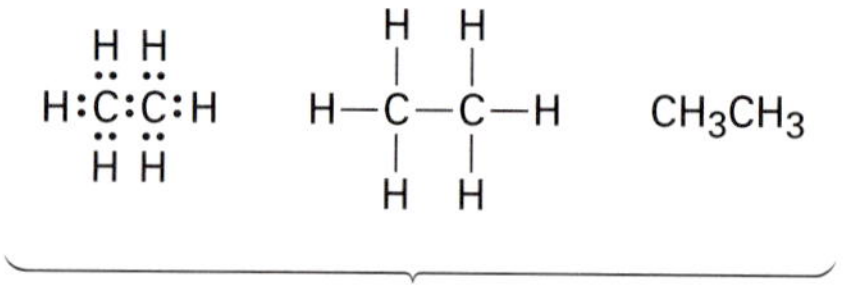

Some representations of ethane

We can picture the ethane molecule by imagining that the two carbon atoms bond to each other by σ overlap of an *sp*^{3} hybrid orbital from each (Figure 1.12). The remaining three *sp*^{3} hybrid orbitals of each carbon overlap with the 1*s* orbitals of three hydrogens to form the six C–H bonds. The C–H bonds in ethane are similar to those in methane, although a bit weaker—423 kJ/mol (101 kcal/mol) for ethane versus 436 kJ/mol for methane. The C–C bond is 154 pm long and has a strength of 376 kJ/mol (90 kcal/mol). All the bond angles of ethane are near, although not exactly at, the tetrahedral value of 109.5°.

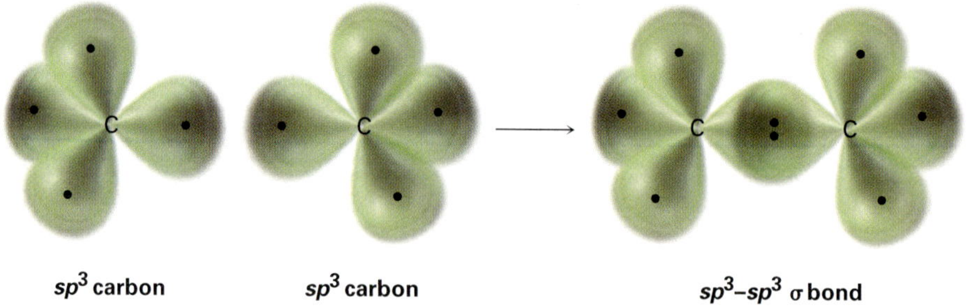

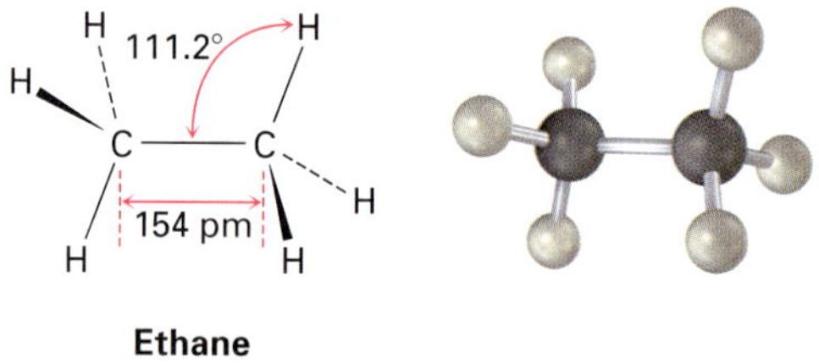

Figure 1.12 The structure of ethane. The carbon–carbon bond is formed by σ overlap of two carbon *sp*^{3} hybrid orbitals. For clarity, the smaller lobes of the *sp*^{3} hybrid orbitals are not shown.

Problem 1.8 Draw a line-bond structure for propane, $CH_3CH_2CH_3$. Predict the value of each bond angle, and indicate the overall shape of the molecule.

Problem 1.9 Convert the following molecular model of hexane, a component of gasoline, into a line-bond structure (gray = C, ivory = H).

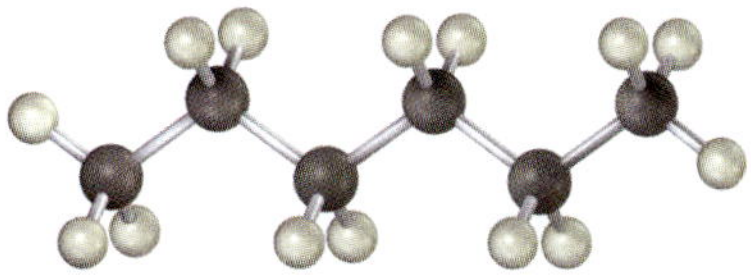

Hexane

1.8 sp^2 Hybrid Orbitals and the Structure of Ethylene

Although sp^3 hybridization is the most common electronic state of carbon, it's not the only possibility. Look at ethylene, C_2H_4, for example. It was recognized more than 100 years ago that ethylene carbons can be tetravalent only if they share *four* electrons and are linked by a *double* bond. Furthermore, ethylene is planar (flat) and has bond angles of approximately 120° rather than 109.5°.

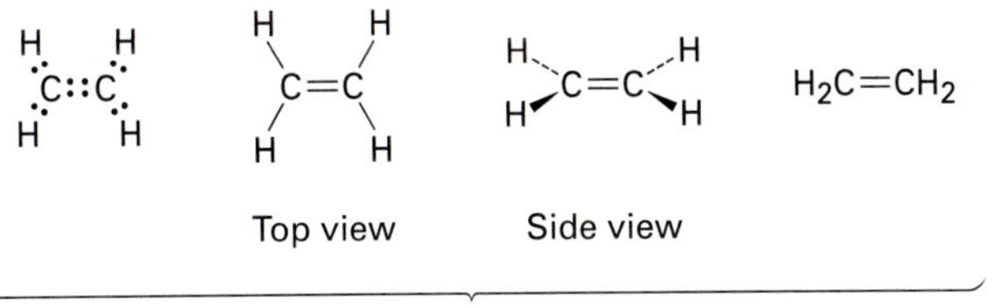

Some representations of ethylene

When we discussed sp^3 hybrid orbitals in Section 1.6, we said that the four valence-shell atomic orbitals of carbon combine to form four equivalent sp^3 hybrids. Imagine instead that the 2*s* orbital combines with only *two* of the three available 2*p* orbitals. Three **sp^2 hybrid orbitals** result, and one 2*p* orbital remains unchanged. The three sp^2 orbitals lie in a plane at angles of 120° to one another, with the remaining *p* orbital perpendicular to the sp^2 plane, as shown in Figure 1.13.

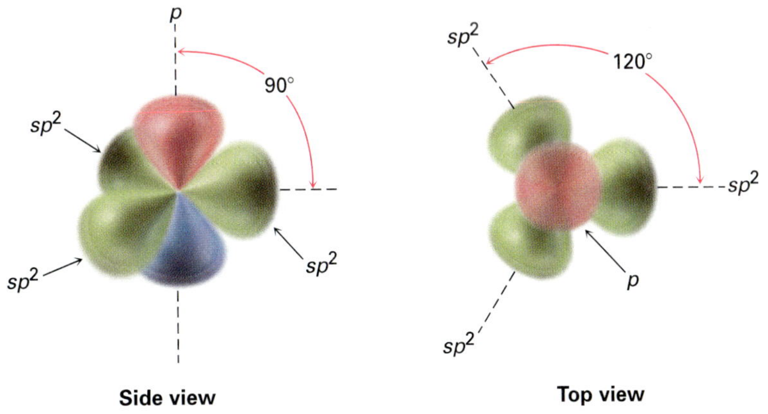

Figure 1.13 An sp^2-hybridized carbon. The three equivalent sp^2 hybrid orbitals (green) lie in a plane at angles of 120° to one another, and a single unhybridized *p* orbital (red/blue) is perpendicular to the sp^2 plane.

When two sp^2-hybridized carbons approach each other, they form a σ bond by sp^2–sp^2 head-on overlap. At the same time, the unhybridized p orbitals approach with the correct geometry for *sideways* overlap, leading to the formation of what is called a **pi (π) bond**. The combination of an sp^2–sp^2 σ bond and a $2p$–$2p$ π bond results in the sharing of four electrons and the formation of a carbon–carbon double bond (Figure 1.14). Note that the electrons in the σ bond occupy the region centered between nuclei, while the electrons in the π bond occupy regions on either side of a line drawn between nuclei.

To complete the structure of ethylene, four hydrogen atoms form σ bonds with the remaining four sp^2 orbitals. Ethylene thus has a planar structure, with H–C–H and H–C–C bond angles of approximately 120°. (The actual values are 117.4° for the H–C–H bond angle and 121.3° for the H–C–C bond angle.) Each C–H bond has a length of 108.7 pm and a strength of 465 kJ/mol (111 kcal/mol).

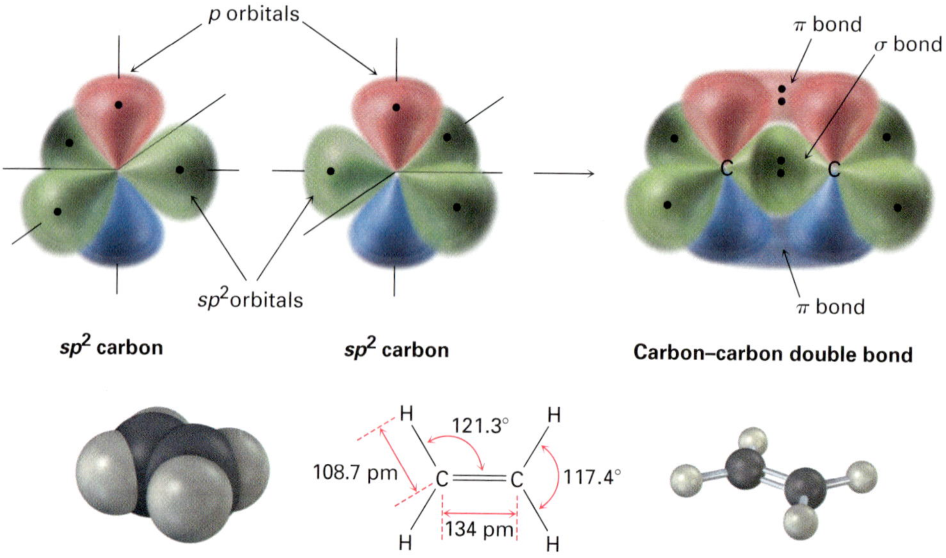

Figure 1.14 The structure of ethylene. Orbital overlap of two sp^2-hybridized carbons forms a carbon–carbon double bond. One part of the double bond results from σ (head-on) overlap of sp^2 orbitals (green), and the other part results from π (sideways) overlap of unhybridized p orbitals (red/blue). The π bond has regions of electron density on either side of a line drawn between nuclei.

As you might expect, the carbon–carbon double bond in ethylene is both shorter and stronger than the single bond in ethane because it has four electrons bonding the nuclei together rather than two. Ethylene has a C=C bond length of 134 pm and a strength of 728 kJ/mol (174 kcal/mol) versus a C–C length of 154 pm and a strength of 376 kJ/mol for ethane. Note that the carbon–carbon double bond is less than twice as strong as a single bond because the overlap in the π part of the double bond is not as effective as the overlap in the σ part.

WORKED EXAMPLE 1.2

Predicting the Structures of Simple Organic Molecules from Their Formulas

Commonly used in biology as a tissue preservative, formaldehyde, CH_2O, contains a carbon–*oxygen* double bond. Draw the line-bond structure of formaldehyde, and indicate the hybridization of the carbon atom.

Strategy We know that hydrogen forms one covalent bond, carbon forms four, and oxygen forms two. Trial and error, combined with intuition, is needed to fit the atoms together.

Solution There is only one way that two hydrogens, one carbon, and one oxygen can combine:

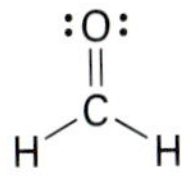

Formaldehyde

Like the carbon atoms in ethylene, the carbon atom in formaldehyde is in a double bond and therefore sp^2-hybridized.

Problem 1.10 Draw a line-bond structure for propene, $CH_3CH{=}CH_2$; indicate the hybridization of each carbon; and predict the value of each bond angle.

Problem 1.11 Draw a line-bond structure for 1,3-butadiene, $H_2C{=}CH{-}CH{=}CH_2$; indicate the hybridization of each carbon; and predict the value of each bond angle.

Problem 1.12 Following is a molecular model of aspirin (acetylsalicylic acid). Identify the hybridization of each carbon atom in aspirin, and tell which atoms have lone pairs of electrons (gray = C, red = O, ivory = H).

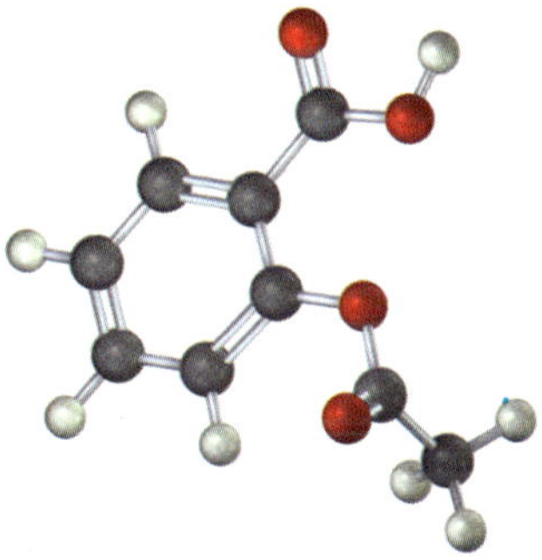

Aspirin (acetylsalicylic acid)

1.9 *sp* Hybrid Orbitals and the Structure of Acetylene

In addition to forming single and double bonds by sharing two and four electrons, respectively, carbon also can form a *triple* bond by sharing six electrons. To account for the triple bond in a molecule such as acetylene, $H{-}C{\equiv}C{-}H$, we need a third kind of hybrid orbital, an ***sp* hybrid**. Imagine that, instead of combining with two or three *p* orbitals, a carbon 2*s* orbital hybridizes with only a single *p* orbital. Two *sp* hybrid orbitals result, and two *p* orbitals remain unchanged. The two *sp* orbitals are oriented 180° apart on the *x*-axis, while the remaining two p orbitals are perpendicular on the y-axis and the z-axis, as shown in Figure 1.15.

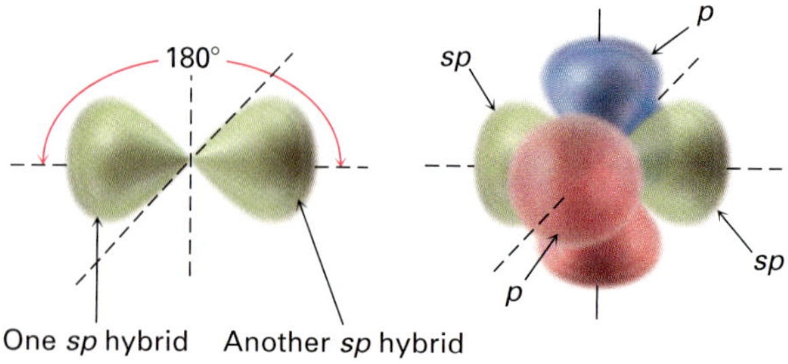

Figure 1.15 An *sp*-hybridized carbon atom. The two *sp* hybrid orbitals (green) are oriented 180° away from each other, perpendicular to the two remaining *p* orbitals (red/blue).

When two *sp*-hybridized carbon atoms approach each other, *sp* hybrid orbitals on each carbon overlap head-on to form a strong *sp–sp* σ bond. In addition, the p_z orbitals from each carbon form a p_z–p_z π bond by sideways overlap and the p_y orbitals overlap similarly to form a p_y–p_y π bond. The net effect is the sharing of six electrons and formation of a carbon–carbon triple bond. The two remaining *sp* hybrid orbitals each form a σ bond with hydrogen to complete the acetylene molecule (Figure 1.16).

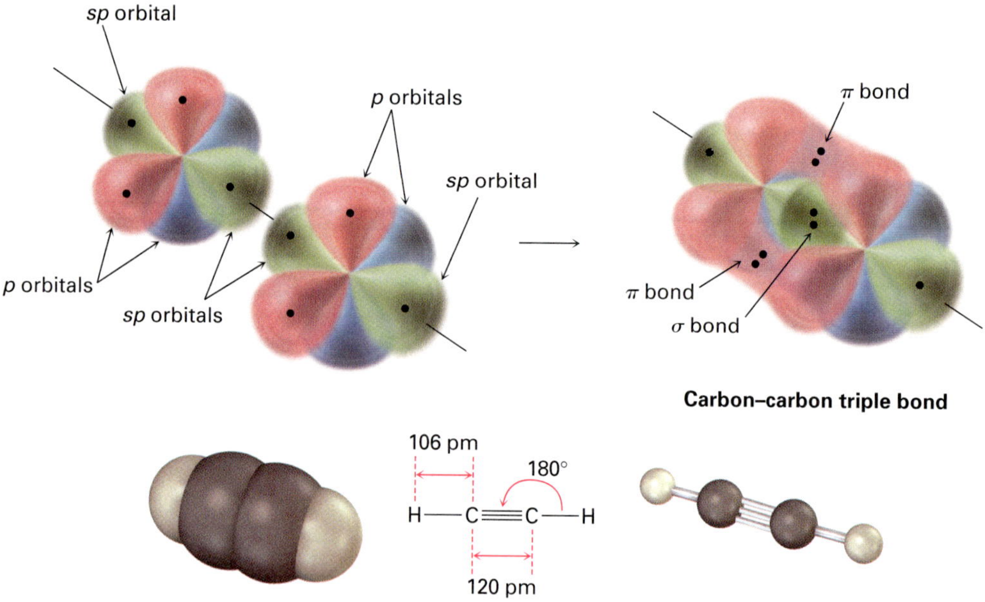

Figure 1.16 The structure of acetylene. The two *sp*-hybridized carbon atoms are joined by one *sp–sp* σ bond and two *p–p* π bonds.

As suggested by *sp* hybridization, acetylene is a linear molecule with H–C–C bond angles of 180°. The C–H bonds have a length of 106 pm and a strength of 556 kJ/mol (133 kcal/mol). The C–C bond length in acetylene is 120 pm, and its strength is about 965 kJ/mol (231 kcal/mol), making it the shortest and strongest of any carbon–carbon bond. A comparison of *sp*, sp^2, and sp^3 hybridization is given in Table 1.2.

Table 1.2 Comparison of C—C and C—H Bonds in Methane, Ethane, Ethylene, and Acetylene

Molecule	Bond	Bond strength (kJ/mol)	Bond strength (kcal/mol)	Bond length (pm)
Methane, CH_4	(sp^3) C—H	436	104	109
Ethane, CH_3CH_3	(sp^3) C—C (sp^3)	376	90	154
	(sp^3) C—H	423	101	109
Ethylene, $H_2C{=}CH_2$	(sp^2) C=C (sp^2)	728	174	134
	(sp^2) C—H	465	111	109
Acetylene, HC≡CH	(sp) C≡C (sp)	965	231	120
	(sp) C—H	556	133	106

Problem 1.13 Draw a line-bond structure for propyne, $CH_3C{\equiv}CH$; indicate the hybridization of each carbon; and predict a value for each bond angle.

1.10 Hybridization of Nitrogen, Oxygen, Phosphorus, and Sulfur

The valence-bond concept of orbital hybridization described in the previous four sections is not limited to carbon compounds. Covalent bonds formed by other elements can also be described using hybrid orbitals. Look, for instance, at the nitrogen atom in methylamine, CH_3NH_2, an organic derivative of ammonia (NH_3) and the substance responsible for the odor of rotting fish.

The experimentally measured H–N–H bond angle in methylamine is 107.1° and the C–N–H bond angle is 110.3°, both of which are close to the 109.5° tetrahedral angle found in methane. We therefore assume that nitrogen hybridizes to form four sp^3 orbitals, just as carbon does. One of the four sp^3 orbitals is occupied by two nonbonding electrons, and the other three hybrid orbitals have one electron each. Overlap of these half-filled nitrogen orbitals with half-filled orbitals from other atoms (C or H) gives methylamine. Note that the unshared lone pair of electrons in the fourth sp^3 hybrid orbital of nitrogen occupies as much space as an N–H bond does and is very important to the chemistry of methylamine and other nitrogen-containing organic molecules.

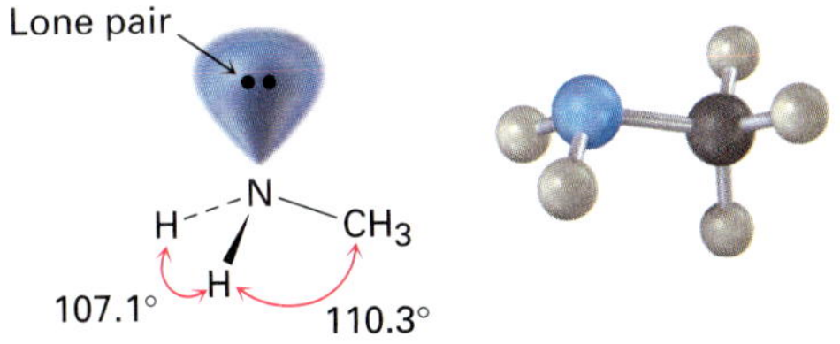

Methylamine

Like the carbon atom in methane and the nitrogen atom in methylamine, the oxygen atom in methanol (methyl alcohol) and many other organic molecules can also be described as sp^3-hybridized. The C–O–H bond angle in methanol is 108.5°, very close to the 109.5° tetrahedral angle. Two of the four sp^3 hybrid orbitals on oxygen are occupied by nonbonding electron lone pairs, and two are used to form bonds.

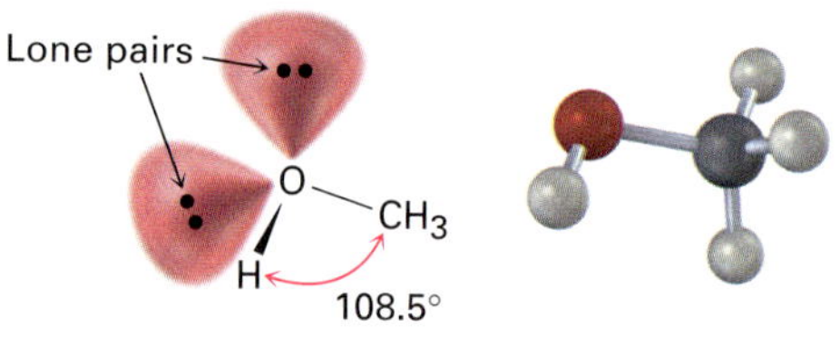

Methanol (methyl alcohol)

Phosphorus and sulfur are the third-row analogs of nitrogen and oxygen, and the bonding in both can be described using hybrid orbitals. Because of their positions in the third row, however, both phosphorus and sulfur can expand their outer-shell octets and form more than the typical number of covalent bonds. Phosphorus, for instance, often forms five covalent bonds, and sulfur occasionally forms four.

Phosphorus is most commonly encountered in biological molecules in *organophosphates,* compounds that contain a phosphorus atom bonded to four oxygens, with one of the oxygens also bonded to carbon. Methyl phosphate, $CH_3OPO_3{}^{2-}$ is the simplest example. The O–P–O bond angle in such compounds is typically in the range 110 to 112°, implying sp^3 hybridization for the phosphorus.

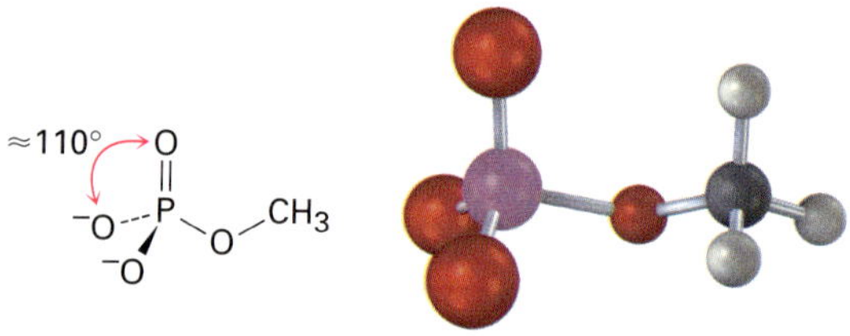

Methyl phosphate (an organophosphate)

Sulfur is most commonly encountered in biological molecules either in compounds called *thiols,* which have a sulfur atom bonded to one hydrogen and one carbon, or in *sulfides,* which have a sulfur atom bonded to two carbons. Produced by some bacteria, methanethiol (CH_3SH) is the simplest example of a thiol, and dimethyl sulfide [$(CH_3)_2S$] is the simplest example of a sulfide. Both can be described by approximate sp^3 hybridization around sulfur, although both have significant deviation from the 109.5° tetrahedral angle.

ThomsonNOW™ Click *Organic Interactive* to learn how to **identify hybridization in a variety of organic molecules.**

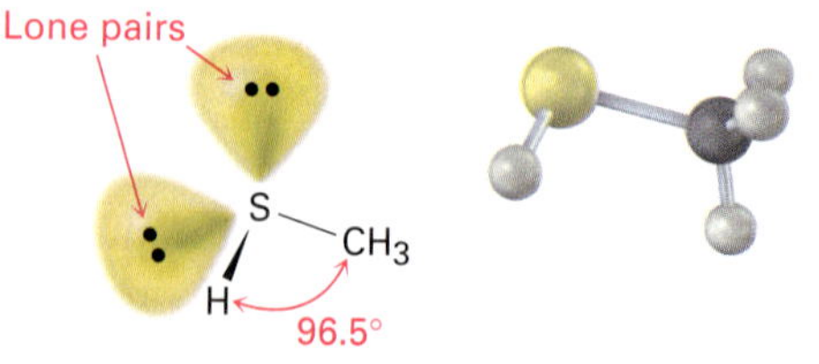

Methanethiol

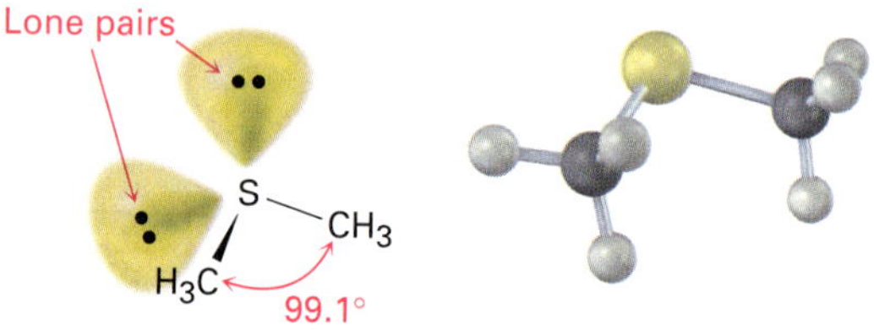

Dimethyl sulfide

Problem 1.14 Identify all nonbonding lone pairs of electrons in the following molecules, and tell what geometry you expect for each of the indicated atoms.

(a) The oxygen atom in dimethyl ether, $CH_3{-}O{-}CH_3$

(b) The nitrogen atom in trimethylamine, $H_3C{-}N{-}CH_3$ (with CH_3 on N)

(c) The phosphorus atom in phosphine, PH_3

(d) The sulfur atom in the amino acid methionine, $CH_3{-}S{-}CH_2CH_2CHCOH$ (with NH_2 on CH and O double-bonded to C)

1.11 The Nature of Chemical Bonds: Molecular Orbital Theory

We said in Section 1.5 that chemists use two models for describing covalent bonds: valence bond theory and molecular orbital theory. Having now seen the valence bond approach, which uses hybrid atomic orbitals to account for geometry and assumes the overlap of atomic orbitals to account for electron sharing, let's look briefly at the molecular orbital approach to bonding. We'll return to the topic in Chapters 14 and 15 for a more in-depth discussion.

Molecular orbital (MO) theory describes covalent bond formation as arising from a mathematical combination of atomic orbitals (wave functions) on different atoms to form *molecular orbitals,* so called because they belong to the entire *molecule* rather than to an individual atom. Just as an *atomic* orbital, whether unhybridized or hybridized, describes a region of space around an *atom* where an electron is likely to be found, so a *molecular* orbital describes a region of space in a *molecule* where electrons are most likely to be found.

Like an atomic orbital, a molecular orbital has a specific size, shape, and energy. In the H_2 molecule, for example, two singly occupied 1*s* atomic orbitals combine to form two molecular orbitals. There are two ways for the orbital combination to occur—an additive way and a subtractive way. The additive combination leads to formation of a molecular orbital that is lower in energy and roughly egg-shaped, while the subtractive combination leads to formation of a molecular orbital that is higher in energy and has a node between nuclei (Figure 1.17). Note that the additive combination is a *single,* egg-shaped, molecular orbital; it is not the same as the two overlapping 1*s* atomic orbitals of the valence bond description. Similarly, the subtractive combination is a single molecular orbital with the shape of an elongated dumbbell.

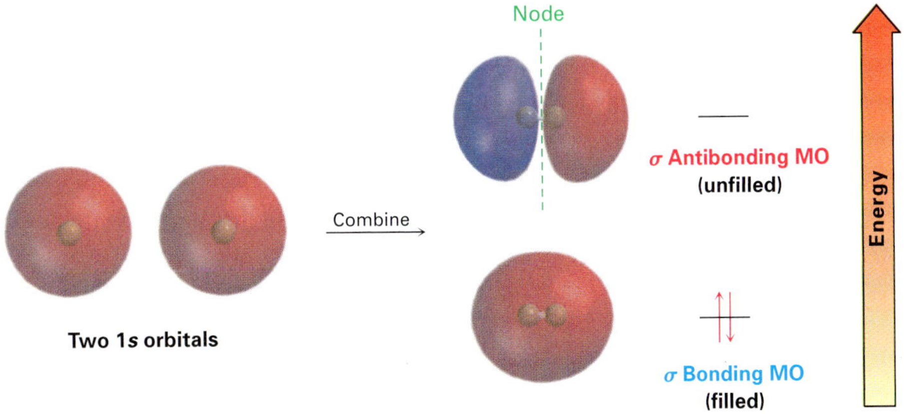

Figure 1.17 Molecular orbitals of H_2. Combination of two hydrogen 1*s* atomic orbitals leads to two H_2 molecular orbitals. The lower-energy, bonding MO is filled, and the higher-energy, antibonding MO is unfilled.

The additive combination is lower in energy than the two hydrogen 1*s* atomic orbitals and is called a **bonding MO** because electrons in this MO spend most of their time in the region between the two nuclei, thereby bonding the atoms together. The subtractive combination is higher in energy than the two hydrogen 1*s* orbitals and is called an **antibonding MO** because any electrons it contains *can't* occupy the central region between the nuclei, where there is a node, and can't contribute to bonding. The two nuclei therefore repel each other.

Just as bonding and antibonding σ molecular orbitals result from the combination of two *s* atomic orbitals in H_2, so bonding and antibonding π molecular orbitals result from the combination of two *p* atomic orbitals in ethylene. As shown in Figure 1.18, the lower-energy, π bonding MO has no node between nuclei and results from combination of *p* orbital lobes with the same algebraic sign. The higher-energy, π antibonding MO has a node between nuclei and results from combination of lobes with opposite algebraic signs. Only the bonding MO is occupied; the higher-energy, antibonding MO is vacant. We'll see in Chapters 14 and 15 that molecular orbital theory is particularly useful for describing π bonds in compounds that have more than one double bond.

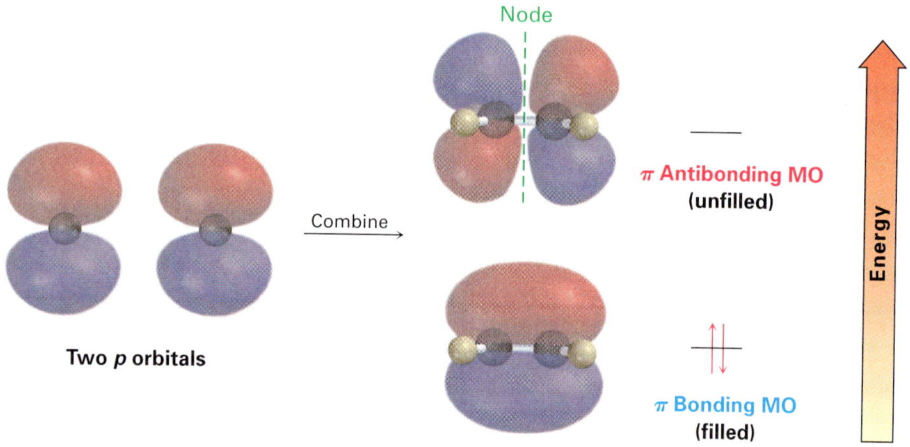

Figure 1.18 A molecular orbital description of the C=C π bond in ethylene. The lower-energy, π bonding MO results from a combination of *p* orbital lobes with the same algebraic sign and is filled. The higher-energy, π antibonding MO results from a combination of *p* orbital lobes with the opposite algebraic signs and is unfilled.

1.12 Drawing Chemical Structures

Let's cover one more point before ending this introductory chapter. In the structures we've been drawing until now, a line between atoms has represented the two electrons in a covalent bond. Drawing every bond and every atom is tedious, however, so chemists have devised several shorthand ways for writing structures. In **condensed structures**, carbon–hydrogen and carbon–carbon single bonds aren't shown; instead, they're understood. If a carbon has three hydrogens bonded to it, we write CH_3; if a carbon has two hydrogens bonded to it, we write CH_2; and so on. The compound called 2-methylbutane, for example, is written as follows:

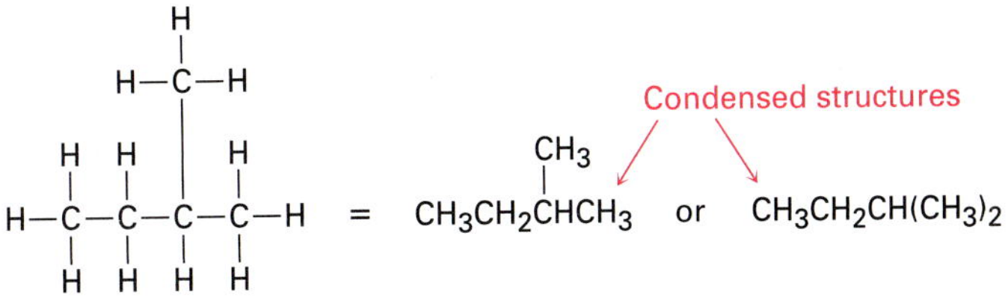

2-Methylbutane

Notice that the horizontal bonds between carbons aren't shown in condensed structures—the CH_3, CH_2, and CH units are simply placed next to each other—but the vertical carbon–carbon bond in the first of the condensed structures drawn above is shown for clarity. Notice also in the second of the condensed structures that the two CH_3 units attached to the CH carbon are grouped together as $(CH_3)_2$.

Even simpler than condensed structures is the use of **skeletal structures** such as those shown in Table 1.3. The rules for drawing skeletal structures are straightforward.

Rule 1 Carbon atoms aren't usually shown. Instead, a carbon atom is assumed to be at each intersection of two lines (bonds) and at the end of each line. Occasionally, a carbon atom might be indicated for emphasis or clarity.

Rule 2 Hydrogen atoms bonded to carbon aren't shown. Since carbon always has a valence of 4, we mentally supply the correct number of hydrogen atoms for each carbon.

Rule 3 Atoms other than carbon and hydrogen *are* shown.

Table 1.3 Kekulé and Skeletal Structures for Some Compounds

Compound	Kekulé structure	Skeletal structure
Isoprene, C_5H_8	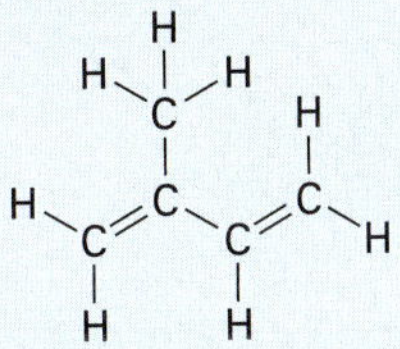	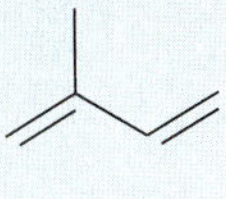
Methylcyclohexane, C_7H_{14}	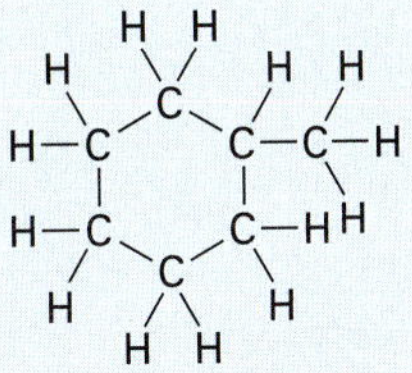	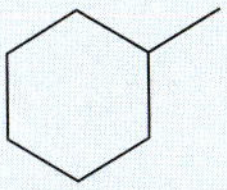
Phenol, C_6H_6O	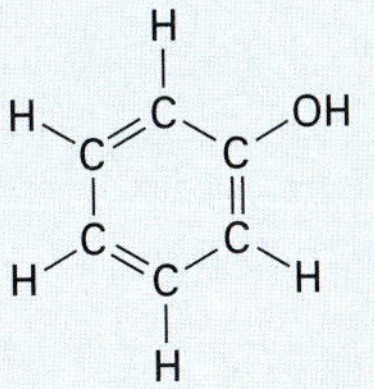	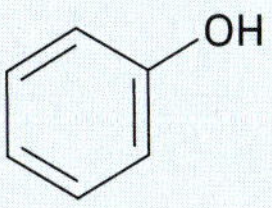

ThomsonNOW™ Click *Organic Interactive* to learn how to **interconvert skeletal structures, condensed structures, and molecular models.**

One further comment: although such groupings as $-CH_3$, $-OH$, and $-NH_2$ are usually written with the C, O, or N atom first and the H atom second, the order of writing is sometimes inverted to H_3C-, $HO-$, and H_2N- if needed to make the bonding connections in a molecule clearer. Larger units such as $-CH_2CH_3$ are not inverted, though; we don't write H_3CH_2C- because it would be confusing. There are, however, no well-defined rules that cover all cases; it's largely a matter of preference.

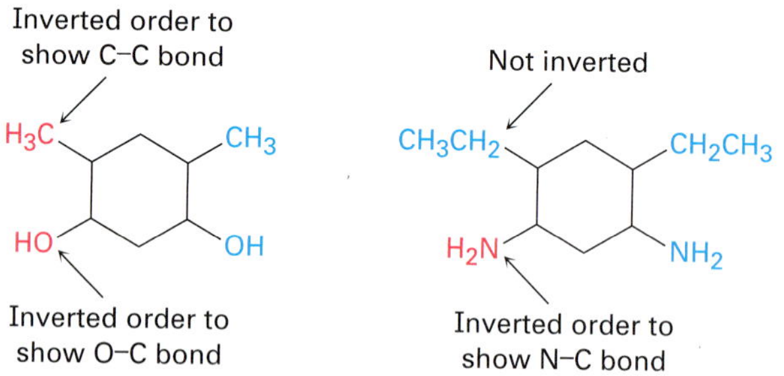

WORKED EXAMPLE 1.3

Interpreting Line-Bond Structures

Carvone, a substance responsible for the odor of spearmint, has the following structure. Tell how many hydrogens are bonded to each carbon, and give the molecular formula of carvone.

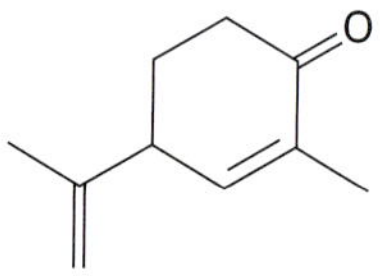

Carvone

Strategy The end of a line represents a carbon atom with 3 hydrogens, CH_3; a two-way intersection is a carbon atom with 2 hydrogens, CH_2; a three-way intersection is a carbon atom with 1 hydrogen, CH; and a four-way intersection is a carbon atom with no attached hydrogens.

Solution

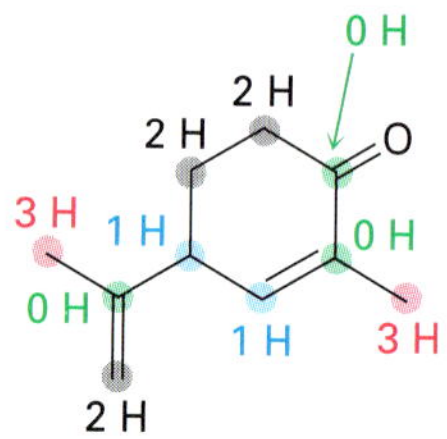

Carvone, $C_{10}H_{14}O$

Problem 1.15 Tell how many hydrogens are bonded to each carbon in the following compounds, and give the molecular formula of each substance:

(a)

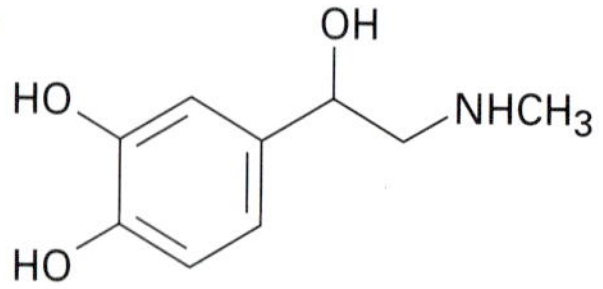

Adrenaline

(b)

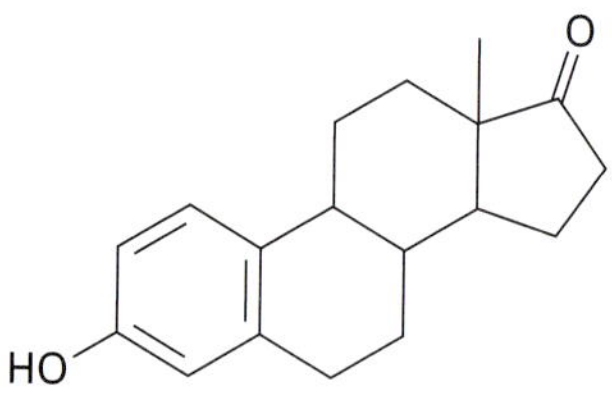

Estrone (a hormone)

Problem 1.16 Propose skeletal structures for compounds that satisfy the following molecular formulas. There is more than one possibility in each case.

(a) C_5H_{12} **(b)** C_2H_7N **(c)** C_3H_6O **(d)** C_4H_9Cl

Problem 1.17 The following molecular model is a representation of *para*-aminobenzoic acid (PABA), the active ingredient in many sunscreens. Indicate the positions of the multiple bonds, and draw a skeletal structure (gray = C, red = O, blue = N, ivory = H).

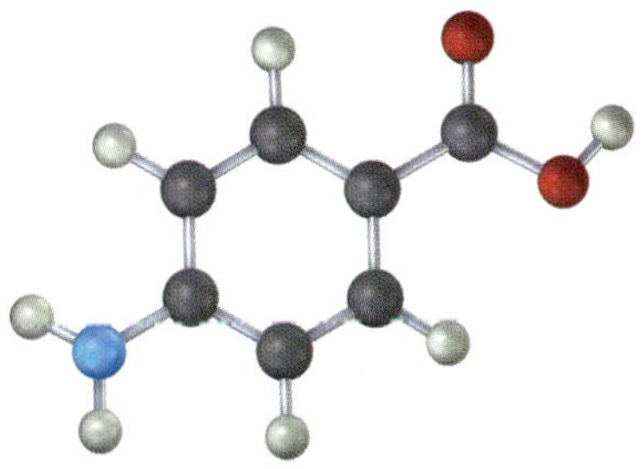

***para*-Aminobenzoic acid (PABA)**

Focus On . . .

Chemicals, Toxicity, and Risk

© Keith Larrett/AP Photo

We all take many risks each day, some much more dangerous than others.

We hear and read a lot these days about the dangers of "chemicals"—about pesticide residues on our food, toxic wastes on our land, unsafe medicines, and so forth. What's a person to believe?

Life is not risk-free; we all take many risks each day. We decide to ride a bike rather than drive, even though there is a ten times greater likelihood per mile of dying in a bicycling accident than in a car. We decide to walk down stairs rather than take an elevator, even though 7000 people die from falls each year in the United States. We decide to smoke cigarettes, even though it increases our chance of getting cancer by 50%. Making decisions that affect our health is something we do routinely without even thinking about it.

What about risks from chemicals? Risk evaluation of chemicals is carried out by exposing test animals (usually rats) to the chemical and then monitoring for signs of harm. To limit the expense and time needed, the amounts administered are hundreds or thousands of times greater than those a person might normally encounter. Data are then reduced to a single number called an LD_{50}, the amount of a substance per kilogram body weight that is lethal to

(continued)

50% of the test animals. The LD_{50}'s of some common substances are shown in Table 1.4. The lower the value, the more toxic the substance.

Table 1.4 Some LD_{50} Values

Substance	LD_{50} (g/kg)	Substance	LD_{50} (g/kg)
Strychnine	0.005	Iron(II) sulfate	1.5
Arsenic trioxide	0.015	Chloroform	3.2
DDT	0.115	Ethyl alcohol	10.6
Aspirin	1.1	Sodium cyclamate	17

Even with animal data available, risk is still hard to assess. If a substance is harmful to animals, is it necessarily harmful to humans? How can a large dose for a small animal be translated into a small dose for a large human? All substances are toxic to some organisms to some extent, and the difference between help and harm is often a matter of degree. Vitamin A, for example, is necessary for vision, yet it can promote cancer at high dosages. Arsenic trioxide is the most classic of poisons, yet recent work has shown it to be effective at inducing remissions in some types of leukemia. Even water can be toxic if drunk in large amounts because it dilutes the salt in body fluids and causes a potentially life-threatening condition called *hyponatremia*. Furthermore, how we evaluate risk is strongly influenced by familiarity. Many foods contain natural ingredients far more toxic than synthetic additives or pesticide residues, but the ingredients are ignored because the foods are familiar.

All decisions involve tradeoffs. Does the benefit of increased food production outweigh possible health risks of a pesticide? Do the beneficial effects of a new drug outweigh a potentially dangerous side effect in a small fraction of users? The answers are rarely obvious, but we should at least try to base our responses on facts.

SUMMARY AND KEY WORDS

antibonding MO, 22
bond angle, 13
bond length, 12
bond strength, 11
bonding MO, 22
condensed structure, 22
covalent bond, 8

Organic chemistry is the study of carbon compounds. Although a division into organic and inorganic chemistry occurred historically, there is no scientific reason for the division.

An atom consists of a positively charged nucleus surrounded by one or more negatively charged electrons. The electronic structure of an atom can be described by a quantum mechanical wave equation, in which electrons are considered to occupy **orbitals** around the nucleus. Different orbitals have different energy levels and different shapes. For example, *s* orbitals are spherical and *p* orbitals are dumbbell-shaped. The **ground-state electron configuration** of an

electron-dot structure, 9
electron shell, 5
ground-state electron configuration, 6
isotope, 4
line-bond structure, 9
lone-pair electrons, 9
molecular orbital (MO) theory, 21
molecule, 8
node, 5
orbital, 4
organic chemistry, 3
pi (π) bond, 16
sigma (σ) bond, 11
skeletal structure, 23
sp hybrid orbital, 17
sp^2 hybrid orbital, 15
sp^3 hybrid orbital, 12
valence bond theory, 11
valence shell, 8

atom can be found by assigning electrons to the proper orbitals, beginning with the lowest-energy ones.

A **covalent bond** is formed when an electron pair is shared between atoms. According to **valence bond theory**, electron sharing occurs by overlap of two atomic orbitals. According to **molecular orbital (MO) theory**, bonds result from the mathematical combination of atomic orbitals to give molecular orbitals, which belong to the entire molecule. Bonds that have a circular cross-section and are formed by head-on interaction are called **sigma (σ) bonds**; bonds formed by sideways interaction of *p* orbitals are called **pi (π) bonds.**

In the valence bond description, carbon uses hybrid orbitals to form bonds in organic molecules. When forming only single bonds with tetrahedral geometry, carbon uses four equivalent **sp^3 hybrid orbitals.** When forming a double bond with planar geometry, carbon uses three equivalent **sp^2 hybrid orbitals** and one unhybridized *p* orbital. When forming a triple bond with linear geometry, carbon uses two equivalent ***sp* hybrid orbitals** and two unhybridized *p* orbitals. Other atoms such as nitrogen, phosphorus, oxygen, and sulfur also use hybrid orbitals to form strong, oriented bonds.

Organic molecules are usually drawn using either condensed structures or skeletal structures. In **condensed structures**, carbon–carbon and carbon–hydrogen bonds aren't shown. In **skeletal structures**, only the bonds and not the atoms are shown. A carbon atom is assumed to be at the ends and at the junctions of lines (bonds), and the correct number of hydrogens is mentally supplied.

Working Problems

There is no surer way to learn organic chemistry than by working problems. Although careful reading and rereading of this text are important, reading alone isn't enough. You must also be able to use the information you've read and be able to apply your knowledge in new situations. Working problems gives you practice at doing this.

Each chapter in this book provides many problems of different sorts. The in-chapter problems are placed for immediate reinforcement of ideas just learned, while end-of-chapter problems provide additional practice and are of several types. They begin with a short section called "Visualizing Chemistry," which helps you "see" the microscopic world of molecules and provides practice for working in three dimensions. After the visualizations are many "Additional Problems." Early problems are primarily of the drill type, providing an opportunity for you to practice your command of the fundamentals. Later problems tend to be more thought-provoking, and some are real challenges.

As you study organic chemistry, take the time to work the problems. Do the ones you can, and ask for help on the ones you can't. If you're stumped by a particular problem, check the accompanying *Study Guide and Solutions Manual* for an explanation that will help clarify the difficulty. Working problems takes effort, but the payoff in knowledge and understanding is immense.

EXERCISES

Organic KNOWLEDGE TOOLS

ThomsonNOW™ Sign in at **www.thomsonedu.com** to assess your knowledge of this chapter's topics by taking a pre-test. The pre-test will link you to interactive organic chemistry resources based on your score in each concept area.

Online homework for this chapter may be assigned in Organic OWL.

■ indicates problems assignable in Organic OWL.

VISUALIZING CHEMISTRY

(Problems 1.1–1.17 appear within the chapter.)

1.18 ■ Convert each of the following molecular models into a skeletal structure, and give the formula of each. Only the connections between atoms are shown; multiple bonds are not indicated (gray = C, red = O, blue = N, ivory = H).

(a)

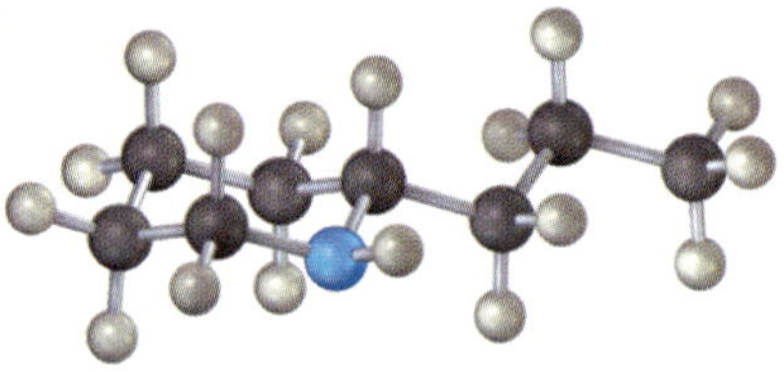

Coniine (the toxic substance in poison hemlock)

(b)

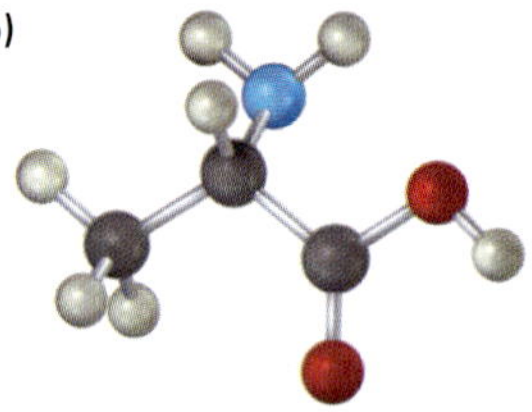

Alanine (an amino acid)

1.19 ■ The following model is a representation of citric acid, the key substance in the so-called citric acid cycle by which food molecules are metabolized in the body. Only the connections between atoms are shown; multiple bonds are not indicated. Complete the structure by indicating the positions of multiple bonds and lone-pair electrons (gray = C, red = O, ivory = H).

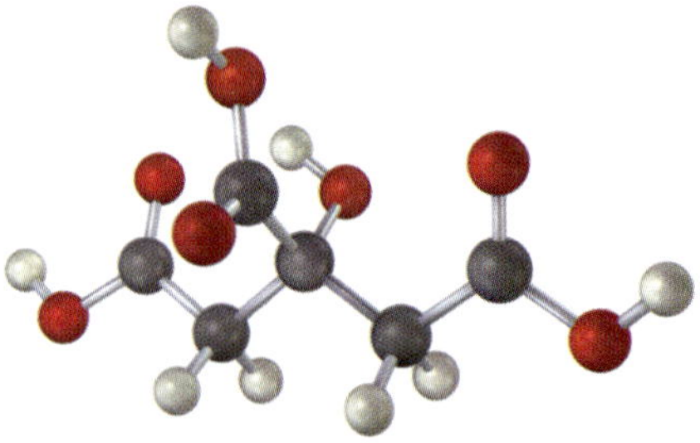

1.20 ■ The following model is a representation of acetaminophen, a pain reliever sold in drugstores as Tylenol. Identify the hybridization of each carbon atom in acetaminophen, and tell which atoms have lone pairs of electrons (gray = C, red = O, blue = N, ivory = H).

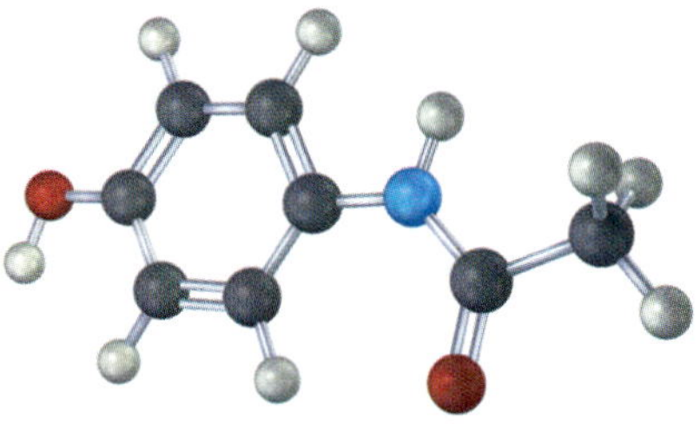

1.21 The following model is a representation of aspartame, $C_{14}H_{18}N_2O_5$, known commercially as NutraSweet. Only the connections between atoms are shown; multiple bonds are not indicated. Complete the structure by indicating the positions of multiple bonds (gray = C, red = O, blue = N, ivory = H).

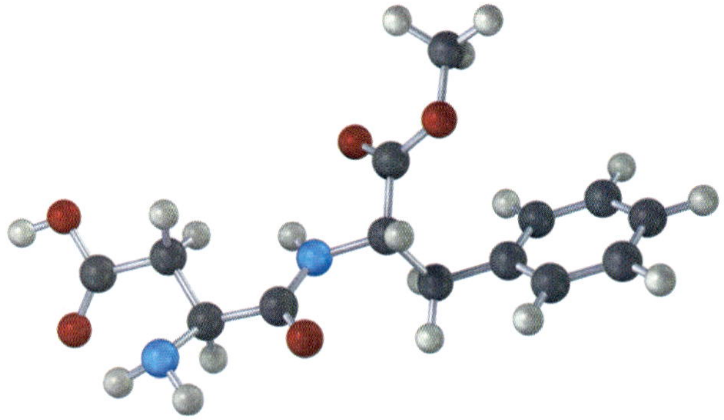

ADDITIONAL PROBLEMS

1.22 ■ How many valence electrons does each of the following dietary trace elements have?
(a) Zinc (b) Iodine (c) Silicon (d) Iron

1.23 ■ Give the ground-state electron configuration for each of the following elements:
(a) Potassium (b) Arsenic (c) Aluminum (d) Germanium

1.24 ■ What are likely formulas for the following molecules?
(a) $NH_?OH$ (b) $AlCl_?$ (c) $CF_2Cl_?$ (d) $CH_?O$

1.25 Draw an electron-dot structure for acetonitrile, C_2H_3N, which contains a carbon–nitrogen triple bond. How many electrons does the nitrogen atom have in its outer shell? How many are bonding, and how many are nonbonding?

1.26 What is the hybridization of each carbon atom in acetonitrile (Problem 1.25)?

1.27 ■ Draw a line-bond structure for vinyl chloride, C_2H_3Cl, the starting material from which PVC [poly(vinyl chloride)] plastic is made.

1.28 ■ Fill in any nonbonding valence electrons that are missing from the following structures:

(a) 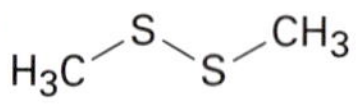

Dimethyl disulfide

(b) 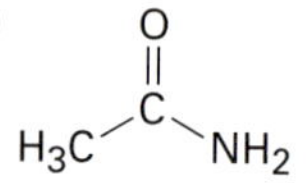

Acetamide

(c) 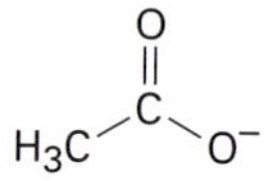

Acetate ion

1.29 Convert the following line-bond structures into molecular formulas:

(a) 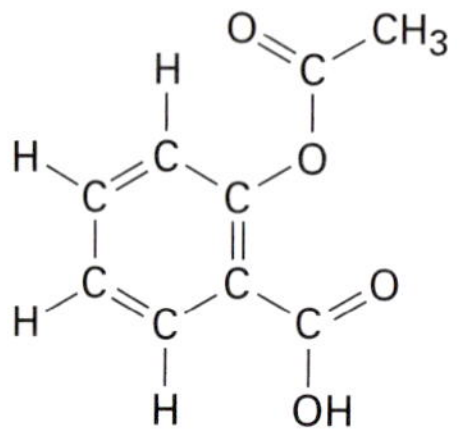

Aspirin (acetylsalicylic acid)

(b) 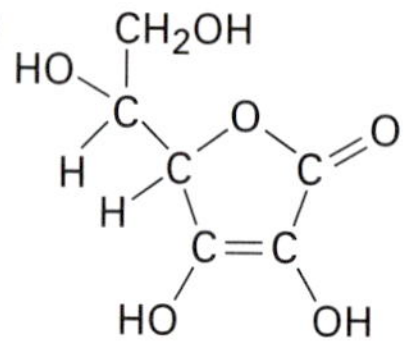

Vitamin C (ascorbic acid)

(c) 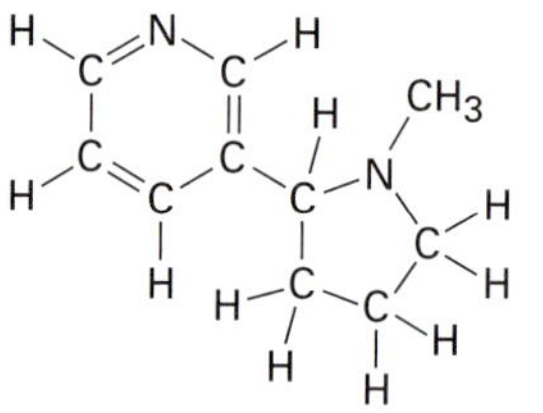

Nicotine

(d) 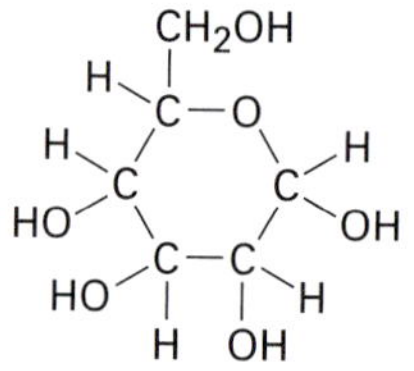

Glucose

1.30 Convert the following molecular formulas into line-bond structures that are consistent with valence rules:

(a) C_3H_8
(b) CH_5N
(c) C_2H_6O (2 possibilities)
(d) C_3H_7Br (2 possibilities)
(e) C_2H_4O (3 possibilities)
(f) C_3H_9N (4 possibilities)

1.31 ■ What kind of hybridization do you expect for each carbon atom in the following molecules?

(a) Propane, $CH_3CH_2CH_3$

(b) 2-Methylpropene, 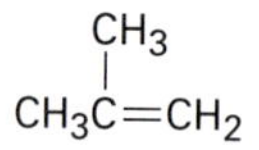

(c) 1-Butene-3-yne, $H_2C{=}CH{-}C{\equiv}CH$

(d) Acetic acid, 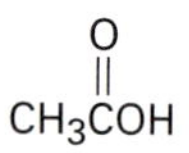

1.32 What is the shape of benzene, and what hybridization do you expect for each carbon?

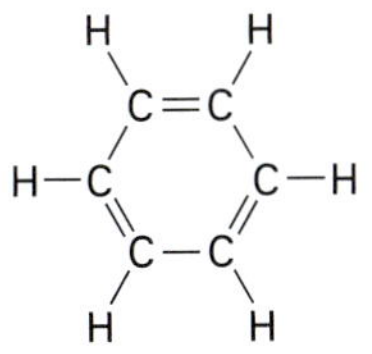

Benzene

1.33 ■ What bond angles do you expect for each of the following, and what kind of hybridization do you expect for the central atom in each?

(a)

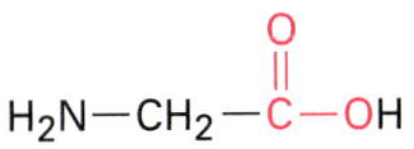

Glycine (an amino acid)

(b)

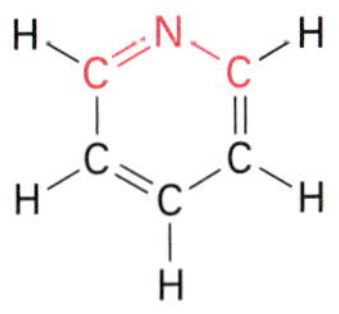

Pyridine

(c)

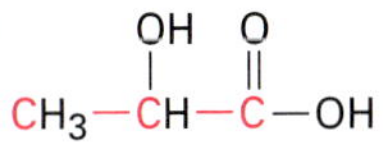

Lactic acid (in sour milk)

1.34 ■ Convert the following structures into skeletal drawings:

(a)

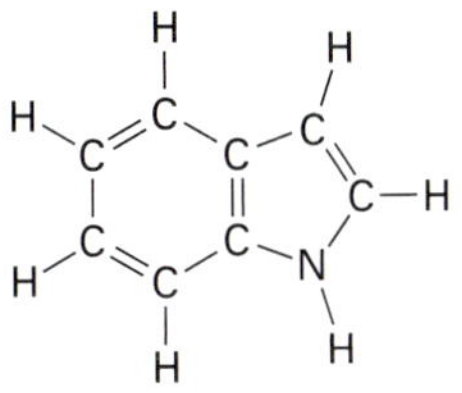

Indole

(b)

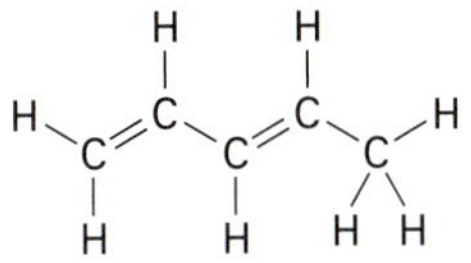

1,3-Pentadiene

(c)

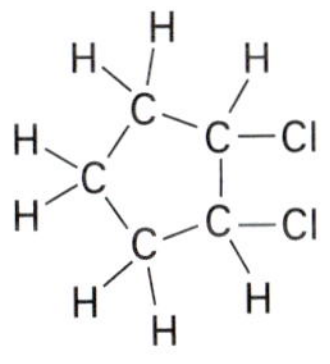

1,2-Dichlorocyclopentane

(d)

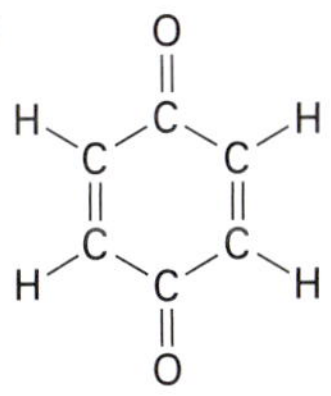

Benzoquinone

1.35 ■ Tell the number of hydrogens bonded to each carbon atom in the following substances, and give the molecular formula of each:

(a)

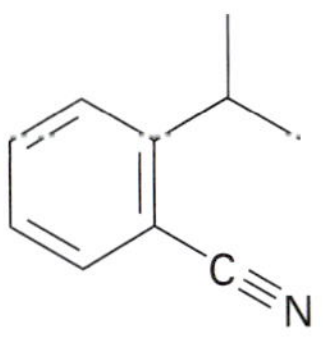

(b)

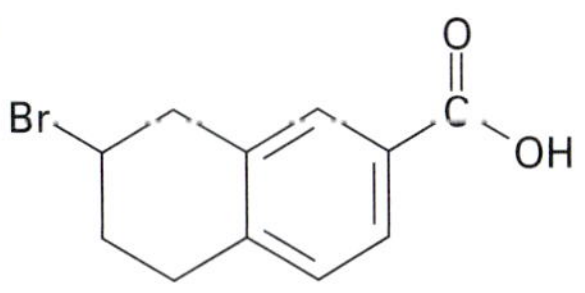

(c)

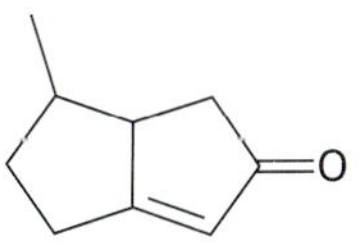

1.36 Propose structures for molecules that meet the following descriptions:
(a) Contains two sp^2-hybridized carbons and two sp^3-hybridized carbons
(b) Contains only four carbons, all of which are sp^2-hybridized
(c) Contains two sp-hybridized carbons and two sp^2-hybridized carbons

1.37 ■ Why can't molecules with the following formulas exist?
(a) CH_5 (b) C_2H_6N (c) $C_3H_5Br_2$

1.38 Draw a three-dimensional representation of the oxygen-bearing carbon atom in ethanol, CH_3CH_2OH, using the standard convention of solid, wedged, and dashed lines.

1.39 Oxaloacetic acid, an important intermediate in food metabolism, has the formula $C_4H_4O_5$ and contains three C=O bonds and two O–H bonds. Propose two possible structures.

1.40 ■ Draw structures for the following molecules, showing lone pairs:
(a) Acrylonitrile, C_3H_3N, which contains a carbon–carbon double bond and a carbon–nitrogen triple bond
(b) Ethyl methyl ether, C_3H_8O, which contains an oxygen atom bonded to two carbons
(c) Butane, C_4H_{10}, which contains a chain of four carbon atoms
(d) Cyclohexene, C_6H_{10}, which contains a ring of six carbon atoms and one carbon–carbon double bond

1.41 Potassium methoxide, $KOCH_3$, contains both covalent and ionic bonds. Which do you think is which?

1.42 What kind of hybridization do you expect for each carbon atom in the following molecules?

(a)

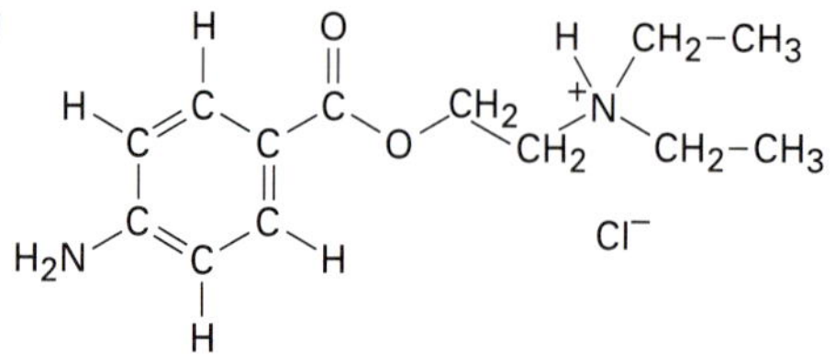

Procaine

(b)

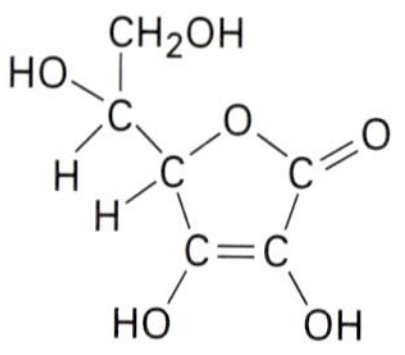

Vitamin C (ascorbic acid)

1.43 Pyridoxal phosphate, a close relative of vitamin B_6, is involved in a large number of metabolic reactions. Tell the hybridization, and predict the bond angles for each nonterminal atom.

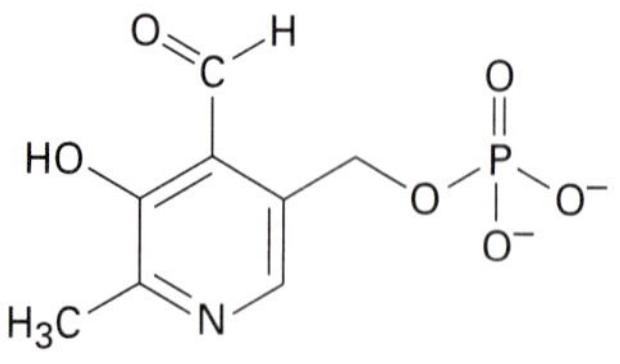

Pyridoxal phosphate

1.44 Why do you suppose no one has ever been able to make cyclopentyne as a stable molecule?

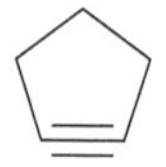

Cyclopentyne

1.45 What is wrong with the following sentence? "The π bonding molecular orbital in ethylene results from sideways overlap of two *p* atomic orbitals."

1.46 Allene, $H_2C{=}C{=}CH_2$, is somewhat unusual in that it has two adjacent double bonds. Draw a picture showing the orbitals involved in the σ and π bonds of allene. Is the central carbon atom sp^2- or *sp*-hybridized? What about the hybridization of the terminal carbons? What shape do you predict for allene?

1.47 Allene (see Problem 1.46) is related structurally to carbon dioxide, CO_2. Draw a picture showing the orbitals involved in the σ and π bonds of CO_2, and identify the likely hybridization of carbon.

1.48 Complete the electron-dot structure of caffeine, showing all lone-pair electrons, and identify the hybridization of the indicated atoms.

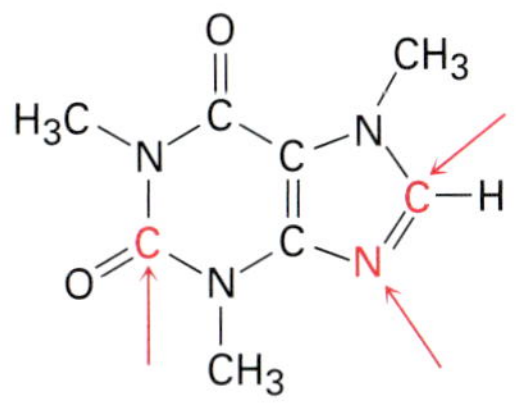

Caffeine

1.49 Almost all stable organic species have tetravalent carbon atoms, but species with trivalent carbon atoms also exist. *Carbocations* are one such class of compounds.

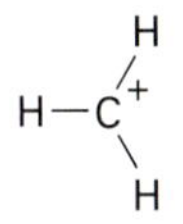

A carbocation

(a) How many valence electrons does the positively charged carbon atom have?
(b) What hybridization do you expect this carbon atom to have?
(c) What geometry is the carbocation likely to have?

1.50 A *carbanion* is a species that contains a negatively charged, trivalent carbon.

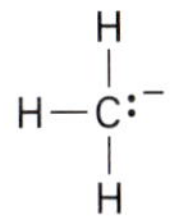

A carbanion

(a) What is the electronic relationship between a carbanion and a trivalent nitrogen compound such as NH_3?
(b) How many valence electrons does the negatively charged carbon atom have?
(c) What hybridization do you expect this carbon atom to have?
(d) What geometry is the carbanion likely to have?

1.51 Divalent carbon species called *carbenes* are capable of fleeting existence. For example, methylene, $:CH_2$, is the simplest carbene. The two unshared electrons in methylene can be either spin-paired in a single orbital or unpaired in different orbitals. Predict the type of hybridization you expect carbon to adopt in singlet (spin-paired) methylene and triplet (spin-unpaired) methylene. Draw a picture of each, and identify the valence orbitals on carbon.

1.52 There are two different substances with the formula C_4H_{10}. Draw both, and tell how they differ.

1.53 There are two different substances with the formula C_3H_6. Draw both, and tell how they differ.

1.54 There are two different substances with the formula C_2H_6O. Draw both, and tell how they differ.

1.55 There are three different substances that contain a carbon–carbon double bond and have the formula C_4H_8. Draw them, and tell how they differ.

1.56 Among the most common over-the-counter drugs you might find in a medicine cabinet are mild pain relievers such ibuprofen (Advil, Motrin), naproxen (Aleve), and acetaminophen (Tylenol).

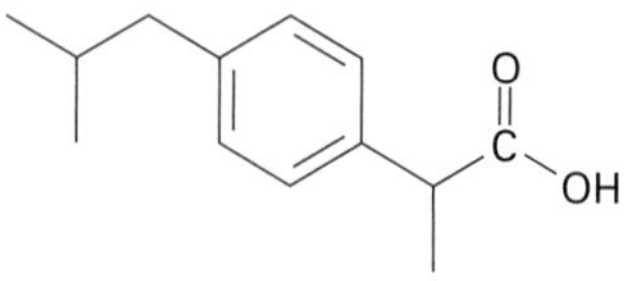

Ibuprofen

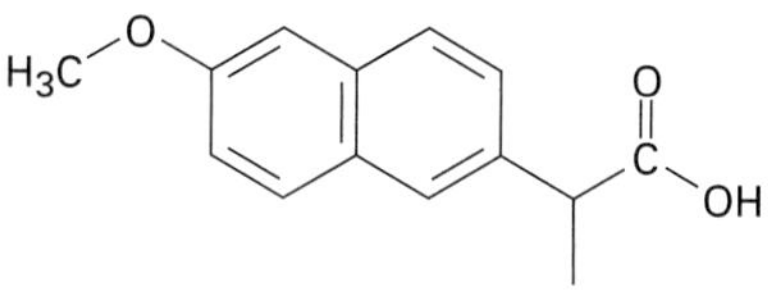

Naproxen

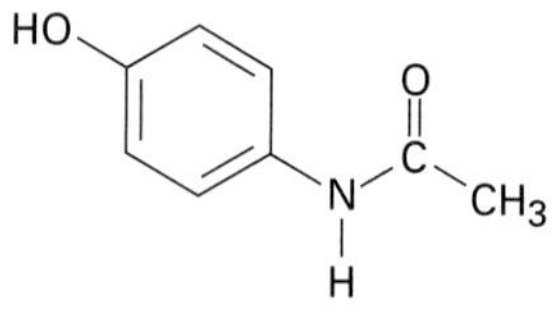

Acetaminophen

(a) How many sp^3-hybridized carbons does each molecule have?
(b) How many sp^2-hybridized carbons does each molecule have?
(c) Can you spot any similarities in their structures?

2 Polar Covalent Bonds; Acids and Bases

Organic KNOWLEDGE TOOLS

ThomsonNOW™ Throughout this chapter, sign in at **www.thomsonedu.com** for online self-study and interactive tutorials based on your level of understanding.

Online homework for this chapter may be assigned in Organic OWL.

We saw in the last chapter how covalent bonds between atoms are described, and we looked at the valence bond model, which uses hybrid orbitals to account for the observed shapes of organic molecules. Before going on to a systematic study of organic chemistry, however, we still need to review a few fundamental topics. In particular, we need to look more closely at how electrons are distributed in covalent bonds and at some of the consequences that arise when the electrons in a bond are not shared equally between atoms.

WHY THIS CHAPTER?

Understanding organic chemistry means knowing not just what happens but also why and how it happens. In this chapter, we'll look some of the basic ways chemists use to describe and account for chemical reactivity, thereby providing a foundation for understanding the specific reactions discussed in subsequent chapters.

2.1 Polar Covalent Bonds: Electronegativity

Up to this point, we've treated chemical bonds as either ionic or covalent. The bond in sodium chloride, for instance, is ionic. Sodium transfers an electron to chlorine to give Na^+ and Cl^- ions, which are held together in the solid by electrostatic attractions. The C–C bond in ethane, however, is covalent. The two bonding electrons are shared equally by the two equivalent carbon atoms, resulting in a symmetrical electron distribution in the bond. Most bonds, however, are neither fully ionic nor fully covalent but are somewhere between the two extremes. Such bonds are called **polar covalent bonds**, meaning that the bonding electrons are attracted more strongly by one atom than the other so that the electron distribution between atoms in not symmetrical (Figure 2.1).

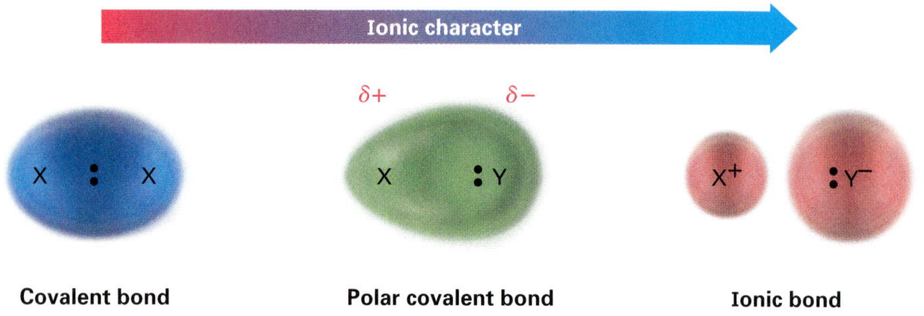

Figure 2.1 The continuum in bonding from covalent to ionic is a result of an unequal distribution of bonding electrons between atoms. The symbol δ (lowercase Greek delta) means *partial* charge, either partial positive ($\delta+$) for the electron-poor atom or partial negative ($\delta-$) for the electron-rich atom.

Bond polarity is due to differences in **electronegativity (EN)**, the intrinsic ability of an atom to attract the shared electrons in a covalent bond. As shown in Figure 2.2, electronegativities are based on an arbitrary scale, with fluorine being the most electronegative (EN = 4.0) and cesium, the least (EN = 0.7). Metals on the left side of the periodic table attract electrons weakly and have lower electronegativities, whereas the halogens and other reactive nonmetals on the right side of the periodic table attract electrons strongly and have higher electronegativities. Carbon, the most important element in organic compounds, has an electronegativity value of 2.5.

Figure 2.2 Electronegativity values and trends. Electronegativity generally increases from left to right across the periodic table and decreases from top to bottom. The values are on an arbitrary scale, with F = 4.0 and Cs = 0.7. Elements in orange are the most electronegative, those in yellow are medium, and those in green are the least electronegative.

H 2.1																	He
Li 1.0	Be 1.6											B 2.0	C 2.5	N 3.0	O 3.5	F 4.0	Ne
Na 0.9	Mg 1.2											Al 1.5	Si 1.8	P 2.1	S 2.5	Cl 3.0	Ar
K 0.8	Ca 1.0	Sc 1.3	Ti 1.5	V 1.6	Cr 1.6	Mn 1.5	Fe 1.8	Co 1.9	Ni 1.9	Cu 1.9	Zn 1.6	Ga 1.6	Ge 1.8	As 2.0	Se 2.4	Br 2.8	Kr
Rb 0.8	Sr 1.0	Y 1.2	Zr 1.4	Nb 1.6	Mo 1.8	Tc 1.9	Ru 2.2	Rh 2.2	Pd 2.2	Ag 1.9	Cd 1.7	In 1.7	Sn 1.8	Sb 1.9	Te 2.1	I 2.5	Xe
Cs 0.7	Ba 0.9	La 1.0	Hf 1.3	Ta 1.5	W 1.7	Re 1.9	Os 2.2	Ir 2.2	Pt 2.2	Au 2.4	Hg 1.9	Tl 1.8	Pb 1.9	Bi 1.9	Po 2.0	At 2.1	Rn

As a loose guide, bonds between atoms whose electronegativities differ by less than 0.5 are nonpolar covalent, bonds between atoms whose electronegativities differ by 0.5 to 2 are polar covalent, and bonds between atoms whose electronegativities differ by more than 2 are largely ionic. Carbon–hydrogen bonds, for example, are relatively nonpolar because carbon (EN = 2.5) and hydrogen (EN = 2.1) have similar electronegativities. Bonds between carbon and *more* electronegative elements such as oxygen (EN = 3.5) and nitrogen (EN = 3.0), by contrast, are polarized so that the bonding electrons are drawn away from carbon toward the electronegative atom. This leaves carbon with a partial positive charge, denoted by $\delta+$, and the electronegative atom with a partial negative charge, $\delta-$. An example is the C–O bond in methanol, CH_3OH (Figure 2.3a). Bonds between carbon and *less* electronegative elements are polarized so that carbon bears a partial negative charge and the other atom bears a partial positive charge. An example is methyllithium, CH_3Li (Figure 2.3b).

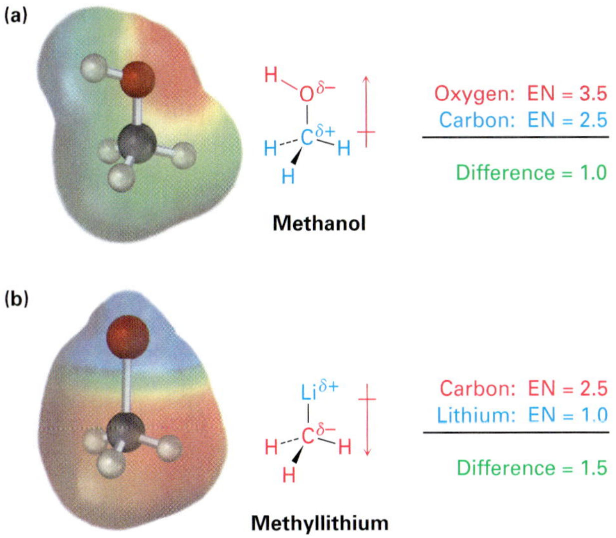

Figure 2.3 **(a)** Methanol, CH_3OH, has a polar covalent C–O bond, and **(b)** methyllithium, CH_3Li, has a polar covalent C–Li bond. The computer-generated representations, called electrostatic potential maps, use color to show calculated charge distributions, ranging from red (electron-rich; $\delta-$) to blue (electron-poor; $\delta+$).

Note in the representations of methanol and methyllithium in Figure 2.3 that a crossed arrow +⟶ is used to indicate the direction of bond polarity. By convention, *electrons are displaced in the direction of the arrow.* The tail of the arrow (which looks like a plus sign) is electron-poor ($\delta+$), and the head of the arrow is electron-rich ($\delta-$).

Note also in Figure 2.3 that calculated charge distributions in molecules can be displayed visually using so-called electrostatic potential maps, which use color to indicate electron-rich (red; $\delta-$) and electron-poor (blue; $\delta+$) regions. In methanol, oxygen carries a partial negative charge and is colored red, while the carbon and hydrogen atoms carry partial positive charges and are colored blue-green. In methyllithium, lithium carries a partial positive charge (blue), while carbon and the hydrogen atoms carry partial negative charges (red). Electrostatic potential maps are useful because they show at a glance the electron-rich and electron-poor atoms in molecules. We'll make frequent use of these maps throughout the text and will see numerous examples of how electronic structure correlates with chemical reactivity.

When speaking of an atom's ability to polarize a bond, we often use the term *inductive effect.* An **inductive effect** is simply the shifting of electrons in a σ bond in response to the electronegativity of nearby atoms. Metals, such as lithium and magnesium, inductively donate electrons, whereas reactive nonmetals, such as oxygen and nitrogen, inductively withdraw electrons. Inductive effects play a major role in understanding chemical reactivity, and we'll use them many times throughout this text to explain a variety of chemical phenomena.

Problem 2.1 Which element in each of the following pairs is more electronegative?

(a) Li or H **(b)** B or Br **(c)** Cl or I **(d)** C or H

Problem 2.2 Use the $\delta+/\delta-$ convention to show the direction of expected polarity for each of the bonds indicated.

(a) $H_3C{-}Cl$ **(b)** $H_3C{-}NH_2$ **(c)** $H_2N{-}H$
(d) $H_3C{-}SH$ **(e)** $H_3C{-}MgBr$ **(f)** $H_3C{-}F$

Problem 2.3 Use the electronegativity values shown in Figure 2.2 to rank the following bonds from least polar to most polar: $H_3C—Li$, $H_3C—K$, $H_3C—F$, $H_3C—MgBr$, $H_3C—OH$.

Problem 2.4 Look at the following electrostatic potential map of chloromethane, and tell the direction of polarization of the C–Cl bond:

Chloromethane

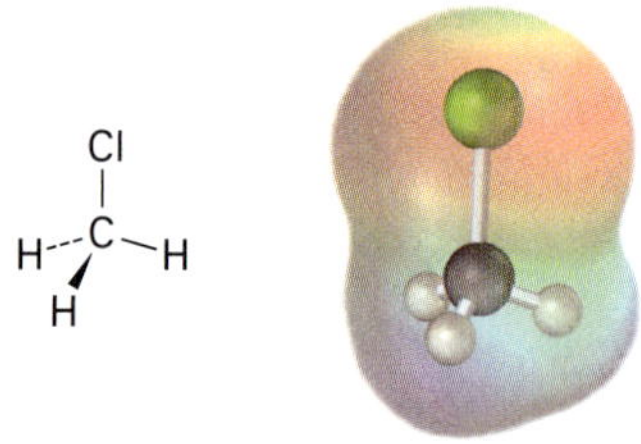

2.2 Polar Covalent Bonds: Dipole Moments

Just as individual bonds are often polar, molecules as a whole are often polar also. Molecular polarity results from the vector summation of all individual bond polarities and lone-pair contributions in the molecule. As a practical matter, strongly polar substances are often soluble in polar solvents like water, whereas nonpolar substances are insoluble in water.

Net molecular polarity is measured by a quantity called the *dipole moment* and can be thought of in the following way: assume that there is a center of mass of all positive charges (nuclei) in a molecule and a center of mass of all negative charges (electrons). If these two centers don't coincide, then the molecule has a net polarity.

The **dipole moment**, μ (Greek mu), is defined as the magnitude of the charge Q at either end of the molecular dipole times the distance *r* between the charges, $\mu = Q \times r$. Dipole moments are expressed in *debyes* (D), where $1\ D = 3.336 \times 10^{-30}$ coulomb meter (C · m) in SI units. For example, the unit charge on an electron is 1.60×10^{-19} C. Thus, if one positive charge and one negative charge were separated by 100 pm (a bit less than the length of a typical covalent bond), the dipole moment would be 1.60×10^{-29} C · m, or 4.80 D.

$$\mu = Q \times r$$

$$\mu = (1.60 \times 10^{-19}\ \text{C})(100 \times 10^{-12}\ \text{m})\left(\frac{1\ \text{D}}{3.336 \times 10^{-30}\ \text{C} \cdot \text{m}}\right) = 4.80\ \text{D}$$

It's relatively easy to measure dipole moments in the laboratory, and values for some common substances are given in Table 2.1. Of the compounds shown in the table, sodium chloride has the largest dipole moment (9.00 D) because it is ionic. Even small molecules like water ($\mu = 1.85$ D), methanol (CH_3OH; $\mu = 1.70$ D), and ammonia ($\mu = 1.47$ D) have substantial dipole moments, however, both because they contain strongly electronegative atoms (oxygen and nitrogen) and because all three molecules have lone-pair electrons. The lone-pair electrons on oxygen and nitrogen atom stick out into space away from the positively charged nuclei, giving rise to a considerable charge separation and making a large contribution to the dipole moment.

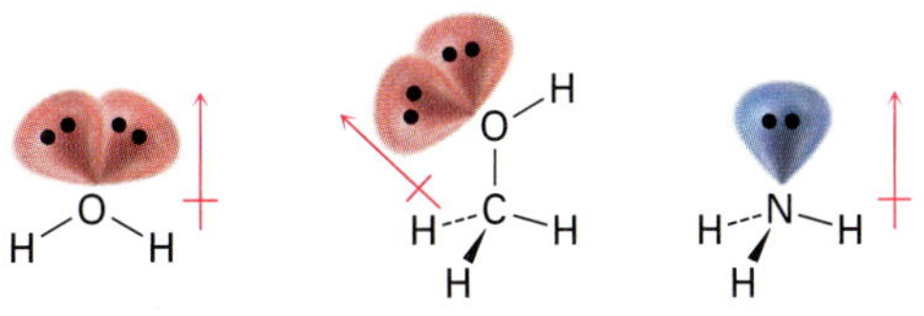

Water (μ = 1.85 D) Methanol (μ = 1.70 D) Ammonia (μ = 1.47 D)

Table 2.1 Dipole Moments of Some Compounds

Compound	Dipole moment (D)	Compound	Dipole moment (D)
NaCl	9.00	NH_3	1.47
CH_2O	2.33	CH_3NH_2	1.31
CH_3Cl	1.87	CO_2	0
H_2O	1.85	CH_4	0
CH_3OH	1.70	CH_3CH_3	0
CH_3CO_2H	1.70	 Benzene	0
CH_3SH	1.52		

In contrast with water, methanol, ammonia, and other substances in Table 2.1, carbon dioxide, methane, ethane, and benzene have zero dipole moments. Because of the symmetrical structures of these molecules, the individual bond polarities and lone-pair contributions exactly cancel.

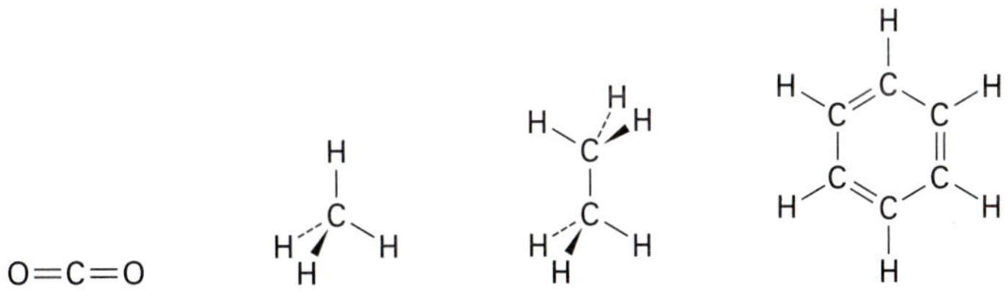

Carbon dioxide (μ = 0) Methane (μ = 0) Ethane (μ = 0) Benzene (μ = 0)

WORKED EXAMPLE 2.1

Predicting the Direction of a Dipole Moment

Make a three-dimensional drawing of methylamine, CH_3NH_2, a substance responsible for the odor of rotting fish, and show the direction of its dipole moment (μ = 1.31).

Strategy Look for any lone-pair electrons, and identify any atom with an electronegativity substantially different from that of carbon. (Usually, this means O, N, F, Cl, or Br.) Electron density will be displaced in the general direction of the electronegative atoms and the lone pairs.

Solution Methylamine contains an electronegative nitrogen atom with two lone-pair electrons. The dipole moment thus points generally from $-CH_3$ toward $-NH_2$.

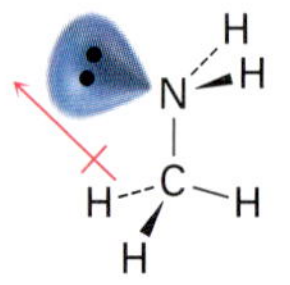

Methylamine
(μ = 1.31)

Problem 2.5 Ethylene glycol, $HOCH_2CH_2OH$, has zero dipole moment even though carbon–oxygen bonds are strongly polarized. Explain.

Problem 2.6 Make three-dimensional drawings of the following molecules, and predict whether each has a dipole moment. If you expect a dipole moment, show its direction.
(a) $H_2C{=}CH_2$ **(b)** $CHCl_3$ **(c)** CH_2Cl_2 **(d)** $H_2C{=}CCl_2$

2.3 Formal Charges

Closely related to the ideas of bond polarity and dipole moment is the concept of assigning *formal charges* to specific atoms within a molecule, particularly atoms that have an apparently "abnormal" number of bonds. Look at dimethyl sulfoxide (CH_3SOCH_3), for instance, a solvent commonly used for preserving biological cell lines at low temperatures. The sulfur atom in dimethyl sulfoxide has three bonds rather than the usual two and has a formal positive charge. The oxygen atom, by contrast, has one bond rather than the usual two and has a formal negative charge. Note that an electrostatic potential map of dimethyl sulfoxide shows the oxygen as negative (red) and the sulfur as relatively positive (blue), in accord with the formal charges.

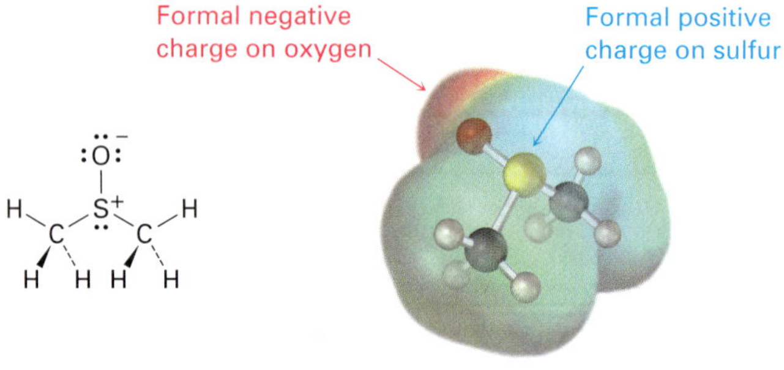

Dimethyl sulfoxide

Formal charges, as the name suggests, are a formalism and don't imply the presence of actual ionic charges in a molecule. Instead, they're a device for electron "bookkeeping" and can be thought of in the following way: a typical covalent bond is formed when each atom donates one electron. Although the bonding electrons are shared by both atoms, each atom can still be considered to "own" one electron for bookkeeping purposes. In methane, for instance, the carbon atom owns one electron in each of the four C–H bonds, for a total of four. Because a neutral, isolated carbon atom has four valence electrons, and because the carbon atom in methane still owns four, the methane carbon atom is neutral and has no formal charge.

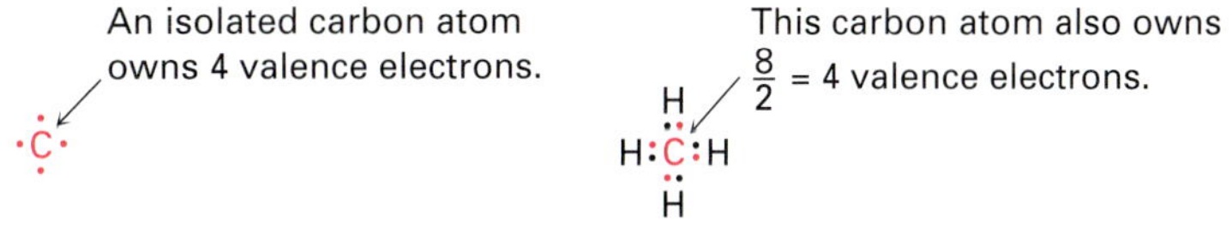

The same is true for the nitrogen atom in ammonia, which has three covalent N–H bonds and two nonbonding electrons (a lone pair). Atomic nitrogen has five valence electrons, and the ammonia nitrogen also has five—one in each of three shared N–H bonds plus two in the lone pair. Thus, the nitrogen atom in ammonia has no formal charge.

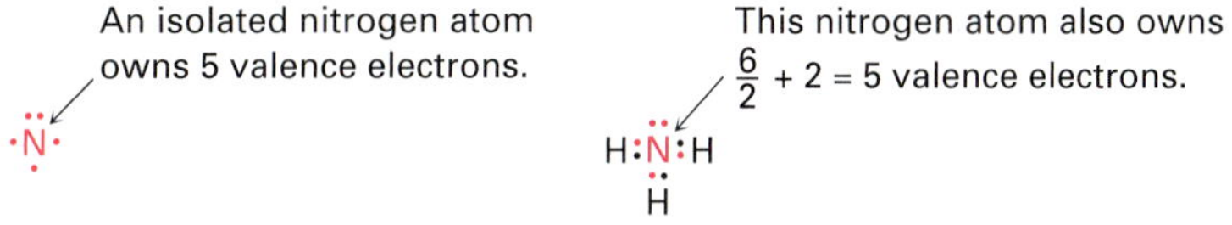

The situation is different in dimethyl sulfoxide. Atomic sulfur has six valence electrons, but the dimethyl sulfoxide sulfur owns only *five*—one in each of the two S–C single bonds, one in the S–O single bond, and two in a lone pair. Thus, the sulfur atom has formally lost an electron and therefore has a positive charge. A similar calculation for the oxygen atom shows that it has formally gained an electron and has a negative charge. Atomic oxygen has six valence electrons, but the oxygen in dimethyl sulfoxide has seven—one in the O–S bond and two in each of three lone pairs.

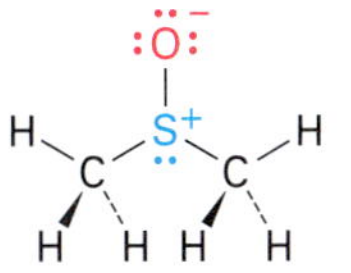

For sulfur:

Sulfur valence electrons = 6
Sulfur bonding electrons = 6
Sulfur nonbonding electrons = 2

Formal charge = 6 − 6/2 − 2 = +1

For oxygen:

Oxygen valence electrons = 6
Oxygen bonding electrons = 2
Oxygen nonbonding electrons = 6

Formal charge = 6 − 2/2 − 6 = −1

To express the calculations in a general way, the **formal charge** on an atom is equal to the number of valence electrons in a neutral, isolated atom minus the number of electrons owned by that atom in a molecule. The number of electrons in the bonded atom, in turn, is equal to half the number of bonding electrons plus the nonbonding, lone-pair electrons.

ThomsonNOW™ Click *Organic Interactive* to learn how to **calculate formal charges in organic molecules.**

$$\textbf{Formal charge} = \begin{pmatrix}\text{Number of}\\ \text{valence electrons}\\ \text{in free atom}\end{pmatrix} - \begin{pmatrix}\text{Number of}\\ \text{valence electrons}\\ \text{in bonded atom}\end{pmatrix}$$

$$= \begin{pmatrix}\text{Number of}\\ \text{valence electrons}\\ \text{in free atom}\end{pmatrix} - \left(\frac{\text{Number of bonding electrons}}{2}\right) - \begin{pmatrix}\text{Number of}\\ \text{nonbonding}\\ \text{electrons}\end{pmatrix}$$

A summary of commonly encountered formal charges and the bonding situations in which they occur is given in Table 2.2. Although only a bookkeeping device, formal charges often give clues about chemical reactivity, so it's helpful to be able to identify and calculate them correctly.

Table 2.2 A Summary of Common Formal Charges

Atom	C			N		O		S		P
Structure	—Ċ— (3 bonds, 1 nonbonding e)	—C⁺—	—C̈⁻—	—N⁺— (4 bonds)	—N̈⁻—	—Ö⁺—	—Ö:⁻	—S̈⁺—	—S̈:⁻	—P⁺— (4 bonds)
Valence electrons	4	4	4	5	5	6	6	6	6	5
Number of bonds	3	3	3	4	2	3	1	3	1	4
Number of nonbonding electrons	1	0	2	0	4	2	6	2	6	0
Formal charge	0	+1	−1	+1	−1	+1	−1	+1	−1	+1

Problem 2.7 Nitromethane has the structure indicated. Explain why it must have formal charges on N and O.

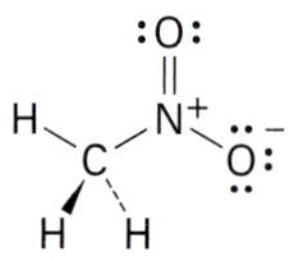

Nitromethane

Problem 2.8 Calculate formal charges for the nonhydrogen atoms in the following molecules:

(a) Diazomethane, $H_2C{=}N{=}\ddot{N}{:}$

(b) Acetonitrile oxide, $H_3C{-}C{\equiv}N{-}\ddot{\underset{\cdot\cdot}{O}}{:}$

(c) Methyl isocyanide, $H_3C{-}N{\equiv}C{:}$

Problem 2.9 Organic phosphate groups occur commonly in biological molecules. Calculate formal charges on the four O atoms in the methyl phosphate dianion.

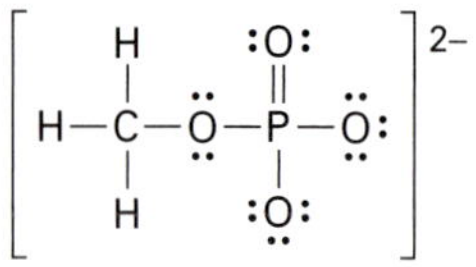

Methyl phosphate

2.4 Resonance

Most substances can be represented without difficulty by the Kekulé line-bond structures we've been using up to this point, but an interesting problem sometimes arises. Look at the acetate ion, for instance. When we draw a line-bond structure for acetate, we need to show a double bond to one oxygen and a single bond to the other. But which oxygen is which? Should we draw a double bond to the "top" oxygen and a single bond to the "bottom" oxygen or vice versa?

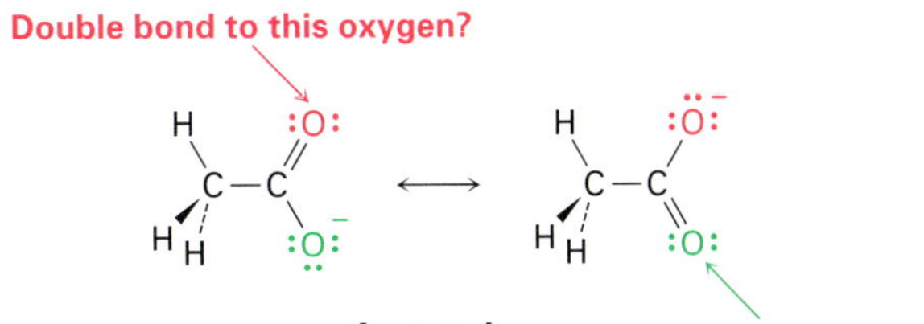

Acetate ion

Although the two oxygen atoms in the acetate ion appear different in line-bond structures, experiments show that they are equivalent. Both carbon–oxygen bonds, for example, are 127 pm in length, midway between the length of a typical C−O bond (135 pm) and a typical C=O bond (120 pm). In other words, *neither* of the two structures for acetate is correct by itself. The true structure is intermediate between the two, and an electrostatic potential map shows that both oxygen atoms share the negative charge and have equal electron densities (red).

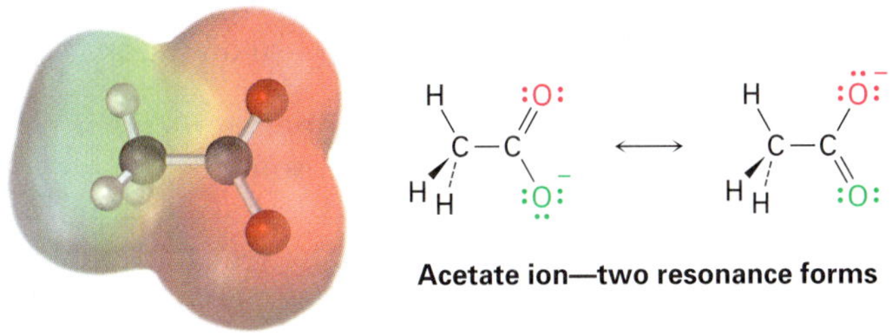

Acetate ion—two resonance forms

The two individual line-bond structures for acetate are called **resonance forms**, and their special resonance relationship is indicated by the double-headed arrow between them. *The only difference between resonance forms is the placement of the π and nonbonding valence electrons.* The atoms themselves occupy exactly the same place in both resonance forms, the connections between atoms are the same, and the three-dimensional shapes of the resonance forms are the same.

A good way to think about resonance forms is to realize that a substance like the acetate ion is no different from any other. Acetate doesn't jump back and forth between two resonance forms, spending part of the time looking like one and part of the time looking like the other. Rather, acetate has a single unchanging structure that is a **resonance hybrid** of the two individual forms and has characteristics of both. The only "problem" with acetate is that we can't draw it accurately using a familiar line-bond structure. Line-bond structures just don't work well for resonance hybrids. The difficulty, however, lies with the *representation* of acetate on paper, not with acetate itself.

Resonance is an extremely useful concept that we'll return to on numerous occasions throughout the rest of this book. We'll see in Chapter 15, for instance, that the six carbon–carbon bonds in so-called *aromatic* compounds, such as benzene, are equivalent and that benzene is best represented as a hybrid of two resonance forms. Although an individual resonance form seems to imply that benzene has alternating single and double bonds, neither form is correct by itself. The true benzene structure is a hybrid of the two individual forms, and all six carbon–carbon bonds are equivalent. This symmetrical distribution of electrons around the molecule is evident in an electrostatic potential map.

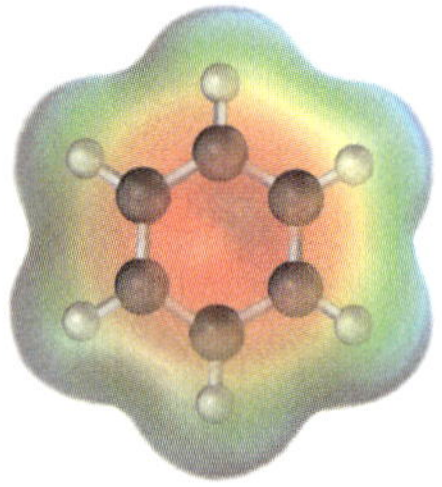

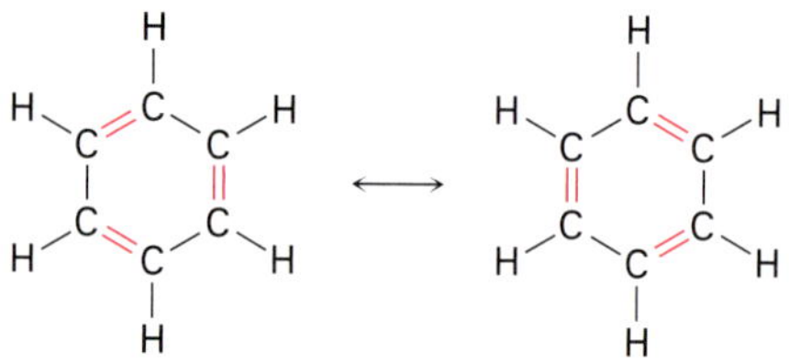

Benzene (two resonance forms)

2.5 Rules for Resonance Forms

When first dealing with resonance forms, it's useful to have a set of guidelines that describe how to draw and interpret them.

Rule 1 **Individual resonance forms are imaginary, not real.** The real structure is a composite, or resonance hybrid, of the different forms. Species such as the acetate ion and benzene are no different from any other. They have single, unchanging structures, and they do not switch back and forth between resonance forms. The only difference between these and other substances is in the way they must be represented on paper.

Rule 2 **Resonance forms differ only in the placement of their π or nonbonding electrons.** Neither the position nor the hybridization of any atom changes from one resonance form to another. In the acetate ion, for example, the carbon atom is sp^2-hybridized and the oxygen atoms remain in exactly the same place in both resonance forms. Only the positions of the π electrons in the C=O bond and the lone-pair electrons on oxygen differ from one form to another. This movement of electrons from one resonance structure to another can be indicated by using curved arrows. *A curved arrow always indicates the movement of electrons, not the movement of atoms.* An arrow shows that a pair of electrons moves *from* the atom or bond at the tail of the arrow *to* the atom or bond at the head of the arrow.

Key **IDEAS**

Test your knowledge of Key Ideas by using resources in ThomsonNOW or by answering end-of-chapter problems marked with ▲.

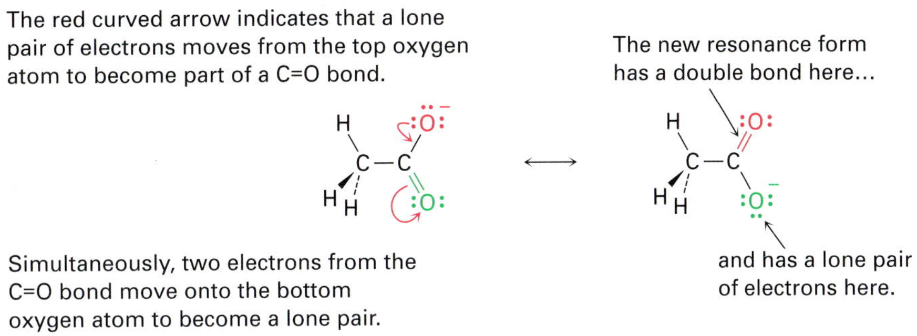

The situation with benzene is similar to that with acetate. The π electrons in the double bonds move, as shown with curved arrows, but the carbon and hydrogen atoms remain in place.

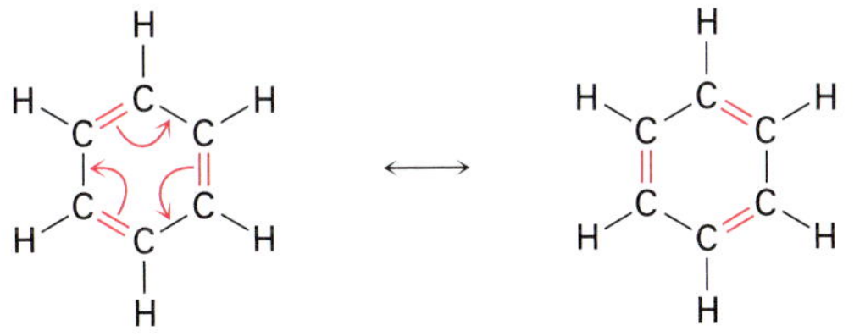

Rule 3 **Different resonance forms of a substance don't have to be equivalent.** For example, we'll see in Chapter 22 that a compound such as acetone, which contains a C=O bond, can be converted into its anion by reaction with a strong base. The resultant anion has two resonance forms. One form contains a carbon–*oxygen* double bond and has a negative charge on *carbon;* the other contains a carbon–*carbon* double bond and has a negative charge on *oxygen.* Even though the two resonance forms aren't equivalent, both contribute to the overall resonance hybrid.

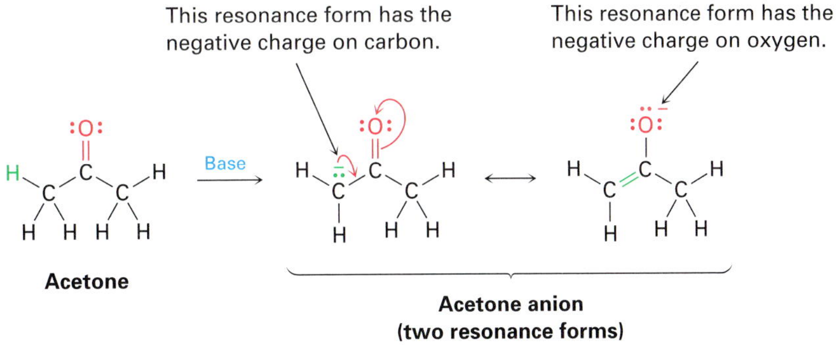

When two resonance forms are nonequivalent, the actual structure of the resonance hybrid is closer to the more stable form than to the less stable form. Thus, we might expect the true structure of the acetone anion to be closer to the resonance form that places the negative charge on an electronegative oxygen atom than to the form that places the charge on a carbon atom.

Rule 4 **Resonance forms obey normal rules of valency.** A resonance form is like any other structure: the octet rule still applies to main-group atoms. For example, one of the following structures for the acetate ion is not a valid resonance form because the carbon atom has five bonds and ten valence electrons:

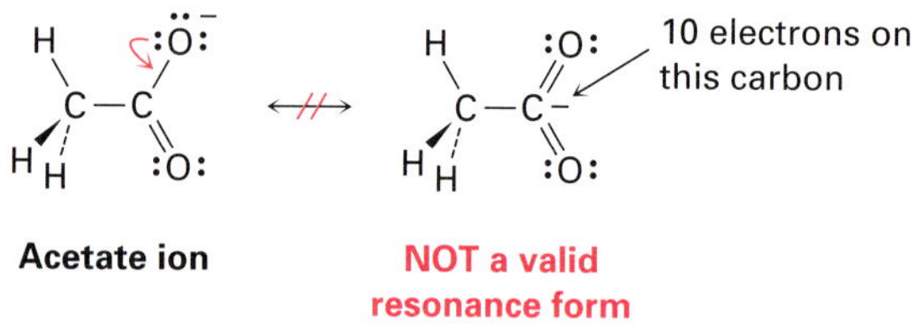

Rule 5 **The resonance hybrid is more stable than any individual resonance form.** In other words, resonance leads to stability. Generally speaking, the larger the number of resonance forms, the more stable a substance is because electrons are spread out over a larger part of the molecule and are closer to more nuclei. We'll see in Chapter 15, for instance, that a benzene ring is more stable because of resonance than might otherwise be expected.

2.6 Drawing Resonance Forms

ThomsonNOW™ Click *Organic Interactive* to **use an online palette to practice drawing resonance forms.**

Look back at the resonance forms of the acetate ion and the acetone anion shown in the previous section. The pattern seen there is a common one that leads to a useful technique for drawing resonance forms. In general, *any three-atom grouping with a p orbital on each atom has two resonance forms.*

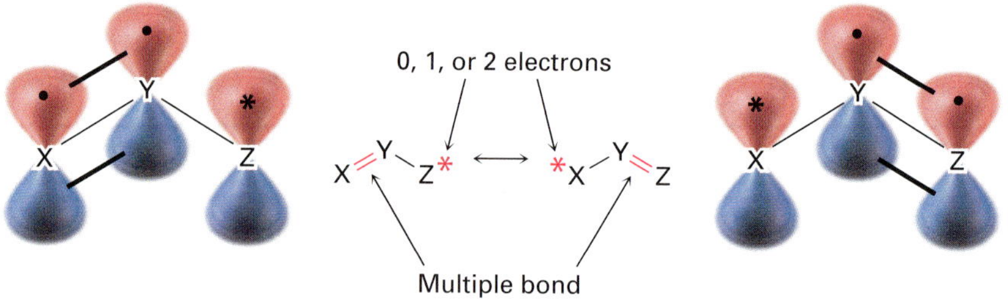

The atoms X, Y, and Z in the general structure might be C, N, O, P, or S, and the asterisk (*) might mean that the *p* orbital on atom Z is vacant, that it contains a single electron, or that it contains a lone pair of electrons. The two resonance forms differ simply by an exchange in position of the multiple bond and the asterisk from one end to the other.

By learning to recognize such three-atom groupings within larger structures, resonance forms can be systematically generated. Look, for instance, at the anion produced when H^+ is removed from 2,4-pentanedione by reaction with a base. How many resonance structures does the resultant anion have?

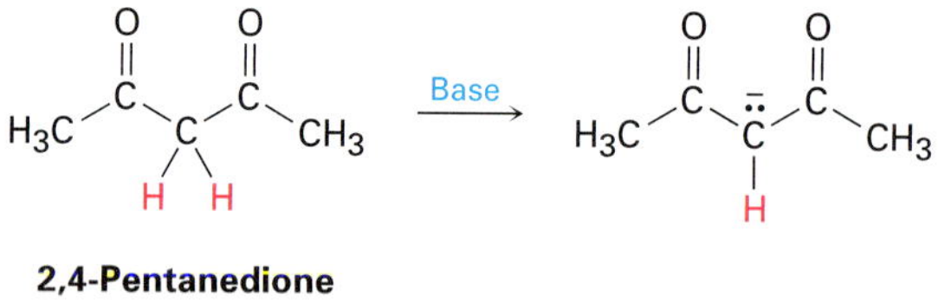

The 2,4-pentanedione anion has a lone pair of electrons and a formal negative charge on the central carbon atom, next to a C=O bond on the left. The O=C–C:⁻ grouping is a typical one for which two resonance structures can be drawn.

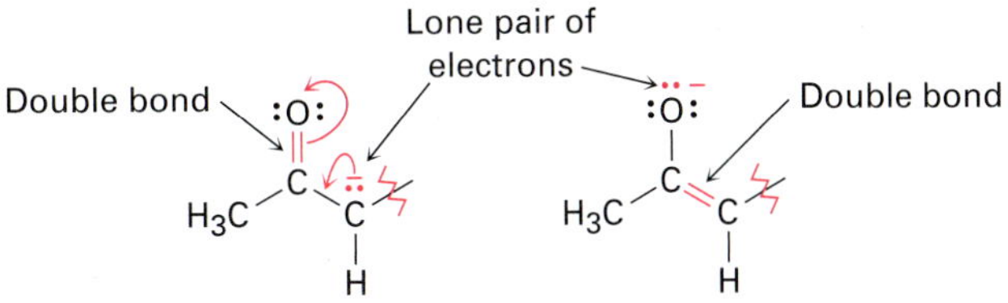

Just as there is a C=O bond to the left of the lone pair, there is a second C=O bond to the right. Thus, we can draw a total of three resonance structures for the 2,4-pentanedione anion.

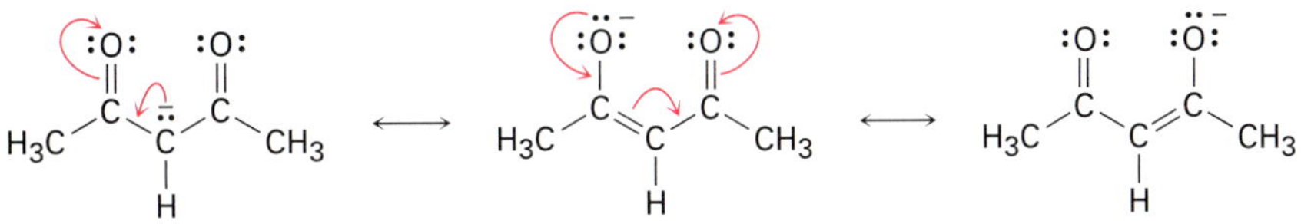

WORKED EXAMPLE 2.2

Drawing Resonance Forms for an Anion

Draw three resonance forms for the carbonate ion, CO_3^{2-}.

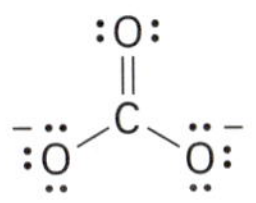

Strategy Look for one or more three-atom groupings that contain a multiple bond next to an atom with a *p* orbital. Then exchange the positions of the multiple bond and the electrons in the *p* orbital. In the carbonate ion, each of the singly bonded oxygen atoms with its lone pairs and negative charge is next to the C=O bond, giving the grouping O=C–O:⁻.

Solution Exchanging the position of the double bond and an electron lone pair in each grouping generates three resonance structures.

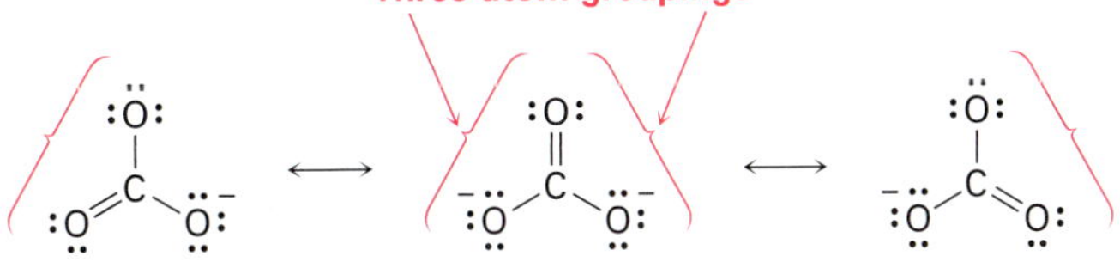

WORKED EXAMPLE 2.3 *Drawing Resonance Forms for a Radical*

Draw three resonance forms for the pentadienyl radical. A *radical* is a substance that contains a single, unpaired electron in one of its orbitals, denoted by a dot (·).

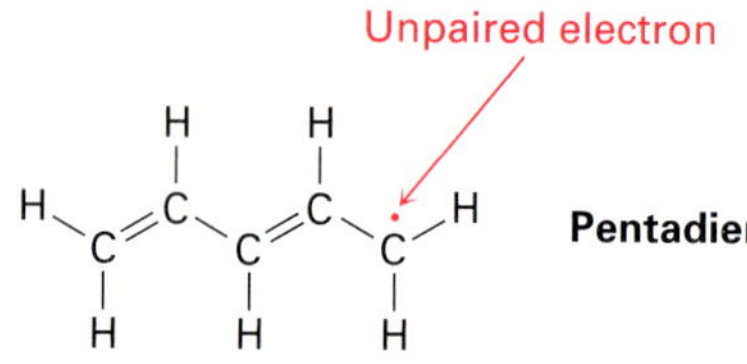

Pentadienyl radical

Strategy Find the three-atom groupings that contain a multiple bond next to a *p* orbital.

Solution The unpaired electron is on a carbon atom next to a C=C bond, giving a typical three-atom grouping that has two resonance forms.

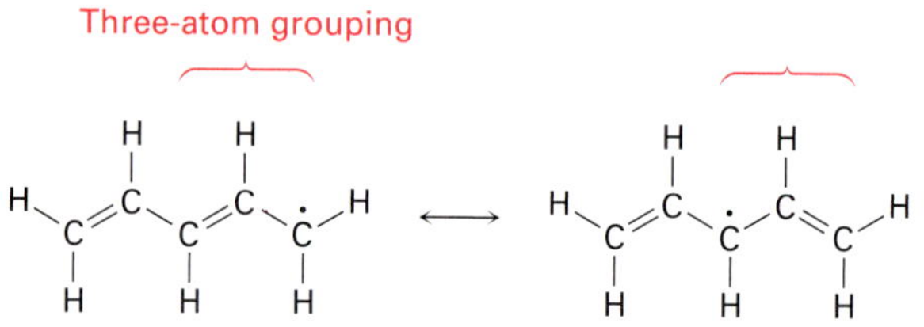

In the second resonance form, the unpaired electron is next to another double bond, giving another three-atom grouping and leading to another resonance form.

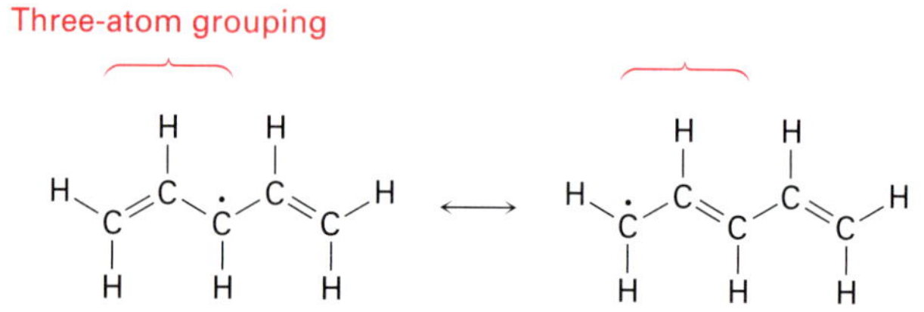

Thus, the three resonance forms for the pentadienyl radical are:

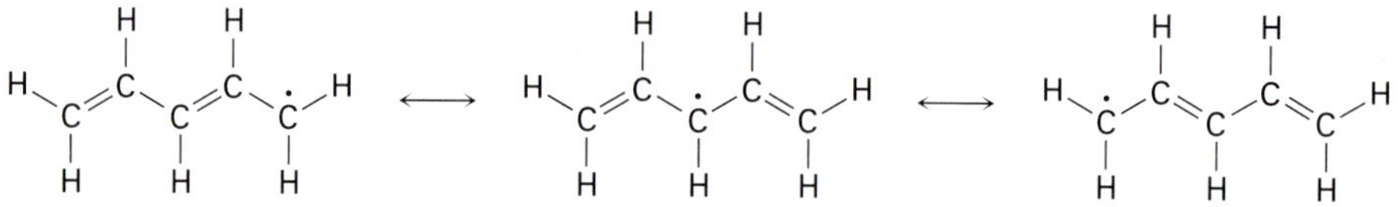

Problem 2.10 Draw the indicated number of resonance forms for each of the following species:

(a) The methyl phosphate anion, $CH_3OPO_3^{2-}$ (3)

(b) The nitrate anion, NO_3^- (3)

(c) The allyl cation, $H_2C{=}CH{-}CH_2^+$ (2)

(d) The benzoate anion (4)

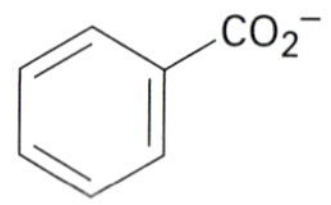

2.7 Acids and Bases: The Brønsted–Lowry Definition

A further important concept related to electronegativity and polarity is that of *acidity* and *basicity*. We'll see, in fact, that much of the chemistry of organic molecules can be explained by their acid–base behavior. You may recall from a course in general chemistry that there are two frequently used definitions of acidity: the *Brønsted–Lowry definition* and the *Lewis definition*. We'll look at the Brønsted–Lowry definition in this and the next three sections and then discuss the Lewis definition in Section 2.11.

A **Brønsted–Lowry acid** is a substance that donates a proton (H^+), and a **Brønsted–Lowry base** is a substance that accepts a proton. (The name *proton* is often used as a synonym for hydrogen ion, H^+, because loss of the valence electron from a neutral hydrogen atom leaves only the hydrogen nucleus—a proton.) When gaseous hydrogen chloride dissolves in water, for example, a polar HCl molecule acts as an acid and donates a proton, while a water molecule acts as a base and accepts the proton, yielding hydronium ion (H_3O^+) and chloride ion (Cl^-).

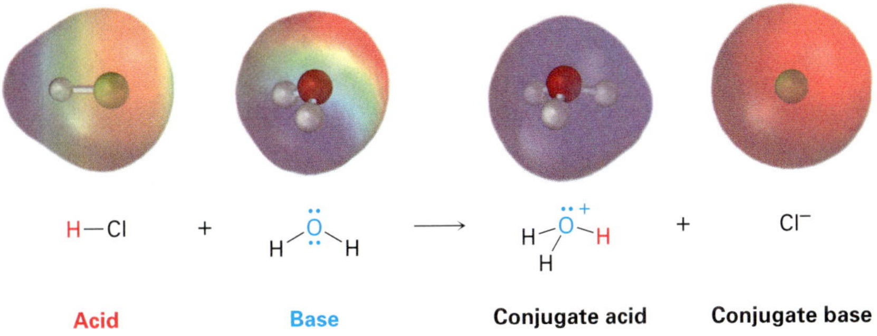

Hydronium ion, the product that results when the base H_2O gains a proton, is called the **conjugate acid** of the base, and chloride ion, the product that results when the acid HCl loses a proton, is called the **conjugate base** of the acid. Other common mineral acids such as H_2SO_4 and HNO_3 behave similarly, as do organic acids such as acetic acid, CH_3CO_2H.

In a general sense,

$$\underset{\text{Acid}}{\text{H—A}} + \underset{\text{Base}}{:\text{B}} \rightleftharpoons \underset{\text{Conjugate base}}{:\text{A}^-} + \underset{\text{Conjugate acid}}{\text{H—B}^+}$$

For example:

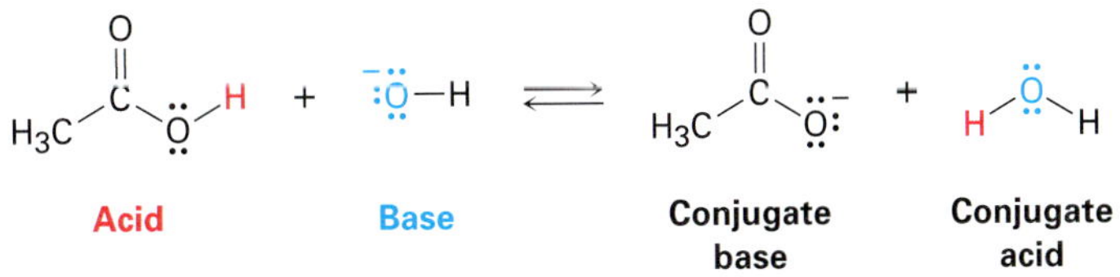

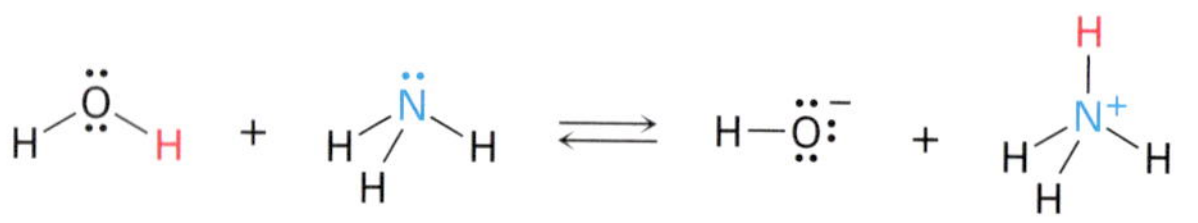

Notice that water can act *either* as an acid or as a base, depending on the circumstances. In its reaction with HCl, water is a base that accepts a proton to give the hydronium ion, H_3O^+. In its reaction with amide ion, $^-NH_2$, however, water is an acid that donates a proton to give ammonia, NH_3, and hydroxide ion, HO^-.

Problem 2.11 Nitric acid (HNO_3) reacts with ammonia (NH_3) to yield ammonium nitrate. Write the reaction, and identify the acid, the base, the conjugate acid product, and the conjugate base product.

2.8 Acid and Base Strength

Acids differ in their ability to donate H^+. Stronger acids such as HCl react almost completely with water, whereas weaker acids such as acetic acid (CH_3CO_2H) react only slightly. The exact strength of a given acid, HA, in water solution is described using the equilibrium constant K_{eq} for the acid-dissociation equilibrium. Remember from general chemistry that brackets [ ] around a substance mean that the concentration of the enclosed species is given in moles per liter, M.

$$HA + H_2O \rightleftharpoons A^- + H_3O^+$$

$$K_{eq} = \frac{[H_3O^+][A^-]}{[HA][H_2O]}$$

In the dilute aqueous solution normally used for measuring acidity, the concentration of water, $[H_2O]$, remains nearly constant at approximately 55.4 M at 25 °C. We can therefore rewrite the equilibrium expression using a new quantity called the **acidity constant, K_a**. The acidity constant for any acid HA is simply the equilibrium constant for the acid dissociation multiplied by the molar concentration of pure water.

$$HA + H_2O \rightleftharpoons A^- + H_3O^+$$

$$K_a = K_{eq}[H_2O] = \frac{[H_3O^+][A^-]}{[HA]}$$

Stronger acids have their equilibria toward the right and thus have larger acidity constants, whereas weaker acids have their equilibria toward the left and have smaller acidity constants. The range of K_a values for different acids is enormous, running from about 10^{15} for the strongest acids to about 10^{-60} for the weakest. The common inorganic acids such as H_2SO_4, HNO_3, and HCl have K_a's in the range of 10^2 to 10^9, while organic acids generally have K_a's in the range of 10^{-5} to 10^{-15}. As you gain more experience, you'll develop a rough feeling for which acids are "strong" and which are "weak" (always remembering that the terms are relative).

For convenience, acid strengths are normally expressed using pK_a values rather than K_a values, where the **pK_a** is the negative common logarithm of the K_a.

$$pK_a = -\log K_a$$

A *stronger* acid (larger K_a) has a *smaller* pK_a, and a *weaker* acid (smaller K_a) has a *larger* pK_a. Table 2.3 lists the pK_a's of some common acids in order of their strength. A more comprehensive table is given in Appendix B.

Table 2.3 Relative Strengths of Some Common Acids and Their Conjugate Bases

Weaker acid ↓ Stronger acid; Stronger base ↑ Weaker base

Acid	Name	pK_a	Conjugate base	Name
CH_3CH_2OH	Ethanol	16.00	$CH_3CH_2O^-$	Ethoxide ion
H_2O	Water	15.74	HO^-	Hydroxide ion
HCN	Hydrocyanic acid	9.31	CN^-	Cyanide ion
$H_2PO_4^-$	Dihydrogen phosphate ion	7.21	HPO_4^{2-}	Hydrogen phosphate ion
CH_3CO_2H	Acetic acid	4.76	$CH_3CO_2^-$	Acetate ion
H_3PO_4	Phosphoric acid	2.16	$H_2PO_4^-$	Dihydrogen phosphate ion
HNO_3	Nitric acid	−1.3	NO_3^-	Nitrate ion
HCl	Hydrochloric acid	−7.0	Cl^-	Chloride ion

Notice that the pK_a value shown in Table 2.3 for water is 15.74, which results from the following calculation: the K_a for any acid in water is the equilibrium constant K_{eq} for the acid dissociation multiplied by 55.4, the molar concentration of pure water. For the acid dissociation of water, we have

$$H_2O + H_2O \rightleftharpoons OH^- + H_3O^+$$

$$K_{eq} = \frac{[H_3O^+][OH^-]}{[H_2O]^2} \quad \text{and} \quad K_a = K_{eq} \times [H_2O] = \frac{[H_3O^+][OH^-]}{[H_2O]}$$

The numerator in this expression, $[H_3O^+][OH^-]$, is the so-called ion-product constant for water, $K_w = 1.00 \times 10^{-14}$, and the denominator is $[H_2O] = 55.4$ M at 25 °C. Thus, we have

$$K_a = \frac{1.0 \times 10^{-14}}{55.4} = 1.8 \times 10^{-16} \quad \text{and} \quad pK_a = 15.74$$

Notice also in Table 2.3 that there is an inverse relationship between the acid strength of an acid and the base strength of its conjugate base. That is, a *strong* acid has a *weak* conjugate base, and a *weak* acid has a *strong* conjugate base. To understand this relationship, think about what happens to the acidic hydrogen in an acid–base reaction. A strong acid is one that loses an H^+ easily, meaning that its conjugate base holds on to the H^+ weakly and is therefore a weak base. A weak acid is one that loses an H^+ with difficulty, meaning that its conjugate base holds on to the H^+ strongly and is therefore a strong base. HCl, for instance, is a strong acid, meaning that Cl^- holds on to the H^+ weakly and is thus a weak base. Water, on the other hand, is a weak acid, meaning that OH^- holds on to the H^+ strongly and is a strong base.

Problem 2.12 The amino acid phenylalanine has $pK_a = 1.83$, and tryptophan has $pK_a = 2.83$. Which is the stronger acid?

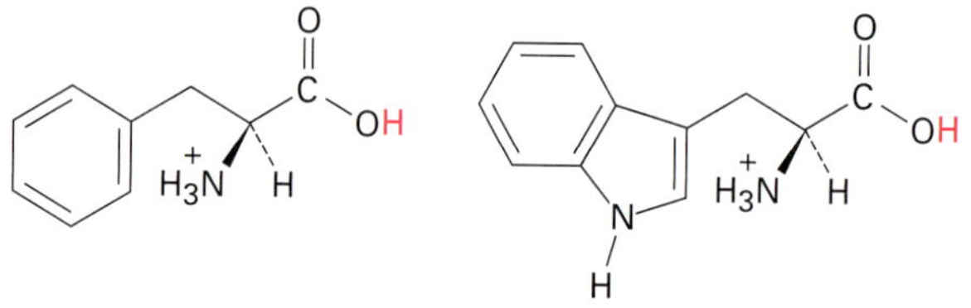

Phenylalanine (pK_a = 1.83) **Tryptophan (pK_a = 2.83)**

Problem 2.13 Amide ion, H_2N^-, is a much stronger base than hydroxide ion, HO^-. Which is the stronger acid, NH_3 or H_2O? Explain.

2.9 Predicting Acid–Base Reactions from pK_a Values

Compilations of pK_a values like those in Table 2.2 and Appendix B are useful for predicting whether a given acid–base reaction will take place because H^+ will always go *from* the stronger acid *to* the stronger base. That is, an acid will donate a proton to the conjugate base of a weaker acid, and the conjugate base of a weaker acid will remove the proton from a stronger acid. For example, since water ($pK_a = 15.74$) is a weaker acid than acetic acid ($pK_a = 4.76$), hydroxide ion holds a proton more tightly than acetate ion does. Hydroxide ion will therefore react with acetic acid, CH_3CO_2H, to yield acetate ion and H_2O.

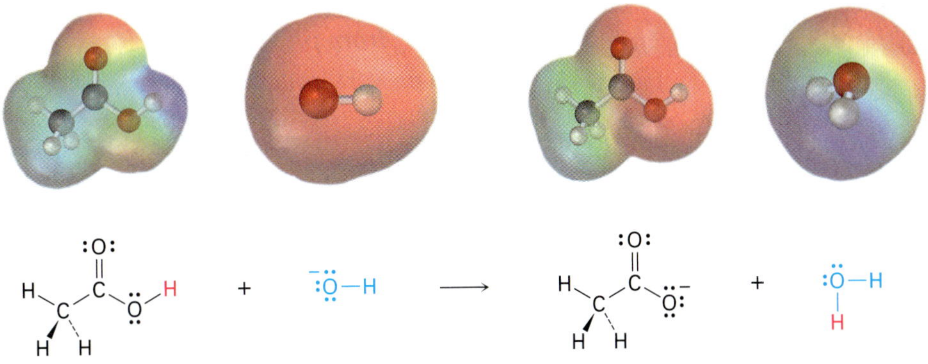

Another way to predict acid–base reactivity is to remember that the product conjugate acid in an acid–base reaction must be weaker and less reactive than the starting acid and the product conjugate base must be weaker and less reactive than the starting base. In the reaction of acetic acid with hydroxide ion, for example, the product conjugate acid (H_2O) is weaker than the starting acid (CH_3CO_2H) and the product conjugate base ($CH_3CO_2^-$) is weaker than the starting base (OH^-).

$$\underset{\text{Stronger acid}}{CH_3\overset{O}{\overset{\|}{C}}OH} + \underset{\text{Stronger base}}{HO^-} \rightleftharpoons \underset{\text{Weaker acid}}{HOH} + \underset{\text{Weaker base}}{CH_3\overset{O}{\overset{\|}{C}}O^-}$$

WORKED EXAMPLE 2.4

Predicting Acid Strengths from pK_a Values

Water has pK_a = 15.74, and acetylene has pK_a = 25. Which is the stronger acid? Does hydroxide ion react with acetylene?

$$\underset{\text{Acetylene}}{H{-}C{\equiv}C{-}H} + OH^- \xrightarrow{?} H{-}C{\equiv}C{:}^- + H_2O$$

Strategy In comparing two acids, the one with the lower pK_a is stronger. Thus, water is a stronger acid than acetylene and gives up H^+ more easily.

Solution Because water is a stronger acid and gives up H^+ more easily than acetylene does, the HO^- ion must have less affinity for H^+ than the $HC{\equiv}C{:}^-$ ion has. In other words, the anion of acetylene is a stronger base than hydroxide ion, and the reaction will not proceed as written.

WORKED EXAMPLE 2.5

Calculating K_a from pK_a

According to the data in Table 2.3, acetic acid has $pK_a = 4.76$. What is its K_a?

Strategy Since pK_a is the negative logarithm of K_a, it's necessary to use a calculator with an ANTILOG or INV LOG function. Enter the value of the pK_a (4.76), change the sign (−4.76), and then find the antilog (1.74×10^{-5}).

Solution $K_a = 1.74 \times 10^{-5}$.

Problem 2.14 Will either of the following reactions take place as written, according to the data in Table 2.3?

(a) $HCN + CH_3CO_2^-\,Na^+ \xrightarrow{?} Na^+\,^-CN + CH_3CO_2H$

(b) $CH_3CH_2OH + Na^+\,^-CN \xrightarrow{?} CH_3CH_2O^-\,Na^+ + HCN$

Problem 2.15 Ammonia, NH_3, has $pK_a \approx 36$, and acetone has pK_a 19. Will the following reaction take place?

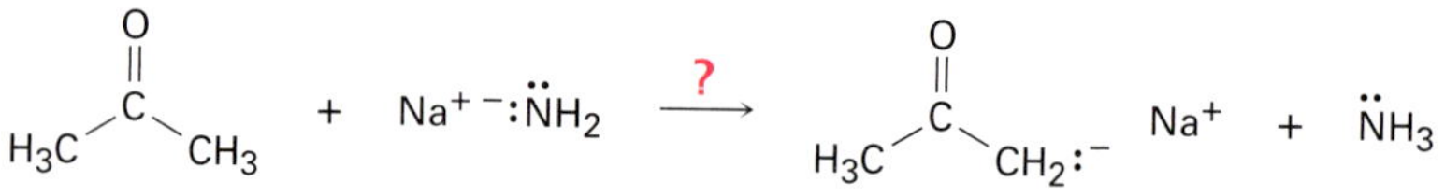

Acetone

Problem 2.16 What is the K_a of HCN if its $pK_a = 9.31$?

2.10 Organic Acids and Organic Bases

Many of the reactions we'll be seeing in future chapters involve organic acids and organic bases. Although it's too early to go into the details of these processes now, you might keep the following generalities in mind as your study progresses.

Organic Acids

Organic acids are characterized by the presence of a positively polarized hydrogen atom (blue in electrostatic potential maps) and are of two main kinds: those acids such as methanol and acetic acid that contain a hydrogen atom bonded to an electronegative oxygen atom (O–H) and those such as acetone (Section 2.5) that contain a hydrogen atom bonded to a carbon atom next to a C=O bond (O=C–C–H).

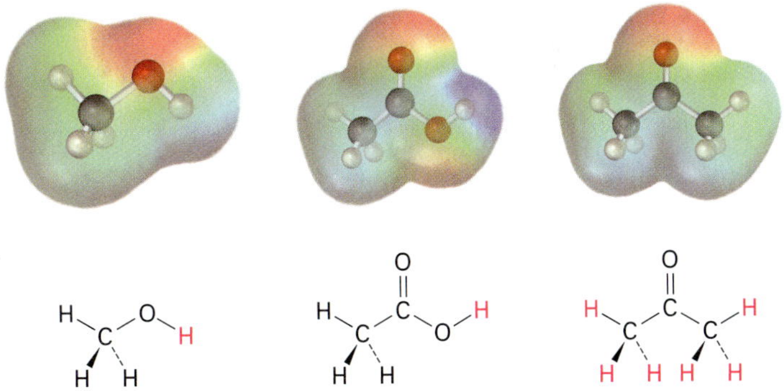

Some organic acids

Methanol **(pK_a = 15.54)**

Acetic acid **(pK_a = 4.76)**

Acetone **(pK_a = 19.3)**

Methanol contains an O–H bond and is a weak acid; acetic acid also contains an O–H bond and is a somewhat stronger acid. In both cases, acidity is due to the fact that the conjugate base resulting from loss of H^+ is stabilized by having its negative charge on a strongly electronegative oxygen atom. In addition, the conjugate base of acetic acid is stabilized by resonance (Sections 2.4 and 2.5).

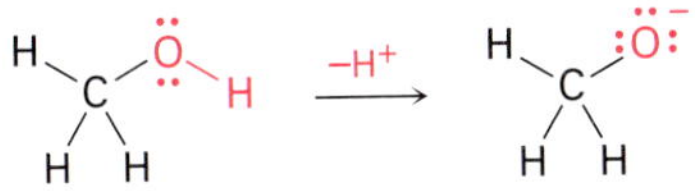

Anion is stabilized by having negative charge on a highly electronegative atom.

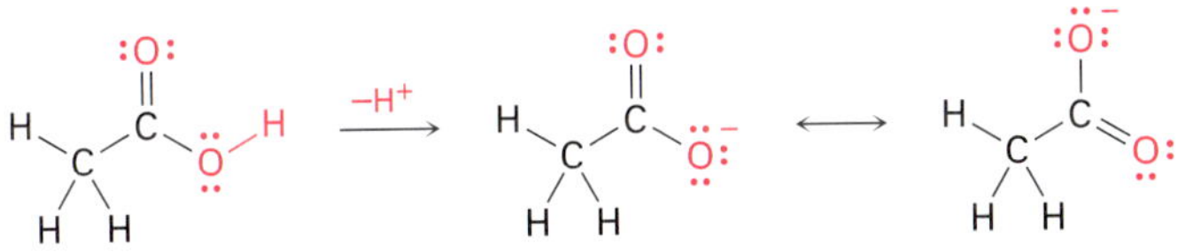

Anion is stabilized both by having negative charge on a highly electronegative atom and by resonance.

The acidity of acetone and other compounds with C=O bonds is due to the fact that the conjugate base resulting from loss of H^+ is stabilized by resonance. In addition, one of the resonance forms stabilizes the negative charge by placing it on an electronegative oxygen atom.

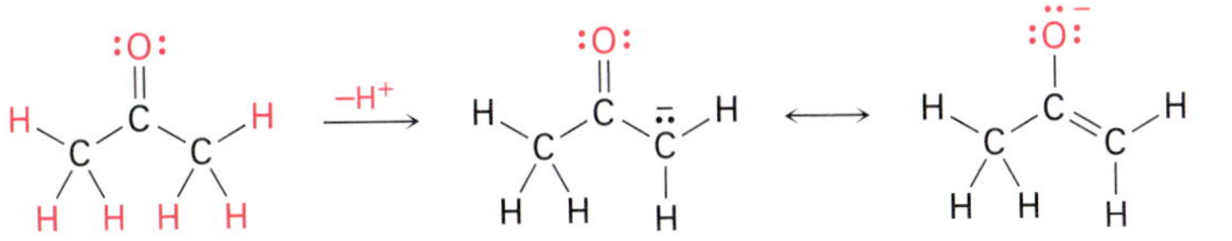

Anion is stabilized both by resonance and by having negative charge on a highly electronegative atom.

Electrostatic potential maps of the conjugate bases from methanol, acetic acid, and acetone are shown in Figure 2.4. As you might expect, all three show a substantial amount of negative charge (red) on oxygen.

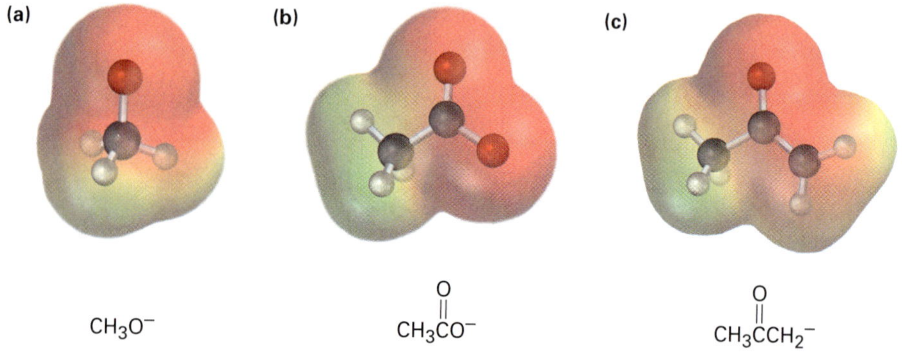

Figure 2.4 Electrostatic potential maps of the conjugate bases of **(a)** methanol, **(b)** acetic acid, and **(c)** acetone. The electronegative oxygen atoms stabilize the negative charge in all three.

Compounds called *carboxylic acids,* which contain the $-CO_2H$ grouping, occur abundantly in all living organisms and are involved in almost all metabolic pathways. Acetic acid, pyruvic acid, and citric acid are examples.

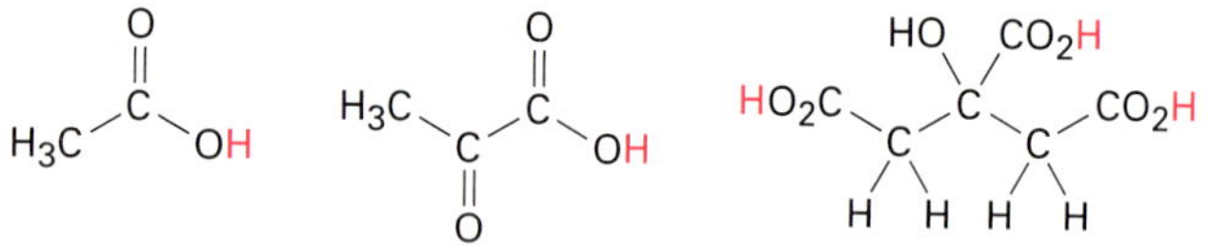

Acetic acid **Pyruvic acid** **Citric acid**

Organic Bases

Organic bases are characterized by the presence of an atom (reddish in electrostatic potential maps) with a lone pair of electrons that can bond to H^+. Nitrogen-containing compounds such as trimethylamine are the most common organic bases, but oxygen-containing compounds can also act as bases when reacting with a sufficiently strong acid. Note that some oxygen-containing compounds can act both as acids and as bases depending on the circumstances, just as water can. Methanol and acetone, for instance, act as *acids* when they donate a proton but as *bases* when their oxygen atom accepts a proton.

Some organic bases

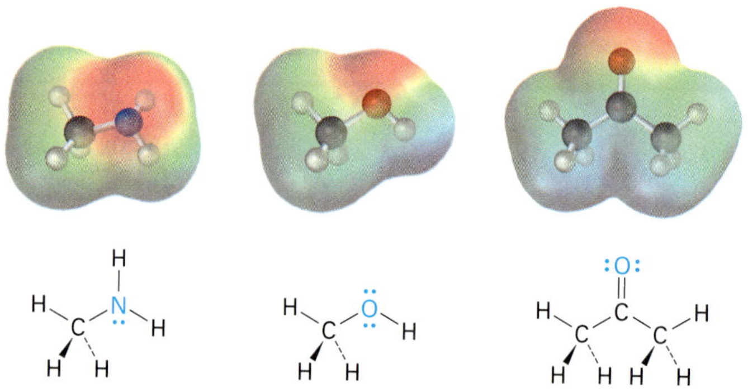

Methylamine **Methanol** **Acetone**

We'll see in Chapter 26 that substances called *amino acids,* so-named because they are both amines ($-NH_2$) and carboxylic acids ($-CO_2H$), are the building blocks from which the proteins present in all living organisms arise. Twenty different amino acids go into making up proteins; alanine is an example.

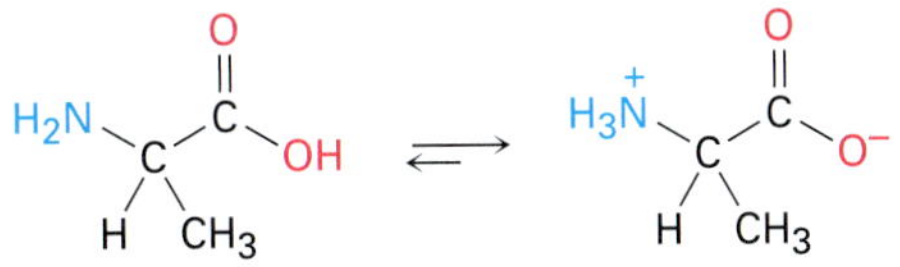

Alanine (uncharged form) **Alanine (zwitterion form)**

Interestingly, alanine and other amino acids exist primarily in a doubly charged form called a *zwitterion* rather than in the uncharged form. The zwitterion form arises because amino acids have both acidic and basic sites within the same molecule and therefore undergo an *internal* acid–base reaction.

2.11 Acids and Bases: The Lewis Definition

The *Lewis definition* of acids and bases is broader and more encompassing than the Brønsted–Lowry definition because it's not limited to substances that donate or accept just protons. A **Lewis acid** is a substance that *accepts an electron pair,* and a **Lewis base** is a substance that *donates an electron pair.* The donated electron pair is shared between the acid and the base in a covalent bond.

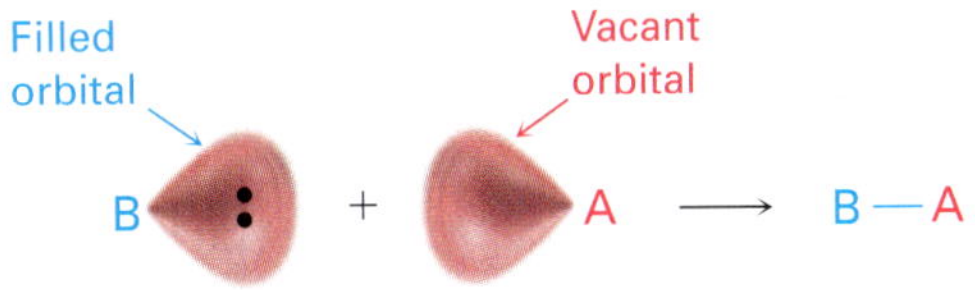

Lewis base **Lewis acid**

Lewis Acids and the Curved Arrow Formalism

The fact that a Lewis acid is able to accept an electron pair means that it must have either a vacant, low-energy orbital or a polar bond to hydrogen so that it can donate H^+ (which has an empty 1*s* orbital). Thus, the Lewis definition of acidity includes many species in addition to H^+. For example, various metal cations, such as Mg^{2+}, are Lewis acids because they accept a pair of electrons when they form a bond to a base. We'll also see in later chapters that certain metabolic reactions begin with an acid–base reaction between Mg^{2+} as a Lewis acid and an organic diphosphate or triphosphate ion as the Lewis base.

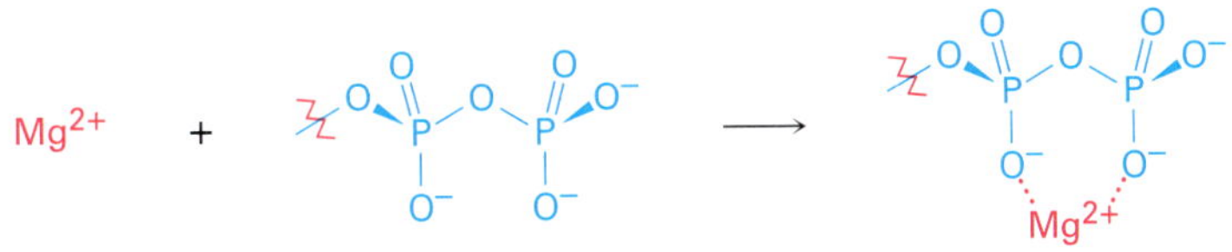

Lewis acid **Lewis base (an organodiphosphate ion)** **Acid–base complex**

In the same way, compounds of group 3A elements, such as BF_3 and $AlCl_3$, are Lewis acids because they have unfilled valence orbitals and can accept electron pairs from Lewis bases, as shown in Figure 2.5. Similarly, many transition-metal compounds, such as $TiCl_4$, $FeCl_3$, $ZnCl_2$, and $SnCl_4$, are Lewis acids.

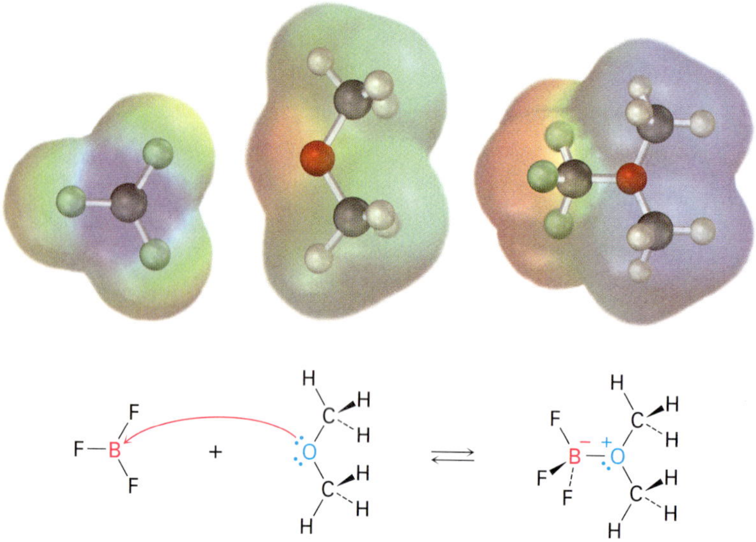

Boron trifluoride (Lewis acid) + Dimethyl ether (Lewis base) ⇌ Acid–base complex

Active Figure 2.5 The reaction of boron trifluoride, a Lewis acid, with dimethyl ether, a Lewis base. The Lewis acid accepts a pair of electrons, and the Lewis base donates a pair of nonbonding electrons. Note how the movement of electrons *from* the Lewis base *to* the Lewis acid is indicated by a curved arrow. Note also how, in electrostatic potential maps, the boron becomes more negative (red) after reaction because it has gained electrons and the oxygen atom becomes more positive (blue) because it has donated electrons. *Sign in at* **www.thomsonedu.com** *to see a simulation based on this figure and to take a short quiz.*

Look closely at the acid–base reaction in Figure 2.5, and note how it is shown. Dimethyl ether, the Lewis base, donates an electron pair to a vacant valence orbital of the boron atom in BF_3, a Lewis acid. The direction of electron-pair flow from the base to acid is shown using curved arrows, just as the direction of electron flow in going from one resonance structure to another was shown using curved arrows in Section 2.5. *A curved arrow always means that a pair of electrons moves* from *the atom at the tail of the arrow* to *the atom at the head of the arrow.* We'll use this curved-arrow notation throughout the remainder of this text to indicate electron flow during reactions.

Some further examples of Lewis acids follow:

Some Lewis acids

Some neutral proton donors:

H_2O HCl HBr HNO_3 H_2SO_4

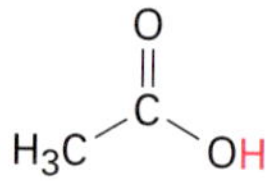

A carboxylic acid

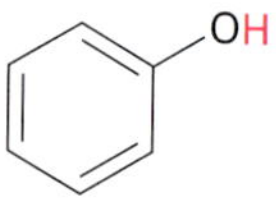

A phenol

CH_3CH_2OH

An alcohol

Some cations:

Li^+ Mg^{2+}

Some metal compounds:

$AlCl_3$ $TiCl_4$ $FeCl_3$ $ZnCl_2$

Lewis Bases

The Lewis definition of a base as a compound with a pair of nonbonding electrons that it can use to bond to a Lewis acid is similar to the Brønsted–Lowry definition. Thus, H_2O, with its two pairs of nonbonding electrons on oxygen, acts as a Lewis base by donating an electron pair to an H^+ in forming the hydronium ion, H_3O^+.

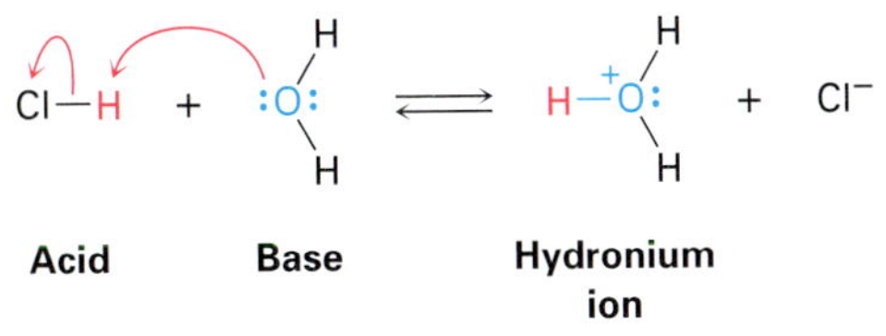

In a more general sense, most oxygen- and nitrogen-containing organic compounds can act as Lewis bases because they have lone pairs of electrons. A divalent oxygen compound has two lone pairs of electrons, and a trivalent nitrogen compound has one lone pair. Note in the following examples that some compounds can act as both acids and bases, just as water can. Alcohols and carboxylic acids, for instance, act as acids when they donate an H^+ but as bases when their oxygen atom accepts an H^+.

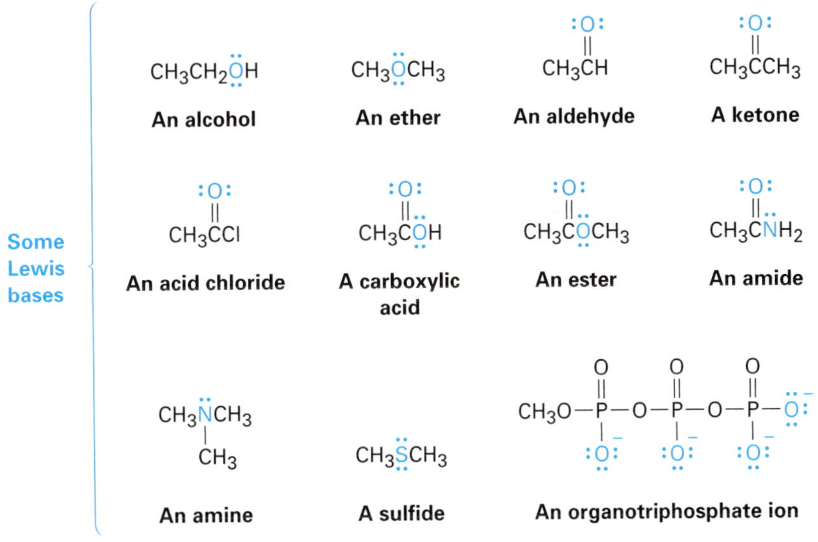

Notice in the list of Lewis bases just given that some compounds, such as carboxylic acids, esters, and amides, have more than one atom with a lone pair of electrons and can therefore react at more than one site. Acetic acid, for example, can be protonated either on the doubly bonded oxygen atom or on the singly bonded oxygen atom. Reaction normally occurs only once in such instances, and the more stable of the two possible protonation products is formed. For acetic acid, protonation by reaction with sulfuric acid occurs on the doubly bonded oxygen because that product is stabilized by two resonance forms.

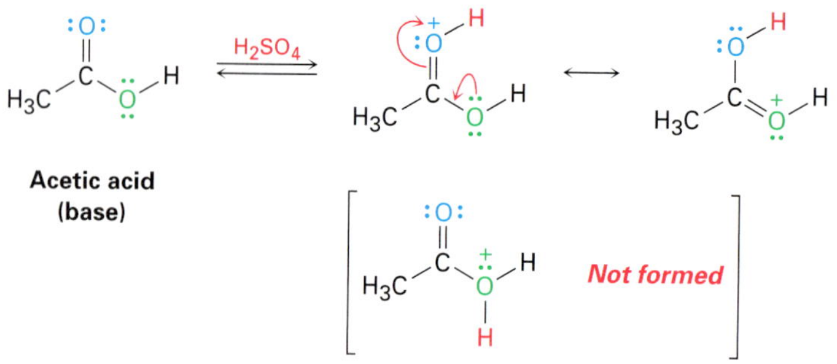

WORKED EXAMPLE 2.6 *Using Curved Arrows to Show Electron Flow*

Using curved arrows, show how acetaldehyde, CH_3CHO, can act as a Lewis base.

Strategy A Lewis base donates an electron pair to a Lewis acid. We therefore need to locate the electron lone pairs on acetaldehyde and use a curved arrow to show the movement of a pair toward the H atom of the acid.

Solution

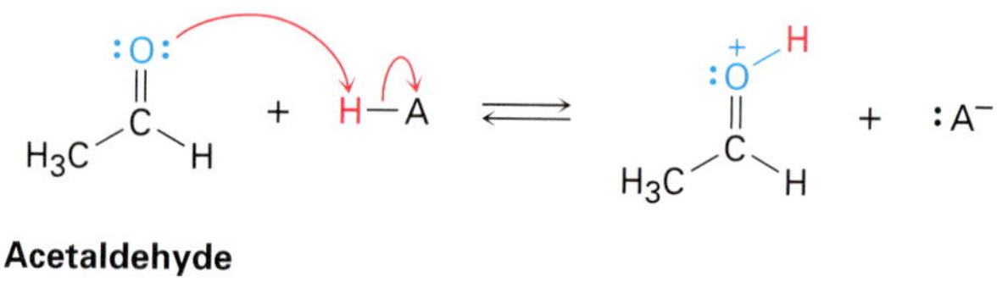

Problem 2.17 Using curved arrows, show how the species in part (a) can act as Lewis bases in their reactions with HCl, and show how the species in part (b) can act as Lewis acids in their reaction with OH^-.

(a) CH_3CH_2OH, $HN(CH_3)_2$, $P(CH_3)_3$ **(b)** H_3C^+, $B(CH_3)_3$, $MgBr_2$

Problem 2.18 Imidazole forms part of the structure of the amino acid histidine and can act as both an acid and a base.

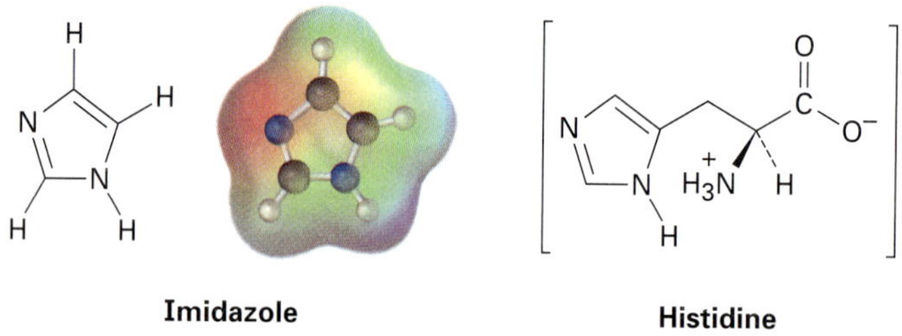

(a) Look at the electrostatic potential map of imidazole, and identify the most acidic hydrogen atom and the most basic nitrogen atom.

(b) Draw structures for the resonance forms of the products that result when imidazole is protonated by an acid and deprotonated by a base.

2.12 Molecular Models

Because organic chemistry is a three-dimensional science, molecular shape is often critical in determining the chemistry a compound undergoes, both in the laboratory and in living organisms. Learning to visualize molecular shapes is therefore an important skill to develop. One helpful technique, particularly when dealing with large biomolecules, is to use one of the many computer programs that are available for rotating and manipulating molecules on the screen. Another technique is to use molecular models. With practice, you can learn to see many spatial relationships even when viewing two-dimensional drawings, but there's no substitute for building a molecular model and turning it in your hands to get different perspectives.

Many kinds of models are available, some at relatively modest cost, and it's a good idea to have access to a set of models while studying this book. *Space-filling models* are better for examining the crowding within a molecule, but *ball-and-stick models* are generally the least expensive and most durable for student use. Figure 2.6 shows two kinds of models of acetic acid, CH_3CO_2H.

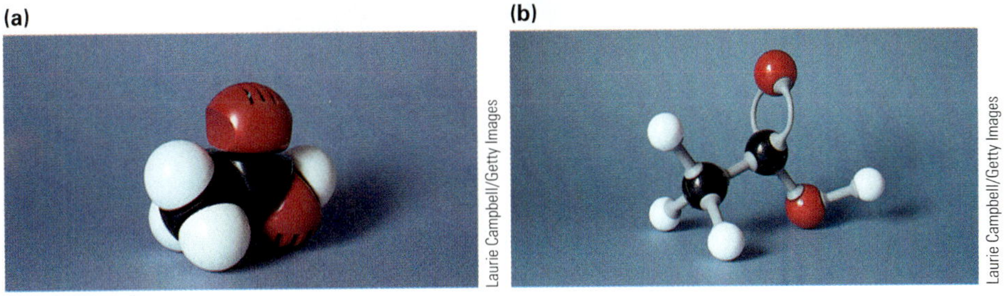

Figure 2.6 Molecular models of acetic acid, CH_3CO_2H. **(a)** Space-filling; **(b)** ball-and-stick.

2.13 Noncovalent Interactions

When thinking about chemical reactivity, chemists usually focus their attention on bonds, the covalent interactions between atoms *within* individual molecules. Also important, however, particularly in large biomolecules like proteins and nucleic acids, are a variety of interactions *between* molecules that strongly affect molecular properties. Collectively called either *intermolecular forces, van der Waals forces,* or **noncovalent interactions**, they are of several different types: dipole–dipole forces, dispersion forces, and hydrogen bonds.

Dipole–dipole forces occur between polar molecules as a result of electrostatic interactions among dipoles. The forces can be either attractive or repulsive depending on the orientation of the molecules—attractive when unlike charges are together and repulsive when like charges are together. The attractive geometry is lower in energy and therefore predominates (Figure 2.7).

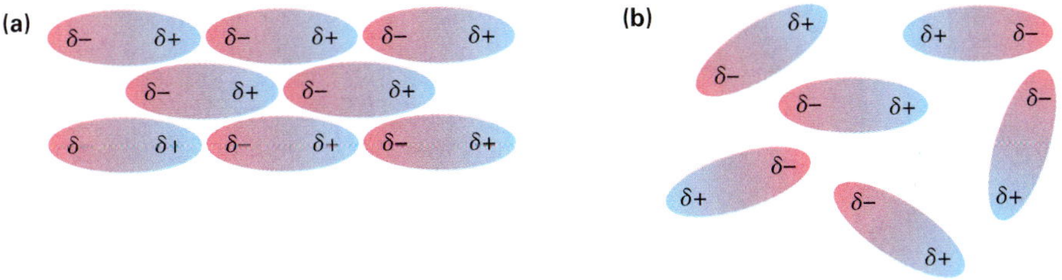

Figure 2.7 Dipole–dipole forces cause polar molecules **(a)** to attract one another when they orient with unlike charges together but **(b)** to repel one another when they orient with like charges together.

Dispersion forces occur between all neighboring molecules and arise because the electron distribution within molecules is constantly changing. Although uniform on a time-averaged basis, the electron distribution even in nonpolar molecules is likely to be nonuniform at any given instant. One side of a molecule may, by chance, have a slight excess of electrons relative to the opposite side, giving the molecule a temporary dipole. This temporary dipole in one molecule causes a nearby molecule to adopt a temporarily opposite dipole, with the result that a tiny attraction is induced between the two (Figure 2.8). Temporary molecular dipoles have only a fleeting existence and are constantly changing, but their cumulative effect is often strong enough to cause a substance to be liquid or solid rather than gaseous.

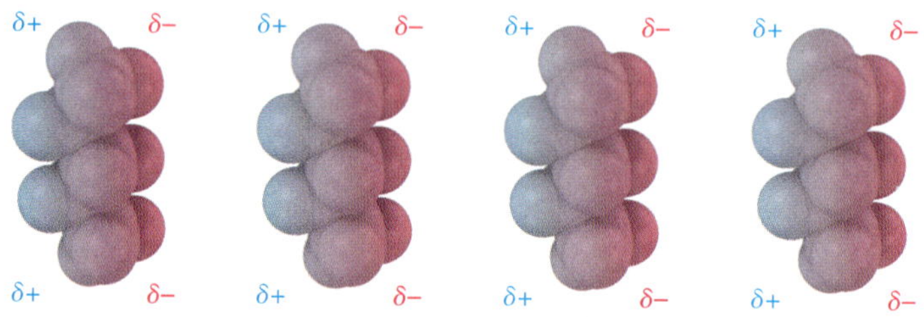

Figure 2.8 Attractive dispersion forces in nonpolar molecules are caused by temporary dipoles, as shown in these models of pentane, C_5H_{12}.

Perhaps the most important noncovalent interaction in biological molecules is the **hydrogen bond**, an attractive interaction between a hydrogen bonded to an electronegative O or N atom and an unshared electron pair on another O or N atom. In essence, a hydrogen bond is a strong dipole–dipole interaction involving polarized O–H and N–H bonds. Electrostatic potential maps of water and ammonia clearly show the positively polarized hydrogens (blue) and the negatively polarized oxygens and nitrogens (red).

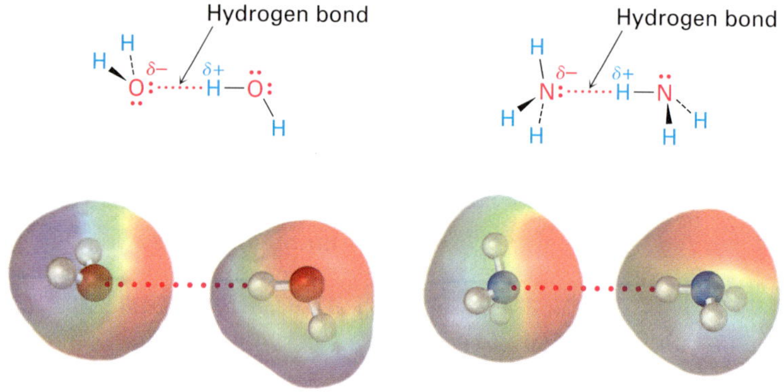

Hydrogen-bonding has enormous consequences for living organisms. Hydrogen bonds cause water to be a liquid rather than a gas at ordinary temperatures, they hold enzymes in the shapes necessary for catalyzing biological reactions, and they cause strands of deoxyribonucleic acid (DNA) to pair up and coil into the double helix that stores genetic information.

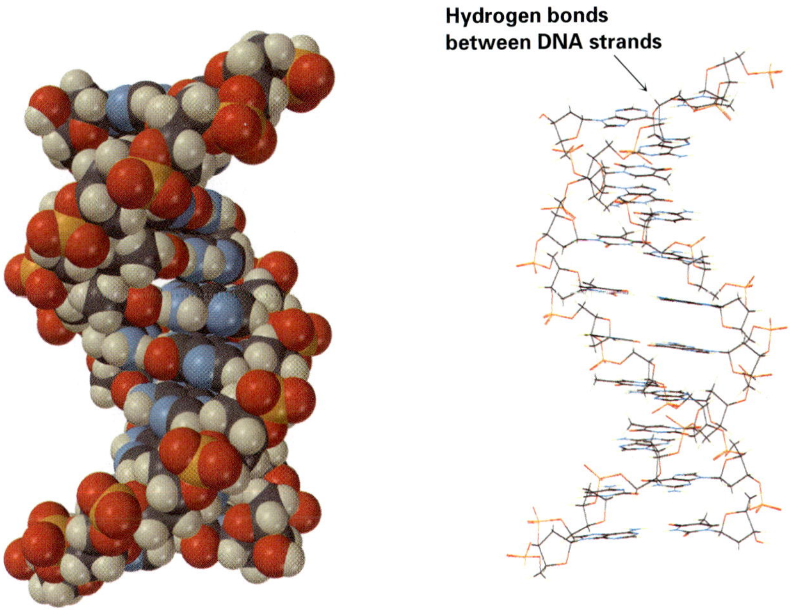

A deoxyribonucleic acid segment

One further point before leaving the subject of noncovalent interactions: chemists frequently use the terms **hydrophilic**, meaning "water-loving," to describe a substance that dissolves in water and **hydrophobic**, meaning "water-fearing," to describe a substance that does not dissolve in water. Hydrophilic substances, such as table sugar, usually have a number of ionic charges or polar –OH groups in their structure, so they are strongly attracted to water. Hydrophobic substances, such as vegetable oil, do not have groups that form hydrogen bonds, so their attraction to water is weak.

Problem 2.19 Of the two vitamins A and C, one is hydrophilic and water-soluble while the other is hydrophobic and fat-soluble. Which is which?

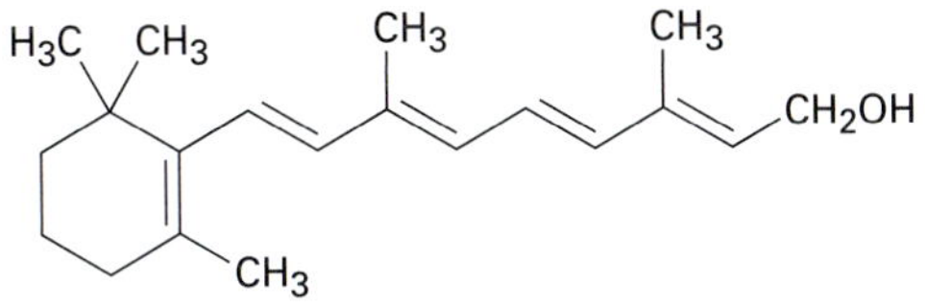

Vitamin A
(retinol)

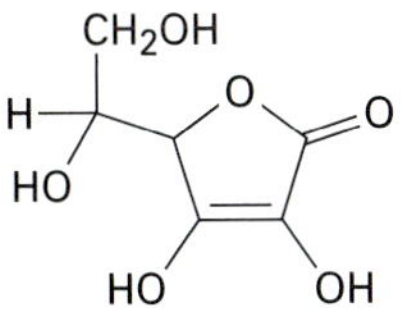

Vitamin C
(ascorbic acid)

Focus On . . .

Alkaloids: Naturally Occurring Bases

The coca bush *Erythroxylon coca,* native to upland rain forest areas of Colombia, Ecuador, Peru, Bolivia, and western Brazil, is the source of the alkaloid cocaine.

Just as ammonia is a weak base, there are a large number of nitrogen-containing organic compounds called *amines* that are also weak bases. In the early days of organic chemistry, basic amines derived from natural sources were known as *vegetable alkali,* but they are now called *alkaloids.* The study of alkaloids provided much of the impetus for the growth of organic chemistry in the 19th century and remains today an active and fascinating area of research.

Alkaloids vary widely in structure, from the simple to the enormously complex. The odor of rotting fish, for example, is caused largely by methylamine, CH_3NH_2, a simple relative of ammonia in which one of the NH_3 hydrogens has been replaced by an organic CH_3 group. In fact, the use of lemon juice to mask fish odors is simply an acid–base reaction of the citric acid in lemons with methylamine base in the fish.

Many alkaloids have pronounced biological properties, and a substantial number of the pharmaceutical agents used today are derived from naturally occurring amines. As a few examples, morphine, an analgesic agent, is obtained from the opium poppy *Papaver somniferum.* Cocaine, both an anesthetic and a central nervous system stimulant, is obtained from the coca bush *Erythroxylon coca,* endemic to upland rain forest areas of Colombia, Ecuador, Peru, Bolivia, and western Brazil. Reserpine, a tranquilizer and antihypertensive, comes from powdered roots of the semitropical plant *Rauwolfia serpentina.* Ephedrine, a bronchodilator and decongestant, is obtained from the Chinese plant *Ephedra sinica.*

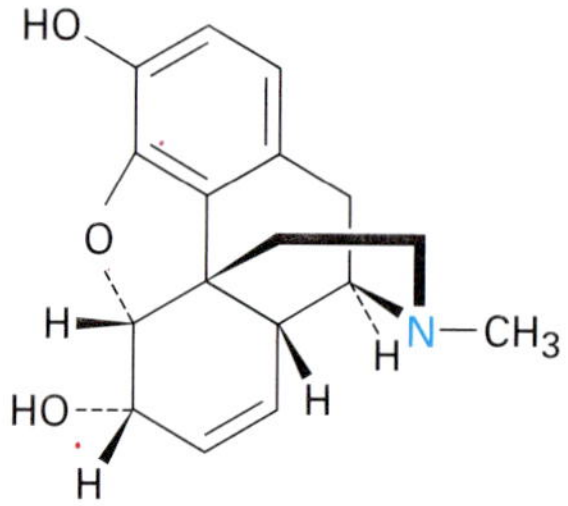

Morphine

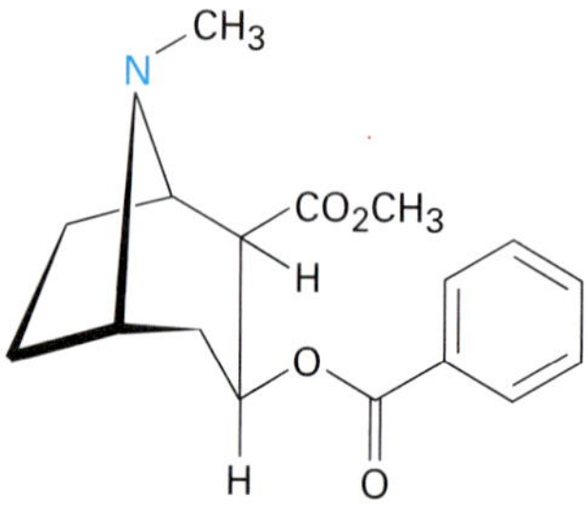

Cocaine

(continued)

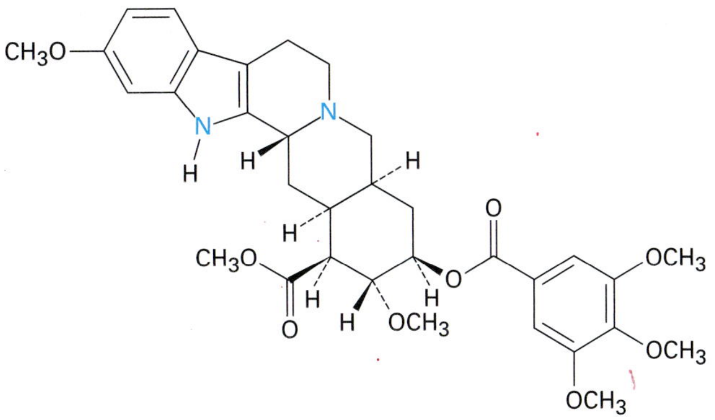

Reserpine

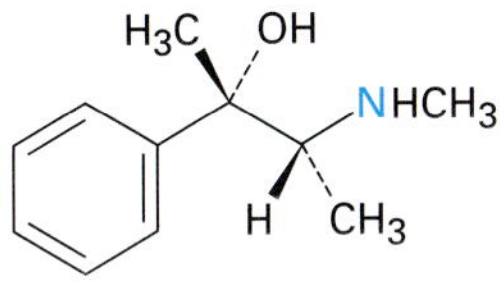

Ephedrine

A recent report from the U.S. National Academy of Sciences estimates than less than 1% of all living species have been characterized. Thus, alkaloid chemistry remains today an active area of research, and innumerable substances with potentially useful properties remain to be discovered.

SUMMARY AND KEY WORDS

acidity constant (K_a), 50
Brønsted–Lowry acid, 49
Brønsted–Lowry base, 49
conjugate acid, 49
conjugate base, 49
dipole moment (μ), 38
electronegativity (EN), 36
formal charge, 41
hydrogen bond, 62
hydrophilic, 63
hydrophobic, 63
inductive effect, 37
Lewis acid, 57
Lewis base, 57
noncovalent interaction, 61

Organic molecules often have **polar covalent bonds** as a result of unsymmetrical electron sharing caused by differences in the **electronegativity** of atoms. A carbon–oxygen bond is polar, for example, because oxygen attracts the shared electrons more strongly than carbon does. Carbon–hydrogen bonds are relatively nonpolar. Many molecules as a whole are also polar owing to the vector summation of individual polar bonds and electron lone pairs. The polarity of a molecule is measured by its **dipole moment**, μ.

Plus (+) and minus (−) signs are often used to indicate the presence of **formal charges** on atoms in molecules. Assigning formal charges to specific atoms is a bookkeeping technique that makes it possible to keep track of the valence electrons around an atom and offers some clues about chemical reactivity.

Some substances, such as acetate ion and benzene, can't be represented by a single line-bond structure and must be considered as a **resonance hybrid** of two or more structures, neither of which is correct by itself. The only difference between two **resonance forms** is in the location of their π and nonbonding electrons. The nuclei remain in the same places in both structures, and the hybridization of the atoms remains the same.

Acidity and basicity are closely related to the ideas of polarity and electronegativity. A **Brønsted–Lowry acid** is a compound that can donate a proton (hydrogen ion, H^+), and a **Brønsted–Lowry base** is a compound that can accept a proton. The strength of a Brønsted–Lowry acid or base is expressed by its **acidity constant, K_a**, or by the negative logarithm of the acidity constant, **pK_a**. The larger the pK_a, the weaker the acid. More useful is the Lewis definition of acids and bases. A **Lewis acid** is a compound that has a low-energy empty orbital that can accept an electron pair; Mg^{2+}, BF_3, $AlCl_3$, and H^+ are examples. A **Lewis base** is a compound that can donate an unshared electron pair; NH_3 and H_2O are examples. Most organic molecules that contain oxygen and nitrogen can act as Lewis bases toward sufficiently strong acids.

A variety of **noncovalent interactions** have a significant effect on the properties of large biomolecules. **Hydrogen-bonding**—the attractive interaction between a positively polarized hydrogen atom bonded to an oxygen or nitrogen atom with an unshared electron pair on another O or N atom, is particularly important in giving proteins and nucleic acids their shapes.

EXERCISES

Organic KNOWLEDGE TOOLS

ThomsonNOW™ Sign in at **www.thomsonedu.com** to assess your knowledge of this chapter's topics by taking a pre-test. The pre-test will link you to interactive organic chemistry resources based on your score in each concept area.

Online homework for this chapter may be assigned in Organic OWL.

■ indicates problems assignable in Organic OWL.

▲ denotes problems linked to Key Ideas of this chapter and testable in ThomsonNOW.

VISUALIZING CHEMISTRY

(Problems 2.1–2.19 appear within the chapter.)

2.20 Fill in the multiple bonds in the following model of naphthalene, $C_{10}H_8$ (gray = C, ivory = H). How many resonance structures does naphthalene have?

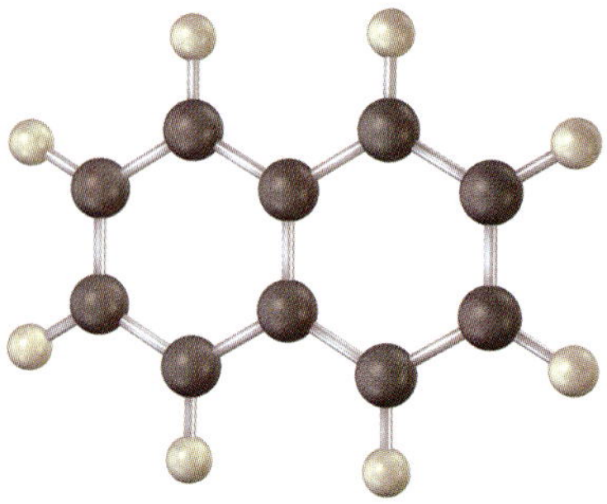

2.21 ■ The following model is a representation of ibuprofen, a common over-the-counter pain reliever. Indicate the positions of the multiple bonds, and draw a skeletal structure (gray = C, red = O, ivory = H).

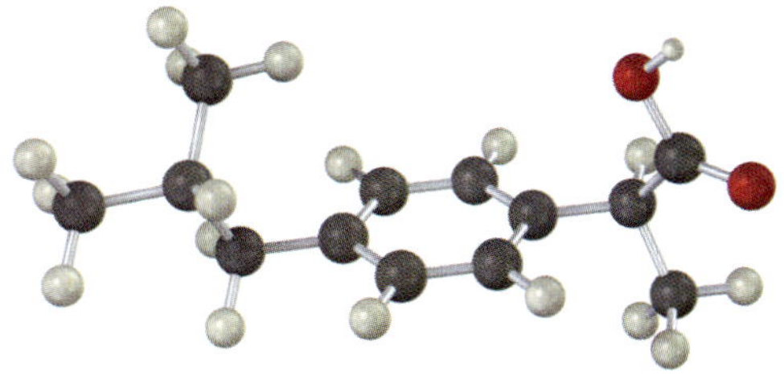

2.22 *cis*-1,2-Dichloroethylene and *trans*-dichloroethylene are *isomers,* compounds with the same formula but different chemical structures. Look at the following electrostatic potential maps, and tell whether either compound has a dipole moment.

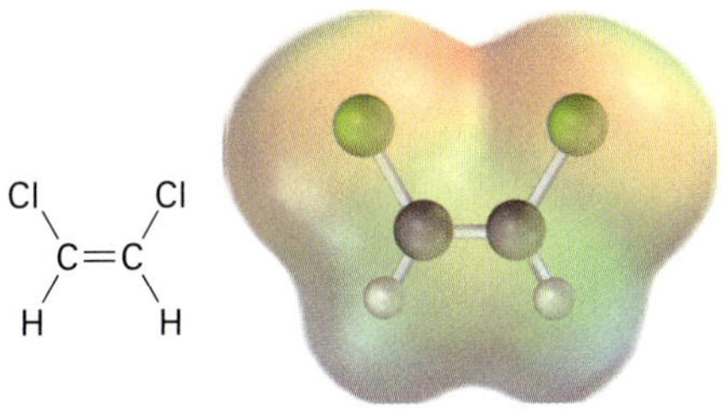

***cis*-1,2-Dichloroethylene**

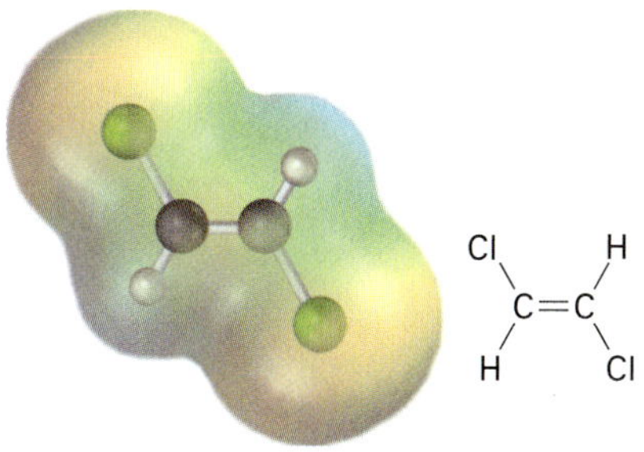

***trans*-1,2-Dichloroethylene**

2.23 ■ The following molecular models are representations of **(a)** adenine and **(b)** cytosine, constituents of DNA. Indicate the positions of multiple bonds and lone pairs for both, and draw skeletal structures (gray = C, red = O, blue = N, ivory = H).

(a)

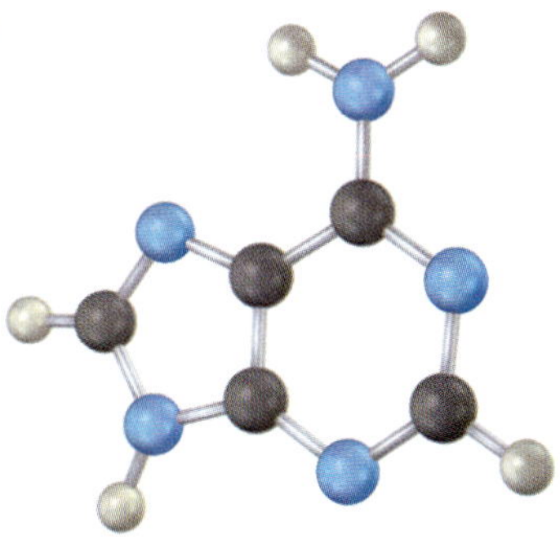

Adenine

(b)

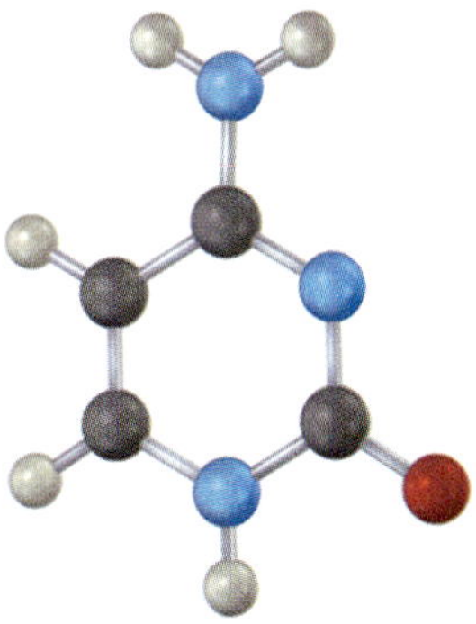

Cytosine

ADDITIONAL PROBLEMS

2.24 Tell the number of hydrogens bonded to each carbon atom in the following substances, and give the molecular formula of each:

(a)

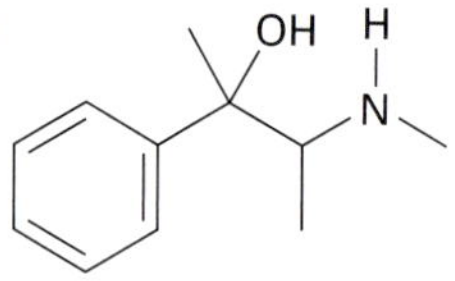

Ephedrine

(b)

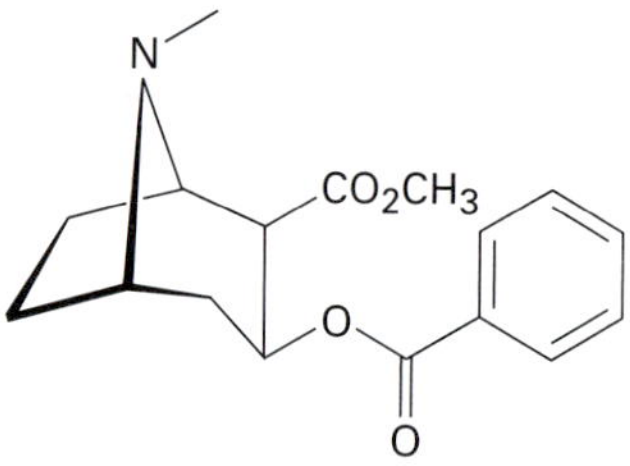

Cocaine

2.25 ■ Identify the most electronegative element in each of the following molecules:
(a) CH_2FCl (b) $FCH_2CH_2CH_2Br$
(c) $HOCH_2CH_2NH_2$ (d) CH_3OCH_2Li

2.26 ■ Use the electronegativity table (Figure 2.2) to predict which bond in each of the following sets is more polar, and indicate the direction of bond polarity for each compound.
(a) $H_3C{-}Cl$ or $Cl{-}Cl$ (b) $H_3C{-}H$ or $H{-}Cl$
(c) $HO{-}CH_3$ or $(CH_3)_3Si{-}CH_3$ (d) $H_3C{-}Li$ or $Li{-}OH$

2.27 ■ Which of the following molecules has a dipole moment? Indicate the expected direction of each.

(a)

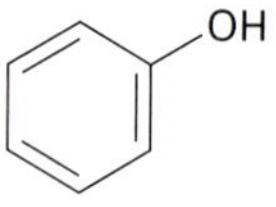

(b)

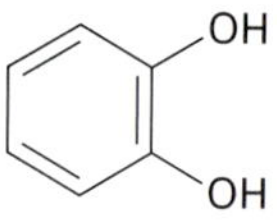

(c)

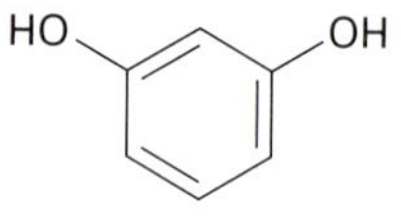

(d)

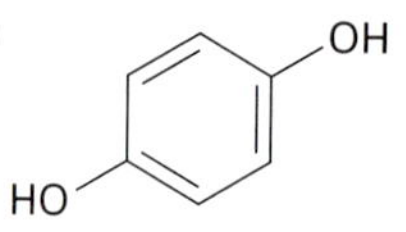

2.28 (a) The H–Cl bond length is 136 pm. What would the dipole moment of HCl be if the molecule were 100% ionic, $H^+ Cl^-$?
(b) The actual dipole moment of HCl is 1.08 D. What is the percent ionic character of the H–Cl bond?

2.29 Phosgene, $Cl_2C{=}O$, has a smaller dipole moment than formaldehyde, $H_2C{=}O$, even though it contains electronegative chlorine atoms in place of hydrogen. Explain.

2.30 Fluoromethane (CH_3F, $\mu = 1.81$ D) has a smaller dipole moment than chloromethane (CH_3Cl, $\mu = 1.87$ D) even though fluorine is more electronegative than chlorine. Explain.

2.31 Methanethiol, CH_3SH, has a substantial dipole moment ($\mu = 1.52$) even though carbon and sulfur have identical electronegativities. Explain.

2.32 ■ Calculate the formal charges on the atoms shown in red.

(a) $(CH_3)_2\ddot{O}BF_3$ (b) $H_2\ddot{C}{-}N{\equiv}N{:}$ (c) $H_2C{=}N{=}\ddot{N}{:}$

(d) $:\ddot{O}{=}\ddot{O}{-}\ddot{O}:$ (e)

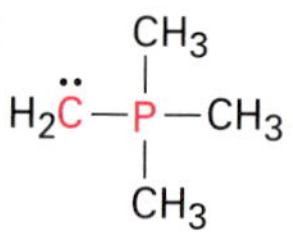

(f)

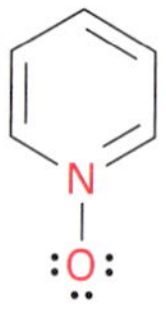

2.33 ■ Which of the following pairs of structures represent resonance forms?

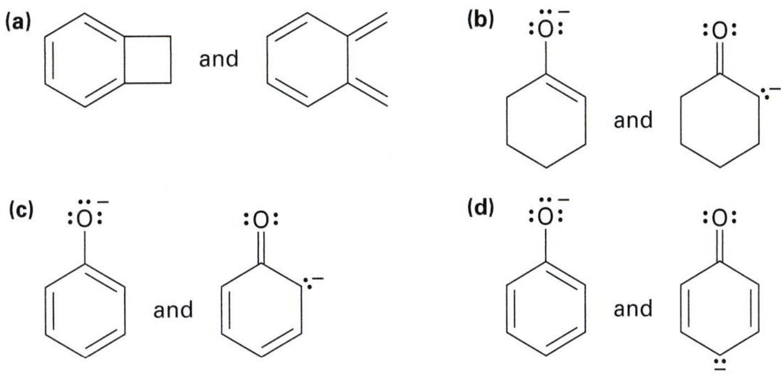

2.34 ■ ▲ Draw as many resonance structures as you can for the following species:

(a)
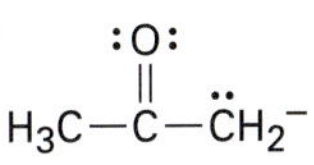

(b)
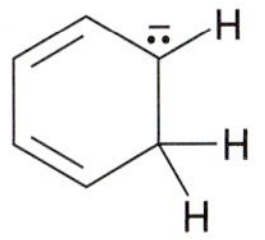

(c)
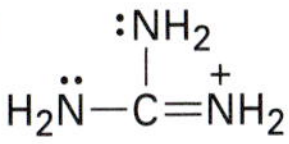

(d) $H_3C-\ddot{S}-\overset{+}{C}H_2$

(e) $H_2C=CH-CH=CH-\overset{+}{C}H-CH_3$

2.35 Cyclobutadiene is a rectangular molecule with two shorter double bonds and two longer single bonds. Why do the following structures *not* represent resonance forms?

2.36 Alcohols can act either as weak acids or as weak bases, just as water can. Show the reaction of methanol, CH_3OH, with a strong acid such as HCl and with a strong base such as $Na^+\ ^-NH_2$.

2.37 ▲ The O–H hydrogen in acetic acid is much more acidic than any of the C–H hydrogens. Explain this result using resonance structures.

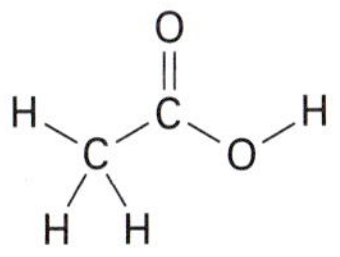
Acetic acid

2.38 ■ Which of the following are likely to act as Lewis acids and which as Lewis bases?

(a) $AlBr_3$ (b) $CH_3CH_2NH_2$ (c) BH_3
(d) HF (e) CH_3SCH_3 (f) $TiCl_4$

2.39 Draw an electron-dot structure for each of the molecules in Problem 2.38, indicating any unshared electron pairs.

2.40 ■ Write the products of the following acid–base reactions:

(a) $CH_3OH + H_2SO_4 \rightleftarrows ?$
(b) $CH_3OH + NaNH_2 \rightleftarrows ?$
(c) $CH_3NH_3^+\ Cl^- + NaOH \rightleftarrows ?$

2.41 ■ Assign formal charges to the atoms in each of the following molecules:

(a)
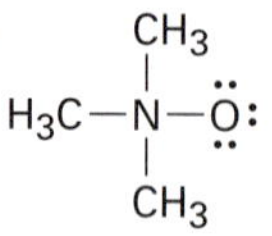
(b)
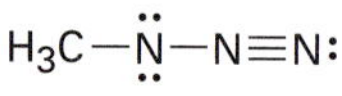
(c)
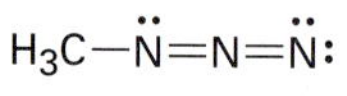

2.42 Maleic acid has a dipole moment, but the closely related fumaric acid, a substance involved in the citric acid cycle by which food molecules are metabolized, does not. Explain.

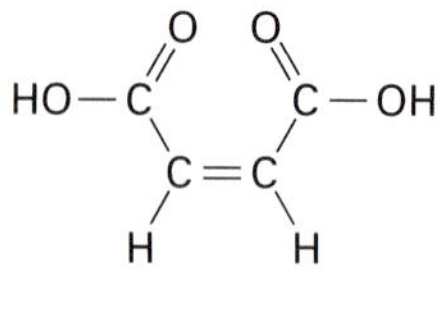
Maleic acid

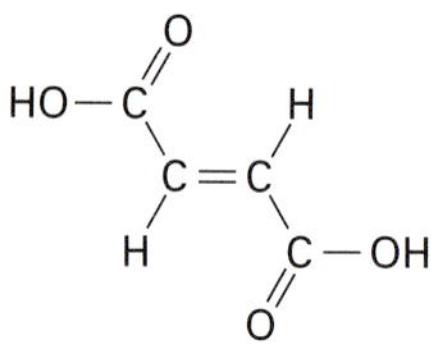
Fumaric acid

2.43 ■ Rank the following substances in order of increasing acidity:

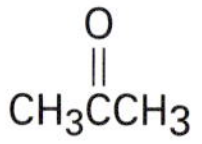
Acetone
(pK_a = 19.3)

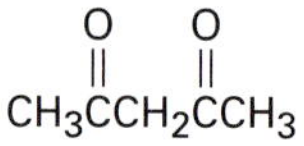
2,4-Pentanedione
(pK_a = 9)

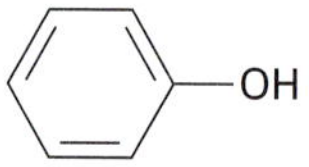
Phenol
(pK_a = 9.9)

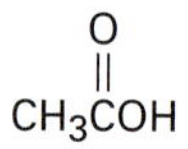
Acetic acid
(pK_a = 4.76)

2.44 Which, if any, of the four substances in Problem 2.43 is a strong enough acid to react almost completely with NaOH? (The pK_a of H_2O is 15.74.)

2.45 The ammonium ion (NH_4^+, pK_a = 9.25) has a lower pK_a than the methylammonium ion ($CH_3NH_3^+$, pK_a = 10.66). Which is the stronger base, ammonia (NH_3) or methylamine (CH_3NH_2)? Explain.

2.46 Is *tert*-butoxide anion a strong enough base to react with water? In other words, can a solution of potassium *tert*-butoxide be prepared in water? The pK_a of *tert*-butyl alcohol is approximately 18.

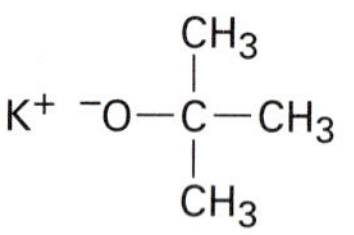
Potassium *tert*-butoxide

2.47 Predict the structure of the product formed in the reaction of the organic base pyridine with the organic acid acetic acid, and use curved arrows to indicate the direction of electron flow.

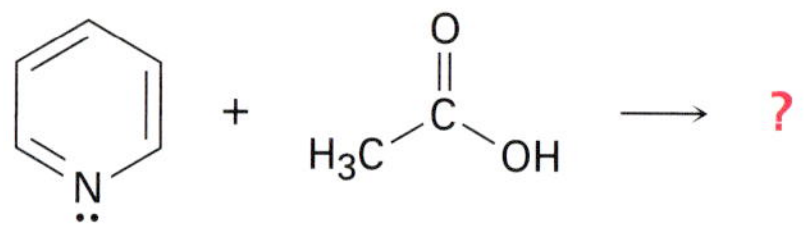
Pyridine **Acetic acid**

■ Assignable in OWL ▲ Key Idea Problems

2.48 ■ Calculate K_a values from the following pK_a's:

(a) Acetone, pK_a = 19.3 **(b)** Formic acid, pK_a = 3.75

2.49 ■ Calculate pK_a values from the following K_a's:

(a) Nitromethane, $K_a = 5.0 \times 10^{-11}$ **(b)** Acrylic acid, $K_a = 5.6 \times 10^{-5}$

2.50 What is the pH of a 0.050 M solution of formic acid, pK_a = 3.75?

2.51 Sodium bicarbonate, $NaHCO_3$, is the sodium salt of carbonic acid (H_2CO_3), pK_a = 6.37. Which of the substances shown in Problem 2.43 will react with sodium bicarbonate?

2.52 Assume that you have two unlabeled bottles, one of which contains phenol (pK_a = 9.9) and one of which contains acetic acid (pK_a = 4.76). In light of your answer to Problem 2.51, suggest a simple way to determine what is in each bottle.

2.53 ■ Identify the acids and bases in the following reactions:

(a) $CH_3OH + H^+ \longrightarrow CH_3\overset{+}{O}H_2$

(b)

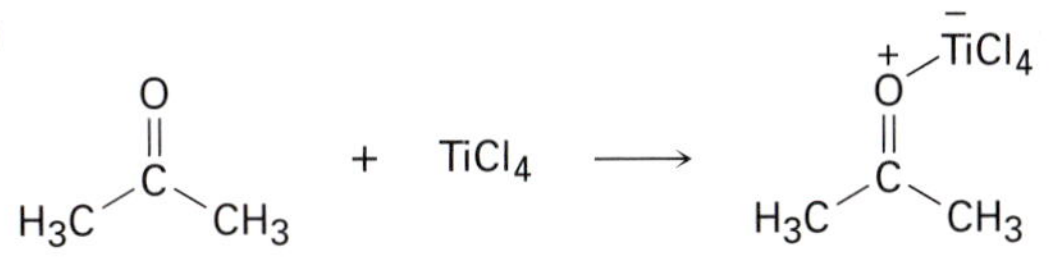

(c)

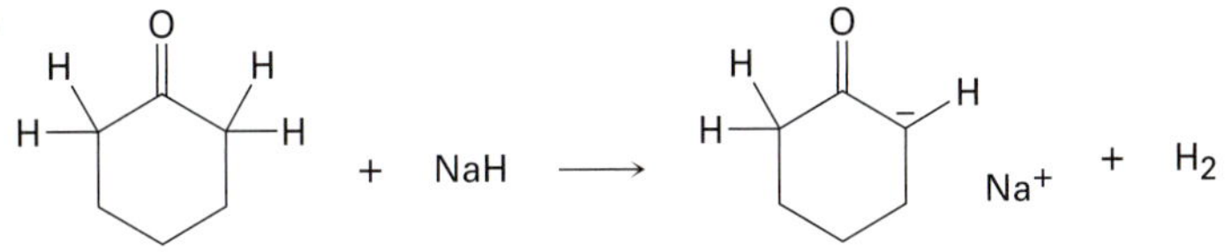

(d)

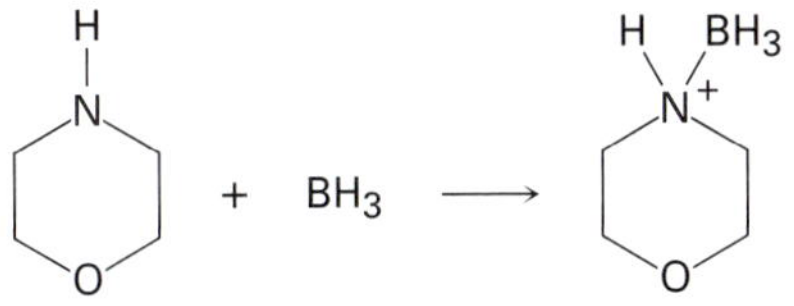

2.54 ■ ▲ Which of the following pairs represent resonance structures?

(a)

(b)

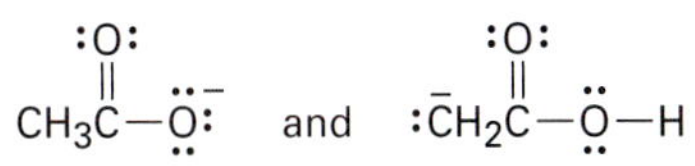

(c)

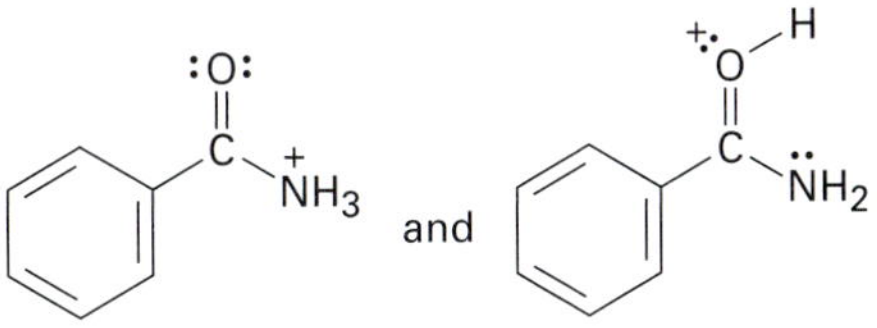

(d)

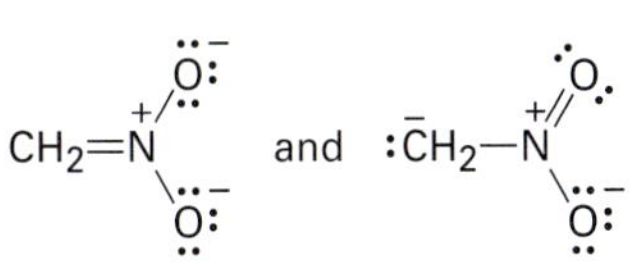

2.55 ▲ Draw as many resonance structures as you can for the following species, adding appropriate formal charges to each:

(a) Nitromethane,

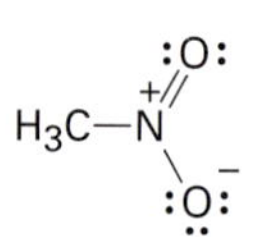

(b) Ozone,

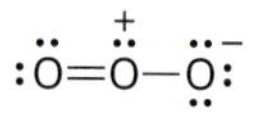

(c) Diazomethane,

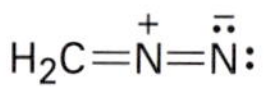

2.56 Carbocations, ions that contain a trivalent, positively charged carbon atom, react with water to give alcohols:

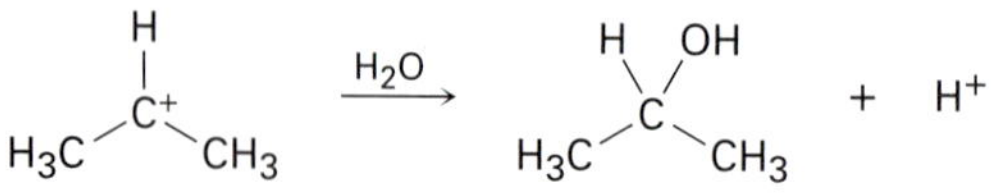

A carbocation **An alcohol**

How can you account for the fact that the following carbocation gives a mixture of *two* alcohols on reaction with water?

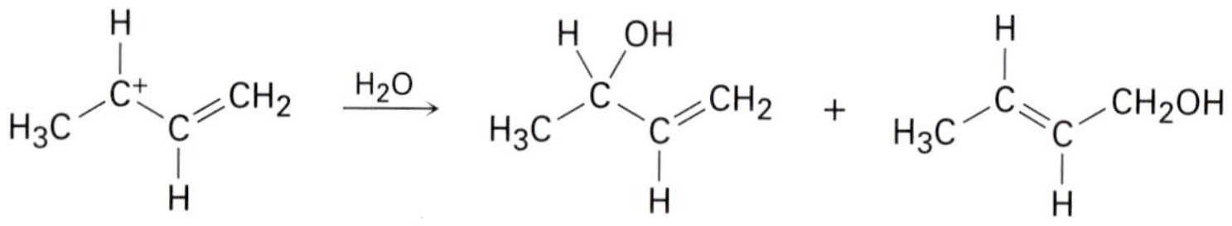

2.57 We'll see in the next chapter that organic molecules can be classified according to the *functional groups* they contain, where a functional group is a collection of atoms with a characteristic chemical reactivity. Use the electronegativity values given in Figure 2.2 to predict the direction of polarization of the following functional groups.

(a)

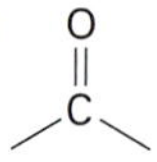

Ketone

(b)

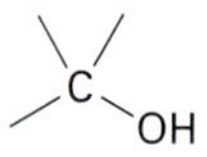

Alcohol

(c)

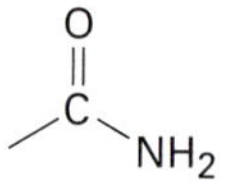

Amide

(d) —C≡N

Nitrile

2.58 Phenol, C_6H_5OH, is a stronger acid than methanol, CH_3OH, even though both contain an O–H bond. Draw the structures of the anions resulting from loss of H^+ from phenol and methanol, and use resonance structures to explain the difference in acidity.

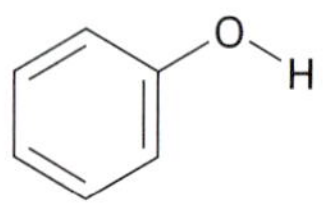

Phenol (pK_a = 9.89)

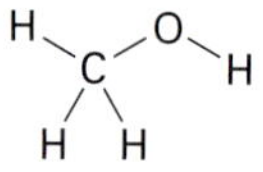

Methanol (pK_a = 15.54)

3 Organic Compounds: Alkanes and Their Stereochemistry

Organic **KNOWLEDGE TOOLS**

ThomsonNOW™ Throughout this chapter, sign in at **www.thomsonedu.com** for online self-study and interactive tutorials based on your level of understanding.

Online homework for this chapter may be assigned in Organic OWL.

According to *Chemical Abstracts,* the publication that abstracts and indexes the chemical literature, there are more than 30 million known organic compounds. Each of these compounds has its own physical properties, such as melting point and boiling point, and each has its own chemical reactivity.

Chemists have learned through many years of experience that organic compounds can be classified into families according to their structural features and that the members of a given family often have similar chemical behavior. Instead of 30 million compounds with random reactivity, there are a few dozen families of organic compounds whose chemistry is reasonably predictable. We'll study the chemistry of specific families throughout much of this book, beginning in this chapter with a look at the simplest family, the *alkanes.*

WHY THIS CHAPTER?

Alkanes are relatively unreactive, but they nevertheless provide a useful vehicle for introducing some important general ideas. In this chapter, we'll use alkanes for discussing the basic approach to naming organic compounds and for taking an initial look at some of the three-dimensional aspects of molecules, a topic of particular importance in understanding biological organic chemistry.

3.1 Functional Groups

ThomsonNOW™ Click *Organic Interactive* to learn how to **recognize functional groups in organic molecules.**

The structural features that make it possible to classify compounds into families are called *functional groups.* A **functional group** is a group of atoms that has a characteristic chemical behavior in every molecule where it occurs. For example, compare ethylene, a plant hormone that causes fruit to ripen, with menthene, a much more complicated molecule. Both substances contain a carbon–carbon double-bond functional group, and both therefore react with Br_2 in the same way to give products in which a Br atom has added to each of the double-bond carbons (Figure 3.1). This example is typical: *the chemistry of every organic molecule, regardless of size and complexity, is determined by the functional groups it contains.*

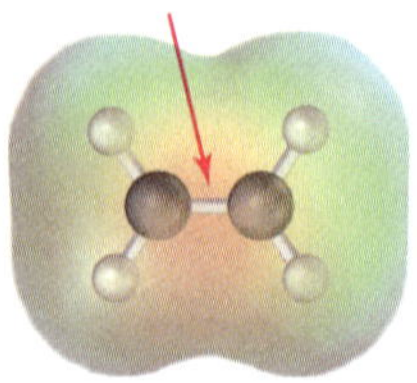

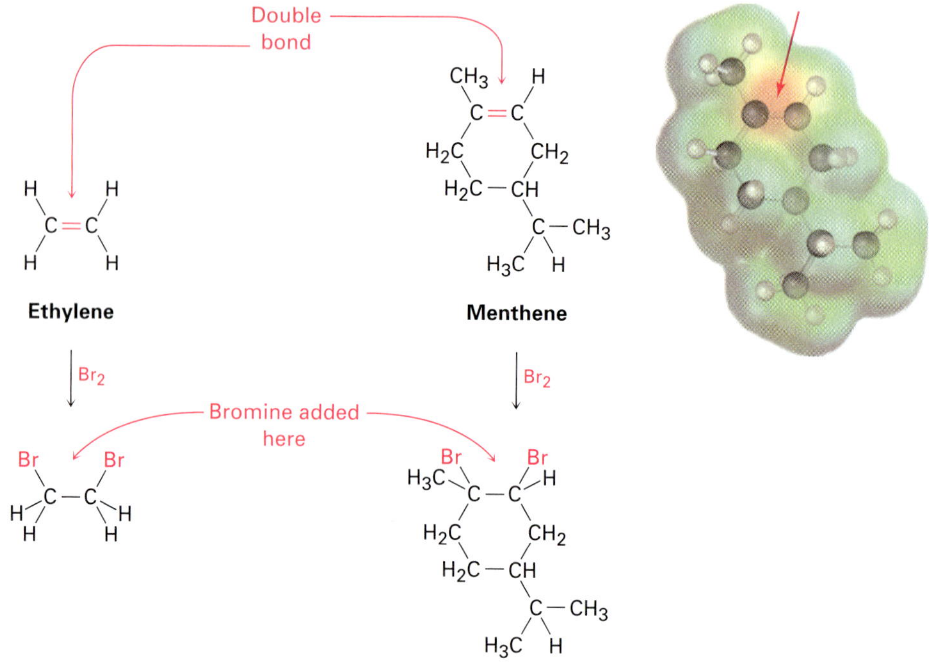

Figure 3.1 The reactions of ethylene and menthene with bromine. In both molecules, the carbon–carbon double-bond functional group has a similar polarity pattern, so both molecules react with Br_2 in the same way. The size and complexity of the remainders of the molecules are not important.

Look carefully at Table 3.1 on pages 76 and 77, which lists many of the common functional groups and gives simple examples of their occurrence. Some functional groups have only carbon–carbon double or triple bonds; others have halogen atoms; and still others contain oxygen, nitrogen, or sulfur. Much of the chemistry you'll be studying is the chemistry of these functional groups.

Functional Groups with Carbon–Carbon Multiple Bonds

Alkenes, alkynes, and arenes (aromatic compounds) all contain carbon–carbon multiple bonds. *Alkenes* have a double bond, *alkynes* have a triple bond, and *arenes* have alternating double and single bonds in a six-membered ring of carbon atoms. Because of their structural similarities, these compounds also have chemical similarities.

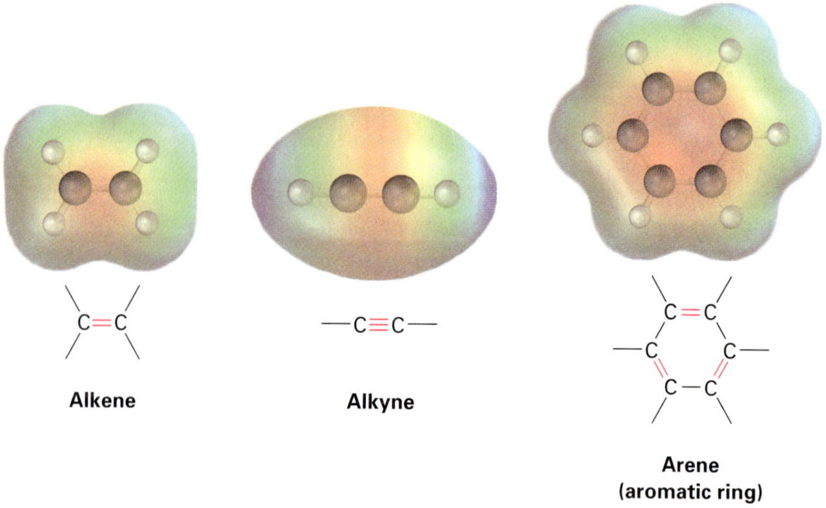

Functional Groups with Carbon Singly Bonded to an Electronegative Atom Alkyl halides (haloalkanes), alcohols, ethers, amines, thiols, sulfides, and disulfides all have a carbon atom singly bonded to an electronegative atom—halogen, oxygen, nitrogen, or sulfur. *Alkyl halides* have a carbon atom bonded to halogen (−X), *alcohols* have a carbon atom bonded to the oxygen of a hydroxyl group (−OH), *ethers* have two carbon atoms bonded to the same oxygen, *organophosphates* have a carbon atom bonded to the oxygen of a phosphate group ($-OPO_3^{2-}$), *amines* have a carbon atom bonded to a nitrogen, *thiols* have a carbon atom bonded to an −SH group, *sulfides* have two carbon atoms bonded to the same sulfur, and *disulfides* have carbon atoms bonded to two sulfurs that are joined together. In all cases, the bonds are polar, with the carbon atom bearing a partial positive charge ($\delta+$) and the electronegative atom bearing a partial negative charge ($\delta-$).

Alkyl halide (haloalkane)

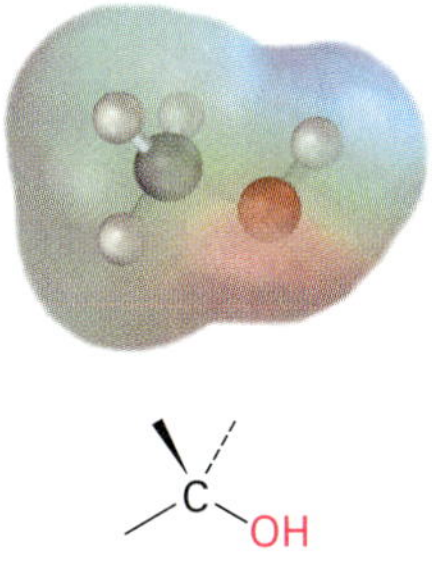

Alcohol

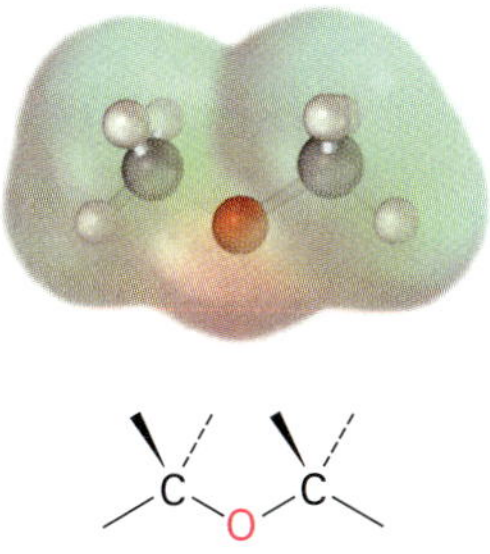

Ether

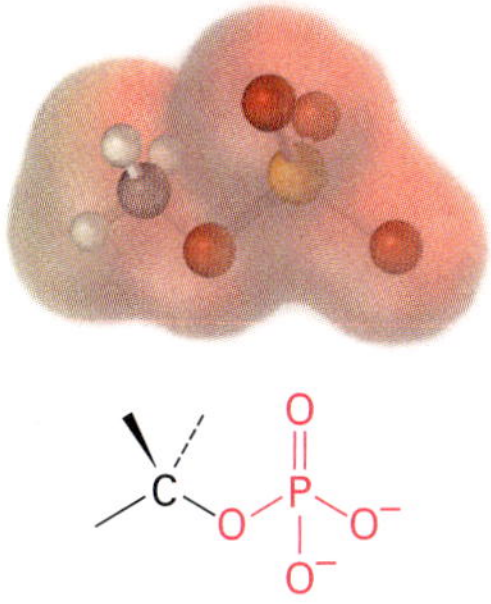

Phosphate

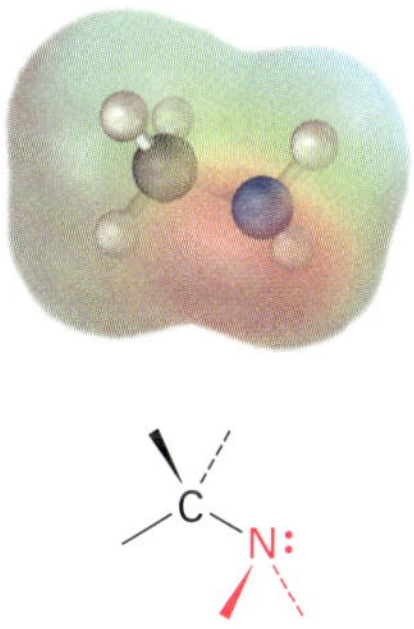

Amine

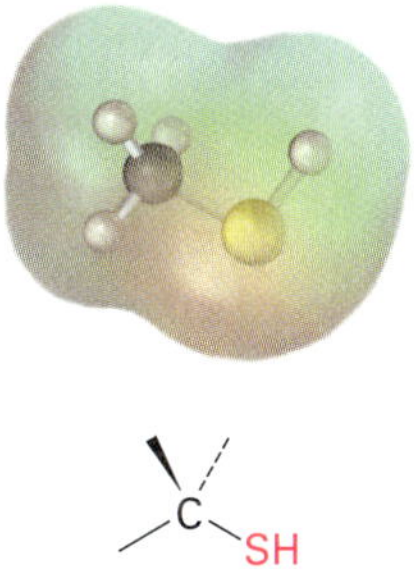

Thiol

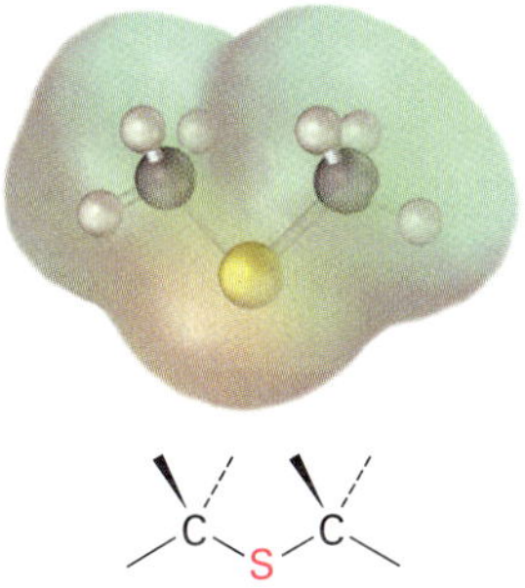

Sulfide

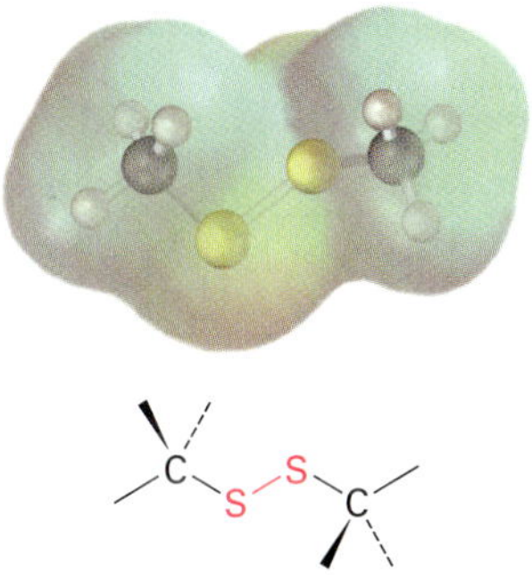

Disulfide

Functional Groups with a Carbon–Oxygen Double Bond (Carbonyl Groups) Note particularly the last seven entries in Table 3.1, which list different families of compounds that contain the *carbonyl group,* C=O (pronounced car-bo-**neel**). Functional groups with a carbon–oxygen double bond are present in the great majority of organic compounds and in practically all biological molecules. These compounds behave similarly in many respects but differ depending on the identity of the atoms bonded to the carbonyl-group carbon. *Aldehydes* have at least one hydrogen bonded to the C=O, *ketones* have two carbons bonded to the C=O, *carboxylic acids* have an −OH group bonded to the C=O, *esters* have an ether-like oxygen bonded to the C=O, *amides* have an amine-like nitrogen

Table 3.1 Structures of Some Common Functional Groups

Name	Structure*	Name ending	Example
Alkene (double bond)	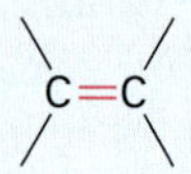	*-ene*	$H_2C{=}CH_2$ Ethene
Alkyne (triple bond)	$—C{\equiv}C—$	*-yne*	$HC{\equiv}CH$ Ethyne
Arene (aromatic ring)	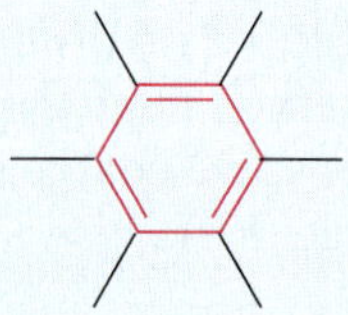	None	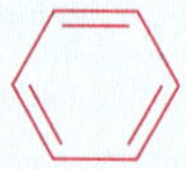 Benzene
Halide	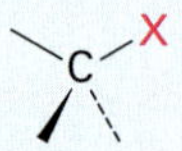 (X = F, Cl, Br, I)	None	CH_3Cl Chloromethane
Alcohol	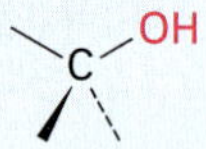	*-ol*	CH_3OH Methanol
Ether	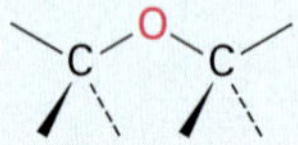	*ether*	CH_3OCH_3 Dimethyl ether
Monophosphate	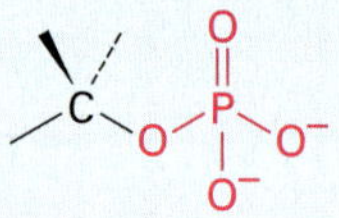	*phosphate*	$CH_3OPO_3{}^{2-}$ Methyl phosphate
Amine	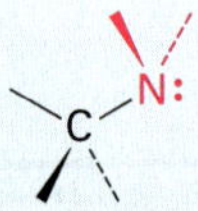	*-amine*	CH_3NH_2 Methylamine
Imine (Schiff base)	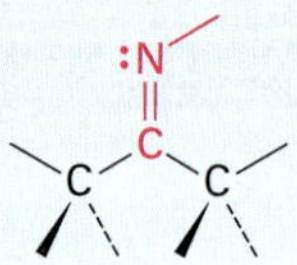	None	NH ‖ CH_3CCH_3 Acetone imine
Nitrile	$—C{\equiv}N$	*-nitrile*	$CH_3C{\equiv}N$ Ethanenitrile
Nitro	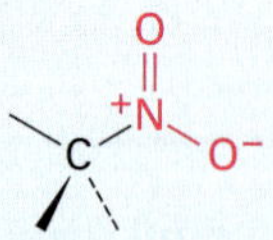	None	CH_3NO_2 Nitromethane
Thiol	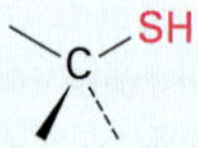	*-thiol*	CH_3SH Methanethiol

*The bonds whose connections aren't specified are assumed to be attached to carbon or hydrogen atoms in the rest of the molecule.

(continued)

Table 3.1 Structures of Some Common Functional Groups *(continued)*

Name	Structure*	Name ending	Example
Sulfide	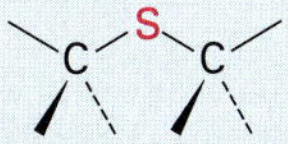	*sulfide*	CH_3SCH_3 Dimethyl sulfide
Disulfide	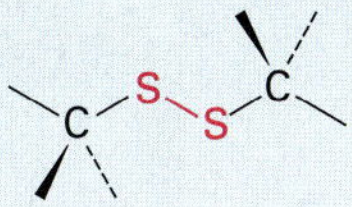	*disulfide*	CH_3SSCH_3 Dimethyl disulfide
Carbonyl	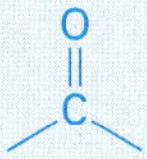		
Aldehyde	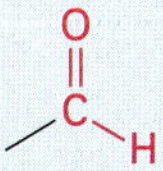	*-al*	$CH_3\overset{O}{\overset{\|}{C}}H$ Ethanal
Ketone	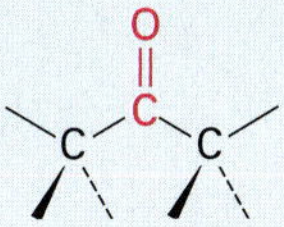	*-one*	$CH_3\overset{O}{\overset{\|}{C}}CH_3$ Propanone
Carboxylic acid	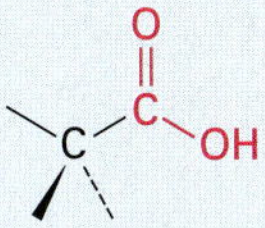	*-oic acid*	$CH_3\overset{O}{\overset{\|}{C}}OH$ Ethanoic acid
Ester	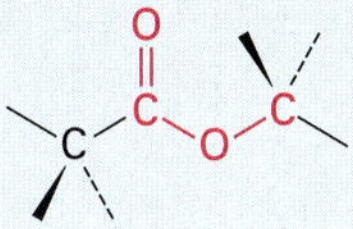	*-oate*	$CH_3\overset{O}{\overset{\|}{C}}OCH_3$ Methyl ethanoate
Amide	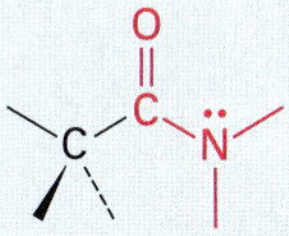	*-amide*	$CH_3\overset{O}{\overset{\|}{C}}NH_2$ Ethanamide
Carboxylic acid anhydride	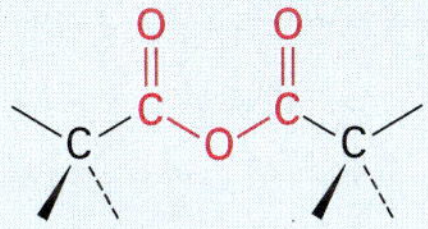	*-oic anhydride*	$CH_3\overset{O}{\overset{\|}{C}}O\overset{O}{\overset{\|}{C}}CH_3$ Ethanoic anhydride
Carboxylic acid chloride	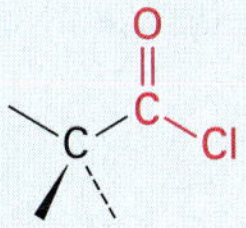	*-oyl chloride*	$CH_3\overset{O}{\overset{\|}{C}}Cl$ Ethanoyl chloride

*The bonds whose connections aren't specified are assumed to be attached to carbon or hydrogen atoms in the rest of the molecule.

bonded to the C=O, *acid chlorides* have a chlorine bonded to the C=O, and so on. The carbonyl carbon atom bears a partial positive charge (δ+), and the oxygen bears a partial negative charge (δ−).

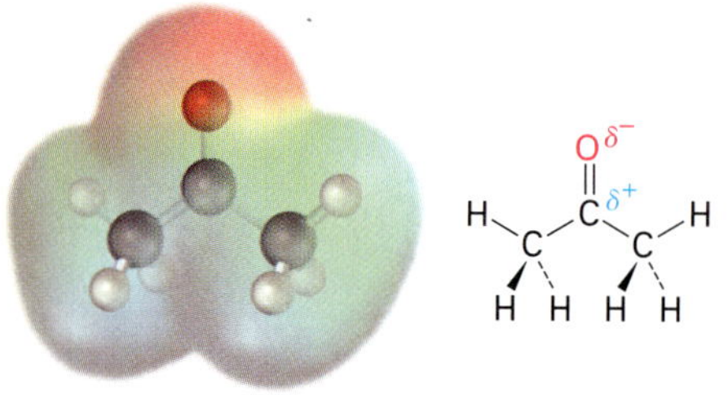

Acetone—a typical carbonyl compound

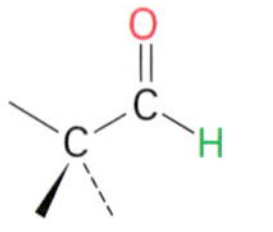

Aldehyde

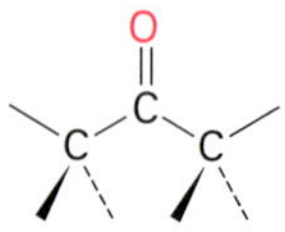

Ketone

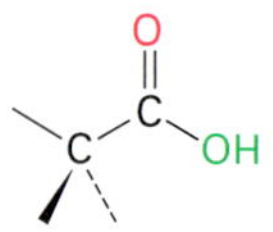

Carboxylic acid

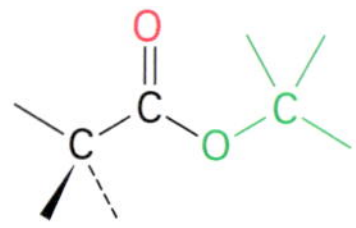

Ester

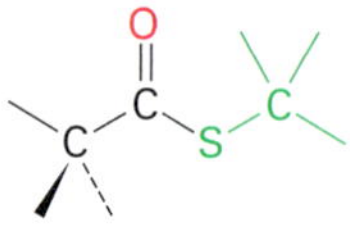

Thioester

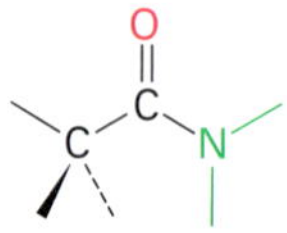

Amide

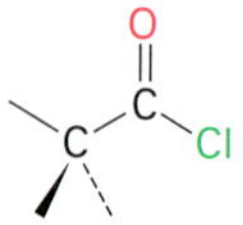

Acid chloride

Problem 3.1 Identify the functional groups in each of the following molecules:

(a) Methionine, an amino acid:

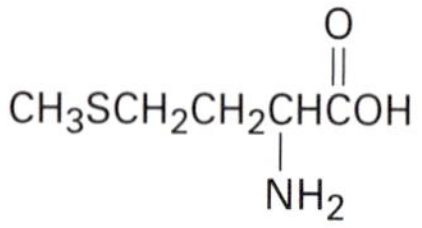

(b) Ibuprofen, a pain reliever:

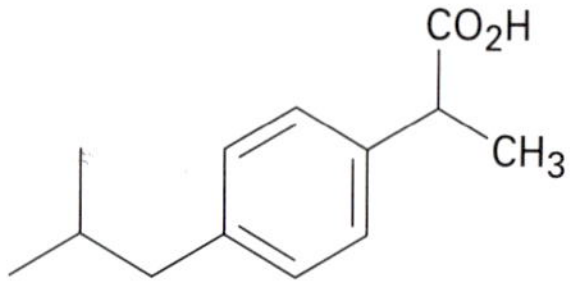

(c) Capsaicin, the pungent substance in chili peppers:

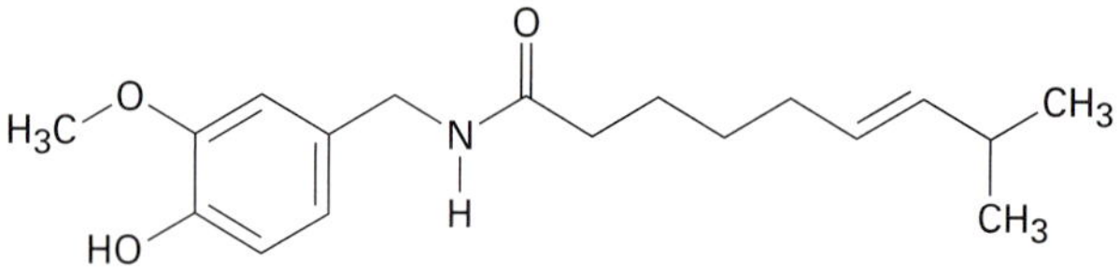

Problem 3.2 Propose structures for simple molecules that contain the following functional groups:

(a) Alcohol (b) Aromatic ring (c) Carboxylic acid
(d) Amine (e) Both ketone and amine (f) Two double bonds

Problem 3.3 Identify the functional groups in the following model of arecoline, a veterinary drug used to control worms in animals. Convert the drawing into a line-bond structure and a molecular formula (red = O, blue = N).

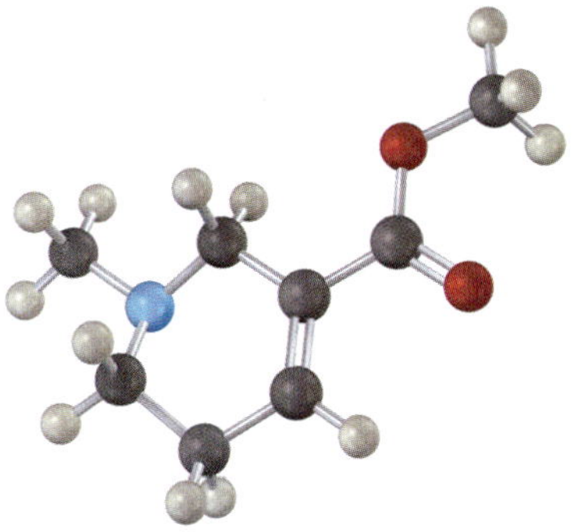

3.2 Alkanes and Alkane Isomers

ThomsonNOW™ Click *Organic Interactive* to learn to **draw and recognize alkane isomers.**

Before beginning a systematic study of the different functional groups, let's look first at the simplest family of molecules—the *alkanes*—to develop some general ideas that apply to all families. We saw in Section 1.7 that the carbon–carbon single bond in ethane results from σ (head-on) overlap of carbon sp^3 orbitals. If we imagine joining three, four, five, or even more carbon atoms by C–C single bonds, we can generate the large family of molecules called **alkanes**.

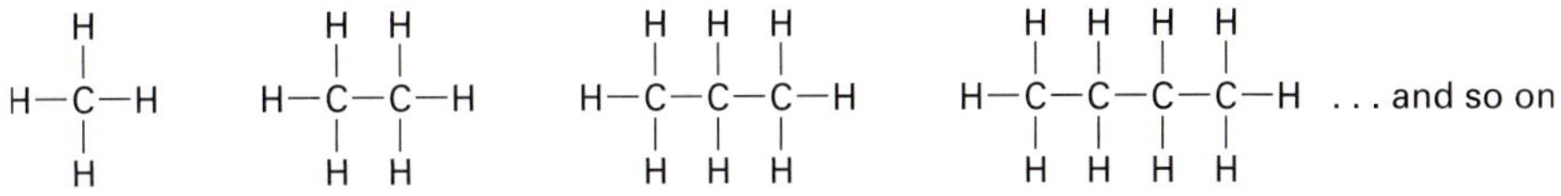

Methane **Ethane** **Propane** **Butane**

Alkanes are often described as *saturated hydrocarbons*—**hydrocarbons** because they contain only carbon and hydrogen; **saturated** because they have only C–C and C–H single bonds and thus contain the maximum possible number of hydrogens per carbon. They have the general formula C_nH_{2n+2}, where n is an integer. Alkanes are also occasionally referred to as **aliphatic** compounds, a name derived from the Greek *aleiphas,* meaning "fat." We'll see in Section 27.1 that many animal fats contain long carbon chains similar to alkanes.

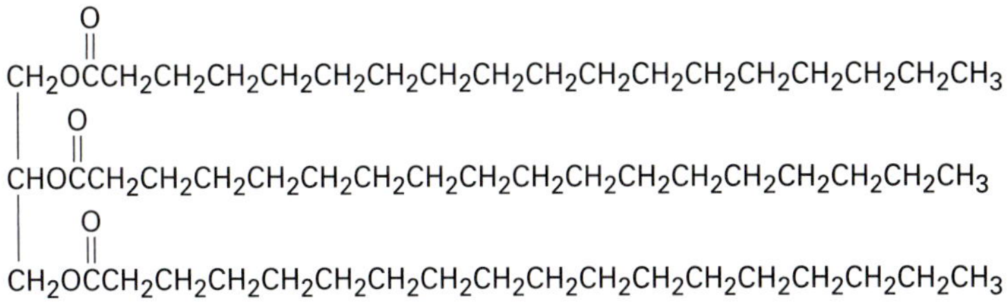

A typical animal fat

Think about the ways that carbon and hydrogen might combine to make alkanes. With one carbon and four hydrogens, only one structure is possible: methane, CH_4. Similarly, there is only one combination of two carbons with six hydrogens (ethane, CH_3CH_3) and only one combination of three carbons with eight hydrogens (propane, $CH_3CH_2CH_3$). If larger numbers of carbons and hydrogens combine, however, more than one structure is possible. For example, there are *two* substances with the formula C_4H_{10}: the four carbons can all be in a row (butane), or they can branch (isobutane). Similarly, there are three C_5H_{12} molecules, and so on for larger alkanes.

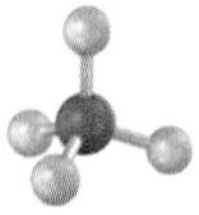

CH_4

Methane, CH_4

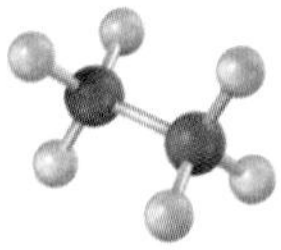

CH_3CH_3

Ethane, C_2H_6

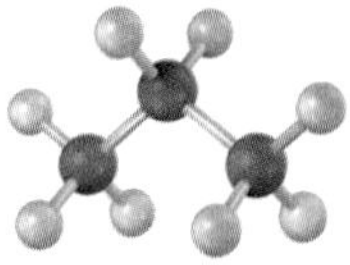

$CH_3CH_2CH_3$

Propane, C_3H_8

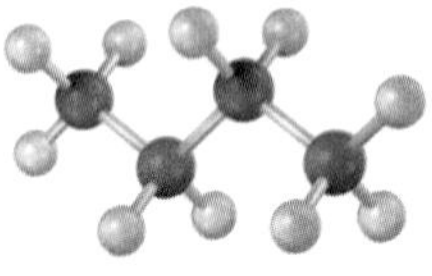

$CH_3CH_2CH_2CH_3$

Butane, C_4H_{10}

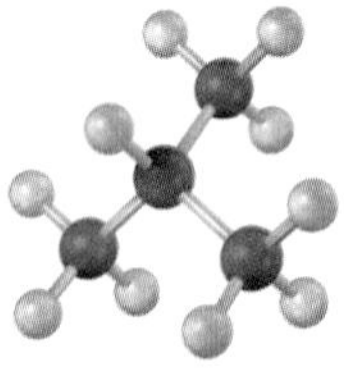

CH_3
CH_3CHCH_3

Isobutane, C_4H_{10}
(2-methylpropane)

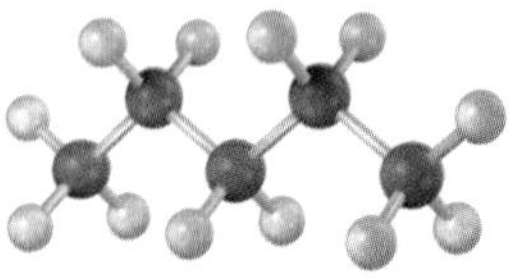

$CH_3CH_2CH_2CH_2CH_3$

Pentane, C_5H_{12}

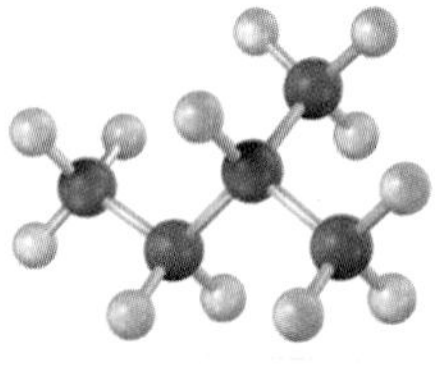

CH_3
$CH_3CH_2CHCH_3$

2-Methylbutane, C_5H_{12}

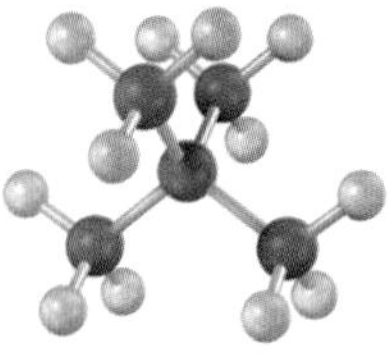

CH_3
CH_3CCH_3
CH_3

2,2-Dimethylpropane, C_5H_{12}

Compounds like butane and pentane, whose carbons are all connected in a row, are called **straight-chain alkanes**, or *normal alkanes*. Compounds like 2-methylpropane (isobutane), 2-methylbutane, and 2,2-dimethylpropane, whose carbon chains branch, are called **branched-chain alkanes**. The difference between the two is that you can draw a line connecting all the carbons of a straight-chain alkane without retracing your path or lifting your pencil from

Table 3.2 Number of Alkane Isomers

Formula	Number of isomers
C_6H_{14}	5
C_7H_{16}	9
C_8H_{18}	18
C_9H_{20}	35
$C_{10}H_{22}$	75
$C_{15}H_{32}$	4,347
$C_{20}H_{42}$	366,319
$C_{30}H_{62}$	4,111,846,763

the paper. For a branched-chain alkane, however, you either have to retrace your path or lift your pencil from the paper to draw a line connecting all the carbons.

Compounds like the two C_4H_{10} molecules and the three C_5H_{12} molecules, which have the same formula but different structures, are called *isomers,* from the Greek *isos* + *meros,* meaning "made of the same parts." **Isomers** are compounds that have the same numbers and kinds of atoms but differ in the way the atoms are arranged. Compounds like butane and isobutane, whose atoms are connected differently, are called **constitutional isomers**. We'll see shortly that other kinds of isomers are also possible, even among compounds whose atoms are connected in the same order. As Table 3.2 shows, the number of possible alkane isomers increases dramatically as the number of carbon atoms increases.

Constitutional isomerism is not limited to alkanes—it occurs widely throughout organic chemistry. Constitutional isomers may have different carbon skeletons (as in isobutane and butane), different functional groups (as in ethanol and dimethyl ether), or different locations of a functional group along the chain (as in isopropylamine and propylamine). Regardless of the reason for the isomerism, constitutional isomers are always different compounds with different properties, but with the same formula.

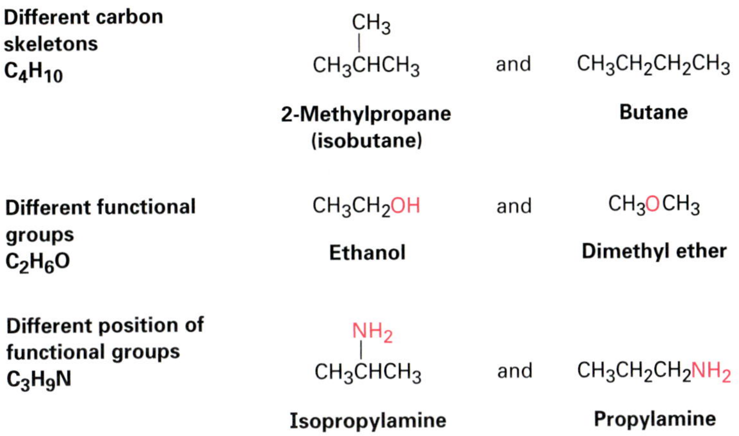

A given alkane can be drawn arbitrarily in many ways. For example, the straight-chain, four-carbon alkane called butane can be represented by any of the structures shown in Figure 3.2. These structures don't imply any particular three-dimensional geometry for butane; they indicate only the connections among atoms. In practice, as noted in Section 1.12, chemists rarely draw all the bonds in a molecule and usually refer to butane by the condensed structure, $CH_3CH_2CH_2CH_3$ or $CH_3(CH_2)_2CH_3$. Still more simply, butane can even be represented as n-C_4H_{10}, where *n* denotes *normal* (straight-chain) butane.

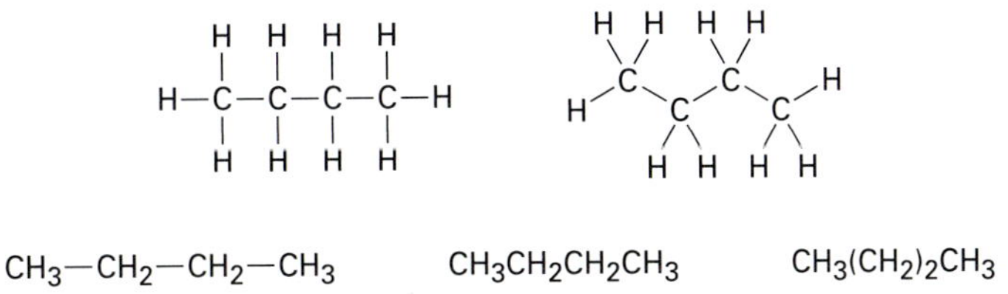

Figure 3.2 Some representations of butane, C_4H_{10}. The molecule is the same regardless of how it's drawn. These structures imply only that butane has a continuous chain of four carbon atoms; they do not imply any specific geometry.

Straight-chain alkanes are named according to the number of carbon atoms they contain, as shown in Table 3.3. With the exception of the first four compounds—methane, ethane, propane, and butane—whose names have historical roots, the alkanes are named based on Greek numbers. The suffix *-ane* is added to the end of each name to indicate that the molecule identified is an alkane. Thus, pent*ane* is the five-carbon alkane, hex*ane* is the six-carbon alkane, and so on. We'll soon see that these alkane names form the basis for naming all other organic compounds, so at least the first ten should be memorized.

Table 3.3 **Names of Straight-Chain Alkanes**

Number of carbons (n)	Name	Formula (C_nH_{2n+2})
1	Methane	CH_4
2	Ethane	C_2H_6
3	Propane	C_3H_8
4	Butane	C_4H_{10}
5	Pentane	C_5H_{12}
6	Hexane	C_6H_{14}
7	Heptane	C_7H_{16}
8	Octane	C_8H_{18}
9	Nonane	C_9H_{20}
10	Decane	$C_{10}H_{22}$
11	Undecane	$C_{11}H_{24}$
12	Dodecane	$C_{12}H_{26}$
13	Tridecane	$C_{13}H_{28}$
20	Icosane	$C_{20}H_{42}$
30	Triacontane	$C_{30}H_{62}$

WORKED EXAMPLE 3.1 *Drawing the Structures of Isomers*

Propose structures for two isomers with the formula C_2H_7N.

Strategy We know that carbon forms four bonds, nitrogen forms three, and hydrogen forms one. Write down the carbon atoms first, and then use a combination of trial and error plus intuition to put the pieces together.

Solution There are two isomeric structures. One has the connection C–C–N, and the other has the connection C–N–C.

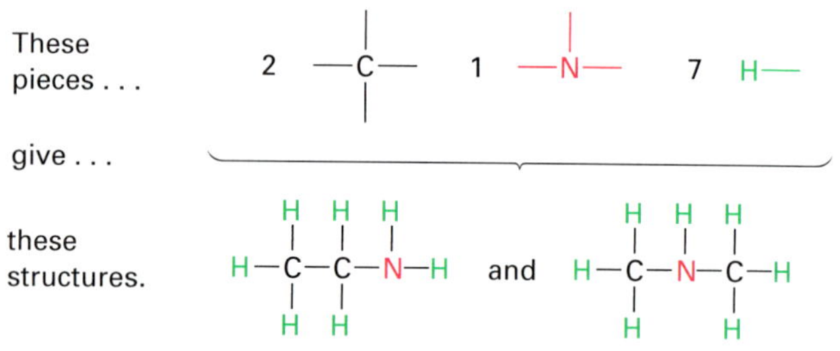

Problem 3.4 Draw structures of the five isomers of C_6H_{14}.

Problem 3.5 Propose structures that meet the following descriptions:
(a) Two isomeric esters with the formula $C_5H_{10}O_2$
(b) Two isomeric nitriles with the formula C_4H_7N
(c) Two isomeric disulfides with the formula $C_4H_{10}S_2$

Problem 3.6 How many isomers are there with the following descriptions?
(a) Alcohols with the formula C_3H_8O
(b) Bromoalkanes with the formula C_4H_9Br

3.3 Alkyl Groups

If you imagine removing a hydrogen atom from an alkane, the partial structure that remains is called an **alkyl group**. Alkyl groups are not stable compounds themselves, they are simply parts of larger compounds. Alkyl groups are named by replacing the *-ane* ending of the parent alkane with an *-yl* ending. For example, removal of a hydrogen from methane, CH_4, generates a *methyl* group, $-CH_3$, and removal of a hydrogen from ethane, CH_3CH_3, generates an *ethyl* group, $-CH_2CH_3$. Similarly, removal of a hydrogen atom from the end carbon of any straight-chain alkane gives the series of straight-chain alkyl groups shown in Table 3.4. Combining an alkyl group with any of the functional groups listed earlier makes it possible to generate and name many thousands of compounds. For example:

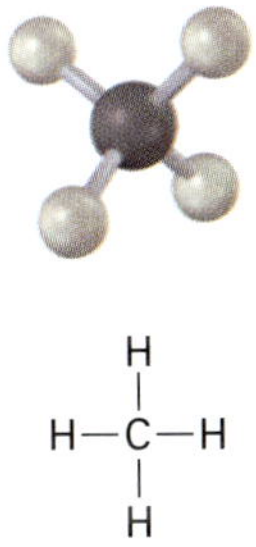
Methane

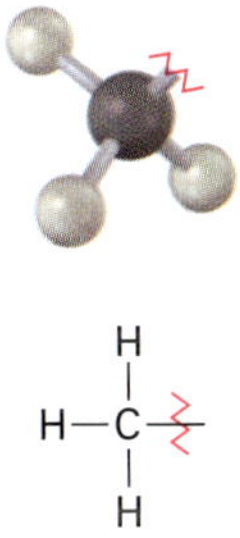
A methyl group

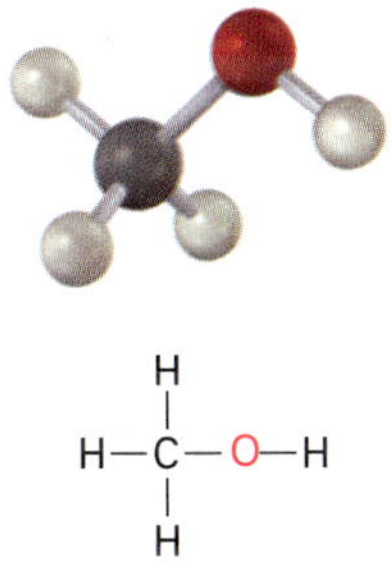
Methyl alcohol (methanol)

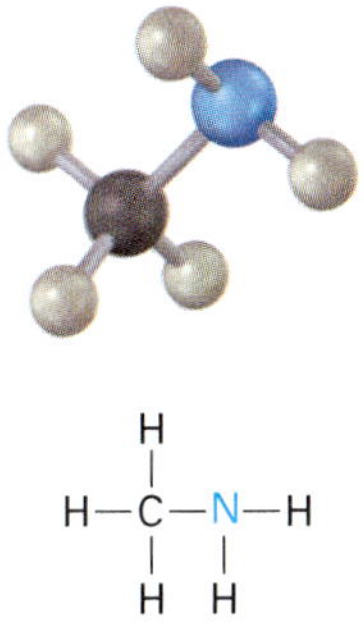
Methylamine

Table 3.4 Some Straight-Chain Alkyl Groups

Alkane	Name	Alkyl group	Name (abbreviation)
CH_4	Methane	$-CH_3$	Methyl (Me)
CH_3CH_3	Ethane	$-CH_2CH_3$	Ethyl (Et)
$CH_3CH_2CH_3$	Propane	$-CH_2CH_2CH_3$	Propyl (Pr)
$CH_3CH_2CH_2CH_3$	Butane	$-CH_2CH_2CH_2CH_3$	Butyl (Bu)
$CH_3CH_2CH_2CH_2CH_3$	Pentane	$-CH_2CH_2CH_2CH_2CH_3$	Pentyl, or amyl

Just as straight-chain alkyl groups are generated by removing a hydrogen from an *end* carbon, branched alkyl groups are generated by removing a hydrogen atom from an *internal* carbon. Two 3-carbon alkyl groups and four 4-carbon alkyl groups are possible (Figure 3.3).

Figure 3.3 Alkyl groups generated from straight-chain alkanes.

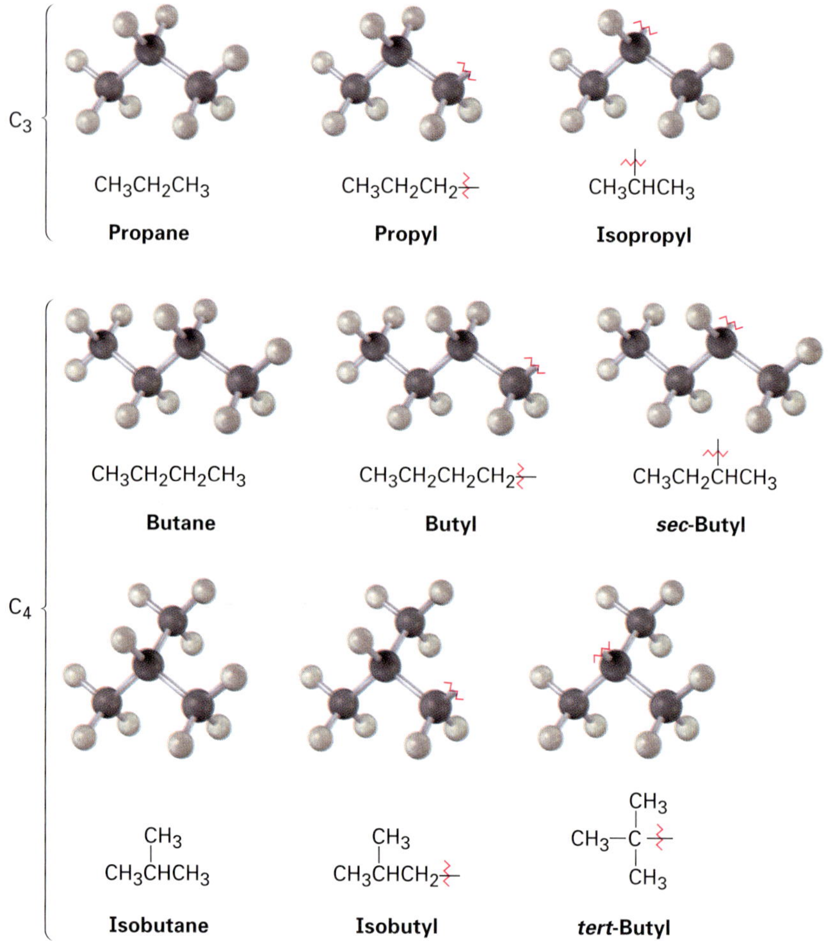

One further word about naming alkyl groups: the prefixes *sec-* (for secondary) and *tert-* (for tertiary) used for the C_4 alkyl groups in Figure 3.3 refer to *the number of other carbon atoms attached to the branching carbon atom.* There are four possibilities: primary (1°), secondary (2°), tertiary (3°), and quaternary (4°).

Primary **carbon (1°) is bonded to one other carbon.**

Secondary **carbon (2°) is bonded to two other carbons.**

Tertiary **carbon (3°) is bonded to three other carbons.**

Quaternary **carbon (4°) is bonded to four other carbons.**

The symbol **R** is used in organic chemistry to represent a *generalized* organic group. The R group can be methyl, ethyl, propyl, or any of a multitude of others.

You might think of **R** as representing the **R**est of the molecule, which we aren't bothering to specify.

The terms *primary, secondary, tertiary,* and *quaternary* are routinely used in organic chemistry, and their meanings need to become second nature. For example, if we were to say, "Citric acid is a tertiary alcohol," we would mean that it has an alcohol functional group (–OH) bonded to a carbon atom that is itself bonded to three other carbons. (These other carbons may in turn connect to other functional groups).

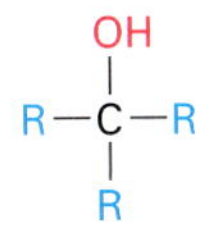

General class of tertiary alcohols, R_3COH

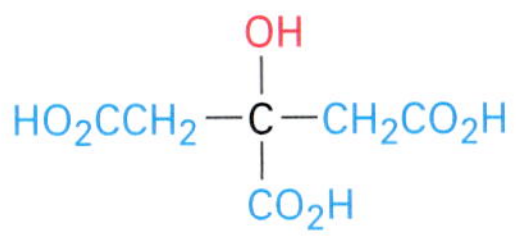

Citric acid—a specific tertiary alcohol

In addition, we also speak about hydrogen atoms as being primary, secondary, or tertiary. Primary hydrogen atoms are attached to primary carbons (RCH_3), secondary hydrogens are attached to secondary carbons (R_2CH_2), and tertiary hydrogens are attached to tertiary carbons (R_3CH). There is, of course, no such thing as a quaternary hydrogen. (Why?)

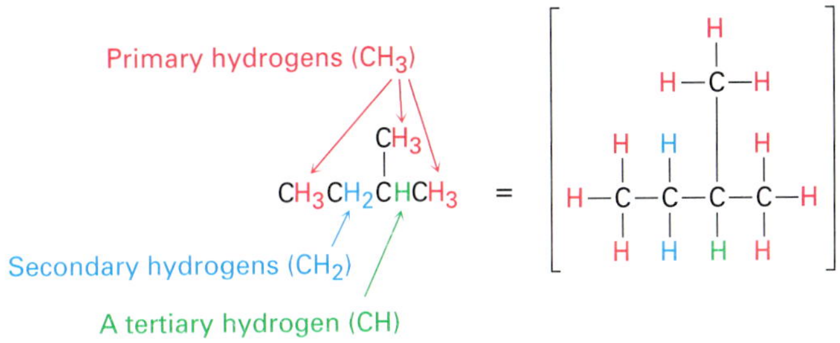

Problem 3.7 Draw the eight 5-carbon alkyl groups (pentyl isomers).

Problem 3.8 Identify the carbon atoms in the following molecules as primary, secondary, tertiary, or quaternary:

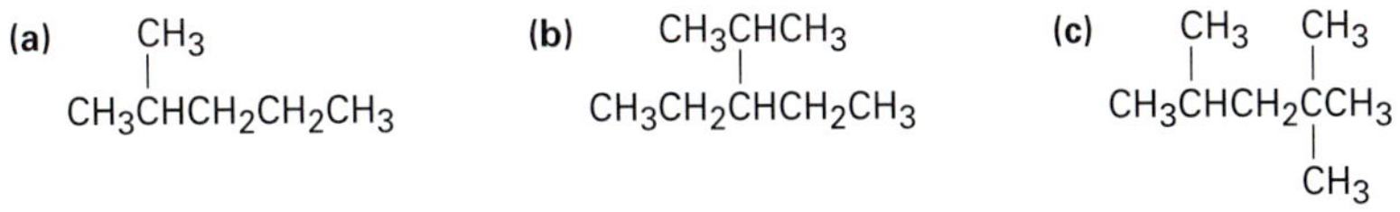

Problem 3.9 Identify the hydrogen atoms on the compounds shown in Problem 3.8 as primary, secondary, or tertiary.

Problem 3.10 Draw structures of alkanes that meet the following descriptions:

(a) An alkane with two tertiary carbons

(b) An alkane that contains an isopropyl group

(c) An alkane that has one quaternary and one secondary carbon

3.4 Naming Alkanes

ThomsonNOW™ Click *Organic Interactive* to learn to **write IUPAC names for simple alkanes.**

In earlier times, when relatively few pure organic chemicals were known, new compounds were named at the whim of their discoverer. Thus, urea (CH_4N_2O) is a crystalline substance isolated from urine; morphine ($C_{17}H_{19}NO_3$) is an analgesic (painkiller) named after Morpheus, the Greek god of dreams; and barbituric acid is a tranquilizing agent said to be named by its discoverer in honor of his friend Barbara.

As the science of organic chemistry slowly grew in the 19th century, so too did the number of known compounds and the need for a systematic method of naming them. The system of nomenclature we'll use in this book is that devised by the International Union of Pure and Applied Chemistry (IUPAC, usually spoken as **eye**-you-pac).

A chemical name typically has four parts in the IUPAC system of nomenclature: prefix, locant, parent, and suffix. The prefix specifies the location and identity of various **substituent** groups in the molecule, the locant gives the location of the primary functional group, the parent selects a main part of the molecule and tells how many carbon atoms are in that part, and the suffix identifies the primary functional group.

Prefix—Locant—Parent—Suffix

Where and what are the substituents? **Where is the primary functional group?** **How many carbons?** **What is the primary functional group?**

As we cover new functional groups in later chapters, the applicable IUPAC rules of nomenclature will be given. In addition, Appendix A at the back of this book gives an overall view of organic nomenclature and shows how compounds that contain more than one functional group are named. For the present, let's see how to name branched-chain alkanes and learn some general naming rules that are applicable to all compounds.

All but the most complex branched-chain alkanes can be named by following four steps. For a very few compounds, a fifth step is needed.

Step 1 **Find the parent hydrocarbon.**

(a) Find the longest continuous chain of carbon atoms in the molecule, and use the name of that chain as the parent name. The longest chain may not always be apparent from the manner of writing; you may have to "turn corners."

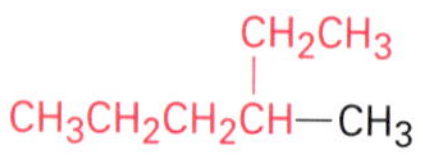

Named as a substituted hexane

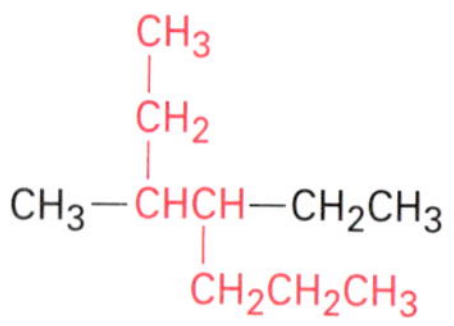

Named as a substituted heptane

(b) If two different chains of equal length are present, choose the one with the larger number of branch points as the parent.

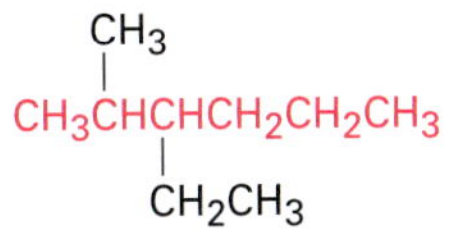

Named as a hexane with *two* substituents

NOT

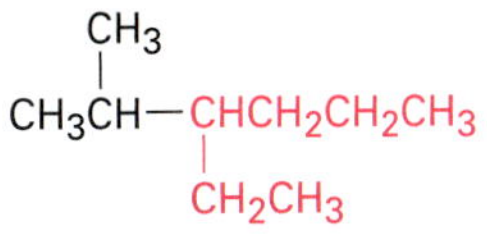

as a hexane with *one* substituent

Step 2 **Number the atoms in the main chain.**

(a) Beginning at the end nearer the first branch point, number each carbon atom in the parent chain.

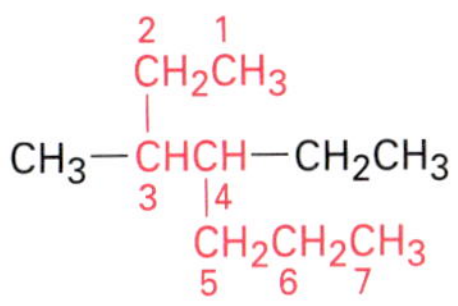

NOT

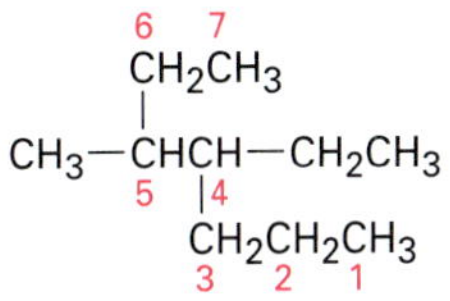

The first branch occurs at C3 in the proper system of numbering, not at C4.

(b) If there is branching an equal distance away from both ends of the parent chain, begin numbering at the end nearer the second branch point.

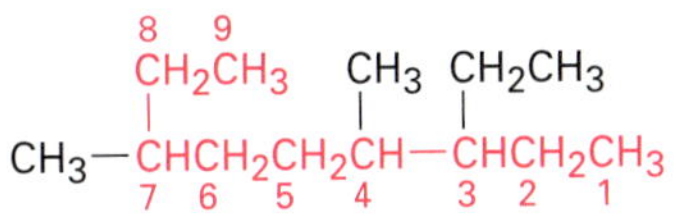

NOT

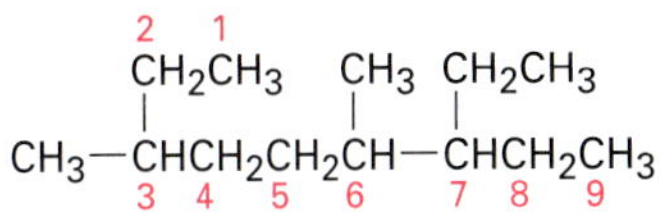

Step 3 **Identify and number the substituents.**

(a) Assign a number, called a *locant,* to each substituent to locate its point of attachment to the parent chain.

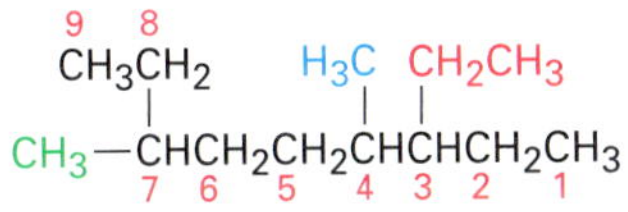

Named as a nonane

Substituents:		
	On C3, CH_2CH_3	(3-ethyl)
	On C4, CH_3	(4-methyl)
	On C7, CH_3	(7-methyl)

(b) If there are two substituents on the same carbon, give both the same number. There must be as many numbers in the name as there are substituents.

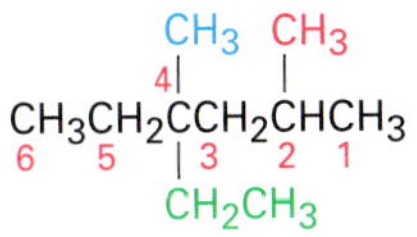

Named as a hexane

Substituents:		
	On C2, CH_3	(2-methyl)
	On C4, CH_3	(4-methyl)
	On C4, CH_2CH_3	(4-ethyl)

Step 4 **Write the name as a single word.**

Use hyphens to separate the different prefixes, and use commas to separate numbers. If two or more different substituents are present, cite them in alphabetical order. If two or more identical substituents are present, use one of the multiplier prefixes *di-*, *tri-*, *tetra-*, and so forth, but don't use these prefixes for alphabetizing. Full names for some of the examples we have been using follow.

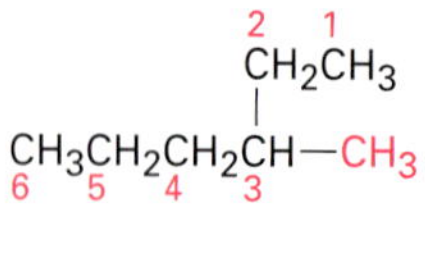

3-Methylhexane

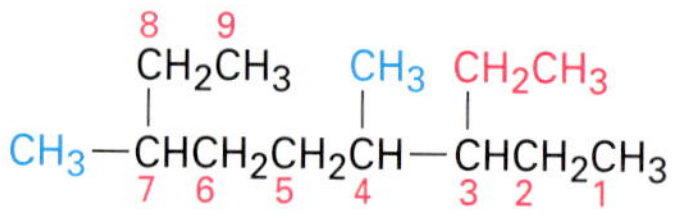

3-Ethyl-4,7-dimethylnonane

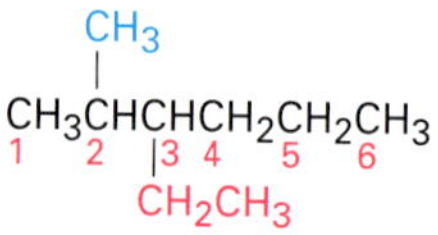

3-Ethyl-2-methylhexane

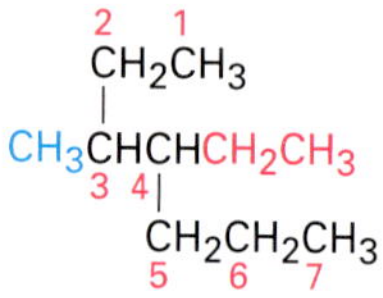

4-Ethyl-3-methylheptane

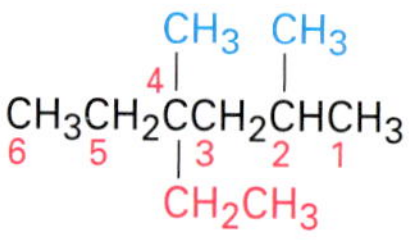

4-Ethyl-2,4-dimethylhexane

Step 5 **Name a complex substituent as though it were itself compound.**

In some particularly complex cases, a fifth step is necessary. It occasionally happens that a substituent on the main chain has sub-branching. In the following case, for instance, the substituent at C6 is a three-carbon chain with a methyl sub-branch. To name the compound fully, the complex substituent must first be named.

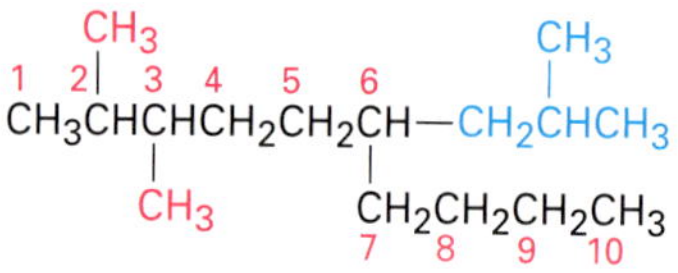

Named as a 2,3,6-trisubstituted decane

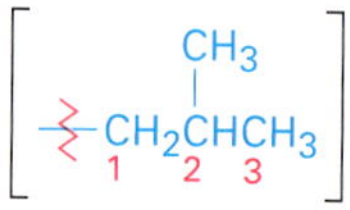

A 2-methylpropyl group

Begin numbering the branched substituent at its point of its attachment to the main chain, and identify it as a 2-methylpropyl group. The substituent is alphabetized according to the first letter of its complete name, including any numerical prefix, and is set off in parentheses when naming the entire molecule.

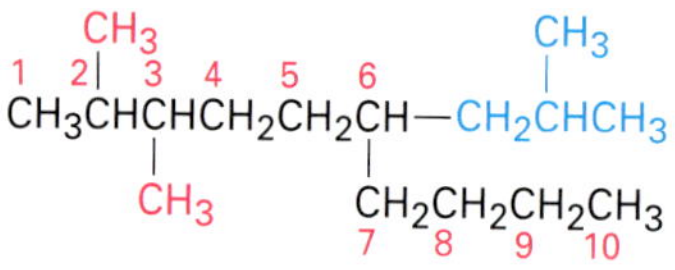

2,3-Dimethyl-6-(2-methylpropyl)decane

As a further example:

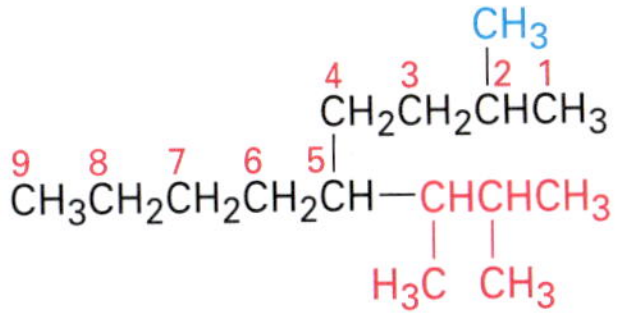

5-(1,2-Dimethylpropyl)-2-methylnonane

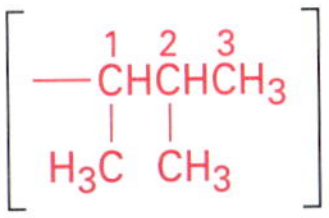

A 1,2-dimethylpropyl group

For historical reasons, some of the simpler branched-chain alkyl groups also have nonsystematic, common names, as noted earlier.

1. Three-carbon alkyl group:

Isopropyl (*i*-Pr)

2. Four-carbon alkyl groups:

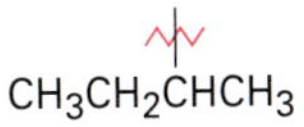

***sec*-Butyl (*sec*-Bu)**

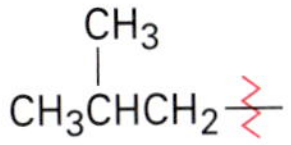

Isobutyl

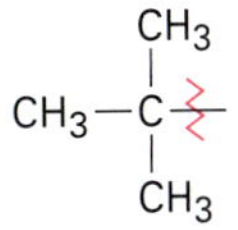

***tert*-Butyl (*t*-butyl or *t*-Bu)**

3. Five-carbon alkyl groups:

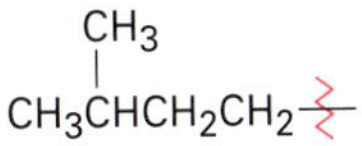

Isopentyl, also called isoamyl (*i*-amyl)

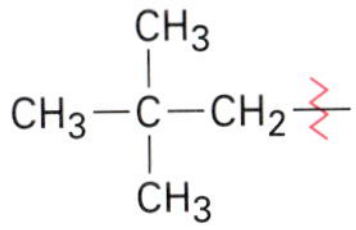

Neopentyl

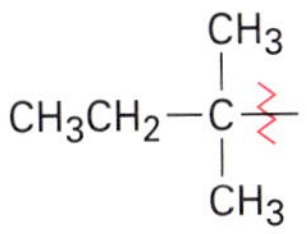

***tert*-Pentyl, also called *tert*-amyl (*t*-amyl)**

The common names of these simple alkyl groups are so well entrenched in the chemical literature that IUPAC rules make allowance for them. Thus, the following compound is properly named either 4-(1-methylethyl)heptane or 4-isopropylheptane. There is no choice but to memorize these common names; fortunately, there are only a few of them.

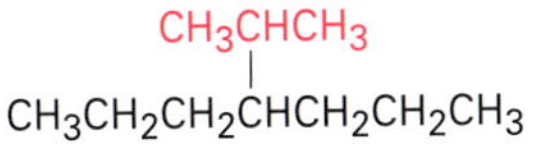

4-(-1-Methylethyl)heptane
or
4-Isopropylheptane

ThomsonNOW™ Click *Organic Interactive* to **use an online palette to draw alkane structures based on IUPAC nomenclature.**

When writing an alkane name, the nonhyphenated prefix iso- is considered part of the alkyl-group name for alphabetizing purposes, but the hyphenated and italicized prefixes *sec-* and *tert-* are not. Thus, isopropyl and isobutyl are listed alphabetically under *i*, but *sec*-butyl and *tert*-butyl are listed under *b*.

WORKED EXAMPLE 3.2 Practice in Naming Alkanes

What is the IUPAC name of the following alkane?

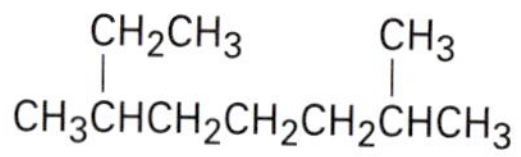

Strategy Find the longest continuous carbon chain in the molecule, and use that as the parent name. This molecule has a chain of eight carbons—octane—with two methyl substituents. (You have to turn corners to see it.) Numbering from the end nearer the first methyl substituent indicates that the methyls are at C2 and C6

Solution

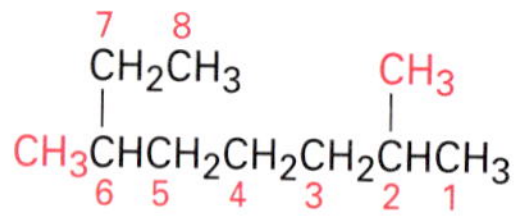

2,6-Dimethyloctane

WORKED EXAMPLE 3.3 Converting a Chemical Name into a Structure

Draw the structure of 3-isopropyl-2-methylhexane.

Strategy This is the reverse of Worked Example 3.2 and uses a reverse strategy. Look at the parent name (hexane), and draw its carbon structure.

C–C–C–C–C–C **Hexane**

Next, find the substituents (3-isopropyl and 2-methyl), and place them on the proper carbons.

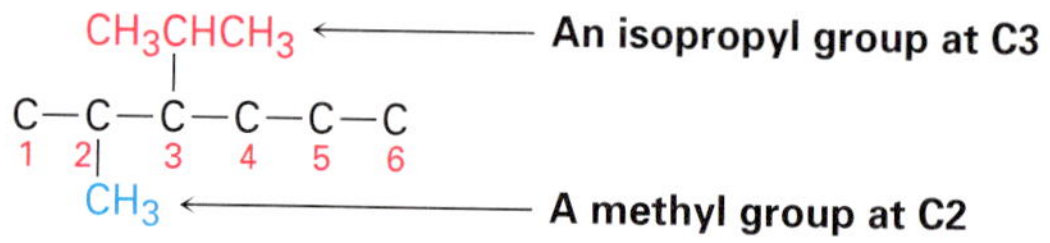

Finally, add hydrogens to complete the structure.

Solution

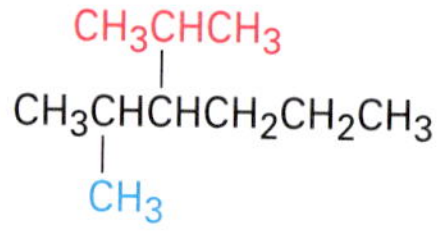

3-Isopropyl-2-methylhexane

Problem 3.11 Give IUPAC names for the following compounds:

(a) The three isomers of C_5H_{12}

(b)

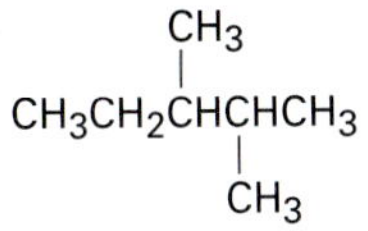

(c)

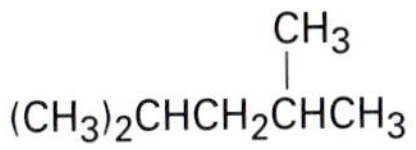

(d)

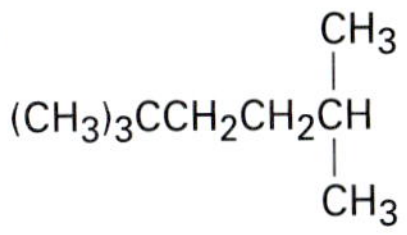

Problem 3.12 Draw structures corresponding to the following IUPAC names:

(a) 3,4-Dimethylnonane
(b) 3-Ethyl-4,4-dimethylheptane
(c) 2,2-Dimethyl-4-propyloctane
(d) 2,2,4-Trimethylpentane

Problem 3.13 Name the eight 5-carbon alkyl groups you drew in Problem 3.7.

Problem 3.14 Give the IUPAC name for the following hydrocarbon, and convert the drawing into a skeletal structure.

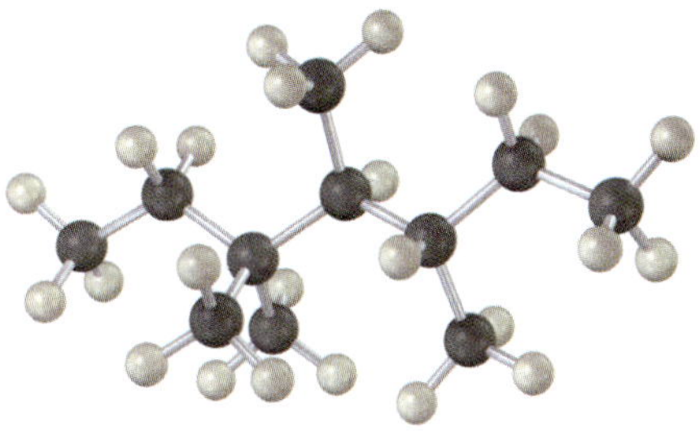

3.5 Properties of Alkanes

Alkanes are sometimes referred to as *paraffins,* a word derived from the Latin *parum affinis,* meaning "little affinity." This term aptly describes their behavior, for alkanes show little chemical affinity for other substances and are chemically inert to most laboratory reagents. They are also relatively inert biologically and are not often involved in the chemistry of living organisms. Alkanes do, however, react with oxygen, halogens, and a few other substances under appropriate conditions.

Reaction with oxygen occurs during combustion in an engine or furnace when the alkane is used as a fuel. Carbon dioxide and water are formed as products, and a large amount of heat is released. For example, methane (natural gas) reacts with oxygen according to the equation

$$CH_4 + 2\,O_2 \longrightarrow CO_2 + 2\,H_2O + 890\text{ kJ/mol } (213\text{ kcal/mol})$$

The reaction of an alkane with Cl_2 occurs when a mixture of the two is irradiated with ultraviolet light (denoted $h\nu$, where ν is the Greek letter nu).

Depending on the relative amounts of the two reactants and on the time allowed, a sequential substitution of the alkane hydrogen atoms by chlorine occurs, leading to a mixture of chlorinated products. Methane, for instance, reacts with Cl_2 to yield a mixture of CH_3Cl, CH_2Cl_2, $CHCl_3$, and CCl_4. We'll look at this reaction in more detail in Section 5.3.

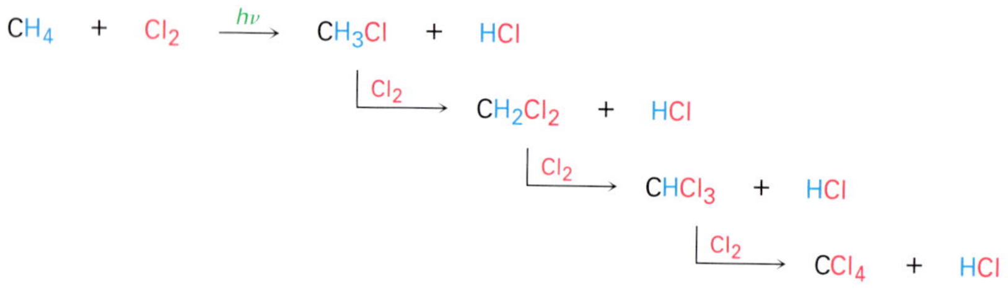

Alkanes show regular increases in both boiling point and melting point as molecular weight increases (Figure 3.4), an effect due to the presence of weak dispersion forces between molecules (Section 2.13). Only when sufficient energy is applied to overcome these forces does the solid melt or liquid boil. As you might expect, dispersion forces increase as molecular size increases, accounting for the higher melting and boiling points of larger alkanes.

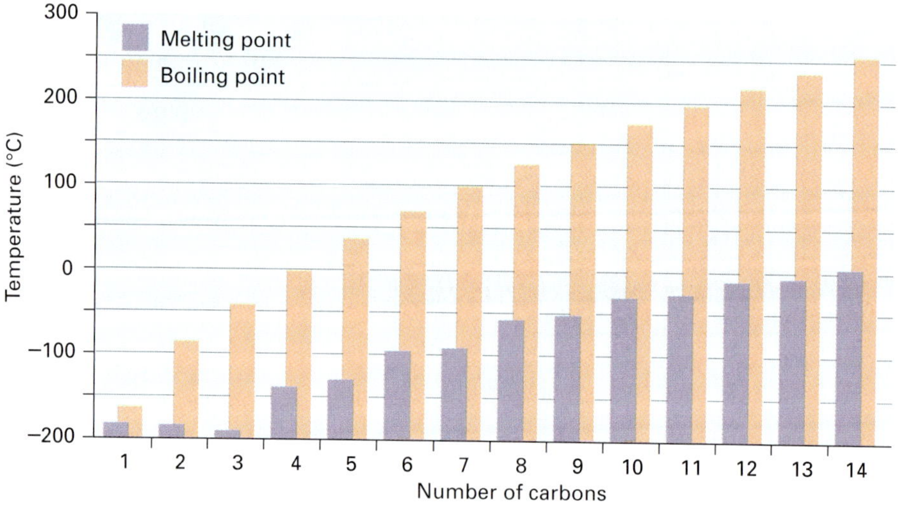

Active Figure 3.4 A plot of melting and boiling points versus number of carbon atoms for the C_1–C_{14} alkanes. There is a regular increase with molecular size. *Sign in at* **www.thomsonedu.com** *to see a simulation based on this figure and to take a short quiz.*

Another interesting effect seen in alkanes is that increased branching lowers an alkane's boiling point. Thus, pentane has no branches and boils at 36.1 °C, isopentane (2-methylbutane) has one branch and boils at 27.85 °C, and neopentane (2,2-dimethylpropane) has two branches and boils at 9.5 °C. Similarly, octane boils at 125.7 °C, whereas isooctane (2,2,4-trimethylpentane) boils at 99.3 °C. Branched-chain alkanes are lower-boiling because they are more nearly spherical than straight-chain alkanes, have smaller surface areas, and consequently have smaller dispersion forces.

3.6 Conformations of Ethane

Up to this point, we've viewed molecules primarily in a two-dimensional way and have given little thought to any consequences that might arise from the spatial arrangement of atoms in molecules. Now it's time to add a third dimension to our study. **Stereochemistry** is the branch of chemistry concerned with the three-dimensional aspects of molecules. We'll see on many occasions in future chapters that the exact three-dimensional structure of a molecule is often crucial to determining its properties and biological behavior.

We know from Section 1.5 that σ bonds are cylindrically symmetrical. In other words, the intersection of a plane cutting through a carbon–carbon single-bond orbital looks like a circle. Because of this cylindrical symmetry, *rotation* is possible around carbon–carbon bonds in open-chain molecules. In ethane, for instance, rotation around the C–C bond occurs freely, constantly changing the spatial relationships between the hydrogens on one carbon and those on the other (Figure 3.5).

Active Figure 3.5 Rotation occurs around the carbon–carbon single bond in ethane because of σ bond cylindrical symmetry. *Sign in at* **www.thomsonedu.com** *to see a simulation based on this figure and to take a short quiz.*

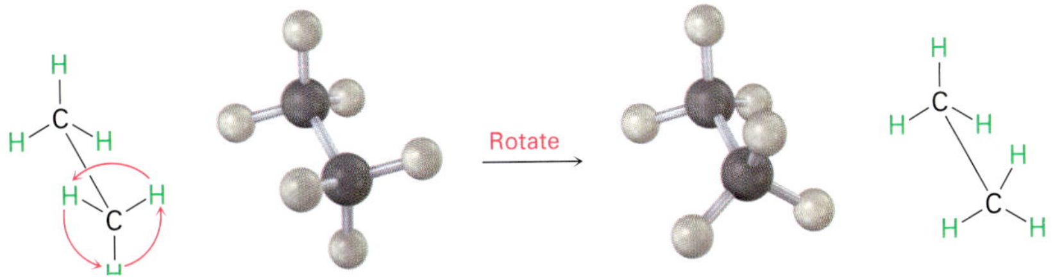

Melvin S. Newman

Melvin S. Newman (1908–1993) was born in New York and received his Ph.D. in 1932 from Yale University. He was professor of chemistry at the Ohio State University (1936–1973), where he was active in both research and chemical education.

The different arrangements of atoms that result from bond rotation are called **conformations**, and molecules that have different arrangements are called conformational isomers, or **conformers**. Unlike constitutional isomers, however, different conformers can't usually be isolated because they interconvert too rapidly.

Conformational isomers are represented in two ways, as shown in Figure 3.6. A *sawhorse representation* views the carbon–carbon bond from an oblique angle and indicates spatial orientation by showing all C–H bonds. A **Newman projection** views the carbon–carbon bond directly end-on and represents the two carbon atoms by a circle. Bonds attached to the front carbon are represented by lines to the center of the circle, and bonds attached to the rear carbon are represented by lines to the edge of the circle.

Figure 3.6 A sawhorse representation and a Newman projection of ethane. The sawhorse representation views the molecule from an oblique angle, while the Newman projection views the molecule end-on. Note that the molecular model of the Newman projection appears at first to have six atoms attached to a single carbon. Actually, the front carbon, with three attached green atoms, is directly in front of the rear carbon, with three attached red atoms.

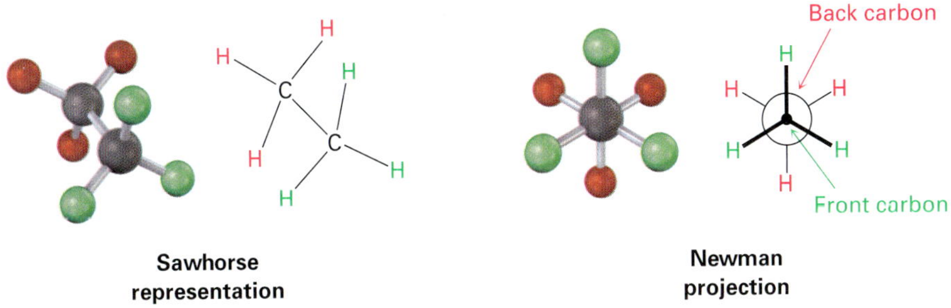

Despite what we've just said, we actually don't observe *perfectly* free rotation in ethane. Experiments show that there is a small (12 kJ/mol; 2.9 kcal/mol) barrier to rotation and that some conformers are more stable than others. The lowest-energy, most stable conformer is the one in which all six C–H bonds are as far away from one another as possible—**staggered** when viewed end-on in a Newman projection. The highest-energy, least stable conformer is the one in which the six C–H bonds are as close as possible—**eclipsed** in a Newman projection. At any given instant, about 99% of ethane molecules have an approximately staggered conformation and only about 1% are near the eclipsed conformation.

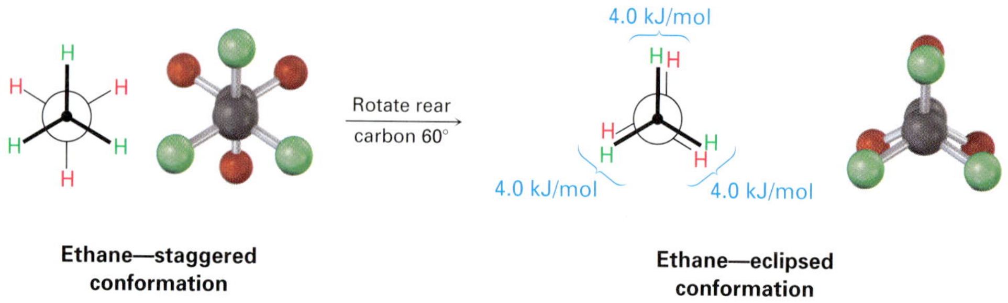

The extra 12 kJ/mol of energy present in the eclipsed conformer of ethane is called **torsional strain**. Its cause has been the subject of controversy, but the major factor is an interaction between C–H bonding orbitals on one carbon with antibonding orbitals on the adjacent carbon, which stabilizes the staggered conformer relative to the eclipsed conformer. Because the total strain of 12 kJ/mol arises from three equal hydrogen–hydrogen eclipsing interactions, we can assign a value of approximately 4.0 kJ/mol (1.0 kcal/mol) to each single interaction. The barrier to rotation that results can be represented on a graph of potential energy versus degree of rotation in which the angle between C–H bonds on front and back carbons as viewed end-on (the *dihedral angle*) goes full circle from 0° to 360°. Energy minima occur at staggered conformations, and energy maxima occur at eclipsed conformations, as shown in Figure 3.7.

Figure 3.7 A graph of potential energy versus bond rotation in ethane. The staggered conformers are 12 kJ/mol lower in energy than the eclipsed conformers.

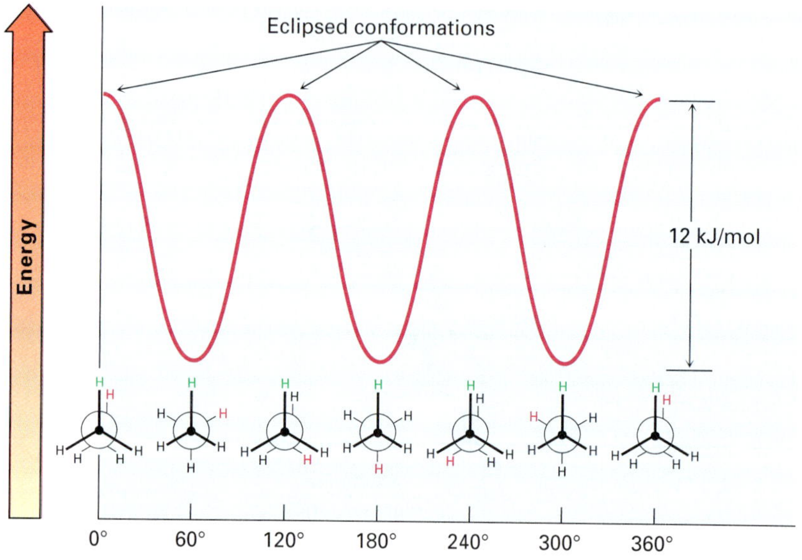

3.7 Conformations of Other Alkanes

Propane, the next higher member in the alkane series, also has a torsional barrier that results in hindered rotation around the carbon–carbon bonds. The barrier is slightly higher in propane than in ethane—a total of 14 kJ/mol (3.4 kcal/mol) versus 12 kJ/mol.

The eclipsed conformer of propane has three interactions—two ethane-type hydrogen–hydrogen interactions and one additional hydrogen–methyl interaction. Since each eclipsing H⟷H interaction is the same as that in ethane and thus has an energy "cost" of 4.0 kJ/mol, we can assign a value of $14 - (2 \times 4.0) = 6.0$ kJ/mol (1.4 kcal/mol) to the eclipsing H⟷CH_3 interaction (Figure 3.8).

Figure 3.8 Newman projections of propane showing staggered and eclipsed conformations. The staggered conformer is lower in energy by 14 kJ/mol.

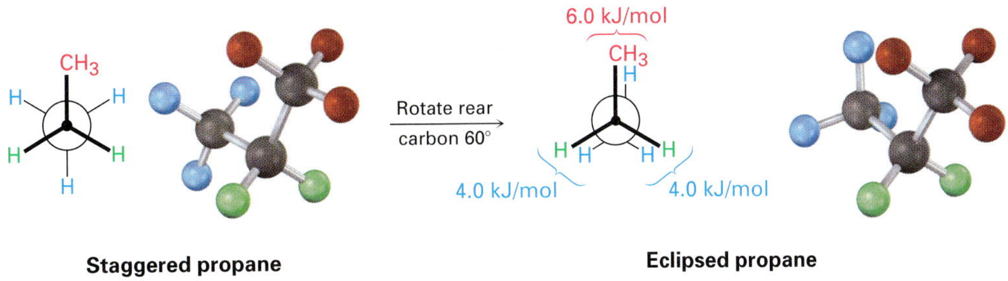

The conformational situation becomes more complex for larger alkanes because not all staggered conformations have the same energy and not all eclipsed conformations have the same energy. In butane, for instance, the lowest-energy arrangement, called the **anti conformation**, is the one in which the two methyl groups are as far apart as possible—180° away from each other. As rotation around the C2–C3 bond occurs, an eclipsed conformation is reached in which there are two CH_3⟷H interactions and one H⟷H interaction. Using the energy values derived previously from ethane and propane, this eclipsed conformation is more strained than the anti conformation by 2×6.0 kJ/mol + 4.0 kJ/mol (two CH_3⟷H interactions plus one H⟷H interaction), for a total of 16 kJ/mol (3.8 kcal/mol).

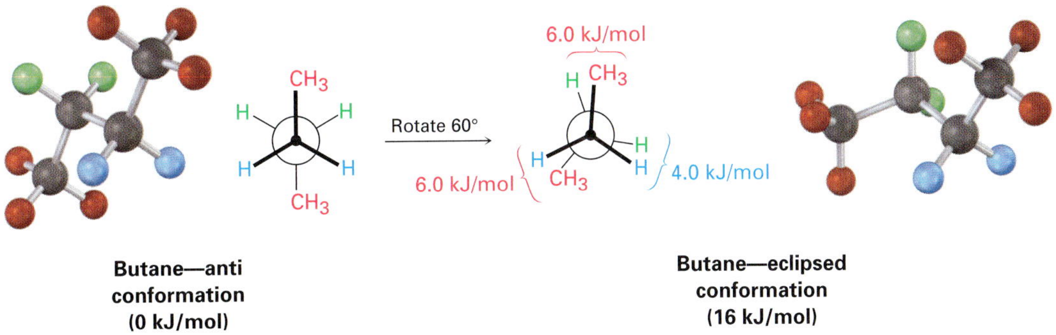

As bond rotation continues, an energy minimum is reached at the staggered conformation where the methyl groups are 60° apart. Called the **gauche**

conformation, it lies 3.8 kJ/mol (0.9 kcal/mol) higher in energy than the anti conformation *even though it has no eclipsing interactions*. This energy difference occurs because the hydrogen atoms of the methyl groups are near one another in the gauche conformation, resulting in what is called *steric strain*. **Steric strain** is the repulsive interaction that occurs when atoms are forced closer together than their atomic radii allow. It's the result of trying to force two atoms to occupy the same space.

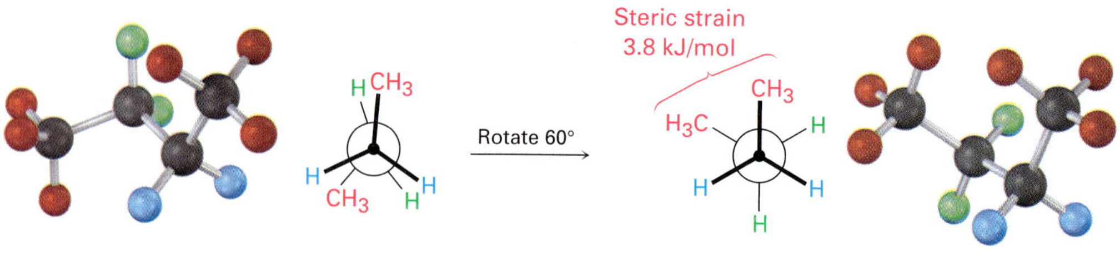

Butane—eclipsed conformation (16 kJ/mol)

Butane—gauche conformation (3.8 kJ/mol)

As the dihedral angle between the methyl groups approaches 0°, an energy maximum is reached at a second eclipsed conformation. Because the methyl groups are forced even closer together than in the gauche conformation, both torsional strain and steric strain are present. A total strain energy of 19 kJ/mol (4.5 kcal/mol) has been estimated for this conformation, making it possible to calculate a value of 11 kJ/mol (2.6 kcal/mol) for the $CH_3 \longleftrightarrow CH_3$ eclipsing interaction: total strain of 19 kJ/mol less the strain of two $H \longleftrightarrow H$ eclipsing interactions (2 × 4.0 kcal/mol) equals 11 kJ/mol.

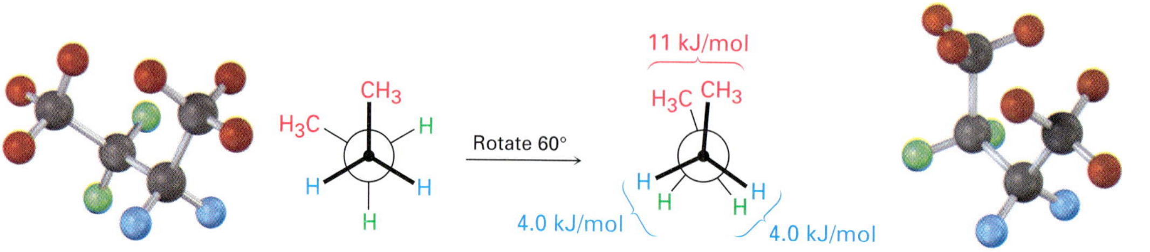

Butane—gauche conformation (3.8 kJ/mol)

Butane—eclipsed conformation (19 kJ/mol)

After 0°, the rotation becomes a mirror image of what we've already seen: another gauche conformation is reached, another eclipsed conformation, and finally a return to the anti conformation. A plot of potential energy versus rotation about the C2–C3 bond is shown in Figure 3.9.

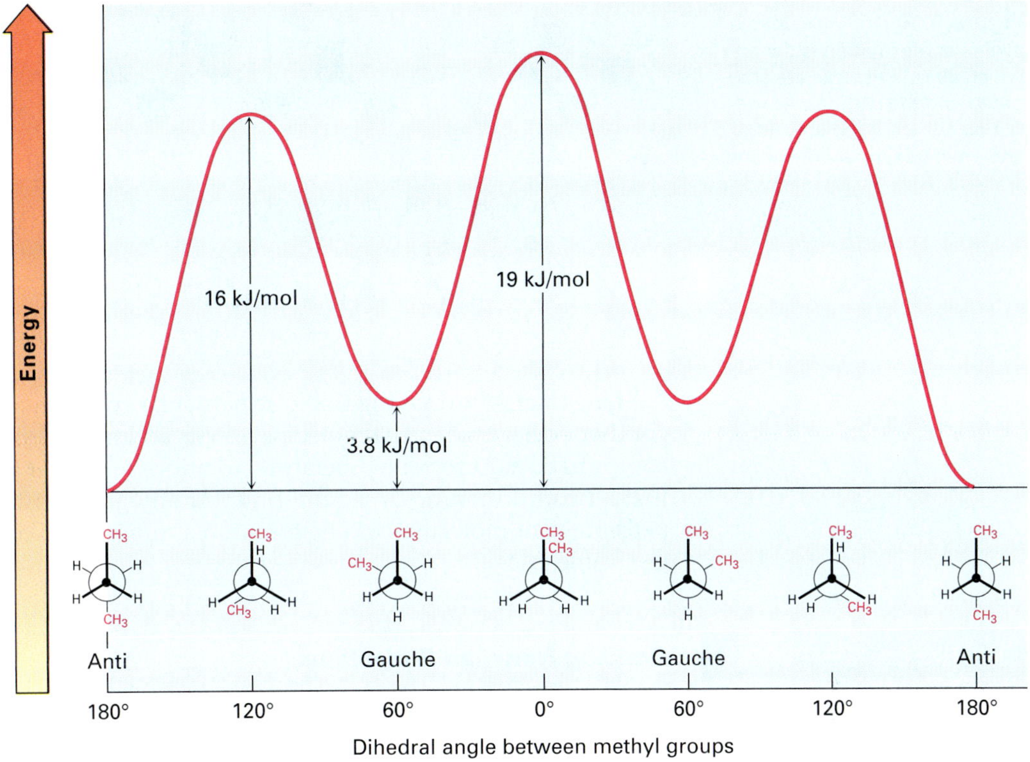

Figure 3.9 A plot of potential energy versus rotation for the C2–C3 bond in butane. The energy maximum occurs when the two methyl groups eclipse each other, and the energy minimum occurs when the two methyl groups are 180° apart (anti).

The notion of assigning definite energy values to specific interactions within a molecule is a very useful one that we'll return to in the next chapter. A summary of what we've seen thus far is given in Table 3.5.

Table 3.5 Energy Costs for Interactions in Alkane Conformers

Interaction	Cause	Energy cost (kJ/mol)	Energy cost (kcal/mol)
H⟷H eclipsed	Torsional strain	4.0	1.0
H⟷CH_3 eclipsed	Mostly torsional strain	6.0	1.4
CH_3⟷CH_3 eclipsed	Torsional and steric strain	11	2.6
CH_3⟷CH_3 gauche	Steric strain	3.8	0.9

The same principles just developed for butane apply to pentane, hexane, and all higher alkanes. The most favorable conformation for any alkane has the carbon–carbon bonds in staggered arrangements, with large substituents arranged anti to one another. A generalized alkane structure is shown in Figure 3.10.

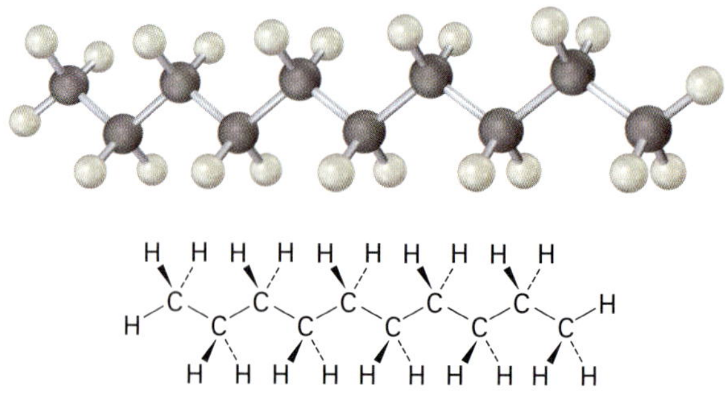

Figure 3.10 The most stable alkane conformation is the one in which all substituents are staggered and the carbon–carbon bonds are arranged anti, as shown in this model of decane.

One final point: saying that one particular conformer is "more stable" than another doesn't mean the molecule adopts and maintains only the more stable conformation. At room temperature, rotations around σ bonds occur so rapidly that all conformers are in equilibrium. At any given instant, however, a larger percentage of molecules will be found in a more stable conformation than in a less stable one.

WORKED EXAMPLE 3.4 *Drawing Newman Projections*

Sighting along the C1–C2 bond of 1-chloropropane, draw Newman projections of the most stable and least stable conformations.

Strategy The most stable conformation of a substituted alkane is generally a staggered one in which large groups have an anti relationship. The least stable conformation is generally an eclipsed one in which large groups are as close as possible.

Solution

Most stable (staggered) **Least stable (eclipsed)**

Problem 3.15 Make a graph of potential energy versus angle of bond rotation for propane, and assign values to the energy maxima.

Problem 3.16 Consider 2-methylpropane (isobutane). Sighting along the C2–C1 bond:

(a) Draw a Newman projection of the most stable conformation.

(b) Draw a Newman projection of the least stable conformation.

(c) Make a graph of energy versus angle of rotation around the C2–C1 bond.

(d) Since an H⟷H eclipsing interaction costs 4.0 kJ/mol and an H⟷CH_3 eclipsing interaction costs 6.0 kJ/mol, assign relative values to the maxima and minima in your graph.

Problem 3.17 Sight along the C2–C3 bond of 2,3-dimethylbutane, and draw a Newman projection of the most stable conformation.

Problem 3.18 Draw a Newman projection along the C2–C3 bond of the following conformation of 2,3-dimethylbutane, and calculate a total strain energy:

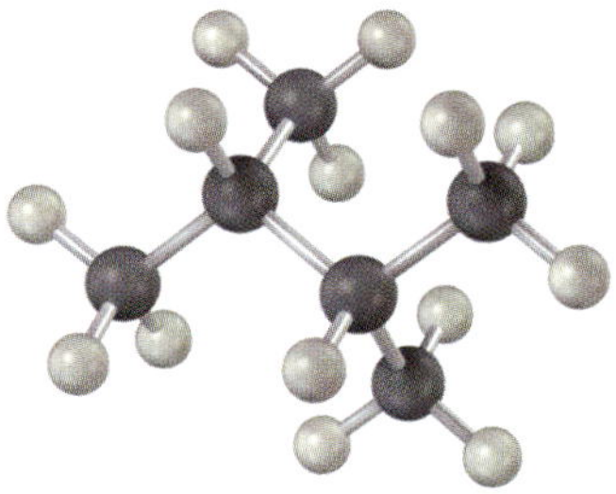

Focus On . . .

Gasoline

© Sascha Burkard

Gasoline is a finite resource; it won't be around forever.

British Foreign Minister Ernest Bevin once said that "The Kingdom of Heaven runs on righteousness, but the Kingdom of Earth runs on alkanes." Well, actually he said "runs on oil" not "runs on alkanes," but they're essentially the same. By far, the major sources of alkanes are the world's natural gas and petroleum deposits. Laid down eons ago, these deposits are thought to be derived from the decomposition of plant and animal matter, primarily of marine origin. *Natural gas* consists chiefly of methane but also contains ethane, propane, and butane. *Petroleum* is a complex mixture of hydrocarbons that must be separated into fractions and then further refined before it can be used.

The petroleum era began in August 1859, when the world's first oil well was drilled near Titusville, Pennsylvania. The petroleum was distilled into fractions according to boiling point, but it was high-boiling kerosene, or lamp oil, rather than gasoline that was primarily sought. Literacy was becoming widespread at the time, and people wanted better light for reading than was available from candles. Gasoline was too volatile for use in lamps and was initially considered a waste by-product. The world has changed greatly since those early days, however, and it is now gasoline rather than lamp oil that is prized.

Petroleum refining begins by fractional distillation of crude oil into three principal cuts according to boiling point (bp): straight-run gasoline (bp 30–200 °C), kerosene (bp 175–300 °C), and heating oil, or diesel fuel (bp 275–400 °C). Further distillation under reduced pressure then yields

(continued)

lubricating oils and waxes and leaves a tarry residue of asphalt. The distillation of crude oil is only the first step in gasoline production, however. Straight-run gasoline turns out to be a poor fuel in automobiles because of *engine knock,* an uncontrolled combustion that can occur in a hot engine.

The *octane number* of a fuel is the measure by which its antiknock properties are judged. It was recognized long ago that straight-chain hydrocarbons are far more prone to induce engine knock than are highly branched compounds. Heptane, a particularly bad fuel, is assigned a base value of 0 octane number, and 2,2,4-trimethylpentane, commonly known as isooctane, has a rating of 100.

$CH_3CH_2CH_2CH_2CH_2CH_2CH_3$

Heptane
(octane number = 0)

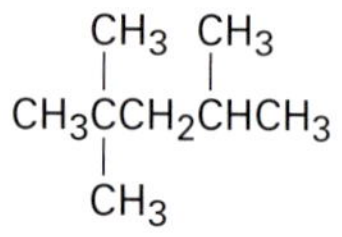

2,2,4-Trimethylpentane
(octane number = 100)

Because straight-run gasoline burns so poorly in engines, petroleum chemists have devised numerous methods for producing higher-quality fuels. One of these methods, *catalytic cracking,* involves taking the high-boiling kerosene cut (C_{11}–C_{14}) and "cracking" it into smaller branched molecules suitable for use in gasoline. Another process, called *reforming,* is used to convert C_6–C_8 alkanes to aromatic compounds such as benzene and toluene, which have substantially higher octane numbers than alkanes. The final product that goes in your tank has an approximate composition of 15% C_4–C_8 straight-chain alkanes, 25% to 40% C_4–C_{10} branched-chain alkanes, 10% cyclic alkanes, 10% straight-chain and cyclic alkenes, and 25% arenes (aromatics).

SUMMARY AND KEY WORDS

aliphatic, 79
alkane, 79
alkyl group, 83
anti conformation, 95
branched-chain alkane, 80
conformation, 93
conformers, 93
constitutional isomers, 81
eclipsed conformation, 94
functional group, 73
gauche conformation, 95

A **functional group** is a group of atoms within a larger molecule that has a characteristic chemical reactivity. Because functional groups behave in approximately the same way in all molecules where they occur, the chemical reactions of an organic molecule are largely determined by its functional groups.

Alkanes are a class of **saturated hydrocarbons** with the general formula C_nH_{2n+2}. They contain no functional groups, are relatively inert, and can be either **straight-chain** *(normal)* or **branched**. Alkanes are named by a series of IUPAC rules of nomenclature. Compounds that have the same chemical formula but different structures are called **isomers**. More specifically, compounds such as butane and isobutane, which differ in their connections between atoms, are called **constitutional isomers**.

Carbon–carbon single bonds in alkanes are formed by σ overlap of carbon sp^3 hybrid orbitals. Rotation is possible around σ bonds because of their cylindrical symmetry, and alkanes therefore exist in a large number of rapidly interconverting **conformations**. **Newman projections** make it possible to visualize the spatial consequences of bond rotation by sighting directly along a carbon–carbon bond axis. Not all alkane conformations are equally stable. The **staggered** conformation of ethane is 12 kJ/mol (2.9 kcal/mol) more stable than the **eclipsed** conformation because of **torsional strain**. In general, any alkane is most stable when all its bonds are staggered.

EXERCISES

Organic KNOWLEDGE TOOLS

ThomsonNOW™ Sign in at **www.thomsonedu.com** to assess your knowledge of this chapter's topics by taking a pre-test. The pre-test will link you to interactive organic chemistry resources based on your score in each concept area.

Online homework for this chapter may be assigned in Organic OWL.

■ indicates problems assignable in Organic OWL.

VISUALIZING CHEMISTRY

(Problems 3.1–3.18 appear within the chapter.)

3.19 ■ Identify the functional groups in the following substances, and convert each drawing into a molecular formula (red = O, blue = N):

(a)

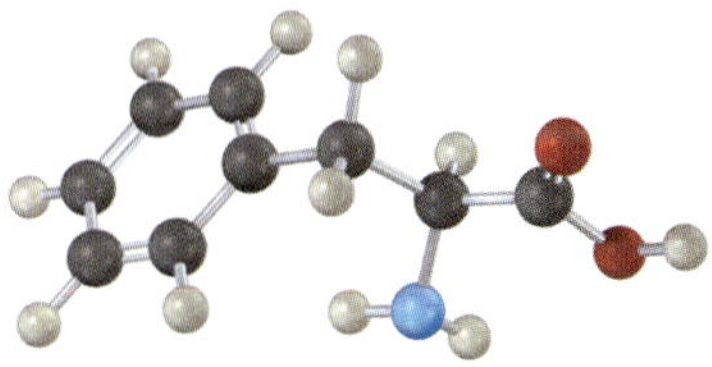

Phenylalanine

(b)

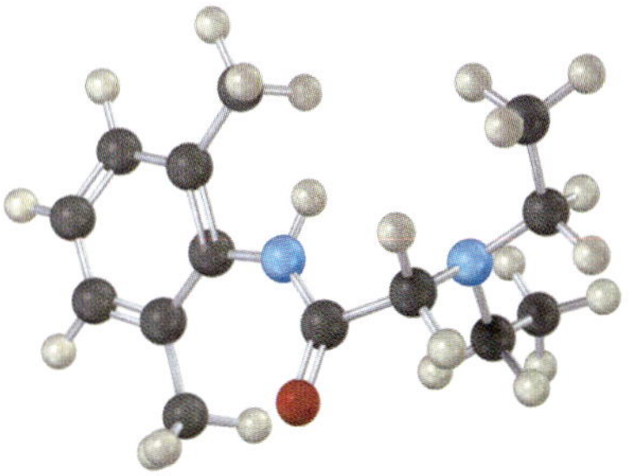

Lidocaine

3.20 ■ Give IUPAC names for the following alkanes, and convert each drawing into a skeletal structure:

(a)

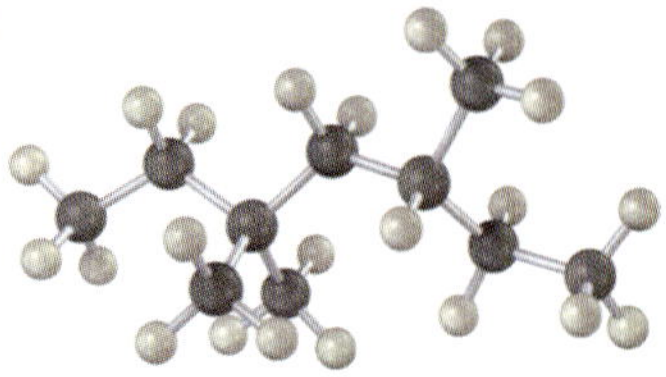

(b)

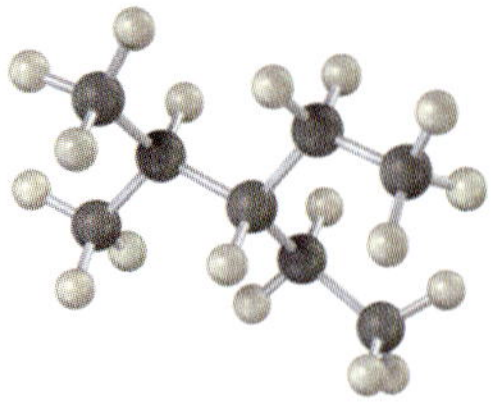

(c)

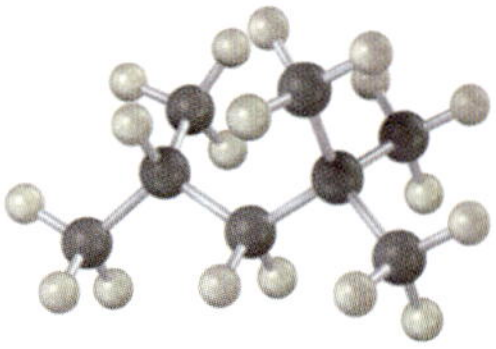

(d)

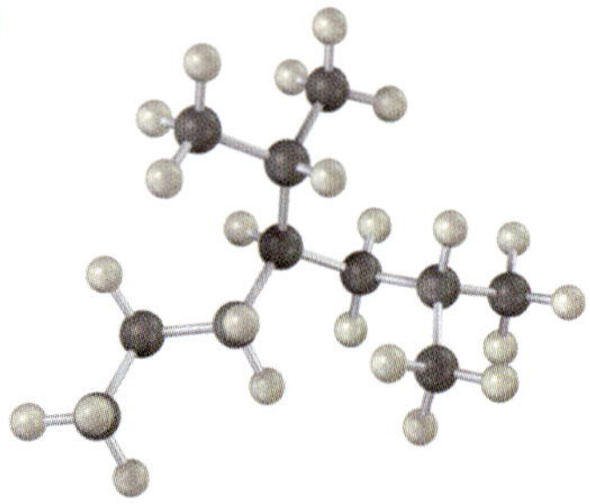

ADDITIONAL PROBLEMS

3.21 ■ Locate and identify the functional groups in the following molecules. In these representations, each intersection of lines and the end of each line represents a carbon atom with the appropriate number of hydrogens attached.

(a)

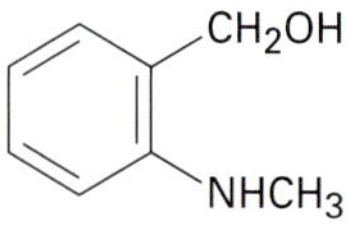

(b)

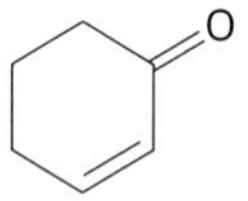

(c)

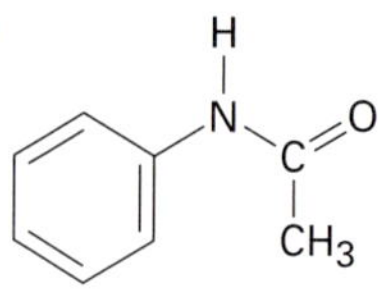

(d)

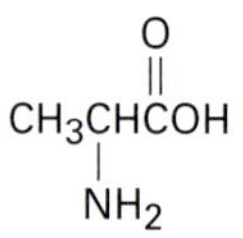

(e)

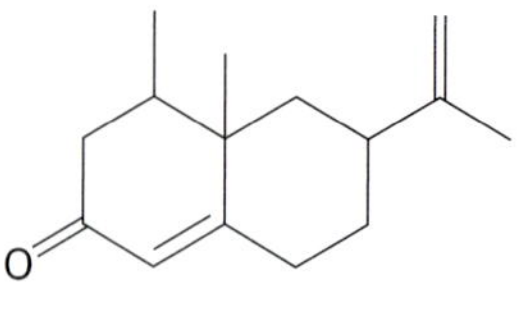

(f)

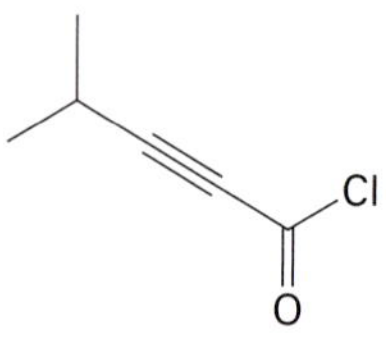

3.22 Draw structures that meet the following descriptions (there are many possibilities):

(a) Three isomers with the formula C_8H_{18}

(b) Two isomers with the formula $C_4H_8O_2$

3.23 Draw structures of the nine isomers of C_7H_{16}.

3.24 ■ In each of the following sets, which structures represent the same compound and which represent different compounds?

(a)

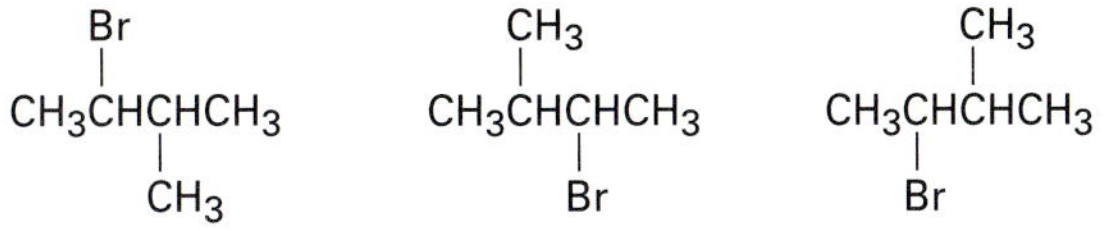

(b)

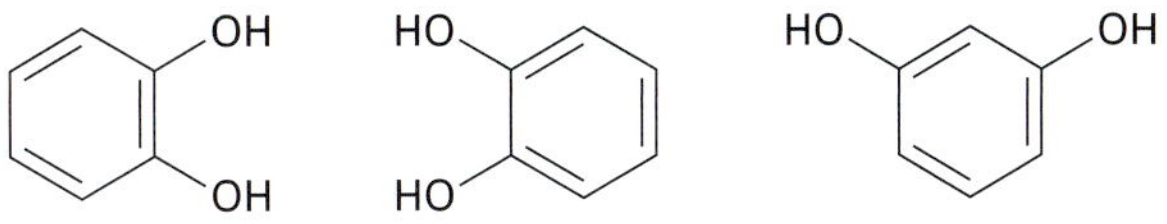

(c)

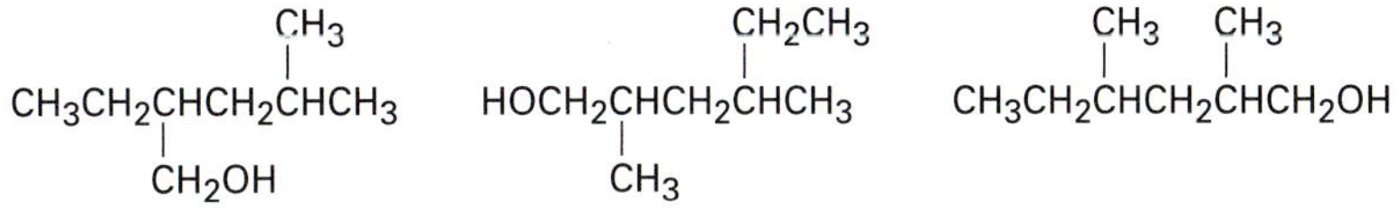

3.25 There are seven constitutional isomers with the formula $C_4H_{10}O$. Draw as many as you can.

3.26 ■ Propose structures that meet the following descriptions:

(a) A ketone with five carbons
(b) A four-carbon amide
(c) A five-carbon ester
(d) An aromatic aldehyde
(e) A keto ester
(f) An amino alcohol

3.27 ■ Propose structures for the following:

(a) A ketone, C_4H_8O
(b) A nitrile, C_5H_9N
(c) A dialdehyde, $C_4H_6O_2$
(d) A bromoalkene, $C_6H_{11}Br$
(e) An alkane, C_6H_{14}
(f) A *cyclic* saturated hydrocarbon, C_6H_{12}
(g) A diene (dialkene), C_5H_8
(h) A keto alkene, C_5H_8O

3.28 Draw as many compounds as you can that fit the following descriptions:

(a) Alcohols with formula $C_4H_{10}O$
(b) Amines with formula $C_5H_{13}N$
(c) Ketones with formula $C_5H_{10}O$
(d) Aldehydes with formula $C_5H_{10}O$
(e) Esters with formula $C_4H_8O_2$
(f) Ethers with formula $C_4H_{10}O$

3.29 ■ Draw compounds that contain the following:

(a) A primary alcohol
(b) A tertiary nitrile
(c) A secondary thiol
(d) Both primary and secondary alcohols
(e) An isopropyl group
(f) A quaternary carbon

3.30 Draw and name all monobromo derivatives of pentane, $C_5H_{11}Br$.

3.31 Draw and name all monochloro derivatives of 2,5-dimethylhexane, $C_8H_{17}Cl$.

3.32 Predict the hybridization of the carbon atom in each of the following functional groups:

(a) Ketone (b) Nitrile (c) Carboxylic acid

3.33 ■ Draw the structures of the following molecules:

(a) *Biacetyl*, $C_4H_6O_2$, a substance with the aroma of butter; it contains no rings or carbon–carbon multiple bonds.
(b) *Ethylenimine*, C_2H_5N, a substance used in the synthesis of melamine polymers; it contains no multiple bonds.
(c) *Glycerol*, $C_3H_8O_3$, a substance isolated from fat and used in cosmetics; it has an $-OH$ group on each carbon.

■ Assignable in OWL

3.34 ■ Draw structures for the following:

(a) 2-Methylheptane
(b) 4-Ethyl-2,2-dimethylhexane
(c) 4-Ethyl-3,4-dimethyloctane
(d) 2,4,4-Trimethylheptane
(e) 3,3-Diethyl-2,5-dimethylnonane
(f) 4-Isopropyl-3-methylheptane

3.35 Draw a compound that:

(a) Has only primary and tertiary carbons
(b) Has no secondary or tertiary carbons
(c) Has four secondary carbons

3.36 Draw a compound that:

(a) Has nine primary hydrogens
(b) Has only primary hydrogens

3.37 For each of the following compounds, draw an isomer that has the same functional groups. Each intersection of lines represents a carbon atom with the appropriate number of hydrogens attached.

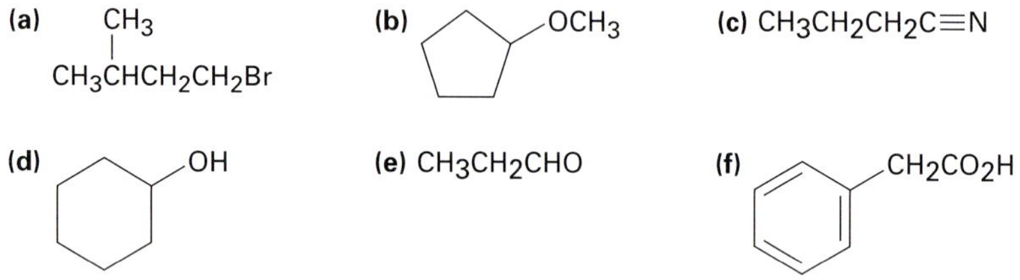

3.38 ■ Give IUPAC names for the following compounds:

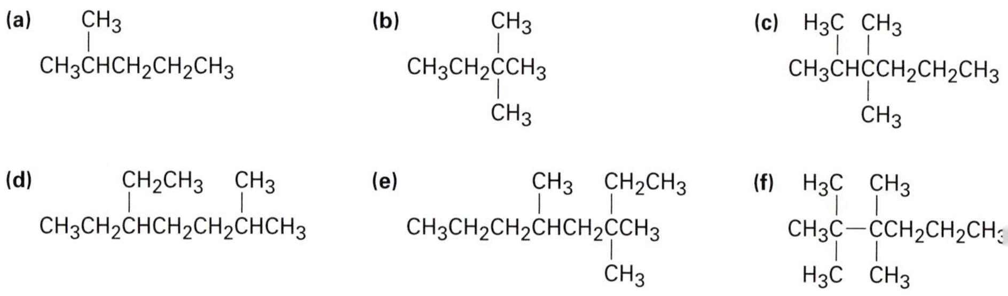

3.39 Name the five isomers of C_6H_{14}.

3.40 Explain why each of the following names is incorrect:

(a) 2,2-Dimethyl-6-ethylheptane
(b) 4-Ethyl-5,5-dimethylpentane
(c) 3-Ethyl-4,4-dimethylhexane
(d) 5,5,6-Trimethyloctane
(e) 2-Isopropyl-4-methylheptane

3.41 Propose structures and give IUPAC names for the following:

(a) A diethyldimethylhexane
(b) A (3-methylbutyl)-substituted alkane

3.42 ■ Consider 2-methylbutane (isopentane). Sighting along the C2–C3 bond:

(a) Draw a Newman projection of the most stable conformation.
(b) Draw a Newman projection of the least stable conformation.
(c) Since a $CH_3 \leftrightarrow CH_3$ eclipsing interaction costs 11 kJ/mol (2.5 kcal/mol) and a $CH_3 \leftrightarrow CH_3$ gauche interaction costs 3.8 kJ/mol (0.9 kcal/mol), make a quantitative plot of energy versus rotation about the C2–C3 bond.

3.43 ■ What are the relative energies of the three possible staggered conformations around the C2–C3 bond in 2,3-dimethylbutane? (See Problem 3.42.)

■ Assignable in OWL

3.44 Construct a qualitative potential-energy diagram for rotation about the C–C bond of 1,2-dibromoethane. Which conformation would you expect to be more stable? Label the anti and gauche conformations of 1,2-dibromoethane.

3.45 Which conformation of 1,2-dibromoethane (Problem 3.44) would you expect to have the larger dipole moment? The observed dipole moment of 1,2-dibromoethane is $\mu = 1.0$ D. What does this tell you about the actual structure of the molecule?

3.46 ■ The barrier to rotation about the C–C bond in bromoethane is 15 kJ/mol (3.6 kcal/mol).
(a) What energy value can you assign to an H–Br eclipsing interaction?
(b) Construct a quantitative diagram of potential energy versus bond rotation for bromoethane.

3.47 Draw the most stable conformation of pentane, using wedges and dashes to represent bonds coming out of the paper and going behind the paper, respectively.

3.48 Draw the most stable conformation of 1,4-dichlorobutane, using wedges and dashes to represent bonds coming out of the paper and going behind the paper, respectively.

3.49 Malic acid, $C_4H_6O_5$, has been isolated from apples. Because this compound reacts with 2 molar equivalents of base, it is a dicarboxylic acid.
(a) Draw at least five possible structures.
(b) If malic acid is a secondary alcohol, what is its structure?

3.50 ■ Formaldehyde, $H_2C{=}O$, is known to all biologists because of its usefulness as a tissue preservative. When pure, formaldehyde *trimerizes* to give trioxane, $C_3H_6O_3$, which, surprisingly enough, has no carbonyl groups. Only one monobromo derivative ($C_3H_5BrO_3$) of trioxane is possible. Propose a structure for trioxane.

3.51 ■ Increased substitution around a bond leads to increased strain. Take the four substituted butanes listed below, for example. For each compound, sight along the C2–C3 bond and draw Newman projections of the most stable and least stable conformations. Use the data in Table 3.5 to assign strain energy values to each conformation. Which of the eight conformations is most strained? Which is least strained?
(a) 2-Methylbutane **(b)** 2,2-Dimethylbutane
(c) 2,3-Dimethylbutane **(d)** 2,2,3-Trimethylbutane

3.52 The cholesterol-lowering agents called *statins,* such as simvastatin (Zocor) and pravastatin (Pravachol), are among the most widely prescribed drugs in the world. Identify the functional groups in both, and tell how the two substances differ.

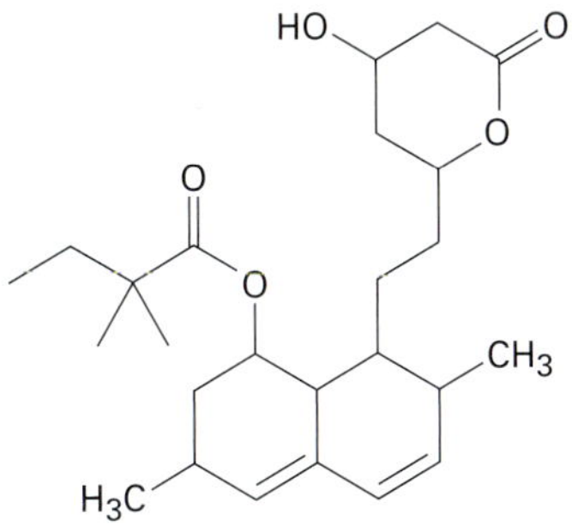

Simvastatin (Zocor)

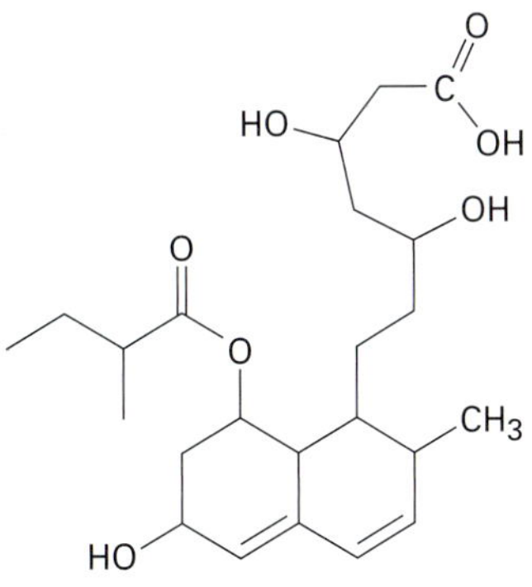

Pravastatin (Pravachol)

3.53 We'll look in the next chapter at *cycloalkanes*—saturated cyclic hydrocarbons– and we'll see that the molecules generally adopt puckered, nonplanar confo mations. Cyclohexane, for instance, has a puckered shape like a lounge cha rather than a flat shape. Why?

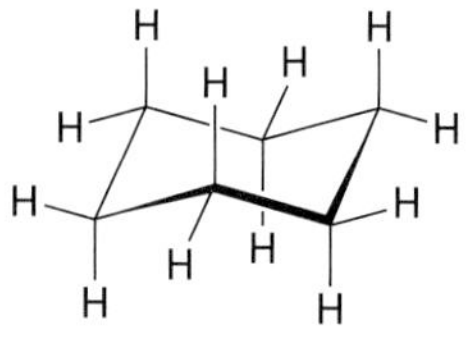

Nonplanar cyclohexane

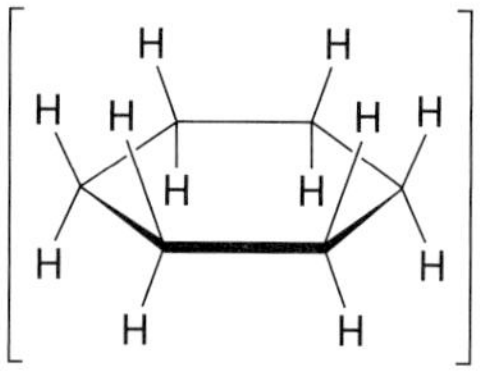

Planar cyclohexane

3.54 We'll see in the next chapter that there are two isomeric substances bot named 1,2-dimethylcyclohexane. Explain.

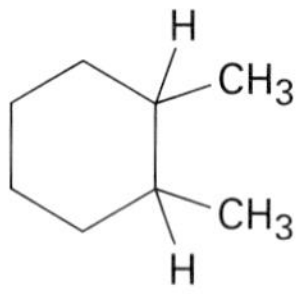

1,2-Dimethylcyclohexane

4 Organic Compounds: Cycloalkanes and Their Stereochemistry

Organic **KNOWLEDGE TOOLS**

ThomsonNOW™ Throughout this chapter, sign in at **www.thomsonedu.com** for online self-study and interactive tutorials based on your level of understanding.

Online homework for this chapter may be assigned in Organic OWL.

We've discussed only open-chain compounds up to this point, but most organic compounds contain *rings* of carbon atoms. Chrysanthemic acid, for instance, whose esters occur naturally as the active insecticidal constituents of chrysanthemum flowers, contains a three-membered (cyclopropane) ring.

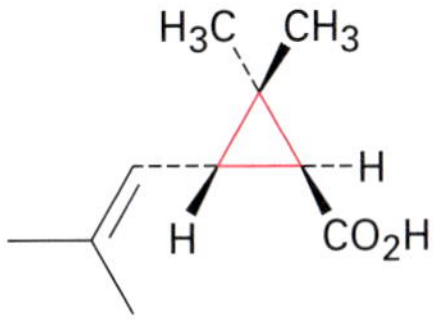

Chrysanthemic acid

Prostaglandins, potent hormones that control an extraordinary variety of physiological functions in humans, contain a five-membered (cyclopentane) ring.

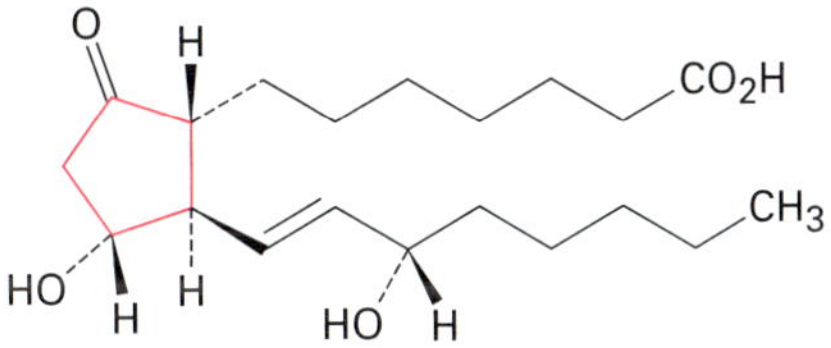

Prostaglandin E_1

Steroids, such as cortisone, contain four rings joined together—3 six-membered (cyclohexane) and 1 five-membered. We'll discuss steroids and their properties in more detail in Sections 27.6 and 27.7.

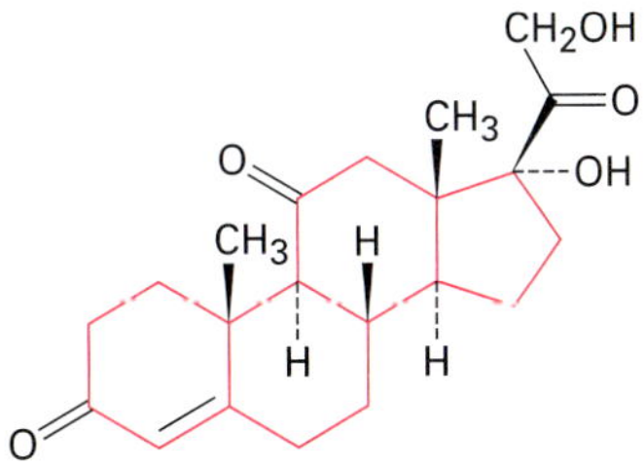

Cortisone

WHY THIS CHAPTER?

We'll see numerous instances in future chapters where the chemistry of a given functional group is strongly affected by being in a ring rather than an open chain. Because cyclic molecules are so commonly encountered in all classes of biomolecules, including proteins, lipids, carbohydrates, and nucleic acids, it's important that the effects of their cyclic structures be understood.

4.1 Naming Cycloalkanes

Saturated cyclic hydrocarbons are called **cycloalkanes**, or **alicyclic** compounds (**ali**phatic **cyclic**). Because cycloalkanes consist of rings of $-CH_2-$ units, they have the general formula $(CH_2)_n$, or C_nH_{2n}, and can be represented by polygons in skeletal drawings.

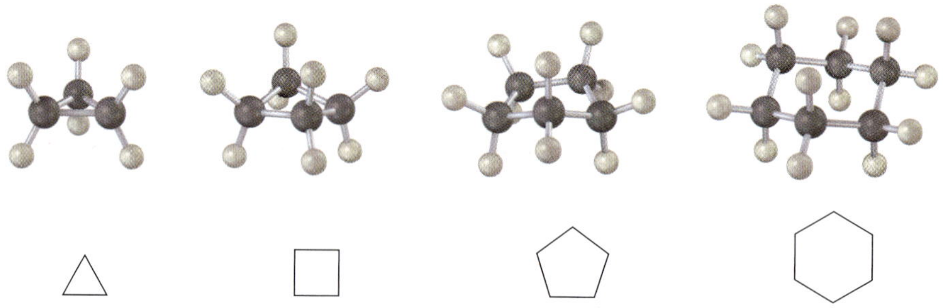

Cyclopropane **Cyclobutane** **Cyclopentane** **Cyclohexane**

Substituted cycloalkanes are named by rules similar to those we saw in the previous chapter for open-chain alkanes (Section 3.4). For most compounds, there are only two steps.

Rule 1 **Find the parent.**

Count the number of carbon atoms in the ring and the number in the largest substituent chain. If the number of carbon atoms in the ring is equal to or greater than the number in the substituent, the compound is named as an alkyl-substituted cycloalkane. If the number of carbon atoms in the largest substituent is greater than the number in the ring, the compound is named as a cycloalkyl-substituted alkane. For example:

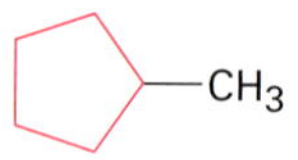

Methylcyclopentane

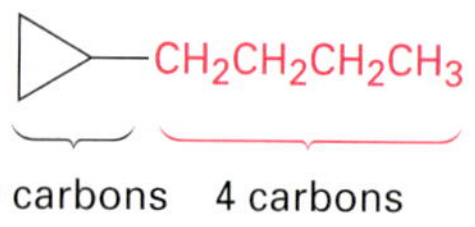

1-Cyclopropylbutane

Rule 2 **Number the substituents, and write the name.**

For an alkyl- or halo-substituted cycloalkane, choose a point of attachment as carbon 1 and number the substituents on the ring so that the *second* substituent has as low a number as possible. If ambiguity still exists, number so that the third or fourth substituent has as low a number as possible, until a point of difference is found.

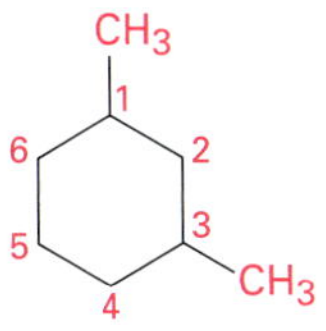

1,3-Dimethylcyclohexane
↑
Lower

NOT

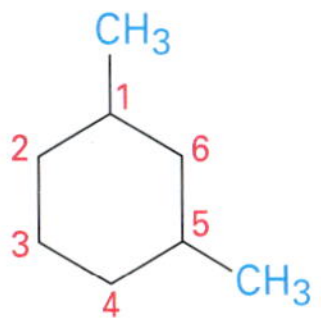

1,5-Dimethylcyclohexane
↑
Higher

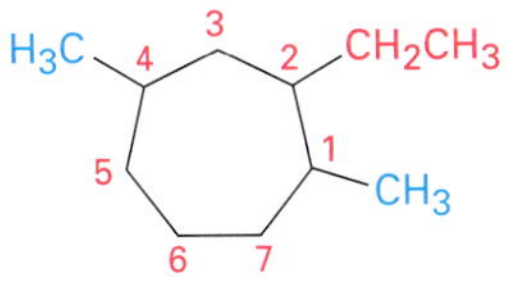

2-Ethyl-1,4-dimethylcycloheptane
↑ ↑
Lower Lower

NOT

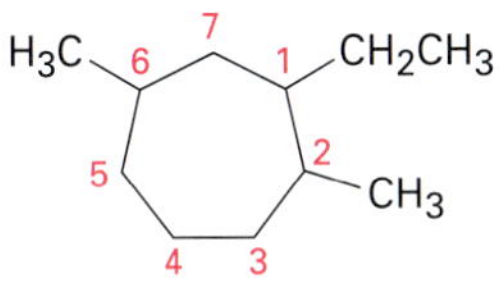

1-Ethyl-2,6-dimethylcycloheptane
↑
Higher

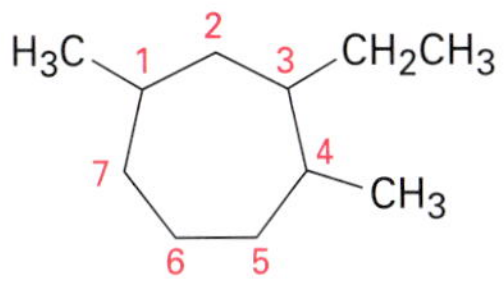

3-Ethyl-1,4-dimethylcycloheptane
↑
Higher

(a) When two or more different alkyl groups that could potentially receive the same numbers are present, number them by alphabetical priority.

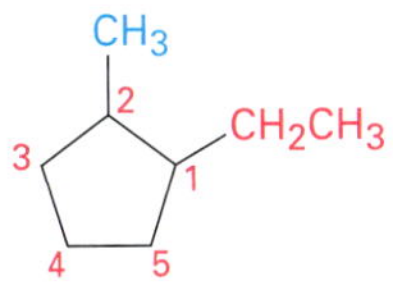

1-Ethyl-2-methylcyclopentane

NOT

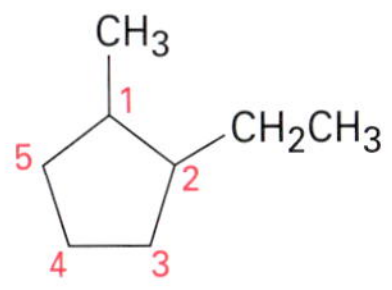

2-Ethyl-1-methylcyclopentane

(b) If halogens are present, treat them just like alkyl groups.

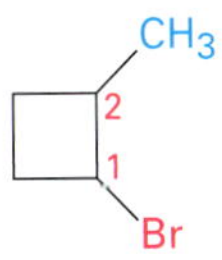

1-Bromo-2-methylcyclobutane

NOT

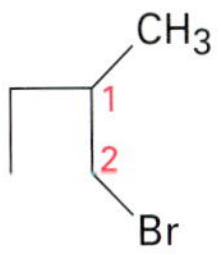

2-Bromo-1-methylcyclobutane

Some additional examples follow:

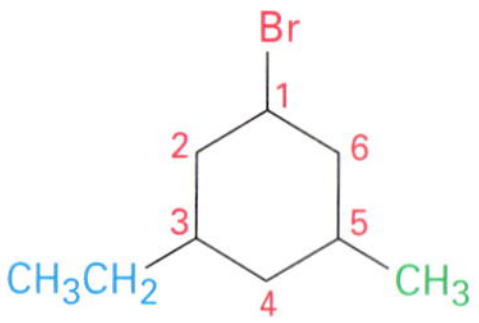

1-Bromo-3-ethyl-5-methylcyclohexane

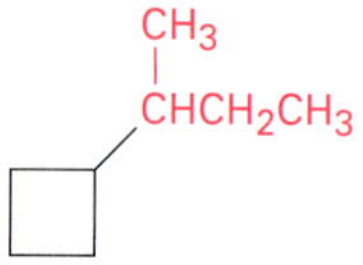

(1-Methylpropyl)cyclobutane or *sec*-butylcyclobutane

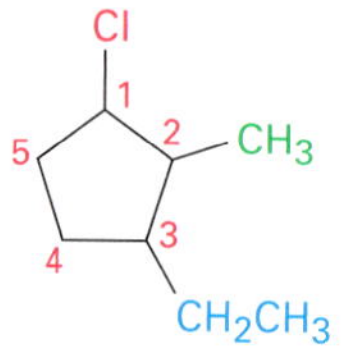

1-Chloro-3-ethyl-2-methylcyclopentane

Problem 4.1 Give IUPAC names for the following cycloalkanes:

(a)

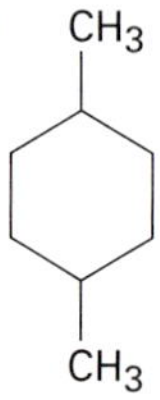

(b)

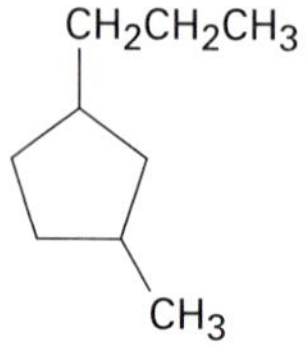

(c)

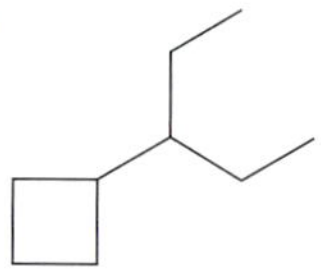

(d)

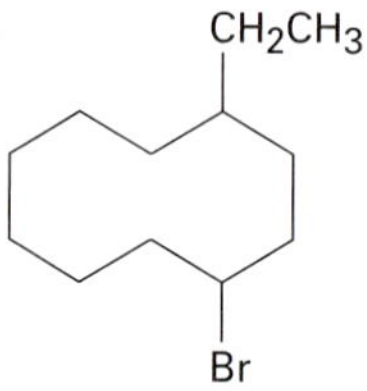

(e)

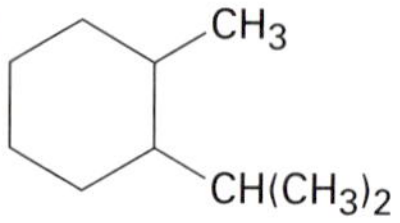

(f)

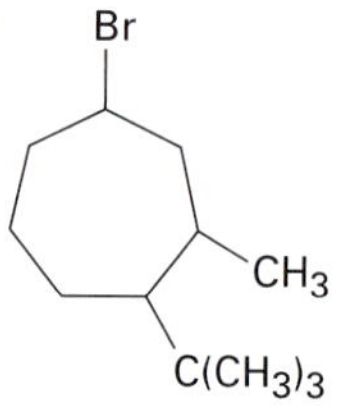

Problem 4.2 Draw structures corresponding to the following IUPAC names:

(a) 1,1-Dimethylcyclooctane **(b)** 3-Cyclobutylhexane

(c) 1,2-Dichlorocyclopentane **(d)** 1,3-Dibromo-5-methylcyclohexane

Problem 4.3 Name the following cycloalkane:

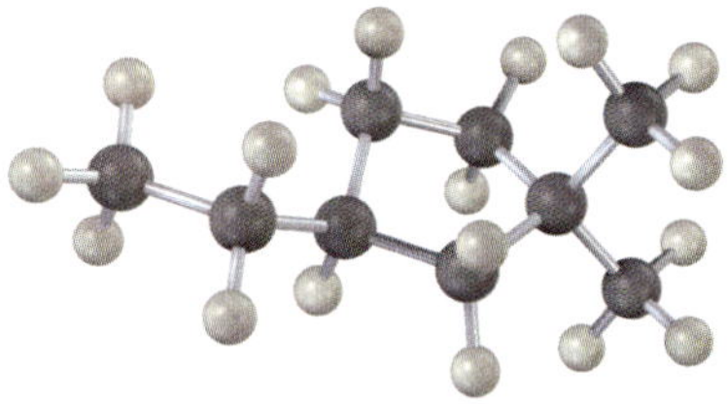

4.2 Cis–Trans Isomerism in Cycloalkanes

In many respects, the chemistry of cycloalkanes is like that of open-chain alkanes: both are nonpolar and fairly inert. There are, however, some important differences. One difference is that cycloalkanes are less flexible than open-chain alkanes. In contrast with the relatively free rotation around single bonds in open-chain alkanes (Sections 3.6 and 3.7), there is much less freedom in cycloalkanes.

Cyclopropane, for example, must be a rigid, planar molecule because three points (the carbon atoms) define a plane. No bond rotation can take place around a cyclopropane carbon–carbon bond without breaking open the ring (Figure 4.1).

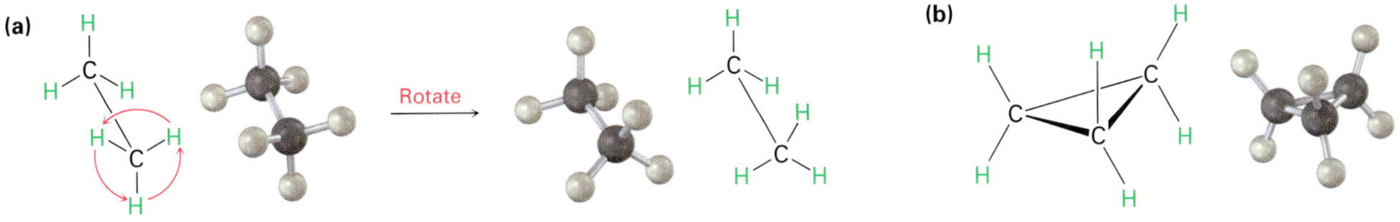

Figure 4.1 **(a)** Rotation occurs around the carbon–carbon bond in ethane, but **(b)** no rotation is possible around the carbon–carbon bonds in cyclopropane without breaking open the ring.

Larger cycloalkanes have increasing rotational freedom, and the very large rings (C_{25} and up) are so floppy that they are nearly indistinguishable from open-chain alkanes. The common ring sizes (C_3–C_7), however, are severely restricted in their molecular motions.

Because of their cyclic structures, cycloalkanes have two faces as viewed edge-on, a "top" face and a "bottom" face. As a result, isomerism is possible in substituted cycloalkanes. For example, there are two different 1,2-dimethylcyclopropane isomers, one with the two methyl groups on the same face of the ring and one with the methyls on opposite faces (Figure 4.2). Both isomers are stable compounds, and neither can be converted into the other without breaking and reforming chemical bonds. Make molecular models to prove this to yourself.

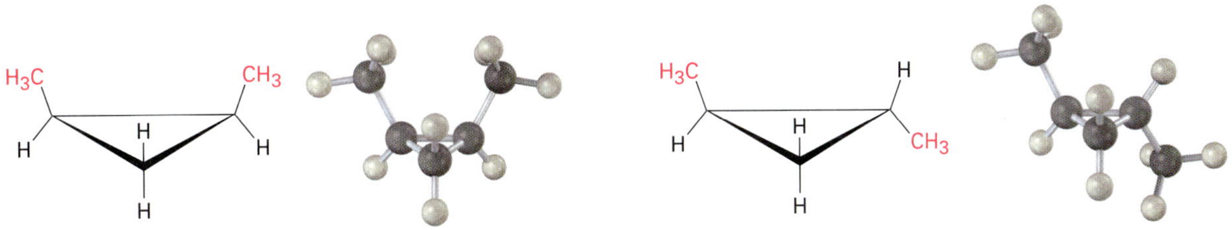

***cis*-1,2-Dimethylcyclopropane** ***trans*-1,2-Dimethylcyclopropane**

Figure 4.2 There are two different 1,2-dimethylcyclopropane isomers, one with the methyl groups on the same face of the ring (cis) and the other with the methyl groups on opposite faces of the ring (trans). The two isomers do not interconvert.

Unlike the constitutional isomers butane and isobutane (Section 3.2), which have their atoms connected in a different order, the two 1,2-dimethylcyclopropanes have the same order of connections but differ in the spatial orientation of the atoms. Such compounds, which have their atoms connected in the same order but differ in three-dimensional orientation, are called stereochemical isomers, or **stereoisomers**.

Constitutional isomers **(different connections between atoms)**

Stereoisomers **(same connections but different three-dimensional geometry)**

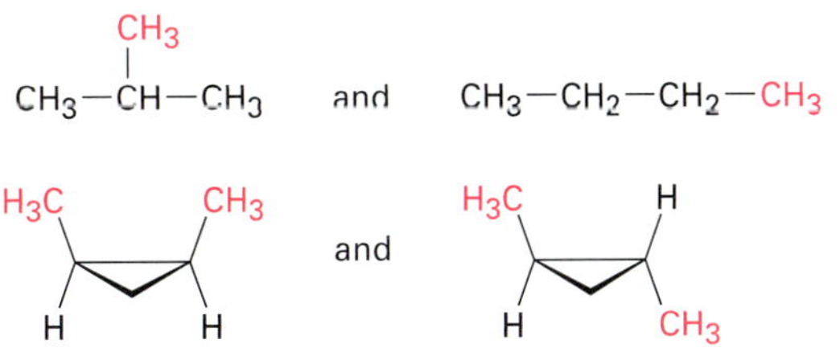

The 1,2-dimethylcyclopropanes are members of a subclass of stereoisomers called **cis–trans isomers.** The prefixes *cis-* (Latin "on the same side") and *trans-* (Latin "across") are used to distinguish between them. Cis–trans isomerism is a common occurrence in substituted cycloalkanes.

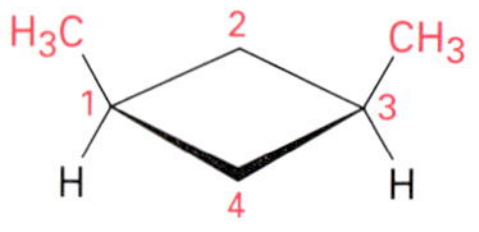

cis-**1,3-Dimethylcyclobutane**

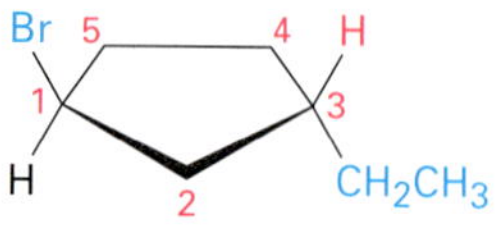

trans-**1-Bromo-3-ethylcyclopentane**

WORKED EXAMPLE 4.1 *Naming Cycloalkanes*

ThomsonNOW™ Click *Organic Interactive* to learn to **write IUPAC names for simple cycloalkanes.**

ThomsonNOW™ Click *Organic Interactive* to **use an online palette to draw cycloalkane structures from their IUPAC names.**

Name the following substances, including the *cis-* or *trans-* prefix:

(a)

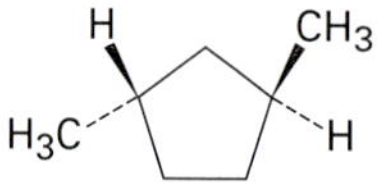

(b)

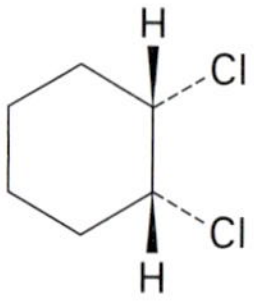

Strategy In these views, the ring is roughly in the plane of the page, a wedged bond protrudes out of the page, and a dashed bond recedes into the page. Two substituents are cis if they are both out of or both into the page, and they are trans if one is out of and one is into.

Solution **(a)** *trans*-1,3-Dimethylcyclopentane **(b)** *cis*-1,2-Dichlorocyclohexane

Problem 4.4 Name the following substances, including the *cis-* or *trans-* prefix:

(a)

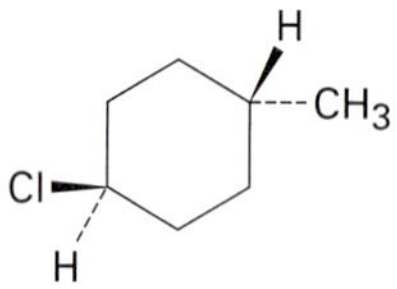

(b)

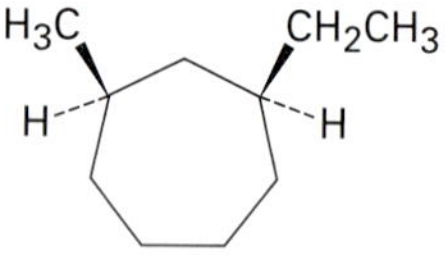

Problem 4.5 Draw the structures of the following molecules:
(a) *trans*-1-Bromo-3-methylcyclohexane **(b)** *cis*-1,2-Dimethylcyclobutane
(c) *trans*-1-*tert*-Butyl-2-ethylcyclohexane

Problem 4.6 Prostaglandin $F_{2\alpha}$, a hormone that causes uterine contraction during childbirth, has the following structure. Are the two hydroxyl groups (–OH) on the cyclopentane ring cis or trans to each other? What about the two carbon chains attached to the ring?

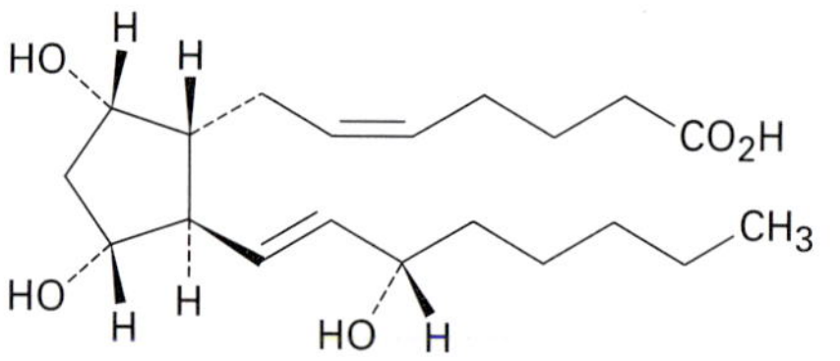

Prostaglandin $F_{2\alpha}$

Problem 4.7 Name the following substances, including the *cis*- or *trans*- prefix (red-brown = Br):

(a)

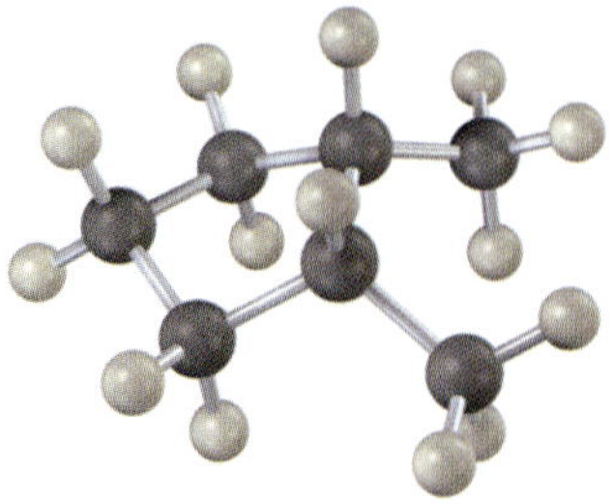

(b)

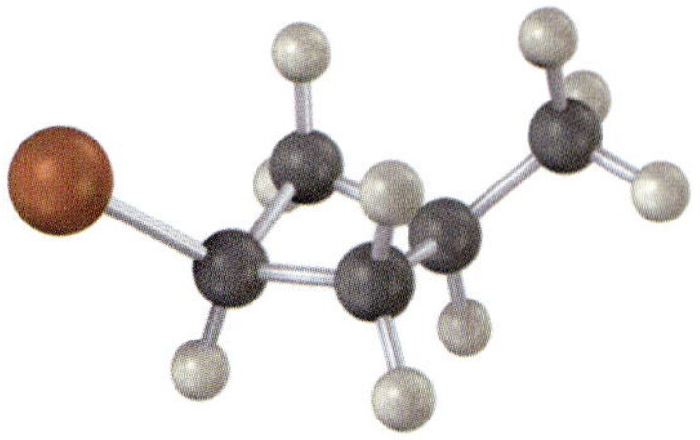

4.3 Stability of Cycloalkanes: Ring Strain

Adolf von Baeyer

Adolf von Baeyer (1835–1917) was born in Berlin, Germany, and received his Ph.D. at the University of Berlin in 1858, working with Robert Bunsen and August Kekulé. After holding positions at Berlin and Strasbourg, he was a professor at Munich from 1875 to 1917. He was the first to synthesize the blue dye indigo and was also discoverer of the first barbiturate sedative, which he named after his friend Barbara. Baeyer was awarded the Nobel Prize in chemistry in 1905.

Chemists in the late 1800s knew that cyclic molecules existed, but the limitations on ring size were unclear. Although numerous compounds containing five- and six-membered rings were known, smaller and larger ring sizes had not been prepared, despite many efforts.

A theoretical interpretation of this observation was proposed in 1885 by Adolf von Baeyer, who suggested that small and large rings might be unstable due to **angle strain**—the strain induced in a molecule when bond angles are forced to deviate from the ideal 109° tetrahedral value. Baeyer based his suggestion on the simple geometric notion that a three-membered ring (cyclopropane) should be an equilateral triangle with bond angles of 60° rather than 109°, a four-membered ring (cyclobutane) should be a square with bond angles of 90°, a five-membered ring should be a regular pentagon with bond angles of 108°, and so on. Continuing this argument, large rings should be strained by having bond angles that are much greater than 109°.

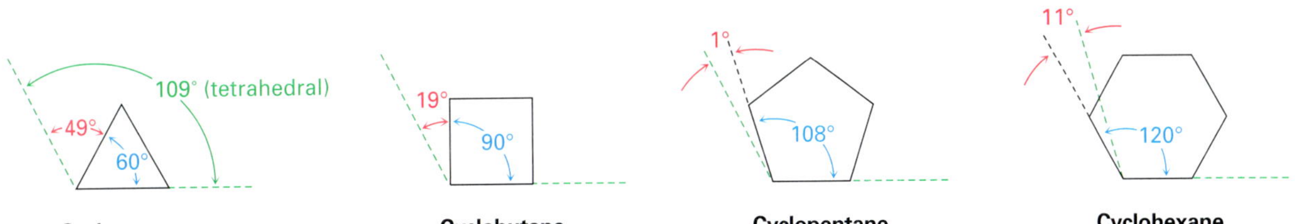

What are the facts? To measure the amount of strain in a compound, we have to measure the total energy of the compound and then subtract the energy of a strain-free reference compound. The difference between the two values should represent the amount of extra energy in the molecule due to strain. The simplest way to do this for a cycloalkane is to measure its *heat of combustion,* the amount of heat released when the compound burns completely with oxygen. The more energy (strain) the compound contains, the more energy (heat) is released on combustion.

$$(CH_2)_n + 3n/2\ O_2 \longrightarrow n\ CO_2 + n\ H_2O + \text{Heat}$$

Because the heat of combustion of a cycloalkane depends on size, we need to look at heats of combustion per CH_2 unit. Subtracting a reference value derived from a strain-free acyclic alkane and then multiplying by the number of CH_2 units in the ring gives the overall strain energy. Figure 4.3 shows the results.

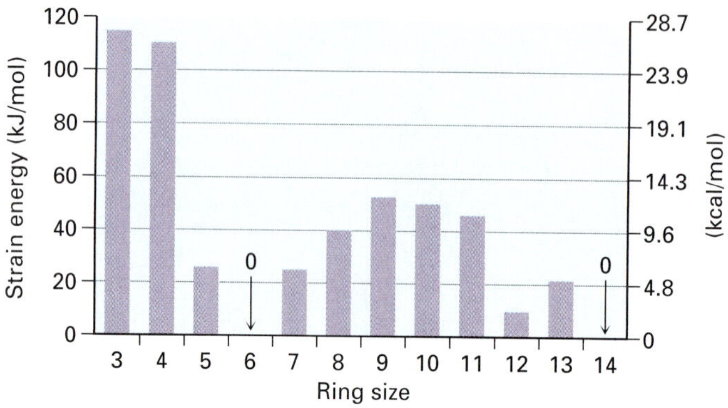

Figure 4.3 Cycloalkane strain energies, calculated by taking the difference between cycloalkane heat of combustion per CH_2 and acyclic alkane heat of combustion per CH_2, and multiplying by the number of CH_2 units in a ring. Small and medium rings are strained, but cyclohexane rings are strain-free.

The data in Figure 4.3 show that Baeyer's theory is only partially correct. Cyclopropane and cyclobutane are indeed strained, just as predicted, but cyclopentane is more strained than predicted, and cyclohexane is strain-free. Cycloalkanes of intermediate size have only modest strain, and rings of 14 carbons or more are strain-free. Why is Baeyer's theory wrong?

Baeyer's theory is wrong for the simple reason that he assumed all cycloalkanes to be flat. In fact, as we'll see shortly, most cycloalkanes are *not* flat; they adopt puckered three-dimensional conformations that allow bond angles to be nearly tetrahedral. As a result, angle strain occurs only in three- and four-membered rings that have little flexibility. For most ring sizes, particularly the medium-ring (C_7–C_{11}) cycloalkanes, torsional strain caused by H⟷H eclipsing interactions on adjacent carbons (Section 3.6) and steric strain caused by the repulsion between nonbonded atoms that approach too closely (Section 3.7) are the most important factors. Thus, three kinds of strain contribute to the overall energy of a cycloalkane.

- **Angle strain**—the strain due to expansion or compression of bond angles
- **Torsional strain**—the strain due to eclipsing of bonds on neighboring atoms
- **Steric strain**—the strain due to repulsive interactions when atoms approach each other too closely

Problem 4.8 Each H⟷H eclipsing interaction in ethane costs about 4.0 kJ/mol. How many such interactions are present in cyclopropane? What fraction of the overall 115 kJ/mol (27.5 kcal/mol) strain energy of cyclopropane is due to torsional strain?

Problem 4.9 *cis*-1,2-Dimethylcyclopropane has more strain than *trans*-1,2-dimethylcyclopropane. How can you account for this difference? Which of the two compounds is more stable?

4.4 Conformations of Cycloalkanes

Cyclopropane

Cyclopropane is the most strained of all rings, primarily because of the angle strain caused by its 60° C–C–C bond angles. In addition, cyclopropane also has considerable torsional strain because the C–H bonds on neighboring carbon atoms are eclipsed (Figure 4.4).

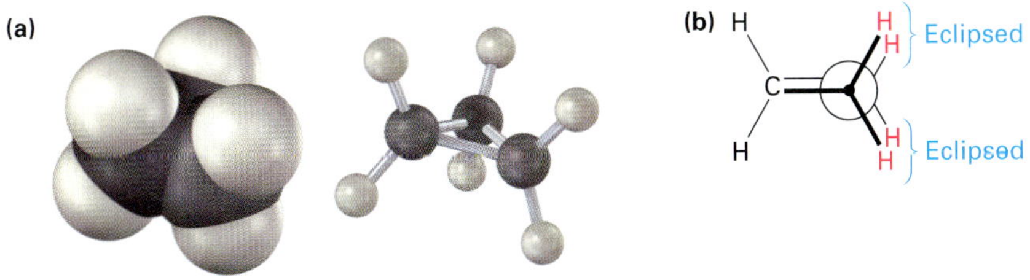

Figure 4.4 The structure of cyclopropane, showing the eclipsing of neighboring C–H bonds that gives rise to torsional strain. Part **(b)** is a Newman projection along a C–C bond.

How can the hybrid-orbital model of bonding account for the large distortion of bond angles from the normal 109° tetrahedral value to 60° in cyclopropane? The answer is that cyclopropane has *bent bonds.* In an unstrained alkane, maximum bonding is achieved when two atoms have their overlapping orbitals pointing directly toward each other. In cyclopropane, though, the orbitals can't point directly toward each other; rather, they overlap at an angle. The result is that cyclopropane bonds are weaker and more reactive than typical alkane bonds—255 kJ/mol (61 kcal/mol) for a C–C bond in cyclopropane versus 355 kJ/mol (85 kcal/mol) for a C–C bond in open-chain propane.

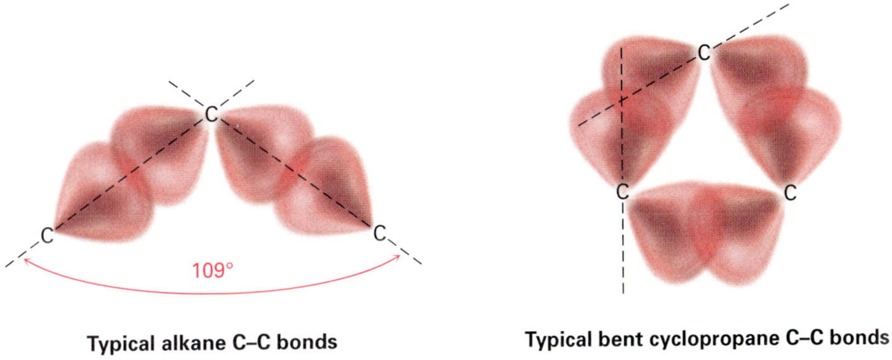

Cyclobutane

Cyclobutane has less angle strain than cyclopropane but has more torsional strain because of its larger number of ring hydrogens. As a result, the total strain for the two compounds is nearly the same—110 kJ/mol (26.4 kcal/mol) for cyclobutane versus 115 kJ/mol (27.5 kcal/mol) for cyclopropane. Experiments show that cyclobutane is not quite flat but is slightly bent so that one carbon atom lies about 25° above the plane of the other three (Figure 4.5). The effect of this slight bend is to *increase* angle strain but to *decrease* torsional strain, until a minimum-energy balance between the two opposing effects is achieved.

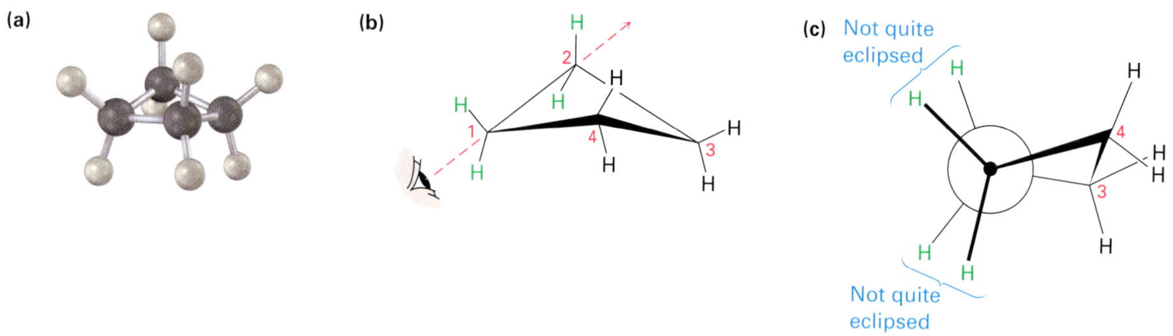

Figure 4.5 The conformation of cyclobutane. Part **(c)** is a Newman projection along the C1–C2 bond, showing that neighboring C–H bonds are not quite eclipsed.

Cyclopentane

Cyclopentane was predicted by Baeyer to be nearly strain-free but in fact has a total strain energy of 26 kJ/mol (6.2 kcal/mol). Although planar cyclopentane has practically no angle strain, it has a large amount of torsional strain. Cyclopentane therefore twists to adopt a puckered, nonplanar conformation that strikes a balance between increased angle strain and decreased torsional strain. Four of the cyclopentane carbon atoms are in approximately the same plane, with the fifth carbon atom bent out of the plane. Most of the hydrogens are nearly staggered with respect to their neighbors (Figure 4.6).

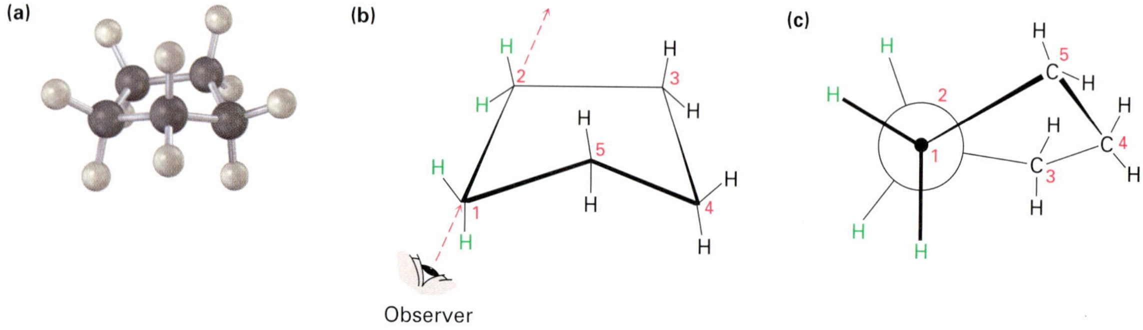

Figure 4.6 The conformation of cyclopentane. Carbons 1, 2, 3, and 4 are nearly planar, but carbon 5 is out of the plane. Part **(c)** is a Newman projection along the C1–C2 bond, showing that neighboring C–H bonds are nearly staggered.

Problem 4.10 How many H⟷H eclipsing interactions would be present if cyclopentane were planar? Assuming an energy cost of 4.0 kJ/mol for each eclipsing interaction, how much torsional strain would planar cyclopentane have? Since the measured total strain of cyclopentane is 26 kJ/mol, how much of the torsional strain is relieved by puckering?

Problem 4.11 Two conformations of *cis*-1,3-dimethylcyclobutane are shown. What is the difference between them, and which do you think is likely to be more stable?

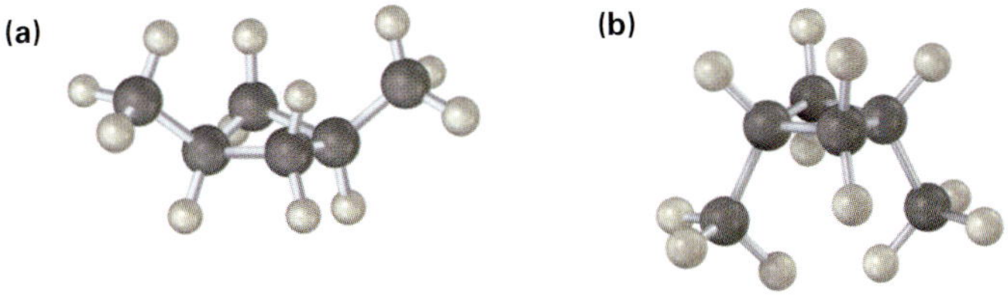

4.5 Conformations of Cyclohexane

Substituted cyclohexanes are the most common cycloalkanes and occur widely in nature. A large number of compounds, including steroids and many pharmaceutical agents, have cyclohexane rings. The flavoring agent menthol, for instance, has three substituents on a six-membered ring.

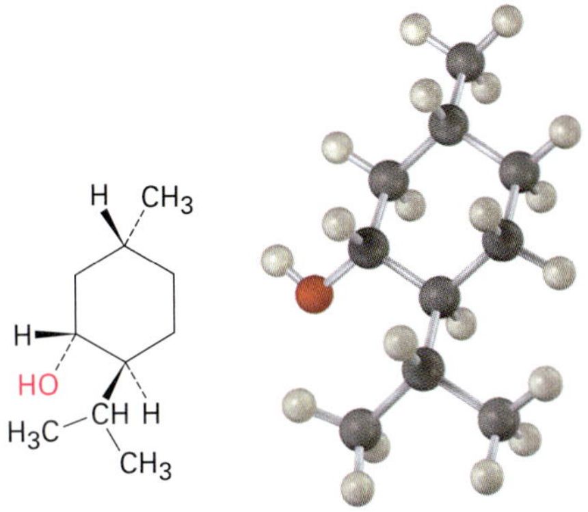

Menthol

Cyclohexane adopts a strain-free, three-dimensional shape, called a **chair conformation** because of its similarity to a lounge chair, with a back, a seat, and a footrest (Figure 4.7). Chair cyclohexane has neither angle strain nor torsional strain—all C–C–C bond angles are near 109°, and all neighboring C–H bonds are staggered.

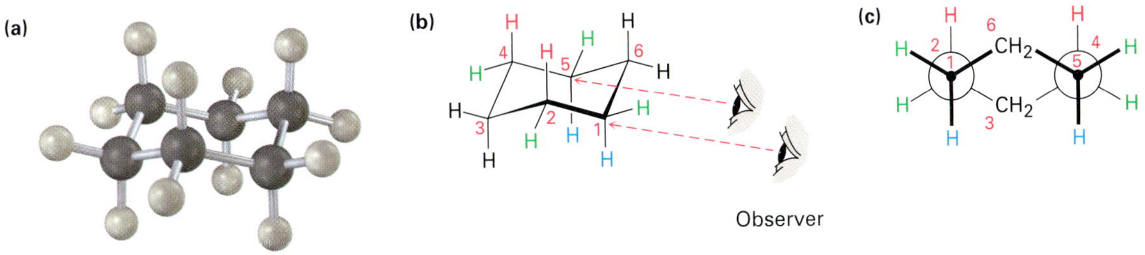

Figure 4.7 The strain-free chair conformation of cyclohexane. All C–C–C bond angles are 111.5°, close to the ideal 109.5° tetrahedral angle, and all neighboring C–H bonds are staggered.

The easiest way to visualize chair cyclohexane is to build a molecular model. (In fact, do it now.) Two-dimensional drawings like that in Figure 4.7 are useful, but there's no substitute for holding, twisting, and turning a three-dimensional model in your own hands. The chair conformation of cyclohexane can be drawn in three steps.

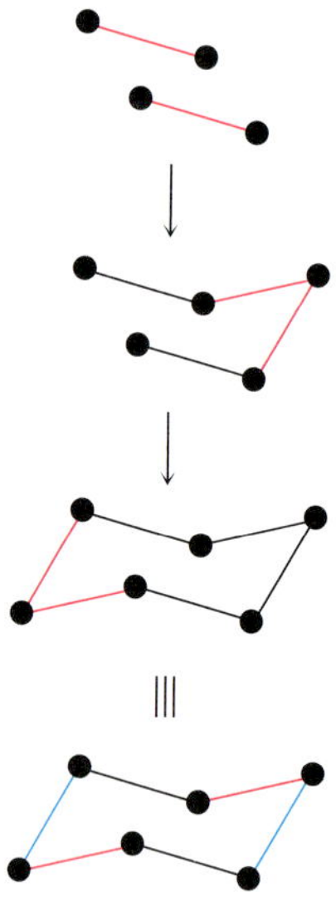

Step 1 Draw two parallel lines, slanted downward and slightly offset from each other. This means that four of the cyclohexane carbons lie in a plane.

Step 2 Place the topmost carbon atom above and to the right of the plane of the other four, and connect the bonds.

Step 3 Place the bottommost carbon atom below and to the left of the plane of the middle four, and connect the bonds. Note that the bonds to the bottommost carbon atom are parallel to the bonds to the topmost carbon.

When viewing cyclohexane, it's helpful to remember that the lower bond is in front and the upper bond is in back. If this convention is not defined, an optical illusion can make it appear that the reverse is true. For clarity, all cyclohexane rings drawn in this book will have the front (lower) bond heavily shaded to indicate nearness to the viewer.

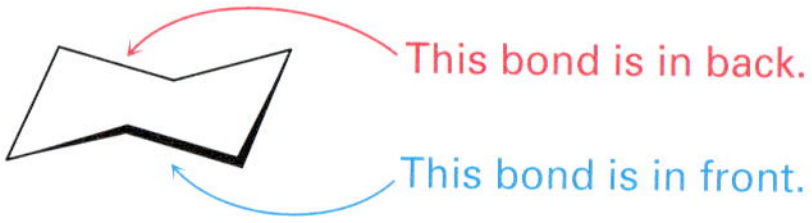

In addition to the chair conformation of cyclohexane, a second arrangement called the **twist-boat conformation** is also nearly free of angle strain. It does, however, have both steric strain and torsional strain and is about 23 kJ/mol (5.5 kcal/mol) higher in energy than the chair conformation. As a result, molecules adopt the twist-boat geometry only under special circumstances.

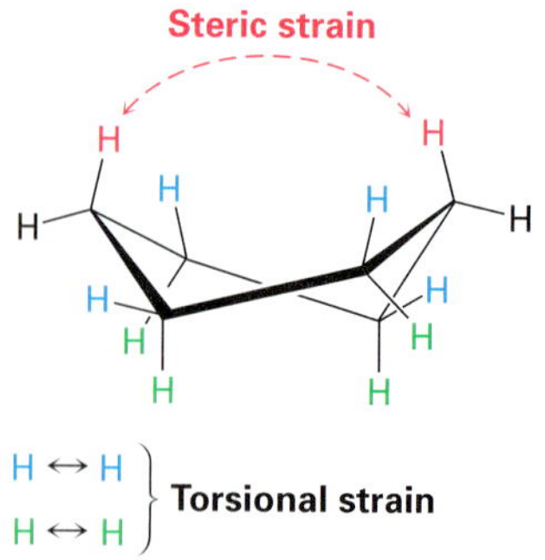

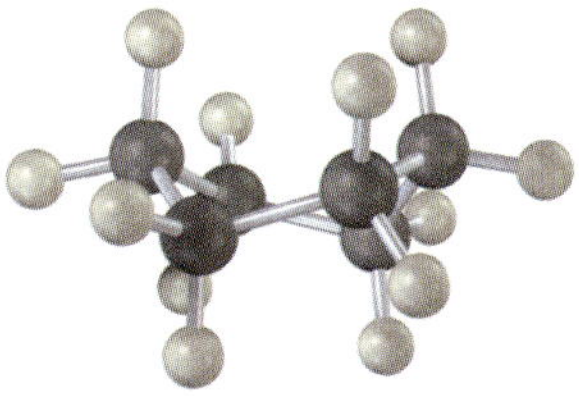

Twist-boat cyclohexane
(23 kJ/mol strain)

4.6 Axial and Equatorial Bonds in Cyclohexane

The chair conformation of cyclohexane has many consequences. We'll see in Section 11.9, for instance, that the chemical behavior of many substituted cyclohexanes is influenced by their conformation. In addition, we'll see in Section 25.5 that simple carbohydrates such as glucose adopt a conformation based on the cyclohexane chair and that their chemistry is directly affected as a result.

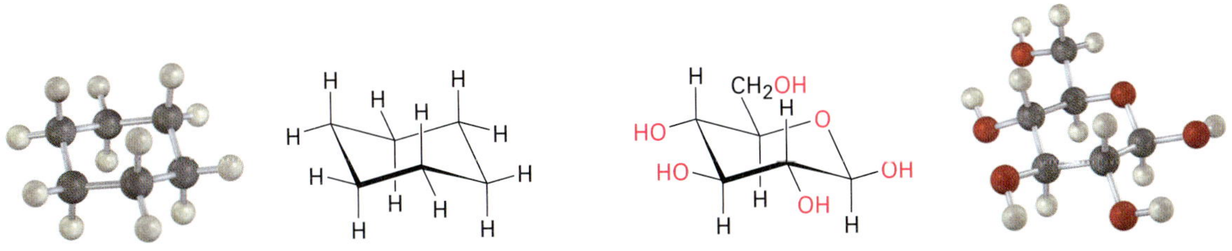

Cyclohexane (chair conformation)

Glucose (chair conformation)

Another consequence of the chair conformation is that there are two kinds of positions for substituents on the cyclohexane ring: *axial* positions and *equatorial* positions (Figure 4.8). The six **axial** positions are perpendicular to the ring, parallel to the ring axis, and the six **equatorial** positions are in the rough plane of the ring, around the ring equator.

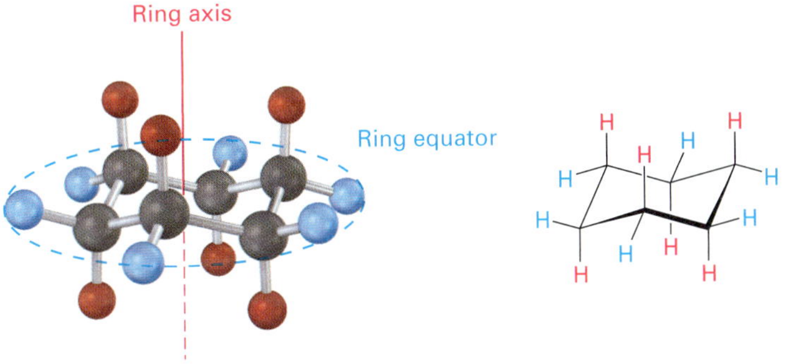

Figure 4.8 Axial (red) and equatorial (blue) positions in chair cyclohexane. The six axial hydrogens are parallel to the ring axis, and the six equatorial hydrogens are in a band around the ring equator.

As shown in Figure 4.8, each carbon atom in cyclohexane has one axial and one equatorial hydrogen. Furthermore, each face of the ring has three axial and three equatorial hydrogens in an alternating arrangement. For example, if the top face of the ring has axial hydrogens on carbons 1, 3, and 5, then it has equatorial hydrogens on carbons 2, 4, and 6. Exactly the reverse is true for the bottom face: carbons 1, 3, and 5 have equatorial hydrogens, but carbons 2, 4, and 6 have axial hydrogens (Figure 4.9).

Note that we haven't used the words *cis* and *trans* in this discussion of cyclohexane conformation. Two hydrogens on the same face of the ring are always cis, regardless of whether they're axial or equatorial and regardless of whether they're adjacent. Similarly, two hydrogens on opposite faces of the ring are always trans.

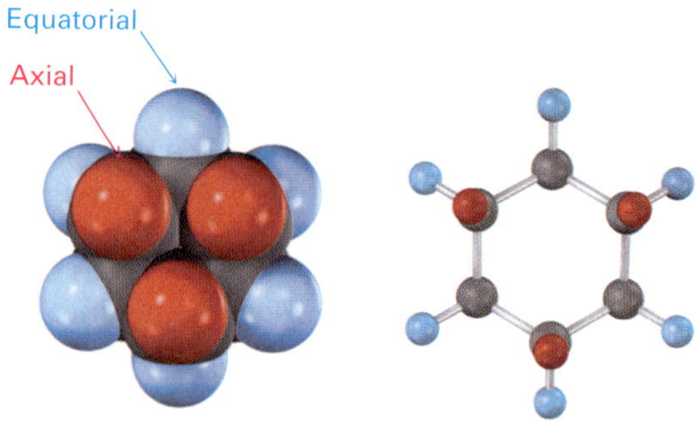

Figure 4.9 Alternating axial and equatorial positions in chair cyclohexane, as shown in a view looking directly down the ring axis. Each carbon atom has one axial and one equatorial position, and each face has alternating axial and equatorial positions.

Axial and equatorial bonds can be drawn following the procedure in Figure 4.10. Look at a molecular model as you practice.

Axial bonds: The six axial bonds, one on each carbon, are parallel and alternate up–down.

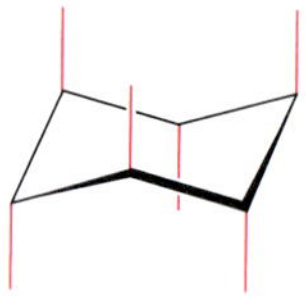

Equatorial bonds: The six equatorial bonds, one on each carbon, come in three sets of two parallel lines. Each set is also parallel to two ring bonds. Equatorial bonds alternate between sides around the ring.

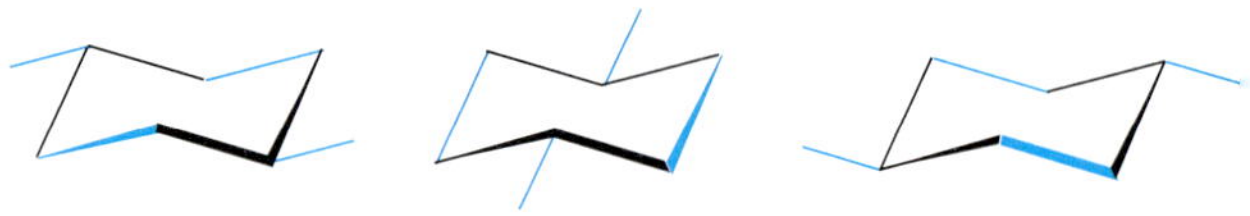

Completed cyclohexane

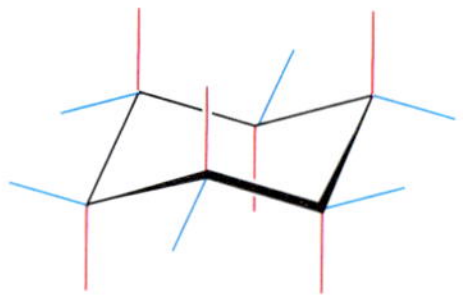

Figure 4.10 A procedure for drawing axial and equatorial bonds in chair cyclohexane.

Because chair cyclohexane has two kinds of positions, axial and equatorial, we might expect to find two isomeric forms of a monosubstituted cyclohexane. In fact, we don't. There is only *one* methylcyclohexane, *one* bromocyclohexane, *one* cyclohexanol (hydroxycyclohexane), and so on, because cyclohexane rings are *conformationally mobile* at room temperature. Different chair conformations readily interconvert, exchanging axial and equatorial positions. This interconversion, usually called a **ring-flip**, is shown in Figure 4.11.

As shown in Figure 4.11, a chair cyclohexane can be ring-flipped by keeping the middle four carbon atoms in place while folding the two end carbons in opposite directions. In so doing, an axial substituent in one chair form becomes an equatorial substituent in the ring-flipped chair form and vice versa. For example, axial bromocyclohexane becomes equatorial bromocyclohexane after ring-flip. Since the energy barrier to chair–chair interconversion is only

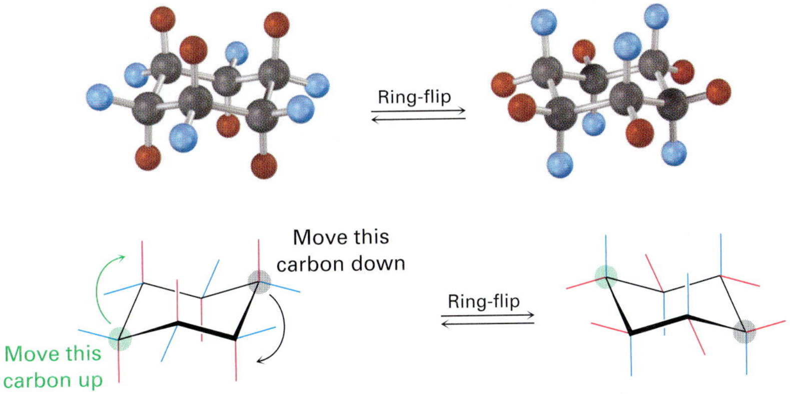

Figure 4.11 A ring-flip in chair cyclohexane interconverts axial and equatorial positions. What is axial (red) in the starting structure becomes equatorial in the ring-flipped structure, and what is equatorial (blue) in the starting structure is axial after ring-flip.

about 45 kJ/mol (10.8 kcal/mol), the process is rapid at room temperature and we see what appears to be a single structure rather than distinct axial and equatorial isomers.

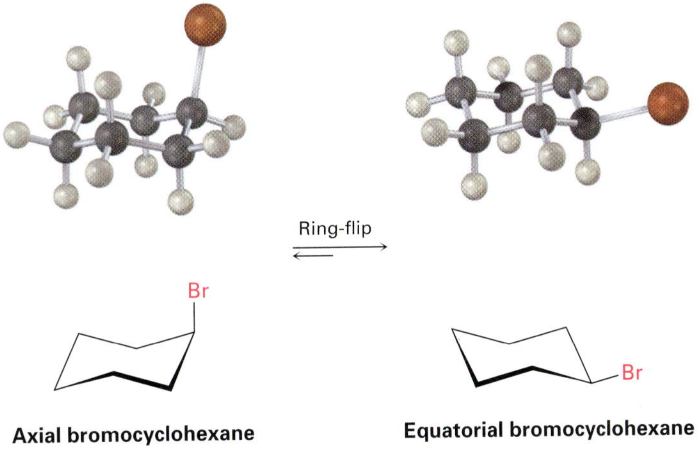

WORKED EXAMPLE 4.2

Drawing the Chair Conformation of a Substituted Cyclohexane

Draw 1,1-dimethylcyclohexane in a chair conformation, indicating which methyl group in your drawing is axial and which is equatorial.

Strategy Draw a chair cyclohexane ring using the procedure in Figure 4.9, and then put two methyl groups on the same carbon. The methyl group in the rough plane of the ring is equatorial, and the other (directly above or below the ring) is axial.

Solution

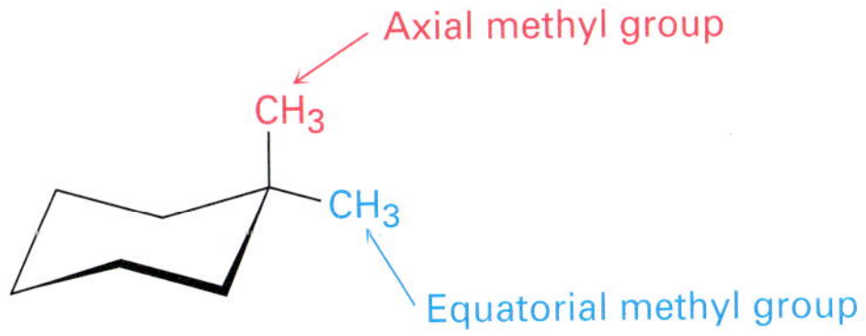

Problem 4.12 Draw two different chair conformations of cyclohexanol (hydroxycyclohexane), showing all hydrogen atoms. Identify each position as axial or equatorial.

Problem 4.13 Draw two different chair conformations of *trans*-1,4-dimethylcyclohexane, and label all positions as axial or equatorial.

Problem 4.14 Identify each of the colored positions—red, blue, and green—as axial or equatorial. Then carry out a ring-flip, and show the new positions occupied by each color.

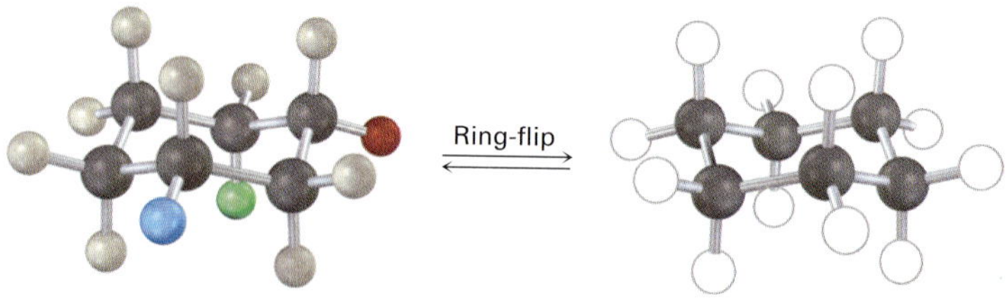

4.7 Conformations of Monosubstituted Cyclohexanes

Key **IDEAS**

Test your knowledge of Key Ideas by using resources in ThomsonNOW or by answering end-of-chapter problems marked with ▲.

Even though cyclohexane rings rapidly flip between chair conformations at room temperature, the two conformations of a monosubstituted cyclohexane aren't equally stable. In methylcyclohexane, for instance, the equatorial conformation is more stable than the axial conformation by 7.6 kJ/mol (1.8 kcal/mol). The same is true of other monosubstituted cyclohexanes: a substituent is almost always more stable in an equatorial position than in an axial position.

You might recall from your general chemistry course that it's possible to calculate the percentages of two isomers at equilibrium using the equation $\Delta E = -RT \ln K$, where ΔE is the energy difference between isomers, R is the gas constant [8.315 J/(K · mol)], T is the Kelvin temperature, and K is the equilibrium constant between isomers. For example, an energy difference of 7.6 kJ/mol means that about 95% of methylcyclohexane molecules have the methyl group equatorial at any given instant and only 5% have the methyl group axial. Figure 4.12 plots the relationship between energy and isomer percentages.

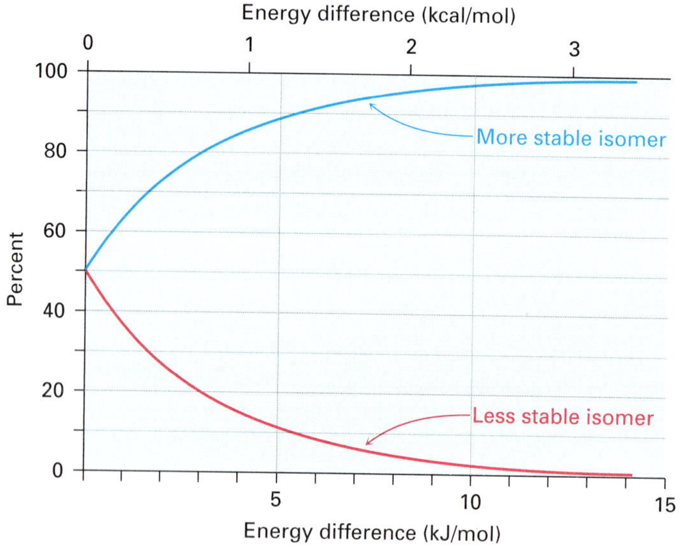

Figure 4.12 A plot of the percentages of two isomers at equilibrium versus the energy difference between them. The curves are calculated using the equation $\Delta E = -RT \ln K$.

The energy difference between axial and equatorial conformations is due to steric strain caused by **1,3-diaxial interactions**. The axial methyl group on C1 is too close to the axial hydrogens three carbons away on C3 and C5, resulting in 7.6 kJ/mol of steric strain (Figure 4.13).

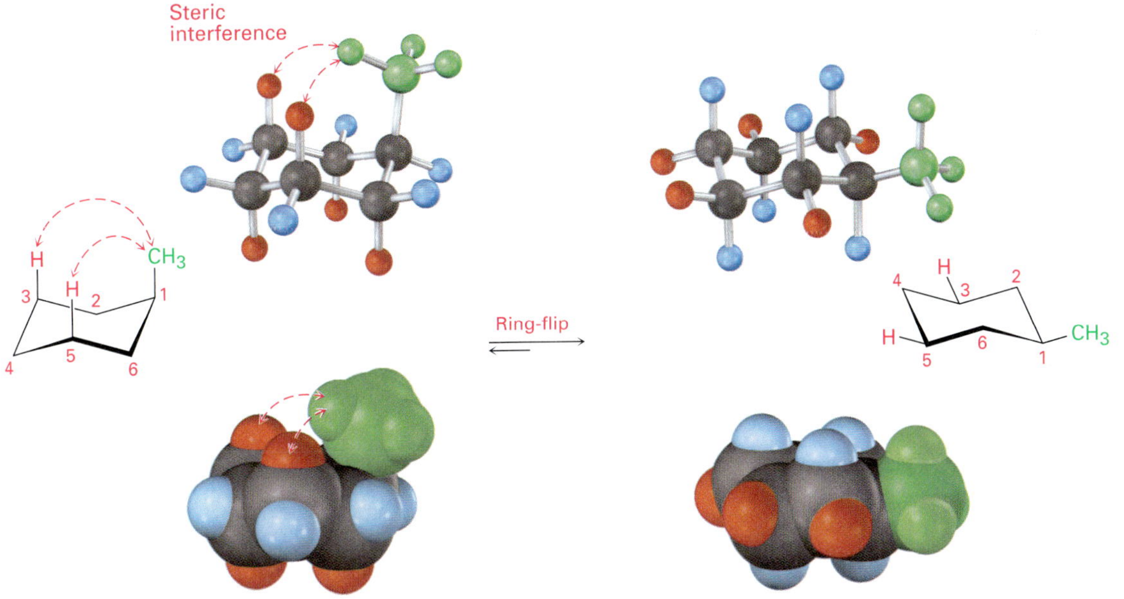

Figure 4.13 Interconversion of axial and equatorial methylcyclohexane, as represented in several formats. The equatorial conformation is more stable than the axial conformation by 7.6 kJ/mol.

The 1,3-diaxial steric strain in substituted methylcyclohexane is already familiar—we saw it previously as the steric strain between methyl groups in gauche butane. Recall from Section 3.7 that gauche butane is less stable than anti butane by 3.8 kJ/mol (0.9 kcal/mol) because of steric interference between hydrogen atoms on the two methyl groups. Comparing a four-carbon fragment of axial methylcyclohexane with gauche butane shows that the steric interaction is the same in both cases (Figure 4.14). Because axial methylcyclohexane has two such interactions, though, it has $2 \times 3.8 = 7.6$ kJ/mol of steric strain. Equatorial methylcyclohexane, however, has no such interactions and is therefore more stable.

Figure 4.14 The origin of 1,3-diaxial interactions in methylcyclohexane. The steric strain between an axial methyl group and an axial hydrogen atom three carbons away is identical to the steric strain in gauche butane. Note that the $-CH_3$ group in methylcyclohexane moves slightly away from a true axial position to minimize the strain.

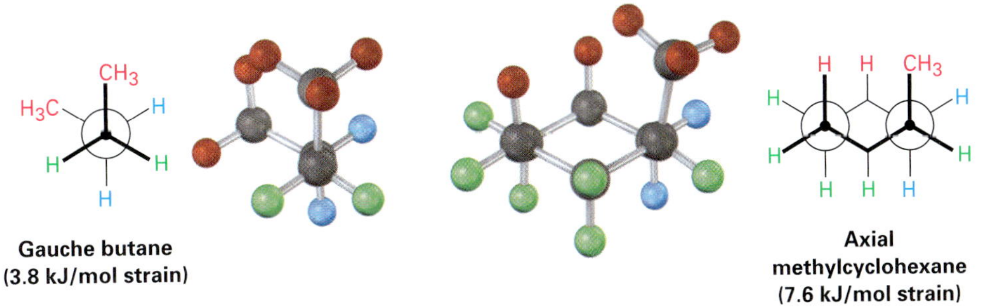

What is true for methylcyclohexane is also true for other monosubstitute cyclohexanes: a substituent is almost always more stable in an equatorial posi tion than in an axial position. The exact amount of 1,3-diaxial steric strain in given substituted cyclohexane depends on the nature and size of the sub stituent, as indicated in Table 4.1. Not surprisingly, the amount of steric strai increases through the series $H_3C– < CH_3CH_2– < (CH_3)_2CH– << (CH_3)_3C–$ paralleling the increasing bulk of the alkyl groups. Note that the values in Tabl 4.1 refer to 1,3-diaxial interactions of the substituent with a *single* hydroger atom. These values must be doubled to arrive at the amount of strain in a mono substituted cyclohexane.

Table 4.1 Steric Strain in Monosubstituted Cyclohexanes

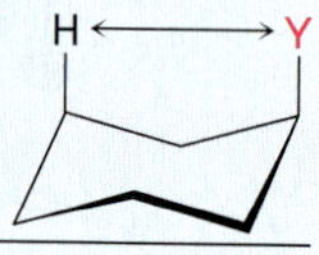

Y	1,3-Diaxial strain (kJ/mol)	(kcal/mol)
F	0.5	0.12
Cl, Br	1.0	0.25
OH	2.1	0.5
CH_3	3.8	0.9
CH_2CH_3	4.0	0.95
$CH(CH_3)_2$	4.6	1.1
$C(CH_3)_3$	11.4	2.7
C_6H_5	6.3	1.5
CO_2H	2.9	0.7
CN	0.4	0.1

Problem 4.15 What is the energy difference between the axial and equatorial conformations of cyclohexanol (hydroxycyclohexane)?

Problem 4.16 Why do you suppose an axial cyano (–CN) substituent causes practically no 1,3-diaxial steric strain (0.4 kJ/mol)? Use molecular models to help with your answer.

Problem 4.17 Look at Figure 4.12, and estimate the percentages of axial and equatorial conformers present at equilibrium in bromocyclohexane.

4.8 Conformations of Disubstituted Cyclohexanes

Monosubstituted cyclohexanes are more stable with their substituent in an equatorial position, but the situation in disubstituted cyclohexanes is more complex because the steric effects of both substituents must be taken into account. All steric interactions in both possible chair conformations must be analyzed before deciding which conformation is favored.

Let's look at 1,2-dimethylcyclohexane as an example. There are two isomers, *cis*-1,2-dimethylcyclohexane and *trans*-1,2-dimethylcyclohexane, which must be considered separately. In the cis isomer, both methyl groups are on the same face of the ring, and the compound can exist in either of the two chair conformations shown in Figure 4.15. (It may be easier for you to see whether a compound is cis- or trans-disubstituted by first drawing the ring as a flat representation and then converting to a chair conformation.) Both chair conformations have one axial methyl group and one equatorial methyl group. The top conformation in Figure 4.15 has an axial methyl group at C2, which has 1,3-diaxial interactions with hydrogens on C4 and C6. The ring-flipped conformation has an axial methyl group at C1, which has 1,3-diaxial interactions with hydrogens on C3 and C5. In addition, both conformations have gauche butane interactions between the two methyl groups. *The two conformations are equal in energy,* with a total steric strain of 3 × 3.8 kJ/mol = 11.4 kJ/mol (2.7 kcal/mol).

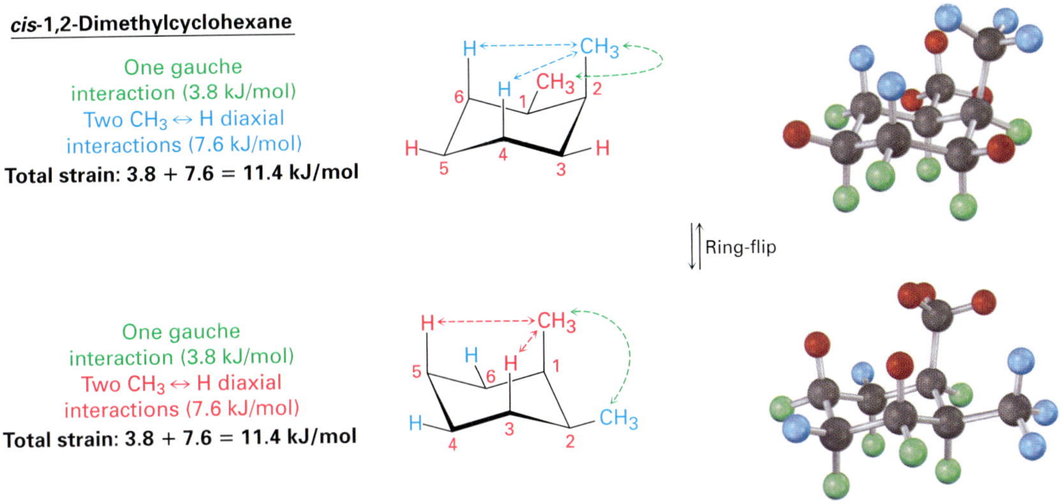

Active Figure 4.15 Conformations of *cis*-1,2-dimethylcyclohexane. The two chair conformations are equal in energy because each has one axial methyl group and one equatorial methyl group. *Sign in at* **www.thomsonedu.com** *to see a simulation based on this figure and to take a short quiz.*

In *trans*-1,2-dimethylcyclohexane, the two methyl groups are on opposite faces of the ring and the compound can exist in either of the two chair conformations shown in Figure 4.16. The situation here is quite different from that of the cis isomer. The top trans conformation in Figure 4.16 has both methyl groups equatorial and therefore has only a gauche butane interaction between methyls (3.8 kJ/mol) but no 1,3-diaxial interactions. The ring-flipped conformation, however, has both methyl groups axial. The axial methyl group at C1 interacts with axial hydrogens at C3 and C5, and the axial methyl group at C2 interacts with axial hydrogens at C4 and C6. These four 1,3-diaxial interactions produce a steric strain of 4 × 3.8 kJ/mol = 15.2 kJ/mol and make the diaxial conformation 15.2 − 3.8 = 11.4 kJ/mol less favorable than the diequatorial conformation. We therefore predict that *trans*-1,2-dimethylcyclohexane will exist almost exclusively in the diequatorial conformation.

ThomsonNOW™ Click *Organic Interactive* to learn to **draw and assess the stability of substituted cyclohexanes.**

The same kind of **conformational analysis** just carried out for *cis*- and *trans*-1,2-dimethylcyclohexane can be done for any substituted cyclohexane, such as *cis*-1-*tert*-butyl-4-chlorocyclohexane (see Worked Example 4.3). As you might imagine, though, the situation becomes more complex as the number of

***trans*-1,2-Dimethylcyclohexane**

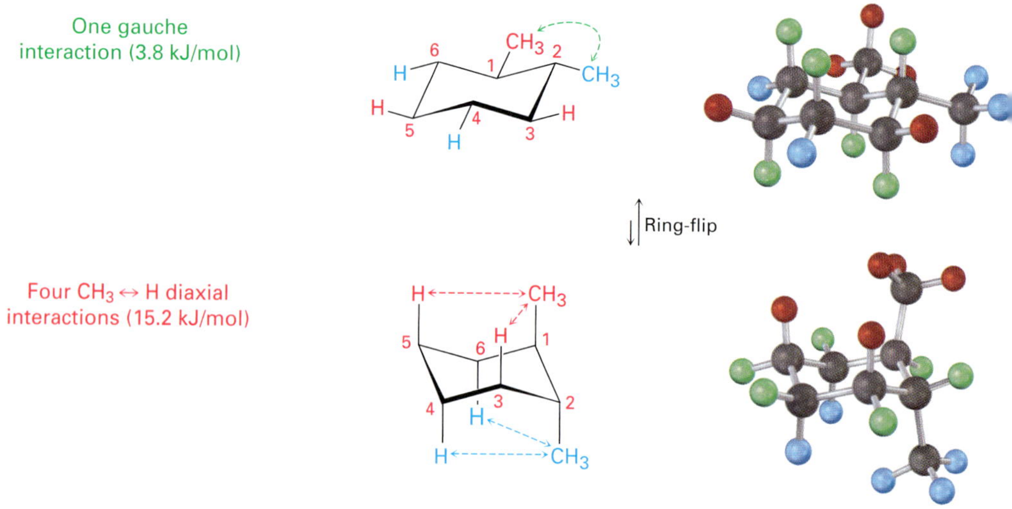

Figure 4.16 Conformations of *trans*-1,2-dimethylcyclohexane. The conformation with both methyl groups equatorial is favored by 11.4 kJ/mol (2.7 kcal/mol) over the conformation with both methyl groups axial.

ThomsonNOW™ Click *Organic Interactive* to learn to **recognize the most stable conformations of cyclohexanes following ring-flips.**

substituents increases. For instance, compare glucose with mannose, a carbohydrate present in seaweed. Which do you think is more strained? In glucose, all substituents on the six-membered ring are equatorial, while in mannose, one of the –OH groups is axial, making mannose more strained.

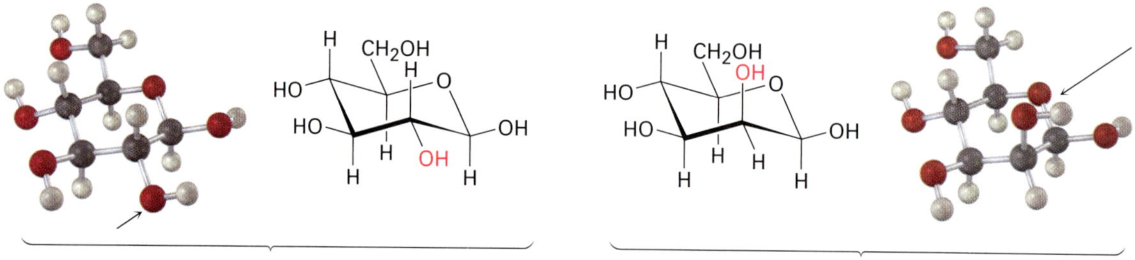

Glucose **Mannose**

ThomsonNOW™ Click *Organic Interactive* to **use an online palette to draw and interconvert cyclohexane structures.**

A summary of the various axial and equatorial relationships among substituent groups in the different possible cis and trans substitution patterns for disubstituted cyclohexanes is given in Table 4.2.

Table 4.2 Axial and Equatorial Relationships in Cis- and Trans-Disubstituted Cyclohexanes

Cis/trans substitution pattern	Axial/equatorial relationships		
1,2-Cis disubstituted	a,e	or	e,a
1,2-Trans disubstituted	a,a	or	e,e
1,3-Cis disubstituted	a,a	or	e,e
1,3-Trans disubstituted	a,e	or	e,a
1,4-Cis disubstituted	a,e	or	e,a
1,4-Trans disubstituted	a,a	or	e,e

WORKED EXAMPLE 4.3

Drawing the Most Stable Conformation of a Substituted Cyclohexane

Draw the most stable conformation of *cis*-1-*tert*-butyl-4-chlorocyclohexane. By how much is it favored?

Strategy Draw the possible conformations, and calculate the strain energy in each. Remember that equatorial substituents cause less strain than axial substituents.

Solution First draw the two chair conformations of the molecule:

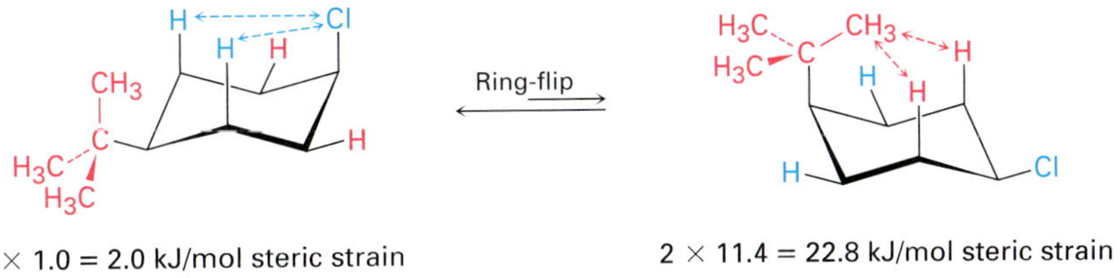

In the left-hand conformation, the *tert*-butyl group is equatorial and the chlorine is axial. In the right-hand conformation, the *tert*-butyl group is axial and the chlorine is equatorial. These conformations aren't of equal energy because an axial *tert*-butyl substituent and an axial chloro substituent produce different amounts of steric strain. Table 4.1 shows that the 1,3-diaxial interaction between a hydrogen and a *tert*-butyl group costs 11.4 kJ/mol (2.7 kcal/mol), whereas the interaction between a hydrogen and a chlorine costs only 1.0 kJ/mol (0.25 kcal/mol). An axial *tert*-butyl group therefore produces $(2 \times 11.4\ \text{kJ/mol}) - (2 \times 1.0\ \text{kJ/mol}) = 20.8$ kJ/mol (4.9 kcal/mol) more steric strain than does an axial chlorine, and the compound preferentially adopts the conformation with the chlorine axial and the *tert*-butyl equatorial.

Problem 4.18 Draw the most stable chair conformation of the following molecules, and estimate the amount of strain in each:

(a) *trans*-1-Chloro-3-methylcyclohexane
(b) *cis*-1-Ethyl-2-methylcyclohexane
(c) *cis*-1-Bromo-4-ethylcyclohexane
(d) *cis*-1-*tert*-Butyl-4-ethylcyclohexane

Problem 4.19 Identify each substituent in the following compound as axial or equatorial, and tell whether the conformation shown is the more stable or less stable chair form (yellow-green = Cl):

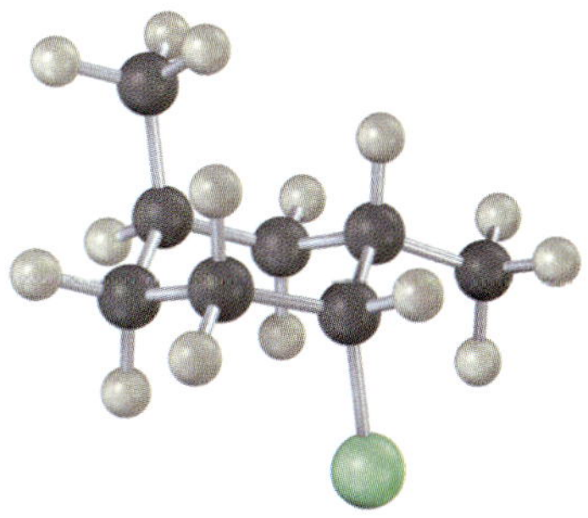

4.9 Conformations of Polycyclic Molecules

The last point we'll consider about cycloalkane stereochemistry is to see wha happens when two or more cycloalkane rings are fused together along a com mon bond to construct a **polycyclic** molecule—for example, decalin.

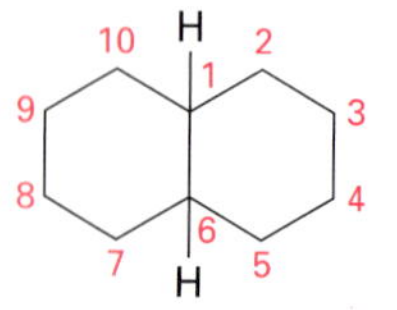

Decalin—two fused cyclohexane rings

Decalin consists of two cyclohexane rings joined to share two carbon atoms (the *bridgehead* carbons, C1 and C6) and a common bond. Decalin can exist in either of two isomeric forms, depending on whether the rings are trans fused or cis fused. In *cis*-decalin, the hydrogen atoms at the bridgehead carbons are on the same face of the rings; in *trans*-decalin, the bridgehead hydrogens are on opposite faces. Figure 4.17 shows how both compounds can be represented using chair cyclohexane conformations. Note that *cis*- and *trans*-decalin are not interconvertible by ring-flips or other rotations. They are cis–trans stereoisomers and have the same relationship to each other that *cis*- and *trans*-1,2-dimethyl-cyclohexane have.

Figure 4.17 Representations of *cis*- and *trans*-decalin. The red hydrogen atoms at the bridgehead carbons are on the same face of the rings in the cis isomer but on opposite faces in the trans isomer.

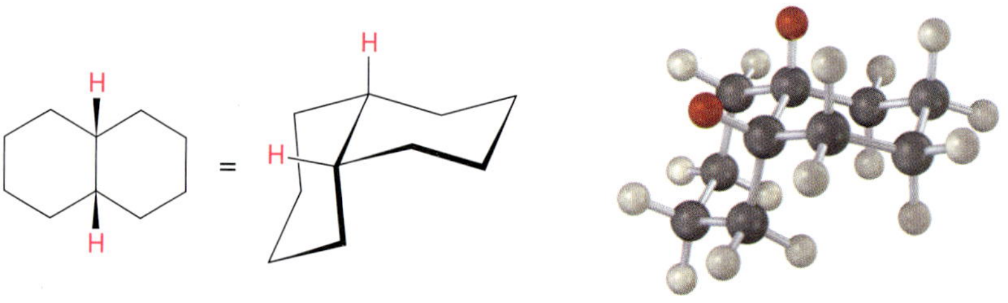

***cis*-Decalin**

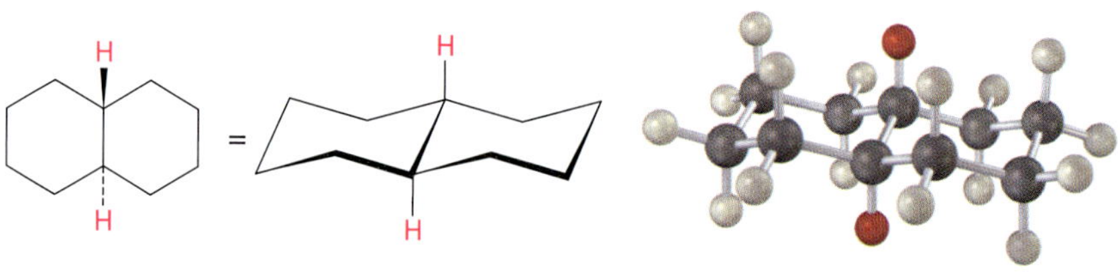

***trans*-Decalin**

Polycyclic compounds are common in nature, and many valuable substances have fused-ring structures. For example, steroids, such as the male hormone testosterone, have 3 six-membered rings and 1 five-membered ring fused together. Although steroids look complicated compared with cyclohexane or decalin, the same principles that apply to the conformational analysis of simple cyclohexane rings apply equally well (and often better) to steroids.

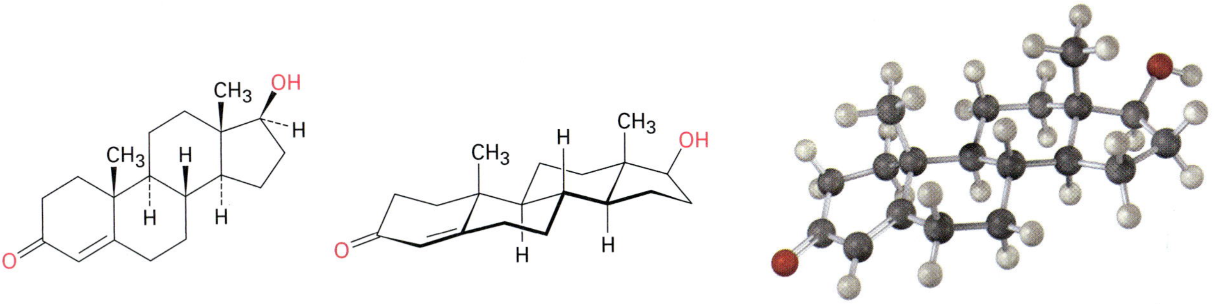

Testosterone (a steroid)

Another common ring system is the norbornane, or bicyclo[2.2.1]heptane, structure. Like decalin, norbornane is a *bicycloalkane,* so called because *two* rings would have to be broken open to generate an acyclic structure. Its systematic name, bicyclo[2.2.1]heptane, reflects the fact that the molecule has seven carbons, is bicyclic, and has three "bridges" of 2, 2, and 1 carbon atoms connecting the two bridgehead carbons.

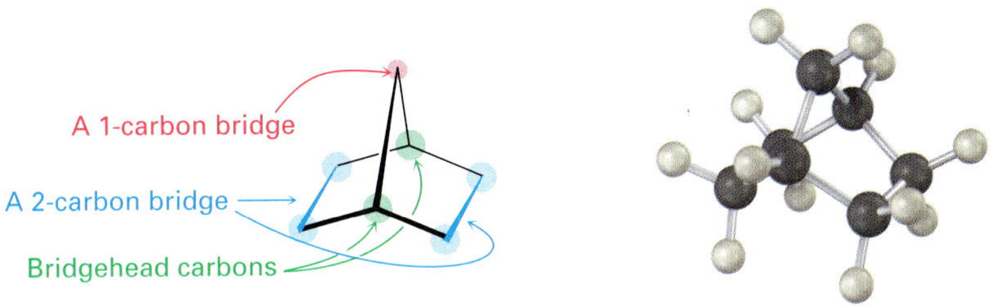

Norbornane
(bicyclo[2.2.1]heptane)

Norbornane has a conformationally locked boat cyclohexane ring (Section 4.5) in which carbons 1 and 4 are joined by an additional CH_2 group. Note how, in drawing this structure, a break in the rear bond indicates that the vertical bond crosses in front of it. Making a molecular model is particularly helpful when trying to see the three-dimensionality of norbornane.

Substituted norbornanes, such as camphor, are found widely in nature, and many have been important historically in developing organic structural theories.

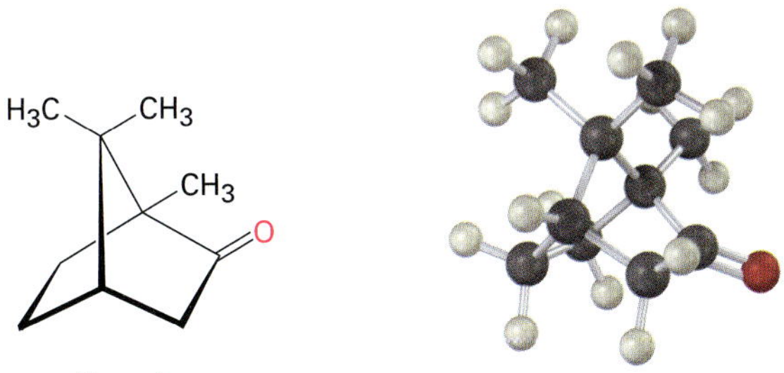

Camphor

Problem 4.20 | Which isomer is more stable, *cis*-decalin or *trans*-decalin? Explain.

Focus On . . .

Molecular Mechanics

Computer programs make it possible to portray accurate representations of molecular geometry.

All the structural models in this book are computer-drawn. To make sure they accurately portray bond angles, bond lengths, torsional interactions, and steric interactions, the most stable geometry of each molecule has been calculated on a desktop computer using a commercially available *molecular mechanics* program based on work by N. L. Allinger of the University of Georgia.

The idea behind molecular mechanics is to begin with a rough geometry for a molecule and then calculate a total strain energy for that starting geometry, using mathematical equations that assign values to specific kinds of molecular interactions. Bond angles that are too large or too small cause angle strain; bond lengths that are too short or too long cause stretching or compressing strain; unfavorable eclipsing interactions around single bonds cause torsional strain; and nonbonded atoms that approach each other too closely cause steric, or *van der Waals,* strain.

$$E_{\text{total}} = E_{\text{bond stretching}} + E_{\text{angle strain}} + E_{\text{torsional strain}} + E_{\text{van der Waals}}$$

After calculating a total strain energy for the starting geometry, the program automatically changes the geometry slightly in an attempt to lower strain—perhaps by lengthening a bond that is too short or decreasing an angle that is too large. Strain is recalculated for the new geometry, more changes are made, and more calculations are done. After dozens or hundreds of iterations, the calculation ultimately converges on a minimum energy that corresponds to the most favorable, least strained conformation of the molecule.

Molecular mechanics calculations have proved to be enormously useful in pharmaceutical research, where the complementary fit between a drug molecule and a receptor molecule in the body is often a key to designing new pharmaceutical agents (Figure 4.18).

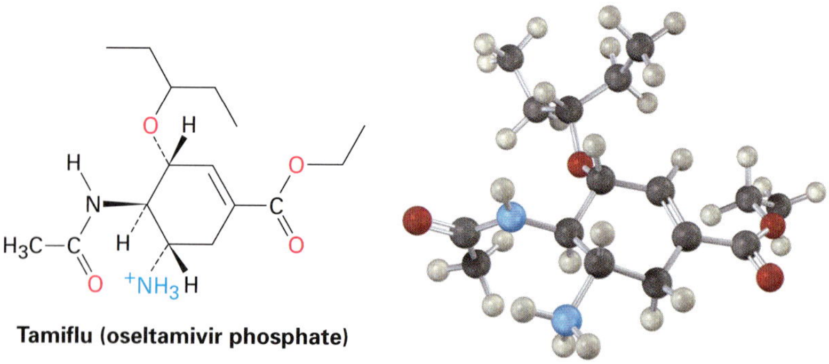

Tamiflu (oseltamivir phosphate)

Figure 4.18 The structure of Tamiflu (oseltamivir phosphate), an antiviral agent active against type A influenza, and a molecular model of its minimum-energy conformation, as calculated by molecular mechanics.

SUMMARY AND KEY WORDS

alicyclic, 108
angle strain, 113
axial position, 119
chair conformation, 117
cis–trans isomers, 112
conformational analysis, 125
cycloalkane, 108
1,3-diaxial interaction, 123
equatorial position, 119
polycyclic compound, 128
ring-flip (cyclohexane), 120
stereoisomers, 111
twist-boat conformation, 118

A **cycloalkane** is a saturated cyclic hydrocarbon with the general formula C_nH_{2n}. In contrast to open-chain alkanes, where nearly free rotation occurs around C–C bonds, rotation is greatly reduced in cycloalkanes. Disubstituted cycloalkanes can therefore exist as **cis–trans isomers**. The cis isomer has both substituents on the same face of the ring; the trans isomer has substituents on opposite faces. Cis–trans isomers are just one kind of **stereoisomers**—isomers that have the same connections between atoms but different three-dimensional arrangements.

Not all cycloalkanes are equally stable. Three kinds of strain contribute to the overall energy of a cycloalkane: (1) **angle strain** is the resistance of a bond angle to compression or expansion from the normal 109° tetrahedral value, (2) *torsional strain* is the energy cost of having neighboring C–H bonds eclipsed rather than staggered, and (3) *steric strain* is the repulsive interaction that arises when two groups attempt to occupy the same space.

Cyclopropane (115 kJ/mol strain) and cyclobutane (110.4 kJ/mol strain) have both angle strain and torsional strain. Cyclopentane is free of angle strain but has a substantial torsional strain due to its large number of eclipsing interactions. Both cyclobutane and cyclopentane pucker slightly away from planarity to relieve torsional strain.

Cyclohexane is strain-free because it adopts a puckered **chair conformation**, in which all bond angles are near 109° and all neighboring C–H bonds are staggered. Chair cyclohexane has two kinds of positions: **axial** and **equatorial**. Axial positions are oriented up and down, parallel to the ring axis, whereas equatorial positions lie in a belt around the equator of the ring. Each carbon atom has one axial and one equatorial position.

Chair cyclohexanes are conformationally mobile and can undergo a **ring-flip**, which interconverts axial and equatorial positions. Substituents on the ring are more stable in the equatorial position because axial substituents cause **1,3-diaxial interactions**. The amount of 1,3-diaxial steric strain caused by an axial substituent depends on its bulk.

EXERCISES

Organic KNOWLEDGE TOOLS

ThomsonNOW™ Sign in at **www.thomsonedu.com** to assess your knowledge of this chapter's topics by taking a pre-test. The pre-test will link you to interactive organic chemistry resources based on your score in each concept area.

Online homework for this chapter may be assigned in Organic OWL.

■ indicates problems assignable in Organic OWL.

▲ denotes problems linked to Key Ideas of this chapter and testable in ThomsonNOW.

VISUALIZING CHEMISTRY

(Problems 4.1–4.20 appear within the chapter.)

4.21 ■ Name the following cycloalkanes:

(a)

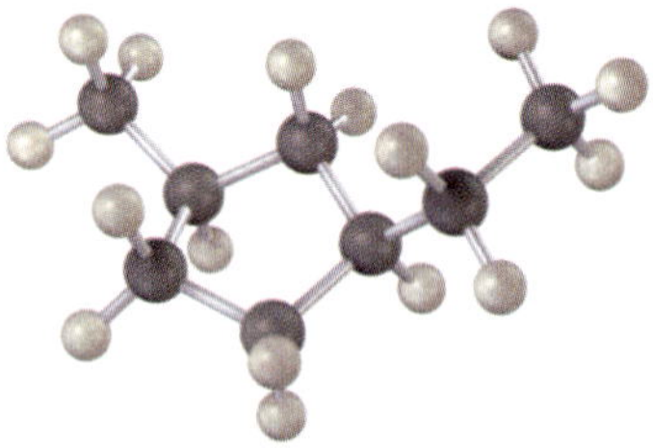

(b)

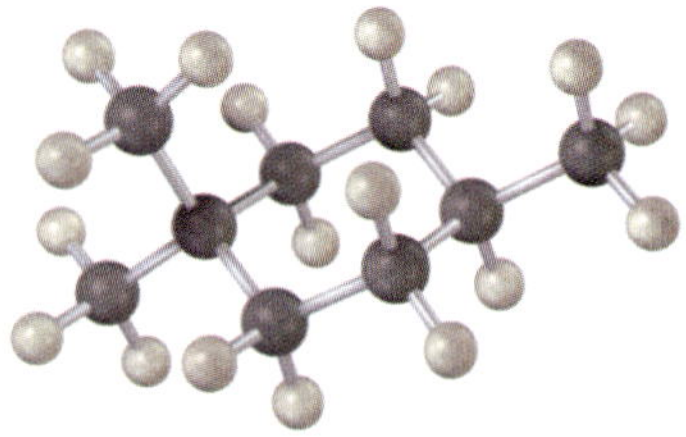

4.22 ■ Name the following compound, identify each substituent as axial or equatorial, and tell whether the conformation shown is the more stable or less stable chair form (yellow-green = Cl):

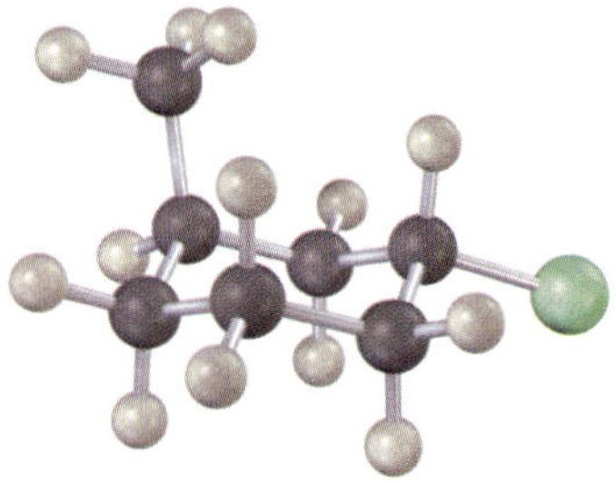

4.23 ▲ A trisubstituted cyclohexane with three substituents—red, yellow, and blue—undergoes a ring-flip to its alternative chair conformation. Identify each substituent as axial or equatorial, and show the positions occupied by the three substituents in the ring-flipped form.

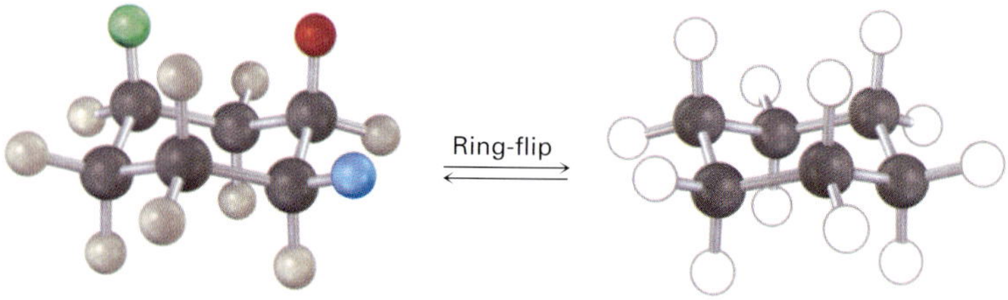

4.24 Glucose exists in two forms having a 36:64 ratio at equilibrium. Draw a skeletal structure of each, describe the difference between them, and tell which of the two you think is more stable (red = O):

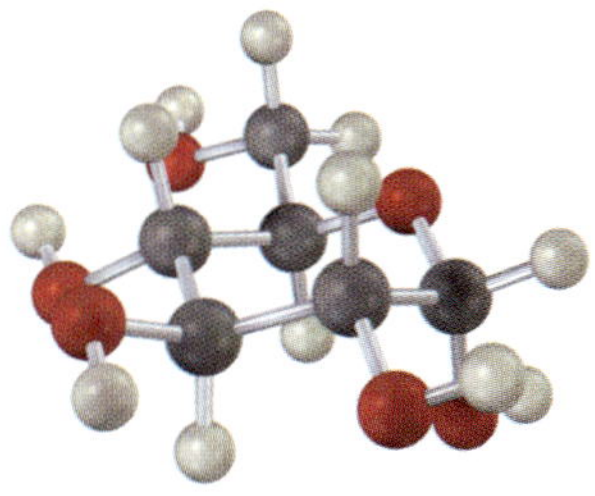

α-Glucose

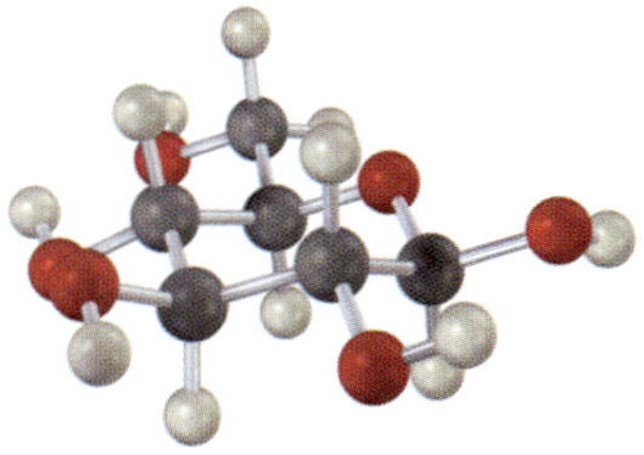

β-Glucose

ADDITIONAL PROBLEMS

4.25 Draw the five cycloalkanes with the formula C_5H_{10}.

4.26 ■ Draw two constitutional isomers of *cis*-1,2-dibromocyclopentane.

4.27 ■ Draw a stereoisomer of *trans*-1,3-dimethylcyclobutane.

4.28 ■ Hydrocortisone, a naturally occurring hormone produced in the adrenal glands, is often used to treat inflammation, severe allergies, and numerous other conditions. Is the indicated –OH group in the molecule axial or equatorial?

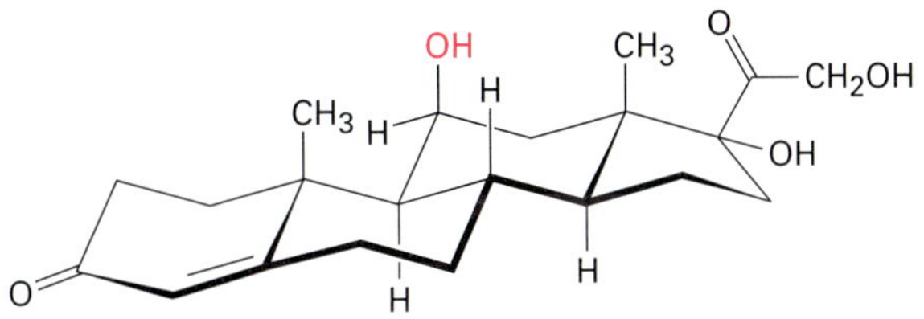

Hydrocortisone

4.29 A 1,2-cis disubstituted cyclohexane, such as *cis*-1,2-dichlorocyclohexane, must have one group axial and one group equatorial. Explain.

4.30 A 1,2-trans disubstituted cyclohexane must have either both groups axial or both groups equatorial. Explain.

4.31 Why is a 1,3-cis disubstituted cyclohexane more stable than its trans isomer?

4.32 ■ Which is more stable, a 1,4-trans disubstituted cyclohexane or its cis isomer?

4.33 *cis*-1,2-Dimethylcyclobutane is less stable than its trans isomer, but *cis*-1,3-dimethylcyclobutane is more stable than its trans isomer. Draw the most stable conformations of both, and explain.

4.34 ■ Draw the two chair conformations of *cis*-1-chloro-2-methylcyclohexane. Which is more stable, and by how much?

4.35 ■ Draw the two chair conformations of *trans*-1-chloro-2-methylcyclohexane. Which is more stable?

4.36 ■ Galactose, a sugar related to glucose, contains a six-membered ring in whic all the substituents except the −OH group indicated below in red are equato rial. Draw galactose in its more stable chair conformation.

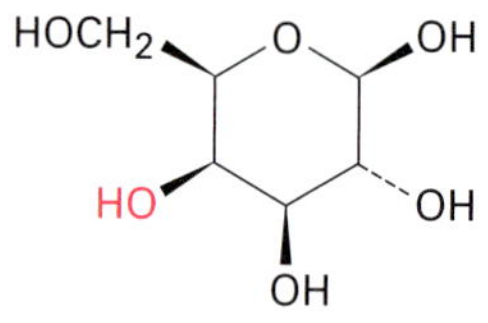

Galactose

4.37 Draw the two chair conformations of menthol, and tell which is more stabl

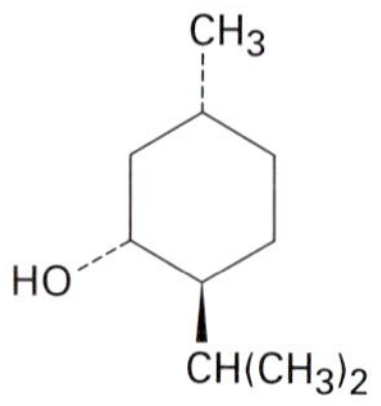

Menthol

4.38 There are four cis–trans isomers of menthol (Problem 4.37), including the on shown. Draw the other three.

4.39 Identify each pair of relationships among the −OH groups in glucose (red–blue red–green, red–black, blue–green, blue–black, green–black) as cis or trans.

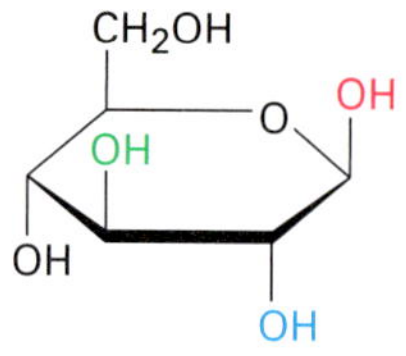

Glucose

4.40 ▲ Draw 1,3,5-trimethylcyclohexane using a hexagon to represent the ring How many cis–trans stereoisomers are possible?

4.41 ■ From the data in Figure 4.12 and Table 4.1, estimate the percentages of molecule that have their substituents in an axial orientation for the following compounds:

(a) Isopropylcyclohexane (b) Fluorocyclohexane

(c) Cyclohexanecarbonitrile, $C_6H_{11}CN$

4.42 ■ ▲ Assume that you have a variety of cyclohexanes substituted in the posi- tions indicated. Identify the substituents as either axial or equatorial. Fo example, a 1,2-cis relationship means that one substituent must be axial an one equatorial, whereas a 1,2-trans relationship means that both substituent are axial or both are equatorial.

(a) 1,3-Trans disubstituted (b) 1,4-Cis disubstituted

(c) 1,3-Cis disubstituted (d) 1,5-Trans disubstituted

(e) 1,5-Cis disubstituted (f) 1,6-Trans disubstituted

4.43 ▲ The diaxial conformation of *cis*-1,3-dimethylcyclohexane is approximately 23 kJ/mol (5.4 kcal/mol) less stable than the diequatorial conformation. Draw the two possible chair conformations, and suggest a reason for the large energy difference.

■ Assignable in OWL ▲ Key Idea Problems

4.44 Approximately how much steric strain does the 1,3-diaxial interaction between the two methyl groups introduce into the diaxial conformation of *cis*-1,3-dimethylcyclohexane? (See Problem 4.43.)

4.45 In light of your answer to Problem 4.44, draw the two chair conformations of 1,1,3-trimethylcyclohexane, and estimate the amount of strain energy in each. Which conformation is favored?

4.46 We saw in Problem 4.20 that *cis*-decalin is less stable than *trans*-decalin. Assume that the 1,3-diaxial interactions in *trans*-decalin are similar to those in axial methylcyclohexane [that is, one $CH_2 \leftrightarrow H$ interaction costs 3.8 kJ/mol (0.9 kcal/mol)], and calculate the magnitude of the energy difference between *cis*- and *trans*-decalin.

4.47 Using molecular models as well as structural drawings, explain why *trans*-decalin is rigid and cannot ring-flip, whereas *cis*-decalin can easily ring-flip.

4.48 *trans*-Decalin is more stable than its cis isomer, but *cis*-bicyclo[4.1.0]heptane is more stable than its trans isomer. Explain.

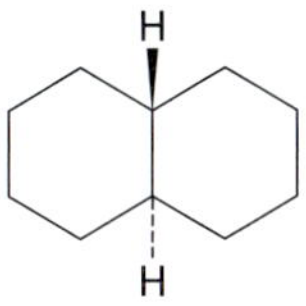

***trans*-Decalin**

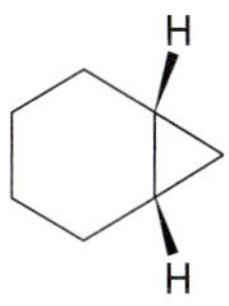

***cis*-Bicyclo[4.1.0]heptane**

4.49 ▲ *myo*-Inositol, one of the isomers of 1,2,3,4,5,6-hexahydroxycyclohexane, acts as a growth factor in both animals and microorganisms. Draw the most stable chair conformation of *myo*-inositol.

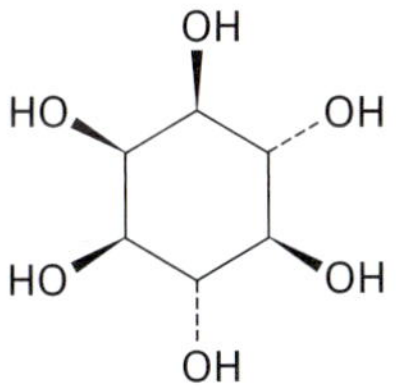

***myo*-Inositol**

4.50 How many cis–trans stereoisomers of *myo*-inositol (Problem 4.49) are there? Draw the structure of the most stable isomer.

4.51 ■ One of the two chair structures of *cis*-1-chloro-3-methylcyclohexane is more stable than the other by 15.5 kJ/mol (3.7 kcal/mol). Which is it? What is the energy cost of a 1,3-diaxial interaction between a chlorine and a methyl group?

4.52 The German chemist J. Bredt proposed in 1935 that bicycloalkenes such as 1-norbornene, which have a double bond to the bridgehead carbon, are too strained to exist. Make a molecular model of 1-norbornene, and explain Bredt's proposal.

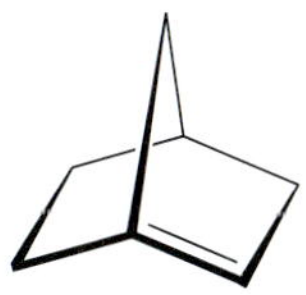

1-Norbornene

4.53 ■ Tell whether each of the following substituents on a steroid is axial or equatorial. (A substituent that is "up" is on the top face of the molecule as drawn, and a substituent that is "down" is on the bottom face.)

(a) Substituent up at C3
(b) Substituent down at C7
(c) Substituent down at C11

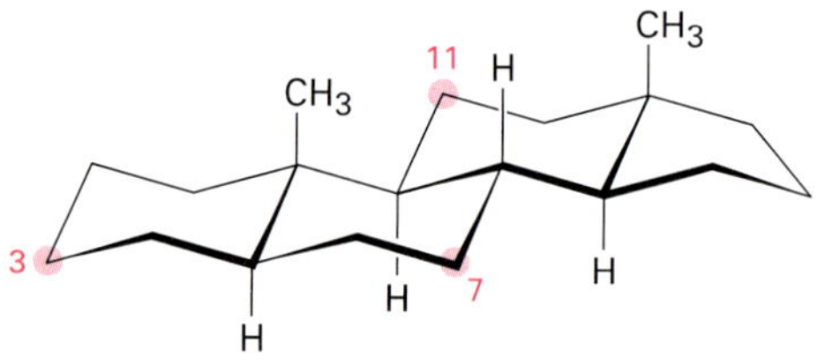

4.54 Amantadine is an antiviral agent that is active against influenza A infection and against some strains of H5N1 avian flu. Draw a three-dimensional representation of amantadine showing the chair cyclohexane rings.

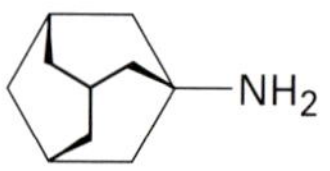

Amantadine

4.55 Ketones react with alcohols to yield products called *acetals*. Why does the all-cis isomer of 4-*tert*-butyl-1,3-cyclohexanediol react readily with acetone and an acid catalyst to form an acetal while other stereoisomers do not react? In formulating your answer, draw the more stable chair conformations of all four stereoisomers and the product acetal. Use molecular models for help.

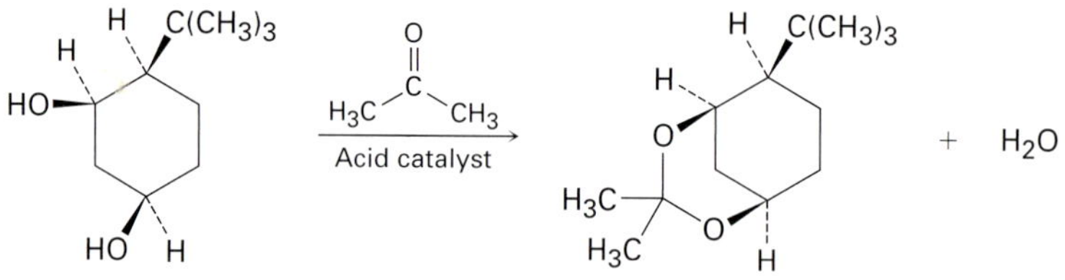

An acetal

4.56 Alcohols undergo an *oxidation* reaction to yield carbonyl compounds on treatment with CrO_3. For example, 2-*tert*-butylcyclohexanol gives 2-*tert*-butylcyclohexanone. If axial –OH groups are generally more reactive than their equatorial isomers, which do you think would react faster, the cis isomer of 2-*tert*-butylcyclohexanol or the trans isomer? Explain.

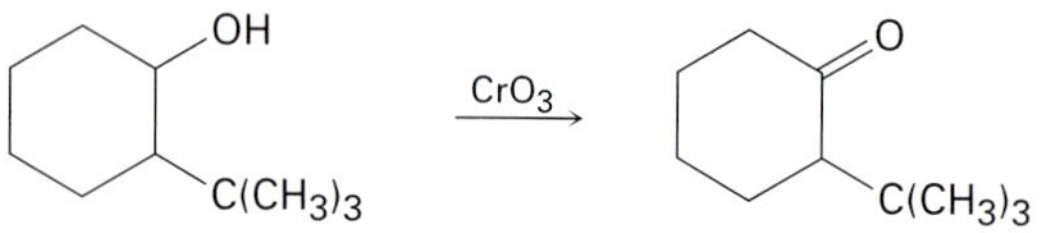

2-*tert*-Butylcyclohexanol **2-*tert*-Butylcyclohexanone**

5

An Overview of Organic Reactions

Organic **KNOWLEDGE TOOLS**

ThomsonNOW™ Throughout this chapter, sign in at **www.thomsonedu.com** for online self-study and interactive tutorials based on your level of understanding.

Online homework for this chapter may be assigned in Organic OWL.

When first approached, organic chemistry can seem overwhelming. It's not so much that any one part is difficult to understand, it's that there are so many parts: literally millions of compounds, dozens of functional groups, and an endless number of reactions. With study, though, it becomes evident that there are only a few fundamental ideas that underlie all organic reactions. Far from being a collection of isolated facts, organic chemistry is a beautifully logical subject that is unified by a few broad themes. When these themes are understood, learning organic chemistry becomes much easier and memorization is minimized. The aim of this book is to describe the themes and clarify the patterns that unify organic chemistry.

WHY THIS CHAPTER?

All chemical reactions, whether in the laboratory or in living organisms, follow the same "rules." Reactions in living organisms often look more complex than laboratory reactions because of the size of the biomolecules and the involvement of biological catalysts called *enzymes,* but the principles governing all reactions are the same.

To understand both organic and biological chemistry, it's necessary to know not just *what* occurs, but also *why* and *how* chemical reactions take place. In this chapter, we'll start with an overview of the fundamental kinds of organic reactions, we'll see why reactions occur, and we'll see how reactions can be described. Once this background is out of the way, we'll then be ready to begin studying the details of organic chemistry.

5.1 Kinds of Organic Reactions

ThomsonNOW™ Click *Organic Interactive* to **classify organic reactions by examining reactants and products.**

Organic chemical reactions can be organized broadly in two ways—by *what kinds* of reactions occur and by *how* those reactions occur. Let's look first at the kinds of reactions that take place. There are four general types of organic reactions: *additions, eliminations, substitutions,* and *rearrangements.*

- **Addition reactions** occur when two reactants add together to form a single product with no atoms "left over." An example that we'll be studying soon is the reaction of an alkene, such as ethylene, with HBr to yield an alkyl bromide.

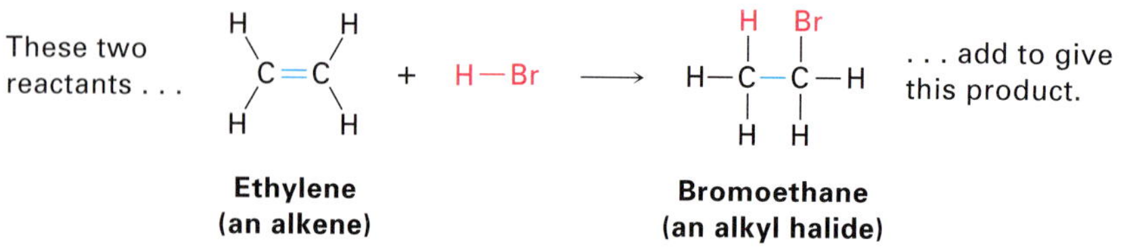

- **Elimination reactions** are, in a sense, the opposite of addition reactions. They occur when a single reactant splits into two products, often with formation of a small molecule such as water or HBr. An example is the acid-catalyzed reaction of an alcohol to yield water and an alkene.

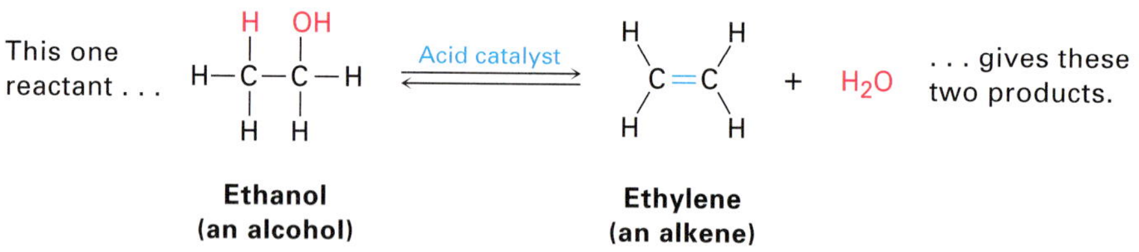

- **Substitution reactions** occur when two reactants exchange parts to give two new products. An example is the reaction of an alkane with Cl_2 in the presence of ultraviolet light to yield an alkyl chloride. A Cl atom from Cl_2 substitutes for an H atom of the alkane, and two new products result.

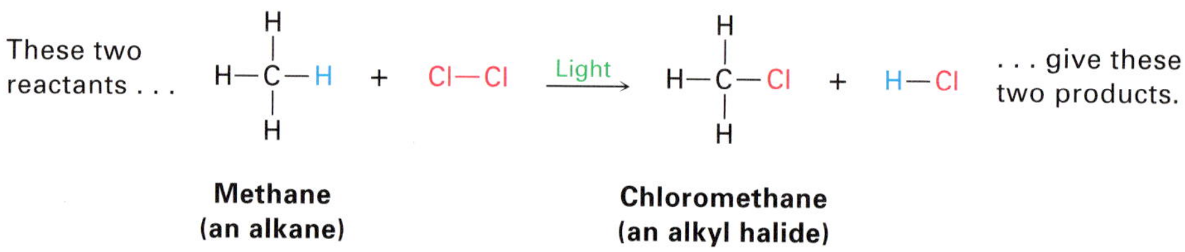

- **Rearrangement reactions** occur when a single reactant undergoes a reorganization of bonds and atoms to yield an isomeric product. An example is the conversion of the alkene 1-butene into its constitutional isomer 2-butene by treatment with an acid catalyst.

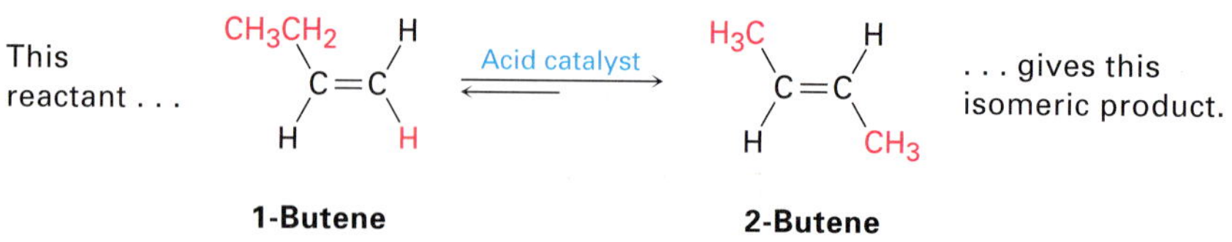

Problem 5.1 Classify each of the following reactions as an addition, elimination, substitution, or rearrangement:

(a) $CH_3Br + KOH \rightarrow CH_3OH + KBr$

(b) $CH_3CH_2Br \rightarrow H_2C{=}CH_2 + HBr$

(c) $H_2C{=}CH_2 + H_2 \rightarrow CH_3CH_3$

5.2 How Organic Reactions Occur: Mechanisms

Having looked at the kinds of reactions that take place, let's now see how reactions occur. An overall description of how a reaction occurs is called a **reaction mechanism**. A mechanism describes in detail exactly what takes place at each stage of a chemical transformation—which bonds are broken and in what order, which bonds are formed and in what order, and what the relative rates of the steps are. A complete mechanism must also account for all reactants used and all products formed.

All chemical reactions involve bond-breaking and bond-making. When two molecules come together, react, and yield products, specific bonds in the reactant molecules are broken and specific bonds in the product molecules are formed. Fundamentally, there are two ways in which a covalent two-electron bond can break. A bond can break in an electronically *symmetrical* way so that one electron remains with each product fragment, or a bond can break in an electronically *unsymmetrical* way so that both bonding electrons remain with one product fragment, leaving the other with a vacant orbital. The symmetrical cleavage is said to be *homolytic,* and the unsymmetrical cleavage is said to be *heterolytic*. We'll develop the point in more detail later, but you might note for now that the movement of *one* electron in the symmetrical process is indicated using a half-headed, or "fishhook," arrow (⌒), whereas the movement of *two* electrons in the unsymmetrical process is indicated using a full-headed curved arrow (⌒).

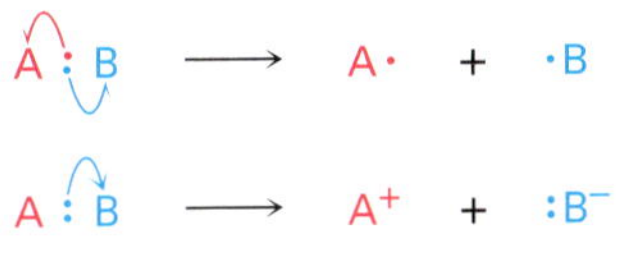

Symmetrical bond-breaking (radical): one bonding electron stays with each product.

Unsymmetrical bond-breaking (polar): two bonding electrons stay with one product.

Just as there are two ways in which a bond can break, there are two ways in which a covalent two-electron bond can form. A bond can form in an electronically symmetrical way if one electron is donated to the new bond by each reactant or in an unsymmetrical way if both bonding electrons are donated by one reactant.

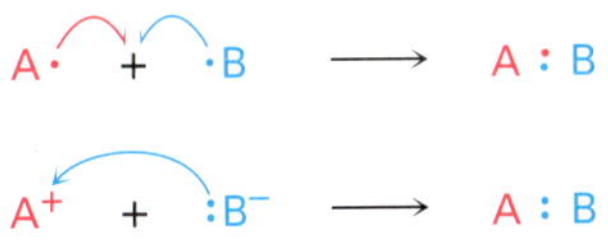

Symmetrical bond-making (radical): one bonding electron is donated by each reactant.

Unsymmetrical bond-making (polar): two bonding electrons are donated by one reactant.

Processes that involve symmetrical bond-breaking and bond-making are called **radical reactions**. A **radical**, often called a *free radical,* is a neutral chemical species that contains an odd number of electrons and thus has a single, unpaired electron in one of its orbitals. Processes that involve unsymmetrical bond-breaking and bond-making are called **polar reactions**. Polar reactions involve species that have an even number of electrons and thus have only electron pairs in their orbitals. Polar processes are by far the more common reaction type in both organic and biological chemistry, and a large part of this book is devoted to their description.

In addition to polar and radical reactions, there is a third, less commonly encountered process called a *pericyclic reaction.* Rather than explain pericyclic reactions now, though, we'll look at them more carefully in Chapter 30.

5.3 Radical Reactions

Radical reactions are not as common as polar reactions but are neverthele
important in some industrial processes and in numerous biological pathway
Let's see briefly how they occur.

A radical is highly reactive because it contains an atom with an odd num
ber of electrons (usually seven) in its valence shell, rather than a stable, noble
gas octet. A radical can achieve a valence-shell octet in several ways. Fo
example, the radical might abstract an atom and one bonding electron fror
another reactant, leaving behind a new radical. The net result is a radical sub
stitution reaction:

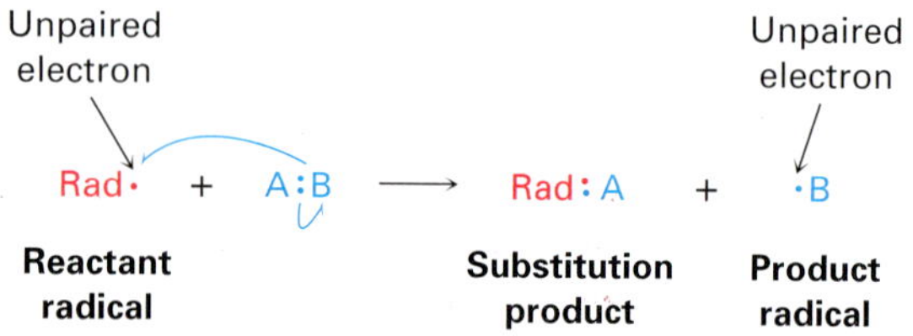

Alternatively, a reactant radical might add to a double bond, taking one electron from the double bond and yielding a new radical. The net result is a radical addi tion reaction:

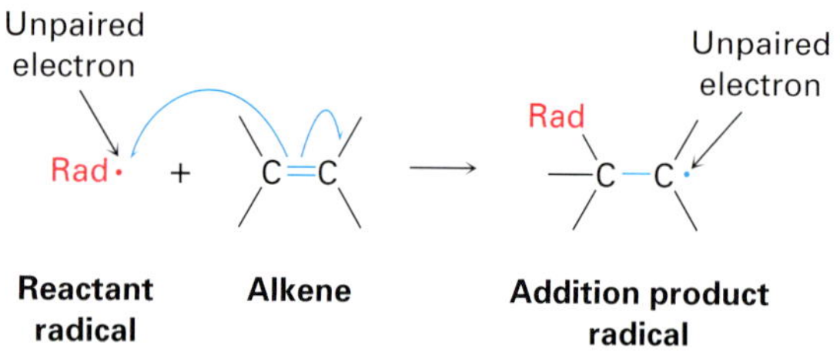

As an example of an industrially useful radical reaction, look at the chlori nation of methane to yield chloromethane. This substitution reaction is the first step in the preparation of the solvents dichloromethane (CH_2Cl_2) and chloro form ($CHCl_3$).

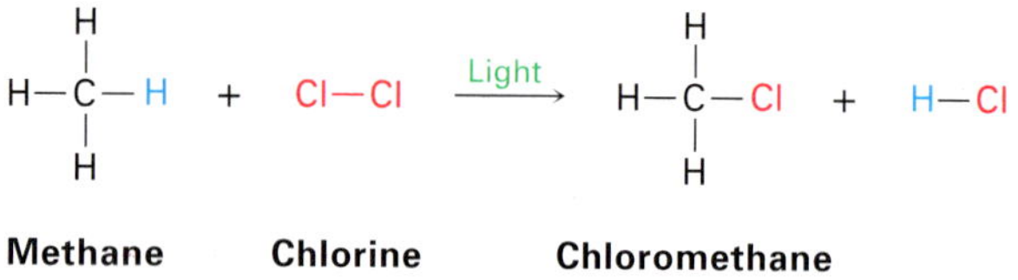

Like many radical reactions in the laboratory, methane chlorination requires three kinds of steps: *initiation*, *propagation*, and *termination*.

Initiation Irradiation with ultraviolet light begins the reaction by break ing the relatively weak Cl–Cl bond of a small number of Cl_2 molecules to give a few reactive chlorine radicals.

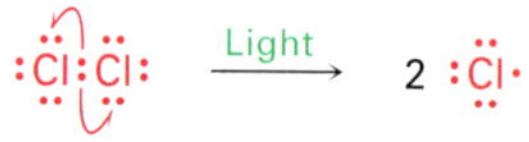

Propagation Once produced, a reactive chlorine radical collides with a methane molecule in a propagation step, abstracting a hydrogen atom to give HCl and a methyl radical (· CH_3). This methyl radical reacts further with Cl_2 in a second propagation step to give the product chloromethane plus a new chlorine radical, which cycles back and repeats the first propagation step. Thus, once the sequence has been initiated, it becomes a self-sustaining cycle of repeating steps (a) and (b), making the overall process a *chain reaction.*

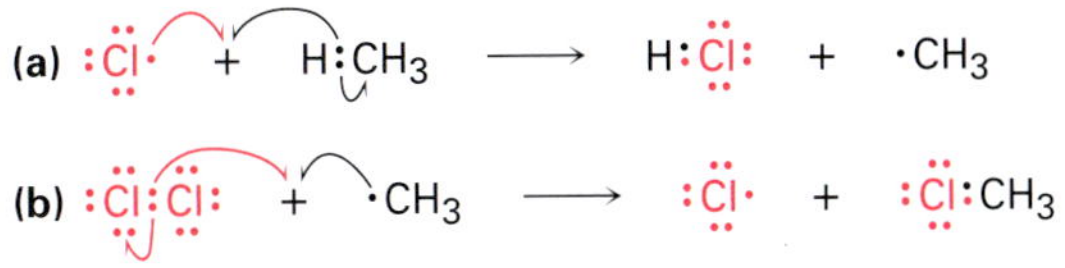

Termination Occasionally, two radicals might collide and combine to form a stable product. When that happens, the reaction cycle is broken and the chain is ended. Such termination steps occur infrequently, however, because the concentration of radicals in the reaction at any given moment is very small. Thus, the likelihood that two radicals will collide is also small.

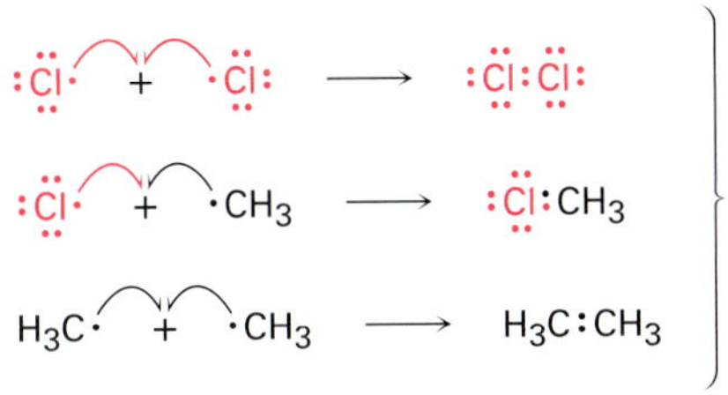

Possible termination steps

As a biological example of a radical reaction, let's look at the biosynthesis of *prostaglandins,* a large class of molecules found in virtually all body tissues and fluids. A number of pharmaceuticals are based on or derived from prostaglandins, including medicines that induce labor during childbirth, reduce intraocular pressure in glaucoma, control bronchial asthma, and help treat congenital heart defects.

Prostaglandin biosynthesis is initiated by abstraction of a hydrogen atom from arachidonic acid by an iron–oxygen radical, thereby generating a new, carbon radical in a substitution reaction. Don't be intimidated by the size of the molecules; focus only on the changes occurring in each step. (To help you do that, the unchanged part of the molecule is "ghosted," with only the reactive part clearly visible.)

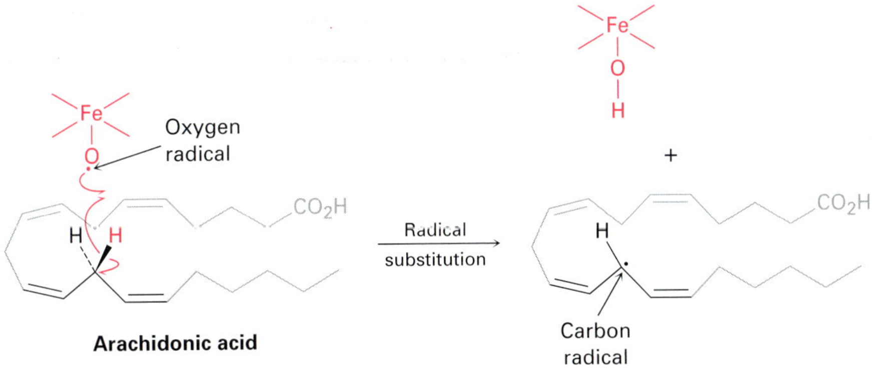

Following the initial abstraction of a hydrogen atom, the carbon radical then reacts with O_2 to give an oxygen radical, which reacts with a C=C bond within the same molecule in an addition reaction. Several further transformations ultimately yield prostaglandin H_2.

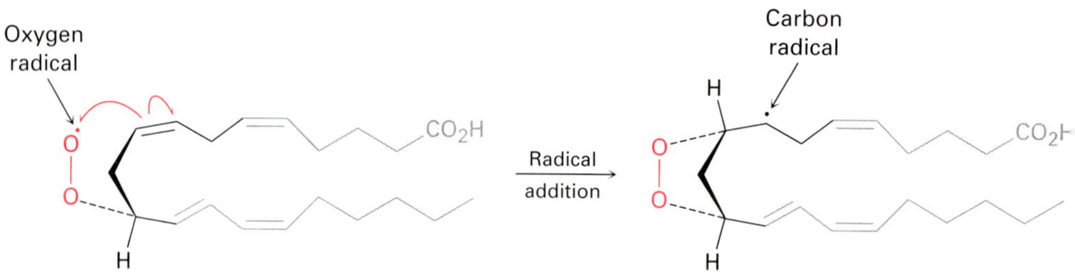

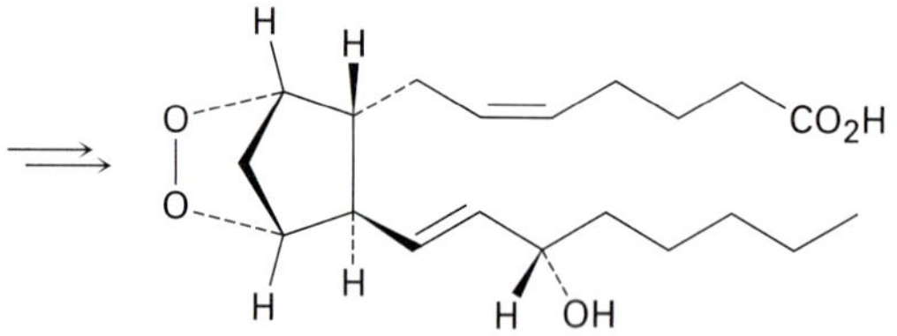

Prostaglandin H_2 (PGH_2)

Problem 5.2 Radical chlorination of alkanes is not generally useful because mixtures of products often result when more than one kind of C–H bond is present in the substrate. Draw and name all monochloro substitution products $C_6H_{13}Cl$ you might obtain by reaction of 2-methylpentane with Cl_2.

Problem 5.3 Using a curved fishhook arrow, propose a mechanism for formation of the cyclopentane ring of prostaglandin H_2. What kind of reaction is occurring?

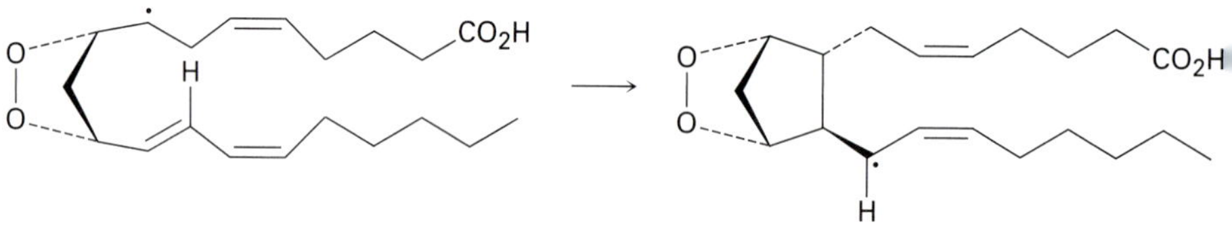

5.4 Polar Reactions

Polar reactions occur because of the electrical attraction between positive and negative centers on functional groups in molecules. To see how these reactions take place, let's first recall the discussion of polar covalent bonds in Section 2.1 and then look more deeply into the effects of bond polarity on organic molecules.

Most organic compounds are electrically neutral; they have no net charge, either positive or negative. We saw in Section 2.1, however, that certain bonds within a molecule, particularly the bonds in functional groups, are polar. Bond polarity is a consequence of an unsymmetrical electron distribution in a bond and is due to the difference in electronegativity of the bonded atoms.

Elements such as oxygen, nitrogen, fluorine, and chlorine are more electronegative than carbon, so a carbon atom bonded to one of these atoms has a partial positive charge ($\delta+$). Conversely, metals are less electronegative than carbon, so a carbon atom bonded to a metal has a partial negative charge (δ−). Electrostatic potential maps of chloromethane and methyllithium illustrate these charge distributions, showing that the carbon atom in chloromethane is electron-poor (blue) while the carbon in methyllithium is electron-rich (red).

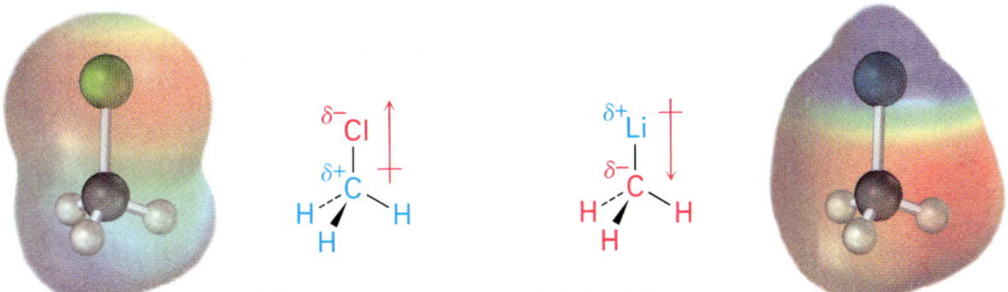

Chloromethane **Methyllithium**

The polarity patterns of some common functional groups are shown in Table 5.1. Carbon is always positively polarized except when bonded to a metal.

Table 5.1 Polarity Patterns in Some Common Functional Groups

Alcohol	$\overset{\delta+}{C}-\overset{\delta-}{OH}$	Carbonyl	$\overset{\delta+}{C}=\overset{\delta-}{O}$
Alkene	$C=C$ Symmetrical, nonpolar	Carboxylic acid	$-\overset{\delta+}{C}(=\overset{\delta-}{O})\overset{\delta-}{OH}$
Alkyl halide	$\overset{\delta+}{C}-\overset{\delta-}{X}$	Carboxylic acid chloride	$-\overset{\delta+}{C}(=\overset{\delta-}{O})\overset{\delta-}{Cl}$
Amine	$\overset{\delta+}{C}-\overset{\delta-}{NH_2}$	Aldehyde	$-\overset{\delta+}{C}(=\overset{\delta-}{O})H$
Ether	$\overset{\delta+}{C}-\overset{\delta-}{O}-\overset{\delta+}{C}$	Ester	$-\overset{\delta+}{C}(=\overset{\delta-}{O})\overset{\delta-}{O}-C$
Thiol	$\overset{\delta+}{C}-\overset{\delta-}{SH}$	Ketone	$-\overset{\delta+}{C}(=\overset{\delta-}{O})C$
Nitrile	$-\overset{\delta+}{C}\equiv\overset{\delta-}{N}$		
Grignard reagent	$\overset{\delta-}{C}-\overset{\delta+}{Mg}Br$		
Alkyllithium	$\overset{\delta-}{C}-\overset{\delta+}{Li}$		

This discussion of bond polarity is oversimplified in that we've considered only bonds that are inherently polar due to differences in electronegativity. Polar bonds can also result from the interaction of functional groups with acids or bases. Take an alcohol such as methanol, for example. In neutral methanol, the carbon atom is somewhat electron-poor because the electronegative oxygen attracts the electrons in the C–O bond. On protonation of the methanol oxygen by an acid, however, a full positive charge on oxygen attracts the electrons in the C–O bond much more strongly and makes the carbon much more electron-poor. We'll see numerous examples throughout this book of reactions that are catalyzed by acids because of the resultant increase in bond polarity.

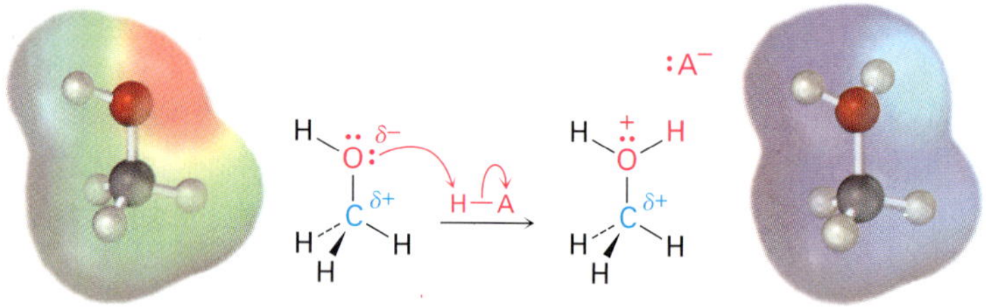

Methanol—weakly electron-poor carbon

Protonated methanol—strongly electron-poor carbon

Yet a further consideration is the *polarizability* (as opposed to polarity) of atoms in a molecule. As the electric field around a given atom changes because of changing interactions with solvent or other polar molecules nearby, the electron distribution around that atom also changes. The measure of this response to an external electrical influence is called the polarizability of the atom. Larger atoms with more, loosely held electrons are more polarizable, and smaller atoms with fewer, tightly held electrons are less polarizable. Thus, sulfur is more polarizable than oxygen, and iodine is more polarizable than chlorine. The effect of this higher polarizability for sulfur and iodine is that carbon–sulfur and carbon–iodine bonds, although nonpolar according to electronegativity values (Figure 2.2), nevertheless usually react as if they were polar.

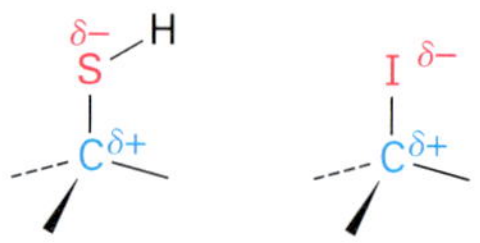

What does functional-group polarity mean with respect to chemical reactivity? Because unlike charges attract, the fundamental characteristic of all polar organic reactions is that electron-rich sites react with electron-poor sites. Bonds are made when an electron-rich atom shares a pair of electrons with an electron-poor atom, and bonds are broken when one atom leaves with both electrons from the former bond.

As we saw in Section 2.11, chemists indicate the movement of an electron pair during a polar reaction by using a curved, full-headed arrow. A curved arrow shows where electrons move when reactant bonds are broken and product bonds are formed. It means that an electron pair moves *from* the atom (or bond) at the tail of the arrow *to* the atom at the head of the arrow during the reaction.

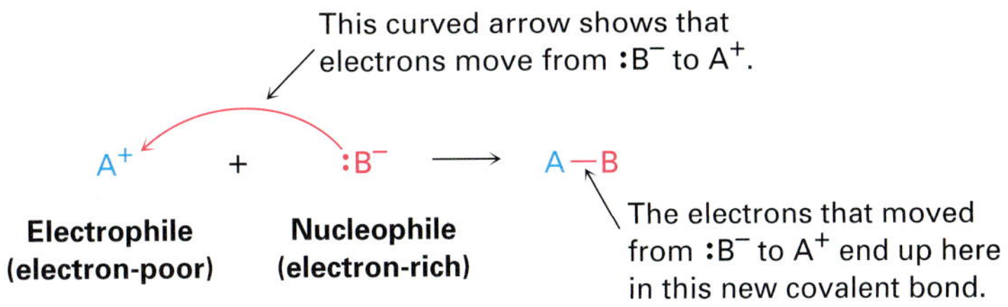

In referring to the electron-rich and electron-poor species involved in polar reactions, chemists use the words *nucleophile* and *electrophile*. A **nucleophile** is a substance that is "nucleus-loving." (Remember that a nucleus is positively charged.) A nucleophile has a negatively polarized, electron-rich atom and can form a bond by donating a pair of electrons to a positively polarized, electron-poor atom. Nucleophiles may be either neutral or negatively charged; ammonia, water, hydroxide ion, and chloride ion are examples. An **electrophile**, by contrast, is "electron-loving." An electrophile has a positively polarized, electron-poor atom and can form a bond by accepting a pair of electrons from a nucleophile. Electrophiles can be either neutral or positively charged. Acids (H^+ donors), alkyl halides, and carbonyl compounds are examples (Figure 5.1).

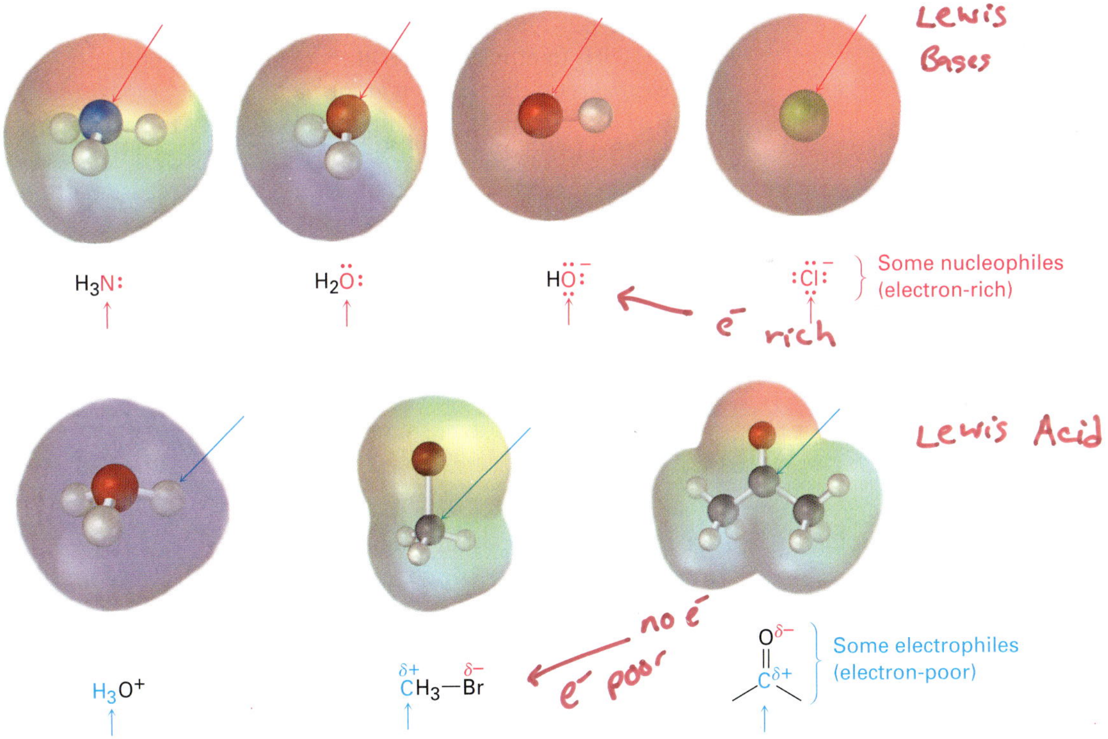

Figure 5.1 Some nucleophiles and electrophiles. Electrostatic potential maps identify the nucleophilic (red; negative) and electrophilic (blue; positive) atoms.

If the definitions of nucleophiles and electrophiles sound similar to those given in Section 2.11 for Lewis acids and Lewis bases, that's because there is indeed a correlation. Lewis bases are electron donors and behave as nucleophiles, whereas Lewis acids are electron acceptors and behave as electrophiles. Thus, much of organic chemistry is explainable in terms of acid–base reactions. The main difference is that the words *acid* and *base* are used broadly, while *nucleophile* and *electrophile* are used primarily when bonds to carbon are involved.

WORKED EXAMPLE 5.1 Identifying Electrophiles and Nucleophiles

ThomsonNOW™ Click *Organic Interactive* to **identify and characterize nucleophiles and electrophiles in organic reactions.**

Which of the following species is likely to behave as a nucleophile and which as an electrophile?

(a) NO_2^+ **(b)** CN^- **(c)** CH_3NH_2 **(d)** $(CH_3)_3S^+$

Strategy Nucleophiles have an electron-rich site, either because they are negatively charged or because they have a functional group containing an atom that has a lone pair of electrons. Electrophiles have an electron-poor site, either because they are positively charged or because they have a functional group containing an atom that is positively polarized.

Solution

(a) NO_2^+ (nitronium ion) is likely to be an electrophile because it is positively charged.

(b) $:C{\equiv}N^-$ (cyanide ion) is likely to be a nucleophile because it is negatively charged.

(c) CH_3NH_2 (methylamine) is likely to be a nucleophile because it has a lone pair of electrons on the nitrogen atom.

(d) $(CH_3)_3S^+$ (trimethylsulfonium ion) is likely to be an electrophile because it is positively charged.

Problem 5.4 Which of the following species is likely to be a nucleophile and which an electrophile?

(a) CH_3Cl **(b)** CH_3S^- **(c)**

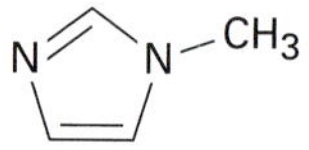

(d)

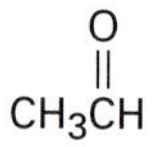

Problem 5.5 An electrostatic potential map of boron trifluoride is shown. Is BF_3 likely to be a nucleophile or an electrophile? Draw a Lewis structure for BF_3, and explain your answer.

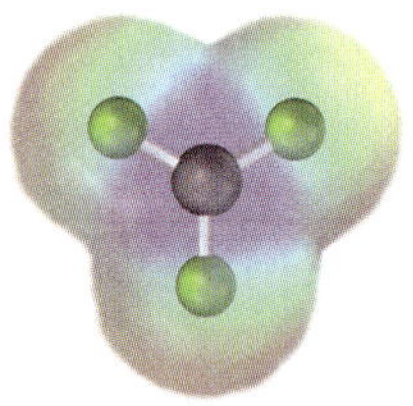

BF_3

5.5 An Example of a Polar Reaction: Addition of HBr to Ethylene

homsonNOW™ Click *Organic* *'rocesses* **to view an animation f the addition of HBr to an lkene.**

Let's look at a typical polar process—the addition reaction of an alkene, such as ethylene, with hydrogen bromide. When ethylene is treated with HBr at room temperature, bromoethane is produced. Overall, the reaction can be formulated as

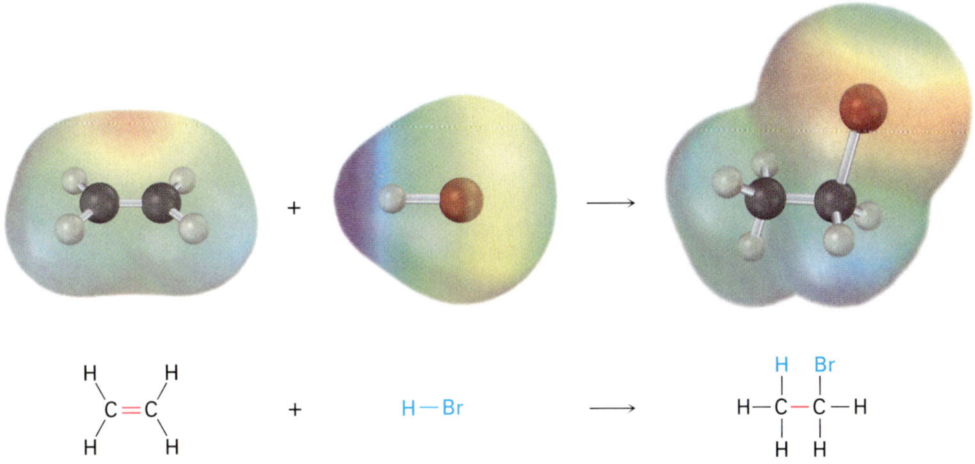

Ethylene (nucleophile) **Hydrogen bromide (electrophile)** **Bromoethane**

The reaction is an example of a polar reaction type known as an *electrophilic addition reaction* and can be understood using the general ideas discussed in the previous section. Let's begin by looking at the two reactants.

What do we know about ethylene? We know from Section 1.8 that a carbon–carbon double bond results from orbital overlap of two sp^2-hybridized carbon atoms. The σ part of the double bond results from sp^2–sp^2 overlap, and the π part results from *p–p* overlap.

What kind of chemical reactivity might we expect of a C=C bond? We know that *alkanes,* such as ethane, are relatively inert because all valence electrons are tied up in strong, nonpolar C–C and C–H bonds. Furthermore, the bonding electrons in alkanes are relatively inaccessible to approaching reactants because they are sheltered in σ bonds between nuclei. The electronic situation in *alkenes* is quite different, however. For one thing, double bonds have a greater electron density than single bonds—four electrons in a double bond versus only two in a single bond. Furthermore, the electrons in the π bond are accessible to approaching reactants because they are located above and below the plane of the double bond rather than being sheltered between the nuclei (Figure 5.2). As a result, the double bond is nucleophilic and the chemistry of alkenes is dominated by reactions with electrophiles.

What about the second reactant, HBr? As a strong acid, HBr is a powerful proton (H^+) donor and electrophile. Thus, the reaction between HBr and ethylene is a typical electrophile–nucleophile combination, characteristic of all polar reactions.

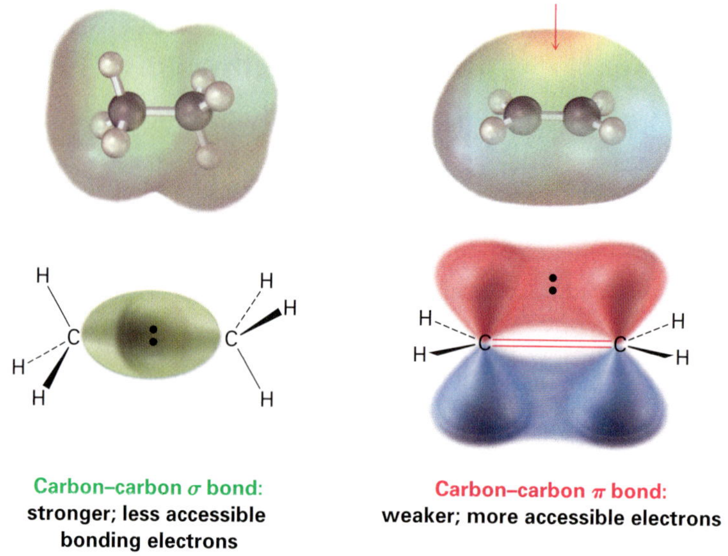

Figure 5.2 A comparison of carbon–carbon single and double bonds. A double bond is both more accessible to approaching reactants than a single bond and more electron-rich (more nucleophilic). An electrostatic potential map of ethylene indicates that the double bond is the region of highest negative charge (red).

We'll see more details about alkene electrophilic addition reactions shortly, but for the present we can imagine the reaction as taking place in two steps by the pathway shown in Figure 5.3. The reaction begins when the alkene donates a pair of electrons from its C=C bond to HBr to form a new C–H bond plus Br^-, as indicated by the path of the curved arrows in the first step of Figure 5.3. One curved arrow begins at the middle of the double bond (the source of the electron pair) and points to the hydrogen atom in HBr (the atom to which a bond will form). This arrow indicates that a new C–H bond forms using electrons from the former C=C bond. A second curved arrow begins in the middle of the H–Br bond and points to the Br, indicating that the H–Br bond breaks and the electrons remain with the Br atom, giving Br^-.

When one of the alkene carbon atoms bonds to the incoming hydrogen, the other carbon atom, having lost its share of the double-bond electrons, now has only six valence electrons and is left with a positive charge. This positively charged species—a carbon-cation, or **carbocation**—is itself an electrophile that can accept an electron pair from nucleophilic Br^- anion in a second step, forming a C–Br bond and yielding the observed addition product. Once again, a curved arrow in Figure 5.3 shows the electron-pair movement from Br^- to the positively charged carbon.

The electrophilic addition of HBr to ethylene is only one example of a polar process; there are many others that we'll study in detail in later chapters. But regardless of the details of individual reactions, all polar reactions take place between an electron-poor site and an electron-rich site and involve the donation of an electron pair from a nucleophile to an electrophile.

Problem 5.6 What product would you expect from reaction of cyclohexene with HBr? With HCl?

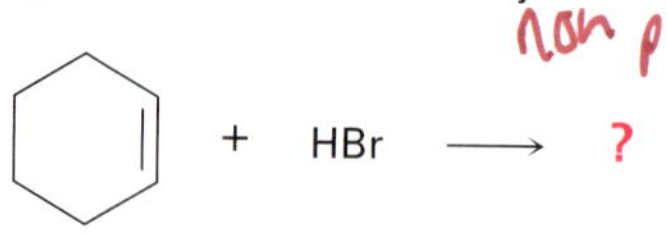

igure 5.3 MECHANISM: The ectrophilic addition reaction of :hylene and HBr. The reaction kes place in two steps, both f which involve electrophile–ucleophile interactions.

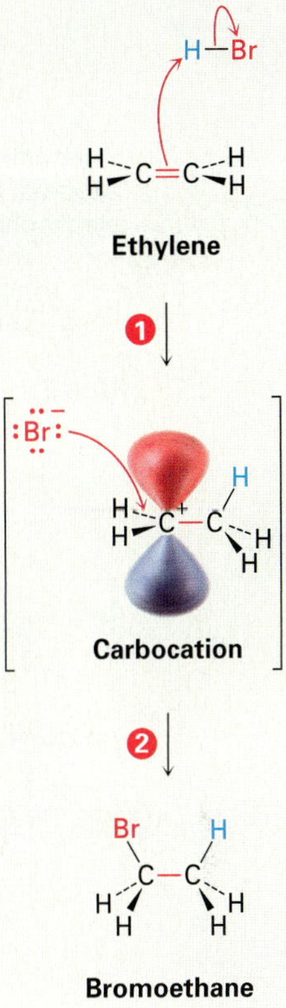

❶ A hydrogen atom on the electrophile HBr is attacked by π electrons from the nucleophilic double bond, forming a new C–H bond. This leaves the other carbon atom with a + charge and a vacant *p* orbital. Simultaneously, two electrons from the H–Br bond move onto bromine, giving bromide anion.

❷ Bromide ion donates an electron pair to the positively charged carbon atom, forming a C–Br bond and yielding the neutral addition product.

Problem 5.7 Reaction of HBr with 2-methylpropene yields 2-bromo-2-methylpropane. What is the structure of the carbocation formed during the reaction? Show the mechanism of the reaction.

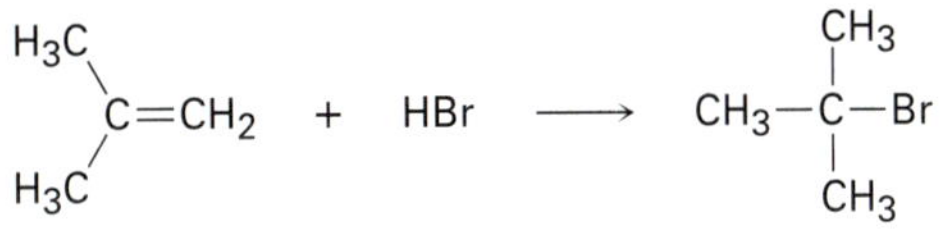

2-Methylpropene **2-Bromo-2-methylpropane**

5.6 Using Curved Arrows in Polar Reaction Mechanisms

It takes practice to use curved arrows properly in reaction mechanisms, but there are a few rules and a few common patterns you should look for that will help you become more proficient.

Rule 1 **Electrons move *from* a nucleophilic source (Nu: or Nu:$^-$) *to* an electrophilic sink (E or E$^+$).** The nucleophilic source must have an electron pair available, usually either in a lone pair or in a multiple bond. For example:

Key **IDEAS**

Test your knowledge of Key Ideas by using resources in ThomsonNOW or by answering end-of-chapter problems marked with ▲.

Electrons usually flow *from* one of these nucleophiles.

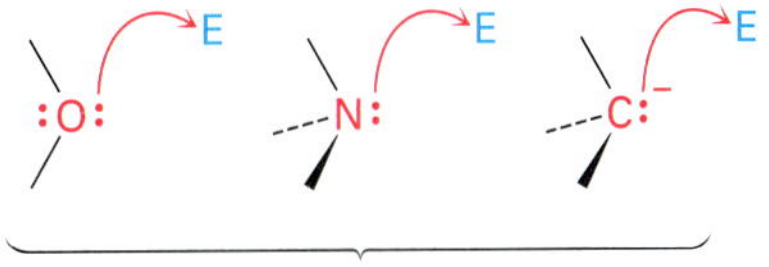

The electrophilic sink must be able to accept an electron pair, usually because it has either a positively charged atom or a positively polarized atom in a functional group. For example:

Electrons usually flow *to* one of these electrophiles.

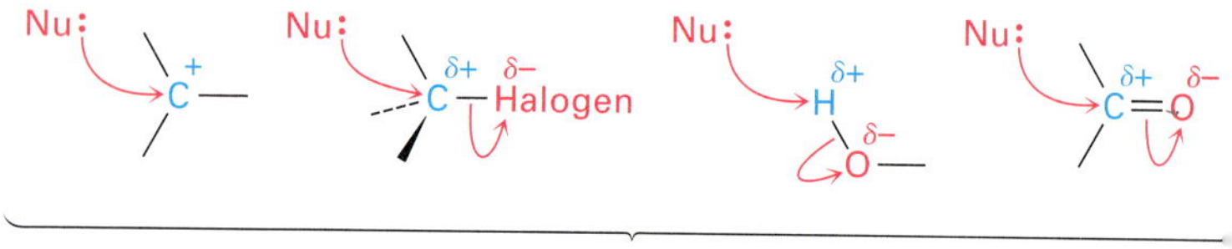

Rule 2 **The nucleophile can be either negatively charged or neutral.** If the nucleophile is negatively charged, the atom that gives away an electron pair becomes neutral. For example:

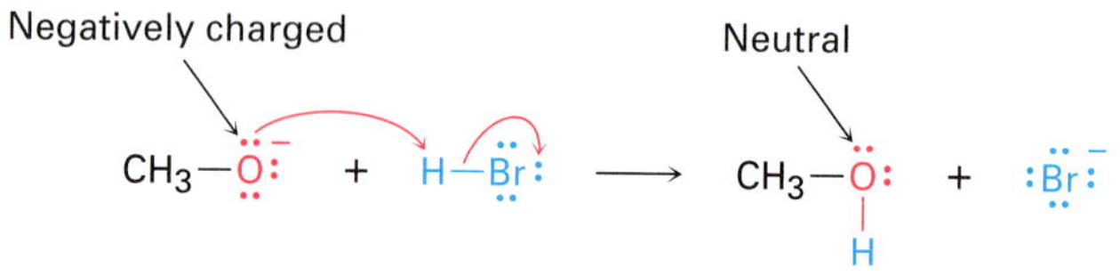

If the nucleophile is neutral, the atom that gives away an electron pair acquires a positive charge. For example:

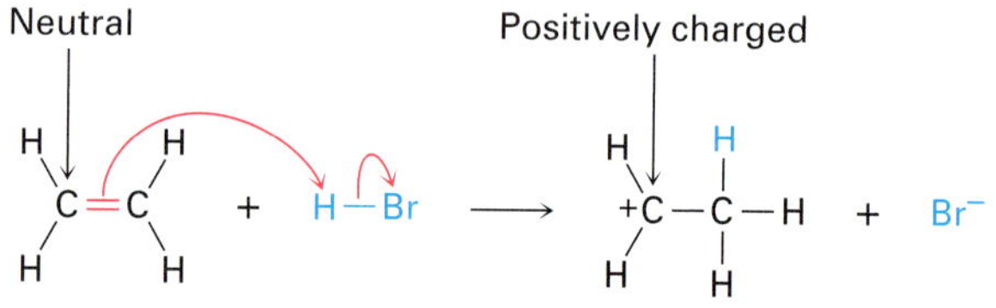

Rule 3 **The electrophile can be either positively charged or neutral.** If the electrophile is positively charged, the atom bearing that charge becomes neutral after accepting an electron pair. For example:

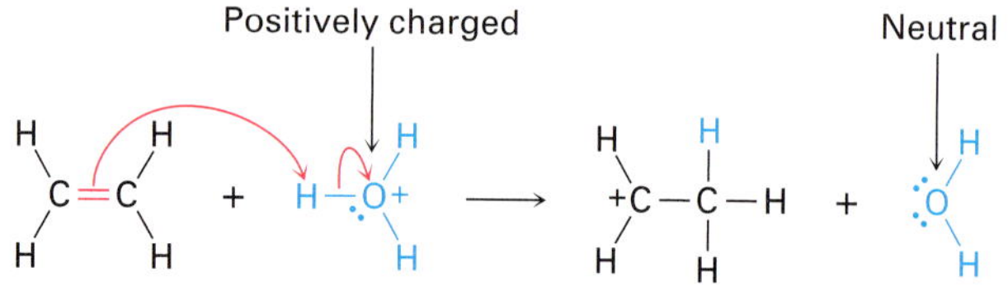

If the electrophile is neutral, the atom that ultimately accepts the electron pair acquires a negative charge. For this to happen, however, the negative charge must be stabilized by being on an electronegative atom such as oxygen, nitrogen, or a halogen. For example:

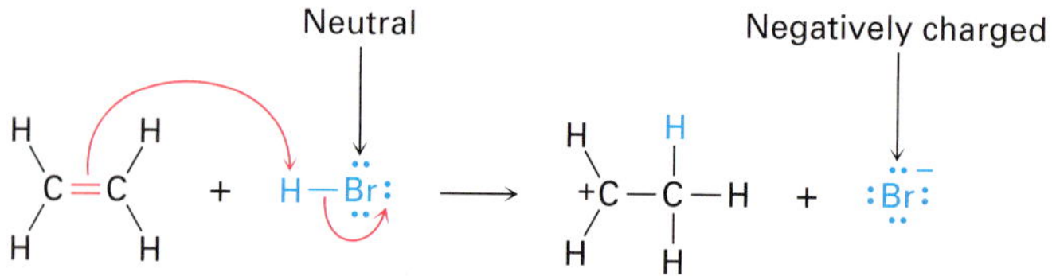

The result of Rules 2 and 3 together is that charge is conserved during the reaction. A negative charge in one of the reactants gives a negative charge in one of the products, and a positive charge in one of the reactants gives a positive charge in one of the products.

Rule 4 **The octet rule must be followed.** That is, no second-row atom can be left with ten electrons (or four for hydrogen). If an electron pair moves *to* an atom that already has an octet (or two for hydrogen), another electron pair must simultaneously move *from* that atom to maintain the octet. When two electrons move from the C=C bond of ethylene to the hydrogen atom of H_3O^+, for instance, two electrons must leave that hydrogen. This means that the H–O bond must break and the electrons must stay with the oxygen, giving neutral water.

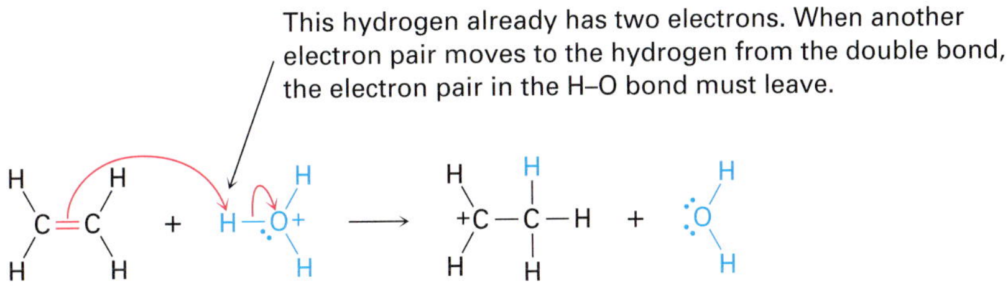

ThomsonNOW™ Click *Organic Interactive* to **practice writing organic mechanisms using curved arrows.**

Worked Example 5.2 gives another example of drawing curved arrows.

WORKED EXAMPLE 5.2 *Using Curved Arrows in Reaction Mechanisms*

Add curved arrows to the following polar reaction to show the flow of electrons:

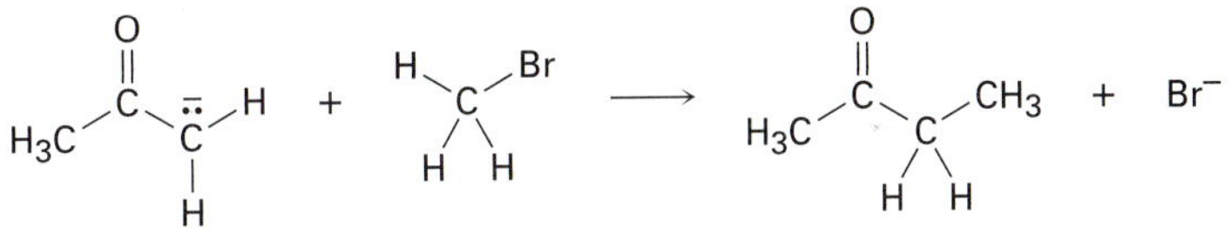

Strategy First, look at the reaction and identify the bonding changes that have occurred. In this case, a C–Br bond has broken and a C–C bond has formed. The formation of the C–C bond involves donation of an electron pair from the nucleophilic carbon atom of the reactant on the left to the electrophilic carbon atom of CH_3Br, so we draw a curved arrow originating from the lone pair on the negatively charged C atom and pointing to the C atom of CH_3Br. At the same time the C–C bond forms, the C–Br bond must break so that the octet rule is not violated. We therefore draw a second curved arrow from the C–Br bond to Br. The bromine is now a stable Br^- ion.

Solution

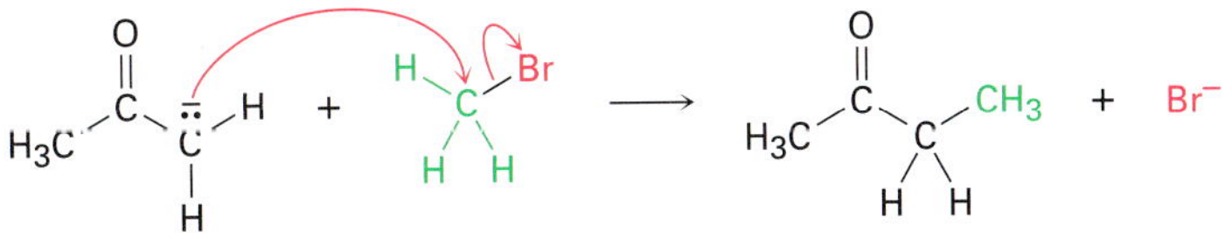

Problem 5.8 Add curved arrows to the following polar reactions to indicate the flow of electron in each:

(a)

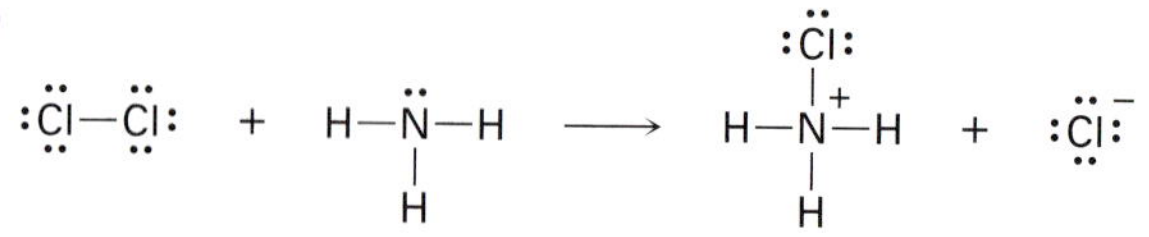

(b)

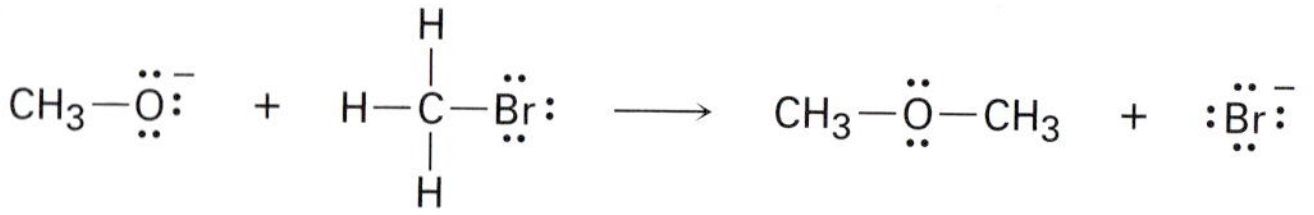

(c)

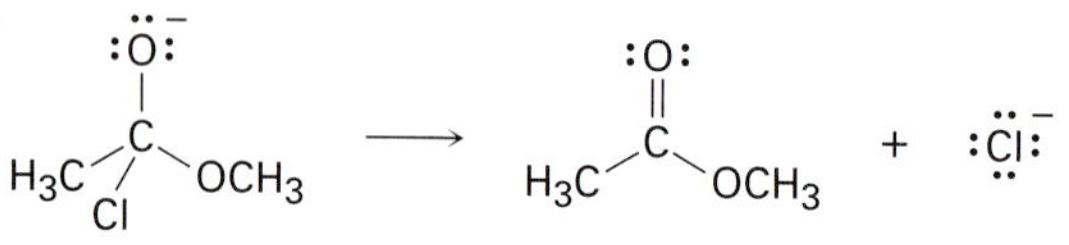

Problem 5.9 Predict the products of the following polar reaction, a step in the citric acid cycle for food metabolism, by interpreting the flow of electrons indicated by the curved arrows:

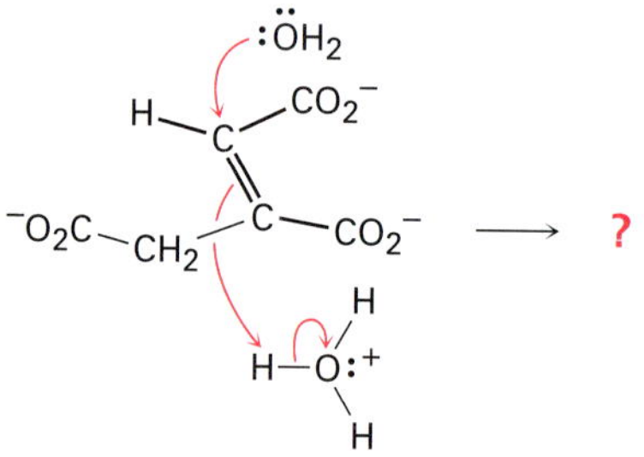

5.7 Describing a Reaction: Equilibria, Rates, and Energy Changes

Every chemical reaction can go in either forward or reverse direction. Reactants can go forward to products, and products can revert to reactants. As you may remember from your general chemistry course, the position of the resulting chemical equilibrium is expressed by an equation in which K_{eq}, the equilibrium constant, is equal to the product concentrations multiplied together, divided by the reactant concentrations multiplied together, with each concentration raised to the power of its coefficient in the balanced equation. For the generalized reaction

$$aA + bB \rightleftharpoons cC + dD$$

we have

$$K_{eq} = \frac{[C]^c\,[D]^d}{[A]^a\,[B]^b}$$

The value of the equilibrium constant tells which side of the reaction arrow is energetically favored. If K_{eq} is much larger than 1, then the product concentration term $[C]^c [D]^d$ is much larger than the reactant concentration term $[A]^a [B]^b$, and the reaction proceeds as written from left to right. If K_{eq} is near 1, appreciable amounts of both reactant and product are present at equilibrium. And if K_{eq} is much smaller than 1, the reaction does not take place as written but instead goes in the reverse direction, from right to left.

In the reaction of ethylene with HBr, for example, we can write the following equilibrium expression, and we can determine experimentally that the equilibrium constant at room temperature is approximately 7.1×10^7:

$$H_2C{=}CH_2 \;+\; HBr \;\rightleftharpoons\; CH_3CH_2Br$$

$$K_{eq} = \frac{[CH_3CH_2Br]}{[HBr]\,[H_2C{=}CH_2]} = 7.1 \times 10^7$$

Because K_{eq} is relatively large, the reaction proceeds as written and greater than 99.999 99% of the ethylene is converted into bromoethane. For practical purposes, an equilibrium constant greater than about 10^3 means that the amount of reactant left over will be barely detectable (less than 0.1%).

What determines the magnitude of the equilibrium constant? For a reaction to have a favorable equilibrium constant and proceed as written, the energy of the products must be lower than the energy of the reactants. In other words, energy must be *released*. The situation is analogous to that of a rock poised precariously in a high-energy position near the top of a hill. When it rolls downhill, the rock releases energy until it reaches a more stable low-energy position at the bottom.

The energy change that occurs during a chemical reaction is called the **Gibbs free-energy change (ΔG).** For a favorable reaction, ΔG has a negative value, meaning that energy is lost by the chemical system and released to the surroundings. Such reactions are said to be **exergonic**. For an unfavorable reaction, ΔG has a positive value, meaning that energy is absorbed by the chemical system *from* the surroundings. Such reactions are said to be **endergonic**. You might also recall from general chemistry that the *standard* free-energy change for a reaction is denoted $\Delta G°$, where the superscript ° means that the reaction is carried out under standard conditions, with pure substances in their most stable form at 1 atm pressure and a specified temperature, usually 298 K.

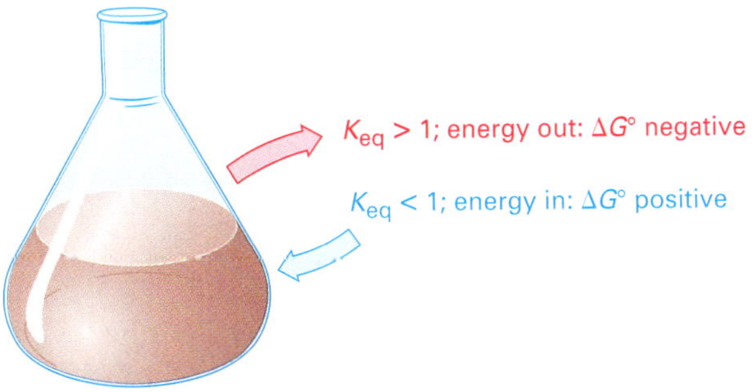

Because the equilibrium constant, K_{eq}, and the standard free-energy change, $\Delta G°$, both measure whether a reaction is favored, they are mathematically related by the equation

$$\Delta G° = -RT \ln K_{eq} \qquad \text{or} \qquad K_{eq} = e^{-\Delta G°/RT}$$

where $R = 8.314\ \text{J/(K} \cdot \text{mol)} = 1.987\ \text{cal/(K} \cdot \text{mol)}$
$T =$ Kelvin temperature
$e = 2.718$
$\ln K_{eq} =$ natural logarithm of K_{eq}

The free-energy change ΔG is made up of two terms, an *enthalpy* term, ΔH, and a temperature-dependent *entropy* term, $T\Delta S$. Of the two terms, the enthalpy term is often larger and more dominant.

$$\Delta G° = \Delta H° - T\Delta S°$$

For the reaction of ethylene with HBr at room temperature (298 K), the approximate values are

$$H_2C{=}CH_2 + HBr \rightleftharpoons CH_3CH_2Br \quad \begin{cases} \Delta G° = -44.8\ \text{kJ/mol} \\ \Delta H° = -84.1\ \text{kJ/mol} \\ \Delta S° = -0.132\ \text{kJ/(K} \cdot \text{mol)} \\ K_{eq} = 7.1 \times 10^7 \end{cases}$$

The **enthalpy change, ΔH,** also called the **heat of reaction**, is a measure of the change in total bonding energy during a reaction. If ΔH is negative, as in the reaction of HBr with ethylene, the bonds in the products are stronger (more stable) than the bonds in the reactants, heat is released, and the reaction is said to be **exothermic**. If ΔH is positive, the bonds in the products are weaker (less stable) than the bonds in the reactants, heat is absorbed, and the reaction is said to be **endothermic**. For example, if a reaction breaks reactant bonds with a total strength of 380 kJ/mol and forms product bonds with a total strength of 400 kJ/mol, then ΔH for the reaction is −20 kJ/mol and the reaction is exothermic.

The **entropy change, ΔS,** is a measure of the change in the amount of molecular randomness, or freedom of motion, that accompanies a reaction. For example, in an elimination reaction of the type

$$\text{A} \longrightarrow \text{B} + \text{C}$$

there is more freedom of movement and molecular randomness in the products than in the reactant because one molecule has split into two. Thus, there is a net increase in entropy during the reaction and ΔS has a positive value.

On the other hand, for an addition reaction of the type

$$\text{A} + \text{B} \longrightarrow \text{C}$$

the opposite is true. Because such reactions restrict the freedom of movement of two molecules by joining them together, the product has less randomness than the reactants and ΔS has a negative value. The reaction of ethylene and

HBr to yield bromoethane, which has $\Delta S° = -0.132$ kJ/(K · mol), is an example. Table 5.2 describes the thermodynamic terms more fully.

Table 5.2 Explanation of Thermodynamic Quantities: $\Delta G° = \Delta H° - T\Delta S°$

Term	Name	Explanation
$\Delta G°$	Gibbs free-energy change	The energy difference between reactants and products. When $\Delta G°$ is negative, the reaction is **exergonic**, has a favorable equilibrium constant, and can occur spontaneously. When $\Delta G°$ is positive, the reaction is **endergonic**, has an unfavorable equilibrium constant, and cannot occur spontaneously.
$\Delta H°$	Enthalpy change	The heat of reaction, or difference in strength between the bonds broken in a reaction and the bonds formed. When $\Delta H°$ is negative, the reaction releases heat and is **exothermic**. When $\Delta H°$ is positive, the reaction absorbs heat and is **endothermic**.
$\Delta S°$	Entropy change	The change in molecular randomness during a reaction. When $\Delta S°$ is negative, randomness decreases; when $\Delta S°$ is positive, randomness increases.

Knowing the value of K_{eq} for a reaction is useful, but it's important to realize the limitations. An equilibrium constant tells only the *position* of the equilibrium, or how much product is theoretically possible. It doesn't tell the *rate* of reaction, or how fast the equilibrium is established. Some reactions are extremely slow even though they have favorable equilibrium constants. Gasoline is stable at room temperature, for instance, because the rate of its reaction with oxygen is slow at 298 K. At higher temperatures, however, such as contact with a lighted match, gasoline reacts rapidly with oxygen and undergoes complete conversion to the equilibrium products water and carbon dioxide. Rates (*how fast* a reaction occurs) and equilibria (*how much* a reaction occurs) are entirely different.

Rate ⟶ **Is the reaction fast or slow?**

Equilibrium ⟶ **In what direction does the reaction proceed?**

Problem 5.10 Which reaction is more energetically favored, one with $\Delta G° = -44$ kJ/mol or one with $\Delta G° = +44$ kJ/mol?

Problem 5.11 Which reaction is likely to be more exergonic, one with $K_{eq} = 1000$ or one with $K_{eq} = 0.001$?

5.8 Describing a Reaction: Bond Dissociation Energies

ThomsonNOW™ Click *Organic Interactive* to **use bond dissociation energies to predict organic reactions and radical stability.**

We've just seen that heat is released (negative ΔH) when a bond is formed and absorbed (positive ΔH) when a bond is broken. The measure of the heat change that occurs on breaking a bond is called the *bond strength,* or **bond dissociation energy (*D*)**, defined as the amount of energy required to break a given bond to produce two radical fragments when the molecule is in the gas phase at 25 °C.

$$\text{A:B} \xrightarrow[\text{energy}]{\text{Bond dissociation}} \text{A}\cdot + \cdot\text{B}$$

Each specific bond has its own characteristic strength, and extensive tables of data are available. For example, a C–H bond in methane has a bond dissociation energy $D = 438.4$ kJ/mol (104.8 kcal/mol), meaning that 438.4 kJ/mol must be added to break a C–H bond of methane to give the two radical fragments $\cdot CH_3$ and $\cdot H$. Conversely, 438.4 kJ/mol of energy is released when a methyl radical and a hydrogen atom combine to form methane. Table 5.3 lists some other bond strengths.

Table 5.3 Some Bond Dissociation Energies, *D*

Bond	*D* (kJ/mol)
H—H	436
H—F	570
H—Cl	432
H—Br	366
H—I	298
Cl—Cl	243
Br—Br	193
I—I	151
CH_3—H	438
CH_3—Cl	351
CH_3—Br	293
CH_3—I	234
CH_3—OH	380
CH_3—NH_2	335
C_2H_5—H	420
C_2H_5—Cl	338
C_2H_5—Br	285
C_2H_5—I	222
C_2H_5—OH	380
$(CH_3)_2CH$—H	401
$(CH_3)_2CH$—Cl	339
$(CH_3)_2CH$—Br	274
$(CH_3)_3C$—H	390
$(CH_3)_3C$—Cl	330
$(CH_3)_3C$—Br	263

Bond	*D* (kJ/mol)
$(CH_3)_3C$—I	209
$H_2C{=}CH$—H	444
$H_2C{=}CH$—Cl	368
$H_2C{=}CHCH_2$—H	361
$H_2C{=}CHCH_2$—Cl	289
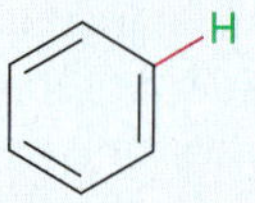	464
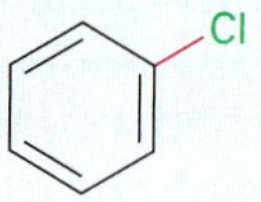	405
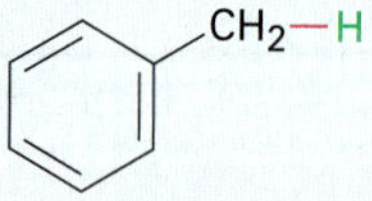	368
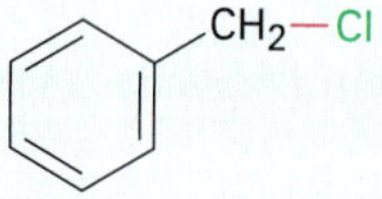	293
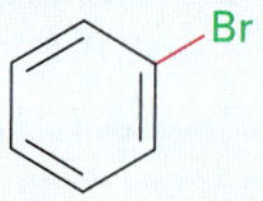	337
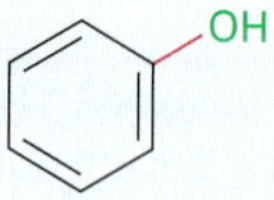	469
HC≡C—H	552
CH_3—CH_3	376

Bond	*D* (kJ/mol)
C_2H_5—CH_3	355
$(CH_3)_2CH$—CH_3	351
$(CH_3)_3C$—CH_3	339
$H_2C{=}CH$—CH_3	406
$H_2C{=}CHCH_2$—CH_3	310
$H_2C{=}CH_2$	611
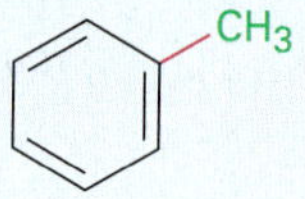	427
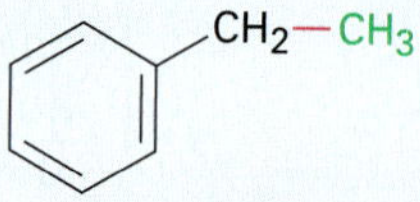	332
$CH_3\overset{\overset{O}{\|}}{C}$—H	368
HO—H	498
HO—OH	213
CH_3O—H	437
CH_3S—H	371
C_2H_5O—H	436
$CH_3\overset{\overset{O}{\|}}{C}$—$CH_3$	322
CH_3CH_2O—CH_3	339
NH_2—H	449
H—CN	518

Think for a moment about the connection between bond strengths and chemical reactivity. In an exothermic reaction, more heat is released than is absorbed. But since making product bonds releases heat and breaking reactant bonds absorbs heat, the bonds in the products must be stronger than the bonds in the reactants. In other words, exothermic reactions are favored by stable products with strong bonds and by reactants with weak, easily broken bonds.

Sometimes, particularly in biochemistry, reactive substances that undergo highly exothermic reactions, such as ATP (adenosine triphosphate), are referred to as "energy-rich" or "high-energy" compounds. Such labels don't mean that ATP is special or different from other compounds; they mean only that ATP has relatively weak bonds that require a smaller amount of heat to break, thus leading to a larger release of heat on reaction. When a typical organic phosphate such as glycerol 3-phosphate reacts with water, for instance, only 9 kJ/mol of heat is released ($\Delta H° = -9$ kJ/mol), but when ATP reacts with water, 30 kJ/mol of heat is released ($\Delta H° = -30$ kJ/mol). The difference between the two reactions is due to the fact that the bond broken in ATP is substantially weaker than the bond broken in glycerol 3-phosphate.

$\Delta H°' = -9$ kJ/mol

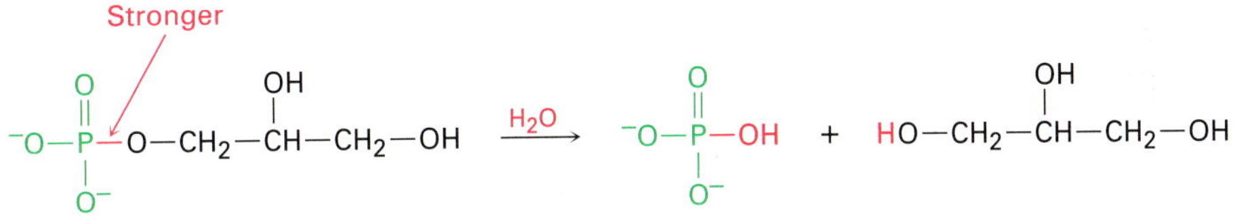

Glycerol 3-phosphate **Glycerol**

$\Delta H°' = -30$ kJ/mol

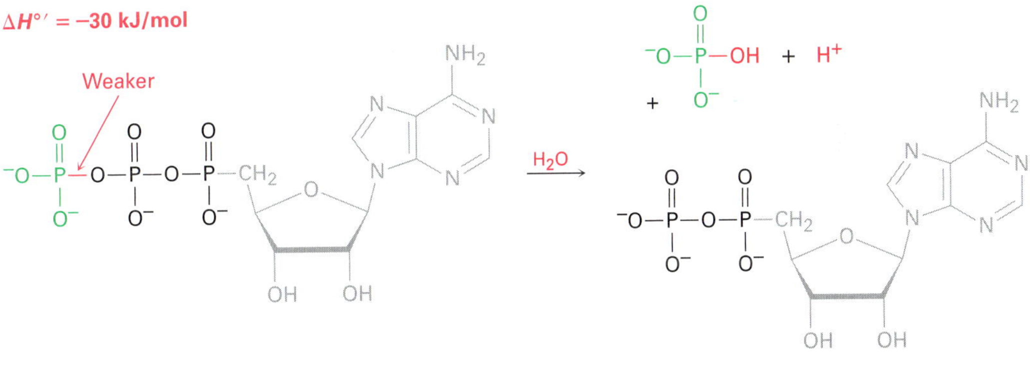

Adenosine triphosphate (ATP) **Adenosine diphosphate (ADP)**

5.9 Describing a Reaction: Energy Diagrams and Transition States

For a reaction to take place, reactant molecules must collide and reorganization of atoms and bonds must occur. Let's again look at the addition reaction of HBr and ethylene, which takes place in two steps.

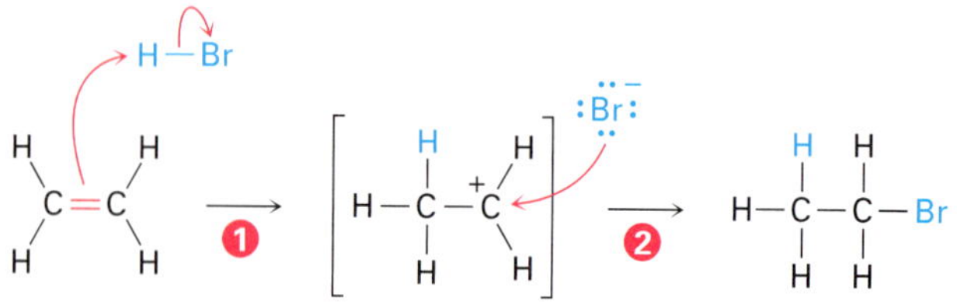

As the reaction proceeds, ethylene and HBr must approach each other, the ethylene π bond and the H–Br bond must break, a new C–H bond must form in the first step, and a new C–Br bond must form in the second step.

To depict graphically the energy changes that occur during a reaction, chemists use reaction energy diagrams, such as that shown in Figure 5.4. The vertical axis of the diagram represents the total energy of all reactants, and the horizontal axis, called the *reaction coordinate,* represents the progress of the reaction from beginning to end. Let's see how the addition of HBr to ethylene can be described in an energy diagram.

Figure 5.4 An energy diagram for the first step in the reaction of ethylene with HBr. The energy difference between reactants and transition state, $\Delta G^{\ddagger}$, defines the reaction rate. The energy difference between reactants and carbocation product, $\Delta G°$, defines the position of the equilibrium.

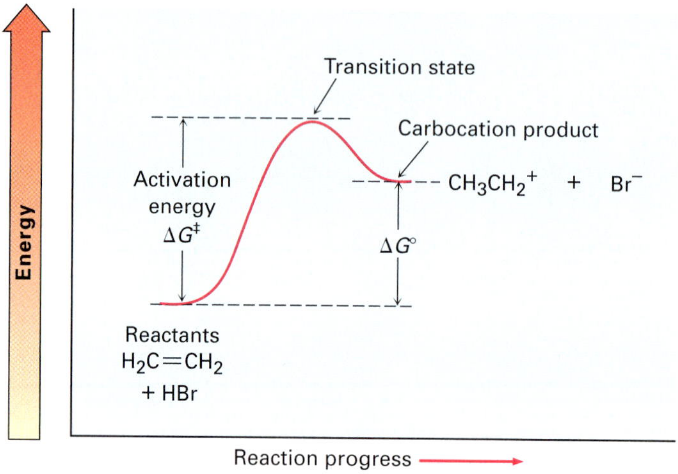

At the beginning of the reaction, ethylene and HBr have the total amount of energy indicated by the reactant level on the left side of the diagram in Figure 5.4. As the two reactants collide and reaction commences, their electron clouds repel each other, causing the energy level to rise. If the collision has occurred with enough force and proper orientation, the reactants continue to approach each other despite the rising repulsion until the new C–H bond starts to form. At some point, a structure of maximum energy is reached, a structure called the *transition state.*

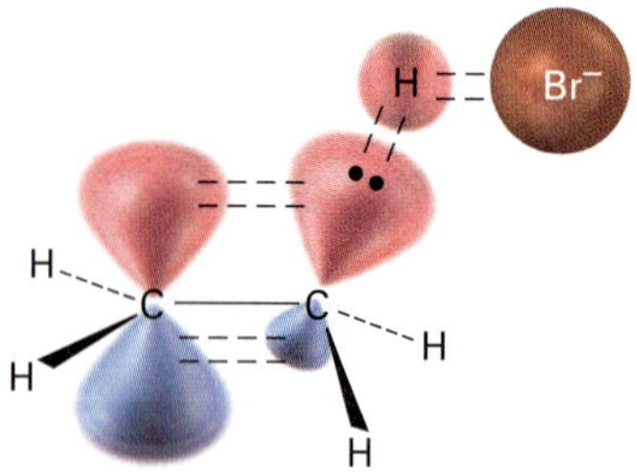

Active Figure 5.5 A hypothetical transition-state structure for the first step of the reaction of ethylene with HBr. The C=C π bond and H–Br bond are just beginning to break, and the C–H bond is just beginning to form. *Sign in at* **www.thomsonedu.com** *to see a simulation based on this figure and to take a short quiz.*

The **transition state** represents the highest-energy structure involved in this step of the reaction. It is unstable and can't be isolated, but we can nevertheless imagine it to be an activated complex of the two reactants in which both the C=C π bond and H–Br bond are partially broken and the new C–H bond is partially formed (Figure 5.5).

The energy difference between reactants and transition state is called the **activation energy,** $\Delta G^{\ddagger}$, and determines how rapidly the reaction occurs at a given temperature. (The double-dagger superscript, ‡, always refers to the transition state.) A large activation energy results in a slow reaction because few collisions occur with enough energy for the reactants to reach the transition state.

A small activation energy results in a rapid reaction because almost all collisions occur with enough energy for the reactants to reach the transition state.

As an analogy, you might think of reactants that need enough energy to climb the activation barrier to the transition state as similar to hikers who need enough energy to climb to the top of a mountain pass. If the pass is a high one, the hikers need a lot of energy and surmount the barrier with difficulty. If the pass is low, however, the hikers need less energy and reach the top easily.

As a rough generalization, many organic reactions have activation energies in the range 40 to 150 kJ/mol (10–35 kcal/mol). The reaction of ethylene with HBr, for example, has an activation energy of approximately 140 kJ/mol (34 kcal/mol). Reactions with activation energies less than 80 kJ/mol take place at or below room temperature, whereas reactions with higher activation energies normally require a higher temperature to give the reactants enough energy to climb the activation barrier.

Once the transition state is reached, the reaction can either continue on to give the carbocation product or revert back to reactant. When reversion to reactant occurs, the transition-state structure comes apart and an amount of free energy corresponding to $-\Delta G^{\ddagger}$ is released. When the reaction continues on to give the carbocation, the new C–H bond forms fully and an amount of energy corresponding to the difference between transition state and carbocation product is released. The net change in energy for the step, $\Delta G°$, is represented in the diagram as the difference in level between reactant and product. Since the carbocation is higher in energy than the starting alkene, the step is endergonic, has a positive value of $\Delta G°$, and absorbs energy.

Not all energy diagrams are like that shown for the reaction of ethylene and HBr. Each reaction has its own energy profile. Some reactions are fast (small $\Delta G^{\ddagger}$) and some are slow (large $\Delta G^{\ddagger}$); some have a negative $\Delta G°$, and some have a positive $\Delta G°$. Figure 5.6 illustrates some different possibilities.

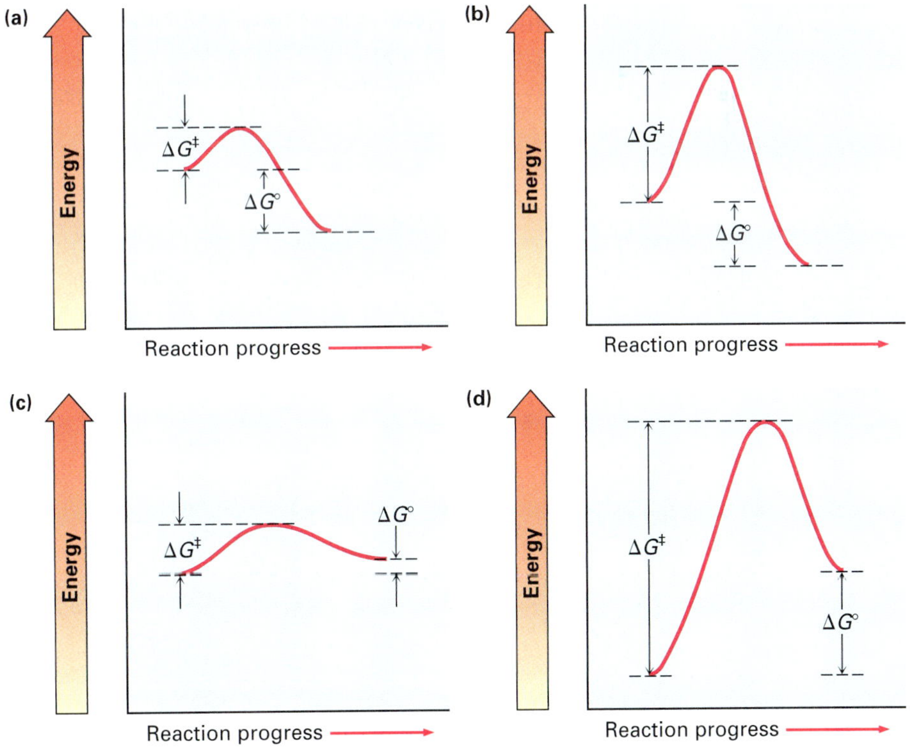

Active Figure 5.6 Some hypothetical energy diagrams: **(a)** a fast exergonic reaction (small $\Delta G^{\ddagger}$, negative $\Delta G°$); **(b)** a slow exergonic reaction (large $\Delta G^{\ddagger}$, negative $\Delta G°$); **(c)** a fast endergonic reaction (small $\Delta G^{\ddagger}$, small positive $\Delta G°$); **(d)** a slow endergonic reaction (large $\Delta G^{\ddagger}$, positive $\Delta G°$). *Sign in at* **www.thomsonedu.com** *to see a simulation based on this figure and to take a short quiz.*

Problem 5.12 | Which reaction is faster, one with $\Delta G^{\ddagger} = +45$ kJ/mol or one with $\Delta G^{\ddagger} = +70$ kJ/mo

5.10 Describing a Reaction: Intermediates

How can we describe the carbocation formed in the first step of the reaction c ethylene with HBr? The carbocation is clearly different from the reactants, yet i isn't a transition state and it isn't a final product.

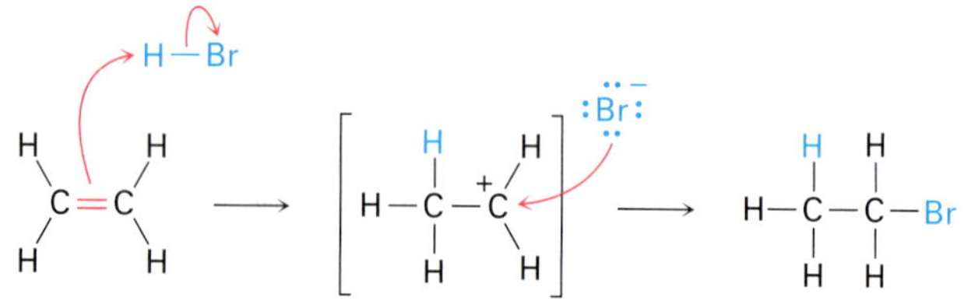

Reaction intermediate

We call the carbocation, which exists only transiently during the cours of the multistep reaction, a **reaction intermediate**. As soon as the interme diate is formed in the first step by reaction of ethylene with H^+, it reacts fur ther with Br^- in a second step to give the final product, bromoethane. Thi second step has its own activation energy ($\Delta G^{\ddagger}$), its own transition state, anc its own energy change ($\Delta G°$). We can picture the second transition state as ar activated complex between the electrophilic carbocation intermediate anc the nucleophilic bromide anion, in which Br^- donates a pair of electrons tc the positively charged carbon atom as the new C−Br bond starts to form.

A complete energy diagram for the overall reaction of ethylene with HBr is shown in Figure 5.7. In essence, we draw a diagram for each of the individua steps and then join them so that the carbocation *product* of step 1 is the *reactan* for step 2. As indicated in Figure 5.7, the reaction intermediate lies at an energy

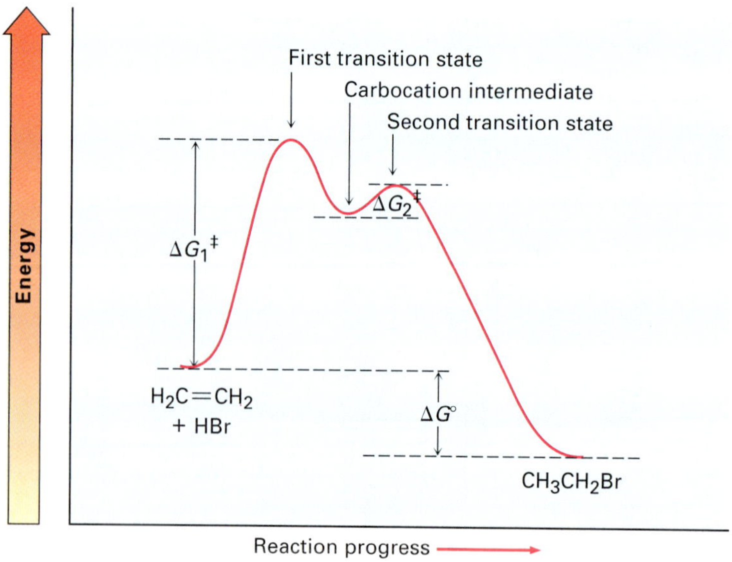

Figure 5.7 An energy diagram for the overall reaction of ethylene with HBr. Two separate steps are involved, each with its own transition state. The energy minimum between the two steps represents the carbocation reaction intermediate.

minimum between steps. Since the energy level of the intermediate is higher than the level of either the reactant that formed it or the product it yields, the intermediate can't normally be isolated. It is, however, more stable than the two transition states that neighbor it.

Each step in a multistep process can always be considered separately. Each step has its own $\Delta G^{\ddagger}$ and its own $\Delta G°$. The *overall* $\Delta G°$ of the reaction, however, is the energy difference between initial reactants and final products.

The biological reactions that take place in living organisms have the same energy requirements as reactions that take place in the laboratory and can be described in similar ways. They are, however, constrained by the fact that they must have low enough activation energies to occur at moderate temperatures, and they must release energy in relatively small amounts to avoid overheating the organism. These constraints are generally met through the use of large, structurally complex, enzyme catalysts that change the mechanism of a reaction to an alternative pathway that proceeds through a series of small steps rather than one or two large steps. Thus, a typical energy diagram for a biological reaction might look like that in Figure 5.8.

Figure 5.8 An energy diagram for a typical, enzyme-catalyzed biological reaction (blue curve) versus an uncatalyzed laboratory reaction (red curve). The biological reaction involves many steps, each of which has a relatively small activation energy and small energy change. The end result is the same, however.

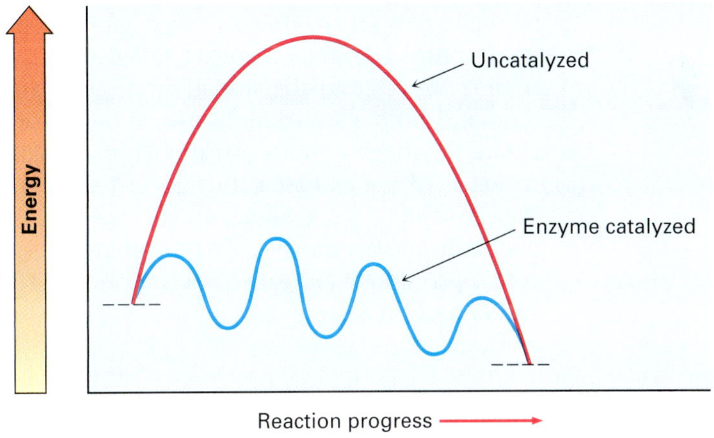

WORKED EXAMPLE 5.3

Drawing Energy Diagrams for Reactions

Sketch an energy diagram for a one-step reaction that is fast and highly exergonic.

Strategy A fast reaction has a small $\Delta G^{\ddagger}$, and a highly exergonic reaction has a large negative $\Delta G°$.

Solution

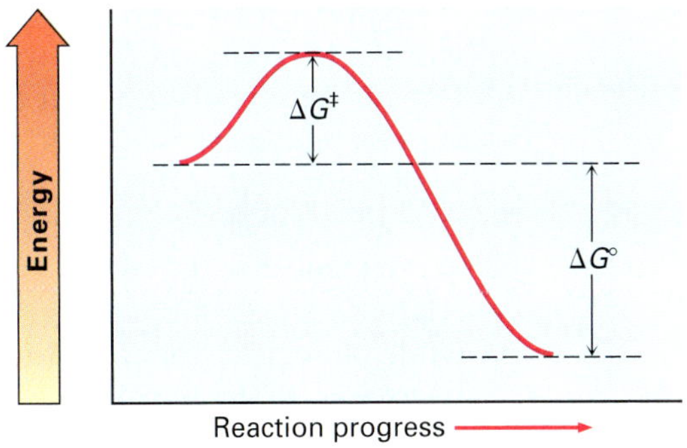

Problem 5.13 Sketch an energy diagram for a two-step reaction with an endergonic first step an an exergonic second step. Label the parts of the diagram corresponding to reactan product, and intermediate.

5.11 A Comparison between Biological Reactions and Laboratory Reactions

In comparing laboratory reactions with biological reactions, several difference are apparent. For one thing, laboratory reactions are usually carried out in a organic solvent such as diethyl ether or dichloromethane to dissolve the reac tants and bring them into contact, whereas biological reactions occur in th aqueous medium inside cells. For another thing, laboratory reactions often tak place over a wide range of temperatures without catalysts, while biological reac tions take place at the temperature of the organism and are catalyzed b *enzymes.*

We'll look at enzymes in more detail in Section 26.10, but you may alread be aware that an enzyme is a large, globular protein molecule that contains i its structure a protected pocket called its *active site.* The active site is lined b acidic or basic groups as needed for catalysis and has precisely the right shape t bind and hold a substrate molecule in the orientation necessary for reaction Figure 5.9 shows a molecular model of hexokinase, along with an X-ray crysta structure of the glucose substrate and adenosine diphosphate (ADP) bound i the active site. Hexokinase is an enzyme that catalyzes the initial step of glucos metabolism—the transfer of a phosphate group from ATP to glucose, giving glu cose 6-phosphate and ADP. The structures of ATP and ADP were shown at th end of Section 5.8.

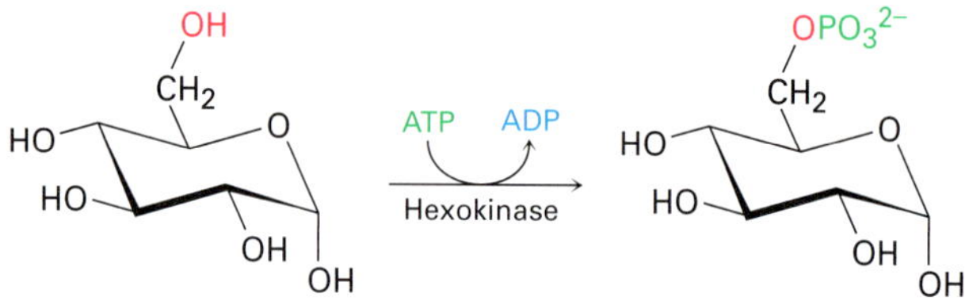

Glucose **Glucose 6-phosphate**

Note how the hexokinase-catalyzed phosphorylation reaction of glucose i written. It's common when writing biological equations to show only the struc ture of the primary reactant and product, while abbreviating the structures o various biological "reagents" and by-products such as ATP and ADP. A curvec arrow intersecting the straight reaction arrow indicates that ATP is also a reac tant and ADP also a product.

Yet another difference is that laboratory reactions are often done using relatively small, simple reagents such as Br_2, HCl, $NaBH_4$, CrO_3, and so forth while biological reactions usually involve relatively complex "reagents" callec *coenzymes.* In the hexokinase-catalyzed phosphorylation of glucose just shown

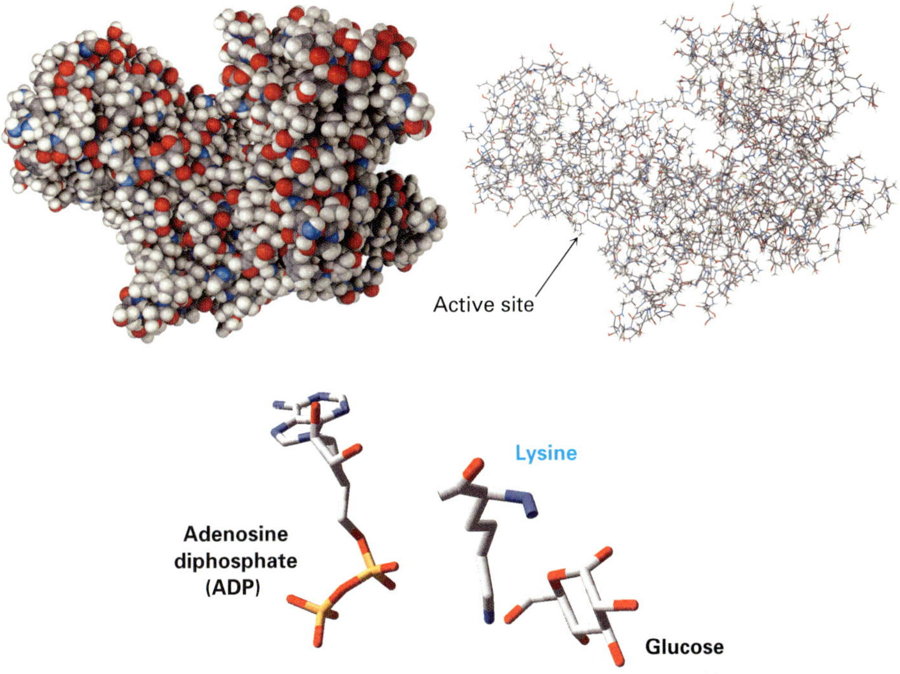

Figure 5.9 Models of hexokinase in space-filling and wire-frame formats, showing the cleft that contains the active site where substrate binding and reaction catalysis occur. At the bottom is an X-ray crystal structure of the enzyme active site, showing the positions of both glucose and ADP as well as a lysine amino acid that acts as a base to deprotonate glucose.

for instance, ATP is the coenzyme. Of all the atoms in the entire coenzyme, only the one phosphate group shown in red is transferred to the glucose substrate.

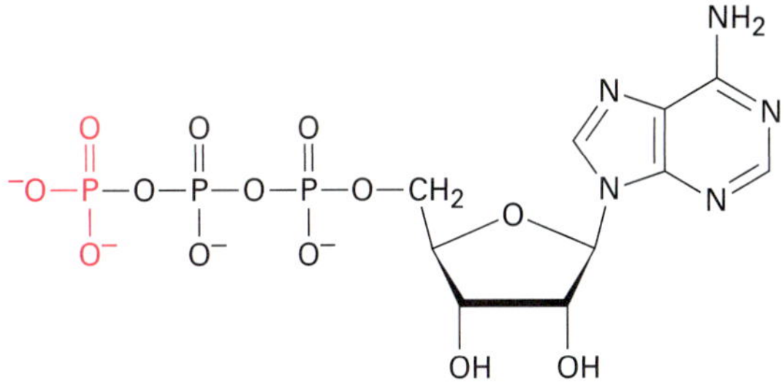

Adenosine triphosphate, ATP
(a coenzyme)

Don't be intimidated by the size of the molecule; most of the structure is there to provide an overall shape for binding to the enzyme and to provide appropriate solubility behavior. When looking at biological molecules, focus on the small part of the molecule where the chemical change takes place.

One final difference between laboratory and biological reactions is in their specificity. A catalyst might be used in the laboratory to catalyze the reaction of thousands of different substances, but an enzyme, because it can bind only a specific substrate molecule having a specific shape, will catalyze only a specific reaction. It's this exquisite specificity that makes biological chemistry so remarkable and that makes life possible. Table 5.4 summarizes some of the differences between laboratory and biological reactions.

Table 5.4 A Comparison of Typical Laboratory and Biological Reactions

	Laboratory reaction	Biological reaction
Solvent	Organic liquid, such as ether	Aqueous environment in cells
Temperature	Wide range; −80 to 150 °C	Temperature of organism
Catalyst	Either none or very simple	Large, complex enzymes needed
Reagent size	Usually small and simple	Large, complex coenzymes
Specificity	Little specificity for substrate	Very high specificity for substrate

Focus On . . .

Where Do Drugs Come From?

Approved for sale in March, 1998, Viagra has been used by more than 16 million men. It is currently undergoing study as a treatment for preeclampsia, a complication of pregnancy that is responsible for as many as 70,000 deaths each year. Where do new drugs like this come from?

It has been estimated that major pharmaceutical companies in the United States spend some $33 billion per year on drug research and development, while government agencies and private foundations spend another $28 billion. What does this money buy? For the period 1981–2004, the money resulted in a total of 912 new molecular entities (NMEs)—new biologically active chemical substances approved for sale as drugs by the U.S. Food and Drug Administration (FDA). That's an average of only 38 new drugs each year spread over all diseases and conditions, and the number has been steadily falling. In 2004, only 23 NMEs were approved.

Where do the new drugs come from? According to a study carried out at the U.S. National Cancer Institute, only 33% of new drugs are entirely synthetic and completely unrelated to any naturally occurring substance. The remaining 67% take their lead, to a greater or lesser extent, from nature. Vaccines and genetically engineered proteins of biological origin account for 15% of NMEs, but most new drugs come from *natural products,* a catchall term generally taken to mean small molecules found in bacteria, plants, and other living organisms. Unmodified natural products isolated directly from the producing organism account for 28% of NMEs, while natural products that have been chemically modified in the laboratory account for the remaining 24%.

Origin of New Drugs 1981–2002

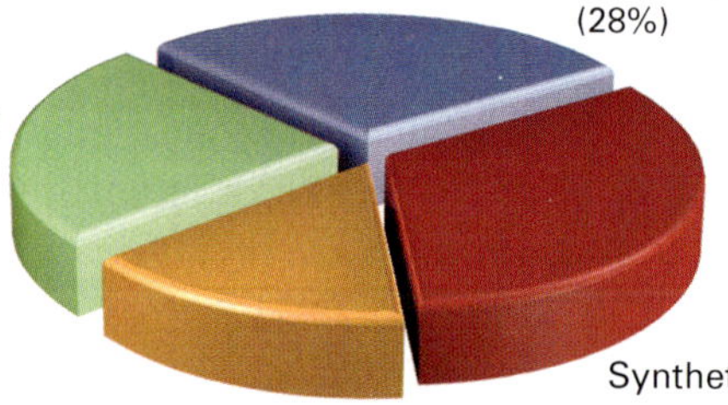

(continued)

Many years of work go into screening many thousands of substances to identify a single compound that might ultimately gain approval as an NME. But after that single compound has been identified, the work has just begun because it takes an average of 9 to 10 years for a drug to make it through the approval process. First, the safety of the drug in animals must be demonstrated and an economical method of manufacture must be devised. With these preliminaries out of the way, an Investigational New Drug (IND) application is submitted to the FDA for permission to begin testing in humans.

Human testing takes 5 to 7 years and is divided into three phases. Phase I clinical trials are carried out on a small group of healthy volunteers to establish safety and look for side effects. Several months to a year are needed, and only about 70% of drugs pass at this point. Phase II clinical trials next test the drug for 1 to 2 years in several hundred patients with the target disease, looking both for safety and for efficacy, and only about 33% of the original group pass. Finally, phase III trials are undertaken on a large sample of patients to document definitively the drug's safety, dosage, and efficacy. If the drug is one of the 25% of the original group that have made it this far, all the data are then gathered into a New Drug Application (NDA) and sent to the FDA for review and approval, which can take another 2 years. Ten years and at least $500 million has now been spent, and only 20% of the drugs that began testing have succeeded. Finally, though, the drug will begin to appear in medicine cabinets. The following timeline shows the process.

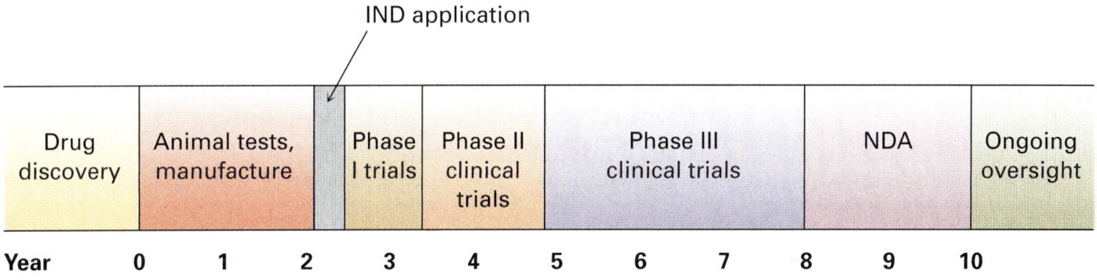

SUMMARY AND KEY WORDS

- activation energy ($\Delta G^{\ddagger}$), 158
- addition reaction, 137
- bond dissociation energy (D), 155
- carbocation, 148
- electrophile, 145
- elimination reaction, 138
- endergonic, 153
- endothermic, 154
- enthalpy change (ΔH), 154
- entropy change (ΔS), 154
- exergonic, 153

There are four common kinds of reactions: **addition reactions** take place when two reactants add together to give a single product; **elimination reactions** take place when one reactant splits apart to give two products; **substitution reactions** take place when two reactants exchange parts to give two new products; and **rearrangement reactions** take place when one reactant undergoes a reorganization of bonds and atoms to give an isomeric product.

A full description of how a reaction occurs is called its **mechanism**. There are two general kinds of mechanisms by which reactions take place: **radical** mechanisms and **polar** mechanisms. Polar reactions, the more common type, occur because of an attractive interaction between a **nucleophilic** (electron-rich) site in one molecule and an **electrophilic** (electron-poor) site in another molecule. A bond is formed in a polar reaction when the nucleophile donates an electron pair to the electrophile. This movement of electrons is indicated by a curved arrow showing the direction of electron travel from the nucleophile to

exothermic, 154
Gibbs free-energy change (ΔG), 153
heat of reaction, 154
nucleophile, 145
polar reaction, 139
radical, 139
radical reaction, 139
reaction intermediate, 160
reaction mechanism, 139
rearrangement reaction, 138
substitution reaction, 138
transition state, 158

the electrophile. Radical reactions involve species that have an odd number of electrons. A bond is formed when each reactant donates one electron.

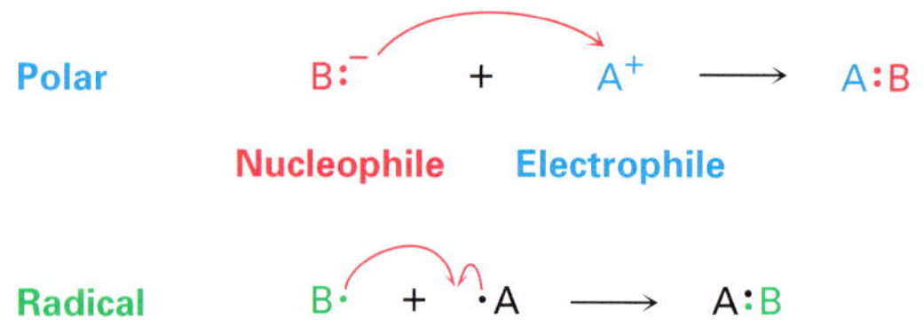

The energy changes that take place during reactions can be described by considering both rates (how fast the reactions occur) and equilibria (how much the reactions occur). The position of a chemical equilibrium is determined by the value of the **free-energy change (ΔG)** for the reaction, where $\Delta G = \Delta H - T\Delta S$. The **enthalpy** term (ΔH) corresponds to the net change in strength of chemical bonds broken and formed during reaction; the **entropy** term (ΔS) corresponds to the change in the amount of randomness during the reaction. Reactions that have negative values of ΔG release energy, are said to be **exergonic**, and have favorable equilibria. Reactions that have positive values of ΔG absorb energy, are said to be **endergonic**, and have unfavorable equilibria.

A reaction can be described pictorially using an energy diagram that follows the reaction course from reactant through transition state to product. The **transition state** is an activated complex occurring at the highest-energy point of a reaction. The amount of energy needed by reactants to reach this high point is the **activation energy, $\Delta G^{\ddagger}$**. The higher the activation energy, the slower the reaction.

Many reactions take place in more than one step and involve the formation of a **reaction intermediate**. An intermediate is a species that lies at an energy minimum between steps on the reaction curve and is formed briefly during the course of a reaction.

EXERCISES

Organic KNOWLEDGE TOOLS

ThomsonNOW™ Sign in at **www.thomsonedu.com** to assess your knowledge of this chapter's topics by taking a pre-test. The pre-test will link you to interactive organic chemistry resources based on your score in each concept area.

Online homework for this chapter may be assigned in Organic OWL.

■ indicates problems assignable in Organic OWL.

▲ denotes problems linked to Key Ideas of this chapter and testable in ThomsonNOW.

VISUALIZING CHEMISTRY

(Problems 5.1–5.13 appear within the chapter.)

5.14 ■ The following alkyl halide can be prepared by addition of HBr to two different alkenes. Draw the structures of both (reddish brown = Br).

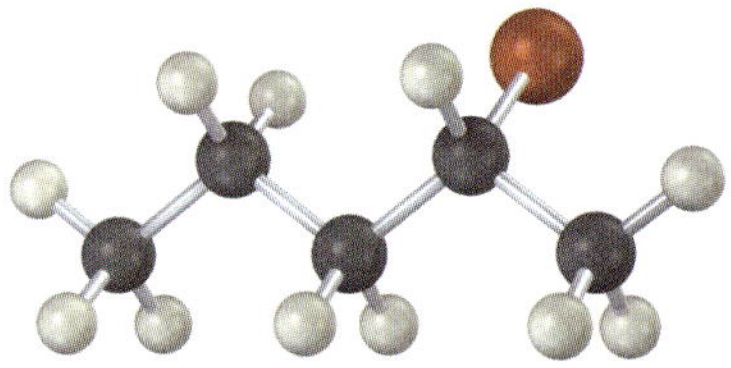

5.15 ■ The following structure represents the carbocation intermediate formed in the addition reaction of HBr to two different alkenes. Draw the structures of both.

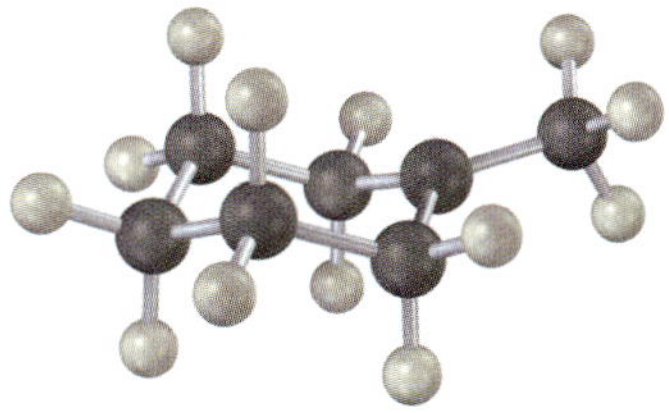

5.16 Electrostatic potential maps of **(a)** formaldehyde (CH_2O) and **(b)** methanethiol (CH_3SH) are shown. Is the formaldehyde carbon atom likely to be electrophilic or nucleophilic? What about the methanethiol sulfur atom? Explain.

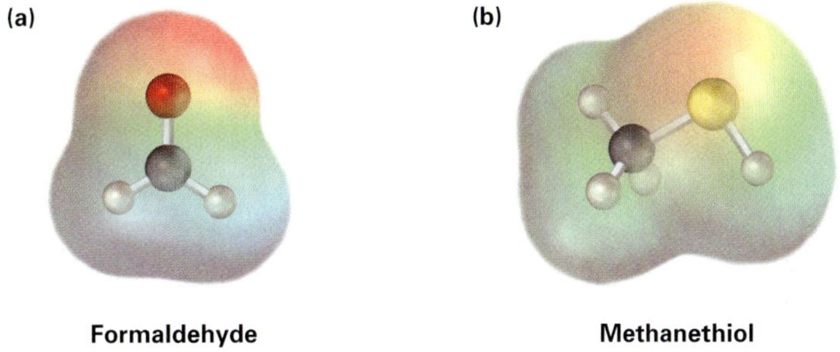

5.17 ■ Look at the following energy diagram:

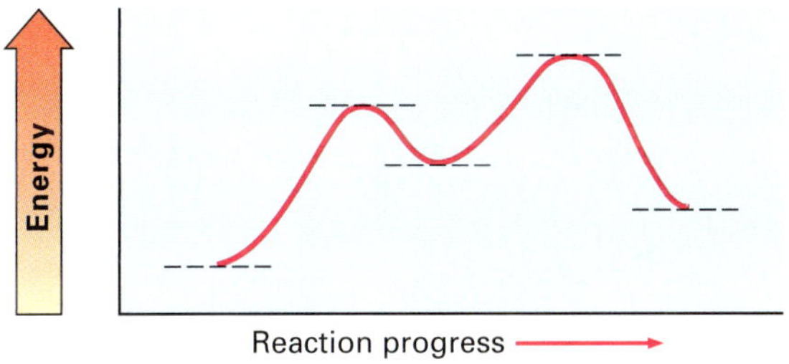

(a) Is $\Delta G°$ for the reaction positive or negative? Label it on the diagram.
(b) How many steps are involved in the reaction?
(c) How many transition states are there? Label them on the diagram.

5.18 Look at the following energy diagram for an enzyme-catalyzed reaction:

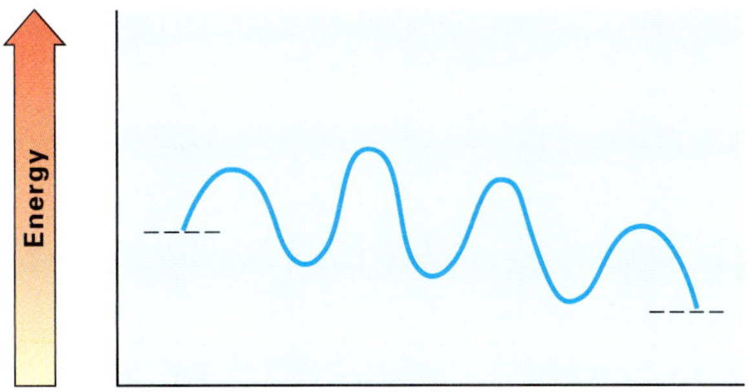

(a) How many steps are involved?
(b) Which step is most exergonic?
(c) Which step is the slowest?

ADDITIONAL PROBLEMS

5.19 ■ Identify the functional groups in the following molecules, and show th polarity of each:

(a) $CH_3CH_2C{\equiv}N$

(b) 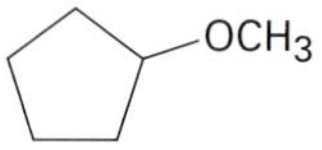

(c) $CH_3\overset{O}{\overset{\|}{C}}CH_2\overset{O}{\overset{\|}{C}}OCH_3$

(d) 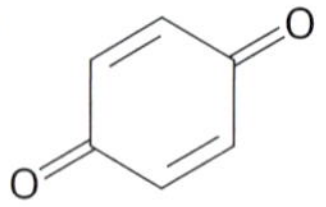

(e) 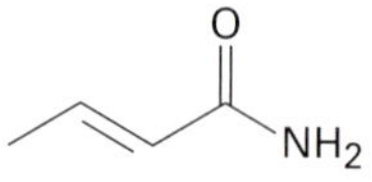

(f) 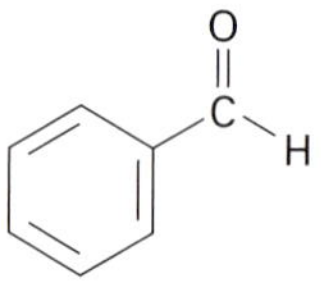

5.20 ■ Identify the following reactions as additions, eliminations, substitutions, o rearrangements:

(a) $CH_3CH_2Br + NaCN \longrightarrow CH_3CH_2CN\ (+\ NaBr)$

(b) 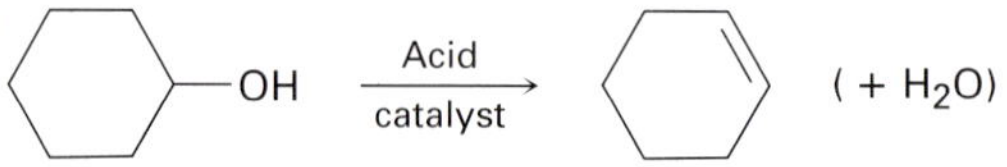

(c) 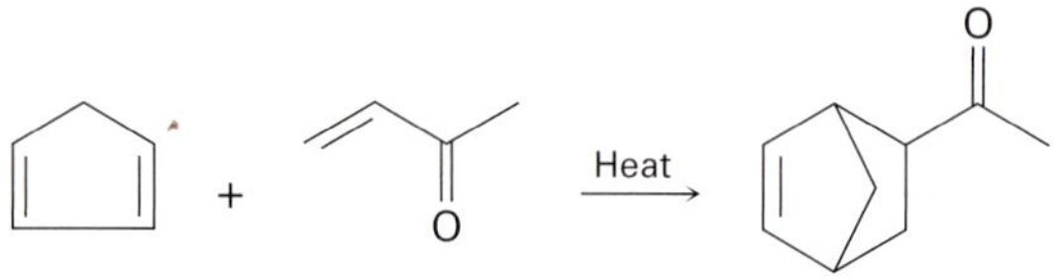

(d) 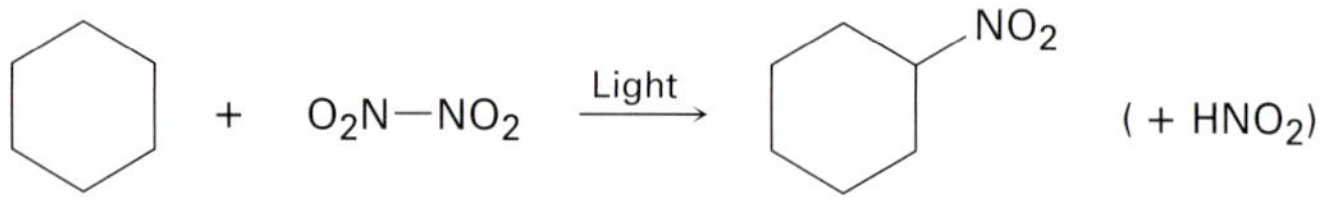

5.21 What is the difference between a transition state and an intermediate?

5.22 Draw an energy diagram for a one-step reaction with $K_{eq} < 1$. Label the part of the diagram corresponding to reactants, products, transition state, $\Delta G°$, an $\Delta G^{\ddagger}$. Is $\Delta G°$ positive or negative?

5.23 Draw an energy diagram for a two-step reaction with $K_{eq} > 1$. Label the ove all $\Delta G°$, transition states, and intermediate. Is $\Delta G°$ positive or negative?

5.24 Draw an energy diagram for a two-step exergonic reaction whose second ste is faster than its first step.

5.25 Draw an energy diagram for a reaction with $K_{eq} = 1$. What is the value of ΔG in this reaction?

5.26 ■ The addition of water to ethylene to yield ethanol has the following the modynamic parameters:

$$H_2C{=}CH_2 + H_2O \rightleftharpoons CH_3CH_2OH \quad \begin{cases} \Delta H° = -44 \text{ kJ/mol} \\ \Delta S° = -0.12 \text{ kJ/(K} \cdot \text{mol)} \\ K_{eq} = 24 \end{cases}$$

(a) Is the reaction exothermic or endothermic?

(b) Is the reaction favorable (spontaneous) or unfavorable (nonspontaneous at room temperature (298 K)?

■ Assignable in OWL ▲ Key Idea Problems

5.27 When a mixture of methane and chlorine is irradiated, reaction commences immediately. When irradiation is stopped, the reaction gradually slows down but does not stop immediately. Explain.

5.28 Radical chlorination of pentane is a poor way to prepare 1-chloropentane, but radical chlorination of neopentane, $(CH_3)_4C$, is a good way to prepare neopentyl chloride, $(CH_3)_3CCH_2Cl$. Explain.

5.29 ■ Despite the limitations of radical chlorination of alkanes, the reaction is still useful for synthesizing certain halogenated compounds. For which of the following compounds does radical chlorination give a single monochloro product?

(a) CH_3CH_3

(b) $CH_3CH_2CH_3$

(c)

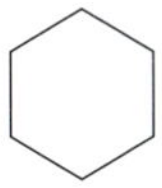

(d)

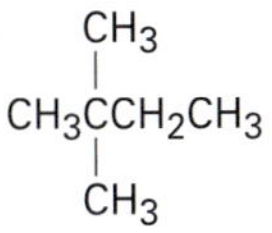

(e) $CH_3C{\equiv}CCH_3$

(f)

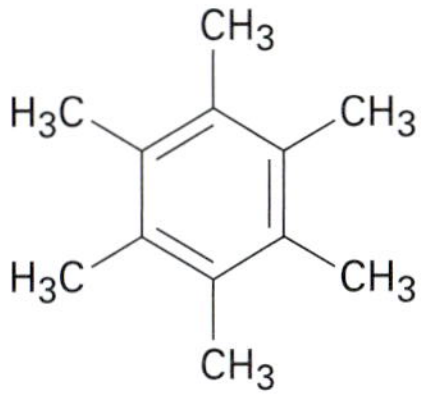

5.30 ■ ▲ Add curved arrows to the following reactions to indicate the flow of electrons in each:

(a)

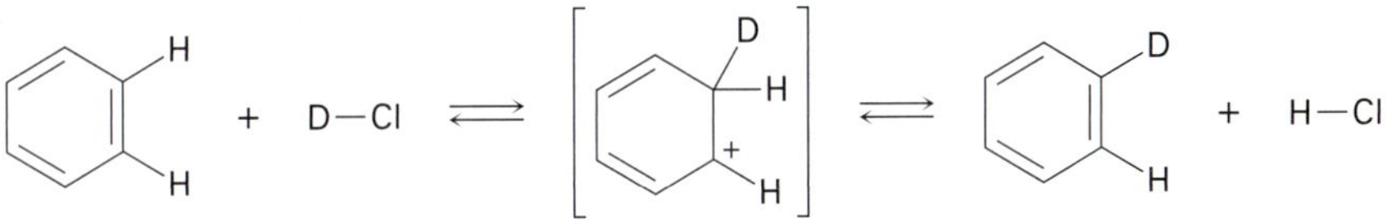

(b)

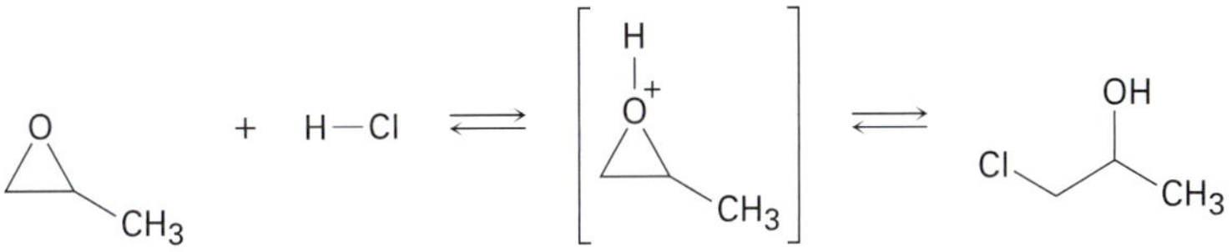

5.31 ■ ▲ Follow the flow of electrons indicated by the curved arrows in each of the following reactions, and predict the products that result:

(a)

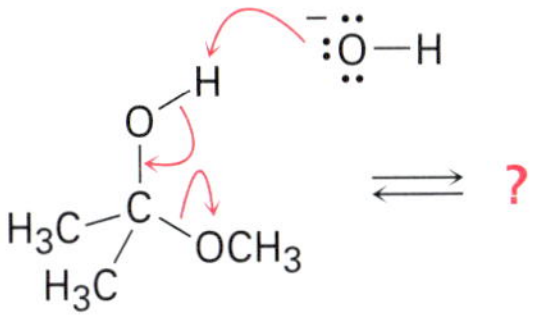

(b)

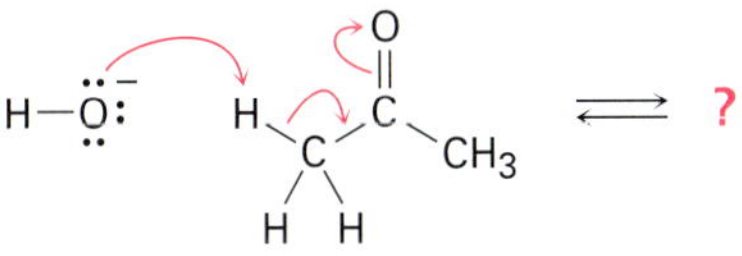

5.32 ■ When isopropylidenecyclohexane is treated with strong acid at room temperature, isomerization occurs by the mechanism shown below to yield 1-isopropylcyclohexene:

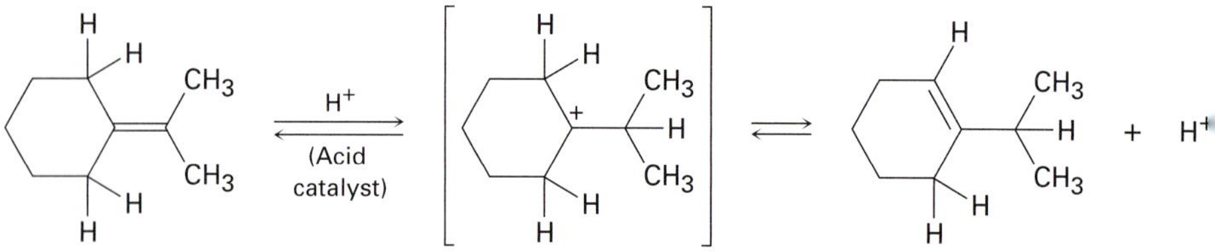

Isopropylidenecyclohexane **1-Isopropylcyclohexene**

At equilibrium, the product mixture contains about 30% isopropylidenecyclohexane and about 70% 1-isopropylcyclohexene.

(a) What is an approximate value of K_{eq} for the reaction?

(b) Since the reaction occurs slowly at room temperature, what is its approximate $\Delta G^{\ddagger}$?

(c) Draw an energy diagram for the reaction.

5.33 ▲ Add curved arrows to the mechanism shown in Problem 5.32 to indicate the electron movement in each step.

5.34 ■ 2-Chloro-2-methylpropane reacts with water in three steps to yield 2-methyl-2-propanol. The first step is slower than the second, which in turn is much slower than the third. The reaction takes place slowly at room temperature, and the equilibrium constant is near 1.

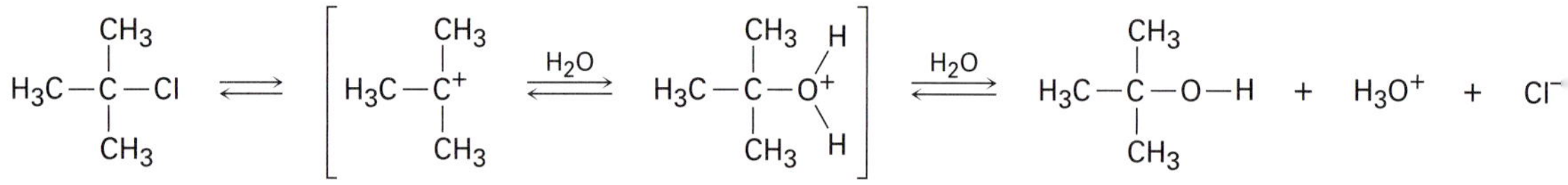

2-Chloro-2-methylpropane **2-Methyl-2-propanol**

(a) Give approximate values for $\Delta G^{\ddagger}$ and $\Delta G°$ that are consistent with the above information.

(b) Draw an energy diagram for the reaction, labeling all points of interest and making sure that the relative energy levels on the diagram are consistent with the information given.

5.35 ▲ Add curved arrows to the mechanism shown in Problem 5.34 to indicate the electron movement in each step.

5.36 ■ The reaction of hydroxide ion with chloromethane to yield methanol and chloride ion is an example of a general reaction type called a *nucleophilic substitution reaction:*

$$HO^- + CH_3Cl \rightleftharpoons CH_3OH + Cl^-$$

The value of $\Delta H°$ for the reaction is −75 kJ/mol, and the value of $\Delta S°$ is +54 J/(K · mol). What is the value of $\Delta G°$ (in kJ/mol) at 298 K? Is the reaction exothermic or endothermic? Is it exergonic or endergonic?

5.37 ■ ▲ Ammonia reacts with acetyl chloride (CH_3COCl) to give acetamide (CH_3CONH_2). Identify the bonds broken and formed in each step of the reaction, and draw curved arrows to represent the flow of electrons in each step.

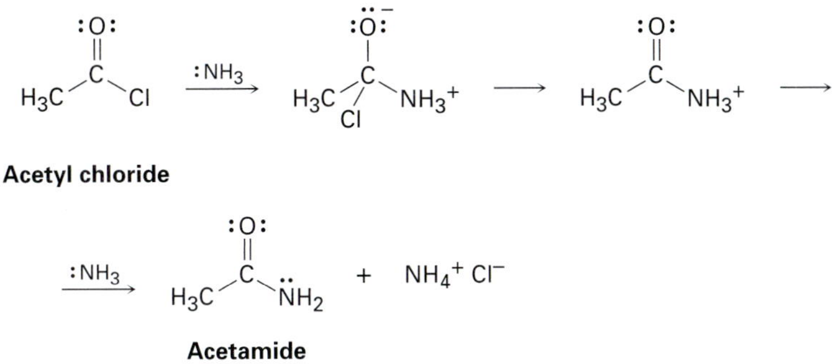

5.38 The naturally occurring molecule α-terpineol is biosynthesized by a route that includes the following step:

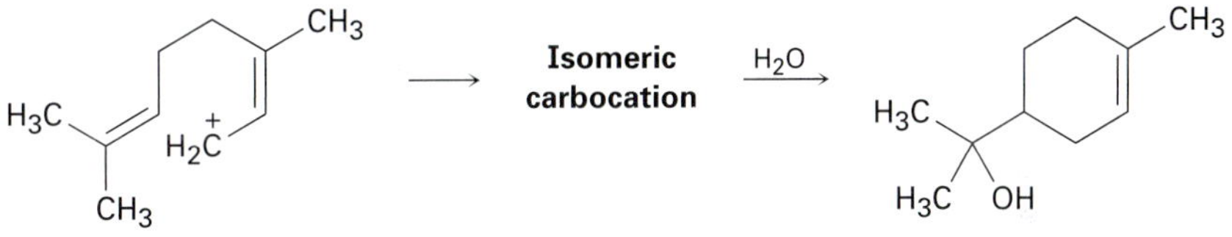

(a) Propose a likely structure for the isomeric carbocation intermediate.
(b) Show the mechanism of each step in the biosynthetic pathway, using curved arrows to indicate electron flow.

5.39 Predict the product(s) of each of the following biological reactions by interpreting the flow of electrons as indicated by the curved arrows:

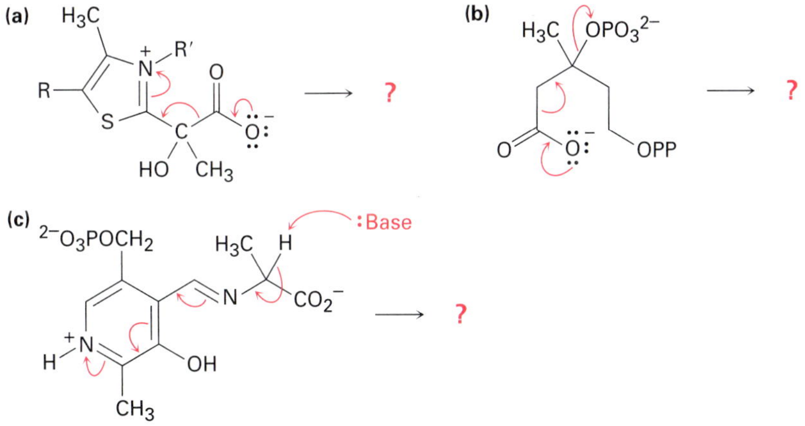

5.40 ■ Reaction of 2-methylpropene with HBr might, in principle, lead to a mixture of two alkyl bromide addition products. Name them, and draw their structures.

5.41 ■ Draw the structures of the two carbocation intermediates that might form during the reaction of 2-methylpropene with HBr (Problem 5.40). We'll see in the next chapter that the stability of carbocations depends on the number of alkyl substituents attached to the positively charged carbon—the more alkyl substituents there are, the more stable the cation. Which of the two carbocation intermediates you drew is more stable?

■ Assignable in OWL ▲ Key Idea Problems

6 Alkenes: Structure and Reactivity

Organic **KNOWLEDGE TOOLS**

ThomsonNOW™ Throughout this chapter, sign in at **www.thomsonedu.com** for online self-study and interactive tutorials based on your level of understanding.

Online homework for this chapter may be assigned in Organic OWL.

An **alkene**, sometimes called an *olefin,* is a hydrocarbon that contains carbon–carbon double bond. Alkenes occur abundantly in nature. Ethylene, fc instance, is a plant hormone that induces ripening in fruit, and α-pinene is th major component of turpentine. Life itself would be impossible without suc alkenes as β-carotene, a compound that contains 11 double bonds. An orang pigment responsible for the color of carrots, β-carotene is a valuable dietar source of vitamin A and is thought to offer some protection against certain type of cancer.

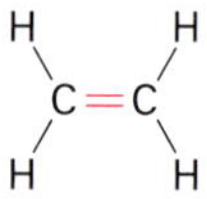

Ethylene

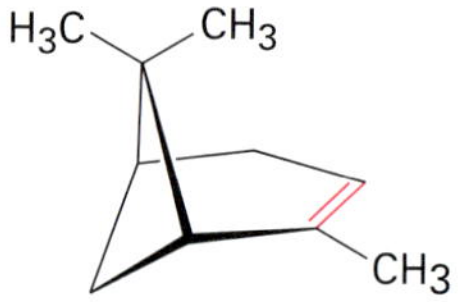

α-Pinene

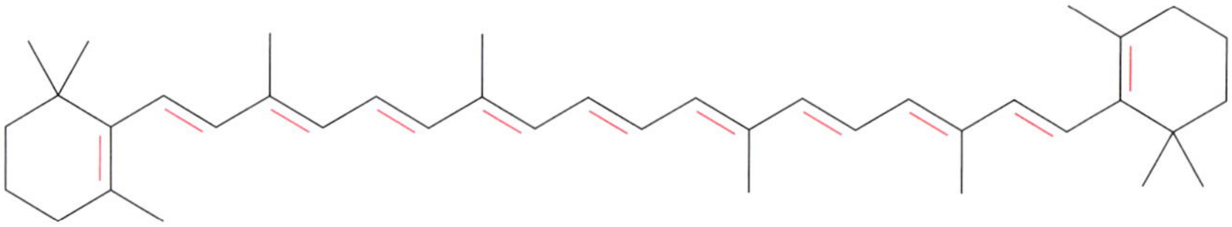

β-Carotene
(orange pigment and vitamin A precursor)

WHY THIS CHAPTER?

Carbon–carbon double bonds are present in most organic and biological mole cules, so a good understanding of their behavior is needed. In this chapter, we'l look at some consequences of alkene stereoisomerism and then focus on th broadest and most general class of alkene reactions, the electrophilic additio reaction.

6.1 Industrial Preparation and Use of Alkenes

Ethylene and propylene, the simplest alkenes, are the two most important organic chemicals produced industrially. Approximately 26 million tons of ethylene and 17 million tons of propylene are produced each year in the United States for use in the synthesis of polyethylene, polypropylene, ethylene glycol, acetic acid, acetaldehyde, and a host of other substances (Figure 6.1).

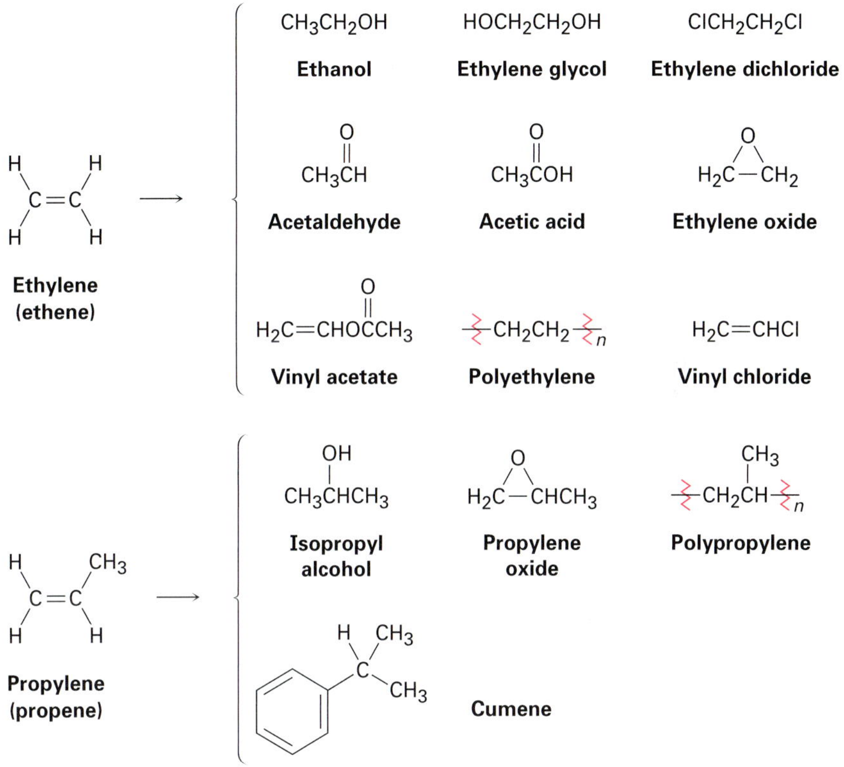

Figure 6.1 Compounds derived industrially from ethylene and propylene.

Ethylene, propylene, and butene are synthesized industrially by thermal cracking of light (C_2–C_8) alkanes.

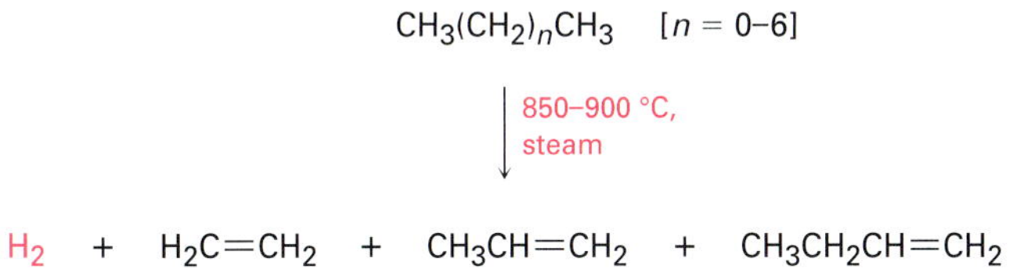

Thermal cracking takes place without a catalyst at temperatures up to 900 °C. The exact processes are complex, although they undoubtedly involve radical reactions. The high-temperature reaction conditions cause spontaneous homolytic breaking of C–C and C–H bonds, with resultant formation of smaller fragments. We might imagine, for instance, that a molecule of butane splits into two ethyl radicals, each of which then loses a hydrogen atom to generate two molecules of ethylene.

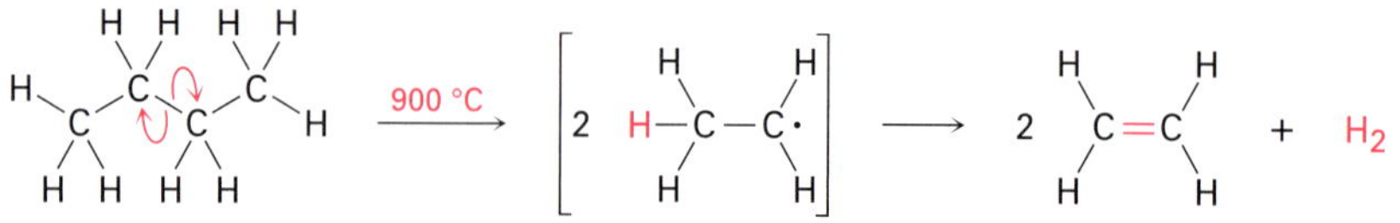

Thermal cracking is an example of a reaction whose energetics are dominated by entropy ($\Delta S°$) rather than by enthalpy ($\Delta H°$) in the free-energy equation $\Delta G° = \Delta H° - T\Delta S°$. Although the bond dissociation energy D for a carbon–carbon single bond is relatively high (about 375 kJ/mol) and cracking is highly endothermic, the large positive entropy change resulting from the fragmentation of one large molecule into several smaller pieces, together with the extremely high temperature, makes the $T\Delta S°$ term larger than the $\Delta H°$ term, thereby favoring the cracking reaction.

6.2 Calculating Degree of Unsaturation

ThomsonNOW™ Click *Organic Interactive* to **practice calculating degrees of unsaturation.**

Because of its double bond, an alkene has fewer hydrogens than an alkane with the same number of carbons—C_nH_{2n} for an alkene versus C_nH_{2n+2} for an alkane—and is therefore referred to as **unsaturated.** Ethylene, for example, has the formula C_2H_4, whereas ethane has the formula C_2H_6.

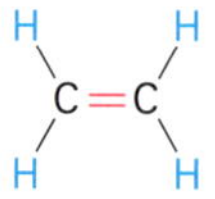

Ethylene: C_2H_4
(fewer hydrogens—*unsaturated*)

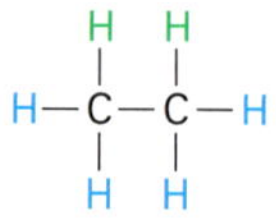

Ethane: C_2H_6
(more hydrogens—*saturated*)

In general, each ring or double bond in a molecule corresponds to a loss of two hydrogens from the alkane formula C_nH_{2n+2}. Knowing this relationship, it's possible to work backward from a molecular formula to calculate a molecule's **degree of unsaturation**—the number of rings and/or multiple bonds present in the molecule.

Let's assume that we want to find the structure of an unknown hydrocarbon. A molecular weight determination on the unknown yields a value of 82, which corresponds to a molecular formula of C_6H_{10}. Since the saturated C_6 alkane (hexane) has the formula C_6H_{14}, the unknown compound has two fewer pairs of hydrogens ($H_{14} - H_{10} = H_4 = 2\ H_2$), and its degree of unsaturation is two. The unknown therefore contains two double bonds, one ring and one double bond, two rings, or one triple bond. There's still a long way to go to establish structure, but the simple calculation has told us a lot about the molecule.

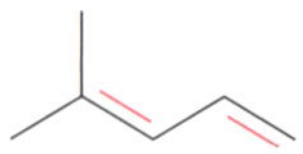

4-Methyl-1,3-pentadiene
(two double bonds)

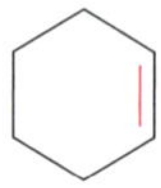

Cyclohexene
(one ring, one double bond)

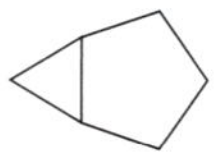

Bicyclo[3.1.0]hexane
(two rings)

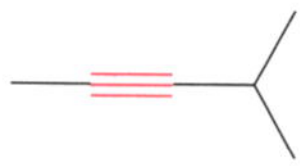

4-Methyl-2-pentyne
(one triple bond)

C_6H_{10}

Similar calculations can be carried out for compounds containing elements other than just carbon and hydrogen.

- **Organohalogen compounds (C, H, X, where X = F, Cl, Br, or I)** A halogen substituent acts simply as a replacement for hydrogen in an organic molecule, so we can add the number of halogens and hydrogens to arrive at an equivalent hydrocarbon formula from which the degree of unsaturation can be found. For example, the organohalogen formula $C_4H_6Br_2$ is equivalent to the hydrocarbon formula C_4H_8 and thus has one degree of unsaturation.

Replace 2 Br by 2 H

$$BrCH_2CH{=}CHCH_2Br \quad = \quad HCH_2CH{=}CHCH_2H$$

$$\underbrace{C_4H_6Br_2}_{\text{Add}} \quad = \quad \text{“}C_4H_8\text{”}$$

One unsaturation: one double bond

- **Organooxygen compounds (C, H, O)** Oxygen forms two bonds, so it doesn't affect the formula of an equivalent hydrocarbon and can be ignored when calculating the degree of unsaturation. You can convince yourself of this by seeing what happens when an oxygen atom is inserted into an alkane bond: C–C becomes C–O–C or C–H becomes C–O–H, and there is no change in the number of hydrogen atoms. For example, the formula C_5H_8O is equivalent to the hydrocarbon formula C_5H_8 and thus has two degrees of unsaturation.

O removed from here

$$H_2C{=}CHCH{=}CHCH_2OH \quad = \quad H_2C{=}CHCH{=}CHCH_2{-}H$$

$$C_5H_8O \quad = \quad \text{“}C_5H_8\text{”}$$

Two unsaturations: two double bonds

- **Organonitrogen compounds (C, H, N)** Nitrogen forms three bonds, so an organonitrogen compound has one more hydrogen than a related hydrocarbon; we therefore *subtract* the number of nitrogens from the number of hydrogens to arrive at the equivalent hydrocarbon formula. Again, you can convince yourself of this by seeing what happens when a nitrogen atom is inserted into an alkane bond: C–C becomes C–NH–C or C–H becomes C–NH_2, meaning that one additional hydrogen atom has been added. We must therefore subtract this extra hydrogen atom to arrive at the equivalent hydrocarbon formula. For example, the formula C_5H_9N is equivalent to C_5H_8 and thus has two degrees of unsaturation.

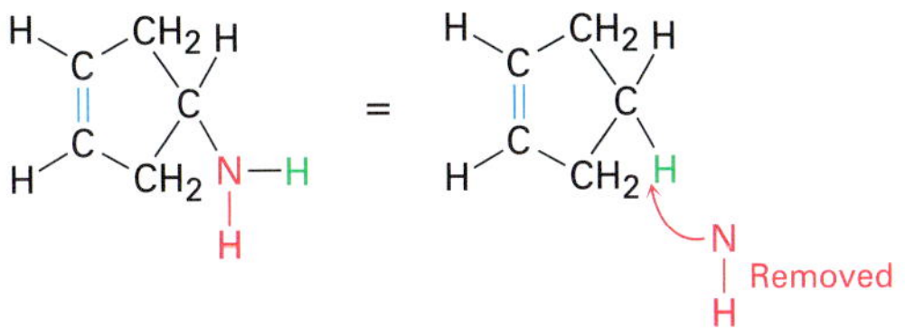

$$C_5H_9N \quad = \quad \text{“}C_5H_8\text{”}$$

Two unsaturations: one ring and one double bond

To summarize:

- **Add** the number of halogens to the number of hydrogens.
- **Ignore** the number of oxygens.
- **Subtract** the number of nitrogens from the number of hydrogens.

Problem 6.1 Calculate the degree of unsaturation in the following formulas, and then draw as many structures as you can for each:

(a) C_4H_8 (b) C_4H_6 (c) C_3H_4

Problem 6.2 Calculate the degree of unsaturation in the following formulas:

(a) C_6H_5N (b) $C_6H_5NO_2$ (c) $C_8H_9Cl_3$
(d) $C_9H_{16}Br_2$ (e) $C_{10}H_{12}N_2O_3$ (f) $C_{20}H_{32}ClN$

Problem 6.3 Diazepam, marketed as an antianxiety medication under the name Valium, has three rings, eight double bonds, and the formula $C_{16}H_?ClN_2O$. How many hydrogens does diazepam have? (Calculate the answer; don't count hydrogens in the structure.)

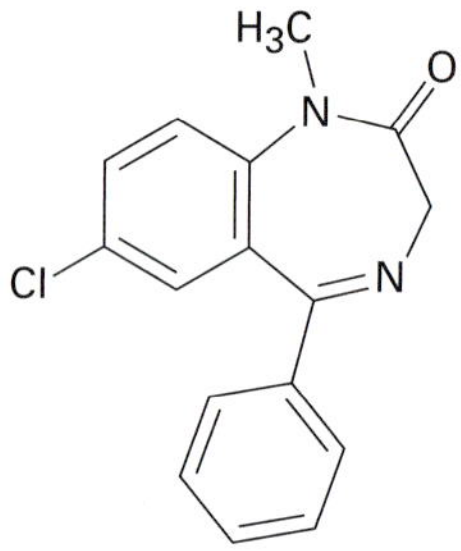

Diazepam

6.3 Naming Alkenes

ThomsonNOW™ Click *Organic Interactive* to **practice naming alkenes in this interactive problem set.**

Alkenes are named using a series of rules similar to those for alkanes (Section 3.4), with the suffix *-ene* used instead of *-ane* to identify the family. There are three steps:

Step 1 **Name the parent hydrocarbon.** Find the longest carbon chain containing the double bond, and name the compound accordingly, using the suffix *-ene*:

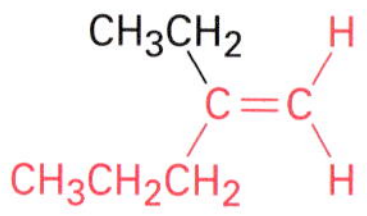

Named as a *pentene*

NOT

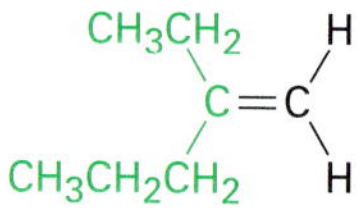

as a hexene, since the double bond is not contained in the six-carbon chain

Step 2 **Number the carbon atoms in the chain.** Begin at the end nearer the double bond or, if the double bond is equidistant from the two ends, begin at the end nearer the first branch point. This rule ensures that the double-bond carbons receive the lowest possible numbers.

$CH_3CH_2CH_2CH{=}CHCH_3$ (carbons numbered 6 5 4 3 2 1)

$CH_3CH(CH_3)CH{=}CHCH_2CH_3$ (carbons numbered 1 2 3 4 5 6)

Step 3 **Write the full name.** Number the substituents according to their positions in the chain, and list them alphabetically. Indicate the position of the double bond by giving the number of the first alkene carbon and placing that number directly before the parent name. If more than one double bond is present, indicate the position of each and use one of the suffixes *-diene*, *-triene*, and so on.

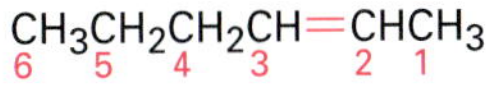

2-Hexene

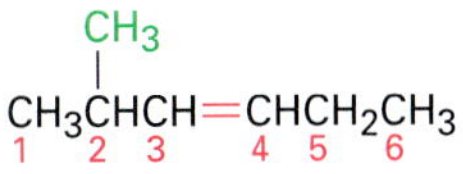

2-Methyl-3-hexene

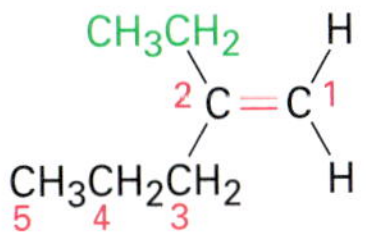

2-Ethyl-1-pentene

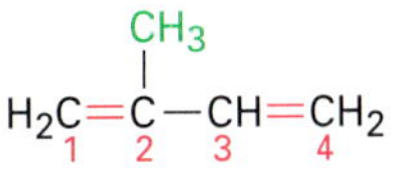

2-Methyl-1,3-butadiene

We should also note that IUPAC changed their naming recommendations in 1993 to place the locant indicating the position of the double bond immediately before the *-ene* suffix rather than before the parent name: but-2-ene rather than 2-butene, for instance. This change has not been widely accepted by the chemical community, however, so we'll stay with the older but more commonly used names. Be aware, though, that you may occasionally encounter the newer system.

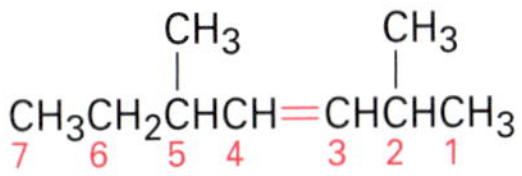

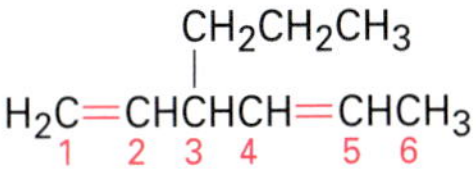

Older naming system:	**2,5-Dimethyl-3-heptene**	**3-Propyl-1,4-hexadiene**
(Newer naming system:	**2,5-Dimethylhept-3-ene**	**3-Propylhexa-1,4-diene)**

Cycloalkenes are named similarly to open-chain alkenes but, because there is no chain end to begin from, we number the cycloalkene so that the double bond is between C1 and C2 and the first substituent has as low a number as possible. Note that it's not necessary to indicate the position of the double bond in the name because it is always between C1 and C2. As with open-chain alkenes, newer but not yet widely accepted naming rules place the locant immediately before the suffix in a diene.

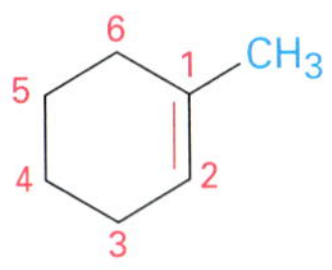

1-Methylcyclohexene

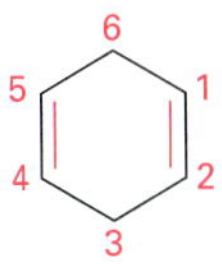

1,4-Cyclohexadiene
(New: Cyclohcxa-1,4-diene)

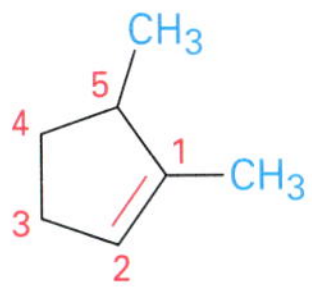

1,5-Dimethylcyclopentene

For historical reasons, there are a few alkenes whose names are firmly entrenched in common usage but don't conform to the rules. For example, the alkene derived from ethane should be called *ethene*, but the name *ethylene* has

ThomsonNOW™ Click *Organic Interactive* to **use a web-based palette to draw alkene structures based on their IUPAC names.**

been used so long that it is accepted by IUPAC. Table 6.1 lists several other con mon names that are often used and are recognized by IUPAC. Note also that $=CH_2$ substituent is called a **methylene group**, a $H_2C{=}CH-$ substituent called a **vinyl group**, and a $H_2C{=}CHCH_2-$ substituent is called an **allyl grou**

$H_2C{=}$	$H_2C{=}CH-$	$H_2C{=}CH-CH_2-$
A methylene group	**A vinyl group**	**An allyl group**

Table 6.1 Common Names of Some Alkenes

Compound	Systematic name	Common name
$H_2C{=}CH_2$	Ethene	Ethylene
$CH_3CH{=}CH_2$	Propene	Propylene
$CH_3C(CH_3){=}CH_2$	2-Methylpropene	Isobutylene
$H_2C{=}C(CH_3)-CH{=}CH_2$	2-Methyl-1,3-butadiene	Isoprene

Problem 6.4 Give IUPAC names for the following compounds:

(a) 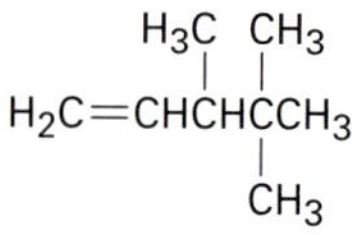

(b) 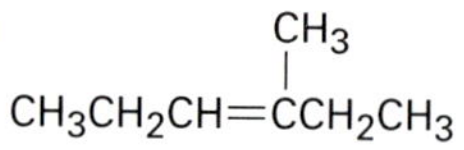

(c) 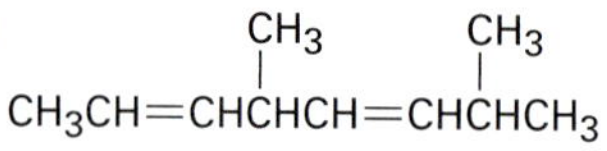

(d) 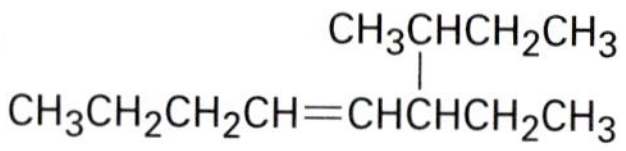

Problem 6.5 Draw structures corresponding to the following IUPAC names:

(a) 2-Methyl-1,5-hexadiene (b) 3-Ethyl-2,2-dimethyl-3-heptene
(c) 2,3,3-Trimethyl-1,4,6-octatriene (d) 3,4-Diisopropyl-2,5-dimethyl-3-hexene

Problem 6.6 Name the following cycloalkenes:

(a) 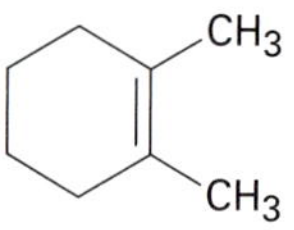

(b) 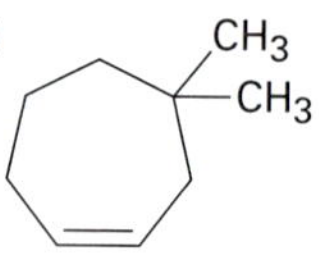

(c) 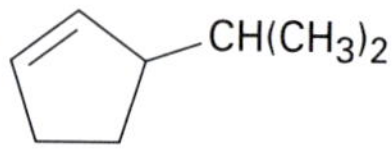

6.4 Cis–Trans Isomerism in Alkenes

We saw in Chapter 1 that the carbon–carbon double bond can be described in two ways. In valence bond language (Section 1.8), the carbons are sp^2-hybridized and have three equivalent hybrid orbitals that lie in a plane at angles of 120° to one another. The carbons form a σ bond by head-on overlap of sp^2 orbitals and a π bond by sideways overlap of unhybridized p orbitals oriented perpendicular to the sp^2 plane, as shown in Figure 1.14 on page 16. In molecular orbital language (Section 1.11), interaction between the p orbitals leads to one bonding and one antibonding π molecular orbital. The π bonding MO has no node between nuclei and results from a combination of p orbital lobes with the same algebraic sign. The π antibonding MO has a node between nuclei and results from a combination of lobes with different algebraic signs, as shown in Figure 1.18, page 22.

Although essentially free rotation is possible around single bonds (Section 3.6), the same is not true of double bonds. For rotation to occur around a double bond, the π bond must break and re-form (Figure 6.2). Thus, the barrier to double-bond rotation must be at least as great as the strength of the π bond itself, an estimated 350 kJ/mol (84 kcal/mol). Recall that the barrier to bond rotation in ethane is only 12 kJ/mol.

Figure 6.2 The π bond must break for rotation to take place around a carbon–carbon double bond.

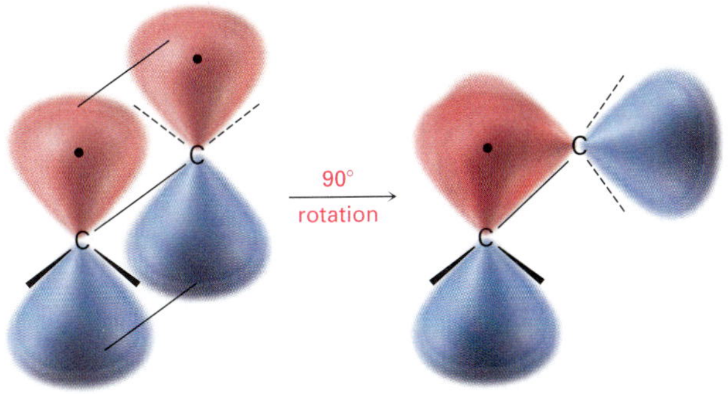

π bond
(p orbitals are parallel)

Broken π bond after rotation
(p orbitals are perpendicular)

The lack of rotation around carbon–carbon double bonds is of more than just theoretical interest; it also has chemical consequences. Imagine the situation for a disubstituted alkene such as 2-butene. (*Disubstituted* means that two substituents other than hydrogen are bonded to the double-bond carbons.) The two methyl groups in 2-butene can be either on the same side of the double bond or on opposite sides, a situation similar to that in disubstituted cycloalkanes (Section 4.2).

Since bond rotation can't occur, the two 2-butenes can't spontaneously interconvert; they are different, isolable compounds. As with disubstituted cycloalkanes, we call such compounds *cis–trans stereoisomers.* The compound with substituents on the same side of the double bond is called *cis*-2-butene, and the isomer with substituents on opposite sides is *trans*-2-butene (Figure 6.3).

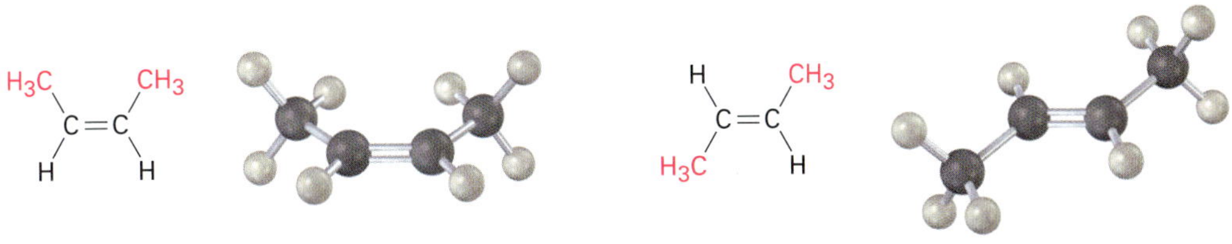

***cis*-2-Butene** ***trans*-2-Butene**

Figure 6.3 Cis and trans isomers of 2-butene. The cis isomer has the two methyl groups on the same side of the double bond, and the trans isomer has the methyl groups on opposite sides.

Cis–trans isomerism is not limited to *di*substituted alkenes. It can occu whenever both double-bond carbons are attached to two different groups. If on of the double-bond carbons is attached to two identical groups, however, the cis–trans isomerism is not possible (Figure 6.4).

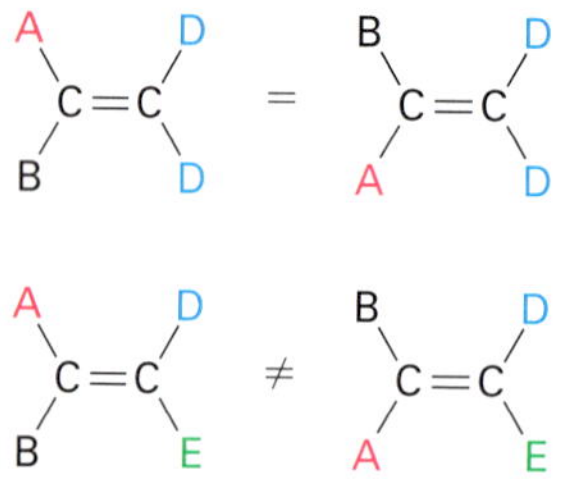

These two compounds are identical; they are not cis–trans isomers.

These two compounds are not identical; they are cis–trans isomers.

Figure 6.4 The requirement for cis–trans isomerism in alkenes. Compounds that have one of their carbons bonded to two identical groups can't exist as cis–trans isomers. Only when both carbons are bonded to two different groups are cis–trans isomers possible.

Problem 6.7 Which of the following compounds can exist as pairs of cis–trans isomers? Drav each cis–trans pair, and indicate the geometry of each isomer.

(a) $CH_3CH{=}CH_2$ (b) $(CH_3)_2C{=}CHCH_3$

(c) $CH_3CH_2CH{=}CHCH_3$ (d) $(CH_3)_2C{=}C(CH_3)CH_2CH_3$

(e) $ClCH{=}CHCl$ (f) $BrCH{=}CHCl$

Problem 6.8 Name the following alkenes, including the cis or trans designation:

(a)

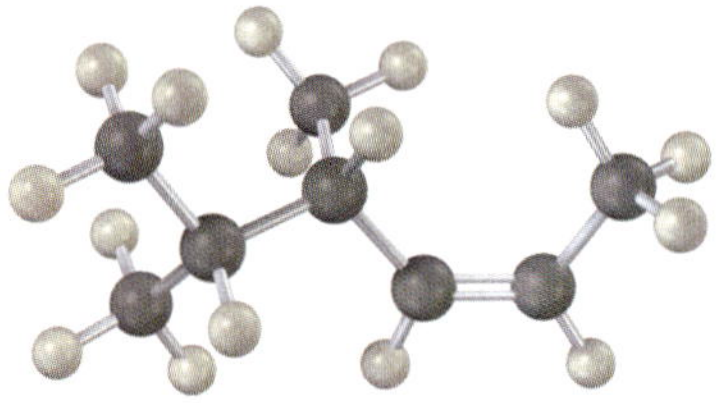

(b)

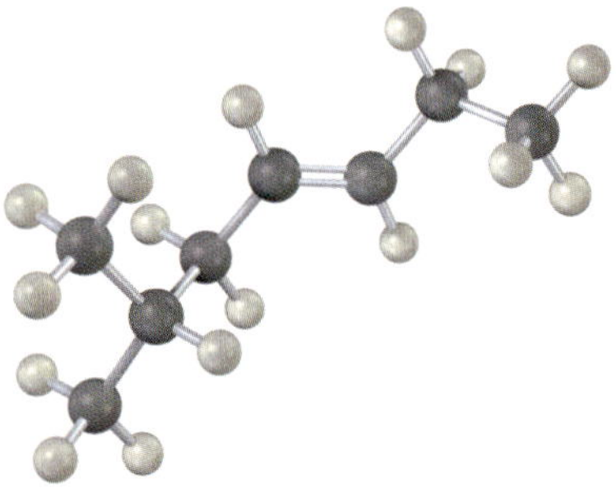

6.5 Sequence Rules: The *E,Z* Designation

Key IDEAS

Test your knowledge of Key Ideas by using resources in ThomsonNOW or by answering end-of-chapter problems marked with ▲.

The cis–trans naming system used in the previous section works only with *disubstituted* alkenes—compounds that have two substituents other than hydrogen on the double bond. With trisubstituted and tetrasubstituted double bonds, a more general method is needed for describing double-bond geometry. (*Trisubstituted* means three substituents other than hydrogen on the double bond; *tetrasubstituted* means four substituents other than hydrogen.)

According to the ***E,Z* system** of nomenclature, a set of *sequence rules* is used to assign priorities to the substituent groups on the double-bond carbons. Considering each doubly bonded carbon atom separately, the sequence rules are used to decide which of the two attached groups is higher in priority. If the higher-priority groups on each carbon are on the same side of the double bond, the alkene is designated ***Z***, for the German *zusammen,* meaning "together." If the higher-priority groups are on opposite sides, the alkene is designated ***E***, for omsonNOW™ Click *Organic* *teractive* to **practice assigning iorities to groups according to e Cahn–Ingold–Prelog rules.**

the German *entgegen,* meaning "opposite." (A simple way to remember which is which is to note that the groups are on "ze zame zide" in the *Z* isomer.)

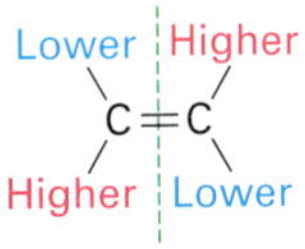

E double bond
(Higher-priority groups are on opposite sides.)

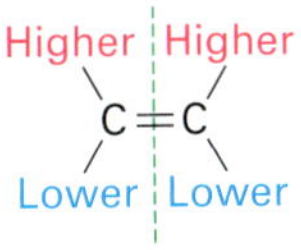

Z double bond
(Higher-priority groups are on the same side.)

Called the *Cahn–Ingold–Prelog rules* after the chemists who proposed them, the sequence rules are as follows:

Rule 1 **Considering the double-bond carbons separately, look at the two atoms directly attached to each and rank them according to atomic number.** An atom with higher atomic number receives higher priority than an atom with lower atomic number. Thus, the atoms commonly found attached to a double bond are assigned the following order. Note that when different isotopes of the same element are compared, such as deuterium (^{2}H) and protium (^{1}H), the heavier isotope receives priority over the lighter isotope.

35		17		16		15		8		7		6		(2)		(1)
Br	>	Cl	>	S	>	P	>	O	>	N	>	C	>	^{2}H	>	^{1}H

Robert Sidney Cahn	Sir Christopher Kelk Ingold	Vladimir Prelog
Robert Sidney Cahn (1899–1981) was born in England and received a doctoral degree in France. Although not specifically trained as a chemist, he became editor of the British *Journal of the Chemical Society.*	**Sir Christopher Kelk Ingold** (1893–1970) was born in Ilford, England, and received his D.Sc. at the University of London. After 6 years as professor at the University of Leeds, he spent his remaining career at University College, London (1930–1961). Ingold published more than 400 scientific papers and, along with Linus Pauling, was instrumental in developing the theory of resonance.	**Vladimir Prelog** (1906–1998) was born in Sarajevo, Bosnia, where, as a young boy, he was close enough to hear the shots that killed Archduke Ferdinand and ignited World War I. After receiving a Dr.Ing. degree in 1929 at the Institute of Technology in Prague, Czechoslovakia, he taught briefly at the University of Zagreb before becoming professor of chemistry at the Swiss Federal Institute of Technology (ETH) in Zürich (1941–1976). He received the 1975 Nobel Prize in chemistry for his lifetime achievements on the stereochemistry of antibiotics, alkaloids, enzymes, and other naturally occurring molecules.

For example:

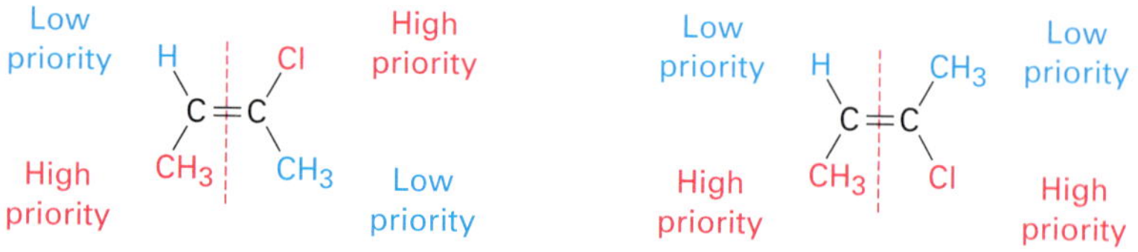

(a) (*E*)-2-Chloro-2-butene **(b) (*Z*)-2-Chloro-2-butene**

Because chlorine has a higher atomic number than carbon, a $-Cl$ substituent receives higher priority than a $-CH_3$ group. Methyl receives higher priority than hydrogen, however, and isomer (a) is assigned *E* geometry because its high-priority groups are on opposite sides of the double bond. Isomer (b) has *Z* geometry because its high-priority groups are on "ze zame zide" of the double bond.

Rule 2 **If a decision can't be reached by ranking the first atoms in the substituent, look at the second, third, or fourth atoms away from the double-bond carbons until the first difference is found.** A $-CH_2CH_3$ substituent and a $-CH_3$ substituent are equivalent by rule 1 because both have carbon as the first atom. By rule 2, however, ethyl receives higher priority than methyl because ethyl has a *carbon* as its highest second atom, while methyl has only *hydrogen* as its second atom. Look at the following examples to see how the rule works:

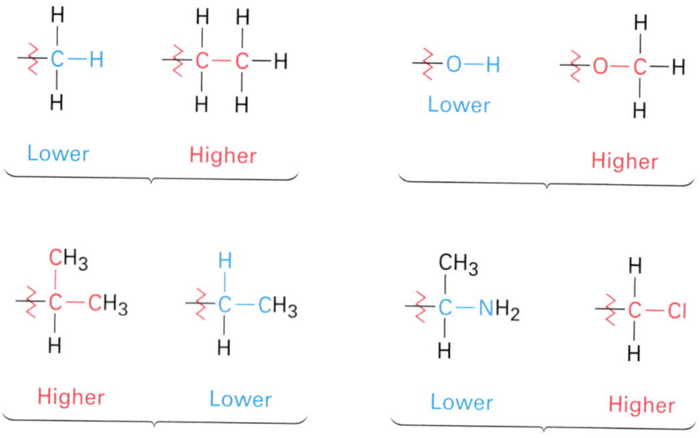

Rule 3 **Multiple-bonded atoms are equivalent to the same number of single-bonded atoms.** For example, an aldehyde substituent ($-CH{=}O$), which has a carbon atom *doubly* bonded to *one* oxygen, is equivalent to a substituent having a carbon atom *singly* bonded to *two* oxygens.

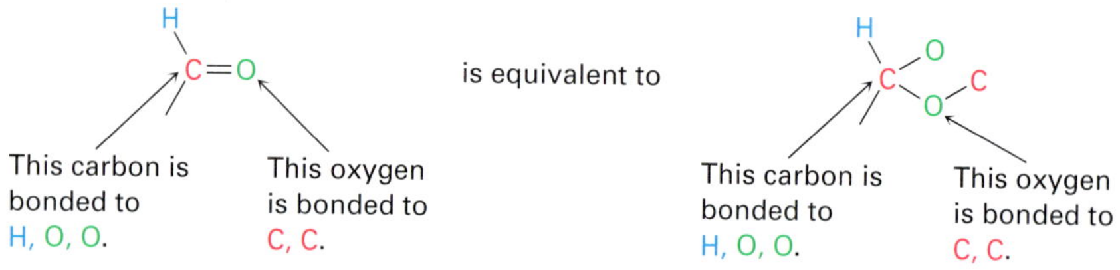

As further examples, the following pairs are equivalent:

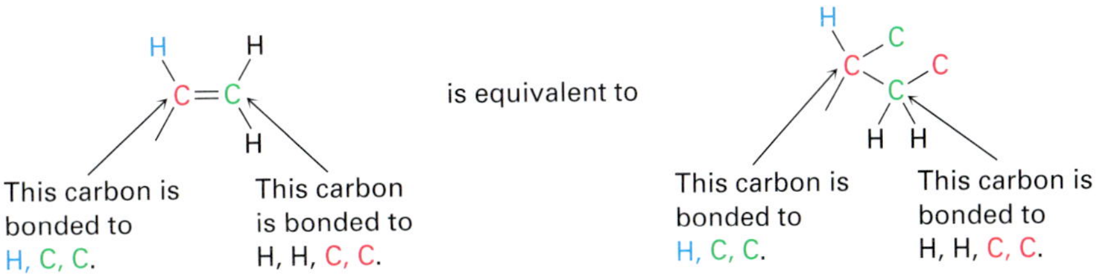

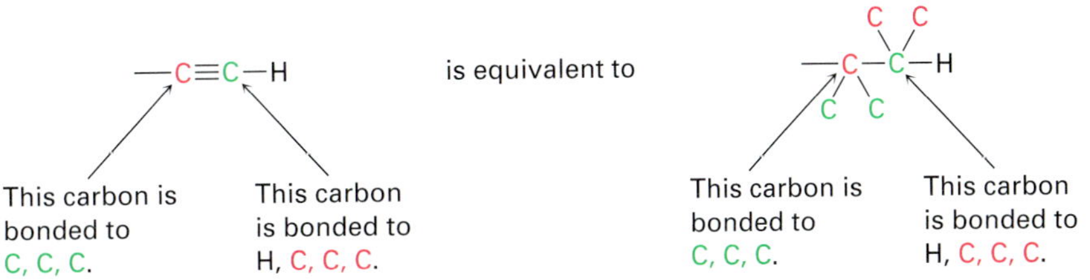

Taking all the sequence rules into account, we can assign the configurations shown in the following examples. Work through each one to convince yourself that the assignments are correct.

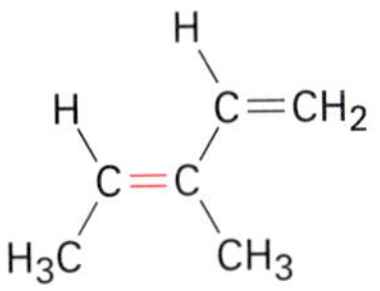

(*E*)-3-Methyl-1,3-pentadiene

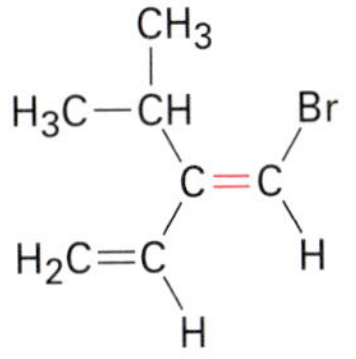

(*E*)-1-Bromo-2-isopropyl-1,3-butadiene

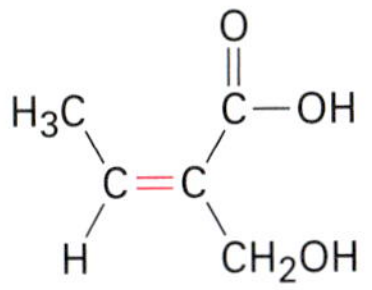

(*Z*)-2-Hydroxymethyl-2-butenoic acid

WORKED EXAMPLE 6.1

Assigning E and Z Configurations to Substituted Alkenes

Assign *E* or *Z* configuration to the double bond in the following compound:

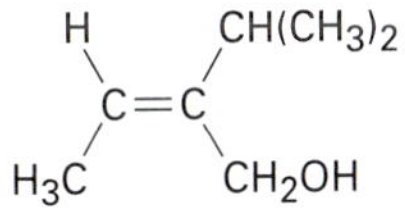

Strategy Look at the two substituents connected to each double-bond carbon, and determine their priorities using the Cahn–Ingold–Prelog rules. Then see whether the two high-priority groups are on the same or opposite sides of the double bond.

Solution The left-hand carbon has $-H$ and $-CH_3$ substituents, of which $-CH_3$ receives higher priority by sequence rule 1. The right-hand carbon has $-CH(CH_3)_2$ and $-CH_2OH$ substituents, which are equivalent by rule 1. By rule 2, however, $-CH_2OH$ receives higher priority than $-CH(CH_3)_2$. The substituent $-CH_2OH$ has an *oxygen* as its highest second atom, but $-CH(CH_3)_2$ has a *carbon* as its highest second atom The two high-priority groups are on the same side of the double bond, so we assign *Z* configuration.

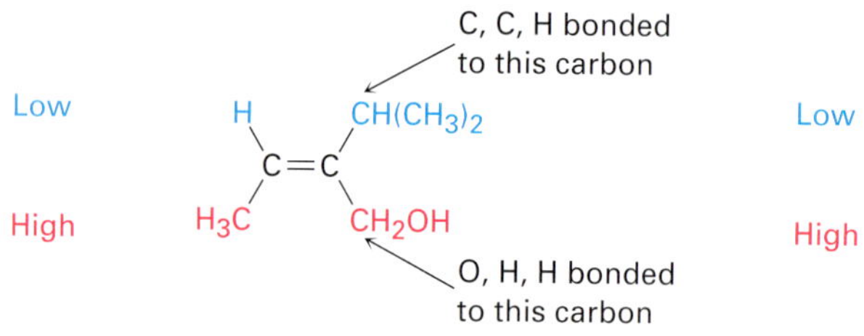

***Z* configuration**

Problem 6.9 Which member in each of the following sets has higher priority?

(a) $-H$ or $-Br$ (b) $-Cl$ or $-Br$ (c) $-CH_3$ or $-CH_2CH_3$
(d) $-NH_2$ or $-OH$ (e) $-CH_2OH$ or $-CH_3$ (f) $-CH_2OH$ or $-CH{=}O$

Problem 6.10 Rank the following sets of substituents in order of Cahn–Ingold–Prelog priorities:

(a) $-CH_3$, $-OH$, $-H$, $-Cl$
(b) $-CH_3$, $-CH_2CH_3$, $-CH{=}CH_2$, $-CH_2OH$
(c) $-CO_2H$, $-CH_2OH$, $-C{\equiv}N$, $-CH_2NH_2$
(d) $-CH_2CH_3$, $-C{\equiv}CH$, $-C{\equiv}N$, $-CH_2OCH_3$

Problem 6.11 Assign *E* or *Z* configuration to the following alkenes:

(a)
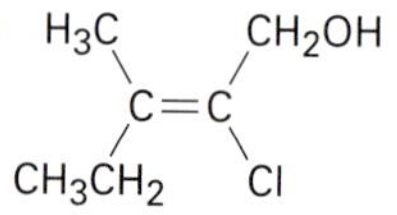

(b)
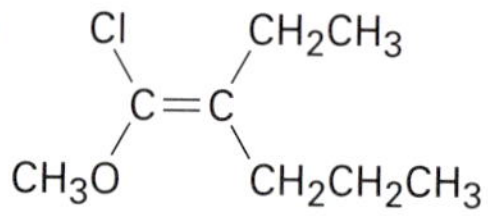

(c)
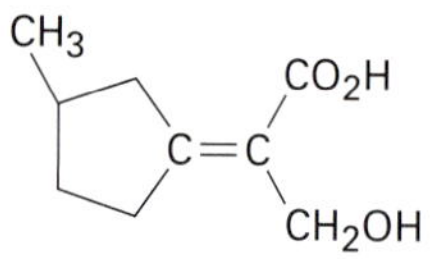

(d)
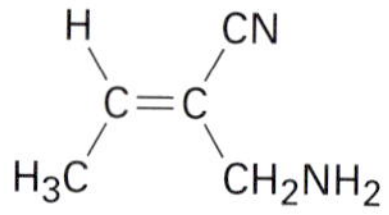

Problem 6.12 Assign stereochemistry (*E* or *Z*) to the double bond in the following compound, and convert the drawing into a skeletal structure (red = O):

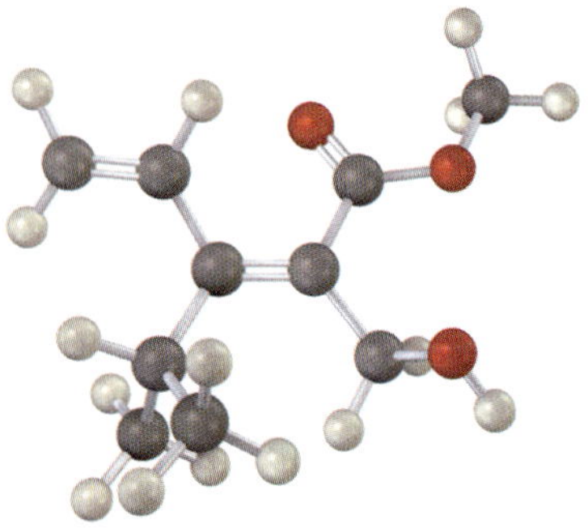

6.6 Stability of Alkenes

Although the cis–trans interconversion of alkene isomers does not occur spontaneously, it can often be brought about by treating the alkene with a strong acid catalyst. If we interconvert *cis*-2-butene with *trans*-2-butene and allow them to reach equilibrium, we find that they aren't of equal stability. The trans isomer is more stable than the cis isomer by 2.8 kJ/mol (0.66 kcal/mol) at room temperature, leading to a 76:24 ratio.

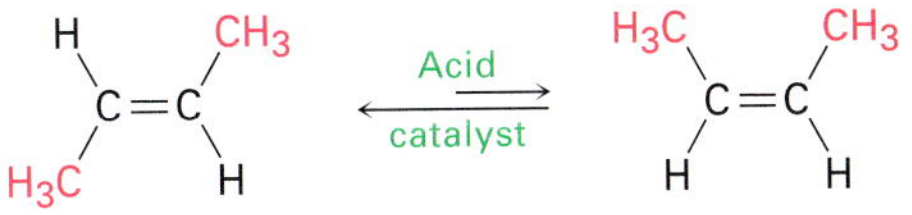

Trans (76%) **Cis (24%)**

Using the relationship between equilibrium constant and free energy shown previously in Figure 4.12, p. 122, we can calculate that *cis*-2-butene is less stable than *trans*-2-butene by 2.8 kJ/mol (0.66 kcal/mol) at room temperature.

Cis alkenes are less stable than their trans isomers because of steric strain between the two larger substituents on the same side of the double bond. This is the same kind of steric interference that we saw previously in the axial conformation of methylcyclohexane (Section 4.7).

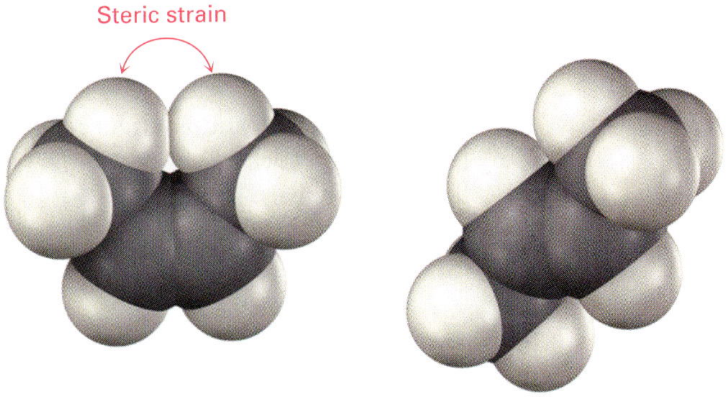

***cis*-2-Butene** ***trans*-2-Butene**

Although it's sometimes possible to find relative stabilities of alkene isomers by establishing a cis–trans equilibrium through treatment with strong acid, a more general method is to take advantage of the fact that alkenes undergo a *hydrogenation* reaction to give the corresponding alkane on treatment with H_2 gas in the presence of a catalyst such as palladium or platinum.

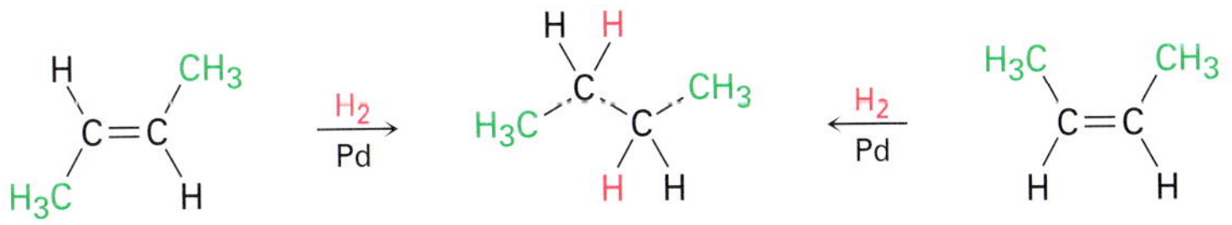

***trans*-2-Butene** **Butane** ***cis*-2-Butene**

Energy diagrams for the hydrogenation reactions of *cis*- and *trans*-2-buter are shown in Figure 6.5. Since *cis*-2-butene is less stable than *trans*-2-butene b 2.8 kJ/mol, the energy diagram shows the cis alkene at a higher energy leve After reaction, however, both curves are at the same energy level (butane). therefore follows that $\Delta G°$ for reaction of the cis isomer must be larger than Δ(for reaction of the trans isomer by 2.8 kJ/mol. In other words, more energy released in the hydrogenation of the cis isomer than the trans isomer becaus the cis isomer has more energy to begin with.

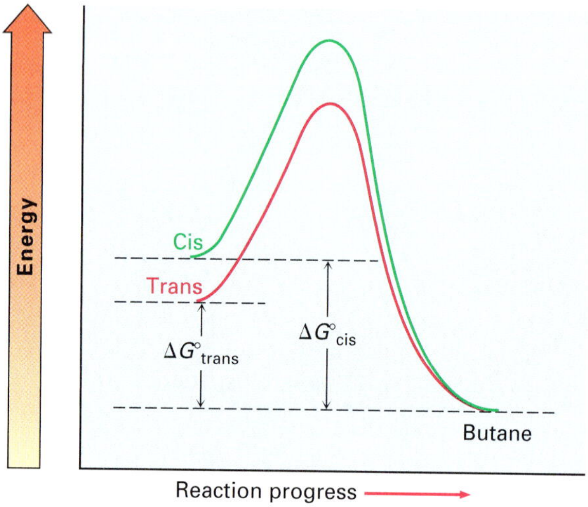

Figure 6.5 Energy diagrams for hydrogenation of *cis*- and *trans*-2-butene. The cis isomer is higher in energy than the trans isomer by about 2.8 kJ/mol and therefore releases more energy in the reaction.

If we were to measure what are called *heats of hydrogenation* ($\Delta H°_{hydrog}$) fo the two double-bond isomers and find their difference, we could determine the relative stabilities of cis and trans isomers without having to measure an equi librium position. In fact, the results bear out our expectation. For *cis*-2-butene $\Delta H°_{hydrog} = -120$ kJ/mol (−28.6 kcal/mol); for the trans isomer, $\Delta H°_{hydrog}$ = −116 kJ/mol (−27.6 kcal/mol).

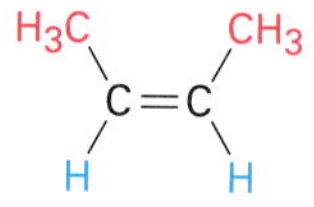

Cis isomer
$\Delta H°_{hydrog}$ = −120 kJ/mol

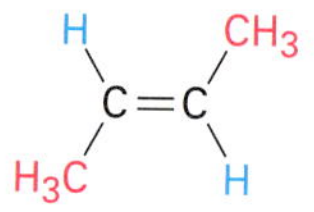

Trans isomer
$\Delta H°_{hydrog}$ = −116 kJ/mol

The energy difference between the 2-butene isomers as calculated from heats of hydrogenation (4 kJ/mol) agrees reasonably well with the energy difference calculated from equilibrium data (2.8 kJ/mol), but the numbers aren't exactly the same for two reasons. First, there is probably some experimental error, since heats of hydrogenation require skill and specialized equipment to measure accurately. Second, heats of reaction and equilibrium constants don't measure exactly the same thing. Heats of reaction measure enthalpy changes, $\Delta H°$, whereas equilibrium constants measure free-energy changes, $\Delta G°$, so we might expect a slight difference between the two.

Table 6.2 lists some representative data for the hydrogenation of different alkenes, showing that alkenes become more stable with increasing substitution.

For example, ethylene has $\Delta H°_{hydrog} = -137$ kJ/mol (-32.8 kcal/mol), but when one alkyl substituent is attached to the double bond, as in 1-butene, the alkene becomes approximately 10 kJ/mol more stable ($\Delta H°_{hydrog} = -126$ kJ/mol). Further increasing the degree of substitution leads to still further stability. As a general rule, alkenes follow the stability order:

Tetrasubstituted > Trisubstituted > Disubstituted > Monosubstituted

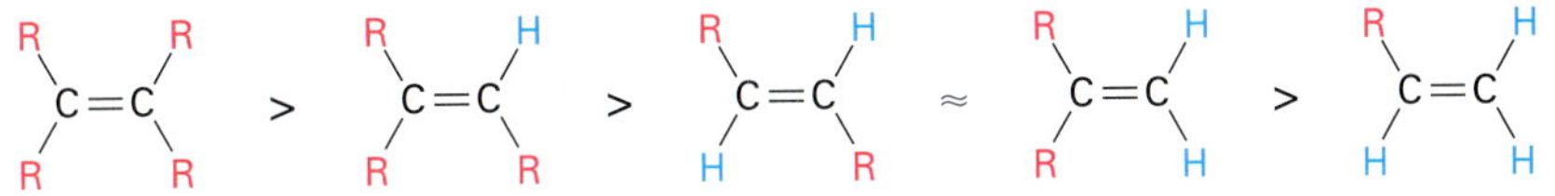

Table 6.2 Heats of Hydrogenation of Some Alkenes

		$\Delta H°_{hydrog}$	
Substitution	**Alkene**	**(kJ/mol)**	**(kcal/mol)**
Ethylene	$H_2C{=}CH_2$	−137	−32.8
Monosubstituted	$CH_3CH{=}CH_2$	−126	−30.1
Disubstituted	$CH_3CH{=}CHCH_3$ (cis)	−120	−28.6
	$CH_3CH{=}CHCH_3$ (trans)	−116	−27.6
	$(CH_3)_2C{=}CH_2$	−119	−28.4
Trisubstituted	$(CH_3)_2C{=}CHCH_3$	−113	−26.9
Tetrasubstituted	$(CH_3)_2C{=}C(CH_3)_2$	−111	−26.6

The stability order of alkenes is due to a combination of two factors. One is a stabilizing interaction between the C=C π bond and adjacent C–H σ bonds on substituents. In valence-bond language, the interaction is called **hyperconjugation**. In a molecular orbital description, there is a bonding MO that extends over the four-atom C=C–C–H grouping, as shown in Figure 6.6. The more substituents that are present on the double bond, the more hyperconjugation there is and the more stable the alkene.

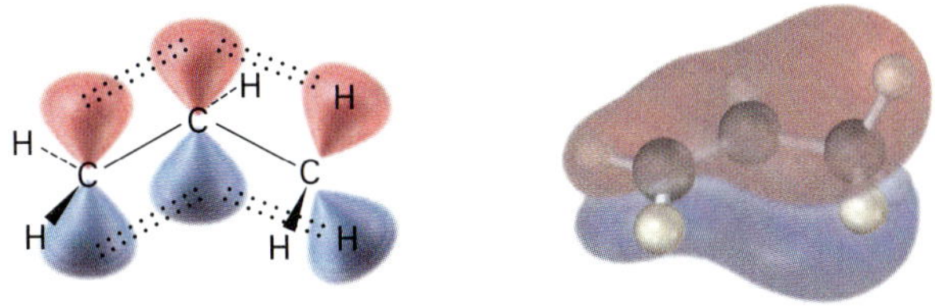

Figure 6.6 Hyperconjugation is a stabilizing interaction between an unfilled π orbital and a neighboring filled C–H σ bond on a substituent. The more substituents there are, the greater the stabilization of the alkene.

A second factor that contributes to alkene stability involves bond strengths. A bond between an sp^2 carbon and an sp^3 carbon is somewhat stronger than a bond between two sp^3 carbons. Thus, in comparing 1-butene and 2-butene, the monosubstituted isomer has one sp^3–sp^3 bond and one sp^3–sp^2 bond, while the disubstituted isomer has two sp^3–sp^2 bonds. More highly substituted alkenes always have a higher ratio of sp^3–sp^2 bonds to sp^3–sp^3 bonds than less highly substituted alkenes and are therefore more stable.

2-Butene (more stable) **1-Butene (less stable)**

Problem 6.13 Name the following alkenes, and tell which compound in each pair is more stable:

(a) **(b)** **(c)**

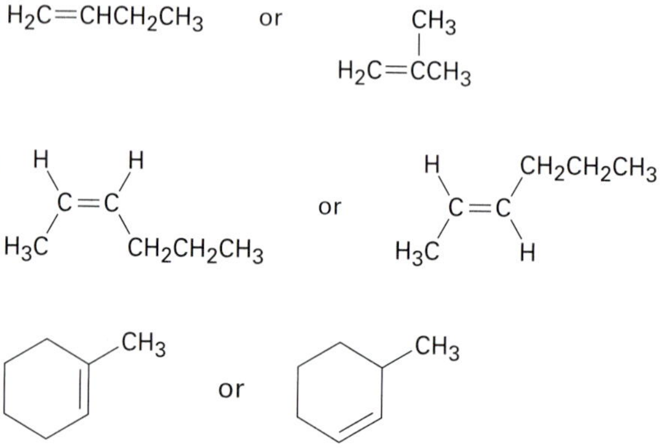

6.7 Electrophilic Addition Reactions of Alkenes

Before beginning a detailed discussion of alkene reactions, let's review briefly some conclusions from the previous chapter. We said in Section 5.5 that alkenes behave as nucleophiles (Lewis bases) in polar reactions. The carbon–carbon double bond is electron-rich and can donate a pair of electrons to an electrophile (Lewis acid). For example, reaction of 2-methylpropene with HBr yields 2-bromo-2-methylpropane. A careful study of this and similar reactions by Christopher Ingold and others in the 1930s led to the generally accepted mechanism shown in Figure 6.7 for **electrophilic addition reactions**.

The reaction begins with an attack on the electrophile, HBr, by the electrons of the nucleophilic π bond. Two electrons from the π bond form a new σ bond between the entering hydrogen and an alkene carbon, as shown by the curved arrow at the top of Figure 6.7. The carbocation intermediate that results is itself an electrophile, which can accept an electron pair from nucleophilic Br^- ion to form a C–Br bond and yield a neutral addition product.

ThomsonNOW™ Click *Organic Process* to **view an animation of this alkene addition reaction.**

The energy diagram for the overall electrophilic addition reaction (Figure 6.8) has two peaks (transition states) separated by a valley (carbocation intermediate). The energy level of the intermediate is higher than that of the starting alkene, but the reaction as a whole is exergonic (negative $\Delta G°$). The first step, protonation of the alkene to yield the intermediate cation, is relatively slow but, once formed, the cation intermediate rapidly reacts further to yield the final alkyl bromide product. The relative rates of the two steps are indicated in Figure 6.8 by the fact that $\Delta G^{\ddagger}_1$ is larger than $\Delta G^{\ddagger}_2$.

igure 6.7 MECHANISM: echanism of the electrophilic ddition of HBr to 2-methyl-ropene. The reaction occurs in vo steps and involves a carbo-ation intermediate.

1 A hydrogen atom on the electrophile HBr is attacked by π electrons from the nucleophilic double bond, forming a new C–H bond. This leaves the other carbon atom with a + charge and a vacant p orbital. Simultaneously, two electrons from the H–Br bond move onto bromine, giving bromide anion.

2 Bromide ion donates an electron pair to the positively charged carbon atom, forming a C–Br bond and yielding the neutral addition product.

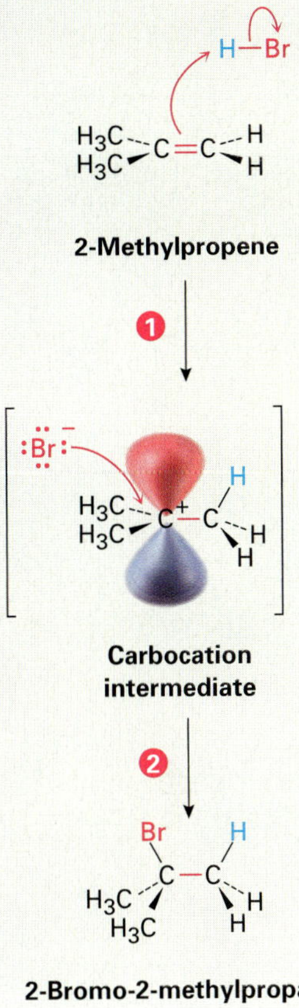

Figure 6.8 Energy diagram for the two-step electrophilic addi-tion of HBr to 2-methylpropene. The first step is slower than the second step.

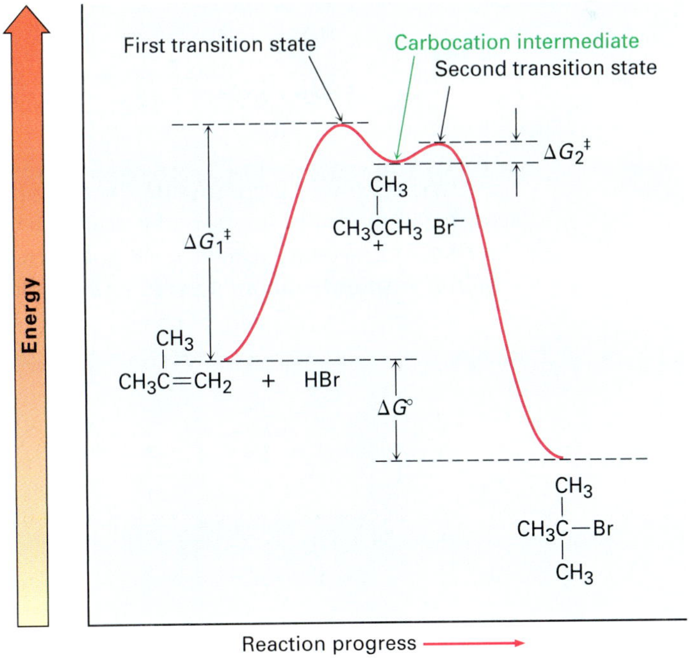

Electrophilic addition of HX to alkenes is successful not only with HBr but with HCl and HI as well. Note that HI is usually generated in the reaction mixture by treating potassium iodide with phosphoric acid.

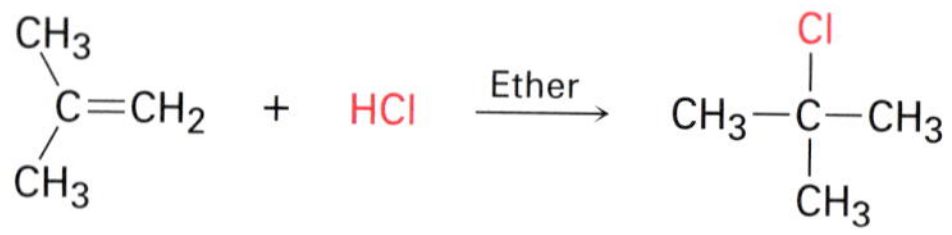

2-Methylpropene **2-Chloro-2-methylpropane (94%)**

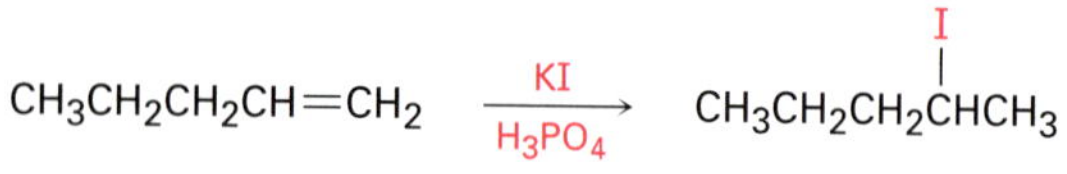

1-Pentene **(HI)** **2-Iodopentane**

Writing Organic Reactions

This is a good time to mention that organic reaction equations are sometimes written in different ways to emphasize different points. In describing a laboratory process, for example, the reaction of 2-methylpropene with HCl just shown might be written in the format A + B → C to emphasize that both reactants are equally important for the purposes of the discussion. The solvent and notes about other reaction conditions, such as temperature, are written either above or below the reaction arrow.

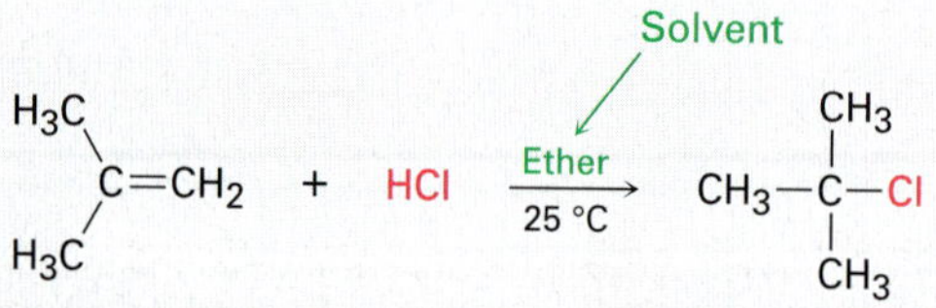

2-Methylpropene **2-Chloro-2-methylpropane**

Alternatively, we might write the same reaction in a format to emphasize that 2-methylpropene is the reactant whose chemistry is of greater interest. The second reactant, HCl, is placed above the reaction arrow together with notes about solvent and reaction conditions.

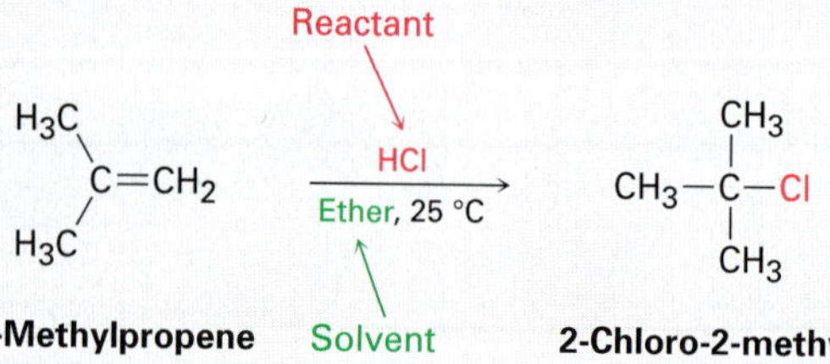

2-Methylpropene **2-Chloro-2-methylpropane**

In describing a biological process, the reaction is usually written to show only the structure of the primary reactant and product, while abbreviating the structures of various biological "reagents" and by-products by using a curved arrow that intersects the straight reaction arrow. As discussed in Section 5.11, the reaction of glucose with ATP to give glucose 6-phosphate plus ADP would be written as

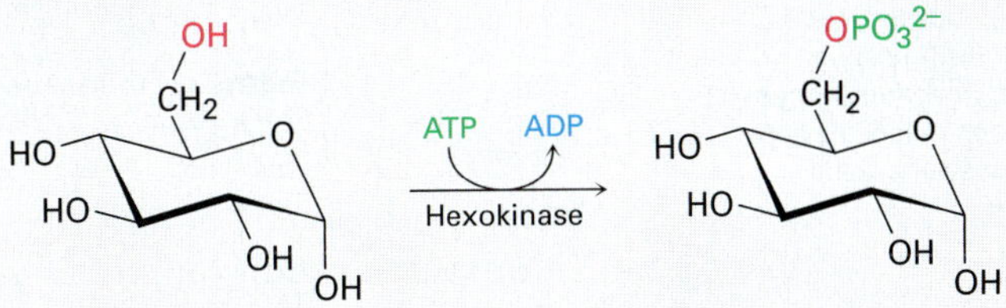

Glucose Glucose 6-phosphate

6.8 Orientation of Electrophilic Additions: Markovnikov's Rule

Key IDEAS

Test your knowledge of Key Ideas by using resources in ThomsonNOW or by answering end-of-chapter problems marked with ▲.

ThomsonNOW™ Click *Organic Interactive* to **use a web-based palette to predict products from the addition of HX to alkenes.**

Look carefully at the reactions shown in the previous section. In each case, an unsymmetrically substituted alkene has given a single addition product, rather than the mixture that might have been expected. As another example, 1-pentene *might* react with HCl to give both 1-chloropentane and 2-chloropentane, but it doesn't. Instead, the reaction gives only 2-chloropentane as the sole product. We say that such reactions are **regiospecific** (**ree**-jee-oh-specific) when only one of two possible orientations of addition occurs.

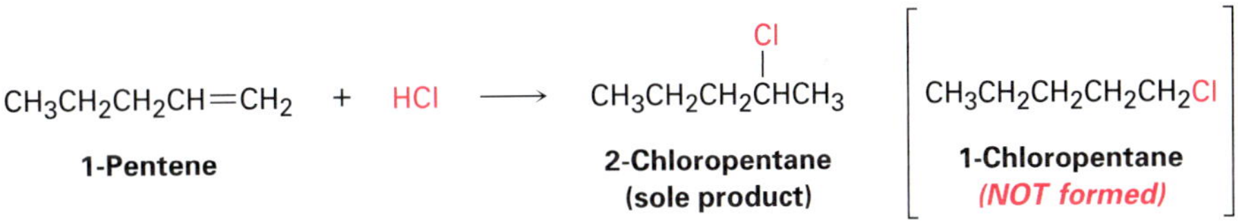

After looking at the results of many such reactions, the Russian chemist Vladimir Markovnikov proposed in 1869 what has become known as **Markovnikov's rule.**

Markovnikov's rule In the addition of HX to an alkene, the H attaches to the carbon with fewer alkyl substituents and the X attaches to the carbon with more alkyl substituents.

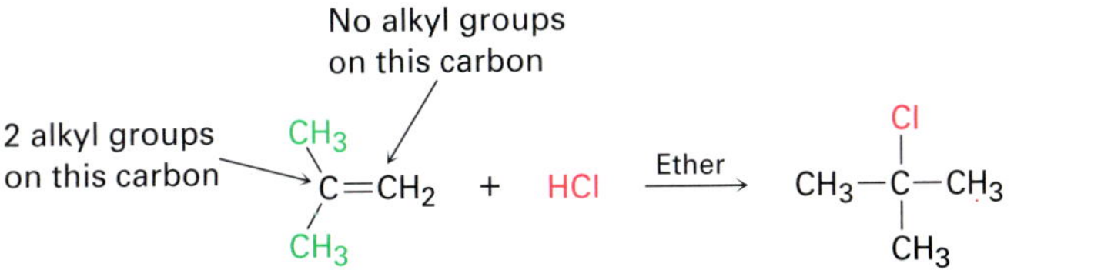

2-Methylpropene 2-Chloro-2-methylpropane

Vladimir Vassilyevich Markovnikov

Vladimir Vassilyevich Markovnikov (1838–1904) was born in Nijni-Novgorod, Russia, and received his Ph.D. working with A. M. Butlerov at the university in Kazan. He was a professor in Kazan (1870), Odessa (1871), and Moscow (1873–1898). In addition to his work on the orientation of addition reactions, he was the first to synthesize a four-membered ring.

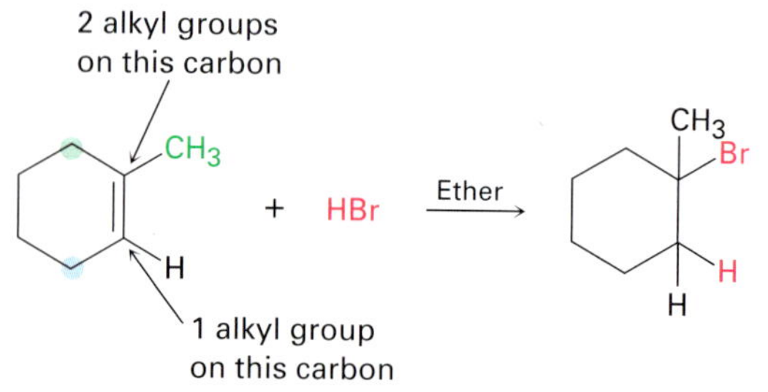

1-Methylcyclohexene **1-Bromo-1-methylcyclohexane**

When both double-bond carbon atoms have the same degree of substitution, a mixture of addition products results.

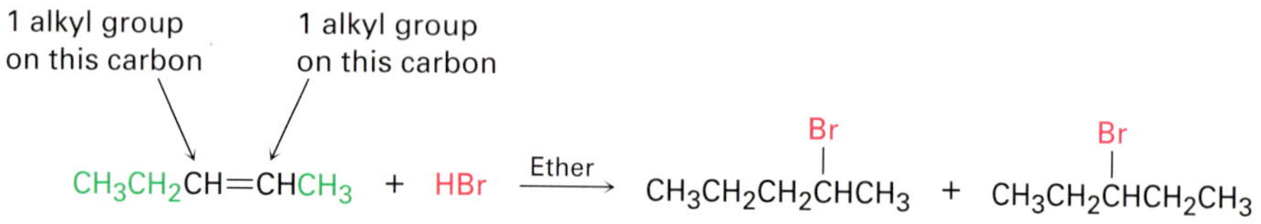

2-Pentene **2-Bromopentane** **3-Bromopentane**

Since carbocations are involved as intermediates in these reactions, Markovnikov's rule can be restated.

Markovnikov's rule (restated) In the addition of HX to an alkene, the more highly substituted carbocation is formed as the intermediate rather than the less highly substituted one.

For example, addition of H^+ to 2-methylpropene yields the intermediate *tertiary* carbocation rather than the alternative primary carbocation, and addition to 1-methylcyclohexene yields a tertiary cation rather than a secondary one. Why should this be?

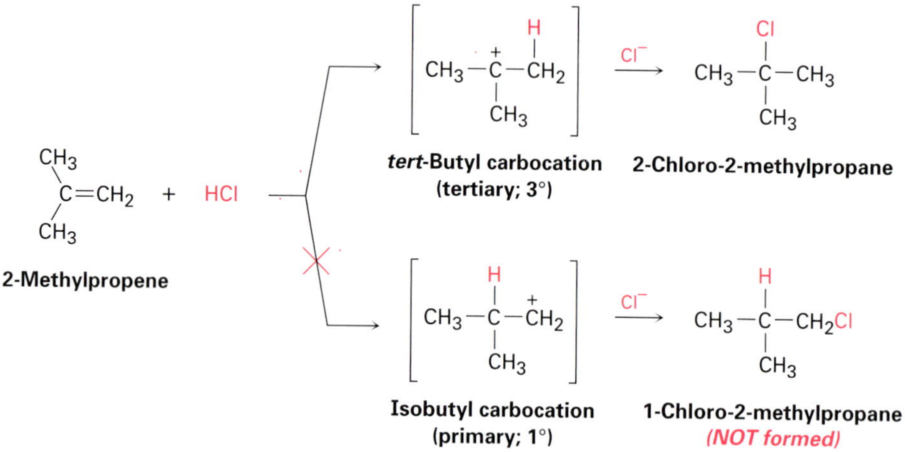

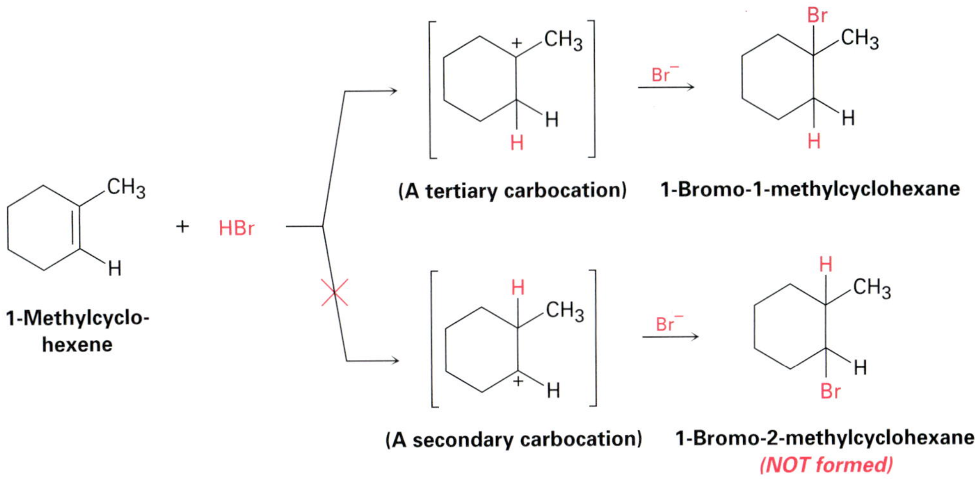

WORKED EXAMPLE 6.2

Predicting the Product of an Electrophilic Addition Reaction

What product would you expect from reaction of HCl with 1-ethylcyclopentene?

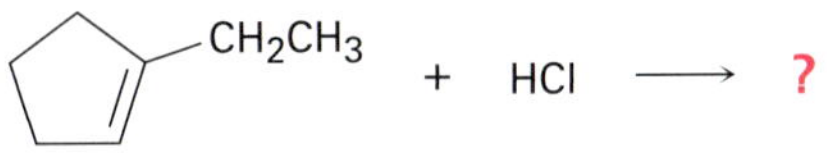

Strategy When solving a problem that asks you to predict a reaction product, begin by looking at the functional group(s) in the reactants and deciding what kind of reaction is likely to occur. In the present instance, the reactant is an alkene that will probably undergo an electrophilic addition reaction with HCl. Next, recall what you know about electrophilic addition reactions, and use your knowledge to predict the product. You know that electrophilic addition reactions follow Markovnikov's rule, so H^+ will add to the double-bond carbon that has one alkyl group (C2 on the ring) and the Cl will add to the double-bond carbon that has two alkyl groups (C1 on the ring).

Solution The expected product is 1-chloro-1-ethylcyclopentane.

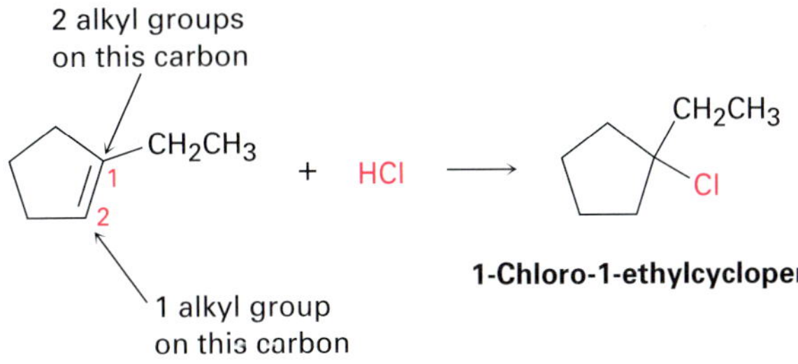

1-Chloro-1-ethylcyclopentane

ThomsonNOW™ Click *Organic Interactive* to **practice predicting products of addition reactions according to Markovnikov's rule.**

WORKED EXAMPLE 6.3 *Synthesizing a Specific Compound*

What alkene would you start with to prepare the following alkyl halide? There may be more than one possibility.

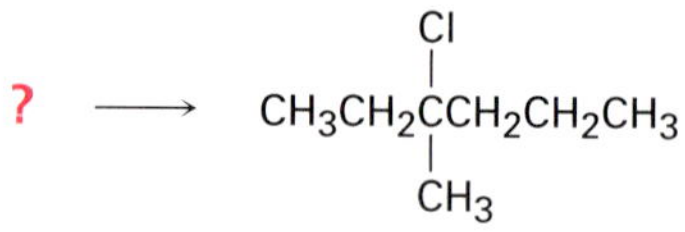

Strategy When solving a problem that asks how to prepare a given product, *always work backward.* Look at the product, identify the functional group(s) it contains, and ask yourself, "How can I prepare that functional group?" In the present instance, the product is a tertiary alkyl chloride, which can be prepared by reaction of an alkene with HCl. The carbon atom bearing the −Cl atom in the product must be one of the double-bond carbons in the reactant. Draw and evaluate all possibilities.

Solution There are three possibilities, any one of which could give the desired product.

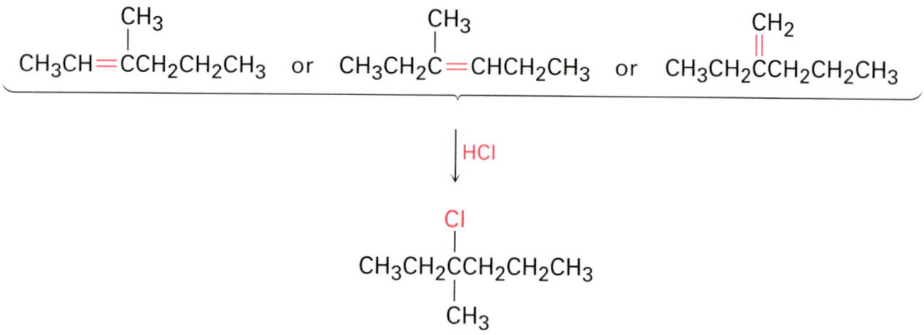

Problem 6.14 Predict the products of the following reactions:

(a)

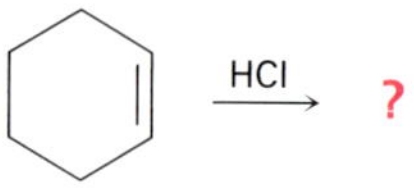

(b)

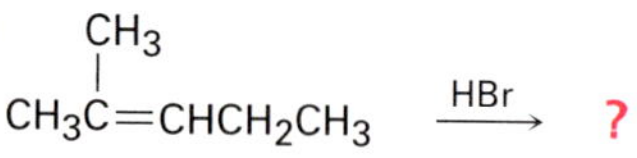

(c)

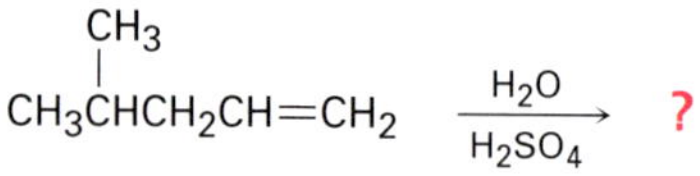

(Addition of H_2O occurs.)

(d)

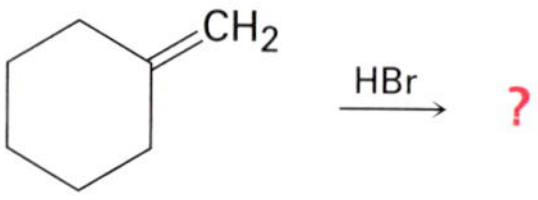

Problem 6.15 What alkenes would you start with to prepare the following alkyl halides?

(a)

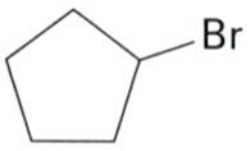

(b)

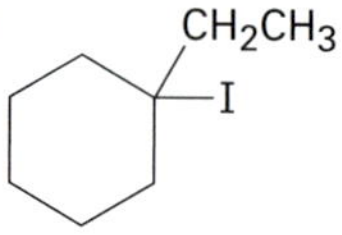

(c)

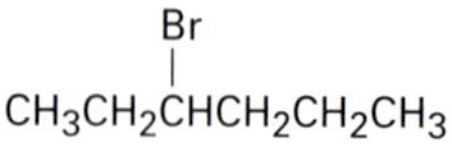

(d)

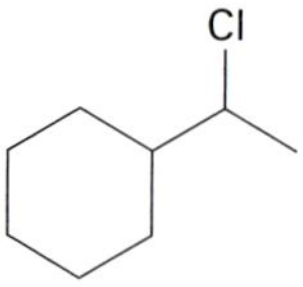

6.9 Carbocation Structure and Stability

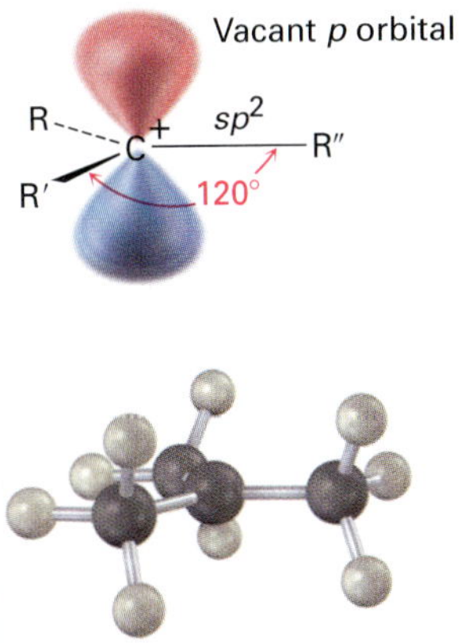

igure 6.9 The structure of a arbocation. The trivalent carbon ; sp^2-hybridized and has a acant *p* orbital perpendicular to ıe plane of the carbon and three ttached groups.

To understand why Markovnikov's rule works, we need to learn more about the structure and stability of carbocations and about the general nature of reactions and transition states. The first point to explore involves structure.

A great deal of evidence has shown that carbocations are *planar.* The trivalent carbon is sp^2-hybridized, and the three substituents are oriented to the corners of an equilateral triangle, as indicated in Figure 6.9. Because there are only six valence electrons on carbon and all six are used in the three σ bonds, the *p* orbital extending above and below the plane is unoccupied.

The second point to explore involves carbocation stability. 2-Methylpropene might react with H^+ to form a carbocation having three alkyl substituents (a tertiary ion, 3°), or it might react to form a carbocation having one alkyl substituent (a primary ion, 1°). Since the tertiary alkyl chloride, 2-chloro-2-methylpropane, is the only product observed, formation of the tertiary cation is evidently favored over formation of the primary cation. Thermodynamic measurements show that, indeed, the stability of carbocations increases with increasing substitution so that the stability order is tertiary > secondary > primary > methyl.

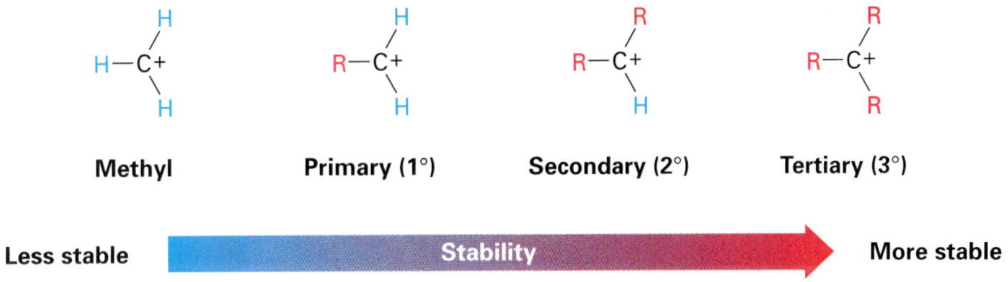

homsonNOW™ Click *Organic* ıteractive to **rank the stability f carbocation intermediates.**

One way of determining carbocation stabilities is to measure the amount of energy required to form the carbocation by dissociation of the corresponding alkyl halide, $R{-}X \rightarrow R^+ + :X^-$. As shown in Figure 6.10, tertiary alkyl halides dissociate to give carbocations more easily than secondary or primary ones. As a result, trisubstituted carbocations are more stable than disubstituted ones, which are more stable than monosubstituted ones. The data in Figure 6.10 are taken from measurements made in the gas phase, but a similar stability order is found for carbocations in solution. The dissociation enthalpies are much lower in solution because polar solvents can stabilize the ions, but the order of carbocation stability remains the same.

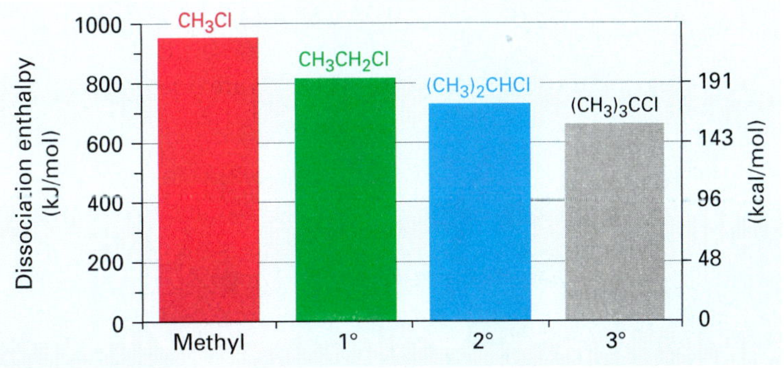

Figure 6.10 A plot of dissociaion enthalpy versus substitution ›attern for the gas-phase dissociıtion of alkyl chlorides to yield :arbocations. More highly substiuted alkyl halides dissociate nore easily than less highly substituted ones.

Why are more highly substituted carbocations more stable than less highl substituted ones? There are at least two reasons. Part of the answer has to do wit inductive effects, and part has to do with hyperconjugation. Inductive effects, di cussed in Section 2.1 in connection with polar covalent bonds, result from th shifting of electrons in a σ bond in response to the electronegativity of nearb atoms. In the present instance, electrons from a relatively larger and more pola izable alkyl group can shift toward a neighboring positive charge more easily tha the electron from a hydrogen. Thus, the more alkyl groups there are attache to the positively charged carbon, the more electron density shifts toward th charge and the more inductive stabilization of the cation occurs (Figure 6.11).

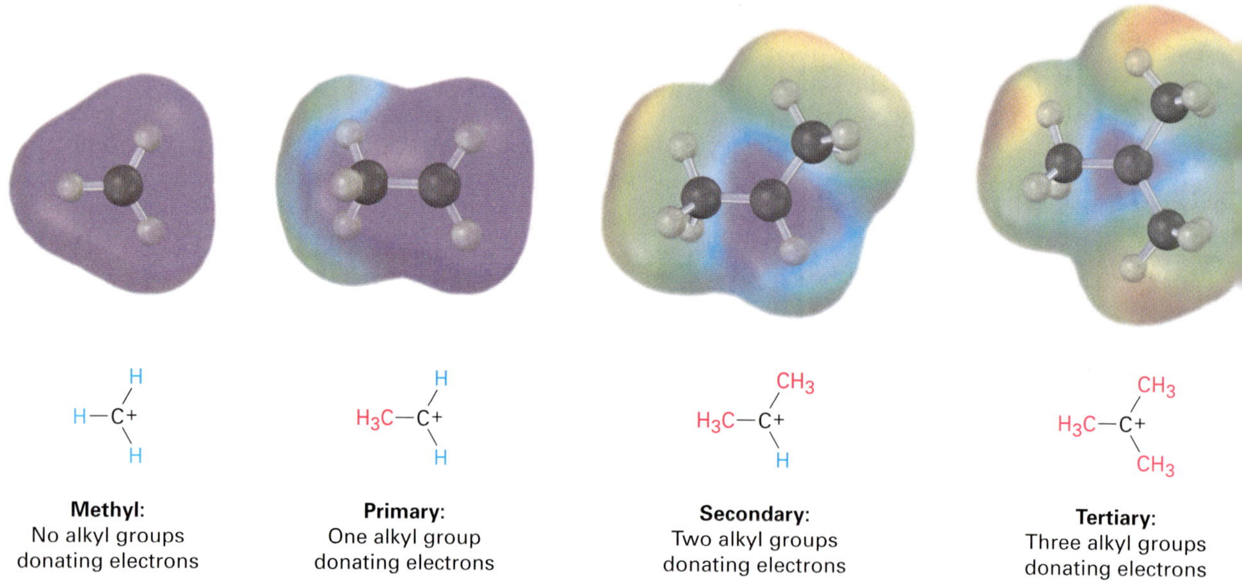

Figure 6.11 A comparison of inductive stabilization for methyl, primary, secondary, and tertiary carbocations. The more alkyl groups there are bonded to the positively charged carbon, the more electron density shifts toward the charge, making the charged carbon less electron-poor (blue in electrostatic potential maps).

Hyperconjugation, discussed in Section 6.6 in connection with the stabilitie of substituted alkenes, is the stabilizing interaction between a vacant *p* orbital an properly oriented C–H σ bonds on neighboring carbons. The more alkyl group there are on the carbocation, the more possibilities there are for hyperconjugatio and the more stable the carbocation. Figure 6.12 shows the molecular orbita involved in hyperconjugation for the ethyl carbocation, $CH_3CH_2^+$, and indicate the difference between the C–H bond perpendicular to the cation *p* orbita and the two C–H bonds more nearly parallel to the cation *p* orbital. Only the roughly parallel C–H bonds are oriented properly to take part in hyperconjugation

Figure 6.12 Stabilization of the ethyl carbocation, $CH_3CH_2^+$, through hyperconjugation. Interaction of neighboring C–H σ bonds with the vacant *p* orbital stabilizes the cation and lowers its energy. The molecular orbital shows that only the two C–H bonds more nearly parallel to the cation *p* orbital are oriented properly for hyper-conjugation. The C–H bond perpendicular to the cation *p* orbital cannot take part.

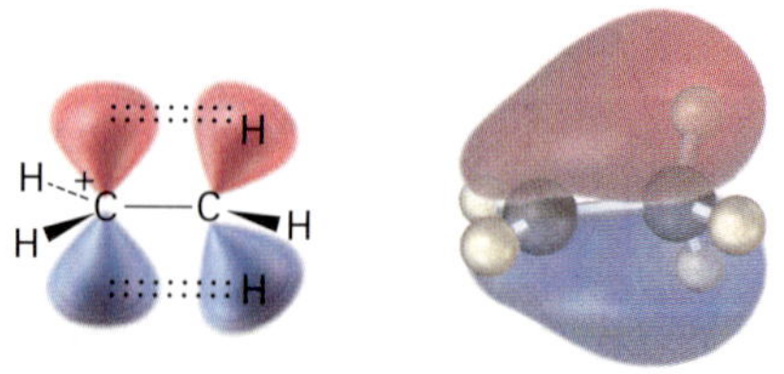

Problem 6.16 Show the structures of the carbocation intermediates you would expect in the following reactions:

(a)

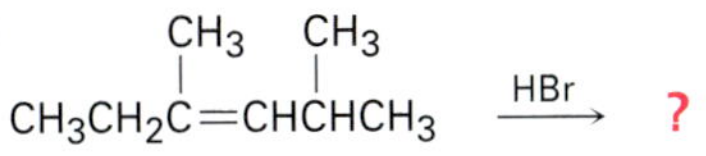

(b)

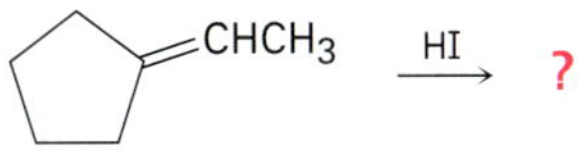

Problem 6.17 Draw a skeletal structure of the following carbocation. Identify it as primary, secondary, or tertiary, and identify the hydrogen atoms that have the proper orientation for hyperconjugation in the conformation shown.

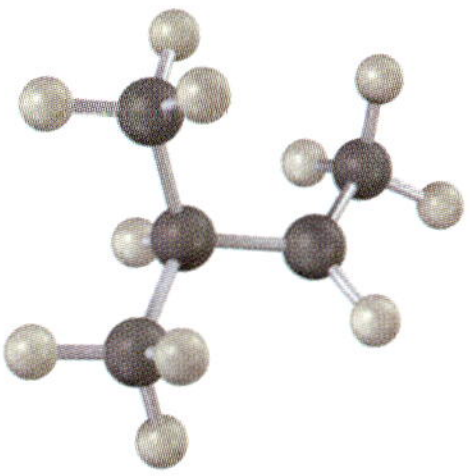

6.10 The Hammond Postulate

Let's summarize our knowledge of electrophilic addition reactions up to this point. We know that:

- **Electrophilic addition to an unsymmetrically substituted alkene gives the more highly substituted carbocation intermediate.** A more highly substituted carbocation forms faster than a less highly substituted one and, once formed, rapidly goes on to give the final product.
- **A more highly substituted carbocation is more stable than a less highly substituted one.** That is, the stability order of carbocations is tertiary > secondary > primary > methyl.

What we have not yet seen is how these two points are related. Why does the *stability* of the carbocation intermediate affect the *rate* at which it's formed and thereby determine the structure of the final product? After all, carbocation stability is determined by the free-energy change $\Delta G°$, but reaction rate is determined by the activation energy $\Delta G^‡$. The two quantities aren't directly related.

Although there is no simple quantitative relationship between the stability of a carbocation intermediate and the rate of its formation, there *is* an intuitive relationship. It's generally true when comparing two similar reactions that the more stable intermediate forms faster than the less stable one. The situation is shown graphically in Figure 6.13, where the reaction energy profile in part (a) represents the typical situation rather than the profile in part (b). That is, the curves for two similar reactions don't cross one another.

An explanation of the relationship between reaction rate and intermediate stability was first advanced in 1955. Known as the **Hammond postulate**, the argument goes like this: transition states represent energy maxima. They are high-energy activated complexes that occur transiently during the course of a reaction and immediately go on to a more stable species. Although we can't

George Simms Hammond

George Simms Hammond (1921–2005) was born on Hardscrabble Road in Auburn, Maine, the son of a dairy farmer. He received his Ph.D. at Harvard University in 1947 and served as professor of chemistry at Iowa State University, California Institute of Technology (1958–1972), and the University of California at Santa Cruz (1972–1978). He was known for his exploratory work on organic photochemistry—the use of light to bring about organic reactions.

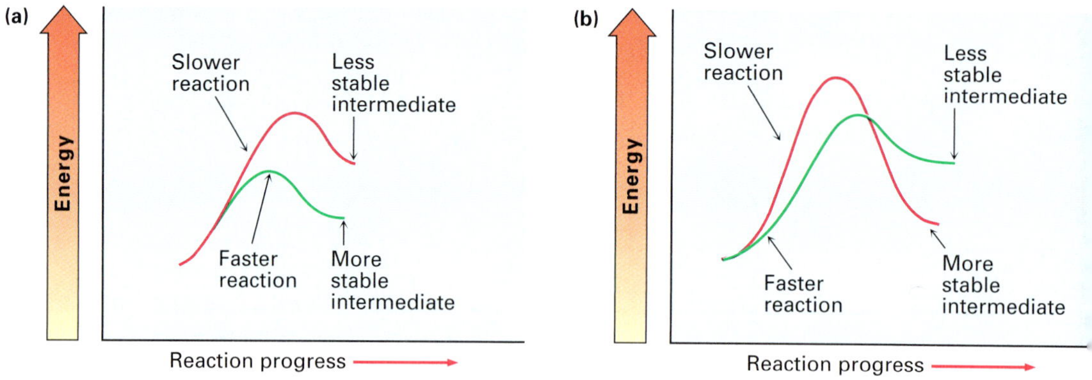

Figure 6.13 Energy diagrams for two similar competing reactions. In **(a)**, the faster reaction yields the more stable intermediate. In **(b)**, the slower reaction yields the more stable intermediate. The curves shown in **(a)** represent the typical situation.

actually observe transition states because they have no finite lifetime, the Hammond postulate says that we can get an idea of a particular transition state's structure by looking at the structure of the nearest stable species. Imagine the two cases shown in Figure 6.14, for example. The reaction profile in part (a) shows the energy curve for an endergonic reaction step, and the profile in part (b) shows the curve for an exergonic step.

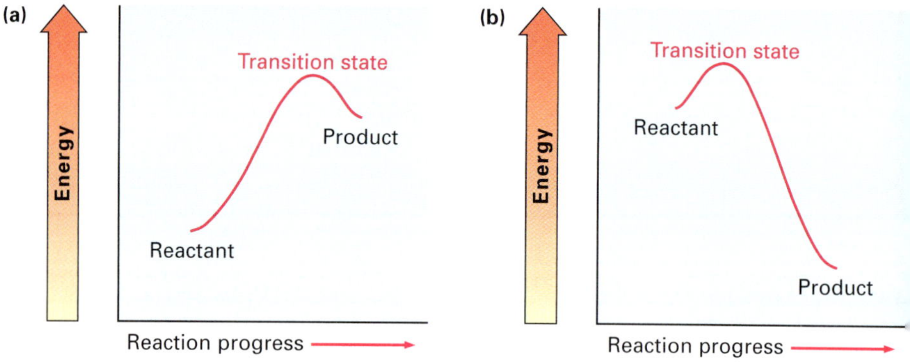

Figure 6.14 Energy diagrams for endergonic and exergonic steps. **(a)** In an endergonic step, the energy levels of transition state and *product* are closer. **(b)** In an exergonic step, the energy levels of transition state and *reactant* are closer.

In an endergonic reaction (Figure 6.14a), the energy level of the transition state is closer to that of the product than to that of the reactant. Since the transition state is closer energetically to the product, we make the natural assumption that it's also closer structurally. In other words, *the transition state for an endergonic reaction step structurally resembles the product of that step.* Conversely, the transition state for an exergonic reaction (Figure 6.14b) is closer energetically, and thus structurally, to the reactant than to the product. We therefore say that *the transition state for an exergonic reaction step structurally resembles the reactant for that step.*

Hammond postulate The structure of a transition state resembles the structure of the nearest stable species. Transition states for endergonic steps structurally resemble products, and transition states for exergonic steps structurally resemble reactants.

How does the Hammond postulate apply to electrophilic addition reactions? The formation of a carbocation by protonation of an alkene is an endergonic step. Thus, the transition state for alkene protonation structurally resembles the carbocation intermediate, and any factor that stabilizes the carbocation will stabilize the nearby transition state. Since increasing alkyl substitution stabilizes carbocations, it also stabilizes the transition states leading to those ions, thus resulting in a faster reaction. More stable carbocations form faster because their greater stability is reflected in the lower-energy transition state leading to them (Figure 6.15).

gure 6.15 Energy diagrams r carbocation formation. The ore stable tertiary carbocation formed faster (green curve) cause its increased stability wers the energy of the transi- n state leading to it.

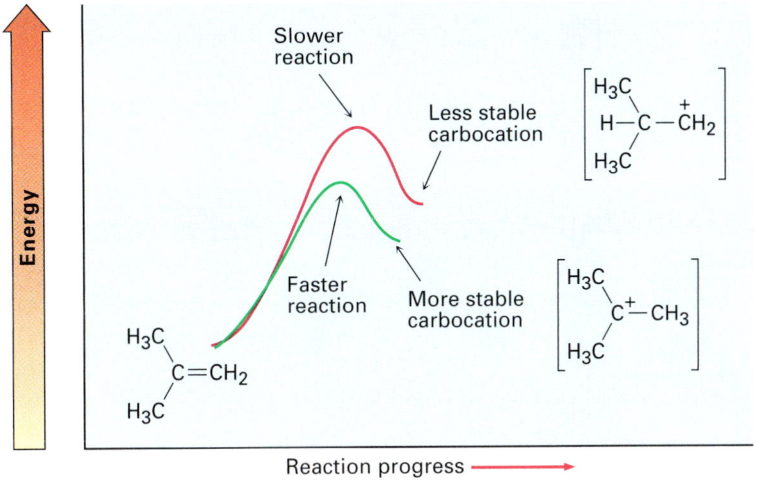

We can imagine the transition state for alkene protonation to be a structure in which one of the alkene carbon atoms has almost completely rehybridized from sp^2 to sp^3 and in which the remaining alkene carbon bears much of the positive charge (Figure 6.16). This transition state is stabilized by hyperconjugation and inductive effects in the same way as the product carbocation. The more alkyl groups that are present, the greater the extent of stabilization and the faster the transition state forms.

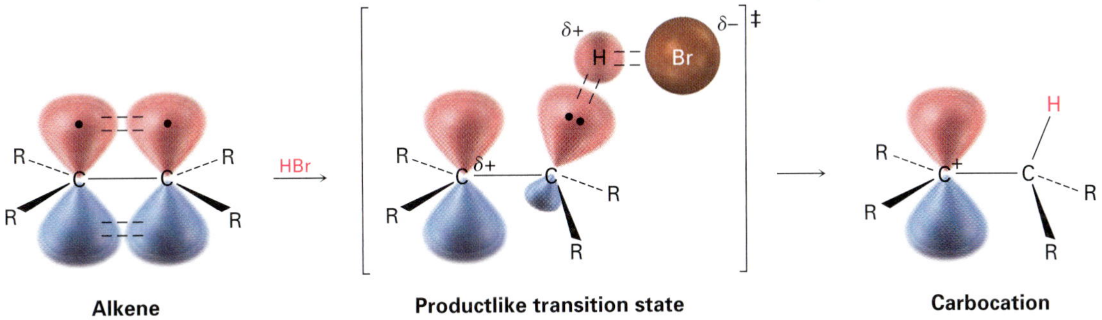

Figure 6.16 The hypothetical structure of a transition state for alkene protonation. The transition state is closer in both energy and structure to the carbocation than to the alkene. Thus, an increase in carbocation stability (lower $\Delta G°$) also causes an increase in transition-state stability (lower $\Delta G^‡$), thereby increasing the rate of its formation.

Problem 6.18 What about the second step in the electrophilic addition of HCl to an alkene—the reaction of chloride ion with the carbocation intermediate? Is this step exergonic or endergonic? Does the transition state for this second step resemble the reactant (carbocation) or product (alkyl chloride)? Make a rough drawing of what the transition-state structure might look like.

6.11 Evidence for the Mechanism of Electrophilic Additions: Carbocation Rearrangements

ThomsonNOW™ Click *Organic Interactive* to **use a web-based palette to predict products from simple carbocation rearrangements.**

Frank C. Whitmore

Frank C. Whitmore (1887–1947) was born in North Attleboro, Massachusetts, and received his Ph.D. at Harvard working with E. L. Jackson. He was professor of chemistry at Minnesota, Northwestern, and the Pennsylvania State University. Nicknamed "Rocky," he wrote an influential advanced textbook in organic chemistry.

How do we know that the carbocation mechanism for electrophilic additic reactions of alkenes is correct? The answer is that we *don't* know it's correc at least we don't know with complete certainty. Although an incorrect rea tion mechanism can be disproved by demonstrating that it doesn't accou for observed data, a correct reaction mechanism can never be entirely prove The best we can do is to show that a proposed mechanism is consistent wi all known facts. If enough facts are accounted for, the mechanism is probab correct.

What evidence is there to support the carbocation mechanism proposed f the electrophilic addition reaction of alkenes? One of the best pieces of eviden was discovered during the 1930s by F. C. Whitmore of the Pennsylvania Sta University, who found that structural rearrangements often occur during th reaction of HX with an alkene. For example, reaction of HCl with 3-methy 1-butene yields a substantial amount of 2-chloro-2-methylbutane in addition the "expected" product, 2-chloro-3-methylbutane.

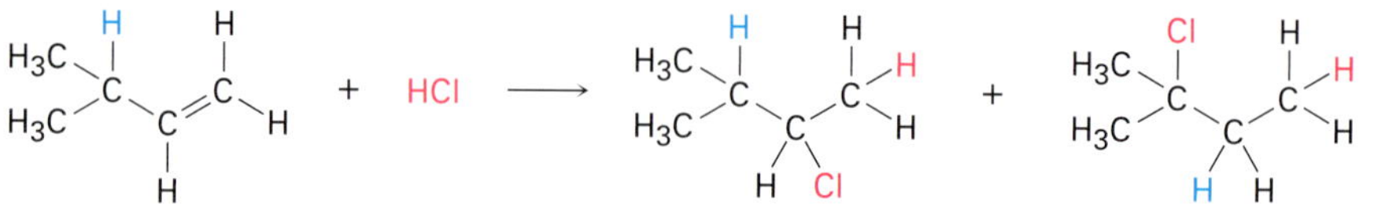

3-Methyl-1- butene **2-Chloro-3-methylbutane (approx. 50%)** **2-Chloro-2-methylbutan (approx. 50%)**

If the reaction takes place in a single step, it would be difficult to account fc rearrangement, but if the reaction takes place in several steps, rearrangement more easily explained. Whitmore suggested that it is a carbocation intermedia that undergoes rearrangement. The secondary carbocation intermediate forme by protonation of 3-methyl-1-butene rearranges to a more stable tertiary carbc cation by a **hydride shift**—the shift of a hydrogen atom and its electron pair (hydride ion, :H$^-$) between neighboring carbons.

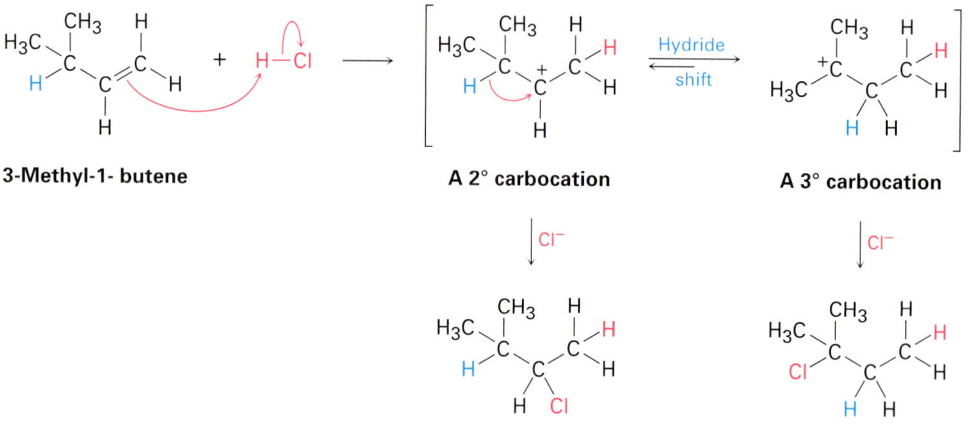

3-Methyl-1- butene **A 2° carbocation** **A 3° carbocation**

2-Chloro-3-methylbutane **2-Chloro-2-methylbutan**

Carbocation rearrangements can also occur by the shift of an alkyl group with its electron pair. For example, reaction of 3,3-dimethyl-1-butene with HCl leads to an equal mixture of unrearranged 2-chloro-3,3-dimethylbutane and rearranged 2-chloro-2,3-dimethylbutane. In this instance, a secondary carbocation rearranges to a more stable tertiary carbocation by the shift of a methyl group.

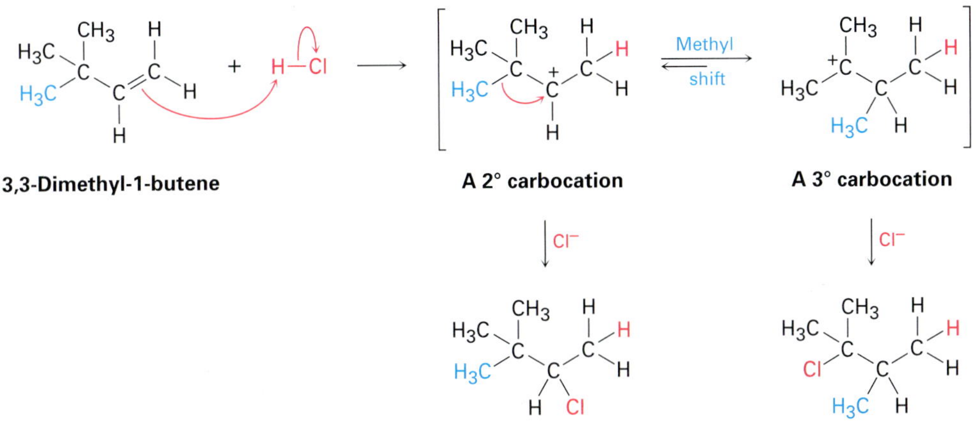

3,3-Dimethyl-1-butene **A 2° carbocation** **A 3° carbocation**

2-Chloro-3,3-dimethylbutane **2-Chloro-2,3-dimethylbutane**

Note the similarities between the two carbocation rearrangements: in both cases, a group (:H^- or :CH_3^-) moves to an adjacent positively charged carbon, taking its bonding electron pair with it. Also in both cases, a less stable carbocation rearranges to a more stable ion. Rearrangements of this kind are a common feature of carbocation chemistry and are particularly important in the biological pathways by which steroids and related substances are synthesized. An example is the following hydride shift that occurs during the biosynthesis of cholesterol.

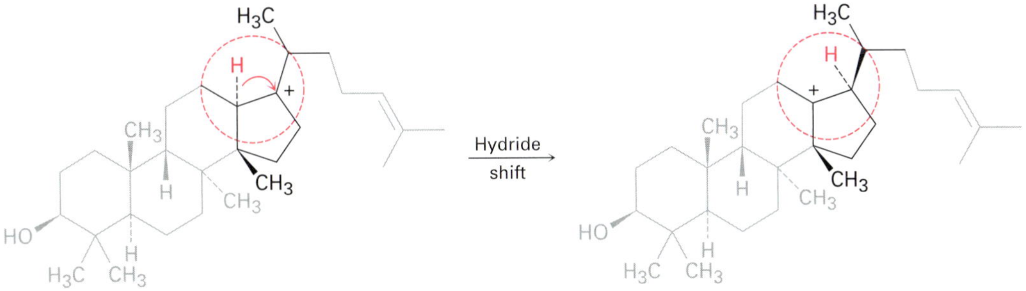

A tertiary carbocation **An isomeric tertiary carbocation**

A word of advice that we'll repeat on occasion: biological molecules are often larger and more complex in appearance than the molecules chemists work with in the laboratory, but don't be intimidated. When looking at *any* chemical transformation, focus only on the part of the molecule where the change is occurring and don't worry about the rest. The tertiary carbocation just pictured looks complicated, but all the chemistry is taking place in the small part of the molecule inside the red circle.

Problem 6.19 On treatment with HBr, vinylcyclohexane undergoes addition and rearrangement to yield 1-bromo-1-ethylcyclohexane. Using curved arrows, propose a mechanism to account for this result.

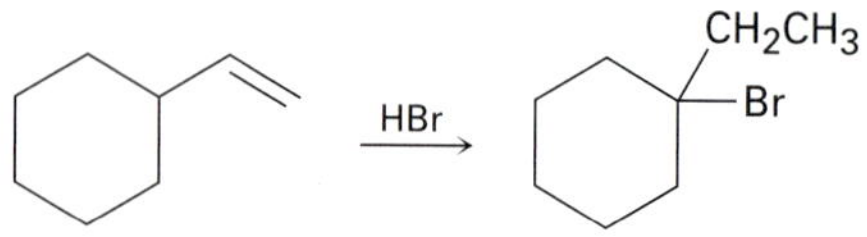

Vinylcyclohexane **1-Bromo-1-ethylcyclohexane**

Focus On . . .

Terpenes: Naturally Occurring Alkenes

The wonderful fragrance of leaves from the California bay tree is due primarily to myrcene, a simple terpene.

It has been known for centuries that codistillation of many plant materials with steam produces a fragrant mixture of liquids called *essential oils*. For hundreds of years, such plant extracts have been used as medicines, spices, and perfumes. The investigation of essential oils also played a major role in the emergence of organic chemistry as a science during the 19th century.

Chemically, plant essential oils consist largely of mixtures of compounds known as *terpenoids*—small organic molecules with an immense diversity of structure. More than 35,000 different terpenoids are known. Some are open-chain molecules, and others contain rings; some are hydrocarbons, and others contain oxygen. Hydrocarbon terpenoids, in particular, are known as *terpenes,* and all contain double bonds. For example:

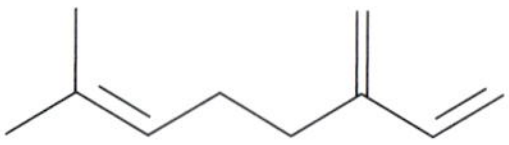

Myrcene
(oil of bay)

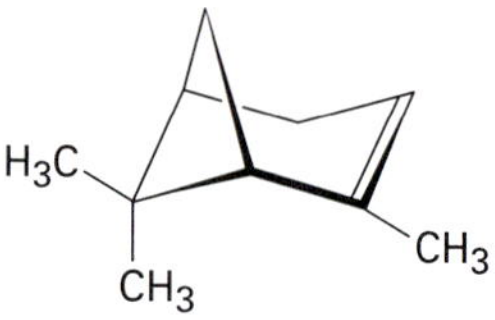

α-Pinene
(turpentine)

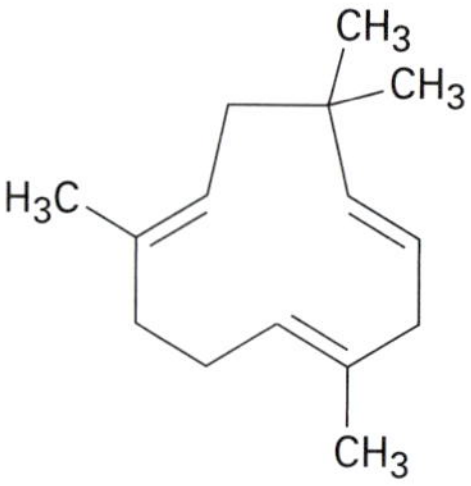

Humulene
(oil of hops)

(continued)

Regardless of their apparent structural differences, all terpenoids are related. According to a formalism called the *isoprene rule,* they can be thought of as arising from head-to-tail joining of 5-carbon isoprene units (2-methyl-1,3-butadiene). Carbon 1 is the head of the isoprene unit, and carbon 4 is the tail. For example, myrcene contains two isoprene units joined head to tail, forming an 8-carbon chain with two 1-carbon branches. α-Pinene similarly contains two isoprene units assembled into a more complex cyclic structure, and humulene contains three isoprene units. See if you can identify the isoprene units in α-pinene and humulene.

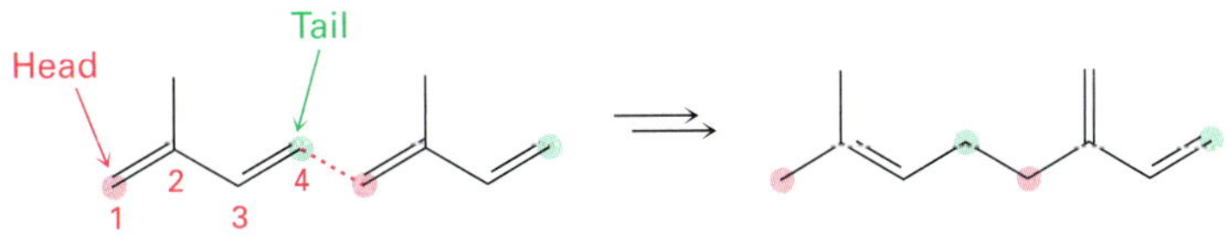

Isoprene **Myrcene**

Terpenes (and terpenoids) are further classified according to the number of 5-carbon units they contain. Thus, *monoterpenes* are 10-carbon substances biosynthesized from two isoprene units, *sesquiterpenes* are 15-carbon molecules from three isoprene units, *diterpenes* are 20-carbon substances from four isoprene units, and so on. Monoterpenes and sesquiterpenes are found primarily in plants, but the higher terpenoids occur in both plants and animals, and many have important biological roles. The triterpenoid lanosterol, for example, is the precursor from which all steroid hormones are made.

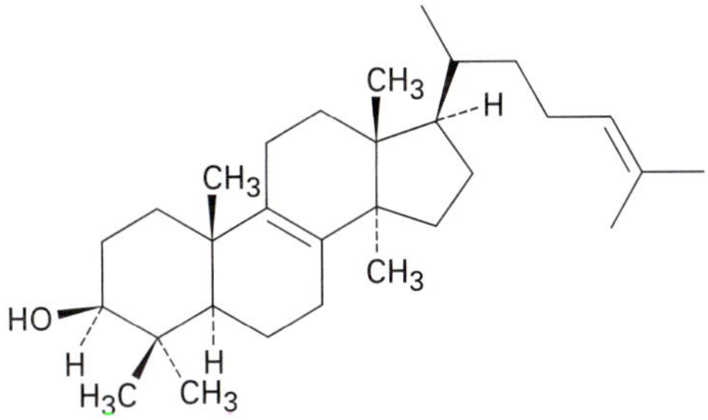

Lanosterol (a triterpene, C_{30})

Isoprene itself is not the true biological precursor of terpenoids. As we'll see in Chapter 27, nature instead uses two "isoprene equivalents"—isopentenyl diphosphate and dimethylallyl diphosphate—which are themselves made by two different routes depending on the organism. Lanosterol, in particular, is biosynthesized from acetic acid by a complex pathway that has been worked out in great detail.

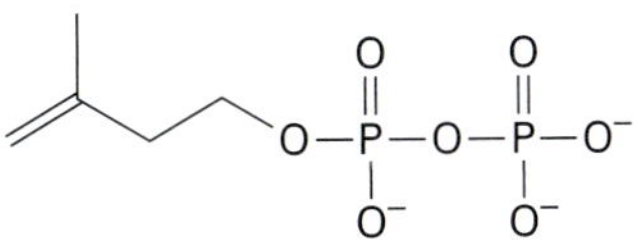

Isopentenyl diphosphate

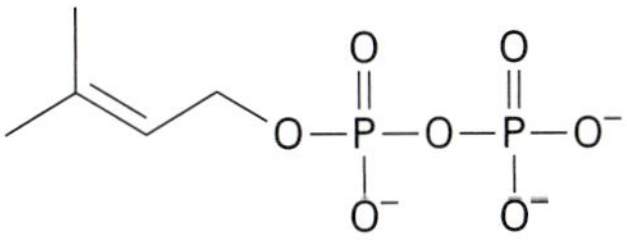

Dimethylallyl diphosphate

SUMMARY AND KEY WORDS

alkene ($R_2C{=}CR_2$), 172
allyl group, 178
degree of unsaturation, 174
E geometry, 180
electrophilic addition reaction, 188
Hammond postulate, 197
hydride shift, 200
hyperconjugation, 187
Markovnikov's rule, 191
methylene group, 178
regiospecific, 191
unsaturated, 174
vinyl group, 178
Z geometry, 180

An **alkene** is a hydrocarbon that contains a carbon–carbon double bonc Because they contain fewer hydrogens than alkanes with the same number c carbons, alkenes are said to be **unsaturated**.

Because rotation around the double bond can't occur, substituted alkene can exist as cis–trans stereoisomers. The geometry of a double bond can be spec ified by application of the Cahn–Ingold–Prelog sequence rules, which assign pr orities to double-bond substituents. If the high-priority groups on each carbo are on the same side of the double bond, the geometry is ***Z*** (*zusammer* "together"); if the high-priority groups on each carbon are on opposite sides c the double bond, the geometry is ***E*** (*entgegen,* "apart").

Alkene chemistry is dominated by **electrophilic addition reactions**. Whe HX reacts with an unsymmetrically substituted alkene, **Markovnikov's rul** predicts that the H will add to the carbon having fewer alkyl substituents an the X group will add to the carbon having more alkyl substituents. Electrophili additions to alkenes take place through carbocation intermediates formed b reaction of the nucleophilic alkene π bond with electrophilic H^+. Carbocatio stability follows the order

Tertiary (3°)	>	Secondary (2°)	>	Primary (1°)	>	Methyl
R_3C^+	>	R_2CH^+	>	RCH_2^+	>	CH_3^+

Markovnikov's rule can be restated by saying that, in the addition of HX tc an alkene, the more stable carbocation intermediate is formed. This result i explained by the **Hammond postulate**, which says that the transition state o an exergonic reaction step structurally resembles the reactant, whereas the tran sition state of an endergonic reaction step structurally resembles the product Since an alkene protonation step is endergonic, the stability of the more highl substituted carbocation is reflected in the stability of the transition state leadin to its formation.

Evidence in support of a carbocation mechanism for electrophilic addition comes from the observation that structural rearrangements often take place dur- ing reaction. Rearrangements occur by shift of either a hydride ion, $:H^-$ (a **hydride shift**), or an alkyl anion, $:R^-$, from a carbon atom to the adjacent pos- itively charged carbon. The result is isomerization of a less stable carbocation tc a more stable one.

EXERCISES

Organic KNOWLEDGE TOOLS

ThomsonNOW™ Sign in at **www.thomsonedu.com** to assess your knowledge of this chapter's topics by taking a pre-test. The pre-test will link you to interactive organic chemistry resources based on your score in each concept area.

Online homework for this chapter may be assigned in Organic OWL.

■ indicates problems assignable in Organic OWL.

▲ denotes problems linked to Key Ideas of this chapter and testable in ThomsonNOW.

VISUALIZING CHEMISTRY

(Problems 6.1–6.19 appear within the chapter.)

6.20 ■ Name the following alkenes, and convert each drawing into a skeletal structure:

(a)

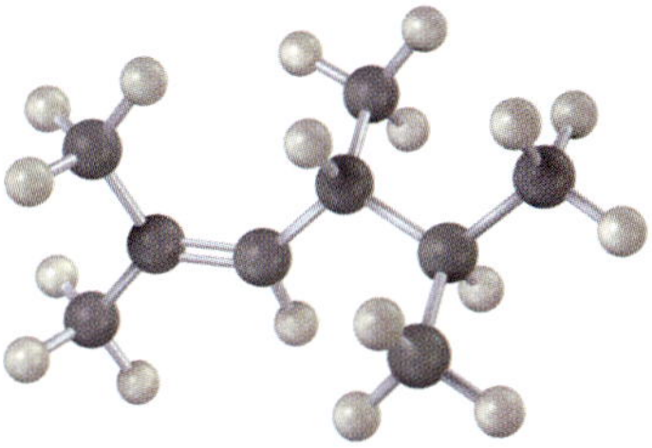

(b)

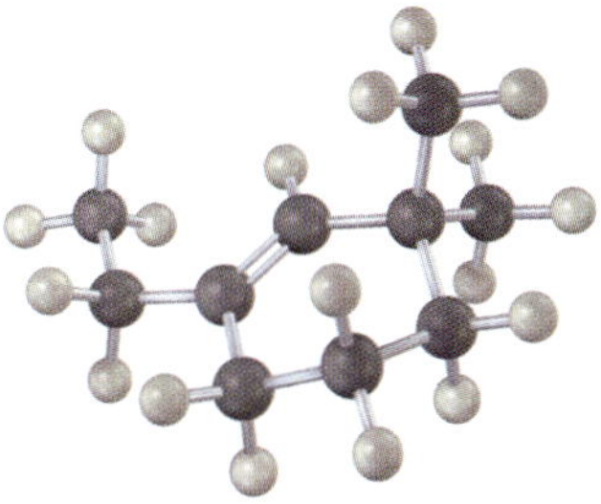

6.21 ■ Assign stereochemistry (*E* or *Z*) to the double bonds in each of the following compounds, and convert each drawing into a skeletal structure (red = O, yellow-green = Cl):

(a)

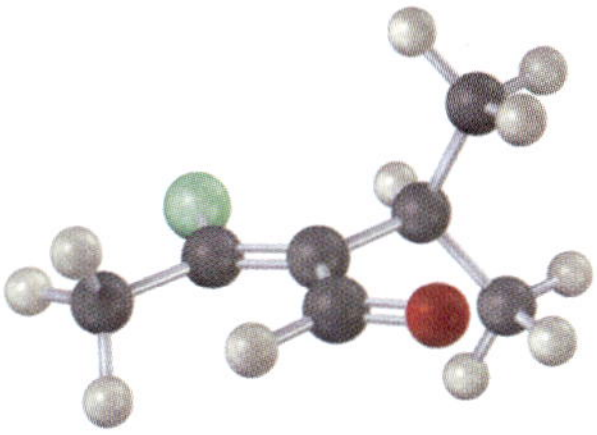

(b)

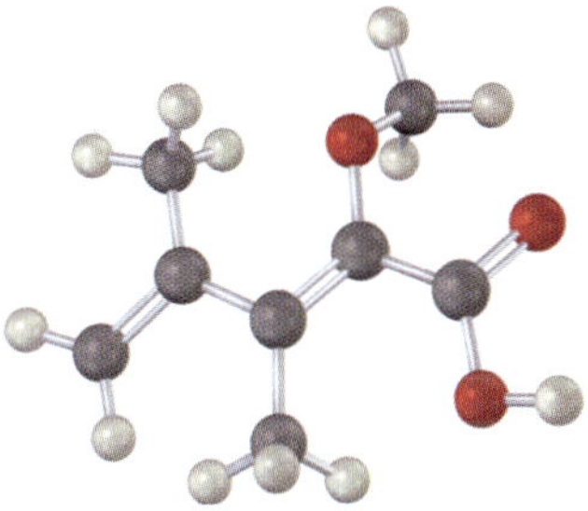

6.22 ■ The following carbocation is an intermediate in the electrophilic addition reaction of HCl with two different alkenes. Identify both, and tell which C–H bonds in the carbocation are aligned for hyperconjugation with the vacant *p* orbital on the positively charged carbon.

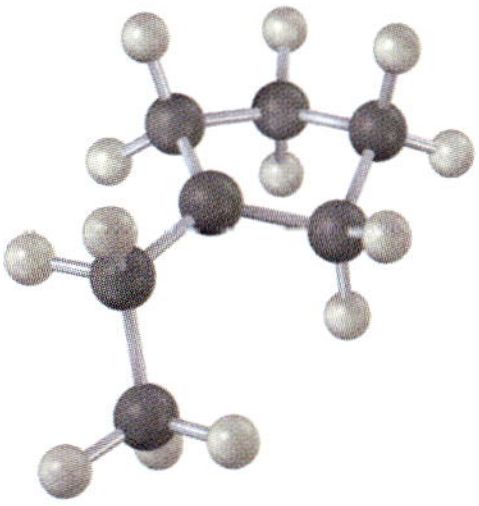

ADDITIONAL PROBLEMS

6.23 ■ Calculate the degree of unsaturation in the following formulas, and dra five possible structures for each:
(a) $C_{10}H_{16}$ (b) C_8H_8O (c) $C_7H_{10}Cl_2$
(d) $C_{10}H_{16}O_2$ (e) $C_5H_9NO_2$ (f) $C_8H_{10}ClNO$

6.24 How many hydrogens does each of the following compounds have?
(a) $C_8H_?O_2$, has two rings and one double bond
(b) $C_7H_?N$, has two double bonds
(c) $C_9H_?NO$, has one ring and three double bonds

6.25 Loratadine, marketed as an antiallergy medication under the name Claritir has four rings, eight double bonds, and the formula $C_{22}H_?ClN_2O_2$. How man hydrogens does loratadine have? (Calculate your answer; don't count hydrc gens in the structure.)

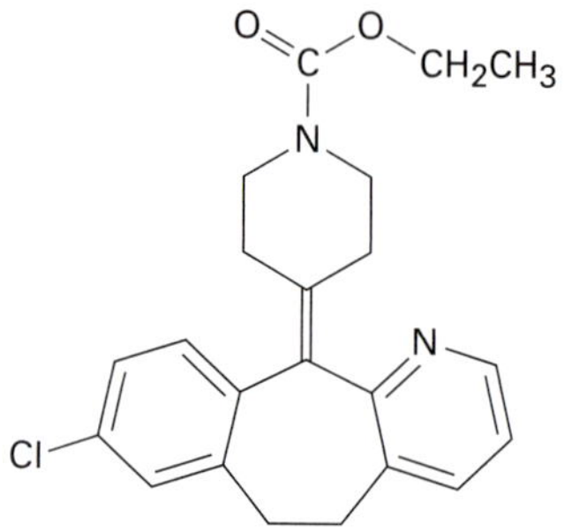

Loratadine

6.26 ■ Name the following alkenes:

(a)
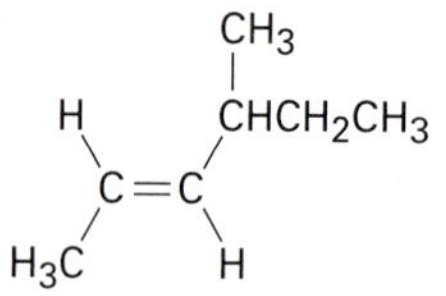

(b)
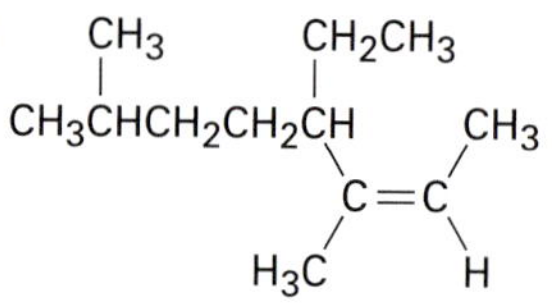

(c)
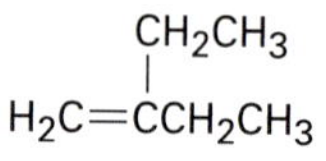

(d)
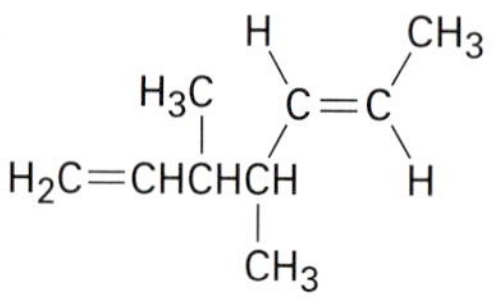

(e)
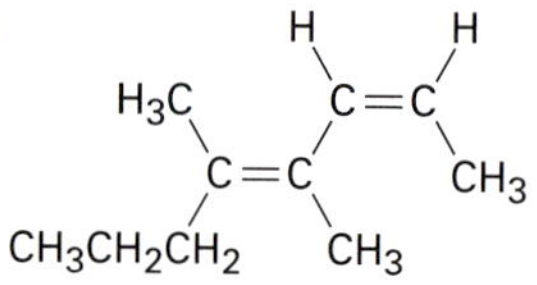

(f) $H_2C{=}C{=}CHCH_3$

6.27 ■ Ocimene is a triene found in the essential oils of many plants. What is its IUPAC name, including stereochemistry?

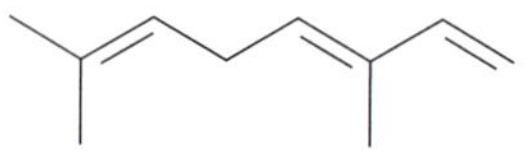

Ocimene

6.28 α-Farnesene is a constituent of the natural wax found on apples. What is its IUPAC name, including stereochemistry?

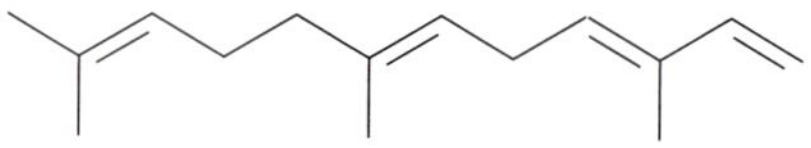

α-Farnesene

6.29 ■ Draw structures corresponding to the following systematic names:
(a) (4*E*)-2,4-Dimethyl-1,4-hexadiene
(b) *cis*-3,3-Dimethyl-4-propyl-1,5-octadiene
(c) 4-Methyl-1,2-pentadiene
(d) (3*E*,5*Z*)-2,6-Dimethyl-1,3,5,7-octatetraene
(e) 3-Butyl-2-heptene
(f) *trans*-2,2,5,5-Tetramethyl-3-hexene

6.30 ■ Menthene, a hydrocarbon found in mint plants, has the systematic name 1-isopropyl-4-methylcyclohexene. Draw its structure.

6.31 Draw and name the 6 pentene isomers, C_5H_{10}, including *E*,*Z* isomers.

6.32 Draw and name the 17 hexene isomers, C_6H_{12}, including *E*,*Z* isomers.

6.33 *trans*-2-Butene is more stable than *cis*-2-butene by only 4 kJ/mol, but *trans*-2,2,5,5-tetramethyl-3-hexene is more stable than its cis isomer by 39 kJ/mol. Explain.

6.34 Cyclodecene can exist in both cis and trans forms, but cyclohexene cannot. Explain. (Making molecular models is helpful.)

6.35 Normally, a trans alkene is *more* stable than its cis isomer. *trans*-Cyclooctene, however, is *less* stable than *cis*-cyclooctene by 38.5 kJ/mol. Explain.

6.36 *trans*-Cyclooctene is less stable than *cis*-cyclooctene by 38.5 kJ/mol, but *trans*-cyclononene is less stable than *cis*-cyclononene by only 12.2 kJ/mol. Explain.

6.37 Allene (1,2-propadiene), $H_2C{=}C{=}CH_2$, has two adjacent double bonds. What kind of hybridization must the central carbon have? Sketch the bonding π orbitals in allene. What shape do you predict for allene?

6.38 The heat of hydrogenation for allene (Problem 6.37) to yield propane is −295 kJ/mol, and the heat of hydrogenation for a typical monosubstituted alkene such as propene is −126 kJ/mol. Is allene more stable or less stable than you might expect for a diene? Explain.

6.39 ■ Predict the major product in each of the following reactions:

(a)
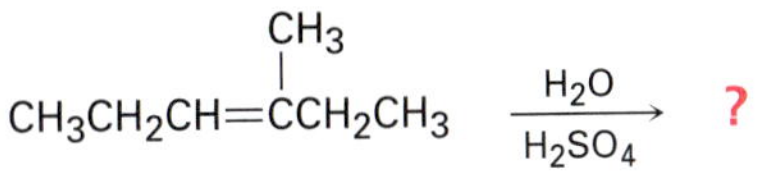

(Addition of H_2O occurs.)

(b)
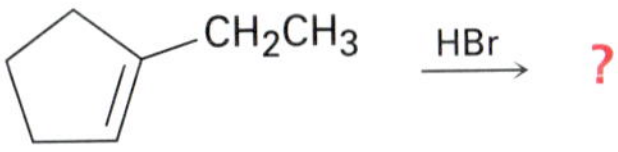

(c)
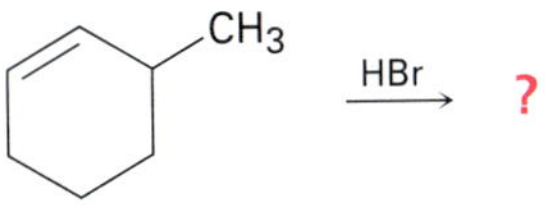

(d) $H_2C{=}CHCH_2CH_2CH_2CH{=}CH_2 \xrightarrow{2\ HCl}$?

6.40 ■ Predict the major product from addition of HBr to each of the following alkenes:

(a)
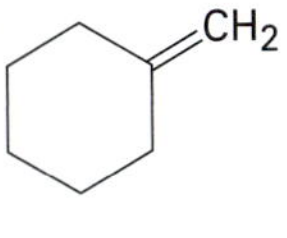

(b)
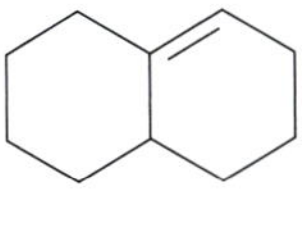

(c)
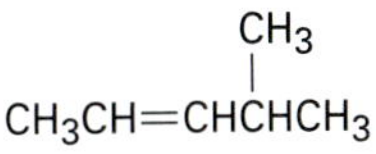

6.41 ■ Rank the following sets of substituents in order of priority according to th Cahn–Ingold–Prelog sequence rules:

(a) $-CH_3$, $-Br$, $-H$, $-I$

(b) $-OH$, $-OCH_3$, $-H$, $-CO_2H$

(c) $-CO_2H$, $-CO_2CH_3$, $-CH_2OH$, $-CH_3$

(d) $-CH_3$, $-CH_2CH_3$, $-CH_2CH_2OH$, $-\overset{\displaystyle O}{\overset{\|}{C}}CH_3$

(e) $-CH{=}CH_2$, $-CN$, $-CH_2NH_2$, $-CH_2Br$

(f) $-CH{=}CH_2$, $-CH_2CH_3$, $-CH_2OCH_3$, $-CH_2OH$

6.42 ▲ Assign *E* or *Z* configuration to each of the following alkenes:

(a)
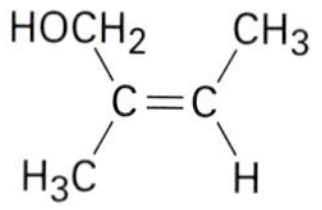

(b)
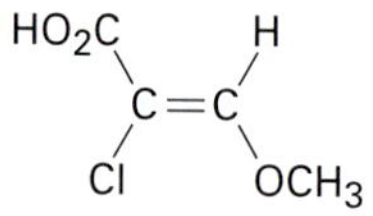

(c)
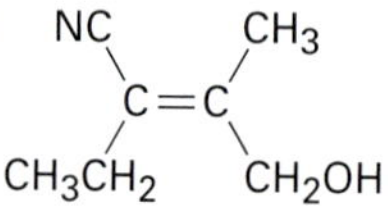

(d)
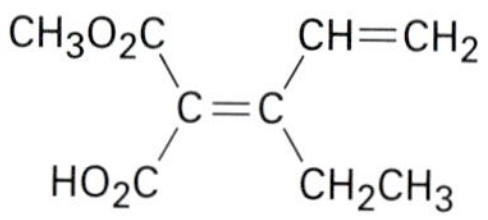

6.43 ■ Name the following cycloalkenes:

(a)
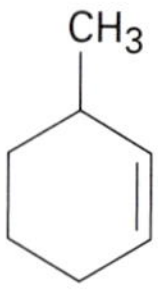

(b)
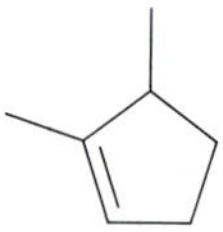

(c)
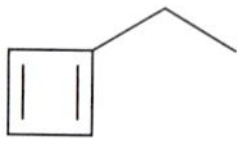

(d)
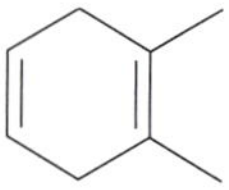

(e)
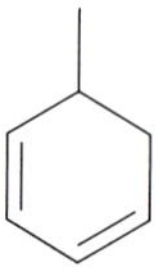

(f)
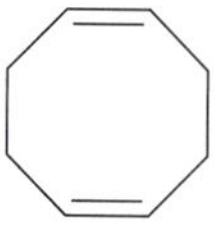

6.44 Fucoserraten, ectocarpen, and multifidene are sex pheromones produced by marine brown algae. What are their systematic names? (The latter two are a bit difficult; make your best guess.)

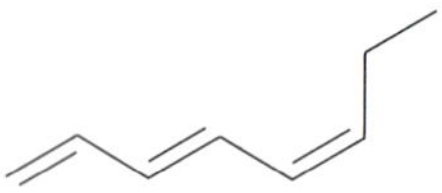

Fucoserraten

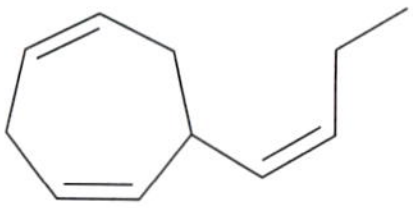

Ectocarpen

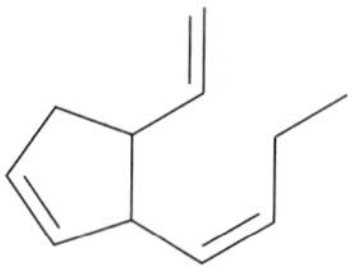

Multifidene

6.45 ▲ Which of the following *E,Z* designations are correct, and which are incorrect?

(a) 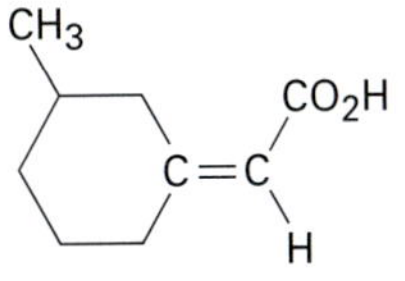

Z

(b) 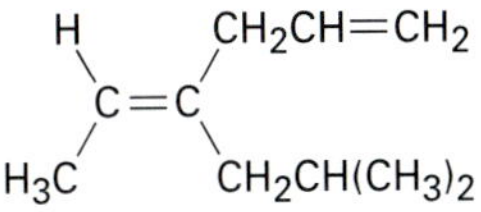

E

(c) 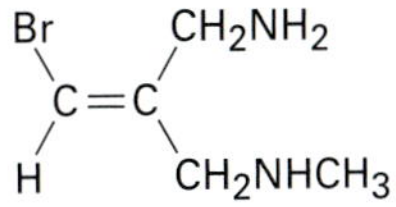

Z

(d) 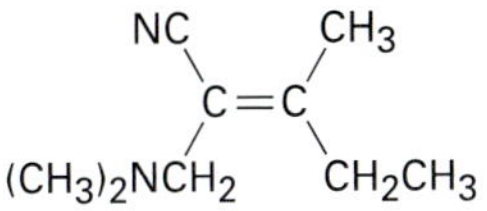

E

(e) 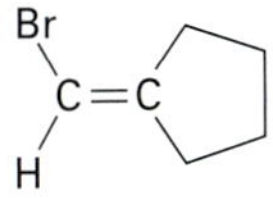

Z

(f) 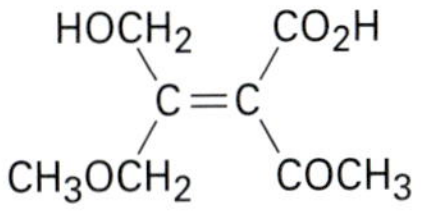

E

6.46 ▲ *tert*-Butyl esters [$RCO_2C(CH_3)_3$] are converted into carboxylic acids (RCO_2H) by reaction with trifluoroacetic acid, a reaction useful in protein synthesis (Section 26.7). Assign *E,Z* designation to the double bonds of both reactant and product in the following scheme, and explain why there is an apparent change of double-bond stereochemistry:

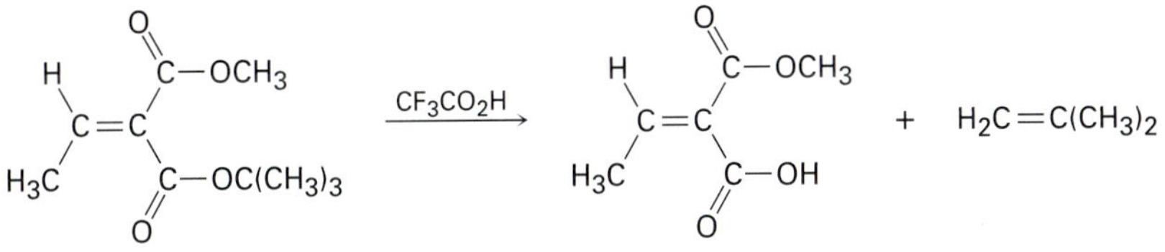

6.47 ■ Each of the following carbocations can rearrange to a more stable ion. Propose structures for the likely rearrangement products.

(a) $CH_3CH_2CH_2CH_2^+$ (b) 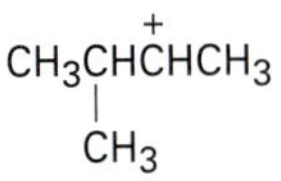 (c) 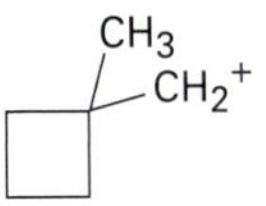

6.48 Addition of HCl to 1-isopropylcyclohexene yields a rearranged product. Propose a mechanism, showing the structures of the intermediates and using curved arrows to indicate electron flow in each step.

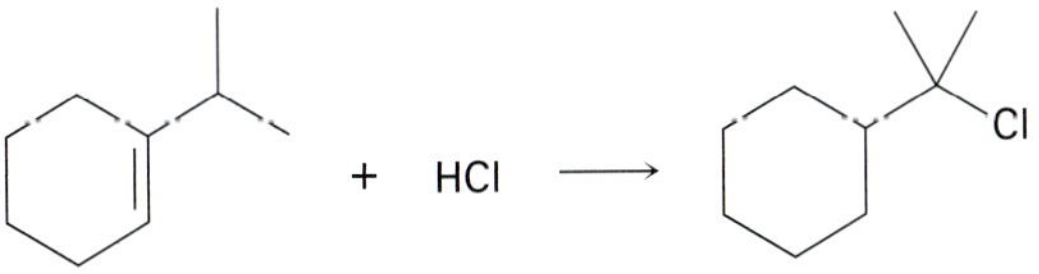

6.49 Addition of HCl to 1-isopropenyl-1-methylcyclopentane yields 1-chloro-1,2,2-trimethylcyclohexane. Propose a mechanism, showing the structures of the intermediates and using curved arrows to indicate electron flow in each step.

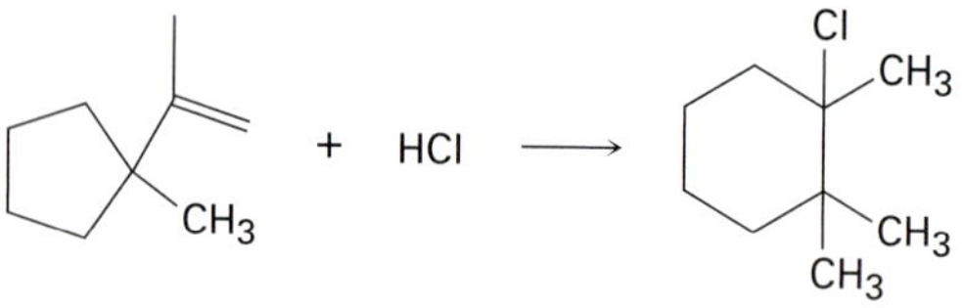

6.50 Vinylcyclopropane reacts with HBr to yield a rearranged alkyl bromide. Follow the flow of electrons as represented by the curved arrows, show the structure of the carbocation intermediate in brackets, and show the structure of the final product.

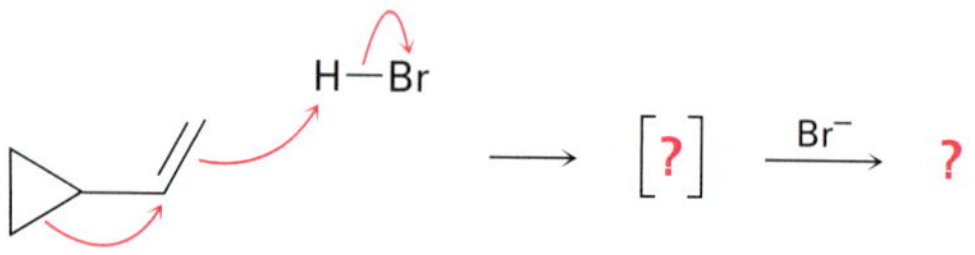

Vinylcyclopropane

6.51 ■ Calculate the degree of unsaturation in each of the following formulas:

(a) Cholesterol, $C_{27}H_{46}O$ **(b)** DDT, $C_{14}H_9Cl_5$

(c) Prostaglandin E_1, $C_{20}H_{34}O_5$ **(d)** Caffeine, $C_8H_{10}N_4O_2$

(e) Cortisone, $C_{21}H_{28}O_5$ **(f)** Atropine, $C_{17}H_{23}NO_3$

6.52 The isobutyl cation spontaneously rearranges to the *tert*-butyl cation by a hydride shift. Is the rearrangement exergonic or endergonic? Draw what you think the transition state for the hydride shift might look like according to the Hammond postulate.

Isobutyl cation ***tert*-Butyl cation**

6.53 Draw an energy diagram for the addition of HBr to 1-pentene. Let one curve on your diagram show the formation of 1-bromopentane product and another curve on the same diagram show the formation of 2-bromopentane product. Label the positions for all reactants, intermediates, and products. Which curve has the higher-energy carbocation intermediate? Which curve has the higher-energy first transition state?

6.54 Make sketches of the transition-state structures involved in the reaction of HBr with 1-pentene (Problem 6.53). Tell whether each structure resembles reactant or product.

6.55 Limonene, a fragrant hydrocarbon found in lemons and oranges, is biosynthesized from geranyl diphosphate by the following pathway. Add curved arrows to show the mechanism of each step. Which step involves an alkene electrophilic addition? (The ion $OP_2O_6^{4-}$ is the diphosphate ion, and "Base" is an unspecified base in the enzyme that catalyzes the reaction.)

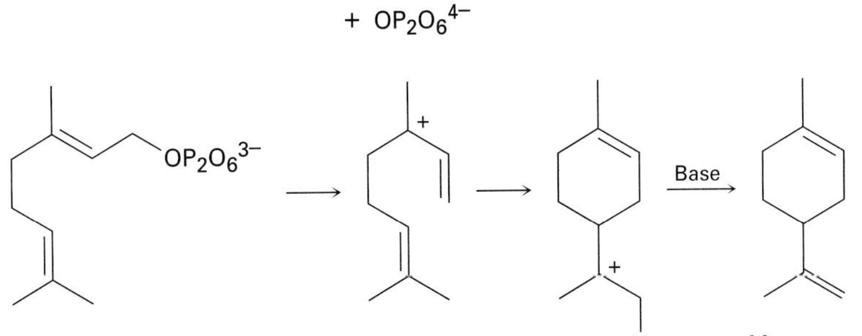

Geranyl diphosphate **Limonene**

6.56 *epi*-Aristolochene, a hydrocarbon found in both pepper and tobacco, is biosynthesized by the following pathway. Add curved arrows to show the mechanism of each step. Which steps involve alkene electrophilic addition(s), and which involve carbocation rearrangement(s)? (The abbreviation H–A stands for an unspecified acid, and "Base" is an unspecified base in the enzyme.)

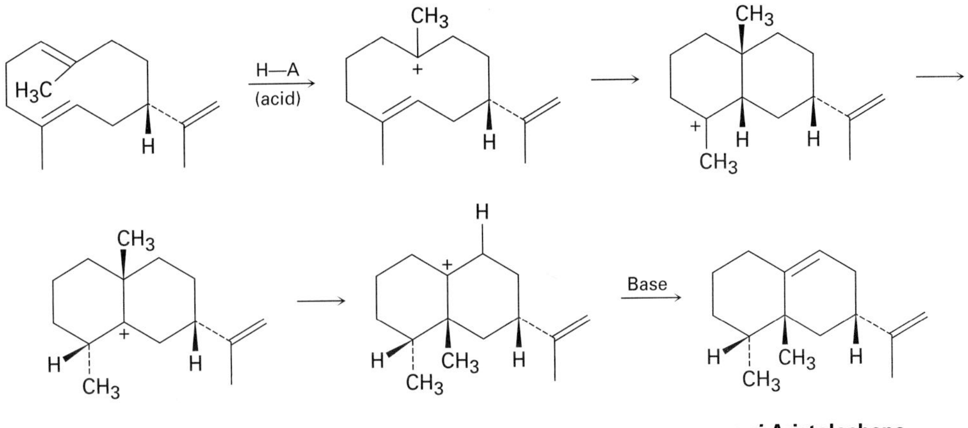

***epi*-Aristolochene**

6.57 Aromatic compounds such as benzene react with alkyl chlorides in the presence of $AlCl_3$ catalyst to yield alkylbenzenes. The reaction occurs through a carbocation intermediate, formed by reaction of the alkyl chloride with $AlCl_3$ ($R-Cl + AlCl_3 \rightarrow R^+ + AlCl_4^-$). How can you explain the observation that reaction of benzene with 1-chloropropane yields isopropylbenzene as the major product?

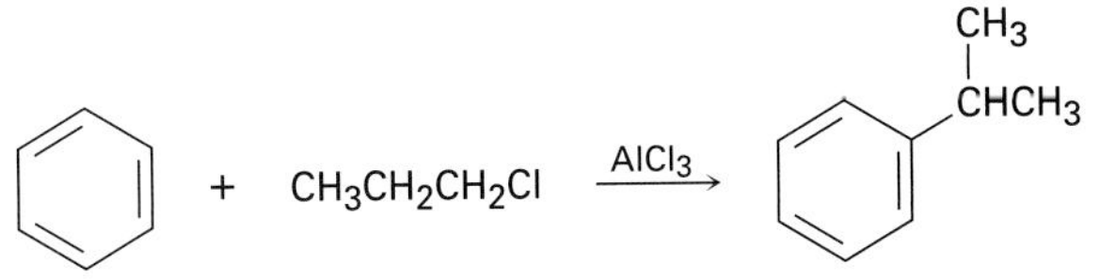

6.58 ▲ Alkenes can be converted into alcohols by acid-catalyzed addition of wate
Assuming that Markovnikov's rule is valid, predict the major alcohol produ
from each of the following alkenes.

(a) 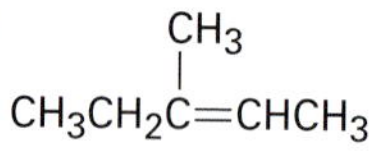

(b) 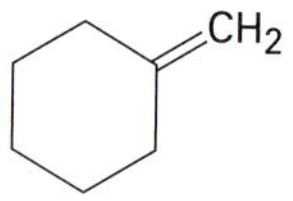

(c) 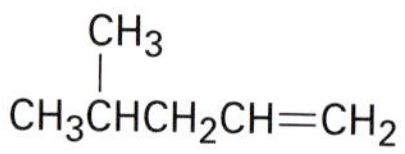

6.59 Reaction of 2,3-dimethyl-1-butene with HBr leads to an alkyl bromid $C_6H_{13}Br$. On treatment of this alkyl bromide with KOH in methanol, elimina tion of HBr to give an alkene occurs and a hydrocarbon that is isomeric wit the starting alkene is formed. What is the structure of this hydrocarbon, an how do you think it is formed from the alkyl bromide?

7

Alkenes: Reactions and Synthesis

Organic KNOWLEDGE TOOLS

ThomsonNOW Throughout this chapter, sign in at **www.thomsonedu.com** for online self-study and interactive tutorials based on your level of understanding.

Online homework for this chapter may be assigned in Organic OWL.

Alkene addition reactions occur widely, both in the laboratory and in living organisms. Although we've studied only the addition of HX thus far, many closely related reactions also take place. In this chapter, we'll see briefly how alkenes are prepared, we'll discuss many further examples of alkene addition reactions, and we'll see the wide variety of compounds that can be made from alkenes.

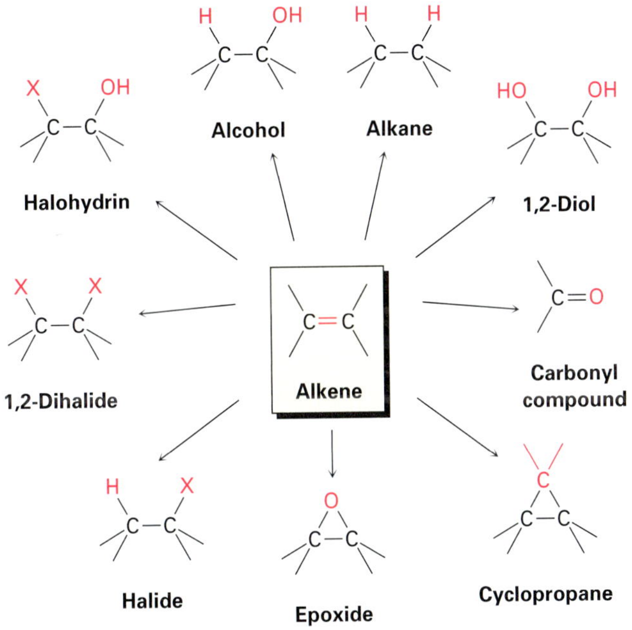

WHY THIS CHAPTER?

Much of the background needed to understand organic reactions has now been covered, and it's time to begin a systematic description of the major functional groups. Both in this chapter on alkenes and in future chapters on other functional groups, we'll discuss a variety of reactions but try to focus on the general principles and patterns of reactivity that tie organic chemistry together. There are no shortcuts: you have to know the reactions to understand organic chemistry.

7.1 Preparation of Alkenes: A Preview of Elimination Reactions

Before getting to the main subject of this chapter—the reactions of alkenes—let's take a brief look at how alkenes are prepared. The subject is a bit complex though, so we'll return in Chapter 11 for a more detailed study. For the present it's enough to realize that alkenes are readily available from simple precursors—usually alcohols in biological systems and either alcohols or alkyl halides in the laboratory.

Just as the chemistry of alkenes is dominated by addition reactions, the preparation of alkenes is dominated by elimination reactions. Additions and eliminations are, in many respects, two sides of the same coin. That is, an addition reaction might involve the addition of HBr or H_2O to an alkene to form an alkyl halide or alcohol, whereas an elimination reaction might involve the loss of HBr or H_2O from an alkyl halide or alcohol to form an alkene.

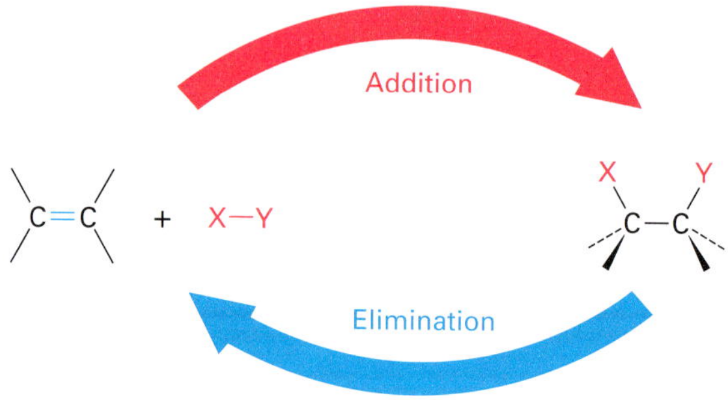

The two most common elimination reactions are *dehydrohalogenation*—the loss of HX from an alkyl halide—and *dehydration*—the loss of water from an alcohol. Dehydrohalogenation usually occurs by reaction of an alkyl halide with strong base such as potassium hydroxide. For example, bromocyclohexane yields cyclohexene when treated with KOH in ethanol solution.

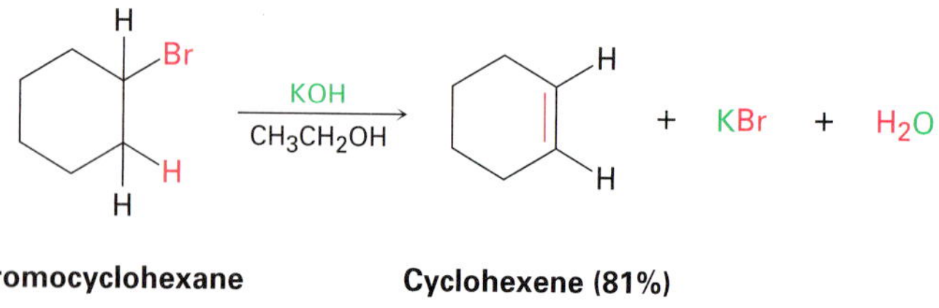

Dehydration is often carried out by treatment of an alcohol with a strong acid. For example, loss of water occurs and 1-methylcyclohexene is formed when 1-methylcyclohexanol is warmed with aqueous sulfuric acid in tetrahydrofuran (THF) solvent.

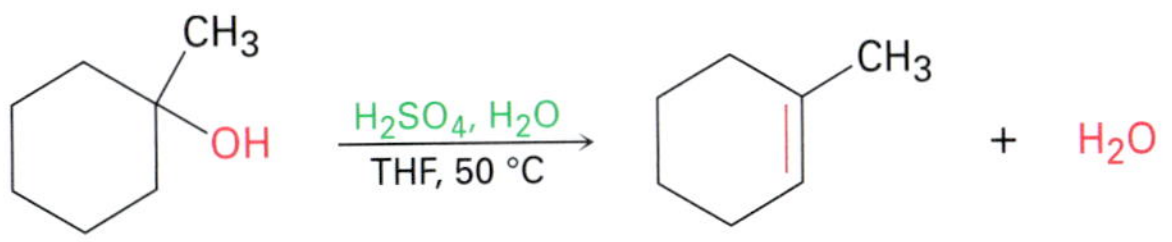

1-Methylcyclohexanol **1-Methylcyclohexene (91%)**

Tetrahydrofuran (THF)—a common solvent

In biological pathways, dehydrations rarely occur with isolated alcohols but instead normally take place on substrates in which the –OH is positioned two carbons away from a carbonyl group. In the biosynthesis of fats, for instance, β-hydroxybutyryl ACP is converted by dehydration to *trans*-crotonyl ACP, where ACP is an abbreviation for *acyl carrier protein*. We'll see the reason for this requirement in Section 11.10.

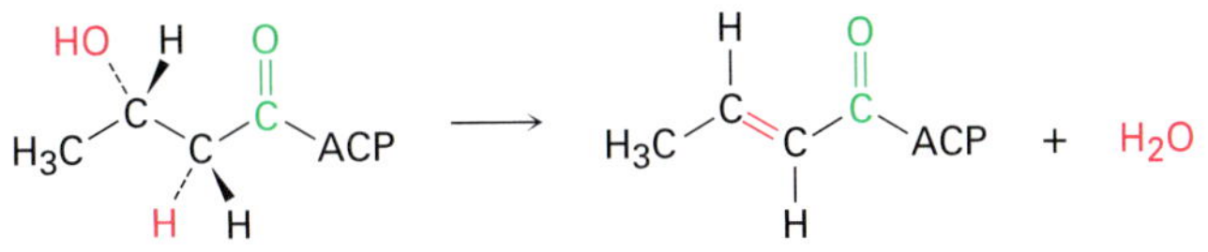

β-Hydroxybutyryl ACP ***trans*-Crotonyl ACP**

Problem 7.1 One problem with elimination reactions is that mixtures of products are often formed. For example, treatment of 2-bromo-2-methylbutane with KOH in ethanol yields a mixture of two alkene products. What are their likely structures?

Problem 7.2 How many alkene products, including *E,Z* isomers, might be obtained by dehydration of 3-methyl-3-hexanol with aqueous sulfuric acid?

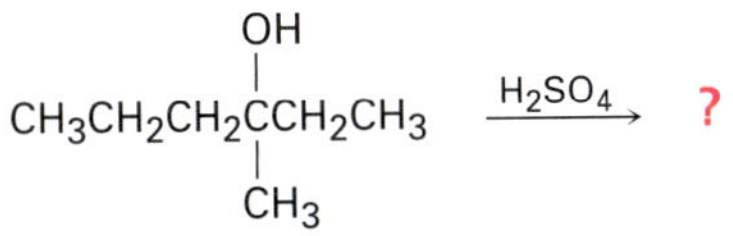

3-Methyl-3-hexanol

7.2 Addition of Halogens to Alkenes

ThomsonNOW™ Click *Organic Interactive* to **use a web-based palette to predict products of the addition of halogens to alkenes.**

Bromine and chlorine add rapidly to alkenes to yield 1,2-dihalides, a process called *halogenation*. For example, approximately 6 million tons per year of 1,2-dichloroethane (ethylene dichloride) are synthesized industrially by addition of Cl_2 to ethylene. The product is used both as a solvent and as starting materia for the manufacture of poly(vinyl chloride), PVC. Fluorine is too reactive and dif ficult to control for most laboratory applications, and iodine does not react with most alkenes.

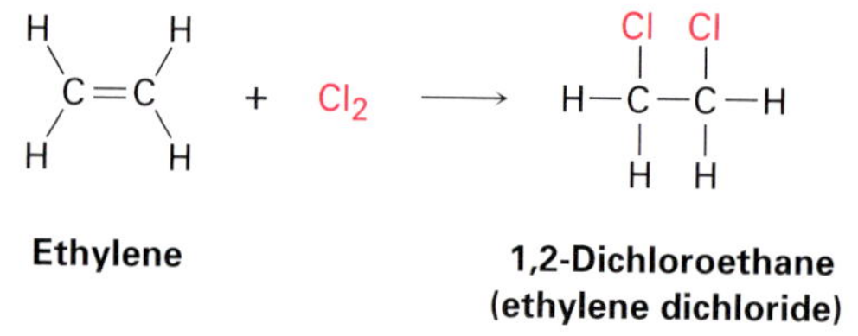

Based on what we've seen thus far, a possible mechanism for the reaction of bromine with alkenes might involve electrophilic addition of Br^+ to the alkene, giving a carbocation that could undergo further reaction with Br^- to yield the dibromo addition product.

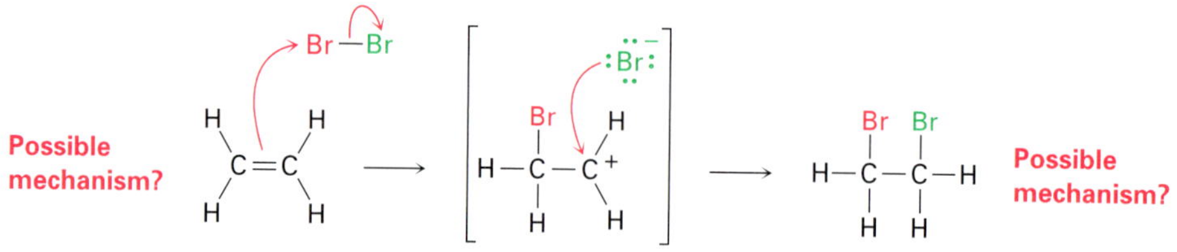

Although this mechanism seems plausible, it's not fully consistent with known facts. In particular, it doesn't explain the *stereochemistry* of the addition reaction. That is, the mechanism doesn't tell which product stereoisomer is formed.

When the halogenation reaction is carried out on a cycloalkene, such as cyclopentene, only the *trans* stereoisomer of the dihalide addition product is formed rather than the mixture of cis and trans isomers that might have been expected if a planar carbocation intermediate were involved. We say that the reaction occurs with **anti stereochemistry**, meaning that the two bromine atoms come from opposite faces of the double bond—one from the top face and one from the bottom face.

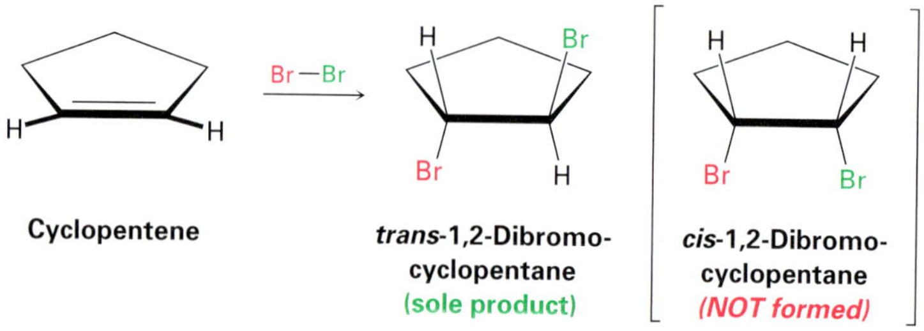

An explanation for the observed anti stereochemistry of addition was suggested in 1937 by George Kimball and Irving Roberts, who proposed that the reaction intermediate is not a carbocation but is instead a **bromonium ion, R_2Br^+**, formed by addition of Br^+ to the alkene. (Similarly, a *chloronium ion* contains a positively charged divalent chlorine, R_2Cl^+.) The bromonium ion is formed in a single step by interaction of the alkene with Br_2 and simultaneous loss of Br^-.

omsonNOW™ Click *Organic ocess* to **view an animation of e bromonium ion intermediate nd product formation in this action.**

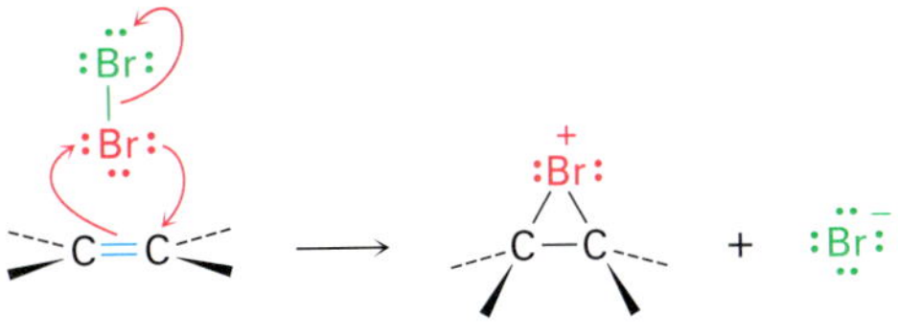

An alkene **A bromonium ion**

How does the formation of a bromonium ion account for the observed anti stereochemistry of addition to cyclopentene? If a bromonium ion is formed as an intermediate, we can imagine that the large bromine atom might "shield" one side of the molecule. Reaction with Br^- ion in the second step could then occur only from the opposite, unshielded side to give trans product.

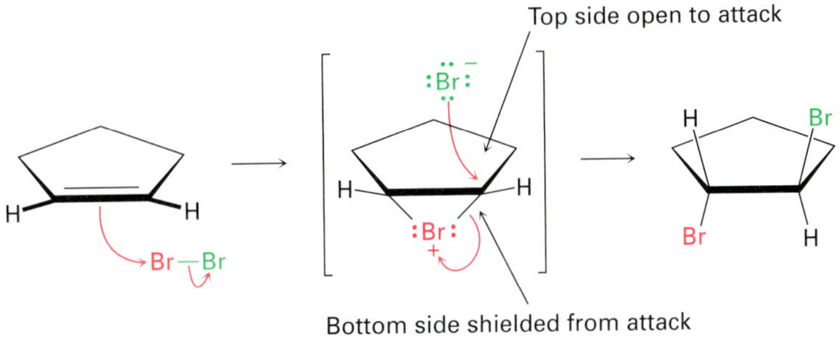

Cyclopentene **Bromonium ion intermediate** ***trans*-1,2-Dibromocyclopentane**

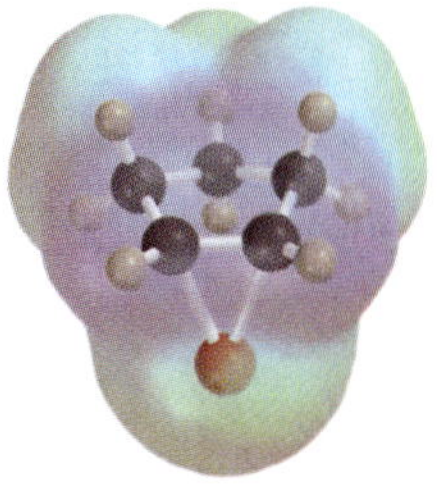

George Andrew Olah

George Andrew Olah (1927–) was born in Budapest, Hungary, and received a doctorate in 1949 at the Technical University of Budapest. During the Hungarian revolution in 1956, he immigrated to Canada and joined the Dow Chemical Company. After moving to the United States, he was professor of chemistry at Case Western Reserve University (1965–1977) and then at the University of Southern California (1977). He received the 1994 Nobel Prize in chemistry for his work on carbocations.

The bromonium ion postulate, made more than 75 years ago to explain the stereochemistry of halogen addition to alkenes, is a remarkable example of deductive logic in chemistry. Arguing from experimental results, chemists were able to make a hypothesis about the intimate mechanistic details of alkene electrophilic reactions. Subsequently, strong evidence supporting the mechanism came from the work of George Olah, who prepared and studied *stable* solutions of cyclic bromonium ions in liquid SO_2. There's no question th bromonium ions exist.

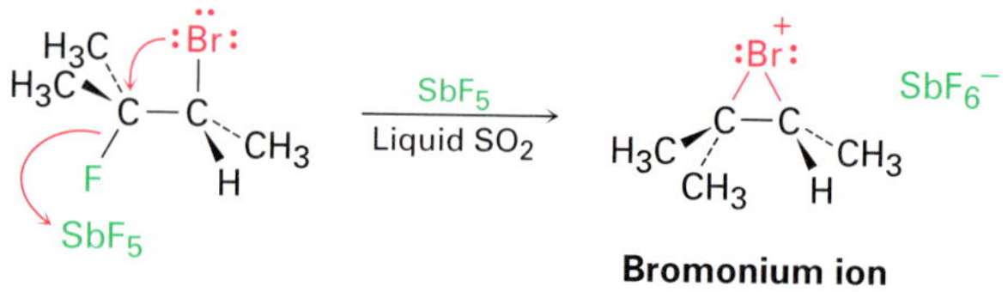

Bromonium ion (stable in SO_2 solution)

Alkene halogenation reactions occur in nature just as they do in the labora tory but are limited primarily to marine organisms, which live in a halide-ric environment. The reactions are carried out by enzymes called *haloperoxidase* which use H_2O_2 to oxidize Br^- or Cl^- ions to a biological equivalent of Br^+ c Cl^+. Electrophilic addition to the double bond of a substrate molecule the yields a bromonium or chloronium ion intermediate just as in the laborator and reaction with another halide ion completes the process. For example, th following tetrahalide, isolated from the red alga *Plocamium cartilagineum,* i thought to arise from β-ocimene by twofold addition of BrCl through the cor responding bromonium ions.

β-Ocimene

Problem 7.3 What product would you expect to obtain from addition of Cl_2 to 1,2-dimethyl cyclohexene? Show the stereochemistry of the product.

Problem 7.4 Addition of HCl to 1,2-dimethylcyclohexene yields a mixture of two products. Show the stereochemistry of each, and explain why a mixture is formed.

7.3 Addition of Hypohalous Acids to Alkenes: Halohydrin Formation

ThomsonNOW™ Click *Organic Interactive* to **use a web-based palette to predict products of the addition of hypohalous acid to alkenes.**

Yet another example of an electrophilic addition is the reaction of alkenes with the hypohalous acids HO–Cl or HO–Br to yield 1,2-halo alcohols, called **halohydrins**. Halohydrin formation doesn't take place by direct reaction of an alkene with HOBr or HOCl, however. Rather, the addition is done indirectly by reaction of the alkene with either Br_2 or Cl_2 in the presence of water.

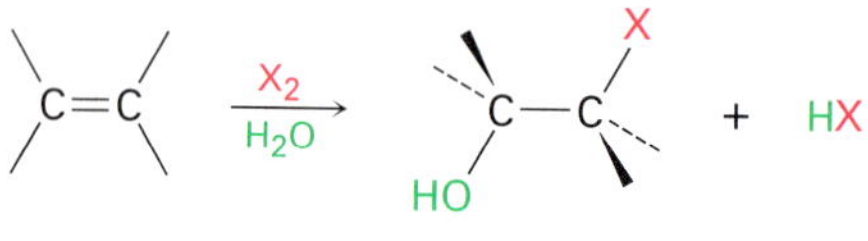

An alkene **A halohydrin**

We saw in the previous section that when Br_2 reacts with an alkene, the cyclic bromonium ion intermediate reacts with the only nucleophile present, Br^- ion. If the reaction is carried out in the presence of an additional nucleophile, however, the intermediate bromonium ion can be intercepted by the added nucleophile and diverted to a different product. In the presence of water, for instance, water competes with Br^- ion as nucleophile and reacts with the bromonium ion intermediate to yield a *bromohydrin*. The net effect is addition of HO–Br to the alkene by the pathway shown in Figure 7.1.

igure 7.1 MECHANISM: Mechanism of bromohydrin formation by reaction of an alkene with Br_2 in the presence of water. Water acts as a nucleophile to react with the intermediate bromonium ion.

1. Reaction of the alkene with Br_2 yields a bromonium ion intermediate, as previously discussed.

2. Water acts as a nucleophile, using a lone pair of electrons to open the bromonium ion ring and form a bond to carbon. Since oxygen donates its electrons in this step, it now has the positive charge.

3. Loss of a proton (H^+) from oxygen then gives H_3O^+ and the neutral bromohydrin addition product.

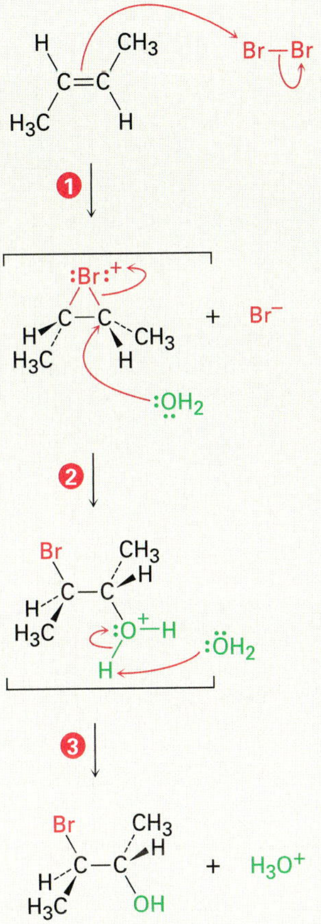

3-Bromo-2-butanol
(a bromohydrin)

In practice, few alkenes are soluble in water, and bromohydrin formation is often carried out in a solvent such as aqueous dimethyl sulfoxide, CH_3SOCH_3 (DMSO), using a reagent called *N*-bromosuccinimide (NBS) as a source of Br_2. NBS is a stable, easily handled compound that slowly decomposes in water to yield Br_2 at a controlled rate. Bromine itself can also be used in the addition reaction, but it is more dangerous and more difficult to handle than NBS.

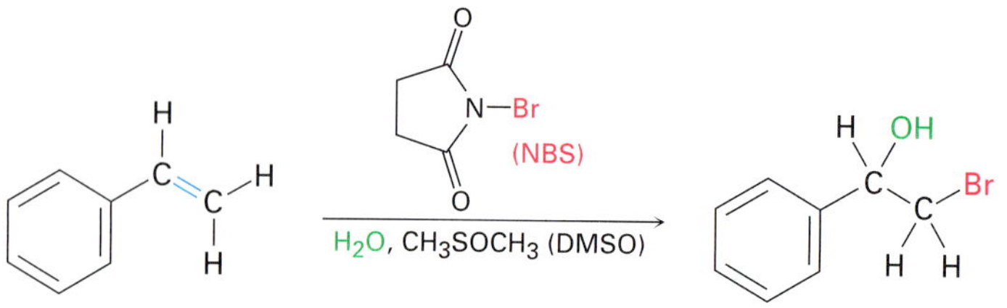

Styrene

2-Bromo-1-phenylethanol (70%)

Note that the aromatic ring in the preceding example does not react with Br_2 under the conditions used, even though it appears to contain three carbon–carbon double bonds. As we'll see in Chapter 15, aromatic rings are a good deal more stable than might be expected.

Problem 7.5 What product would you expect from the reaction of cyclopentene with NBS and water? Show the stereochemistry.

Problem 7.6 When an unsymmetrical alkene such as propene is treated with *N*-bromosuccinimide in aqueous dimethyl sulfoxide, the major product has the bromine atom bonded to the less highly substituted carbon atom. Is this Markovnikov or non-Markovnikov orientation? Explain.

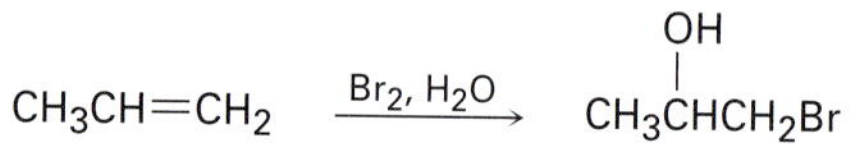

7.4 Addition of Water to Alkenes: Oxymercuration

Water adds to alkenes to yield alcohols, a process called *hydration*. The reaction takes place on treatment of the alkene with water and a strong acid catalyst (HA) by a mechanism similar to that of HX addition. Thus, protonation of an alkene double bond yields a carbocation intermediate, which reacts with water to yield a protonated alcohol product (ROH_2^+). Loss of H^+ from this protonated alcohol gives the neutral alcohol and regenerates the acid catalyst (Figure 7.2).

Acid-catalyzed alkene hydration is particularly suited to large-scale industrial procedures, and approximately 300,000 tons of ethanol are manufactured each year in the United States by hydration of ethylene. The reaction is of little value in the typical laboratory, however, because it requires high temperatures—250 °C in the case of ethylene—and strongly acidic conditions.

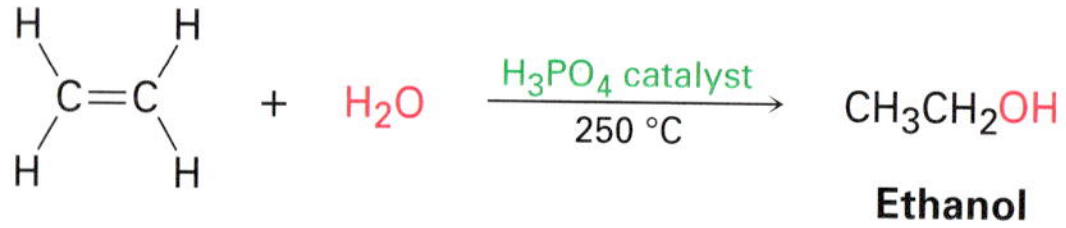

Ethylene

Ethanol

igure 7.2 MECHANISM: echanism of the acid-catalyzed ydration of an alkene to yield an cohol. Protonation of the kene gives a carbocation inter- ediate that reacts with water.

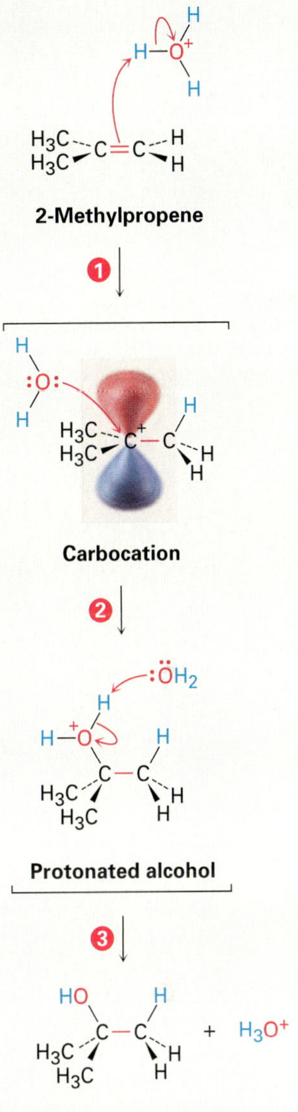

1. A hydrogen atom on the electrophile H_3O^+ is attacked by π electrons from the nucleophilic double bond, forming a new C–H bond. This leaves the other carbon atom with a + charge and a vacant *p* orbital. Simultaneously, two electrons from the H–O bond move onto oxygen, giving neutral water.

2. The nucleophile H_2O donates an electron pair to the positively charged carbon atom, forming a C–O bond and leaving a positive charge on oxygen in the protonated alcohol addition product.

3. Water acts as a base to remove H^+, regenerating H_3O^+ and yielding the neutral alcohol addition product.

Acid-catalyzed hydration of isolated double bonds is also uncommon in biological pathways. More frequently, biological hydrations require that the double bond be adjacent to a carbonyl group for reaction to proceed. Fumarate, for instance, is hydrated to give malate as one step in the citric acid cycle of food metabolism. Note that the requirement for an adjacent carbonyl group in the addition of water is the same as that we saw in Section 7.1 for the elimination of water. We'll see the reason for the requirement in Section 19.13, but might note for now that the reaction is not an electrophilic addition but instead occurs through a mechanism that involves formation of an anion intermediate followed by protonation by an acid HA.

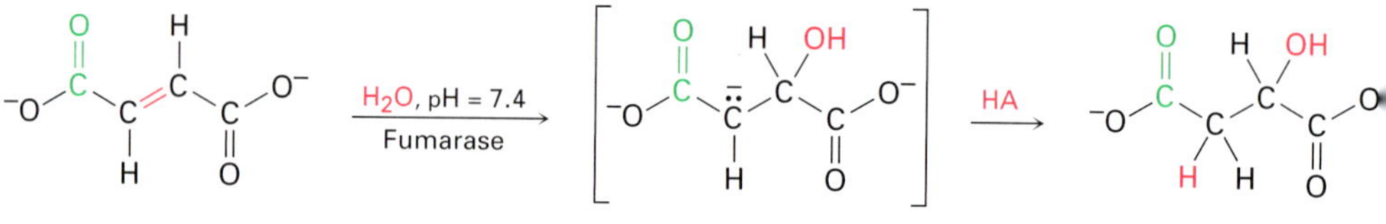

Fumarate **Anion intermediate** **Malate**

ThomsonNOW™ Click *Organic Interactive* to **use a web-based palette to predict products of the oxymercuration of alkenes.**

In the laboratory, alkenes are often hydrated by the **oxymercuration** procedure. When an alkene is treated with mercury(II) acetate [$Hg(O_2CCH_3)_2$, usually abbreviated $Hg(OAc)_2$] in aqueous tetrahydrofuran (THF) solvent, electrophilic addition of Hg^{2+} to the double bond rapidly occurs. The intermediate *organomercury* compound is then treated with sodium borohydride, $NaBH_4$, and an alcohol is produced. For example:

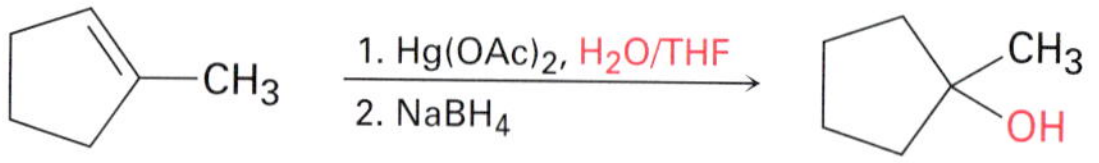

1-Methylcyclopentene **1-Methylcyclopentanol (92%)**

Alkene oxymercuration is closely analogous to halohydrin formation. The reaction is initiated by electrophilic addition of Hg^{2+} (mercuric) ion to the alkene to give an intermediate *mercurinium ion,* whose structure resembles that of a bromonium ion (Figure 7.3). Nucleophilic addition of water as in halohydrin formation, followed by loss of a proton, then yields a stable organomercury product. The final step, reaction of the organomercury compound with sodium borohydride, is complex and appears to involve radicals. Note that the regiochemistry of the reaction corresponds to Markovnikov addition of water; that is, the −OH group attaches to the more highly substituted carbon atom, and the −H attaches to the less highly substituted carbon.

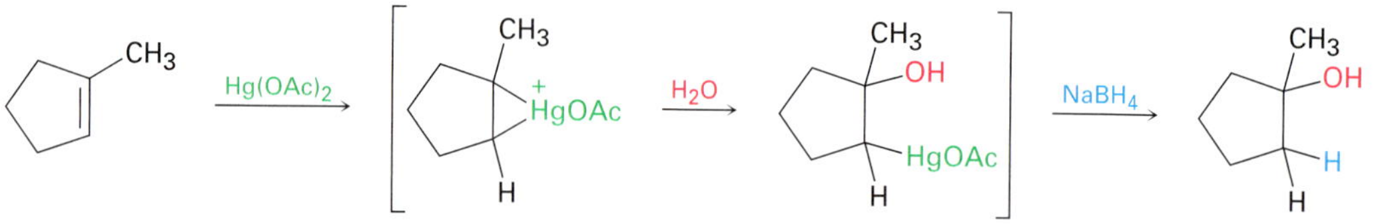

1-Methyl-cyclopentene **Mercurinium ion** **Organomercury compound** **1-Methyl-cyclopentanol (92% yield)**

Figure 7.3 Mechanism of the oxymercuration of an alkene to yield an alcohol. The reaction involves a mercurinium ion intermediate and proceeds by a mechanism similar to that of halohydrin formation. The product of the reaction is the more highly substituted alcohol, corresponding to Markovnikov regiochemistry.

Problem 7.7 What products would you expect from oxymercuration of the following alkenes?

(a) $CH_3CH_2CH_2CH{=}CH_2$

(b) $CH_3C(CH_3){=}CHCH_2CH_3$

Problem 7.8 What alkenes might the following alcohols have been prepared from?

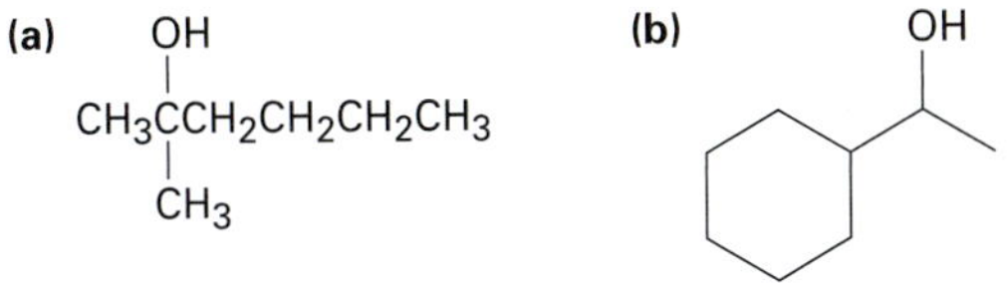

7.5 Addition of Water to Alkenes: Hydroboration

ThomsonNOW™ Click *Organic Interactive* to **use a web-based palette to predict products of the hydroboration/oxidation of alkenes.**

In addition to the oxymercuration method, which yields the Markovnikov product, a complementary method that yields the non-Markovnikov product is also useful. Discovered in 1959 by H. C. Brown and called **hydroboration**, the reaction involves addition of a B–H bond of borane, BH_3, to an alkene to yield an organoborane intermediate, RBH_2. Oxidation of the organoborane by reaction with basic hydrogen peroxide, H_2O_2, then gives an alcohol. For example:

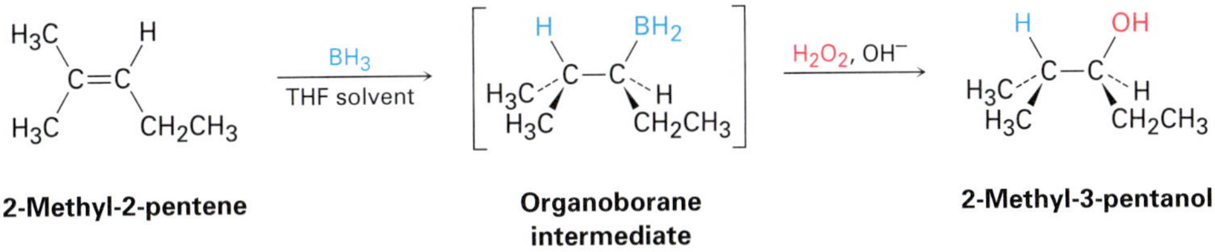

Herbert Charles Brown

Herbert Charles Brown (1912–2004) was born in London to Ukrainian parents and brought to the United States in 1914. Brown received his Ph.D. in 1938 from the University of Chicago, taught at Chicago and at Wayne State University, and then became professor of chemistry at Purdue University. The author of more than 1000 scientific papers, he received the 1979 Nobel Prize in chemistry for his work on organoboranes.

Borane is very reactive because the boron atom has only six electrons in its valence shell. In tetrahydrofuran solution, BH_3 accepts an electron pair from a solvent molecule in a Lewis acid–base reaction to complete its octet and form a stable BH_3–THF complex.

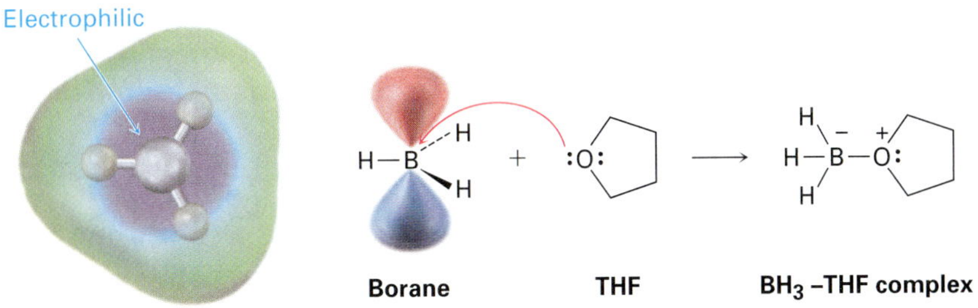

When an alkene reacts with BH_3 in THF solution, rapid addition to the double bond occurs three times and a *trialkylborane*, R_3B, is formed. For example, 1 molar equivalent of BH_3 adds to 3 molar equivalents of cyclohexene to yield tricyclohexylborane. When tricyclohexylborane is then treated with aqueous hydrogen peroxide (H_2O_2) in basic solution, an oxidation takes place. The three C–B bonds are broken, –OH groups bond to the three carbons, and 3 equivalents of cyclohexanol are produced. The net effect of the two-step hydroboration/oxidation sequence is hydration of the alkene dou ble bond.

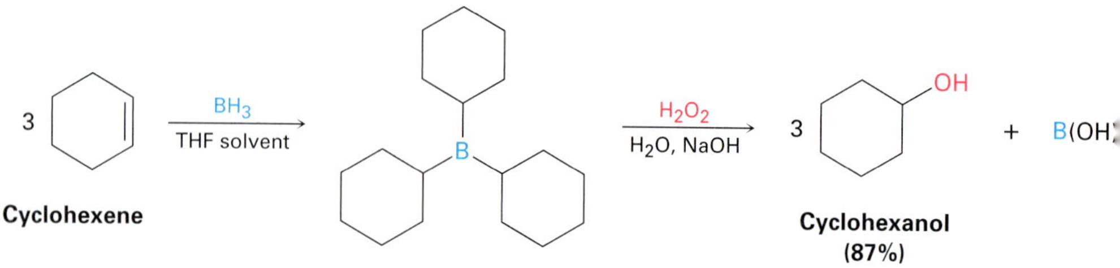

Cyclohexene **Tricyclohexylborane** **Cyclohexanol (87%)**

One of the features that makes the hydroboration reaction so useful is th regiochemistry that results when an unsymmetrical alkene is hydroborated. Fc example, hydroboration/oxidation of 1-methylcyclopentene yields *trans* 2-methylcyclopentanol. Boron and hydrogen both add to the alkene from th same face of the double bond—that is, with **syn stereochemistry**, the opposit of anti—with boron attaching to the less highly substituted carbon. Durin the oxidation step, the boron is replaced by an $-OH$ with the same stereo chemistry, resulting in an overall syn non-Markovnikov addition of water. Thi stereochemical result is particularly useful because it is complementary to th Markovnikov regiochemistry observed for oxymercuration.

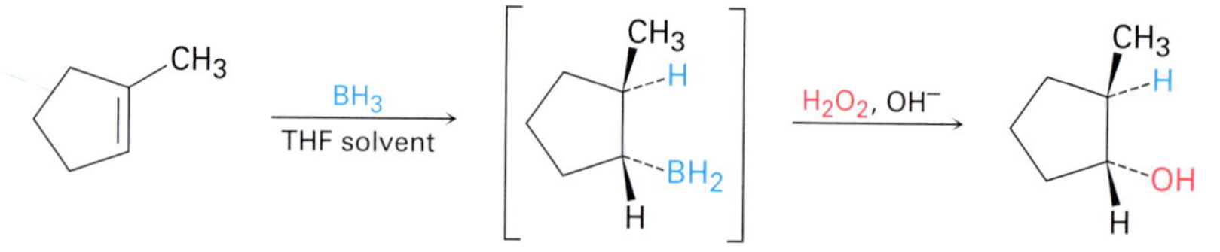

1-Methyl-cyclopentene **Organoborane intermediate** ***trans*-2-Methyl-cyclopentanol (85% yield)**

Why does alkene hydroboration take place with non-Markovnikov regio chemistry, yielding the less highly substituted alcohol? Hydroboration differs from many other alkene addition reactions in that it occurs in a single step through a four-center, cyclic transition state without a carbocation intermediate (Figure 7.4). Because both C–H and C–B bonds form at the same time and from the same face of the alkene, syn stereochemistry results. This mechanism accounts not only for the syn stereochemistry of the reaction but also for the regiochemistry. Attachment of boron is favored at the less sterically hindered carbon atom of the alkene, rather than at the more hindered carbon, because there is less steric crowding in the resultant transition state.

WORKED EXAMPLE 7.1

Predicting the Products Formed in a Reaction

What products would you obtain from reaction of 2,4-dimethyl-2-pentene with:
(a) BH_3, followed by H_2O_2, OH^- **(b)** $Hg(OAc)_2$, followed by $NaBH_4$

Strategy When predicting the product of a reaction, you have to recall what you know about the kind of reaction being carried out and then apply that knowledge to the specific case you're dealing with. In the present instance, recall that the two methods of

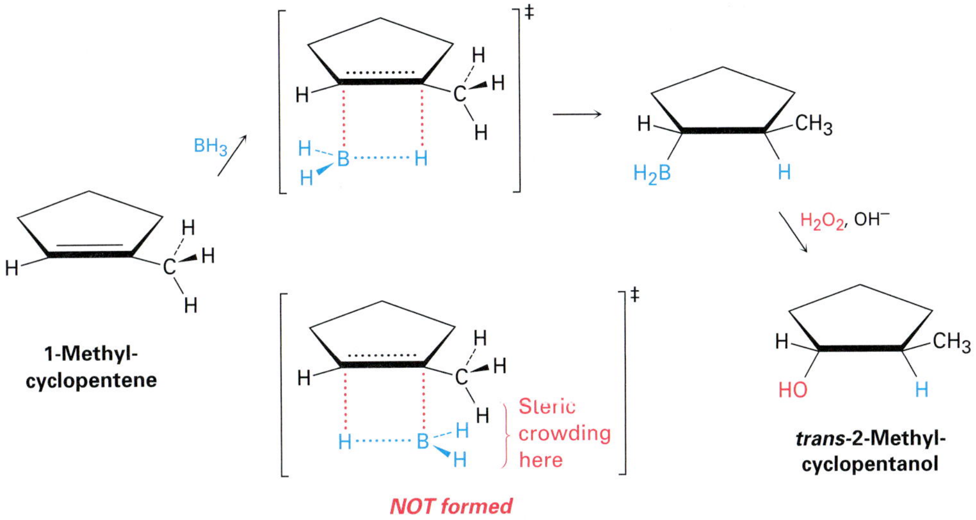

Active Figure 7.4 Mechanism of alkene hydroboration. The reaction occurs in a single step in which both C–H and C–B bonds form at the same time and on the same face of the double bond. The lower energy, more rapidly formed transition state is the one with less steric crowding, leading to non-Markovnikov regiochemistry. *Sign in at* **www.thomsonedu.com** *to see a simulation based on this figure and to take a short quiz.*

hydration—hydroboration/oxidation and oxymercuration—give complementary products. Hydroboration/oxidation occurs with syn stereochemistry and gives the non-Markovnikov addition product; oxymercuration gives the Markovnikov product.

Solution

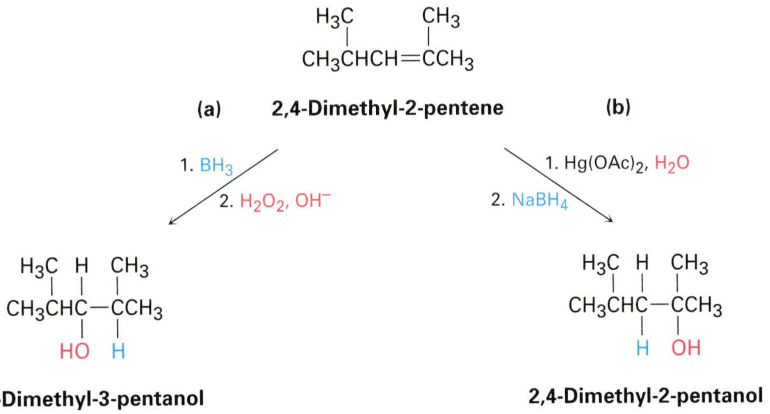

WORKED EXAMPLE 7.2 ***Choosing a Reactant to Synthesize a Specific Compound***

How might you prepare the following alcohol?

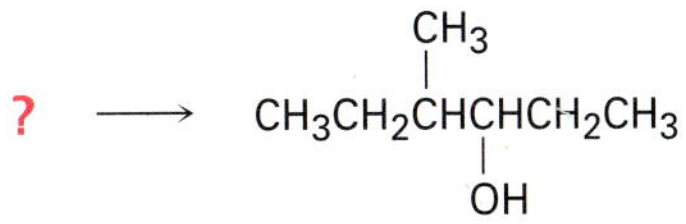

Strategy Problems that require the synthesis of a specific target molecule should always b worked backward. Look at the target, identify its functional group(s), and ask you self "What are the methods for preparing this functional group?" In the prese instance, the target molecule is a secondary alcohol (R_2CHOH), and we've seen th alcohols can be prepared from alkenes by either hydroboration/oxidation o oxymercuration. The –OH bearing carbon in the product must have been double-bond carbon in the alkene reactant, so there are two possibilities: 4-methy 2-hexene and 3-methyl-3-hexene.

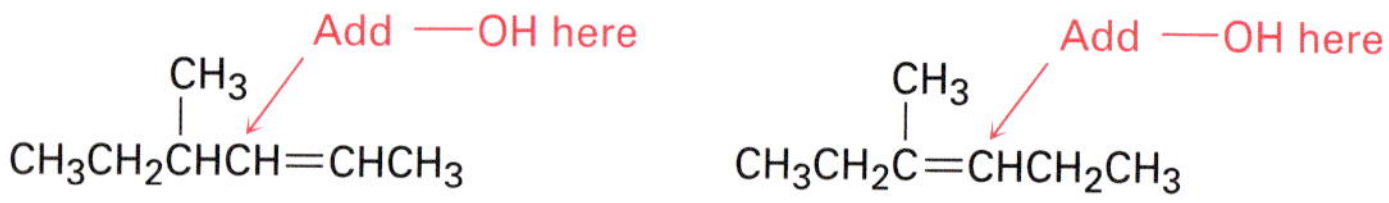

4-Methyl-2-hexene **3-Methyl-3-hexene**

4-Methyl-2-hexene has a disubstituted double bond, RCH=CHR′, and woul probably give a mixture of two alcohols with either hydration method sinc Markovnikov's rule does not apply to symmetrically substituted alkenes. 3-Methy 3-hexene, however, has a trisubstituted double bond, and would give only the desire product on non-Markovnikov hydration using the hydroboration/oxidation metho

Solution

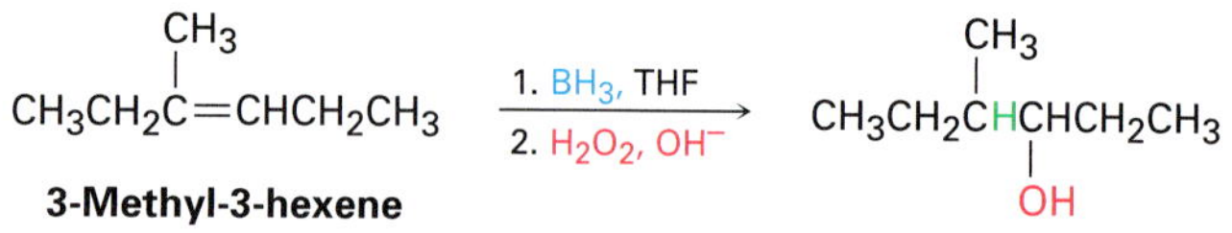

3-Methyl-3-hexene

Problem 7.9 Show the structures of the products you would obtain by hydroboration/oxidatio of the following alkenes:

(a) 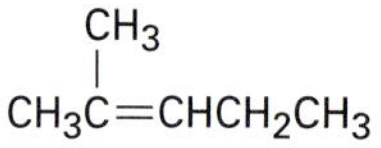

(b) 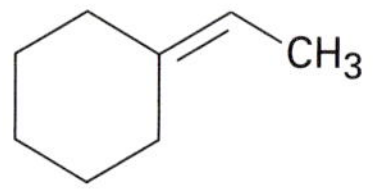

Problem 7.10 What alkenes might be used to prepare the following alcohols by hydroboration oxidation?

(a) 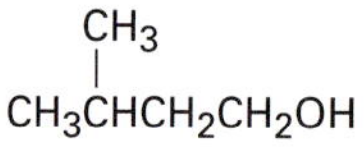

(b) 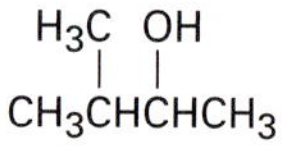

(c) 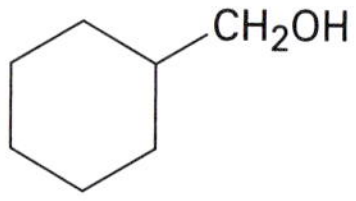

Problem 7.11 The following cycloalkene gives a mixture of two alcohols on hydroboration fol lowed by oxidation. Draw the structures of both, and explain the result.

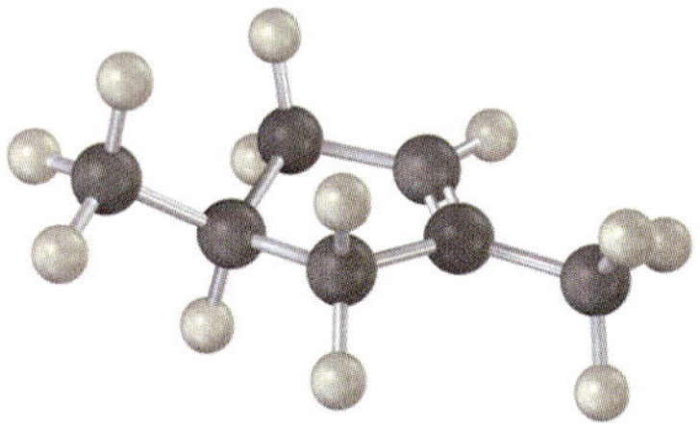

7.6 Addition of Carbenes to Alkenes: Cyclopropane Synthesis

omsonNOW™ Click *Organic* *teractive* to **use a web-based alette to predict products of the ddition of various carbenes to kenes.**

Yet another kind of alkene addition is the reaction of a *carbene* with an alkene to yield a cyclopropane. A **carbene, R_2C:**, is a neutral molecule containing a divalent carbon with only six electrons in its valence shell. It is therefore highly reactive and is generated only as a reaction intermediate, rather than as an isolable molecule. Because they're electron-deficient, carbenes behave as electrophiles and react with nucleophilic C=C bonds. The reaction occurs in a single step without intermediates.

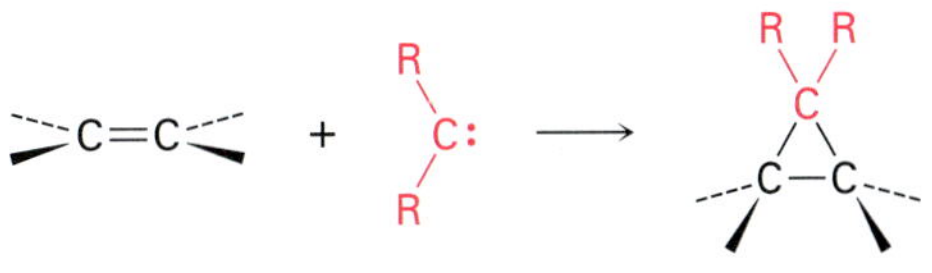

An alkene **A carbene** **A cyclopropane**

One of the simplest methods for generating a substituted carbene is by treatment of chloroform, $CHCl_3$, with a strong base such as KOH. Loss of a proton from $CHCl_3$ gives the trichloromethanide anion, $^-{:}CCl_3$, which expels a Cl^- ion to yield dichlorocarbene, $:CCl_2$ (Figure 7.5).

igure 7.5 **MECHANISM:** Mechanism of the formation of ichlorocarbene by reaction f chloroform with strong base.

homsonNOW™ Click *Organic* *rocess* to **view an animation of he mechanism for the addition f dichlorocarbene to alkenes.**

1 Base abstracts the hydrogen from chloroform, leaving behind the electron pair from the C–H bond and forming the trichloromethanide anion.

2 Spontaneous loss of chloride ion then yields the neutral dichlorocarbene.

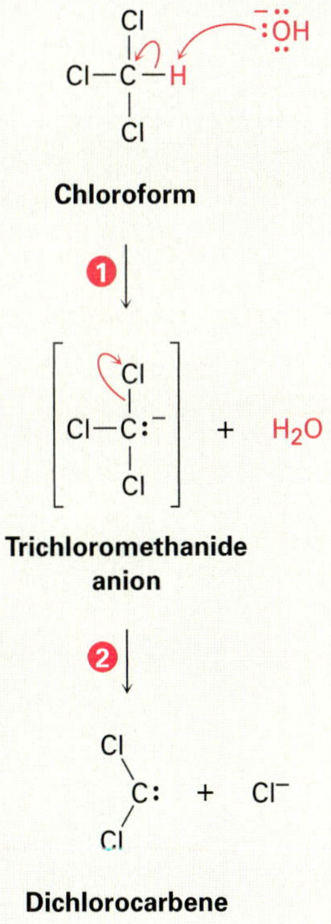

The dichlorocarbene carbon atom is sp^2-hybridized, with a vacant p orbit extending above and below the plane of the three atoms and with an unshare pair of electrons occupying the third sp^2 lobe. Note that this electronic descrip tion of dichlorocarbene is similar to that for a carbocation (Section 6.9) wit respect to both the sp^2 hybridization of carbon and the vacant p orbital. Electro static potential maps further show this similarity (Figure 7.6).

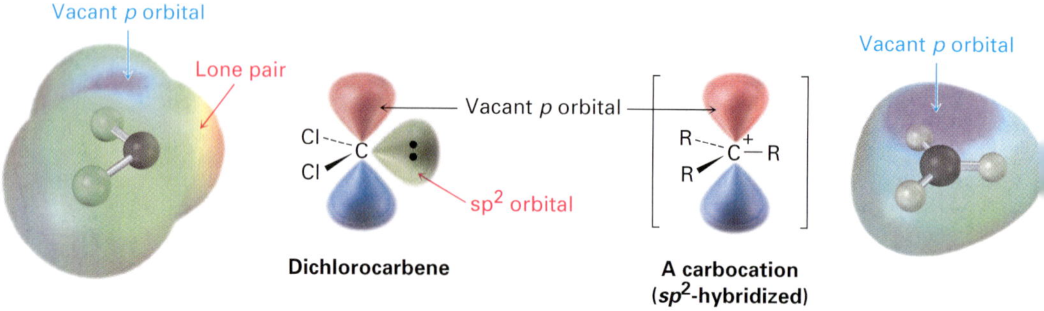

Figure 7.6 The structure of dichlorocarbene. Electrostatic potential maps show how the positive region (blue) coincides with the empty p orbital in both dichlorocarbene and a carbocation (CH_3^+). The negative region (red) in the dichlorocarbene map coincides with th lone-pair electrons.

If dichlorocarbene is generated in the presence of an alkene, addition to th double bond occurs and a dichlorocyclopropane is formed. As the reaction c dichlorocarbene with *cis*-2-pentene demonstrates, the addition is **stereospecific** meaning that only a single stereoisomer is formed as product. Starting from a ci alkene, for instance, only cis-disubstituted cyclopropane is produced; startin from a trans alkene, only trans-disubstituted cyclopropane is produced.

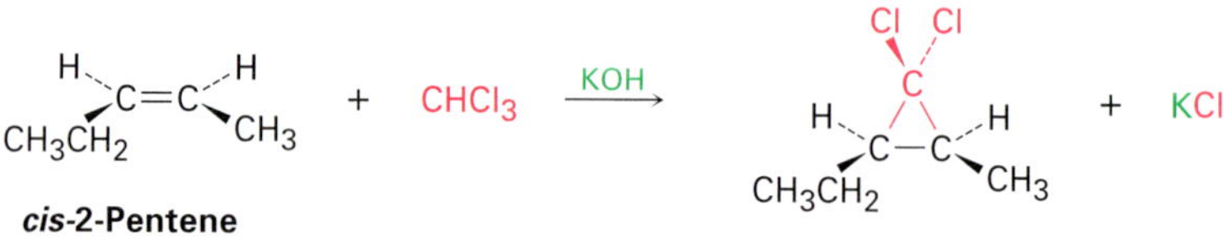

***cis*-2-Pentene**

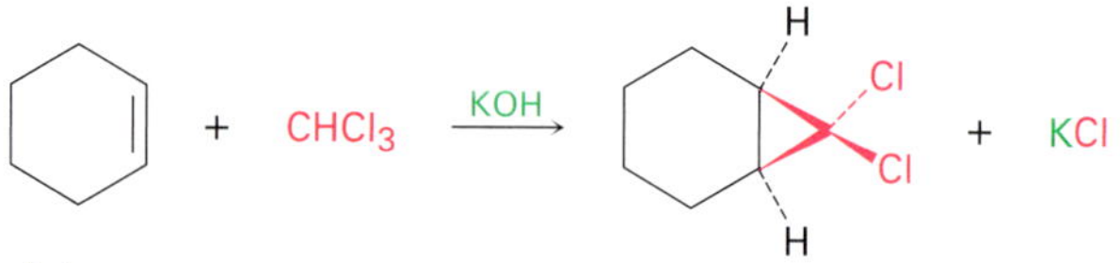

Cyclohexene

The best method for preparing nonhalogenated cyclopropanes is by a proces called the **Simmons–Smith reaction**. First investigated at the DuPont company this reaction does not involve a free carbene. Rather, it utilizes a *carbenoid*—a metal-complexed reagent with carbene-like reactivity. When diiodomethane is treated with a specially prepared zinc–copper mix, (iodomethyl)zinc iodide ICH_2ZnI, is formed. In the presence of an alkene, (iodomethyl)zinc iodide trans fers a CH_2 group to the double bond and yields the cyclopropane. For example cyclohexene reacts cleanly and in good yield to give the corresponding cyclo propane. Although we won't discuss the mechanistic details, carbene addition tc an alkene is one of a general class of reactions called *cycloadditions,* which we'll study more carefully in Chapter 30.

$$CH_2I_2 + Zn(Cu) \longrightarrow ICH_2{-}ZnI \quad [\text{“}{:}CH_2\text{”}]$$

Diiodomethane **(Iodomethyl)zinc iodide (a carbenoid)**

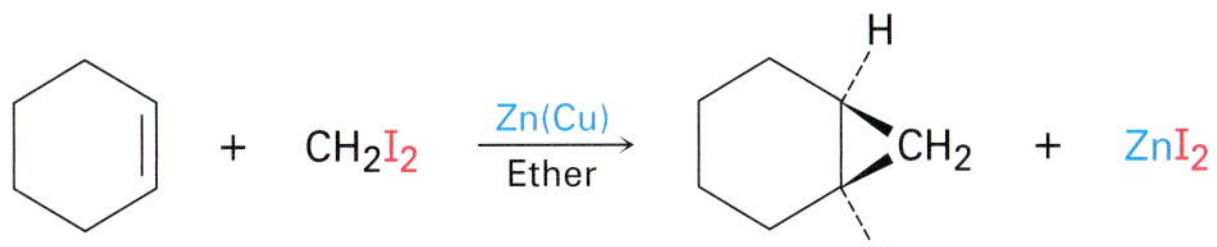

Cyclohexene **Bicyclo[4.1.0]heptane (92%)**

Problem 7.12 What products would you expect from the following reactions?

(a)

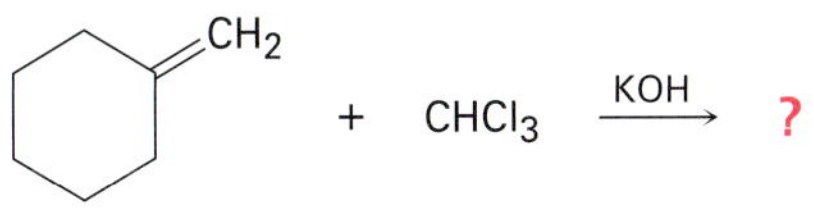

(b)

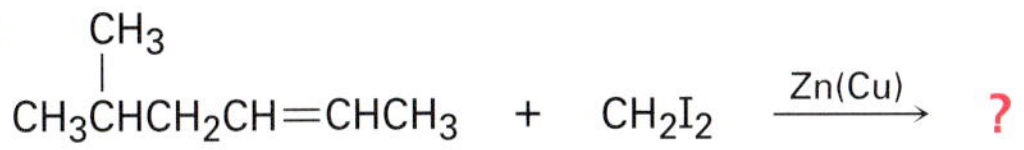

7.7 Reduction of Alkenes: Hydrogenation

ıomsonNOW™ Click *Organic* ıteractive to use a web-based alette to predict products from ıe reduction of alkenes.

Alkenes react with H_2 in the presence of a metal catalyst to yield the corresponding saturated alkane addition products. We describe the result by saying that the double bond has been **hydrogenated**, or *reduced*. Note that the words *oxidation* and *reduction* are used somewhat differently in organic chemistry from what you might have learned previously. In general chemistry, a reduction is defined as the gain of one or more electrons by an atom. In organic chemistry, however, a **reduction** is a reaction that results in a gain of electron density by carbon, caused either by bond formation between carbon and a less electronegative atom or by bond-breaking between carbon and a more electronegative atom. We'll explore the topic in more detail in Section 10.9.

Reduction Increases electron density on carbon by:
- forming this: C–H
- or breaking one of these: C–O C–N C–X

A reduction:

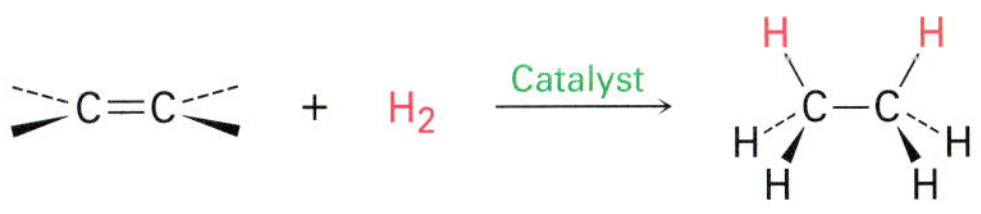

An alkene **An alkane**

Roger Adams

Roger Adams (1889–1971) was born in Boston, Massachusetts, and received his Ph.D. in 1912 at Harvard. He taught at the University of Illinois from 1916 until his retirement in 1957, during which time he had an enormous influence on the development of organic chemistry in the United States. Among many other accomplishments, he established the structure of tetrahydrocannabinol, the active ingredient in marijuana.

Platinum and palladium are the most common catalysts for alkene hydr genations. Palladium is normally used as a very fine powder "supported" on a inert material such as charcoal (Pd/C) to maximize surface area. Platinum normally used as PtO_2, a reagent called *Adams' catalyst* after its discovere Roger Adams.

Catalytic hydrogenation, unlike most other organic reactions, is a *heter geneous* process rather than a homogeneous one. That is, the hydrogenatio reaction does not occur in a homogeneous solution but instead takes plac on the surface of insoluble catalyst particles. Hydrogenation usually occu with syn stereochemistry—both hydrogens add to the double bond from th same face.

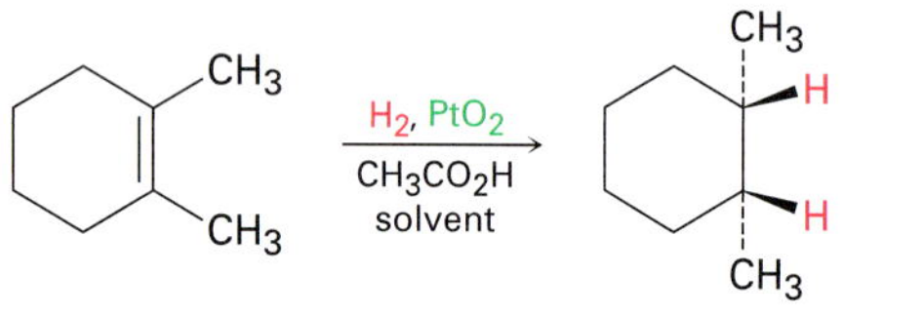

1,2-Dimethylcyclohexene — ***cis*-1,2-Dimethylcyclohexane (82%)**

The first step in the reaction is adsorption of H_2 onto the catalyst surfac Complexation between catalyst and alkene then occurs as a vacant orbital o the metal interacts with the filled alkene π orbital. In the final steps, hydro gen is inserted into the double bond and the saturated product diffuses awa from the catalyst (Figure 7.7). The stereochemistry of hydrogenation is sy because both hydrogens add to the double bond from the same catalys surface.

An interesting feature of catalytic hydrogenation is that the reaction extremely sensitive to the steric environment around the double bond. As result, the catalyst often approaches only the more accessible face of an alken giving rise to a single product. In α-pinene, for example, one of the methy groups attached to the four-membered ring hangs over the top face of the dou ble bond and blocks approach of the hydrogenation catalyst from that side Reduction therefore occurs exclusively from the bottom face to yield the prod uct shown.

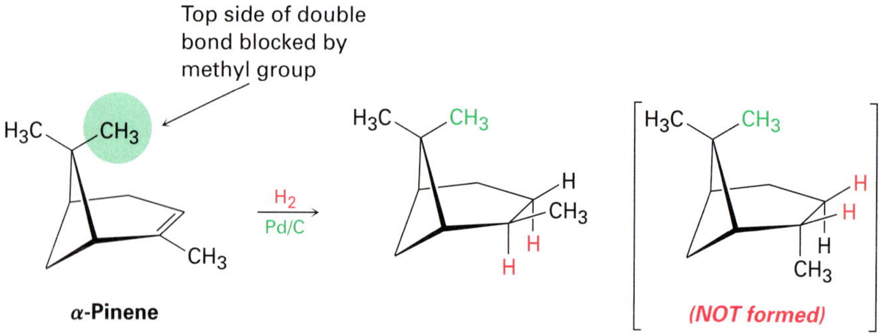

α-Pinene — *(NOT formed)*

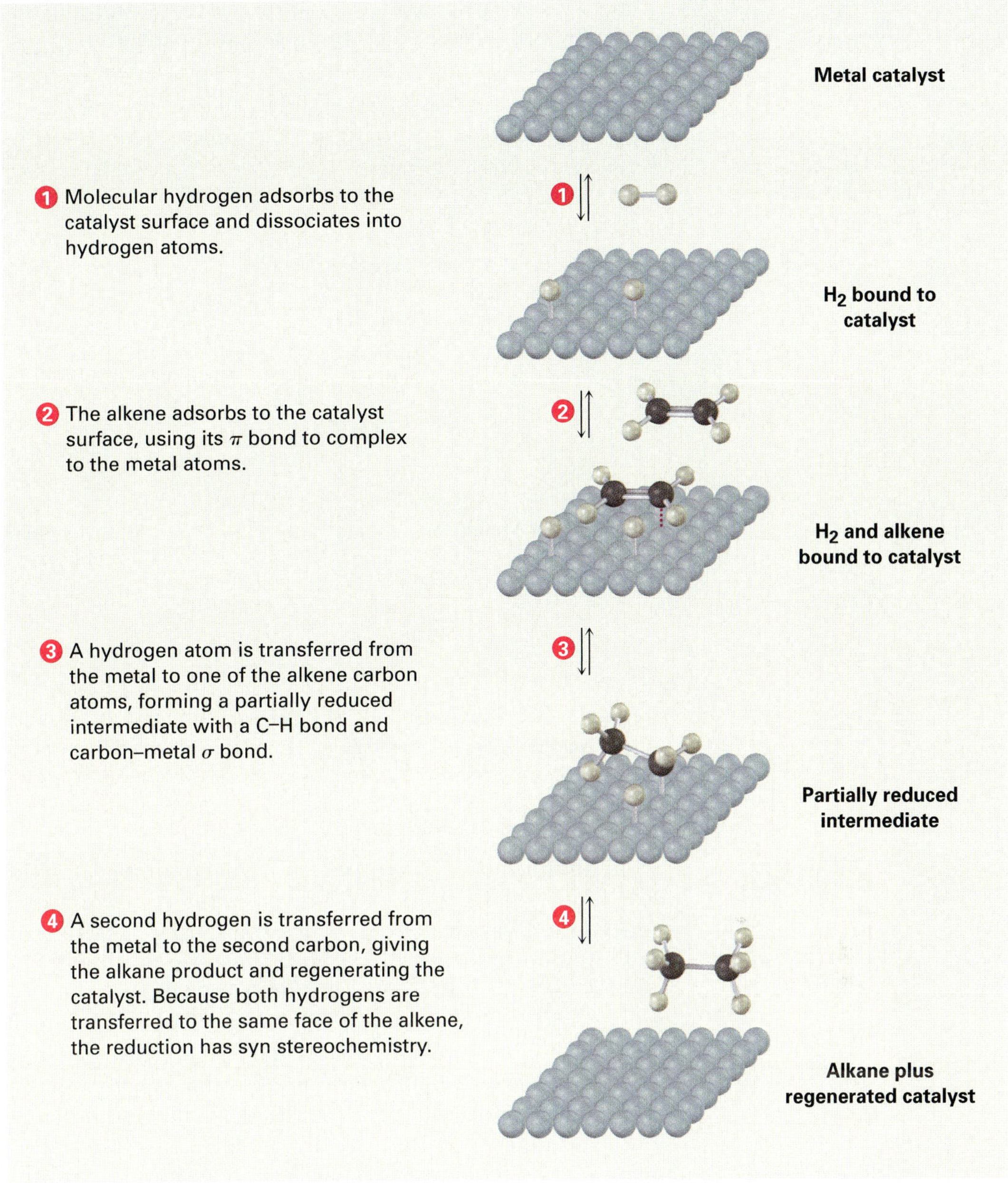

Figure 7.7 **MECHANISM:** Mechanism of alkene hydrogenation. The reaction takes place with syn stereochemistry on the surface of insoluble catalyst particles.

Alkenes are much more reactive than most other unsaturated functional groups toward catalytic hydrogenation, and the reaction is therefore quite selective. Other functional groups such as aldehydes, ketones, esters, and nitriles survive normal alkene hydrogenation conditions unchanged, although reaction with these groups does occur under more vigorous conditions. Note particularly in the hydrogenation of methyl 3-phenylpropenoate shown below that the ar matic ring is not reduced by hydrogen and palladium even though it contai apparent double bonds.

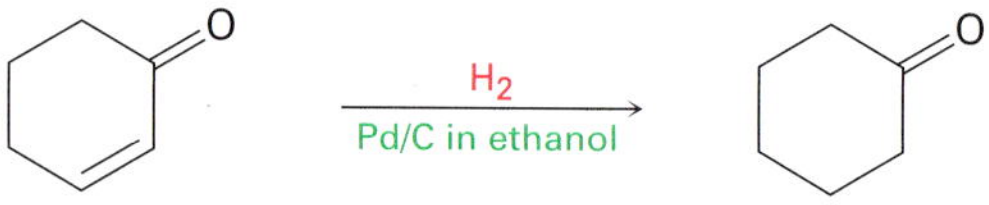

2-Cyclohexenone **Cyclohexanone (ketone *NOT* reduced)**

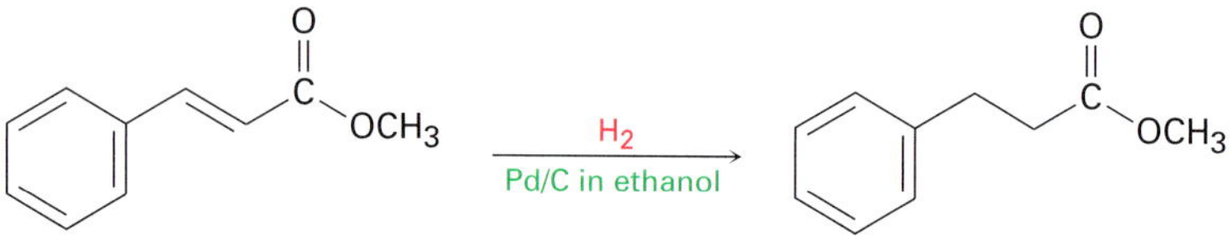

Methyl 3-phenylpropenoate **Methyl 3-phenylpropanoate (aromatic ring *NOT* reduced)**

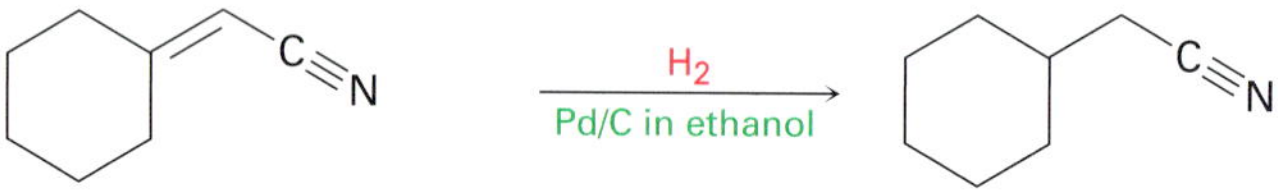

Cyclohexylideneacetonitrile **Cyclohexylacetonitrile (nitrile *NOT* reduced)**

In addition to its usefulness in the laboratory, catalytic hydrogenation also important in the food industry, where unsaturated vegetable oils a reduced on a vast scale to produce the saturated fats used in margarine an cooking products (Figure 7.8). As we'll see in Section 27.1, vegetable oils a triesters of glycerol, $HOCH_2CH(OH)CH_2OH$, with three long-chain ca boxylic acids called *fatty acids*. The fatty acids are generally polyunsaturate and their double bonds invariably have cis stereochemistry. Complete hydr genation yields the corresponding saturated fatty acids, but incomple hydrogenation often results in partial cis–trans isomerization of a remainin double bond. When eaten and digested, the free trans fatty acids are release raising blood cholesterol levels and contributing to potential corona problems.

Problem 7.13 What product would you obtain from catalytic hydrogenation of the followin alkenes?

(a) $CH_3C(CH_3)=CHCH_2CH_3$ (drawn with CH_3 above the C)

(b)

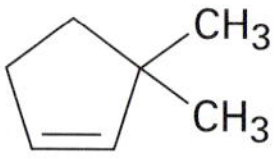

gure 7.8 Catalytic hydro-
nation of polyunsaturated
s leads to saturated prod-
ts, along with a small
nount of isomerized
ns fats.

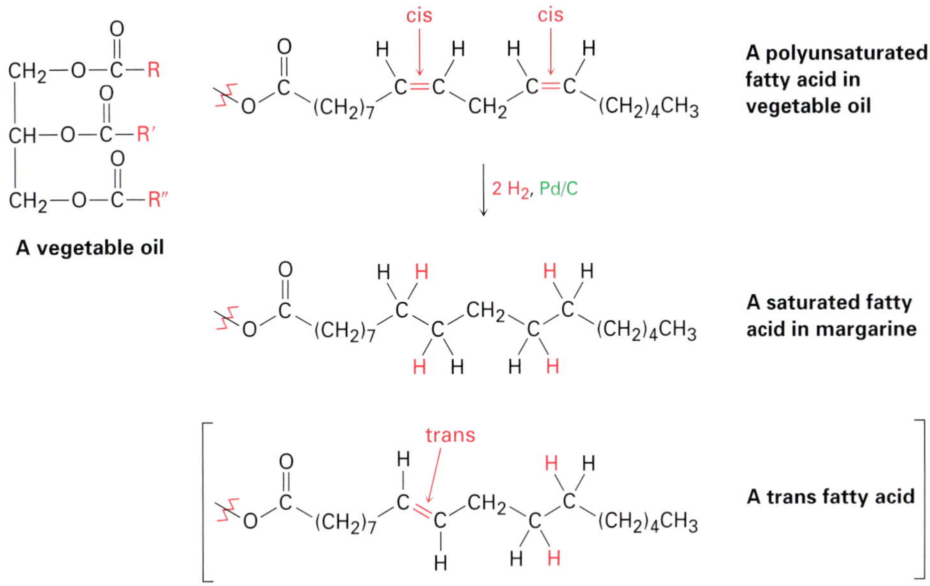

7.8 Oxidation of Alkenes: Epoxidation and Hydroxylation

Like the word *reduction* used in the previous section for addition of hydrogen to a double bond, the word *oxidation* has a slightly different meaning in organic chemistry from what you might have previously learned. In general chemistry, an oxidation is defined as the loss of one or more electrons by an atom. In organic chemistry, however, an **oxidation** is a reaction that results in a loss of electron density by carbon, caused either by bond formation between carbon and a more electronegative atom—usually oxygen, nitrogen, or a halogen—or by bond-breaking between carbon and a less electronegative atom—usually hydrogen. Note that an *oxidation* often adds oxygen, while a *reduction* often adds hydrogen.

Oxidation Decreases electron density on carbon by:
- forming one of these: C–O C–N C–X
- or breaking this: C–H

Alkenes are oxidized to give *epoxides* on treatment with a peroxyacid (RCO_3H), such as *meta*-chloroperoxybenzoic acid. An **epoxide**, also called an *oxirane*, is a cyclic ether with an oxygen atom in a three-membered ring. For example:

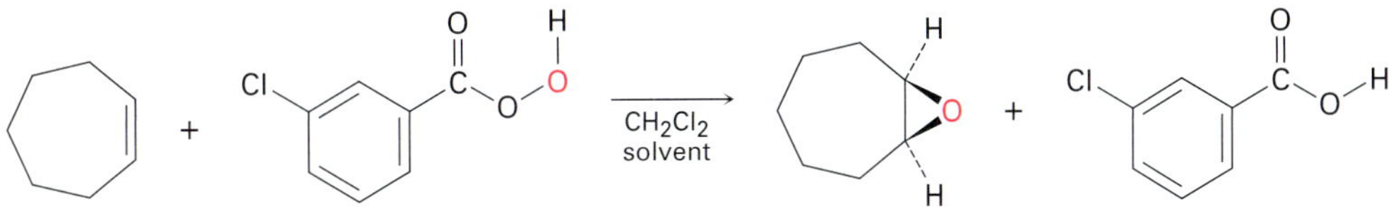

Cycloheptene **meta-Chloroperoxybenzoic acid** **1,2-Epoxycycloheptane** **meta-Chlorobenzoic acid**

Peroxyacids transfer an oxygen atom to the alkene with syn stere chemistry—both C–O bonds form on the same face of the double bonc through a one-step mechanism without intermediates. The oxygen atc farthest from the carbonyl group is the one transferred.

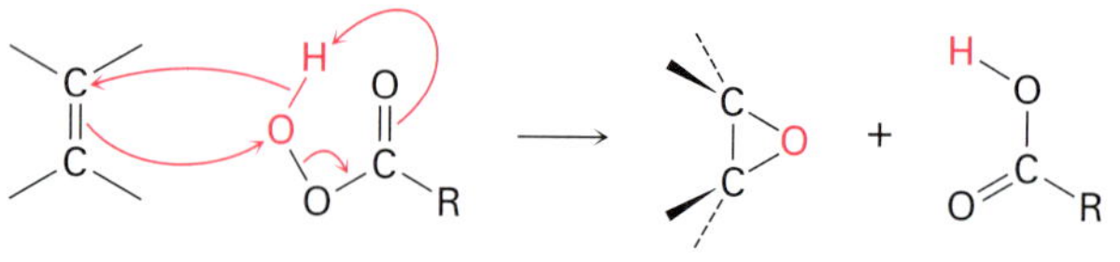

Alkene **Peroxyacid** **Epoxide** **Acid**

Another method for the synthesis of epoxides is through the use of hal hydrins, prepared by electrophilic addition of HO–X to alkenes (Section 7. When a halohydrin is treated with base, HX is eliminated and an epoxide produced.

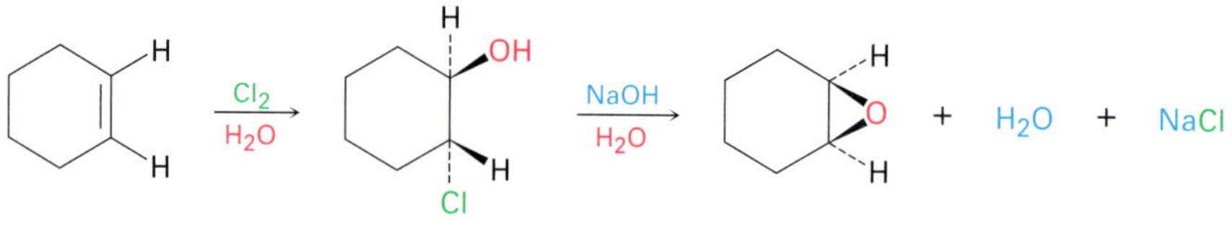

Cyclohexene ***trans*-2-Chloro-cyclohexanol** **1,2-Epoxycyclohexane (73%)**

Epoxides undergo an acid-catalyzed ring-opening reaction with water *hydrolysis*) to give the corresponding dialcohol *(diol),* also called a **glycol**. Thus, tl net result of the two-step alkene epoxidation/hydrolysis is **hydroxylation**- the addition of an –OH group to each of the two double-bond carbons. In fac more than 3 million tons of ethylene glycol, $HOCH_2CH_2OH$, most of it used f automobile antifreeze, is produced each year in the United States by epoxid tion of ethylene followed by hydrolysis.

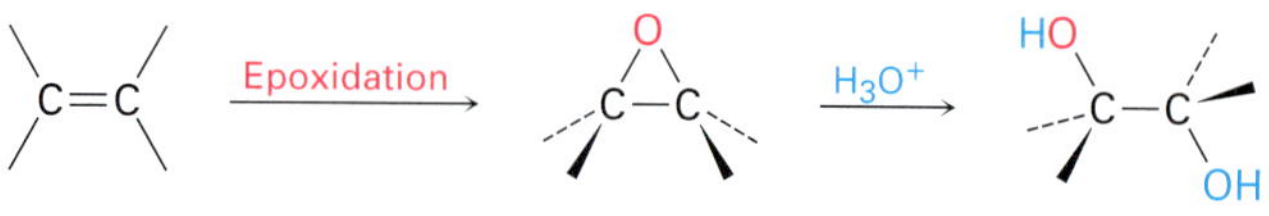

An alkene **An epoxide** **A 1,2-diol**

Acid-catalyzed epoxide opening takes place by protonation of the epoxic to increase its reactivity, followed by nucleophilic addition of water. Th nucleophilic addition is analogous to the final step of alkene bromination, i which a cyclic bromonium ion is opened by a nucleophile (Section 7.2). That i a *trans*-1,2-diol results when an epoxycycloalkane is opened by aqueous acid, just as a *trans*-1,2-dibromide results when a cycloalkene is halogenated. We'll look at epoxide chemistry in more detail in Section 18.6.

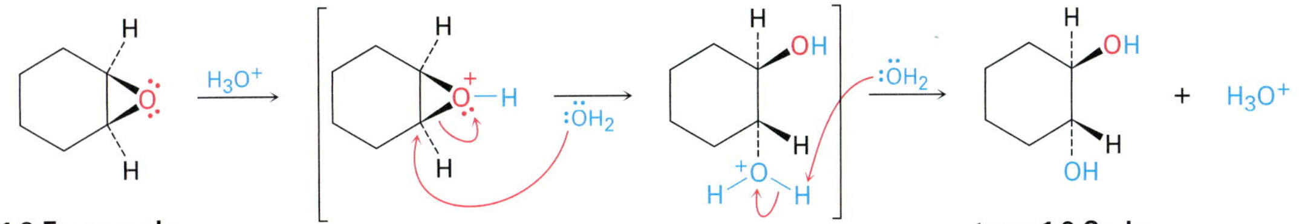

1,2-Epoxycyclohexane → ***trans*-1,2-Cyclohexanediol (86%)**

Recall the following:

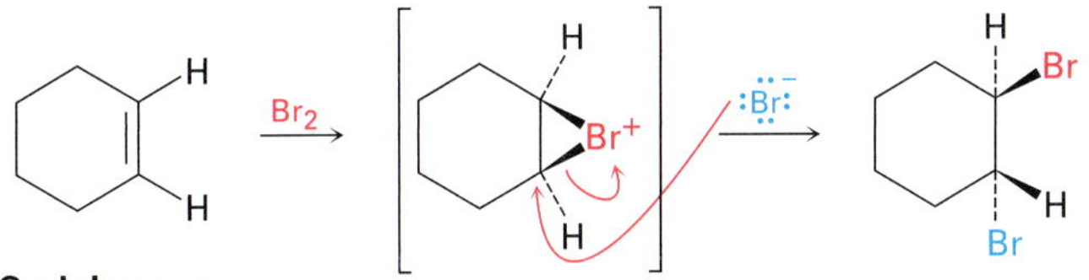

Cyclohexene → ***trans*-1,2-Dibromocyclohexane**

Hydroxylation can be carried out directly without going through the intermediate epoxide by treating an alkene with osmium tetroxide, OsO_4. The reaction occurs with syn stereochemistry and does not involve a carbocation intermediate. Instead, it takes place through an intermediate cyclic *osmate,* which is formed in a single step by addition of OsO_4 to the alkene. This cyclic osmate is then cleaved using aqueous sodium bisulfite, $NaHSO_3$.

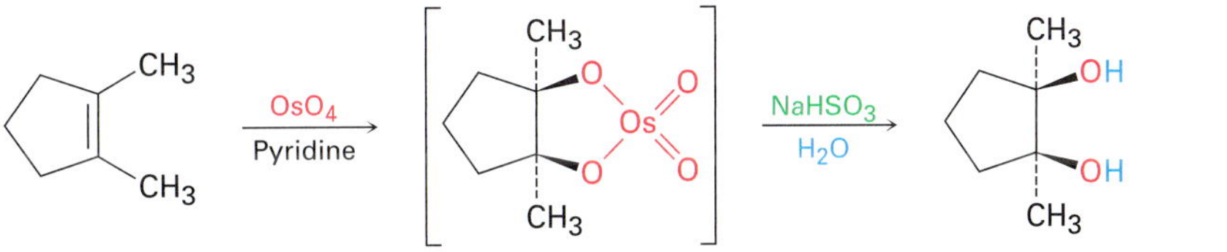

1,2-Dimethylcyclopentene — **A cyclic osmate intermediate** — ***cis*-1,2-Dimethyl-1,2-cyclopentanediol (87%)**

Unfortunately, a serious problem with the osmium tetroxide reaction is that OsO_4 is both very expensive and *very* toxic. As a result, the reaction is usually carried out using only a small, catalytic amount of OsO_4 in the presence of a stoichiometric amount of a safe and inexpensive co-oxidant such as *N*-methylmorpholine *N*-oxide, abbreviated NMO. The initially formed osmate intermediate reacts rapidly with NMO to yield the product diol plus

N-methylmorpholine and reoxidized OsO_4. The OsO_4 then reacts with mor alkene in a catalytic cycle.

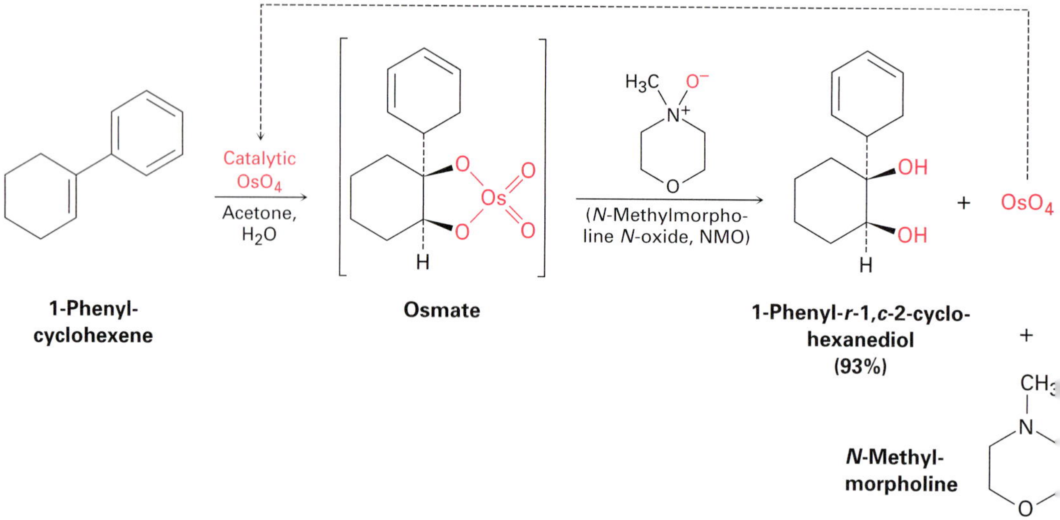

Note that a *cis*- or *trans*- prefix would be ambiguous when naming the dic derived from 1-phenylcyclohexene because the ring has three substituents. In such a case, the substituent with the lowest number is taken as the reference substituent, denoted *r*, and the other substituents are identified as being cis (*c*) or trans (*t*) to that reference. When two substituents share the same lowest number, the one with the highest priority by the Cahn–Ingold–Prelog sequence rule (Section 6.5) is taken as the reference. In the case of 1-phenyl-1,2-cyclohexanediol, the −OH group at C1 is the reference (*r*-1), and the −OH at C2 is either ci (*c*-2) or trans (*t*-2) to that reference. Thus, the diol resulting from cis hydroxylation is named 1-phenyl-*r*-1,*c*-2-cyclohexanediol, and its isomer resulting from trans hydroxylation would be named 1-phenyl-*r*-1,*t*-2-cyclohexanediol.

Problem 7.14 What product would you expect from reaction of *cis*-2-butene with *meta*-chloroperoxybenzoic acid? Show the stereochemistry.

Problem 7.15 How would you prepare each of the following compounds starting with an alkene?

(a)

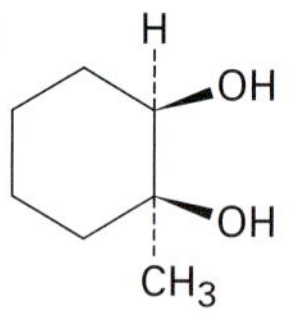

(b)

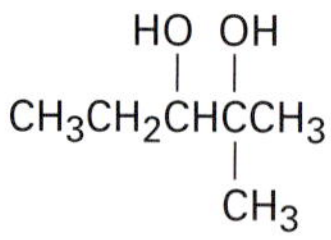

(c)

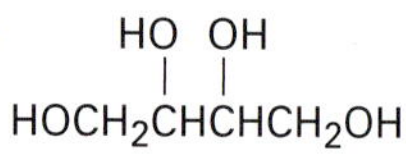

7.9 Oxidation of Alkenes: Cleavage to Carbonyl Compounds

ThomsonNOW™ Click *Organic Interactive* to **use a web-based palette to predict products from the oxidation of alkenes.**

In all the alkene addition reactions we've seen thus far, the carbon–carbon double bond has been converted into a single bond but the carbon skeleton has been left intact. There are, however, powerful oxidizing reagents that will cleave C=C bonds and produce two carbonyl-containing fragments.

Ozone (O_3) is perhaps the most useful double-bond cleavage reagent. Prepared by passing a stream of oxygen through a high-voltage electrical discharge, ozone adds rapidly to an alkene at low temperature to give a cyclic intermediate called a *molozonide*. Once formed, the molozonide then spontaneously rearranges to form an **ozonide**. Although we won't study the mechanism of this rearrangement in detail, it involves the molozonide coming apart into two fragments that then recombine in a different way.

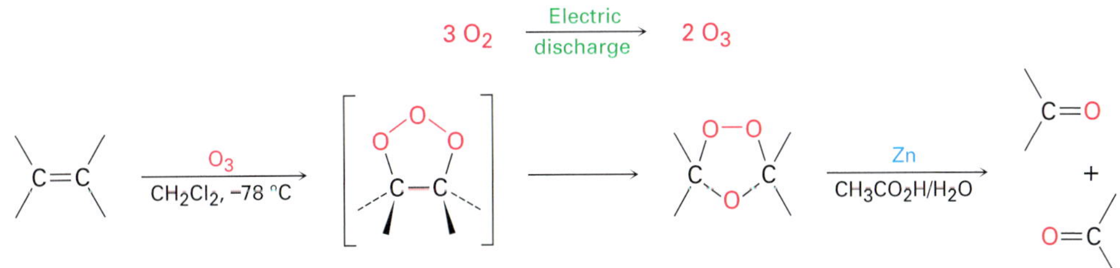

An alkene **A molozonide** **An ozonide**

Low-molecular-weight ozonides are explosive and are therefore not isolated. Instead, the ozonide is immediately treated with a reducing agent such as zinc metal in acetic acid to convert it to carbonyl compounds. The net result of the ozonolysis/reduction sequence is that the C=C bond is cleaved and oxygen becomes doubly bonded to each of the original alkene carbons. If an alkene with a tetrasubstituted double bond is ozonized, two ketone fragments result; if an alkene with a trisubstituted double bond is ozonized, one ketone and one aldehyde result; and so on.

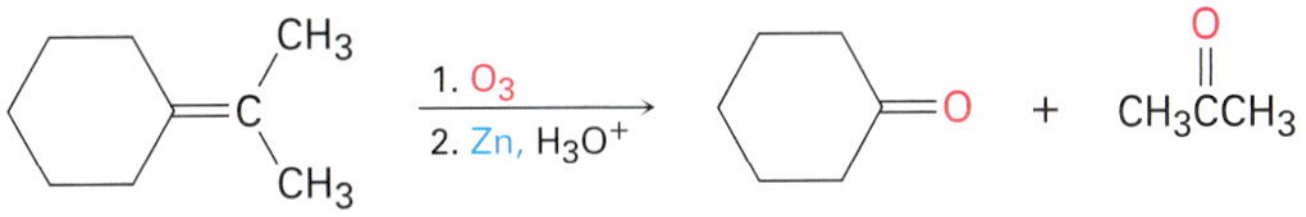

Isopropylidenecyclohexane (tetrasubstituted) **Cyclohexanone** **Acetone**

84%; two ketones

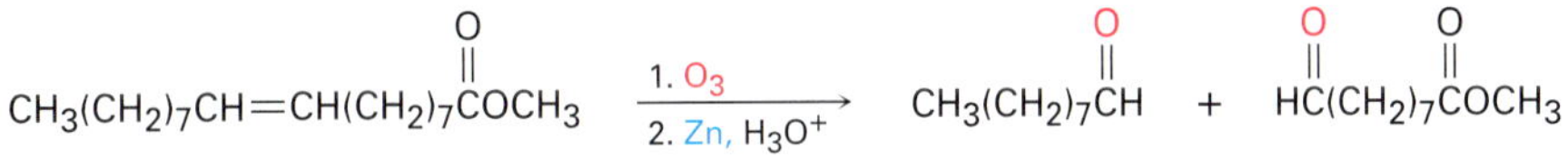

Methyl 9-octadecenoate (disubstituted) **Nonanal** **Methyl 9-oxononanoate**

78%; two aldehydes

Several oxidizing reagents other than ozone also cause double-bond cleavage. For example, potassium permanganate ($KMnO_4$) in neutral or acidic solution cleaves alkenes to give carbonyl-containing products. If hydrogens are present on the double bond, carboxylic acids are produced; if two hydrogens are present on one carbon, CO_2 is formed.

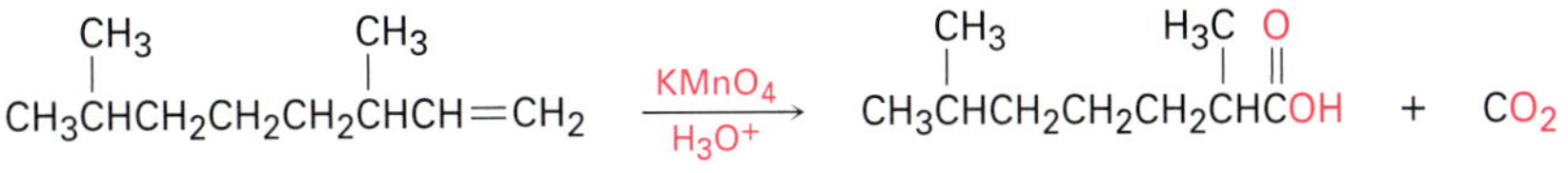

3,7-Dimethyl-1-octene **2,6-Dimethylheptanoic acid (45%)**

In addition to direct cleavage with ozone or $KMnO_4$, an alkene can also cleaved by initial hydroxylation to a 1,2-diol followed by treatment with pe iodic acid, HIO_4. If the two $-OH$ groups are in an open chain, two carbon compounds result. If the two $-OH$ groups are on a ring, a single, open-cha dicarbonyl compound is formed. As indicated in the following examples, th cleavage reaction takes place through a cyclic periodate intermediate.

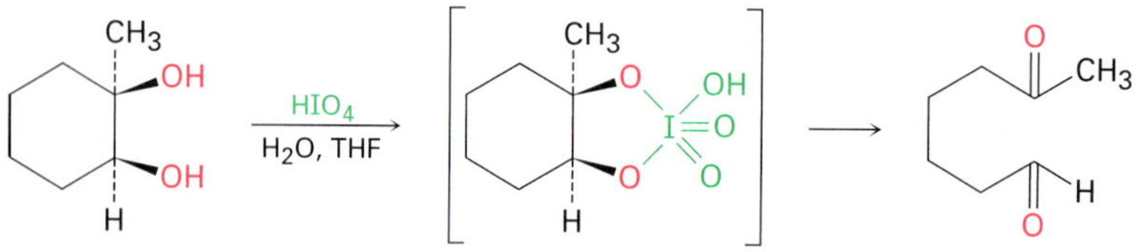

A 1,2-diol — **Cyclic periodate intermediate** — **6-Oxoheptanal (86%)**

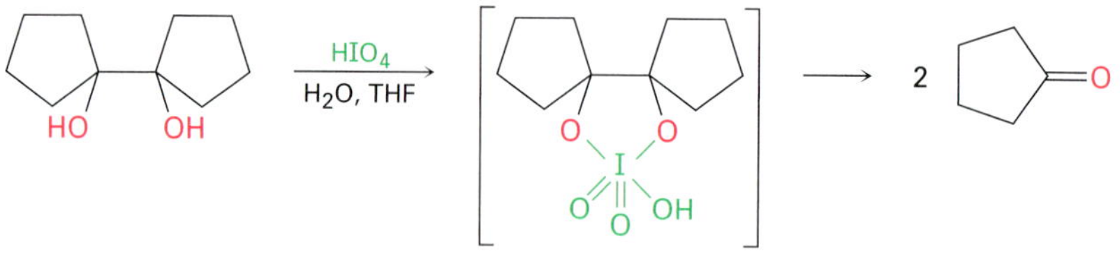

A 1,2-diol — **Cyclic periodate intermediate** — **Cyclopentanone (81%)**

WORKED EXAMPLE 7.3

Predicting the Reactant in an Ozonolysis Reaction

What alkene would yield a mixture of cyclopentanone and propanal on treatmen with ozone followed by reduction with zinc?

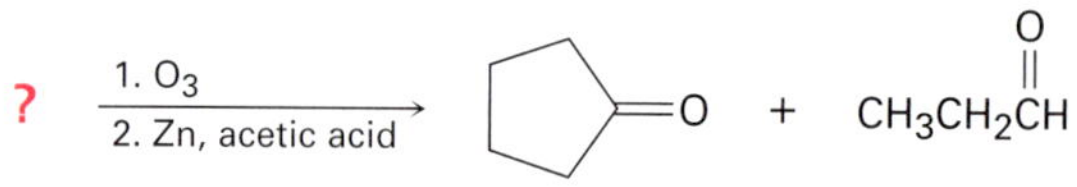

Strategy Reaction of an alkene with ozone, followed by reduction with zinc, cleaves th carbon–carbon double bond and gives two carbonyl-containing fragments. That is the C=C bond becomes two C=O bonds. Working backward from the carbonyl containing products, the alkene precursor can be found by removing the oxyge from each product and joining the two carbon atoms to form a double bond.

Solution

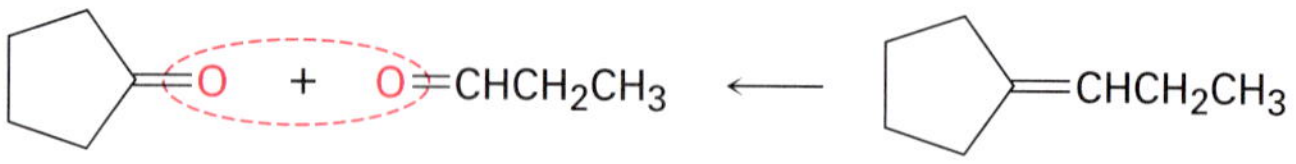

Problem 7.16 What products would you expect from reaction of 1-methylcyclohexene with th following reagents?

(a) Aqueous acidic $KMnO_4$ **(b)** O_3, followed by Zn, CH_3CO_2H

Problem 7.17 Propose structures for alkenes that yield the following products on reaction wit ozone followed by treatment with Zn:

(a) $(CH_3)_2C{=}O + H_2C{=}O$ **(b)** 2 equiv $CH_3CH_2CH{=}O$

7.10 Radical Additions to Alkenes: Polymers

We had a brief introduction to radical reactions in Section 5.3 and said at that time that radicals can add to alkene double bonds, taking one electron from the double bond and leaving one behind to yield a new radical. Let's now look at the process in more detail, focusing on the industrial synthesis of alkene polymers.

A **polymer** is simply a large—sometimes *very* large—molecule built up by repetitive bonding together of many smaller molecules, called **monomers**. Nature makes wide use of biological polymers. Cellulose, for instance, is a polymer built of repeating glucose monomer units; proteins are polymers built of repeating amino acid monomers; and nucleic acids are polymers built of repeating nucleotide monomers. Synthetic polymers, such as polyethylene, are chemically much simpler than biopolymers, but there is still a great diversity to their structures and properties, depending on the identity of the monomers and on the reaction conditions used for polymerization.

Cellulose—a glucose polymer

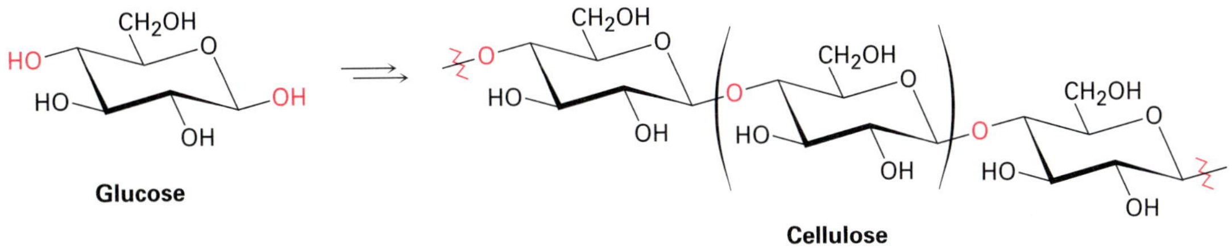

Protein—an amino acid polymer

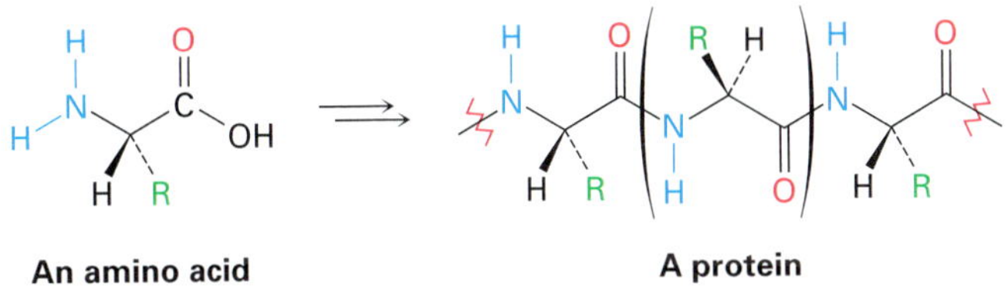

Nucleic acid—a nucleotide polymer

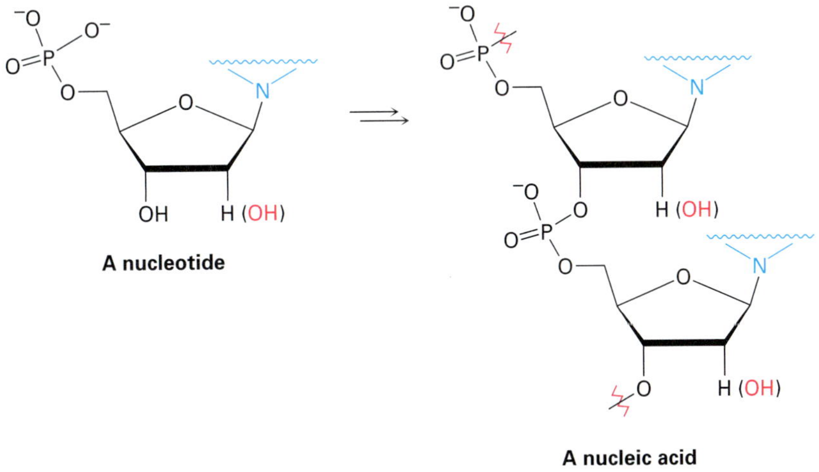

Polyethylene—a synthetic alkene polymer

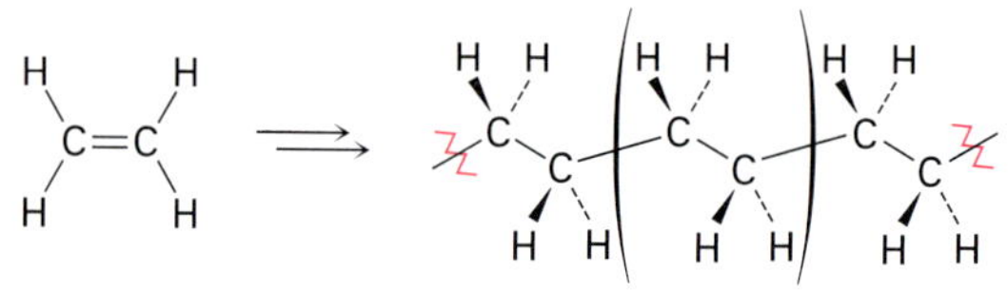

Ethylene **Polyethylene**

The simplest synthetic polymers are those that result when an alkene treated with a small amount of a radical as catalyst. Ethylene, for example, yielc polyethylene, an enormous alkane that may have up to *200,000* monomer unit incorporated into a gigantic hydrocarbon chain. Approximately 14 million tor per year of polyethylene is manufactured in the United States alone.

Historically, ethylene polymerization was carried out at high pressur (1000–3000 atm) and high temperature (100–250 °C) in the presence of a cata lyst such as benzoyl peroxide, although other catalysts and reaction condition are now more often used. The key step is the addition of a radical to the ethylen double bond, a reaction similar in many respects to what takes place in the add tion of an electrophile. In writing the mechanism, recall that a curved hal arrow, or "fishhook" ↶, is used to show the movement of a single electron, a opposed to the full curved arrow used to show the movement of an electron pai in a polar reaction.

- **Initiation** The polymerization reaction is initiated when a few radicals ar generated on heating a small amount of benzoyl peroxide catalyst to break th weak O–O bond. A benzoyloxy radical then adds to the C=C bond o ethylene to generate a carbon radical. One electron from the C=C bond pair up with the odd electron on the benzoyloxy radical to form a C–O bond, an the other electron remains on carbon.

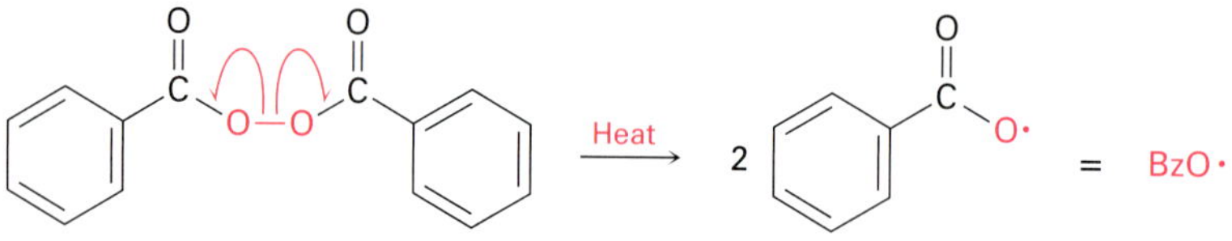

Benzoyl peroxide **Benzoyloxy radical**

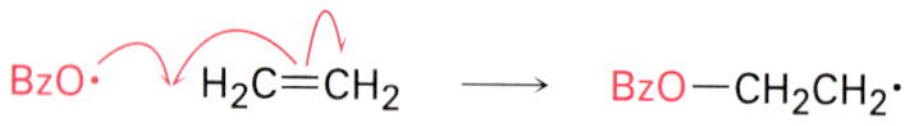

- **Propagation** Polymerization occurs when the carbon radical formed in the initiation step adds to another ethylene molecule to yield another radical.

Repetition of the process for hundreds or thousands of times builds the polymer chain.

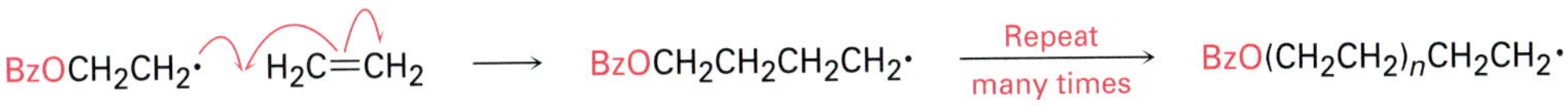

Termination The chain process is eventually ended by a reaction that consumes the radical. Combination of two growing chains is one possible chain-terminating reaction.

$$2\,R{-}CH_2CH_2\cdot \longrightarrow R{-}CH_2CH_2CH_2CH_2{-}R$$

Ethylene is not unique in its ability to form a polymer. Many substituted ethylenes, called *vinyl monomers,* also undergo polymerization to yield polymers with substituent groups regularly spaced on alternating carbon atoms along the chain. Propylene, for example, yields polypropylene, and styrene yields polystyrene.

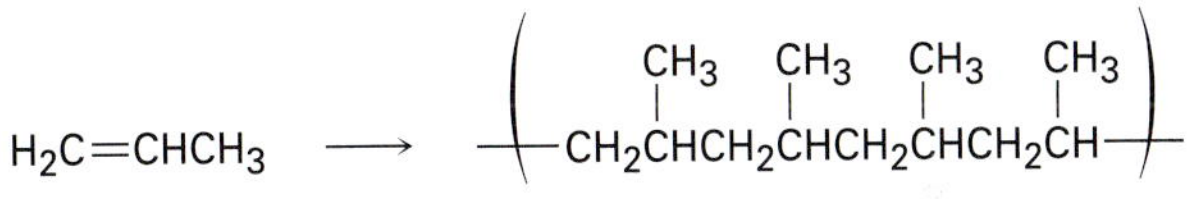

Propylene **Polypropylene**

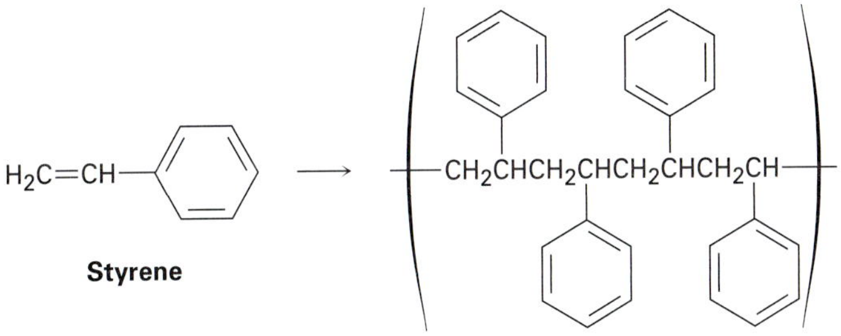

Styrene **Polystyrene**

When an unsymmetrically substituted vinyl monomer such as propylene or styrene is polymerized, the radical addition steps can take place at either end of the double bond to yield either a primary radical intermediate ($RCH_2\cdot$) or a secondary radical ($R_2CH\cdot$). Just as in electrophilic addition reactions, however, we find that only the more highly substituted, secondary radical is formed.

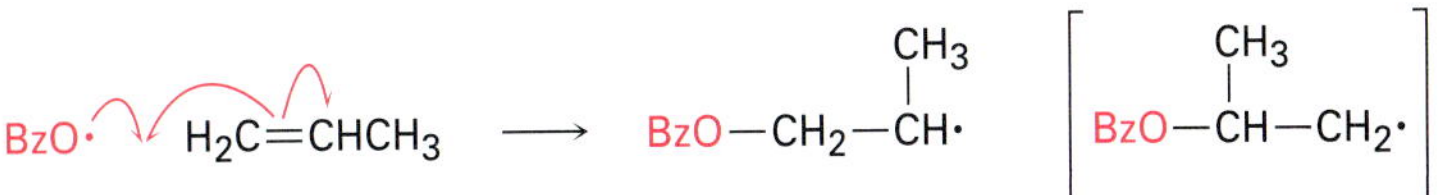

Secondary radical **Primary radical** *(NOT formed)*

Table 7.1 shows some commercially important alkene polymers, their use and the vinyl monomers from which they are made.

Table 7.1 Some Alkene Polymers and Their Uses

Monomer	Formula	Trade or common name of polymer	Uses
Ethylene	$H_2C{=}CH_2$	Polyethylene	Packaging, bottles
Propene (propylene)	$H_2C{=}CHCH_3$	Polypropylene	Moldings, rope, carpets
Chloroethylene (vinyl chloride)	$H_2C{=}CHCl$	Poly(vinyl chloride) Tedlar	Insulation, films, pipes
Styrene	$H_2C{=}CHC_6H_5$	Polystyrene	Foam, moldings
Tetrafluoroethylene	$F_2C{=}CF_2$	Teflon	Gaskets, nonstick coatings
Acrylonitrile	$H_2C{=}CHCN$	Orlon, Acrilan	Fibers
Methyl methacrylate	$H_2C{=}C(CH_3)CO_2CH_3$	Plexiglas, Lucite	Paint, sheets, moldings
Vinyl acetate	$H_2C{=}CHOCOCH_3$	Poly(vinyl acetate)	Paint, adhesives, foams

WORKED EXAMPLE 7.4 ***Predicting the Structure of a Polymer***

Show the structure of poly(vinyl chloride), a polymer made from $H_2C{=}CHCl$, by drawing several repeating units.

Strategy Mentally break the carbon–carbon double bond in the monomer unit, and form single bonds by connecting numerous units together.

Solution The general structure of poly(vinyl chloride) is

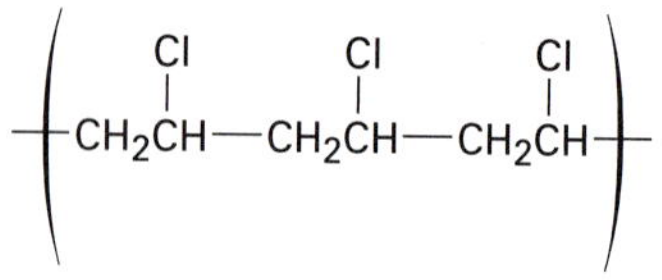

Problem 7.18 Show the monomer units you would use to prepare the following polymers:

(a)

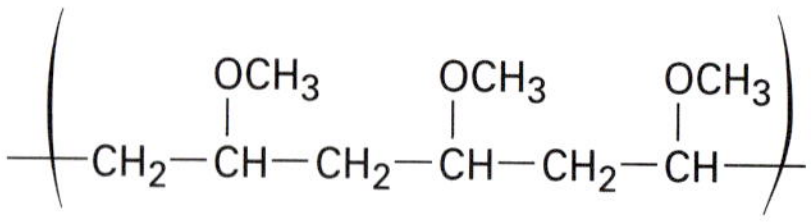

(b)

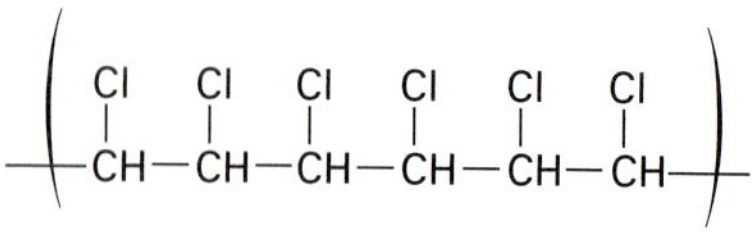

Problem 7.19 One of the chain-termination steps that sometimes occurs to interrupt polymerization is the following reaction between two radicals. Propose a mechanism for the reaction, using fishhook arrows to indicate electron flow.

7.11 Biological Additions of Radicals to Alkenes

The same high reactivity of radicals that makes possible the alkene polymerization we saw in the previous section also makes it difficult to carry out controlled radical reactions on complex molecules. As a result, there are severe limitations on the usefulness of radical addition reactions in the laboratory. In contrast to an *electrophilic* addition, where reaction occurs once and the reactive cation intermediate is rapidly quenched in the presence of a nucleophile, the reactive intermediate in a *radical* reaction is not usually quenched, so it reacts again and again in a largely uncontrollable way.

Electrophilic addition
(Intermediate is quenched,
so reaction stops.)

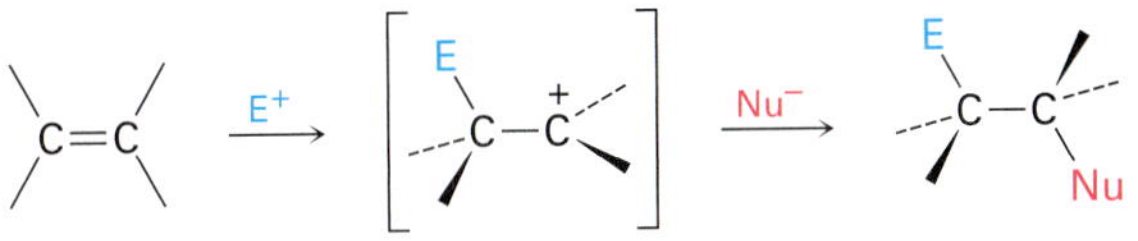

Radical addition
(Intermediate is not quenched,
so reaction does not stop.)

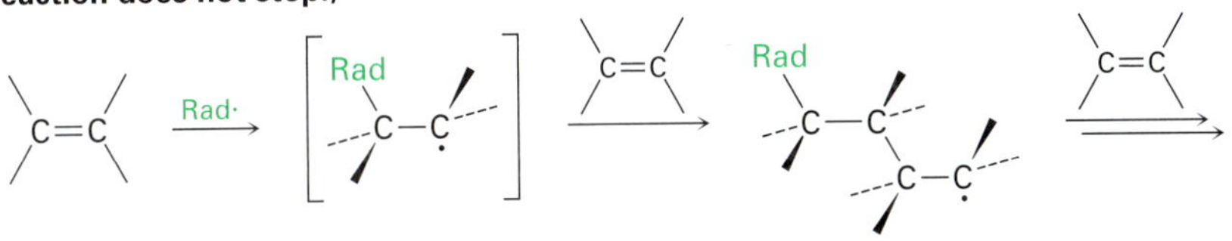

In biological reactions, the situation is different from that in the laboratory. Only one substrate molecule at a time is present in the active site of the enzyme where reaction takes place, and that molecule is held in a precise position, with coenzymes and other necessary reacting groups nearby. As a result, biological radical reactions are both more controlled and more common than laboratory or industrial radical reactions. A particularly impressive example occurs in the biosynthesis of prostaglandins from arachidonic acid, where a sequence of four radical additions take place. The reaction mechanism was discussed briefly in Section 5.3.

Prostaglandin biosynthesis begins with abstraction of a hydrogen atom from C13 of arachidonic acid by an iron–oxy radical (Figure 7.9, step 1) to give a carbon radical that reacts with O_2 at C11 through a resonance form (step 2). The oxygen radical that results adds to the C8–C9 double bond (step 3) to give a carbon radical at C8, which then adds to the C12–C13 double bond ar gives a carbon radical at C13 (step 4). A resonance form of this carbon radic adds at C15 to a second O_2 molecule (step 5), completing the prostaglandi skeleton, and reduction of the O–O bond then gives prostaglandin H_2 (step 6 The pathway looks complicated, but the entire process is catalyzed with exqu site control by just one enzyme.

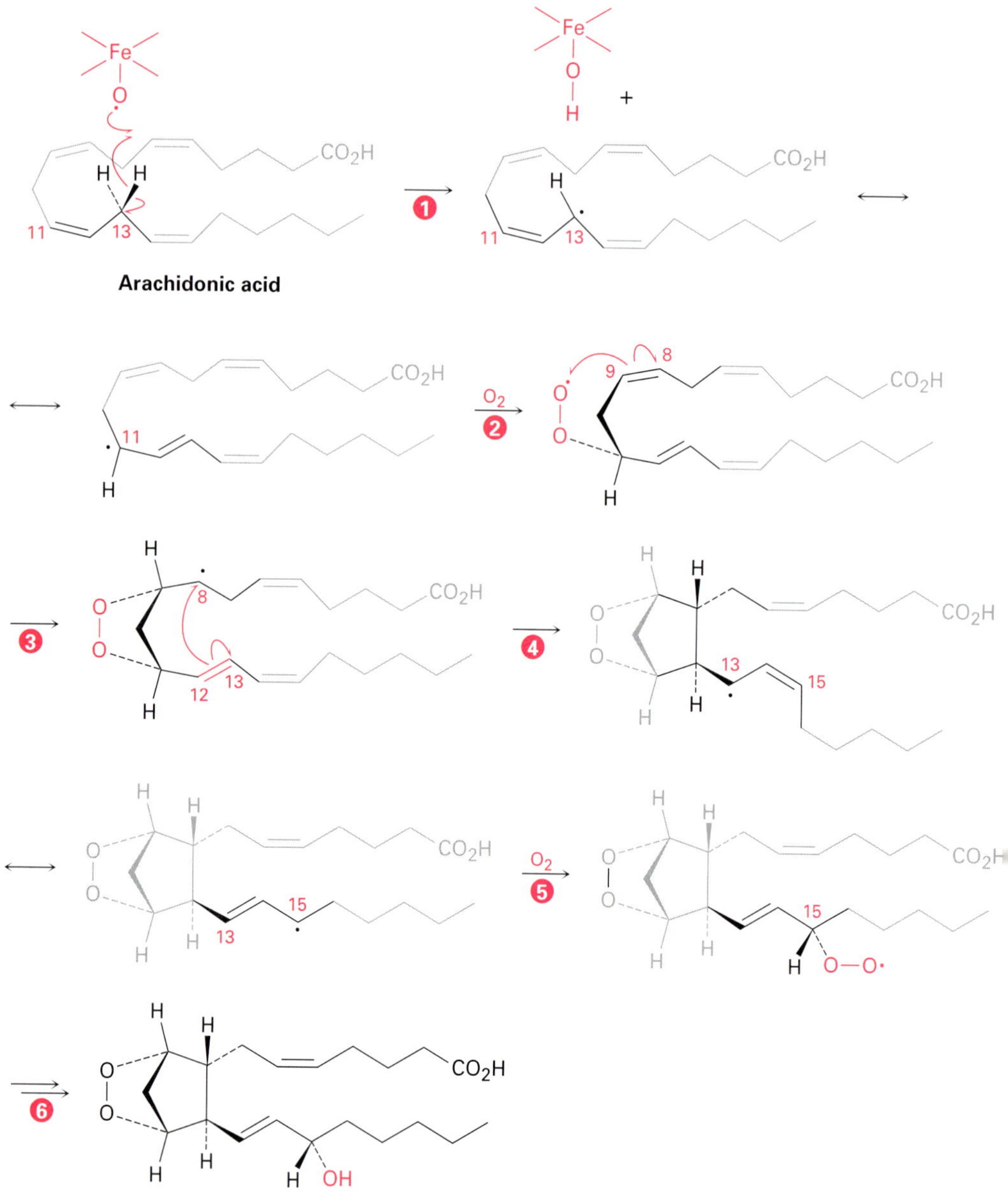

Prostaglandin H_2

Figure 7.9 Pathway for the biosynthesis of prostaglandins from arachidonic acid. Steps 2 and 5 are radical addition reactions to O_2; steps 3 and 4 are radical additions to carbon–carbon double bonds.

Focus On . . .

Natural Rubber

Natural rubber is obtained from the bark of the rubber tree, *Hevea brasiliensis,* grown on enormous plantations in Southeast Asia.

Rubber—an unusual name for an unusual substance—is a naturally occurring alkene polymer produced by more than 400 different plants. The major source is the so-called rubber tree, *Hevea brasiliensis,* from which the crude material is harvested as it drips from a slice made through the bark. The name *rubber* was coined by Joseph Priestley, the discoverer of oxygen and early researcher of rubber chemistry, for the simple reason that one of rubber's early uses was to rub out pencil marks on paper.

Unlike polyethylene and other simple alkene polymers, natural rubber is a polymer of a *diene,* isoprene (2-methyl-1,3-butadiene). The polymerization takes place by addition of isoprene monomer units to the growing chain, leading to formation of a polymer that still contains double bonds spaced regularly at four-carbon intervals. As the following structure shows, these double bonds have *Z* stereochemistry:

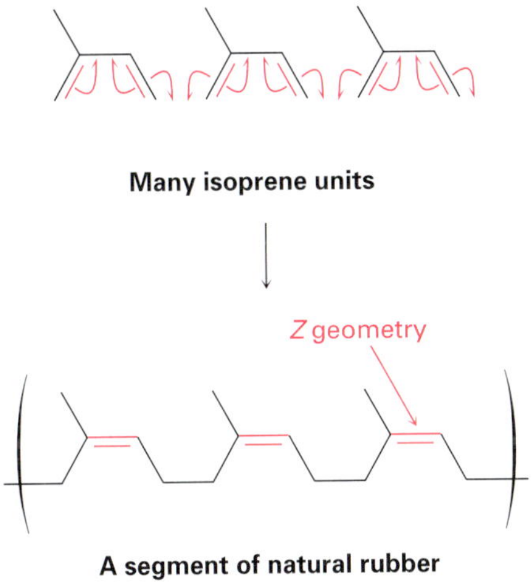

Crude rubber, called *latex,* is collected from the tree as an aqueous dispersion that is washed, dried, and coagulated by warming in air. The resultant polymer has chains that average about 5000 monomer units in length and have molecular weights of 200,000 to 500,000 amu. This crude coagulate is too soft and tacky to be useful until it is hardened by heating with elemental sulfur, a process called *vulcanization.* By mechanisms that are still not fully understood, vulcanization cross-links the rubber chains together by forming

(continued)

carbon–sulfur bonds between them, thereby hardening and stiffening the polymer. The exact degree of hardening can be varied, yielding material soft enough for automobile tires or hard enough for bowling balls *(ebonite)*.

The remarkable ability of rubber to stretch and then contract to its original shape is due to the irregular shapes of the polymer chains caused by the double bonds. These double bonds introduce bends and kinks into the polymer chains, thereby preventing neighboring chains from nestling together. When stretched, the randomly coiled chains straighten out and orient along the direction of the pull but are kept from sliding over one another by the cross-links. When the stretch is released, the polymer reverts to its original random state.

SUMMARY AND KEY WORDS

Alkenes are generally prepared by an *elimination reaction,* such as *dehydrohalogenation,* the elimination of HX from an alkyl halide, or *dehydration,* the elimination of water from an alcohol.

HCl, HBr, and HI add to alkenes by a two-step electrophilic addition mechanism. Initial reaction of the nucleophilic double bond with H^+ gives a carbocation intermediate, which then reacts with halide ion. Bromine and chlorine add to alkenes via three-membered-ring **bromonium ion** or chloronium ion intermediates to give addition products having **anti stereochemistry**. If water is present during the halogen addition reaction, a **halohydrin** is formed.

Hydration of an alkene—the addition of water—is carried out by either of two procedures, depending on the product desired. **Oxymercuration** involves electrophilic addition of Hg^{2+} to an alkene, followed by trapping of the cation intermediate with water and subsequent treatment with $NaBH_4$. **Hydroboration** involves addition of borane (BH_3) followed by oxidation of the intermediate organoborane with alkaline H_2O_2. The two hydration methods are complementary: oxymercuration gives the product of Markovnikov addition, whereas hydroboration/oxidation gives the product with non-Markovnikov **syn stereochemistry**.

A **carbene, R_2C:**, is a neutral molecule containing a divalent carbon with only six valence electrons. Carbenes are highly reactive toward alkenes, adding to give cyclopropanes. Nonhalogenated cyclopropanes are best prepared by treatment of the alkene with CH_2I_2 and zinc–copper, a process called the **Simmons–Smith reaction.**

Alkenes are **reduced** by addition of H_2 in the presence of a catalyst such as platinum or palladium to yield alkanes, a process called **catalytic hydrogenation**. Alkenes are also **oxidized** by reaction with a peroxyacid to give **epoxides,** which can be converted into trans-1,2-diols by acid-catalyzed epoxide hydrolysis. The corresponding cis-1,2-diols can be made directly from alkenes by **hydroxylation** with OsO_4. Alkenes can also be cleaved to produce carbonyl compounds by reaction with ozone, followed by reduction with zinc metal.

Alkene **polymers**—large molecules resulting from repetitive bonding together of many hundreds or thousands of small **monomer** units—are formed by reaction of simple alkenes with a radical initiator at high temperature and pressure. Polyethylene, polypropylene, and polystyrene are common examples. As a general rule, radical addition reactions are not common in the laboratory but occur much more frequently in biological pathways.

Learning Reactions

What's seven times nine? Sixty-three, of course. You didn't have to stop and figure it out; you knew the answer immediately because you long ago learned the multiplication tables. Learning the reactions of organic chemistry requires the same approach: reactions have to be learned for immediate recall if they are to be useful.

Different people take different approaches to learning reactions. Some people make flash cards; others find studying with friends to be helpful. To help guide your study, most chapters in this book end with a summary of the reactions just presented. In addition, the accompanying *Study Guide and Solutions Manual* has several appendixes that organize organic reactions from other viewpoints. Fundamentally, though, there are no shortcuts. Learning organic chemistry does take effort.

SUMMARY OF REACTIONS

Note: No stereochemistry is implied unless specifically indicated with wedged, solid, and dashed lines.

1. Addition reactions of alkenes
   (a) Addition of HCl, HBr, and HI (Sections 6.7 and 6.8)

   Markovnikov regiochemistry occurs, with H adding to the less highly substituted alkene carbon and halogen adding to the more highly substituted carbon.

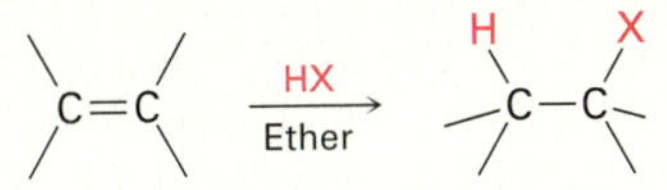

   (b) Addition of halogens Cl_2 and Br_2 (Section 7.2)

   Anti addition is observed through a halonium ion intermediate.

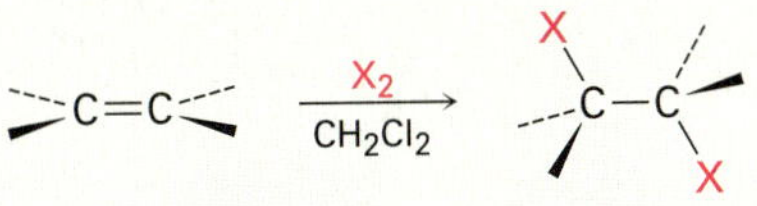

   (c) Halohydrin formation (Section 7.3)

   Markovnikov regiochemistry and anti stereochemistry occur.

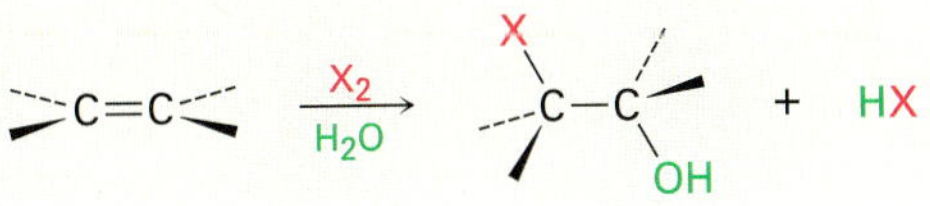

(d) Addition of water by oxymercuration (Section 7.4)
Markovnikov regiochemistry occurs.

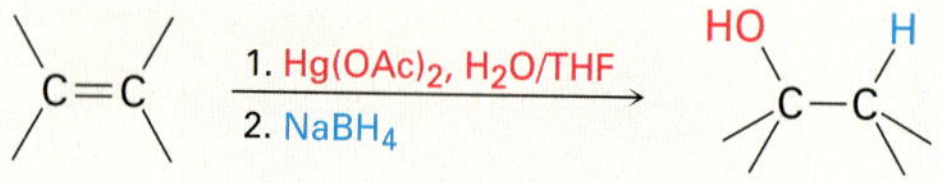

(e) Addition of water by hydroboration/oxidation (Section 7.5)
Non-Markovnikov syn addition occurs.

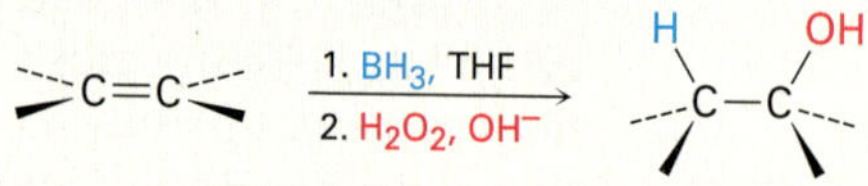

(f) Addition of carbenes to yield cyclopropanes (Section 7.6)
(1) Dichlorocarbene addition

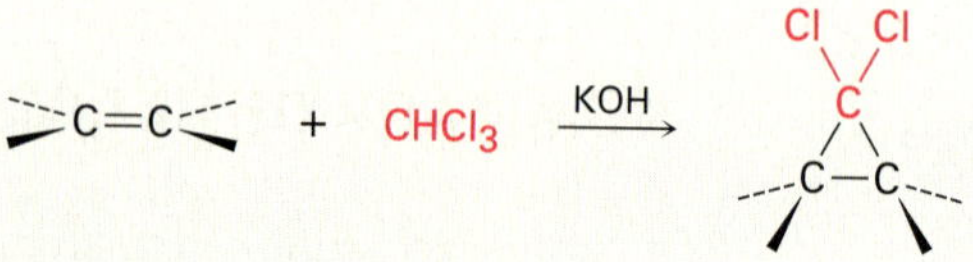

(2) Simmons–Smith reaction

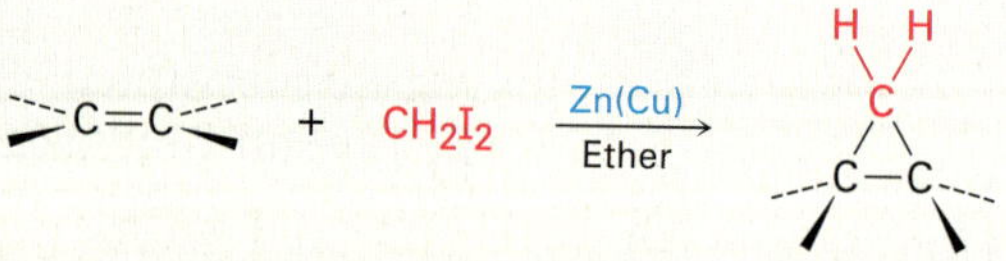

(g) Catalytic hydrogenation (Section 7.7)
Syn addition occurs.

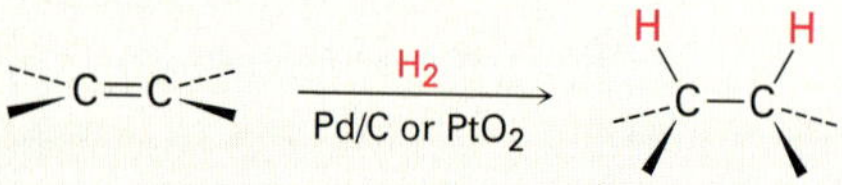

(h) Epoxidation with a peroxyacid (Section 7.8)
Syn addition occurs.

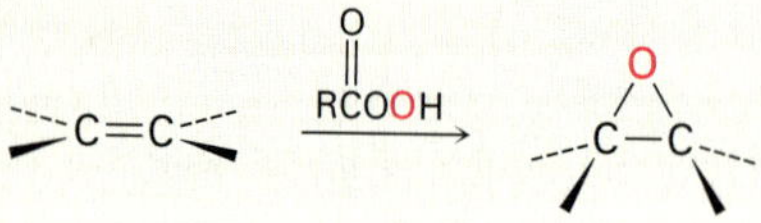

(i) Hydroxylation by acid-catalyzed epoxide hydrolysis (Section 7.8)
Anti stereochemistry occurs.

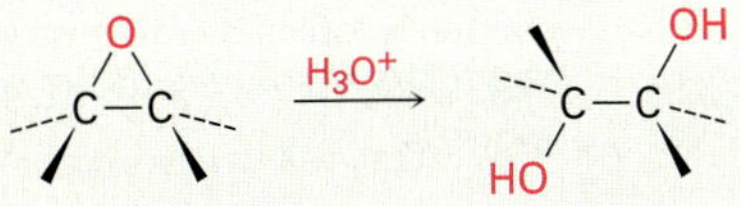

(j) Hydroxylation with OsO_4 (Section 7.8)
Syn addition occurs.

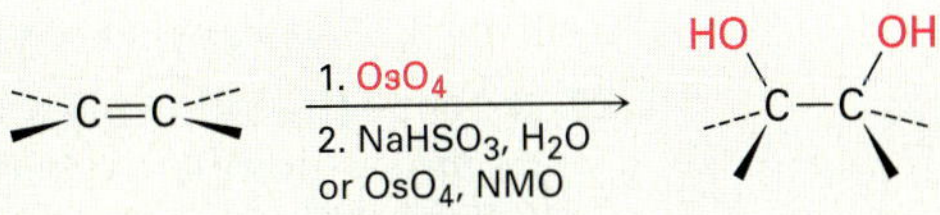

(k) Radical polymerization (Section 7.10)

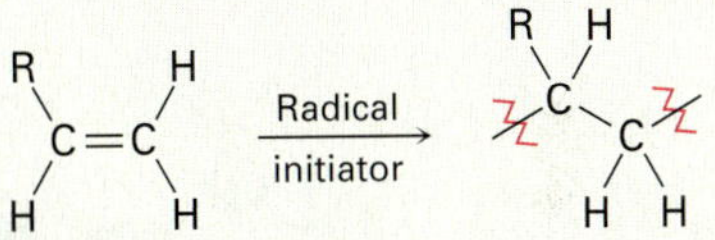

2. Oxidative cleavage of alkenes (Section 7.9)
(a) Reaction with ozone followed by zinc in acetic acid

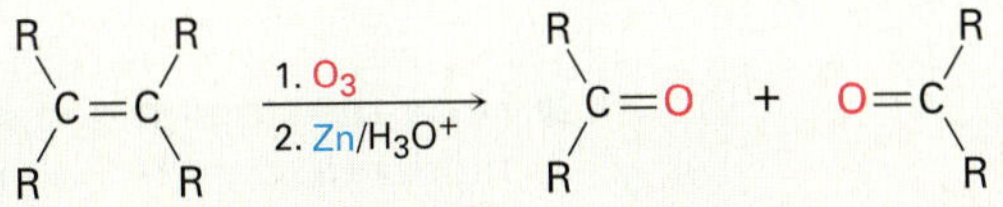

(b) Reaction with $KMnO_4$ in acidic solution

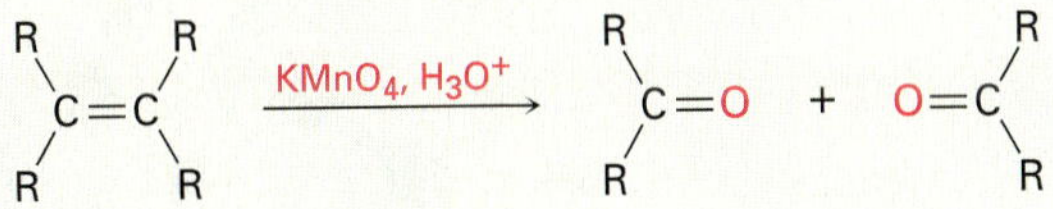

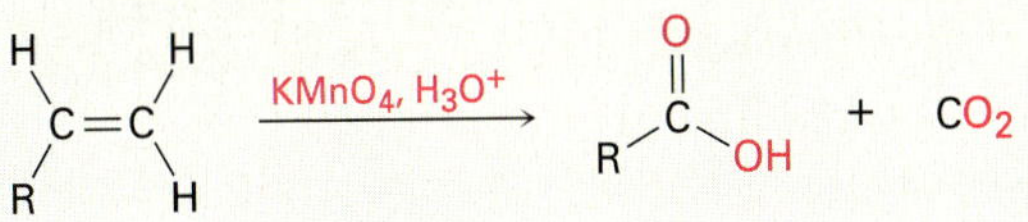

3. Cleavage of 1,2-diols (Section 7.9)

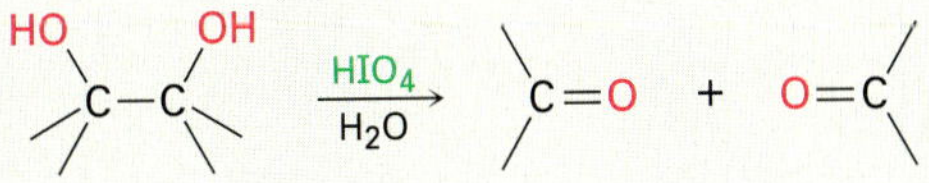

EXERCISES

Organic KNOWLEDGE TOOLS

ThomsonNOW Sign in at **www.thomsonedu.com** to assess your knowledge of this chapter's topics by taking a pre-test. The pre-test will link you to interactive organic chemistry resources based on your score in each concept area.

Online homework for this chapter may be assigned in Organic OWL.

■ indicates problems assignable in Organic OWL.

VISUALIZING CHEMISTRY

(Problems 7.1–7.19 appear within the chapter.)

7.20 ■ Name the following alkenes, and predict the products of their reaction with (i) *meta*-chloroperoxybenzoic acid, (ii) $KMnO_4$ in aqueous acid, and (iii) O_3, followed by Zn in acetic acid:

(a)

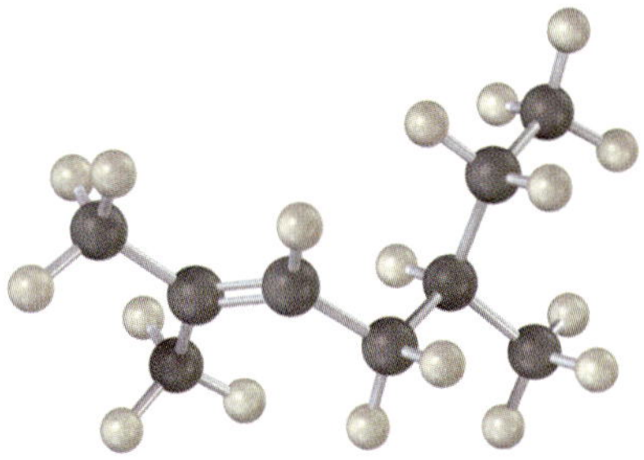

(b)

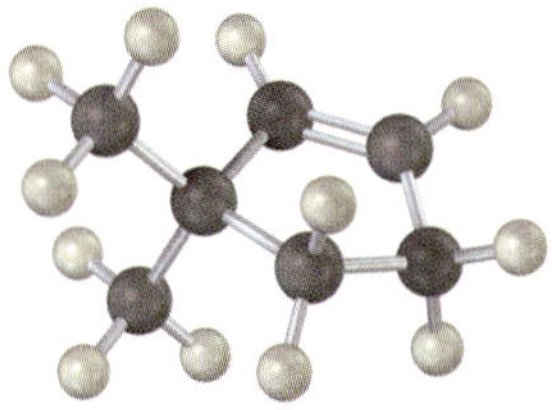

7.21 ■ Draw the structures of alkenes that would yield the following alcohols on hydration (red = O). Tell in each case whether you would use hydroboration/oxidation or oxymercuration.

(a)

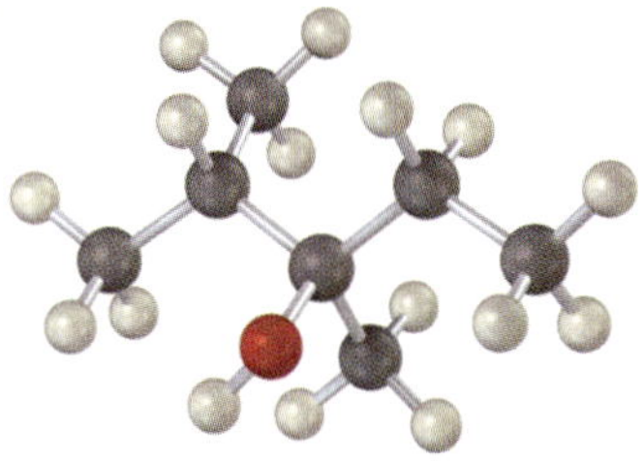

(b)

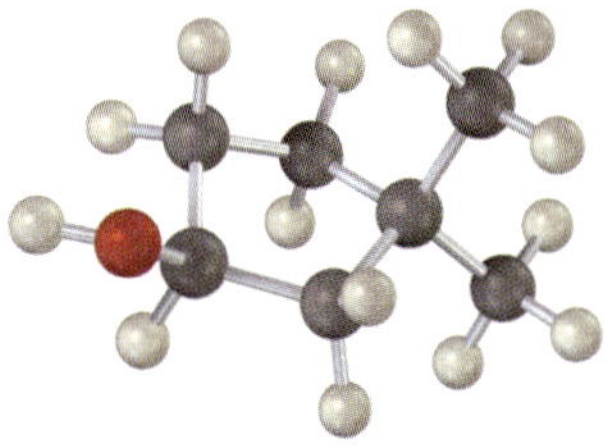

7.22 The following alkene undergoes hydroboration/oxidation to yield a single product rather than a mixture. Explain the result, and draw the product showing its stereochemistry.

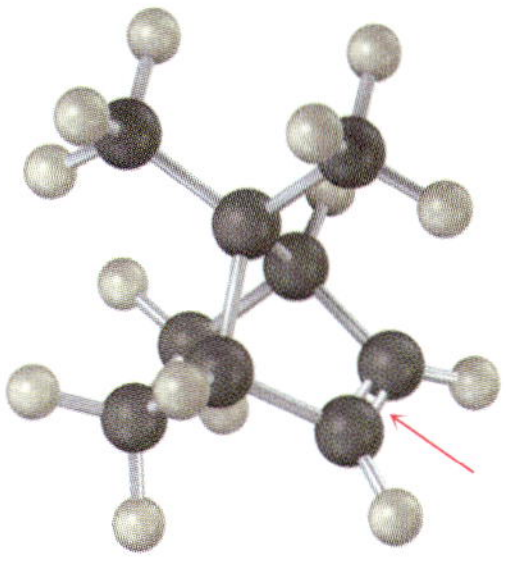

7.23 ■ From what alkene was the following 1,2-diol made, and what method was used, epoxide hydrolysis or OsO_4?

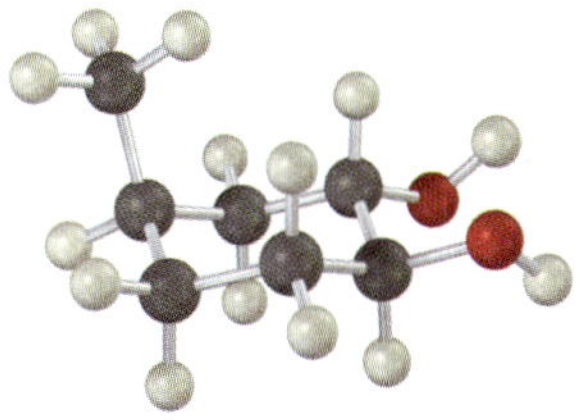

ADDITIONAL PROBLEMS

homsonNOW™ Click *Organic nteractive* to **use a web-based alette to synthesize new unctional groups beginning with alkenes.**

7.24 ■ Predict the products of the following reactions (the aromatic ring is unreactive in all cases). Indicate regiochemistry when relevant.

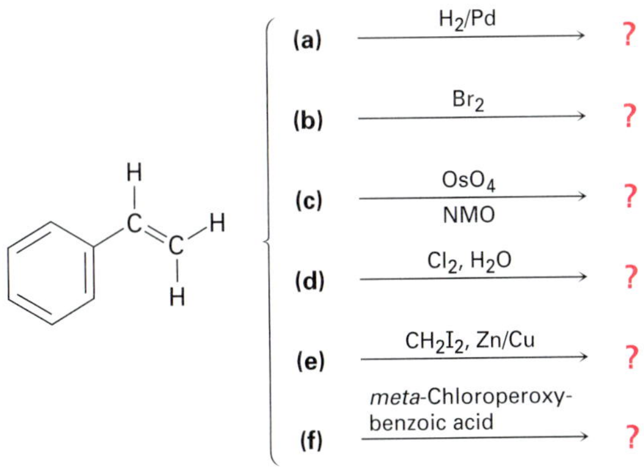

7.25 ■ Suggest structures for alkenes that give the following reaction produc
There may be more than one answer for some cases.

(a)
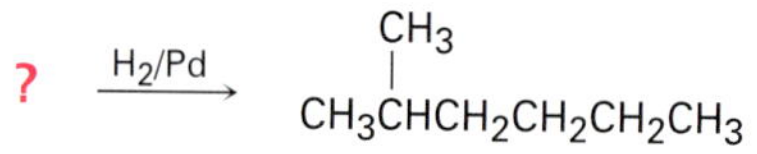

(b)
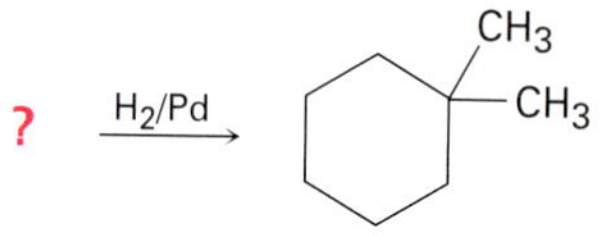

(c)
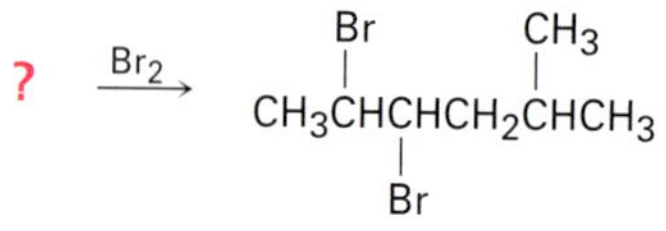

(d)
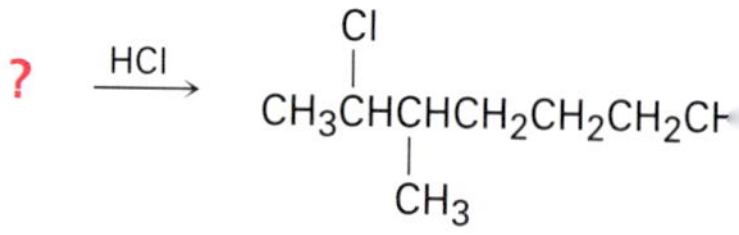

(e)
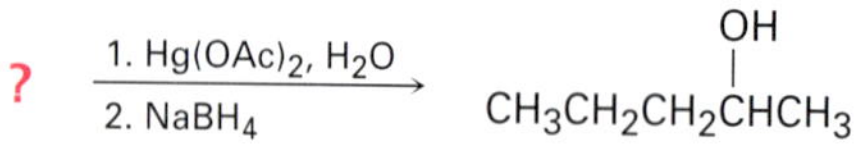

(f)
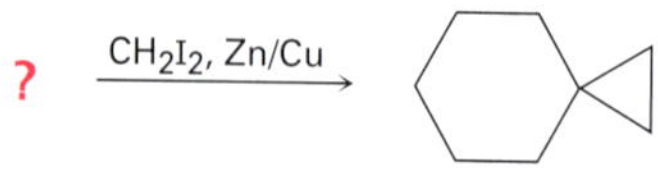

7.26 ■ Predict the products of the following reactions, showing both regiochem
istry and stereochemistry where appropriate:

(a)
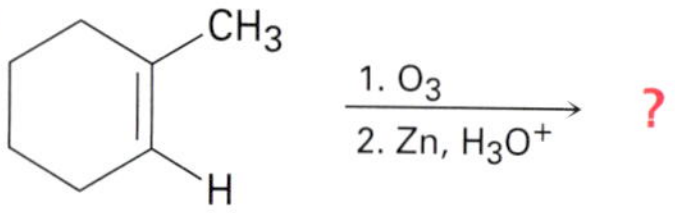

(b)
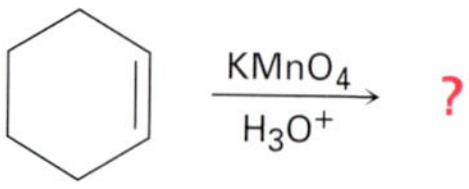

(c)
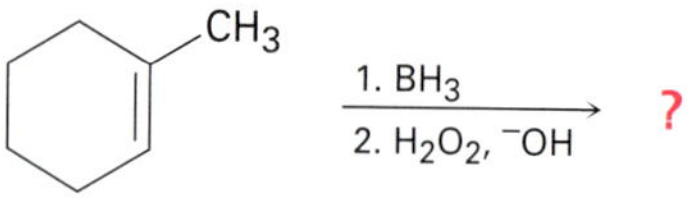

(d)
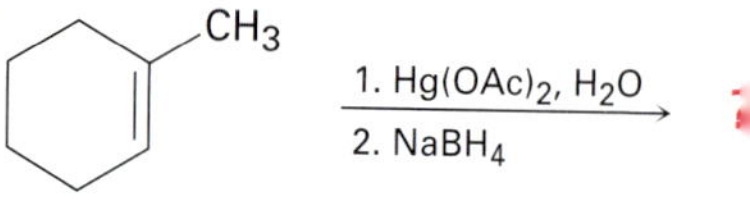

7.27 ■ How would you carry out the following transformations? Tell the reagent
you would use in each case.

(a)
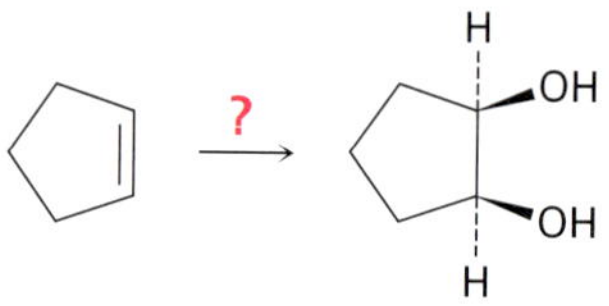

(b)
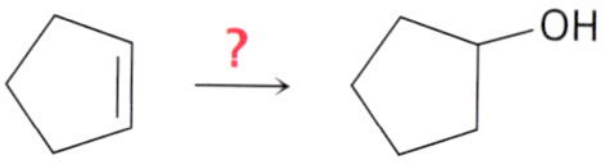

(c)
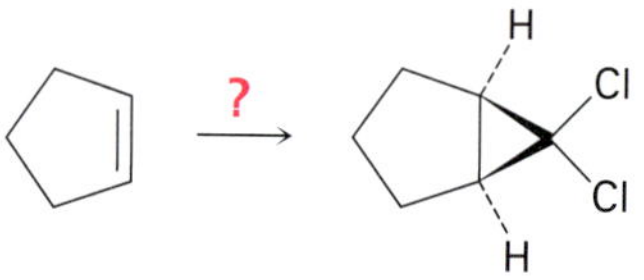

(d)
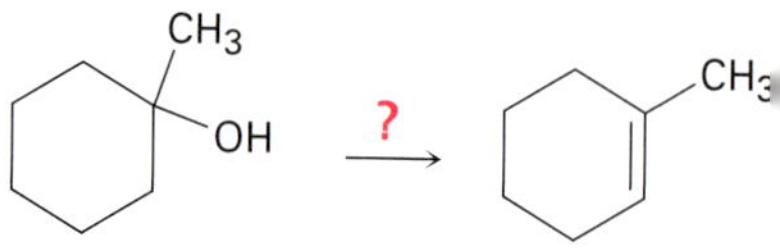

(e)
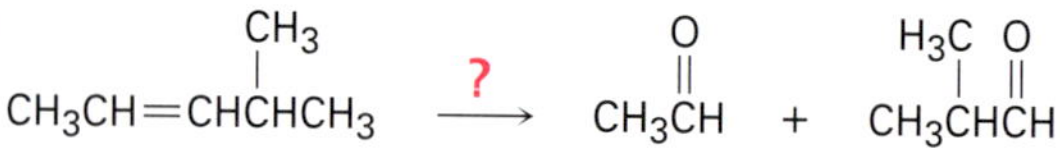

(f)

7.28 Which reaction would you expect to be faster, addition of HBr to cyclohexene or to 1-methylcyclohexene? Explain.

7.29 What product will result from hydroboration/oxidation of 1-methylcyclopentene with deuterated borane, BD_3? Show both the stereochemistry (spatial arrangement) and the regiochemistry (orientation) of the product.

7.30 ■ Draw the structure of an alkene that yields only acetone, $(CH_3)_2C{=}O$, on ozonolysis followed by treatment with Zn.

7.31 ■ Show the structures of alkenes that give the following products on oxidative cleavage with $KMnO_4$ in acidic solution:

(a) $CH_3CH_2CO_2H$ + CO_2

(b) $(CH_3)_2C{=}O$ + $CH_3CH_2CH_2CO_2H$

(c) 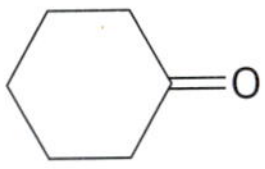 + $(CH_3)_2C{=}O$

(d) $CH_3CH_2\overset{\overset{O}{\|}}{C}CH_2CH_2CH_2CH_2CO_2H$

7.32 ■ Compound A has the formula $C_{10}H_{16}$. On catalytic hydrogenation over palladium, it reacts with only 1 molar equivalent of H_2. Compound A also undergoes reaction with ozone, followed by zinc treatment, to yield a symmetrical diketone, B ($C_{10}H_{16}O_2$).
(a) How many rings does A have?
(b) What are the structures of A and B?
(c) Write the reactions.

7.33 An unknown hydrocarbon A with the formula C_6H_{12} reacts with 1 molar equivalent of H_2 over a palladium catalyst. Hydrocarbon A also reacts with OsO_4 to give diol B. When oxidized with $KMnO_4$ in acidic solution, A gives two fragments. One fragment is propanoic acid, $CH_3CH_2CO_2H$, and the other fragment is ketone C. What are the structures of A, B, and C? Write all reactions, and show your reasoning.

7.34 Using an oxidative cleavage reaction, explain how you would distinguish between the following two isomeric dienes:

 and 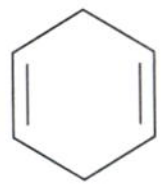

7.35 Compound A, $C_{10}H_{18}O$, undergoes reaction with dilute H_2SO_4 at 50 °C to yield a mixture of two alkenes, $C_{10}H_{16}$. The major alkene product, B, gives only cyclopentanone after ozone treatment followed by reduction with zinc in acetic acid. Identify A and B, and write the reactions.

7.36 The cis and trans isomers of 2-butene give different cyclopropane products in the Simmons–Smith reaction. Show the structure of each, and explain the difference.

$$cis\text{-}CH_3CH{=}CHCH_3 \xrightarrow{CH_2I_2,\ Zn(Cu)} ?$$

$$trans\text{-}CH_3CH{=}CHCH_3 \xrightarrow{CH_2I_2,\ Zn(Cu)} ?$$

7.37 Iodine azide, IN_3, adds to alkenes by an electrophilic mechanism similar t that of bromine. If a monosubstituted alkene such as 1-butene is used, on one product results:

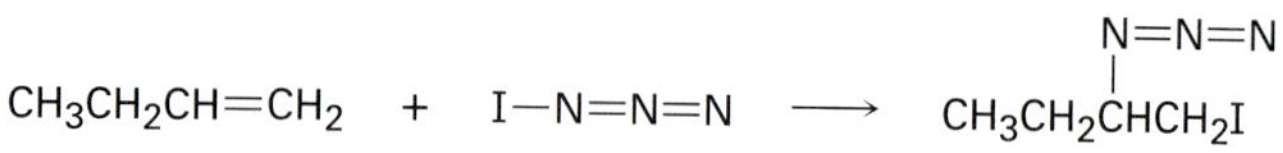

(a) Add lone-pair electrons to the structure shown for IN_3, and draw a secon resonance form for the molecule.
(b) Calculate formal charges for the atoms in both resonance structures yo drew for IN_3 in part **(a)**.
(c) In light of the result observed when IN_3 adds to 1-butene, what is th polarity of the $I{-}N_3$ bond? Propose a mechanism for the reaction usin curved arrows to show the electron flow in each step.

7.38 ■ 10-Bromo-α-chamigrene, a compound isolated from marine algae, i thought to be biosynthesized from γ-bisabolene by the following route:

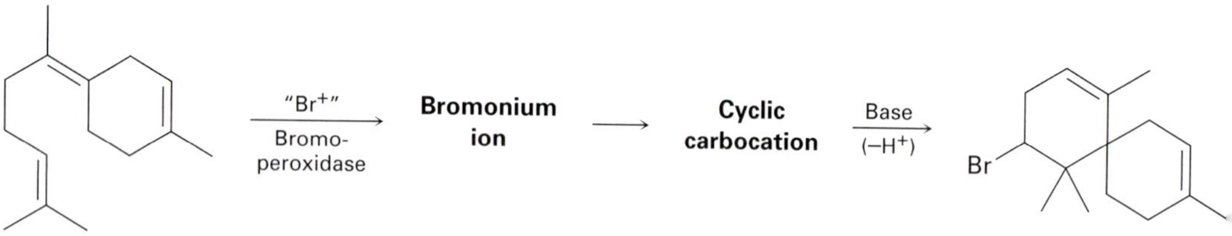

γ-Bisabolene

10-Bromo-α-chamigrene

Draw the structures of the intermediate bromonium and cyclic carbocation, and propose mechanisms for all three steps.

7.39 ■ Draw the structure of a hydrocarbon that absorbs 2 molar equivalents of H_2 on catalytic hydrogenation and gives only butanedial on ozonolysis.

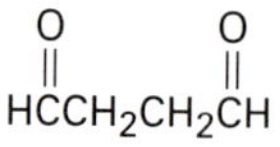

Butanedial

7.40 Simmons–Smith reaction of cyclohexene with diiodomethane gives a single cyclopropane product, but the analogous reaction of cyclohexene with 1,1-diiodoethane gives (in low yield) a mixture of two isomeric methylcyclopropane products. What are the two products, and how do they differ?

7.41 In planning the synthesis of one compound from another, it's just as important to know what *not* to do as to know what to do. The following reactions all have serious drawbacks to them. Explain the potential problems of each.

(a)

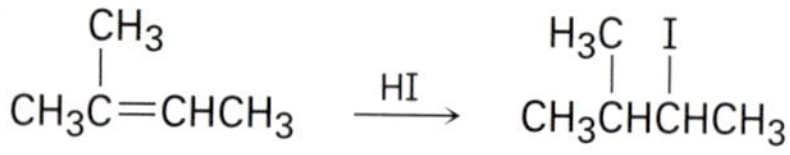

(b)

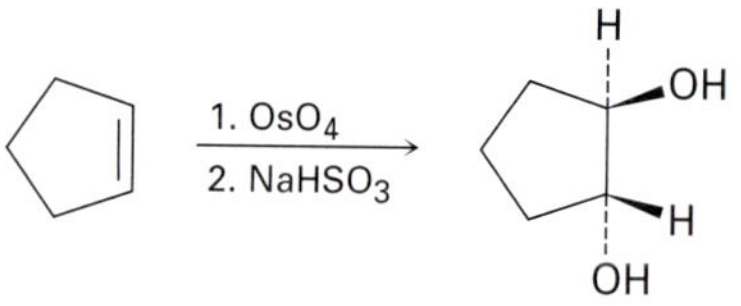

(c)

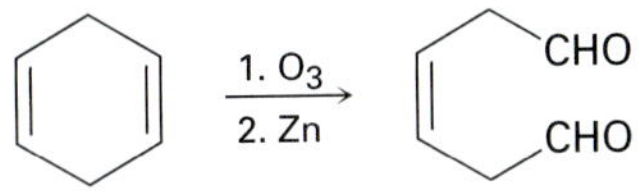

(d)

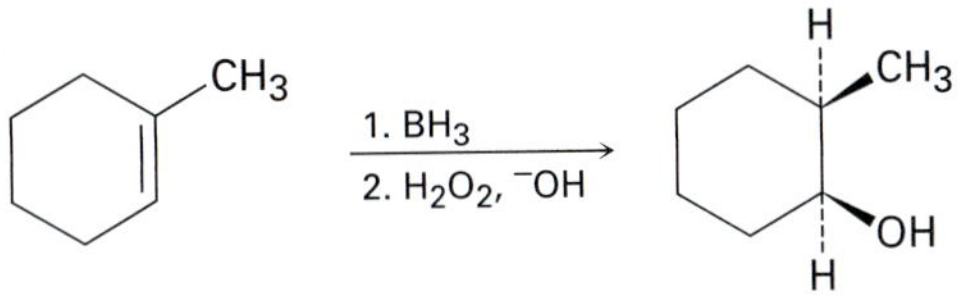

7.42 Which of the following alcohols could *not* be made selectively by hydroboration/oxidation of an alkene? Explain.

(a) $CH_3CH_2CH_2CH(OH)CH_3$

(b) $(CH_3)_2CHC(OH)(CH_3)_2$

(c)

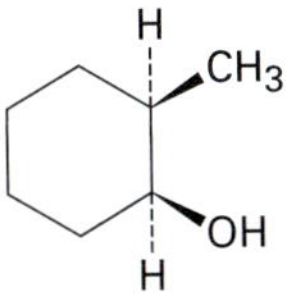

(d)

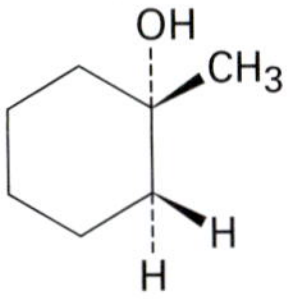

7.43 ■ Predict the products of the following reactions. Don't worry about the size of the molecule; concentrate on the functional groups.

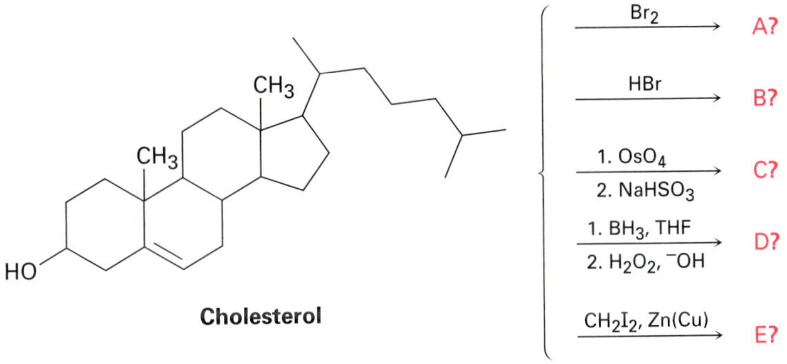

7.44 The sex attractant of the common housefly is a hydrocarbon with the formula $C_{23}H_{46}$. On treatment with aqueous acidic $KMnO_4$, two products are obtained, $CH_3(CH_2)_{12}CO_2H$ and $CH_3(CH_2)_7CO_2H$. Propose a structure.

7.45 Compound A has the formula C_8H_8. It reacts rapidly with $KMnO_4$ to give CO_2 and a carboxylic acid, B ($C_7H_6O_2$), but reacts with only 1 molar equivalent of H_2 on catalytic hydrogenation over a palladium catalyst. On hydrogenation under conditions that reduce aromatic rings, 4 equivalents of H_2 are taken up and hydrocarbon C (C_8H_{16}) is produced. What are the structures of A, B, and C? Write the reactions.

7.46 ■ Plexiglas, a clear plastic used to make many molded articles, is made b polymerization of methyl methacrylate. Draw a representative segment Plexiglas.

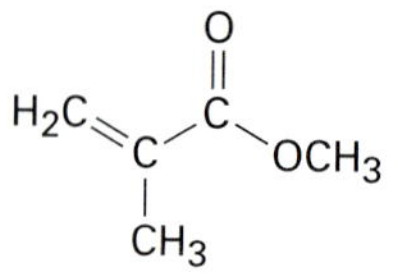

Methyl methacrylate

7.47 ■ Poly(vinyl pyrrolidone), prepared from *N*-vinylpyrrolidone, is used both i cosmetics and as a synthetic blood substitute. Draw a representative segmen of the polymer.

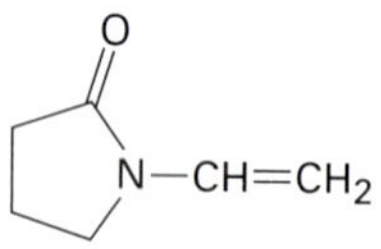

***N*-Vinylpyrrolidone**

7.48 Reaction of 2-methylpropene with CH_3OH in the presence of H_2SO_4 catalys yields methyl *tert*-butyl ether, $CH_3OC(CH_3)_3$, by a mechanism analogous t that of acid-catalyzed alkene hydration. Write the mechanism, using curve arrows for each step.

7.49 ■ Isolated from marine algae, prelaureatin is thought to be biosynthesize from laurediol by the following route. Propose a mechanism.

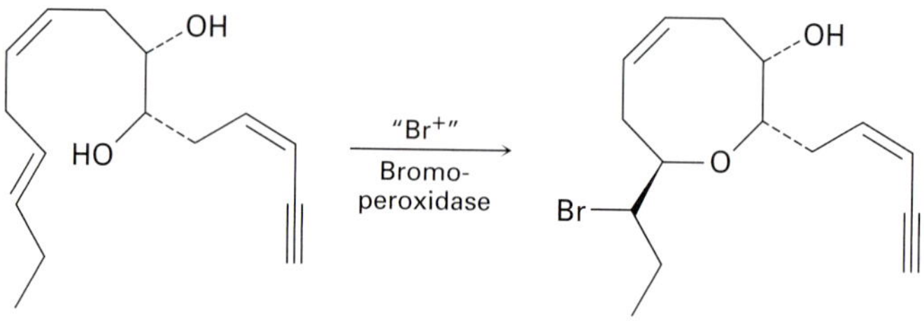

Laurediol **Prelaureatin**

7.50 How would you distinguish between the following pairs of compounds using simple chemical tests? Tell what you would do and what you would see.
(a) Cyclopentene and cyclopentane (b) 2-Hexene and benzene

7.51 Dichlorocarbene can be generated by heating sodium trichloroacetate. Propose a mechanism for the reaction, and use curved arrows to indicate the movement of electrons in each step. What relationship does your mechanism bear to the base-induced elimination of HCl from chloroform?

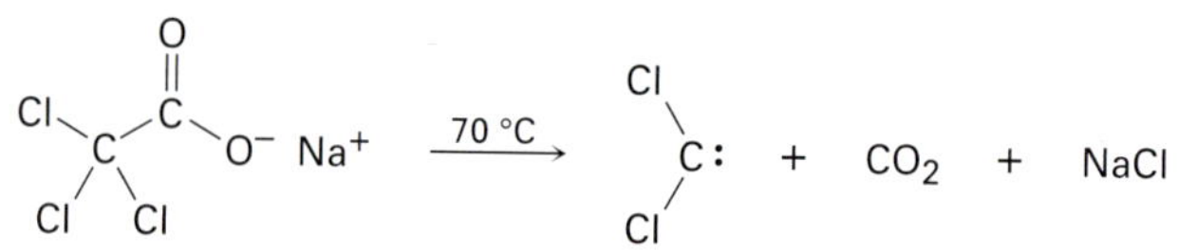

7.52 ■ α-Terpinene, $C_{10}H_{16}$, is a pleasant-smelling hydrocarbon that has been isolated from oil of marjoram. On hydrogenation over a palladium catalyst, α-terpinene reacts with 2 molar equivalents of H_2 to yield a hydrocarbon, $C_{10}H_{20}$. On ozonolysis, followed by reduction with zinc and acetic acid, α-terpinene yields two products, glyoxal and 6-methyl-2,5-heptanedione.

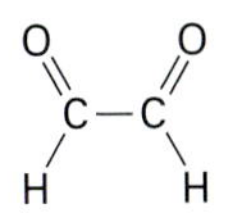

Glyoxal

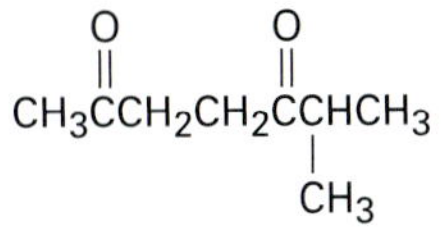

6-Methyl-2,5-heptanedione

(a) How many degrees of unsaturation does α-terpinene have?
(b) How many double bonds and how many rings does it have?
(c) Propose a structure for α-terpinene.

7.53 Evidence that cleavage of 1,2-diols by HIO_4 occurs through a five-membered cyclic periodate intermediate is based on *kinetic data*—the measurement of reaction rates. When diols A and B were prepared and the rates of their reaction with HIO_4 were measured, it was found that diol A cleaved approximately 1 million times faster than diol B. Make molecular models of A and B and of potential cyclic periodate intermediates, and then explain the kinetic results.

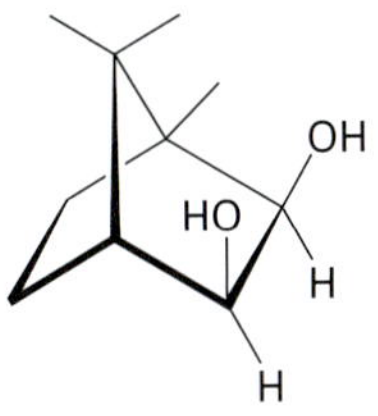

A
(*cis* diol)

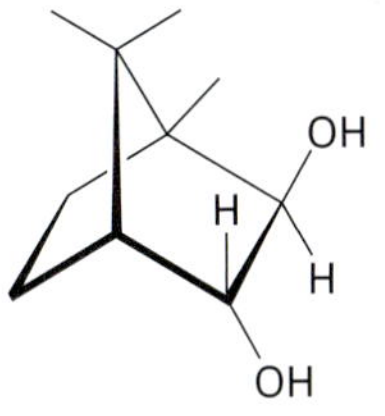

B
(*trans* diol)

7.54 ■ Reaction of HBr with 3-methylcyclohexene yields a mixture of four products: *cis*- and *trans*-1-bromo-3-methylcyclohexane and *cis*- and *trans*-1-bromo-2-methylcyclohexane. The analogous reaction of HBr with 3-bromocyclohexene yields *trans*-1,2-dibromocyclohexane as the sole product. Draw structures of the possible intermediates, and then explain why only a single product is formed in the reaction of HBr with 3-bromocyclohexene.

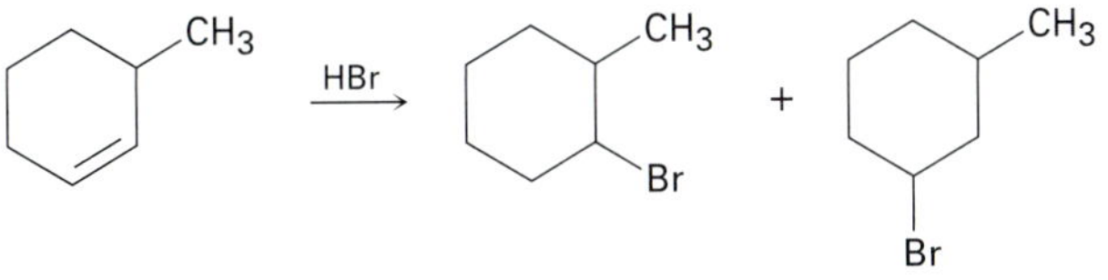

cis, trans **cis, trans**

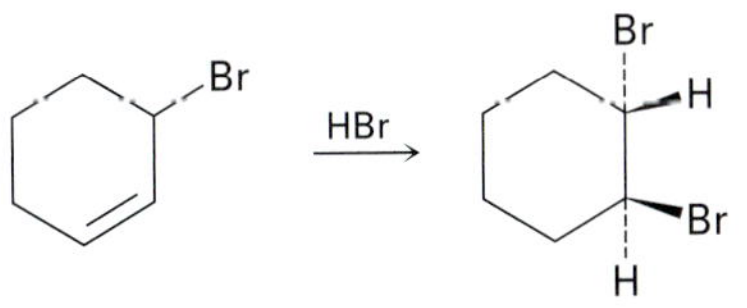

7.55 Reaction of cyclohexene with mercury(II) acetate in CH_3OH rather than H_2O, followed by treatment with $NaBH_4$, yields cyclohexyl methyl ether rather than cyclohexanol. Suggest a mechanism.

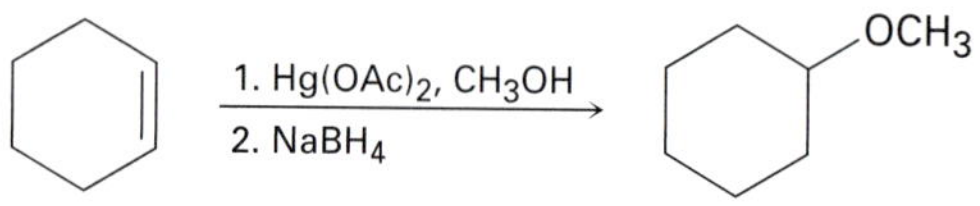

Cyclohexene **Cyclohexyl methyl ether**

7.56 Use your general knowledge of alkene chemistry to suggest a mechanism for the following reaction:

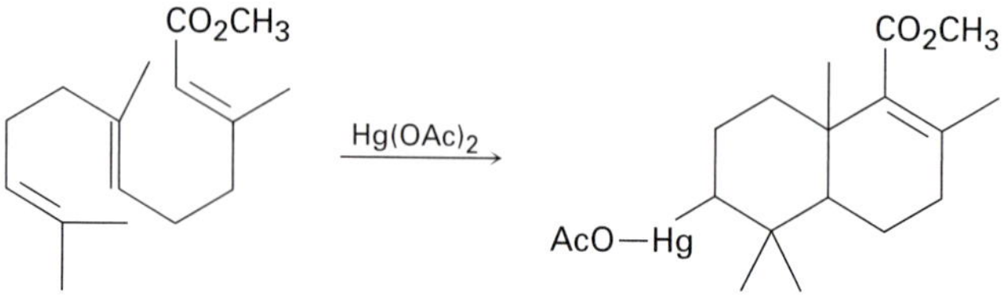

7.57 ■ Treatment of 4-penten-1-ol with aqueous Br_2 yields a cyclic bromo ether rather than the expected bromohydrin. Suggest a mechanism, using curved arrows to show electron movement.

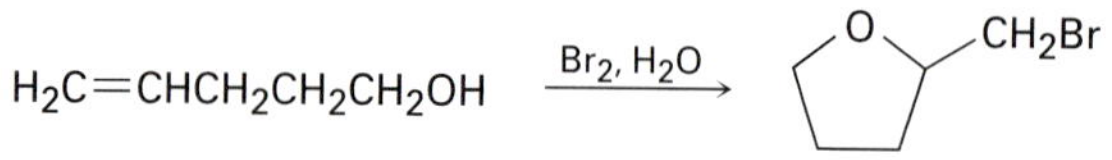

4-Penten-1-ol **2-(Bromomethyl)tetrahydrofuran**

7.58 Hydroboration of 2-methyl-2-pentene at 25 °C followed by oxidation with alkaline H_2O_2 yields 2-methyl-3-pentanol, but hydroboration at 160 °C followed by oxidation yields 4-methyl-1-pentanol. Suggest a mechanism.

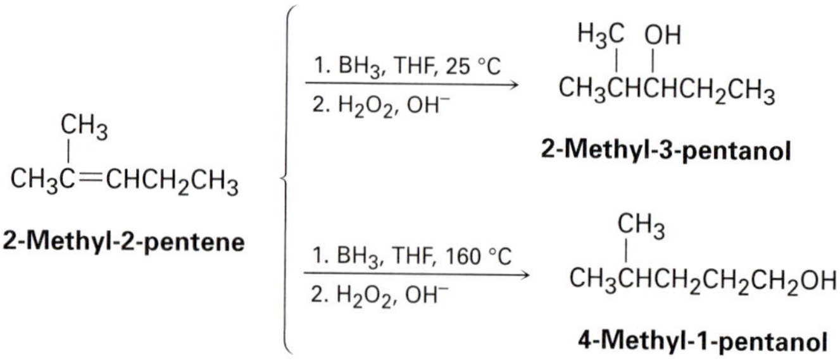

7.59 We'll see in the next chapter that alkynes undergo many of the same reactions that alkenes do. What product might you expect from each of the following reactions?

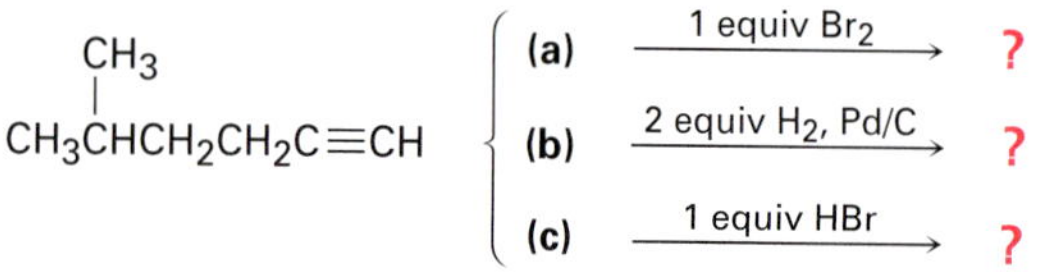

7.60 Hydroxylation of *cis*-2-butene with OsO_4 yields a different product than hydroxylation of *trans*-2-butene. Draw the structure, show the stereochemistry of each product, and explain the difference between them.

■ Assignable in OWL

8

Alkynes: An Introduction to Organic Synthesis

rganic KNOWLEDGE TOOLS

iomsonNOW™ Throughout iis chapter, sign in at vww.thomsonedu.com for nline self-study and interactive itorials based on your level of nderstanding.

Online homework for this chapter may be assigned in Organic OWL.

An **alkyne** is a hydrocarbon that contains a carbon–carbon triple bond. Acetylene, $H-C\equiv C-H$, the simplest alkyne, was once widely used in industry as the starting material for the preparation of acetaldehyde, acetic acid, vinyl chloride, and other high-volume chemicals, but more efficient routes to these substances using ethylene as starting material are now available. Acetylene is still used in the preparation of acrylic polymers but is probably best known as the gas burned in high-temperature oxy–acetylene welding torches.

Much current research is centering on *polyynes*—linear carbon chains of *sp*-hybridized carbon atoms. Polyynes with up to eight triple bonds have been detected in interstellar space, and evidence has been presented for the existence of *carbyne,* an allotrope of carbon consisting of repeating triple bonds in long chains of indefinite length.

$$H-C\equiv C-C\equiv C-C\equiv C-C\equiv C-C\equiv C-C\equiv C-C\equiv C-H$$

A polyyne detected in interstellar space

WHY THIS CHAPTER?

Alkynes are less common than alkenes, both in the laboratory and in living organisms, so we won't cover them in great detail. The real importance of this chapter is that we'll use alkyne chemistry as a vehicle to begin looking at some of the general strategies used in *organic synthesis*—the construction of complex molecules in the laboratory. Without the ability to design and synthesize new molecules in the laboratory, many of the medicines we take for granted would not exist and few new ones would be made.

8.1 Naming Alkynes

Alkyne nomenclature follows the general rules for hydrocarbons discussed in Sections 3.4 and 6.3. The suffix *-yne* is used, and the position of the triple bond is indicated by giving the number of the first alkyne carbon in the chain. Numbering the main chain begins at the end nearer the triple bond so that the triple bond receives as low a number as possible.

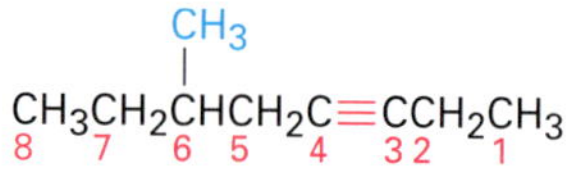

Begin numbering at the end nearer the triple bond.

6-Methyl-3-octyne

(New: 6-Methyloct-3-yne)

Compounds with more than one triple bond are called *diynes*, *triynes*, and so forth; compounds containing both double and triple bonds are called *enynes* (not *ynenes*). Numbering of an enyne chain starts from the end nearer the first multiple bond, whether double or triple. When there is a choice in numbering, double bonds receive lower numbers than triple bonds. For example:

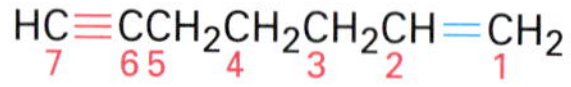

1-Hepten-6-yne

(New: Hept-1-en-6-yne)

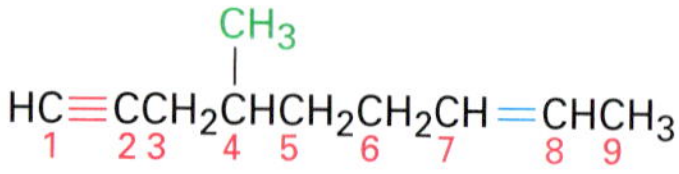

4-Methyl-7-nonen-1-yne

(New: 4-Methylnon-7-en-1-yne)

As with alkyl and alkenyl substituents derived from alkanes and alkenes, respectively, *alkynyl* groups are also possible.

$CH_3CH_2CH_2CH_2$–	$CH_3CH_2CH{=}CH$–	$CH_3CH_2C{\equiv}C$–
Butyl (an alkyl group)	**1-Butenyl (a vinylic group)**	**1-Butynyl (an alkynyl group)**
	(New: But-1-enyl)	**(New: But-1-ynyl)**

Problem 8.1 Name the following compounds:

(a)

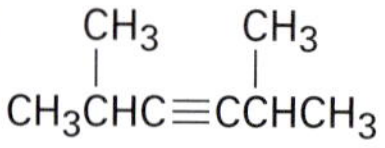

(b)

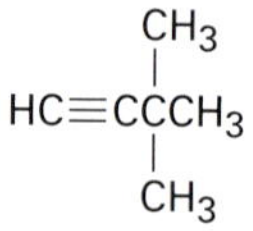

(c)

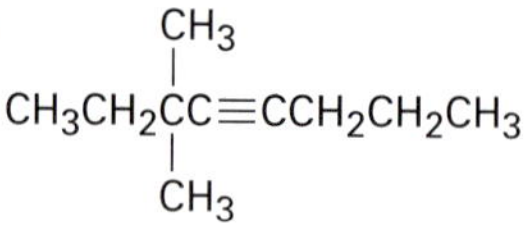

(d)

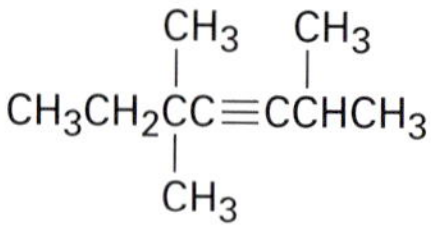

(e)

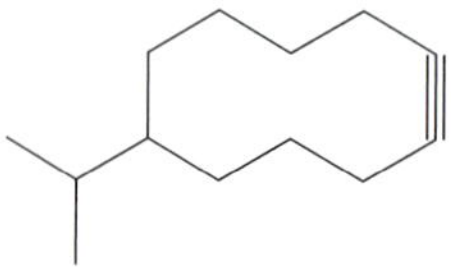

(f) $CH_3CH{=}CHCH{=}CHC{\equiv}CCH_3$

Problem 8.2 There are seven isomeric alkynes with the formula C_6H_{10}. Draw and name them.

8.2 Preparation of Alkynes: Elimination Reactions of Dihalides

Alkynes can be prepared by the elimination of HX from alkyl halides in much the same manner as alkenes (Section 7.1). Treatment of a 1,2-dihaloalkane (a *vicinal* dihalide) with excess strong base such as KOH or $NaNH_2$ results in a twofold elimination of HX and formation of an alkyne. As with the elimination of HX to form an alkene, we'll defer a discussion of the mechanism until Chapter 11.

The necessary vicinal dihalides are themselves readily available by addition of Br_2 or Cl_2 to alkenes. Thus, the overall halogenation/dehydrohalogenation sequence makes it possible to go from an alkene to an alkyne. For example, diphenylethylene is converted into diphenylacetylene by reaction with Br_2 and subsequent base treatment.

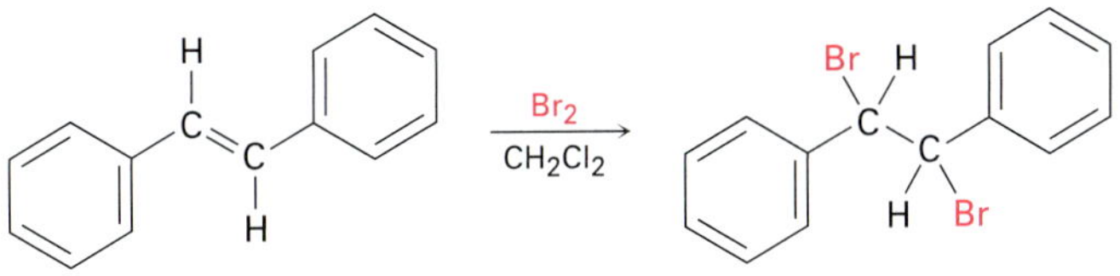

1,2-Diphenylethylene (stilbene)

1,2-Dibromo-1,2-diphenylethane (a vicinal dibromide)

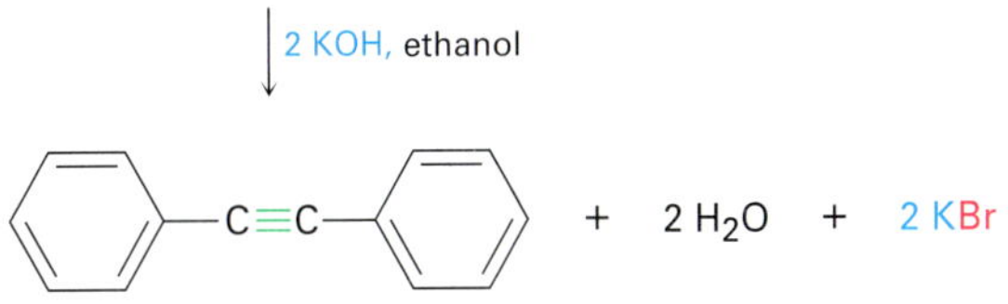

Diphenylacetylene (85%)

The twofold dehydrohalogenation takes place through a vinylic halide intermediate, which suggests that vinylic halides themselves should give alkynes when treated with strong base. (*Recall:* A *vinylic* substituent is one that is attached to a double-bond carbon.) This is indeed the case. For example:

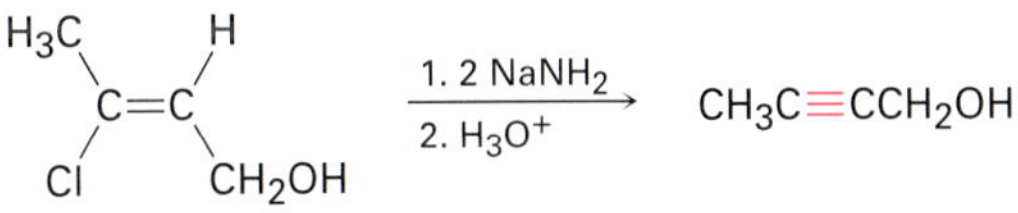

(*Z*)-3-Chloro-2-buten-1-ol

2-Butyn-1-ol

8.3 Reactions of Alkynes: Addition of HX and X_2

You might recall from Section 1.9 that a carbon–carbon triple bond results from the interaction of two *sp*-hybridized carbon atoms. The two *sp* hybrid orbitals of carbon lie at an angle of 180° to each other along an axis perpendicular to the axes of the two unhybridized $2p_y$ and $2p_z$ orbitals. When two *sp*-hybridized carbons approach each other, one *sp*–*sp* σ bond and two *p*–*p* π bonds are formed. The two remaining *sp* orbitals form bonds to other atoms at an angle 180° from the carbon–carbon bond. Thus, acetylene is a linear molecule wi H–C≡C bond angles of 180° (Figure 8.1).

Figure 8.1 The structure of acetylene, H—C≡C—H. The H–C≡C bond angles are 180°, and the C≡C bond length is 120 pm. The electrostatic potential map shows that the π bonds create a negative (red) belt around the molecule.

The length of the carbon–carbon triple bond in acetylene is 120 pm, and tł strength is approximately 835 kJ/mol (200 kcal/mol), making it the shortest ar strongest known carbon–carbon bond. Measurements show that approximate 318 kJ/mol (76 kcal/mol) is needed to break a π bond in acetylene, a value son 50 kJ/mol larger than the 268 kJ/mol needed to break an alkene π bond.

ThomsonNOW™ Click *Organic Interactive* to **use a web-based palette to predict products for alkyne addition reactions.**

As a general rule, electrophiles undergo addition reactions with alkynes muc as they do with alkenes. Take the reaction of alkynes with HX, for instanc The reaction often can be stopped after addition of 1 equivalent of HX, but rea tion with an excess of HX leads to a dihalide product. For example, reaction (1-hexyne with 2 equivalents of HBr yields 2,2-dibromohexane. As the followir examples indicate, the regiochemistry of addition follows Markovnikov's rul halogen adds to the more highly substituted side of the alkyne bond, and hydro gen adds to the less highly substituted side. Trans stereochemistry of H and X no mally, although not always, results in the product.

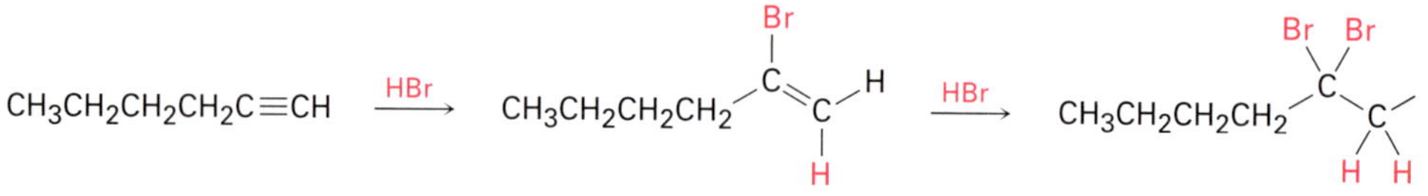

1-Hexyne **2-Bromo-1-hexene** **2,2-Dibromohexane**

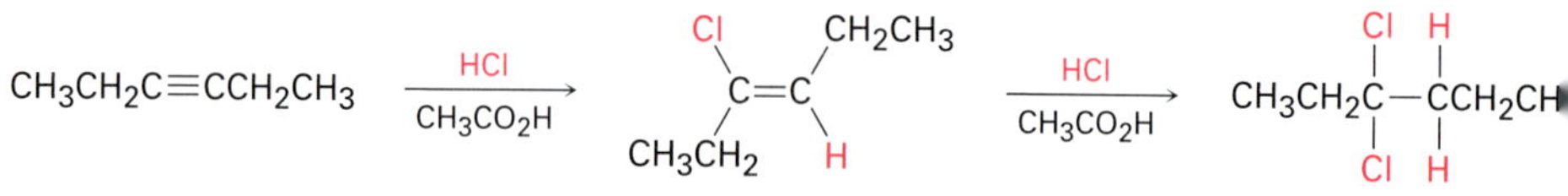

3-Hexyne **(*Z*)-3-Chloro-3-hexene** **3,3-Dichlorohexane**

Bromine and chlorine also add to alkynes to give addition products, an trans stereochemistry again results.

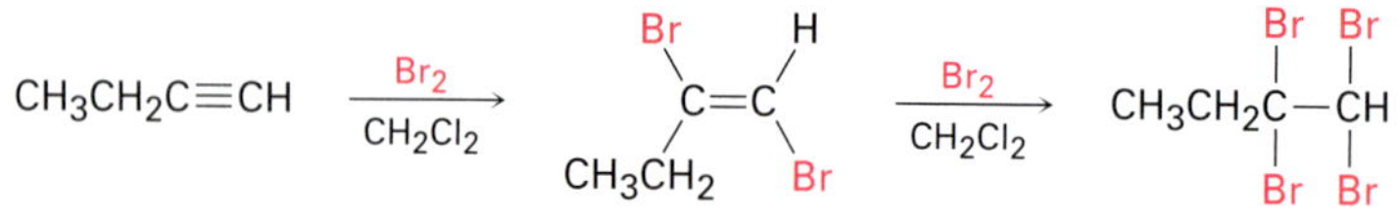

1-Butyne **(*E*)-1,2-Dibromo-1-butene** **1,1,2,2-Tetrabromobutane**

The mechanism of alkyne additions is similar but not identical to that of alkene additions. When an electrophile such as HBr adds to an *alkene* (Sections 6.7 and 6.8), the reaction takes place in two steps and involves an *alkyl* carbocation intermediate. If HBr were to add by the same mechanism to an *alkyne,* an analogous *vinylic* carbocation would be formed as the intermediate.

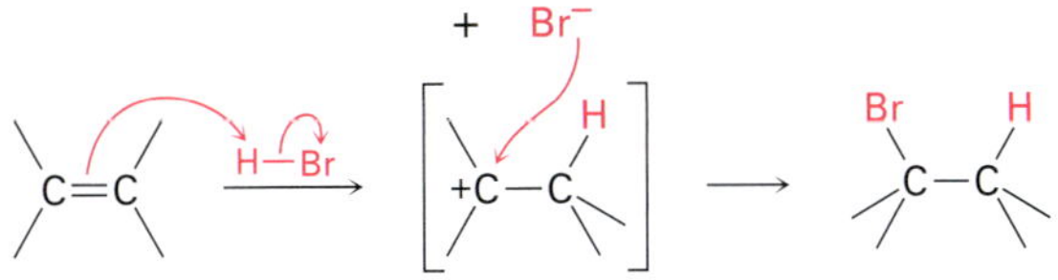

An alkene **An alkyl carbocation** **An alkyl bromide**

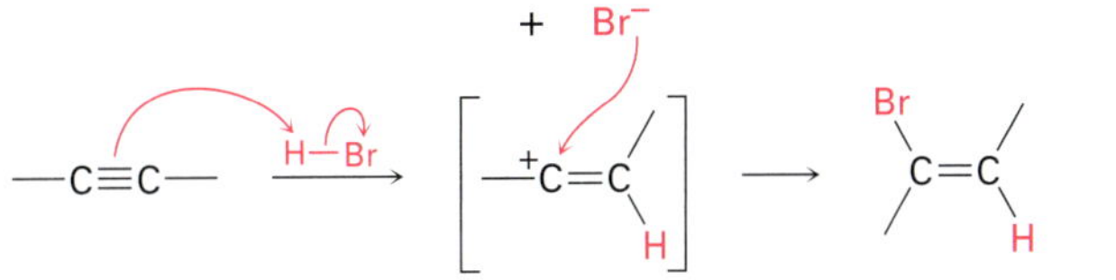

An alkyne **A vinylic carbocation** **A vinylic bromide**

A vinylic carbocation has an *sp*-hybridized carbon and generally forms less readily than an alkyl carbocation (Figure 8.2). As a rule, a *secondary* vinylic carbocation forms about as readily as a *primary* alkyl carbocation, but a *primary* vinylic carbocation is so difficult to form that there is no clear evidence it even exists. Thus, many alkyne additions occur through more complex mechanistic pathways.

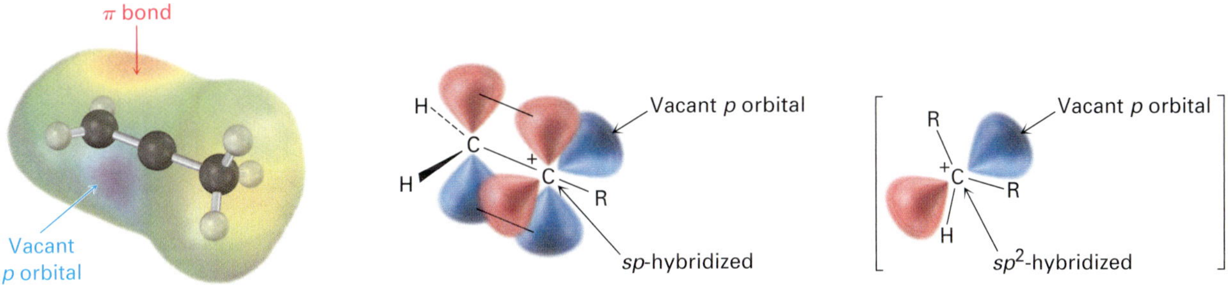

A 2° vinylic carbocation **A 2° alkyl carbocation**

Figure 8.2 The structure of a secondary vinylic carbocation. The cationic carbon atom is *sp*-hybridized and has a vacant *p* orbital perpendicular to the plane of the π bond orbitals. Only one R group is attached to the positively charged carbon rather than two, as in a secondary alkyl carbocation. The electrostatic potential map shows that the most positive (blue) regions coincide with lobes of the vacant *p* orbital and are perpendicular to the most negative (red) regions associated with the π bond.

Problem 8.3 What products would you expect from the following reactions?

(a) $CH_3CH_2CH_2C{\equiv}CH$ + 2 Cl_2 ⟶ ?

(b) 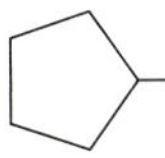 —C≡CH + 1 HBr ⟶ ?

(c) $CH_3CH_2CH_2CH_2C{\equiv}CCH_3$ + 1 HBr ⟶ ?

8.4 Hydration of Alkynes

Like alkenes (Sections 7.4 and 7.5), alkynes can be hydrated by either of tw methods. Direct addition of water catalyzed by mercury(II) ion yields th Markovnikov product, and indirect addition of water by a hydroboration oxidation sequence yields the non-Markovnikov product.

Mercury(II)-Catalyzed Hydration of Alkynes

Alkynes don't react directly with aqueous acid but will undergo hydratio readily in the presence of mercury(II) sulfate as a Lewis acid catalyst. The reac tion occurs with Markovnikov regiochemistry: the −OH group adds to th more highly substituted carbon, and the −H attaches to the less highly sub stituted one.

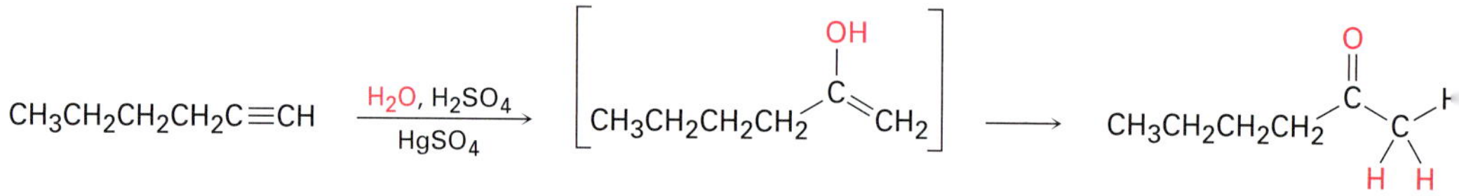

1-Hexyne | An enol | 2-Hexanone (78%)

ThomsonNOW™ Click *Organic Interactive* to **learn to interconvert enol and carbonyl tautomers.**

Interestingly, the product actually isolated from alkyne hydration is not th vinylic alcohol, or **enol** (*ene* + *ol*), but is instead a *ketone*. Although the enol i an intermediate in the reaction, it immediately rearranges to a ketone by a process called *keto–enol tautomerism*. The individual keto and enol forms are saic to be **tautomers**, a word used to describe constitutional isomers that intercon vert rapidly. With few exceptions, the keto–enol tautomeric equilibrium lies or the side of the ketone; enols are almost never isolated. We'll look more closely at this equilibrium in Section 22.1.

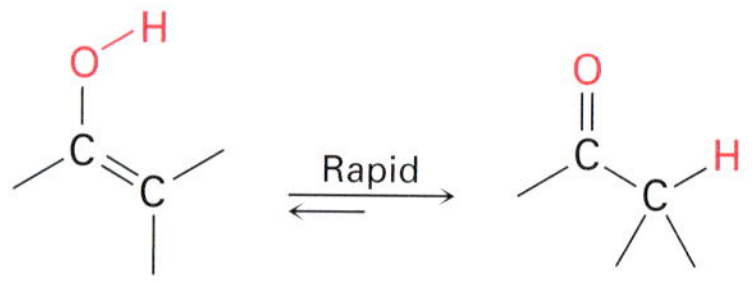

Enol tautomer (less favored) | **Keto tautomer (more favored)**

As shown in Figure 8.3, the mechanism of the mercury(II)-catalyzed alkyne hydration reaction is analogous to the oxymercuration reaction of alkenes (Section 7.4). Electrophilic addition of mercury(II) ion to the alkyne gives a vinylic cation, which reacts with water and loses a proton to yield a mercury-containing enol intermediate. In contrast with alkene oxymercuration, however, no treatment with $NaBH_4$ is necessary to remove the mercury. The acidic reaction conditions alone are sufficient to effect replacement of mercury by hydrogen.

1 The alkyne uses a pair of electrons to attack the electrophilic mercury(II) ion, yielding a mercury-containing vinylic carbocation intermediate.

2 Nucleophilic attack of water on the carbocation forms a C–O bond and yields a protonated mercury-containing enol.

3 Abstraction of H^+ from the protonated enol by water gives an organomercury compound.

4 Replacement of Hg^{2+} by H^+ occurs to give a neutral enol.

5 The enol undergoes tautomerization to give the final ketone product.

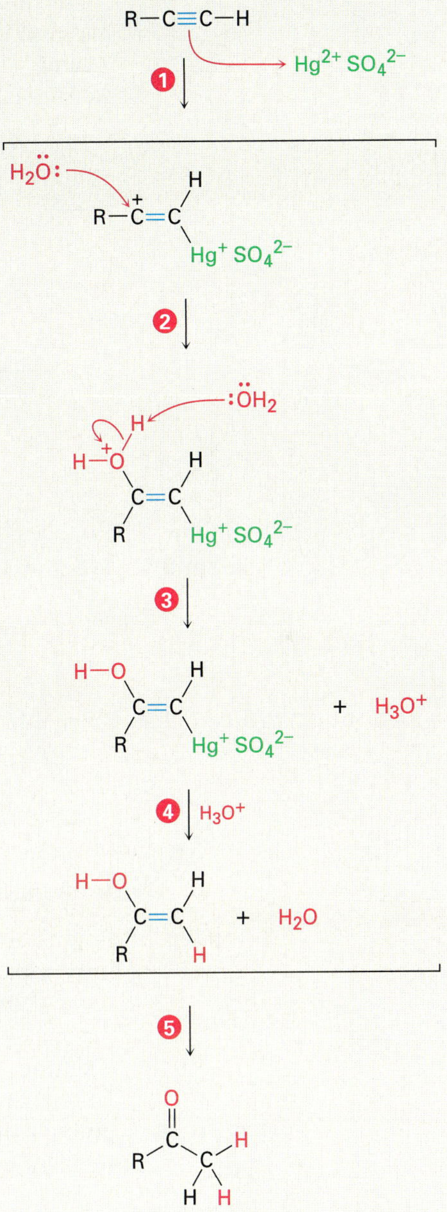

Figure 8.3 **MECHANISM:** Mechanism of the mercury(II)-catalyzed hydration of an alkyne to yield a ketone. The reaction occurs through initial formation of an intermediate enol, which rapidly tautomerizes to the ketone.

A mixture of both possible ketones results when an unsymmetrically su stituted internal alkyne (RC≡CR′) is hydrated. The reaction is therefore mc useful when applied to a terminal alkyne (RC≡CH) because only a meth ketone is formed.

An internal alkyne

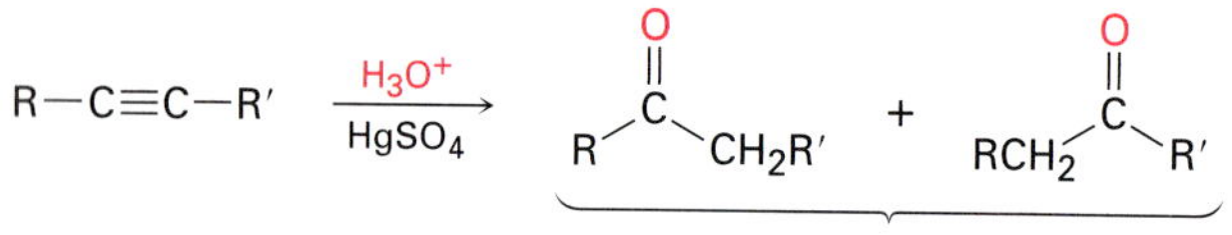

Mixture

A terminal alkyne

A methyl ketone

Problem 8.4 What product would you obtain by hydration of the following alkynes?

(a) $CH_3CH_2CH_2C{\equiv}CCH_2CH_2CH_3$

(b) $CH_3CH(CH_3)CH_2C{\equiv}CCH_2CH_2CH_3$

Problem 8.5 What alkynes would you start with to prepare the following ketones?

(a) $CH_3CH_2CH_2C(=O)CH_3$

(b) $CH_3CH_2C(=O)CH_2CH_3$

Hydroboration/Oxidation of Alkynes

Borane adds rapidly to an alkyne just as it does to an alkene, and the resultin vinylic borane can be oxidized by H_2O_2 to yield an enol. Tautomerization the gives either a ketone or an aldehyde, depending on the structure of the alkyn reactant. Hydroboration/oxidation of an internal alkyne such as 3-hexyne give a ketone, and hydroboration/oxidation of a terminal alkyne gives an aldehyd Note that the relatively unhindered terminal alkyne undergoes *two* addition giving a doubly hydroborated intermediate. Oxidation with H_2O_2 at pH 8 the replaces both boron atoms by oxygen and generates the aldehyde.

An internal alkyne

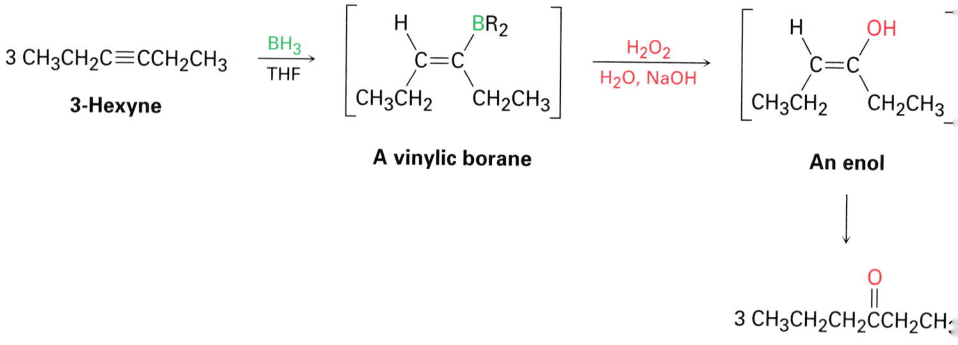

3-Hexyne

A vinylic borane

An enol

3-Hexanone

A terminal alkyne

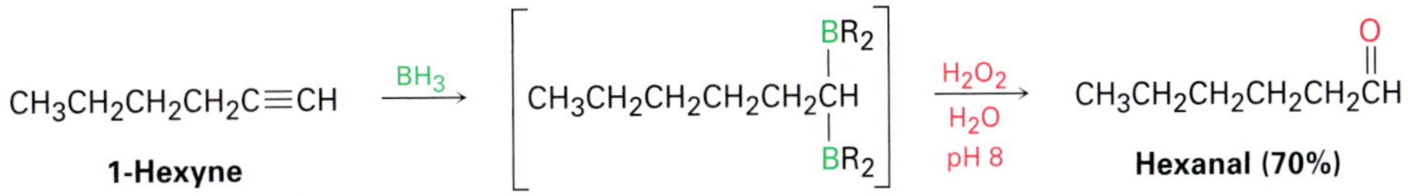

The hydroboration/oxidation sequence is complementary to the direct, mercury(II)-catalyzed hydration reaction of a terminal alkyne because different products result. Direct hydration with aqueous acid and mercury(II) sulfate leads to a methyl ketone, whereas hydroboration/oxidation of the same terminal alkyne leads to an aldehyde.

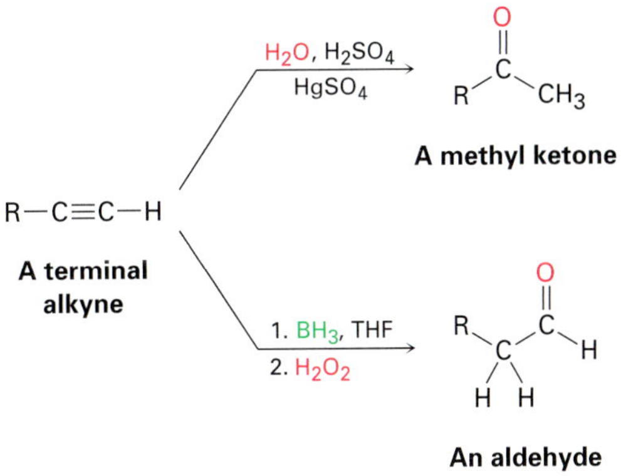

Problem 8.6 What alkyne would you start with to prepare each of the following compounds by a hydroboration/oxidation reaction?

(a)

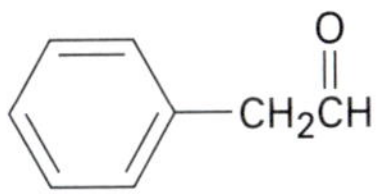

(b)

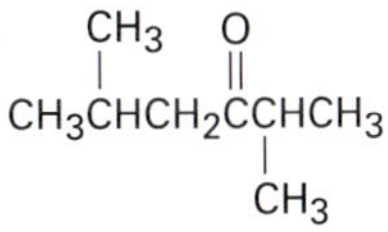

Problem 8.7 How would you prepare the following carbonyl compounds starting from an alkyne (reddish brown = Br)?

(a)

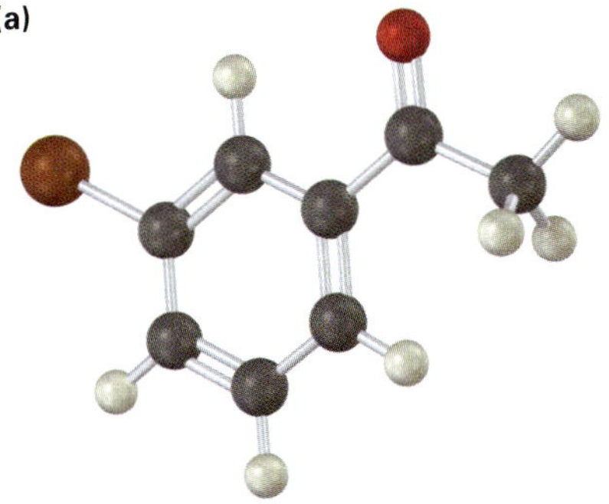

(b)

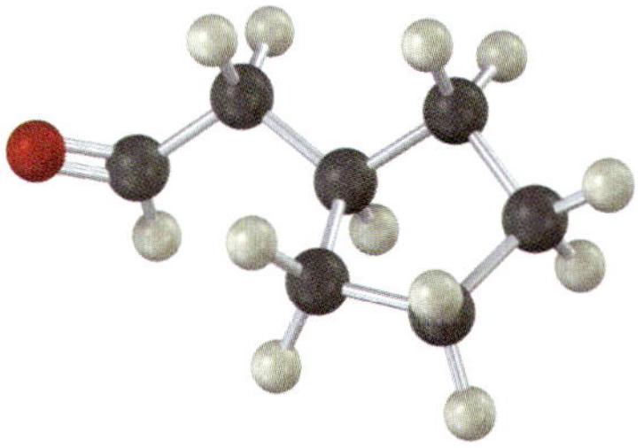

8.5 Reduction of Alkynes

ThomsonNOW™ Click *Organic Interactive* to **use a web-based palette to predict products for alkyne reduction reactions.**

Alkynes are reduced to alkanes by addition of H_2 over a metal catalyst. The reaction occurs in steps through an alkene intermediate, and measurements indicate that the first step in the reaction is more exothermic than the second step.

$$HC{\equiv}CH \xrightarrow[\text{Catalyst}]{H_2} H_2C{=}CH_2 \qquad \Delta H^\circ_{hydrog} = -176\ \text{kJ/mol}\ (-42\ \text{kcal/mol})$$

$$H_2C{=}CH_2 \xrightarrow[\text{Catalyst}]{H_2} CH_3{-}CH_3 \qquad \Delta H^\circ_{hydrog} = -137\ \text{kJ/mol}\ (-33\ \text{kcal/mol})$$

Complete reduction to the alkane occurs when palladium on carbon (Pd/C) is used as catalyst, but hydrogenation can be stopped at the alkene if the less active *Lindlar catalyst* is used. The Lindlar catalyst is a finely divided palladium metal that has been precipitated onto a calcium carbonate support and then deactivated by treatment with lead acetate and quinoline, an aromatic amine. The hydrogenation occurs with syn stereochemistry (Section 7.5), giving a cis alkene product.

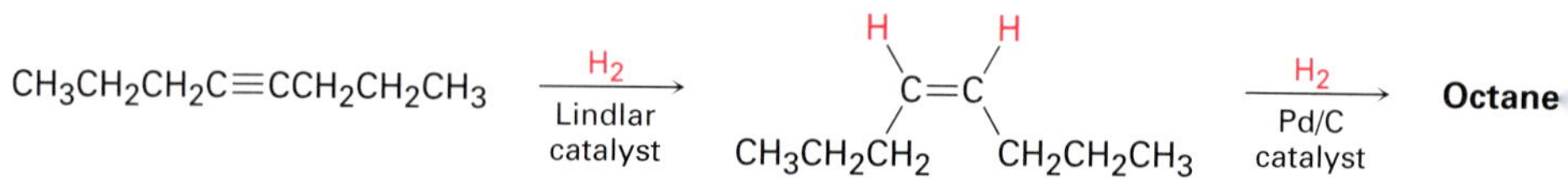

4-Octyne **cis-4-Octene**

The alkyne hydrogenation reaction has been explored extensively by the Hoffmann–La Roche pharmaceutical company, where it is used in the commercial synthesis of vitamin A. The cis isomer of vitamin A produced on hydrogenation is converted to the trans isomer by heating.

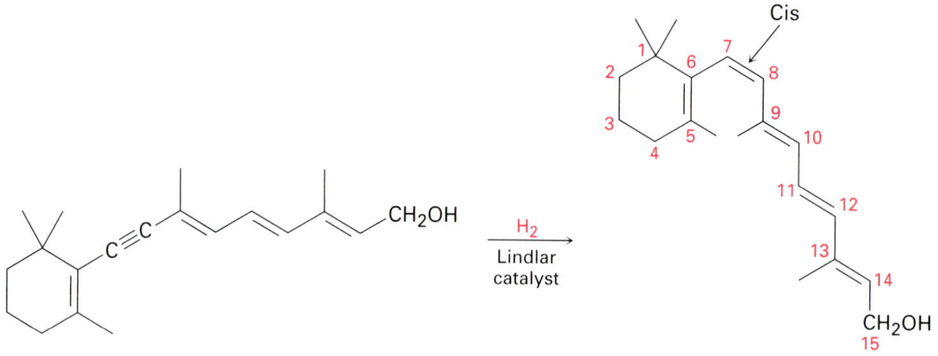

7-*cis*-Retinol
(7-*cis*-vitamin A; vitamin A has a trans double bond at C7)

An alternative method for the conversion of an alkyne to an alkene uses sodium or lithium metal as the reducing agent in liquid ammonia as solvent. This method is complementary to the Lindlar reduction because it produces trans rather than cis alkenes. For example, 5-decyne gives *trans*-5-decene on treatment with lithium in liquid ammonia.

5-Decyne **trans-5-Decene (78%)**

Alkali metals dissolve in liquid ammonia at −33 °C to produce a deep blue solution containing the metal cation and ammonia-solvated electrons. When an alkyne is then added to the solution, an electron adds to the triple bond to yield an intermediate *anion radical*—a species that is both an anion (has a negative charge) and a radical (has an odd number of electrons). This anion radical is a strong base, which removes H^+ from ammonia to give a vinylic radical. Addition of a second electron to the vinylic radical gives a vinylic anion, which abstracts a second H^+ from ammonia to give trans alkene product. The mechanism is shown in Figure 8.4.

igure 8.4 MECHANISM: Mechanism of the lithium/ mmonia reduction of an alkyne o produce a trans alkene.

1 Lithium metal donates an electron to the alkyne to give an anion radical . . .

2 . . . which abstracts a proton from ammonia solvent to yield a vinylic radical.

3 The vinylic radical accepts another electron from a second lithium atom to produce a vinylic anion . . .

4 . . . which abstracts another proton from ammonia solvent to yield the final trans alkene product.

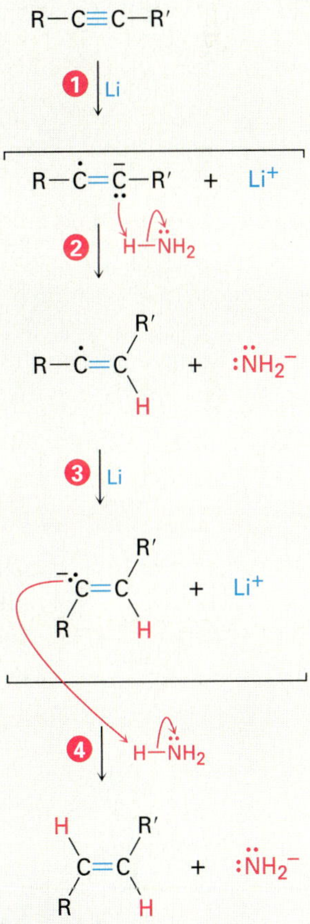

A trans alkene

Trans stereochemistry of the alkene product is established during the se ond reduction step when the less hindered trans vinylic anion is formed fro the vinylic radical. Vinylic radicals undergo rapid cis–trans equilibration, b vinylic anions equilibrate much less rapidly. Thus, the more stable trans vinyl anion is formed rather than the less stable cis anion and is then protonat without equilibration.

Problem 8.8 Using any alkyne needed, how would you prepare the following alkenes?
(a) *trans*-2-Octene **(b)** *cis*-3-Heptene **(c)** 3-Methyl-1-pentene

8.6 Oxidative Cleavage of Alkynes

ThomsonNOW™ Click *Organic Interactive* to **use a web-based palette to predict products for the oxidative cleavage of alkynes.**

Alkynes, like alkenes, can be cleaved by reaction with powerful oxidizing agen such as ozone or $KMnO_4$, although the reaction is of little value and we mer tion it only for completeness. A triple bond is generally less reactive than a dou ble bond and yields of cleavage products are sometimes low. The product obtained from cleavage of an internal alkyne are carboxylic acids; from a term nal alkyne, CO_2 is formed as one product.

An internal alkyne

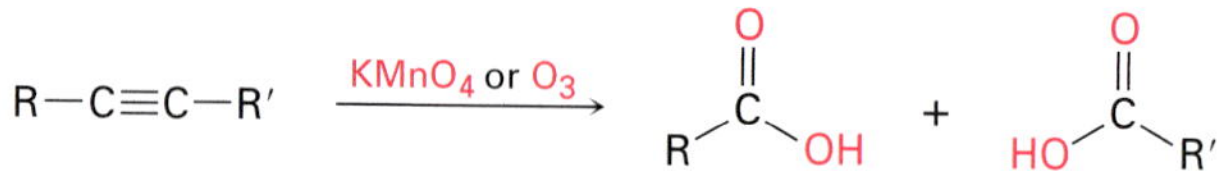

A terminal alkyne

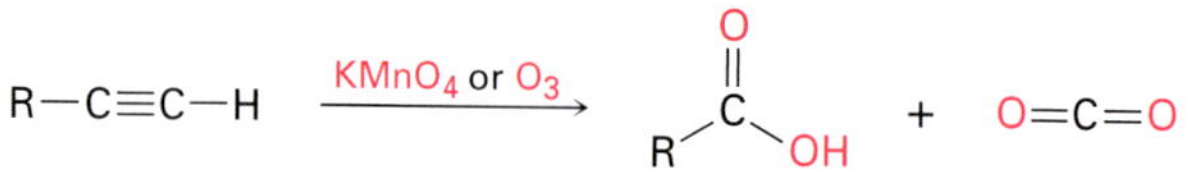

8.7 Alkyne Acidity: Formation of Acetylide Anions

The most striking difference between alkenes and alkynes is that termina alkynes are weakly acidic. When a terminal alkyne is treated with a strong base such as sodium amide, $Na^+ \ ^-NH_2$, the terminal hydrogen is removed and ar **acetylide anion** is formed.

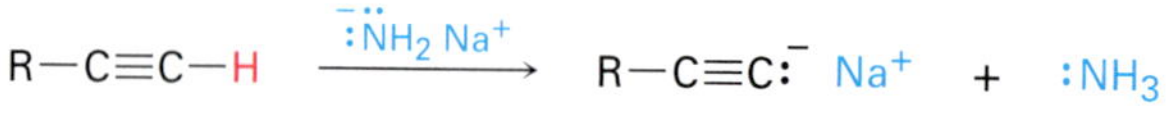

A terminal alkyne **An acetylide anion**

According to the Brønsted–Lowry definition (Section 2.7), an acid is a sub stance that donates H^+. Although we usually think of oxyacids (H_2SO_4, HNO_3) o halogen acids (HCl, HBr) in this context, any compound containing a hydroger atom can be an acid under the right circumstances. By measuring dissociation constants of different acids and expressing the results as pK_a values, an acidity order can be established. Recall from Section 2.8 that a low pK_a corresponds to a strong acid and a high pK_a corresponds to a weak acid.

Where do hydrocarbons lie on the acidity scale? As the data in Table 8.1 show, both methane (p$K_a \approx 60$) and ethylene (p$K_a = 44$) are very weak acids and thus do not react with any of the common bases. Acetylene, however, has p$K_a = 25$ and can be deprotonated by the conjugate base of any acid whose pK_a is greater than 25. Amide ion (NH_2^-), for example, the conjugate base of ammonia (p$K_a = 35$), is often used to deprotonate terminal alkynes.

Table 8.1 Acidity of Simple Hydrocarbons

Family	Example	K_a	pK_a	
Alkyne	HC≡CH	10^{-25}	25	Stronger acid
Alkene	$H_2C{=}CH_2$	10^{-44}	44	↑
Alkane	CH_4	10^{-60}	60	Weaker acid

Why are terminal alkynes more acidic than alkenes or alkanes? In other words, why are acetylide anions more stable than vinylic or alkyl anions? The simplest explanation involves the hybridization of the negatively charged carbon atom. An acetylide anion has an *sp*-hybridized carbon, so the negative charge resides in an orbital that has 50% "*s* character." A vinylic anion has an sp^2-hybridized carbon with 33% *s* character, and an alkyl anion (sp^3) has only 25% *s* character. Because *s* orbitals are nearer the positive nucleus and lower in energy than *p* orbitals, the negative charge is stabilized to a greater extent in an orbital with higher *s* character (Figure 8.5).

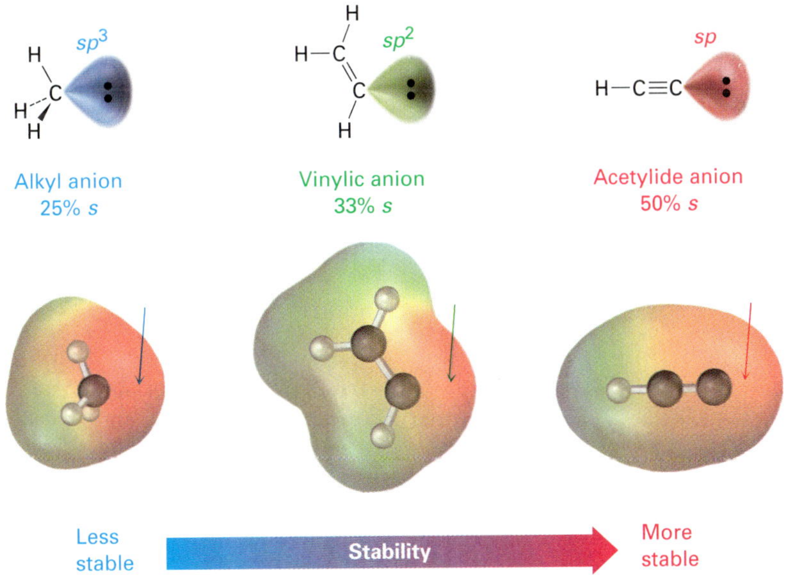

Figure 8.5 A comparison of alkyl, vinylic, and acetylide anions. The acetylide anion, with *sp* hybridization, has more *s* character and is more stable. Electrostatic potential maps show that placing the negative charge closer to the carbon nucleus makes carbon appear less negative (red).

The presence of a negative charge and an unshared electron pair on carbon makes acetylide anions strongly nucleophilic. As a result, they react with many different kinds of electrophiles.

Problem 8.9 The pK_a of acetone, CH_3COCH_3, is 19.3. Which of the following bases is strong enough to deprotonate acetone?

(a) KOH (pK_a of H_2O = 15.7)
(b) $Na^+ \ ^-C{\equiv}CH$ (pK_a of C_2H_2 = 25)
(c) $NaHCO_3$ (pK_a of H_2CO_3 = 6.4)
(d) $NaOCH_3$ (pK_a of CH_3OH = 15.6)

8.8 Alkylation of Acetylide Anions

ThomsonNOW Click *Organic Interactive* to **use a web-based palette to predict products for alkyne alkylation reactions.**

The negative charge and unshared electron pair on carbon make an acetylide anion strongly nucleophilic. As a result, an acetylide anion can react with an alkyl halide such as bromomethane to substitute for the halogen and yield a new alkyne product.

Acetylide anion **Propyne**

We won't study the details of this substitution reaction until Chapter 11 but for now can picture it as happening by the pathway shown in Figure 8.6. The nucleophilic acetylide ion uses an electron pair to form a bond to the positively polarized, electrophilic carbon atom of bromomethane. As the new C–C bond forms, Br^- departs, taking with it the electron pair from the former C–Br bond and yielding propyne as product. We call such a reaction an **alkylation** because a new alkyl group has become attached to the starting alkyne.

Active Figure 8.6 **MECHANISM:** A mechanism for the alkylation reaction of acetylide anion with bromomethane to give propyne. *Sign in at* **www.thomsonedu.com** *to see a simulation based on this figure and to take a short quiz.*

1 The nucleophilic acetylide anion uses its electron lone pair to form a bond to the positively polarized, electrophilic carbon atom of bromomethane. As the new C–C bond begins to form, the C–Br bond begins to break in the transition state.

2 The new C–C bond is fully formed and the old C–Br bond is fully broken at the end of the reaction.

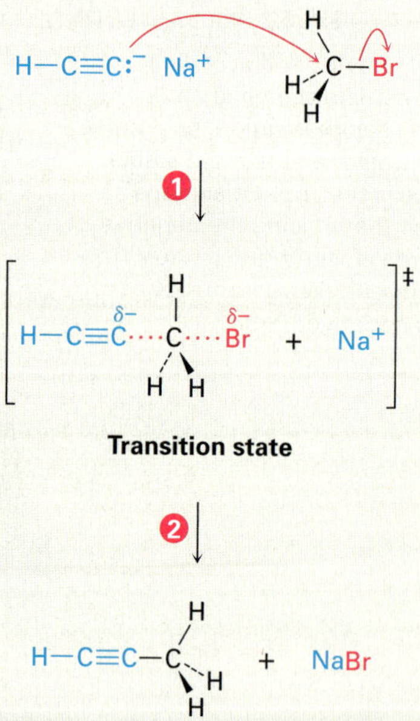

Alkyne alkylation is not limited to acetylene itself. *Any* terminal alkyne can be converted into its corresponding anion and then alkylated by treatment with an alkyl halide, yielding an internal alkyne. For example, conversion of 1-hexyne into its anion, followed by reaction with 1-bromobutane, yields 5-decyne.

$$\underset{\textbf{1-Hexyne}}{CH_3CH_2CH_2CH_2C{\equiv}CH} \xrightarrow[\text{2. } CH_3CH_2CH_2CH_2Br]{\text{1. } NaNH_2,\ NH_3} \underset{\textbf{5-Decyne (76\%)}}{CH_3CH_2CH_2CH_2C{\equiv}CCH_2CH_2CH_2CH_3}$$

Because of its generality, acetylide alkylation is an excellent method for preparing substituted alkynes from simpler precursors. A terminal alkyne can be prepared by alkylation of acetylene itself, and an internal alkyne can be prepared by further alkylation of a terminal alkyne.

$$\underset{\textbf{Acetylene}}{H{-}C{\equiv}C{-}H} \xrightarrow{NaNH_2} \left[H{-}C{\equiv}C{:}^-\ Na^+\right] \xrightarrow{RCH_2Br} \underset{\textbf{A terminal alkyne}}{H{-}C{\equiv}C{-}CH_2R}$$

$$\underset{\textbf{A terminal alkyne}}{R{-}C{\equiv}C{-}H} \xrightarrow{NaNH_2} \left[R{-}C{\equiv}C{:}^-\ Na^+\right] \xrightarrow{R'CH_2Br} \underset{\textbf{An internal alkyne}}{R{-}C{\equiv}C{-}CH_2R'}$$

The alkylation reaction is limited to the use of primary alkyl bromides and alkyl iodides because acetylide ions are sufficiently strong bases to cause dehydrohalogenation instead of substitution when they react with secondary and tertiary alkyl halides. For example, reaction of bromocyclohexane with propyne anion yields the elimination product cyclohexene rather than the substitution product 1-propynylcyclohexane.

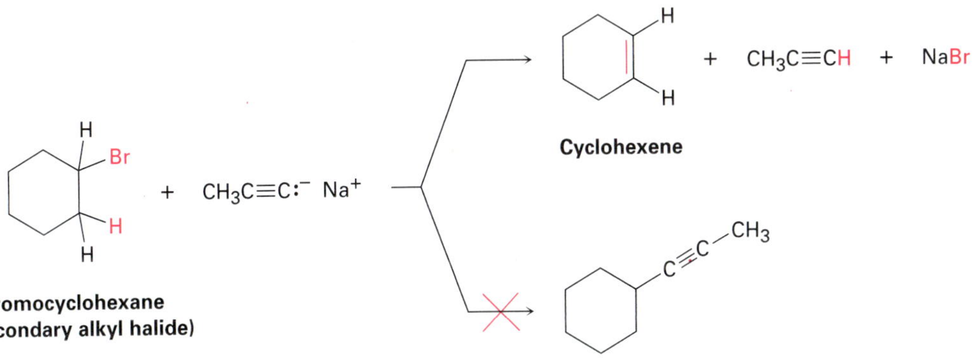

Bromocyclohexane (a secondary alkyl halide)

NOT formed

Problem 8.10 Show the terminal alkyne and alkyl halide from which the following products can be obtained. If two routes look feasible, list both.

(a) $CH_3CH_2CH_2C{\equiv}CCH_3$ **(b)** $(CH_3)_2CHC{\equiv}CCH_2CH_3$ **(c)**

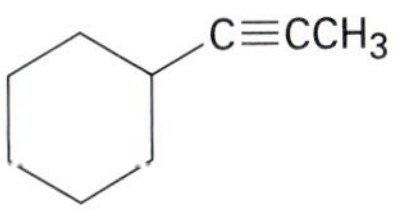

Problem 8.11 How would you prepare *cis*-2-butene starting from propyne, an alkyl halide, and any other reagents needed? This problem can't be worked in a single step. You'll have to carry out more than one reaction.

8.9 An Introduction to Organic Synthesis

There are many reasons for carrying out the laboratory synthesis of an organic compound. In the pharmaceutical industry, new organic molecules are designed and synthesized in the hope that some might be useful new drugs. In the chemical industry, syntheses are done to devise more economical routes to known compounds. In academic laboratories, the synthesis of complex molecules is sometimes done purely for the intellectual challenge involved in mastering so difficult a subject. The successful synthesis route is a highly creative work that is sometimes described by such subjective terms as *elegant* or *beautiful.*

In this book, too, we will often devise syntheses of molecules from simpler precursors. Our purpose, however, is pedagogical. The ability to plan a workable synthetic sequence requires knowledge of a variety of organic reactions. Furthermore, it requires the practical ability to fit together the steps in a sequence such that each reaction does only what is desired without causing changes elsewhere in the molecule. Working synthesis problems is an excellent way to learn organic chemistry.

Some of the syntheses we plan may seem trivial. Here's an example:

WORKED EXAMPLE 8.1

Devising a Synthesis Route

Prepare octane from 1-pentyne.

$$CH_3CH_2CH_2C{\equiv}CH \xrightarrow{?} CH_3CH_2CH_2CH_2CH_2CH_2CH_2CH_3$$

1-Pentyne **Octane**

Strategy Compare the product with the starting material, and catalog the differences. In this case, we need to add three carbons to the chain and reduce the triple bond. Since the starting material is a terminal alkyne that can be alkylated, we might first prepare the acetylide anion of 1-pentyne, let it react with 1-bromopropane, and then reduce the product using catalytic hydrogenation.

Solution

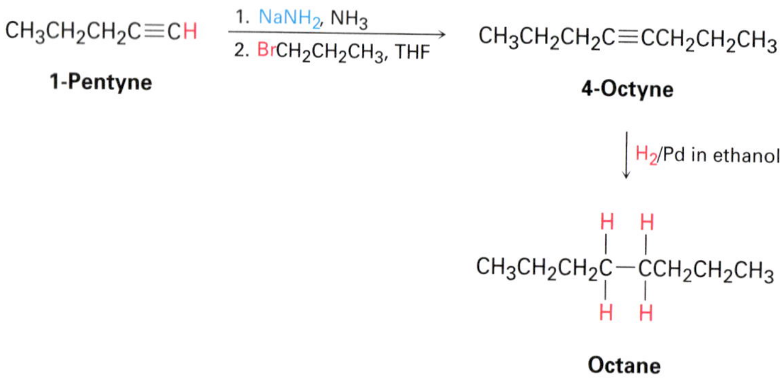

The synthesis route just presented will work perfectly well but has little practical value because you can simply *buy* octane from any of several dozen

Solution The desired product can be synthesized in four steps from acetylene and 1-bromopropane.

$$\underset{\textbf{Acetylene}}{HC{\equiv}CH} \xrightarrow[\text{2. } CH_3CH_2CH_2Br\text{, THF}]{\text{1. } NaNH_2\text{, } NH_3} \underset{\textbf{1-Pentyne}}{CH_3CH_2CH_2C{\equiv}CH} \xrightarrow[\text{Lindlar catalyst}]{H_2} \underset{\textbf{1-Pentene}}{CH_3CH_2CH_2CH{=}CH_2}$$

$$\xrightarrow{HBr\text{, ether}} \underset{\textbf{2-Bromopentane}}{CH_3CH_2CH_2CH(Br)CH_3}$$

WORKED EXAMPLE 8.4

Devising a Synthesis Route

Synthesize 1-hexanol (1-hydroxyhexane) from acetylene and an alkyl halide.

$$\underset{\textbf{Acetylene}}{HC{\equiv}CH} + \underset{\textbf{Alkyl halide}}{RX} \xrightarrow{?} \underset{\textbf{1-Hexanol}}{CH_3CH_2CH_2CH_2CH_2CH_2OH}$$

Strategy "What is an immediate precursor of a primary alcohol?" Perhaps a terminal alkene, which could be hydrated with non-Markovnikov regiochemistry by reaction with borane followed by oxidation with H_2O_2.

$$CH_3CH_2CH_2CH_2CH{=}CH_2 \xrightarrow[\text{2. } H_2O_2\text{, NaOH}]{\text{1. } BH_3} CH_3CH_2CH_2CH_2CH_2CH_2OH$$

"What is an immediate precursor of a terminal alkene?" Perhaps a terminal alkyne, which could be reduced.

$$CH_3CH_2CH_2CH_2C{\equiv}CH \xrightarrow[\text{Lindlar catalyst}]{H_2} CH_3CH_2CH_2CH_2CH{=}CH_2$$

"What is an immediate precursor of 1-hexyne?" Perhaps acetylene and 1-bromobutane.

$$HC{\equiv}CH \xrightarrow{NaNH_2} Na^+ \ {}^-C{\equiv}CH \xrightarrow{CH_3CH_2CH_2CH_2Br} CH_3CH_2CH_2CH_2C{\equiv}CH$$

Solution The synthesis can be completed in four steps from acetylene and 1-bromobutane:

$$\underset{\textbf{Acetylene}}{HC{\equiv}CH} \xrightarrow[\text{2. } CH_3CH_2CH_2CH_2Br]{\text{1. } NaNH_2} \underset{\textbf{1-Hexyne}}{CH_3CH_2CH_2CH_2C{\equiv}CH} \xrightarrow[\text{Lindlar catalyst}]{H_2} \underset{\textbf{1-Hexene}}{CH_3CH_2CH_2CH_2CH{=}CH_2}$$

$$\xrightarrow[\text{2. } H_2O_2\text{, NaOH}]{\text{1. } BH_3} \underset{\textbf{1-Hexanol}}{CH_3CH_2CH_2CH_2CH_2CH_2OH}$$

Problem 8.12 Beginning with 4-octyne as your only source of carbon, and using any inorganic reagents necessary, how would you synthesize the following compounds?

(a) *cis*-4-Octene (b) Butanal (c) 4-Bromooctane
(d) 4-Octanol (e) 4,5-Dichlorooctane (f) Butanoic acid

Problem 8.13 Beginning with acetylene and any alkyl halides needed, how would you synthesize the following compounds?

(a) Decane (b) 2,2-Dimethylhexane
(c) Hexanal (d) 2-Heptanone

Focus On . . .

The Art of Organic Synthesis

Vitamin B_{12} has been synthesized from scratch in the laboratory, but bacteria growing on sludge from municipal sewage plants do a much better job.

If you think some of the synthesis problems at the end of this chapter are hard, try devising a synthesis of vitamin B_{12} starting only from simple substances you can buy in a chemical catalog. This extraordinary achievement was reported in 1973 as the culmination of a collaborative effort headed by Robert B. Woodward of Harvard University and Albert Eschenmoser of the Swiss Federal Institute of Technology in Zürich. More than 100 graduate students and postdoctoral associates contributed to the work, which took more than a decade.

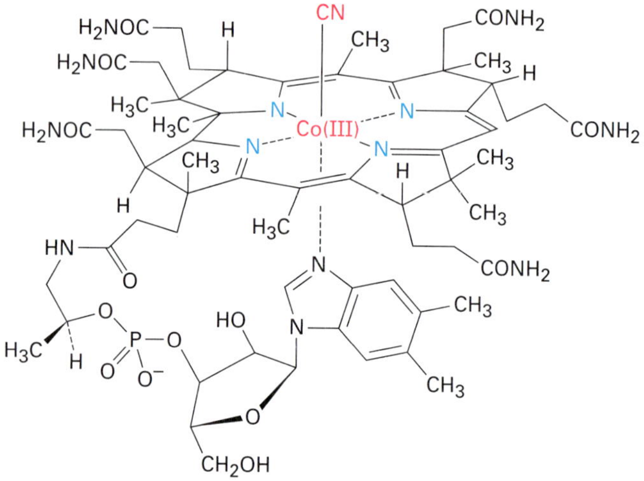

Vitamin B_{12}

(continued)

Why put such extraordinary effort into the laboratory synthesis of a molecule so easily obtained from natural sources? There are many reasons. On a basic human level, a chemist might be motivated primarily by the challenge, much as a climber might be challenged by the ascent of a difficult peak. Beyond the pure challenge, the completion of a difficult synthesis is also valuable for the way in which it establishes new standards and raises the field to a new level. If vitamin B_{12} can be made, then why can't any molecule found in nature be made? Indeed, the three and a half decades that have passed since the work of Woodward and Eschenmoser have seen the laboratory synthesis of many enormously complex and valuable substances. Sometimes these substances—the anticancer compound Taxol, for instance—are not easily available in nature, so laboratory synthesis is the only method for obtaining larger quantities.

But perhaps the most important reason for undertaking a complex synthesis is that, in so doing, new reactions and new chemistry are discovered. It invariably happens in synthesis that a point is reached at which the planned route fails. At such a time, the only alternatives are to quit or to devise a way around the difficulty. New reactions and new principles come from such situations, and it is in this way that the science of organic chemistry grows richer. In the synthesis of vitamin B_{12}, for example, unexpected findings emerged that led to the understanding of an entire new class of reactions—the *pericyclic* reactions that are the subject of Chapter 30 in this book. From synthesizing vitamin B_{12} to understanding pericyclic reactions—no one could have possibly predicted such a link at the beginning of the synthesis, but that is the way of science.

SUMMARY AND KEY WORDS

acetylide anion, 270
alkylation, 272
alkyne ($RC{\equiv}CR$), 259
enol, 264
retrosynthetic, 275
tautomer, 264

An **alkyne** is a hydrocarbon that contains a carbon–carbon triple bond. Alkyne carbon atoms are *sp*-hybridized, and the triple bond consists of one *sp–sp* σ bond and two *p–p* π bonds. There are relatively few general methods of alkyne synthesis. Two good ones are the alkylation of an acetylide anion with a primary alkyl halide and the twofold elimination of HX from a vicinal dihalide.

The chemistry of alkynes is dominated by electrophilic addition reactions, similar to those of alkenes. Alkynes react with HBr and HCl to yield *vinylic* halides and with Br_2 and Cl_2 to yield 1,2-dihalides (*vicinal* dihalides). Alkynes can be hydrated by reaction with aqueous sulfuric acid in the presence of mercury(II) catalyst. The reaction leads to an intermediate **enol** that immediately isomerizes to yield a ketone **tautomer**. Since the addition reaction occurs with Markovnikov regiochemistry, a methyl ketone is produced from a terminal alkyne. Alternatively, hydroboration/oxidation of a terminal alkyne yields an aldehyde.

Alkynes can be reduced to yield alkenes and alkanes. Complete reduction of the triple bond over a palladium hydrogenation catalyst yields an alkane; partial reduction by catalytic hydrogenation over a *Lindlar catalyst* yields a cis alkene. Reduction of the alkyne with lithium in ammonia yields a trans alkene.

Terminal alkynes are weakly acidic. The alkyne hydrogen can be removed by a strong base such as $Na^+ \, {}^-NH_2$ to yield an **acetylide anion**. An acetylide anion acts as a nucleophile and can displace a halide ion from a primary alk halide in an **alkylation** reaction. Acetylide anions are more stable than eith alkyl anions or vinylic anions because their negative charge is in a hybrid orbit with 50% *s* character, allowing the charge to be closer to the nucleus.

SUMMARY OF REACTIONS

1. Preparation of alkynes
   (a) Dehydrohalogenation of vicinal dihalides (Section 8.2)

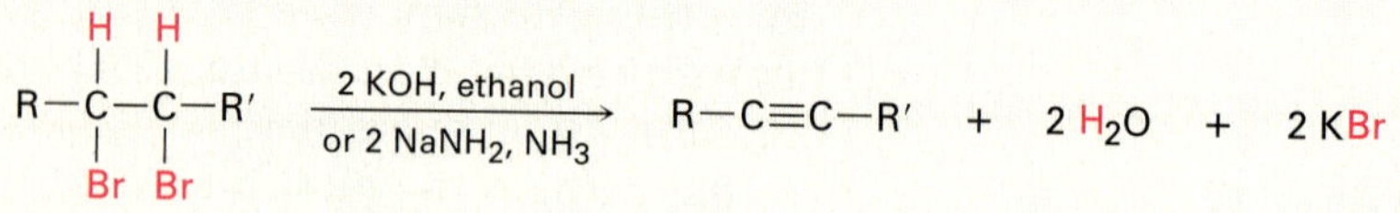

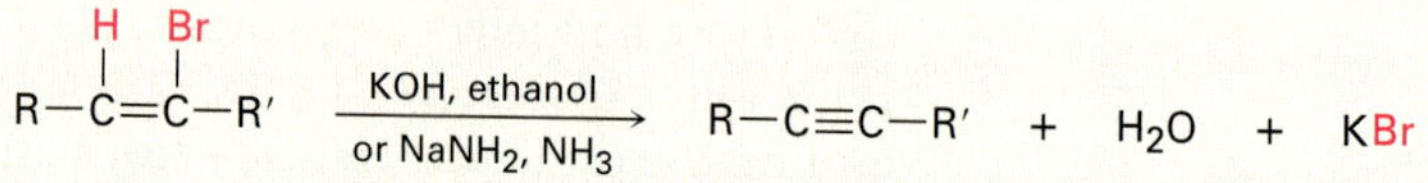

   (b) Alkylation of acetylide anions (Section 8.8)

$$HC{\equiv}CH \xrightarrow{NaNH_2} HC{\equiv}C^- \ Na^+ \xrightarrow{RCH_2Br} HC{\equiv}CCH_2R$$

Acetylene → **A terminal alkyne**

$$RC{\equiv}CH \xrightarrow{NaNH_2} RC{\equiv}C^- \ Na^+ \xrightarrow{R'CH_2Br} RC{\equiv}CCH_2R'$$

A terminal alkyne → **An internal alkyne**

2. Reactions of alkynes
   (a) Addition of HCl and HBr (Section 8.3)

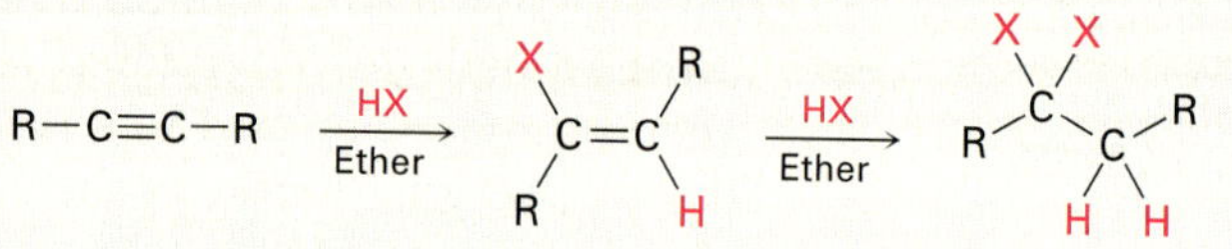

   (b) Addition of Cl_2 and Br_2 (Section 8.3)

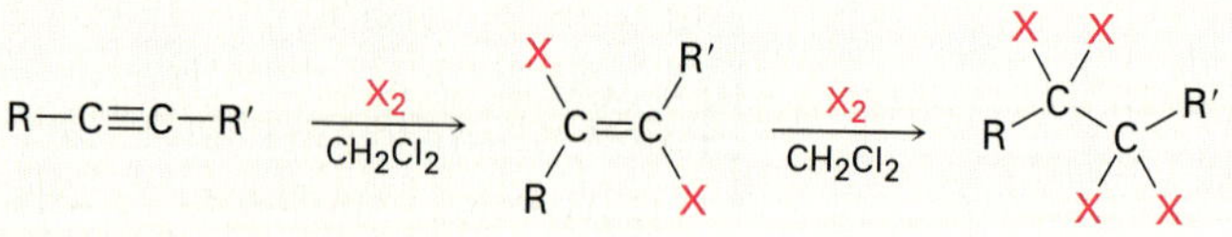

   (c) Hydration (Section 8.4)
      (1) Mercuric sulfate catalyzed

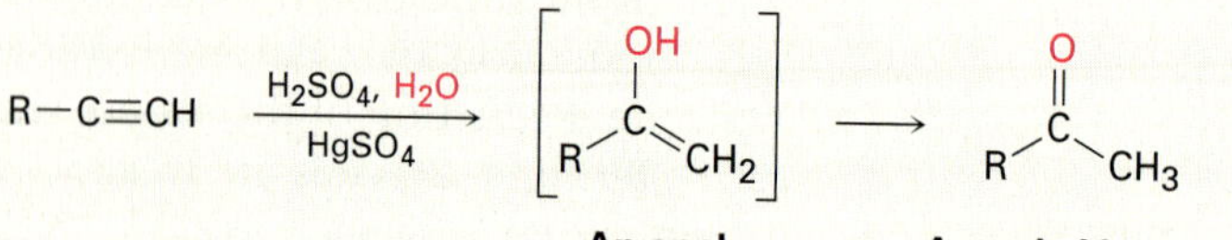

(2) Hydroboration/oxidation

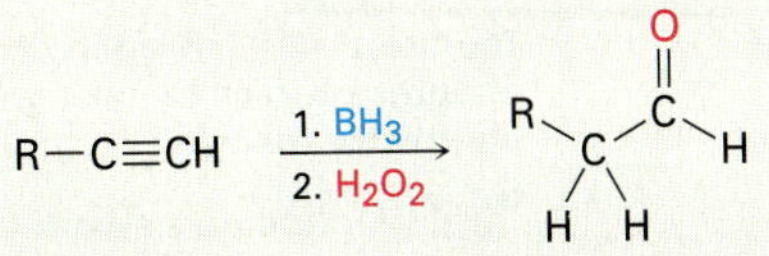

An aldehyde

(d) Reduction (Section 8.5)

(1) Catalytic hydrogenation

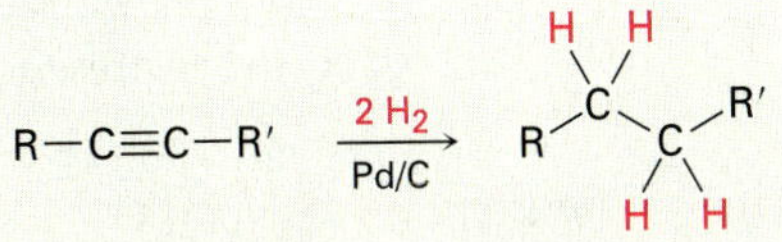

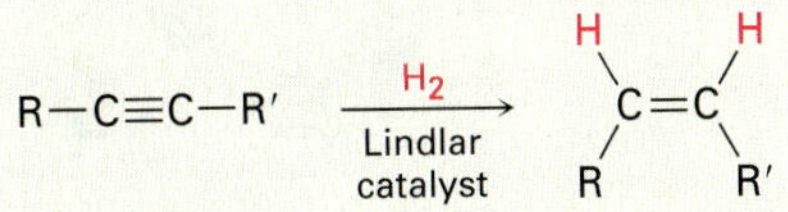

A cis alkene

(2) Lithium in liquid ammonia

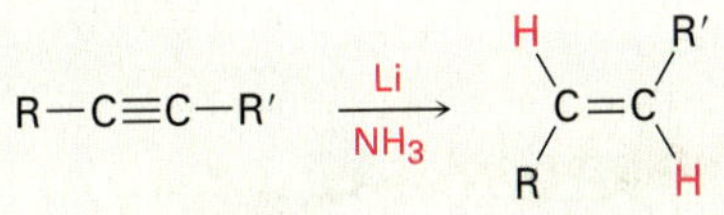

A trans alkene

(e) Conversion into acetylide anions (Section 8.7)

$$R-C\equiv C-H \xrightarrow[NH_3]{NaNH_2} R-C\equiv C:^-\ Na^+ \quad + \quad NH_3$$

(f) Alkylation of acetylide anions (Section 8.8)

$$HC\equiv CH \xrightarrow{NaNH_2} HC\equiv C^-\ Na^+ \xrightarrow{RCH_2Br} HC\equiv CCH_2R$$

Acetylene → **A terminal alkyne**

$$RC\equiv CH \xrightarrow{NaNH_2} RC\equiv C^-\ Na^+ \xrightarrow{R'CH_2Br} RC\equiv CCH_2R'$$

A terminal alkyne → **An internal alkyne**

EXERCISES

Organic KNOWLEDGE TOOLS

ThomsonNOW™ Sign in at **www.thomsonedu.com** to assess your knowledge of this chapter's topics by taking a pre-test. The pre-test will link you to interactive organic chemistry resources based on your score in each concept area.

Online homework for this chapter may be assigned in Organic OWL.

■ indicates problems assignable in Organic OWL.

VISUALIZING CHEMISTRY

(Problems 8.1–8.13 appear within the chapter.)

8.14 ■ Name the following alkynes, and predict the products of their reaction with (i) H_2 in the presence of a Lindlar catalyst and (ii) H_3O^+ in the presence of $HgSO_4$:

(a)

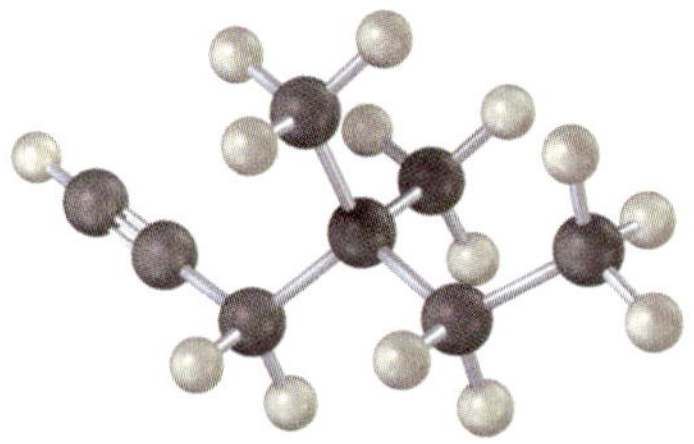

(b)

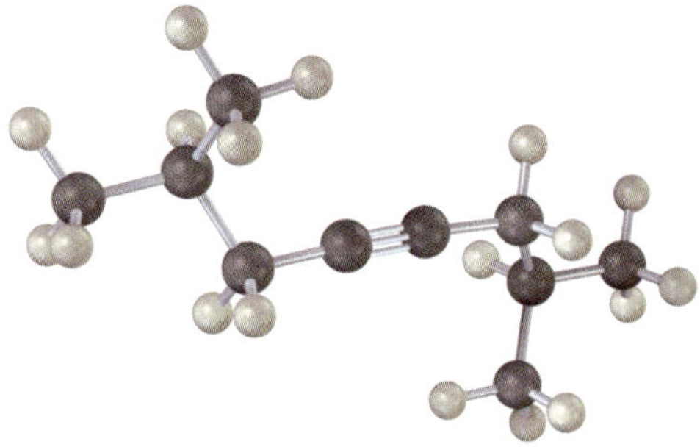

8.15 ■ From what alkyne might each of the following substances have been made? (Yellow-green = Cl.)

(a)

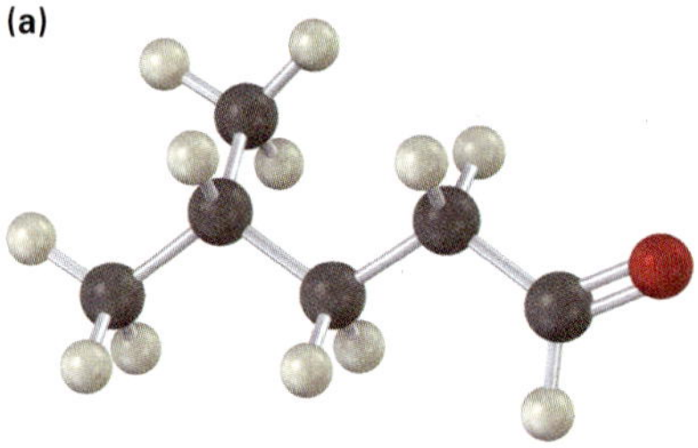

(b)

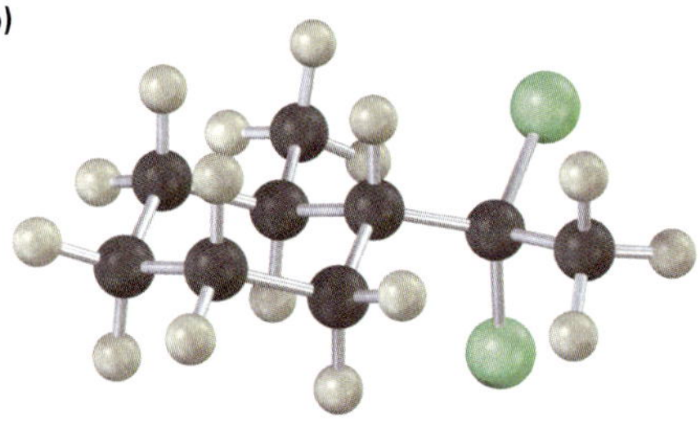

8.16 How would you prepare the following substances, starting from any compounds having four carbons or fewer?

(a)

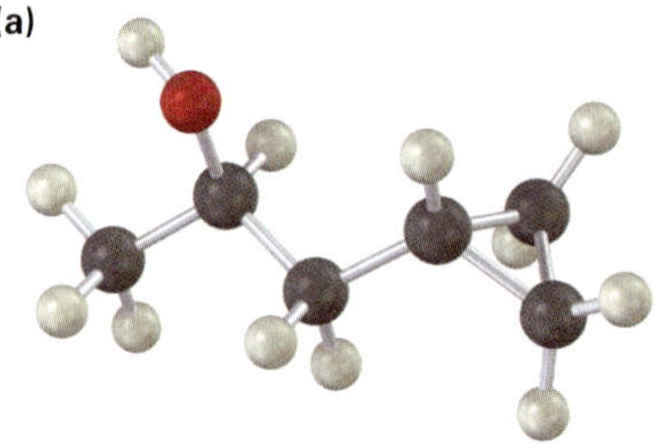

(b)

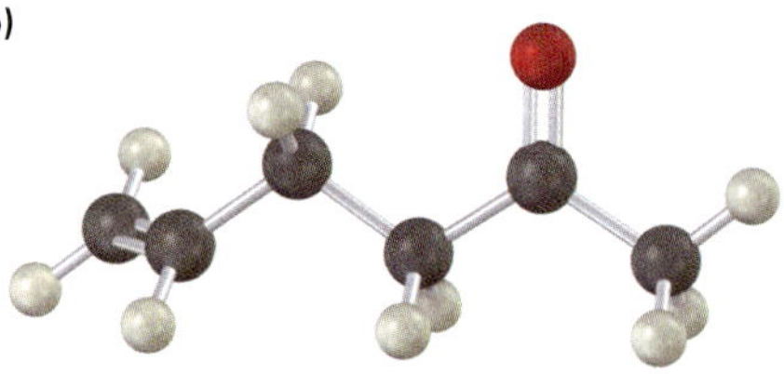

8.17 The following cycloalkyne is too unstable to exist. Explain.

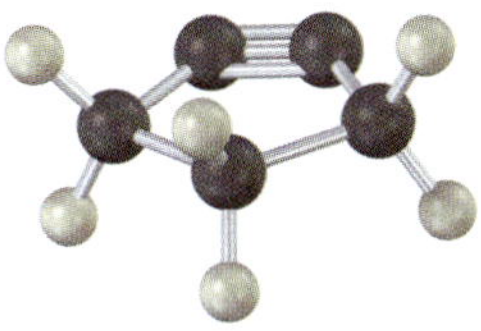

ADDITIONAL PROBLEMS

8.18 Give IUPAC names for the following compounds:

(a)

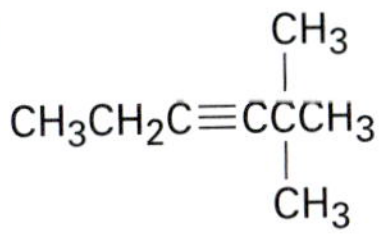

(b) $CH_3C{\equiv}CCH_2C{\equiv}CCH_2CH_3$

(c)

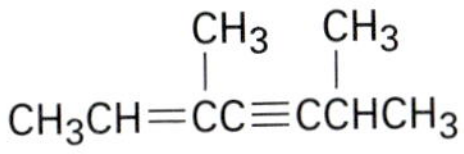

(d)

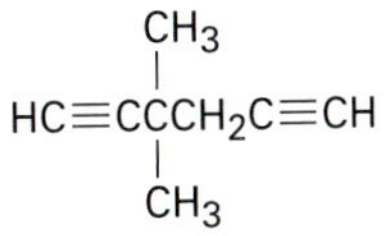

(e) $H_2C{=}CHCH{=}CHC{\equiv}CH$

(f)

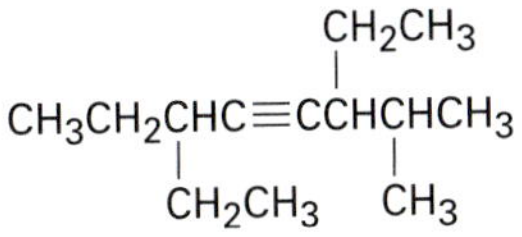

8.19 ■ Draw structures corresponding to the following names:
(a) 3,3-Dimethyl-4-octyne
(b) 3-Ethyl-5-methyl-1,6,8-decatriyne
(c) 2,2,5,5-Tetramethyl-3-hexyne
(d) 3,4-Dimethylcyclodecyne
(e) 3,5-Heptadien-1-yne
(f) 3-Chloro-4,4-dimethyl-1-nonen-6-yne
(g) 3-*sec*-Butyl-1-heptyne
(h) 5-*tert*-Butyl-2-methyl-3-octyne

8.20 The following two hydrocarbons have been isolated from various plants in the sunflower family. Name them according to IUPAC rules.
(a) $CH_3CH{=}CHC{\equiv}CC{\equiv}CCH{=}CHCH{=}CHCH{=}CH_2$ (all trans)
(b) $CH_3C{\equiv}CC{\equiv}CC{\equiv}CC{\equiv}CC{\equiv}CCH{=}CH_2$

8.21 ■ Predict the products of the following reactions:

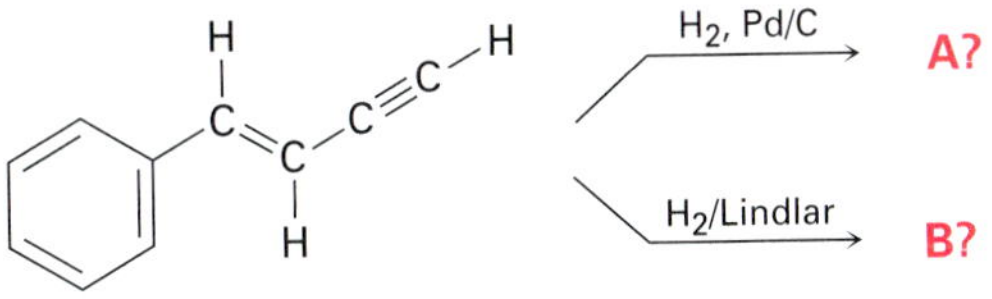

8.22 ■ A hydrocarbon of unknown structure has the formula C_8H_{10}. On catalytic hydrogenation over the Lindlar catalyst, 1 equivalent of H_2 is absorbed. On hydrogenation over a palladium catalyst, 3 equivalents of H_2 are absorbed.
(a) How many degrees of unsaturation are present in the unknown?
(b) How many triple bonds are present?
(c) How many double bonds are present?
(d) How many rings are present?
(e) Draw a structure that fits the data.

8.23 ■ Predict the products from reaction of 1-hexyne with the following reagen
(a) 1 equiv HBr (b) 1 equiv Cl_2
(c) H_2, Lindlar catalyst (d) $NaNH_2$ in NH_3, then CH_3Br
(e) H_2O, H_2SO_4, $HgSO_4$ (f) 2 equiv HCl

8.24 ■ Predict the products from reaction of 5-decyne with the following reagen
(a) H_2, Lindlar catalyst (b) Li in NH_3
(c) 1 equiv Br_2 (d) BH_3 in THF, then H_2O_2, OH^-
(e) H_2O, H_2SO_4, $HgSO_4$ (f) Excess H_2, Pd/C catalyst

8.25 ■ Predict the products from reaction of 2-hexyne with the following reagen
(a) 2 equiv Br_2 (b) 1 equiv HBr (c) Excess HBr
(d) Li in NH_3 (e) H_2O, H_2SO_4, $HgSO_4$

8.26 ■ How would you carry out the following conversions? More than one ste may be needed in some instances.

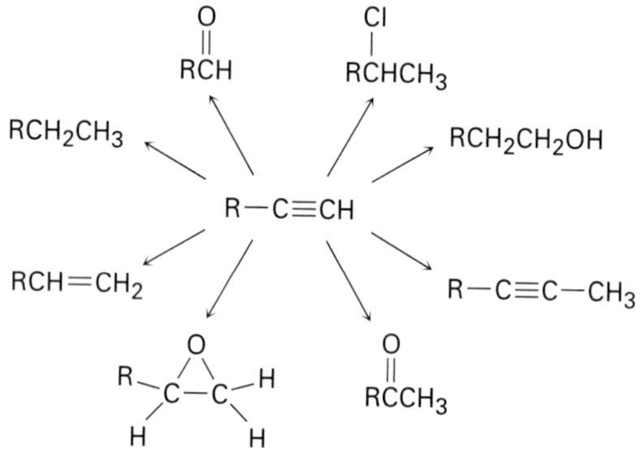

8.27 ■ Hydrocarbon A has the formula C_9H_{12} and absorbs 3 equivalents of H_2 t yield B, C_9H_{18}, when hydrogenated over a Pd/C catalyst. On treatment of A wit aqueous H_2SO_4 in the presence of mercury(II), two isomeric ketones, C and D are produced. Oxidation of A with $KMnO_4$ gives a mixture of acetic aci (CH_3CO_2H) and the tricarboxylic acid E. Propose structures for compounds A–D and write the reactions.

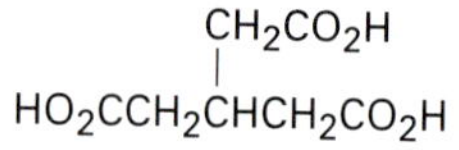

E

8.28 How would you carry out the following reactions?

(a)

(b) $CH_3CH_2C{\equiv}CH \xrightarrow{?} CH_3CH_2CH_2CHO$

(c)

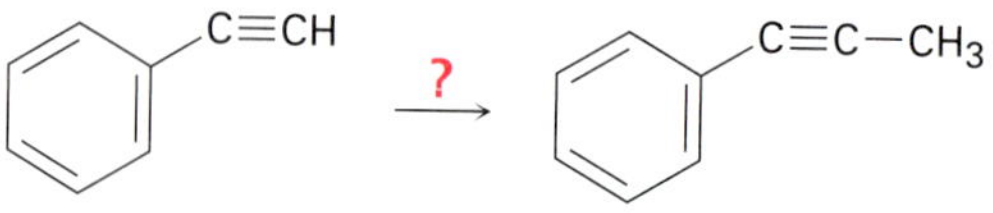

(d)

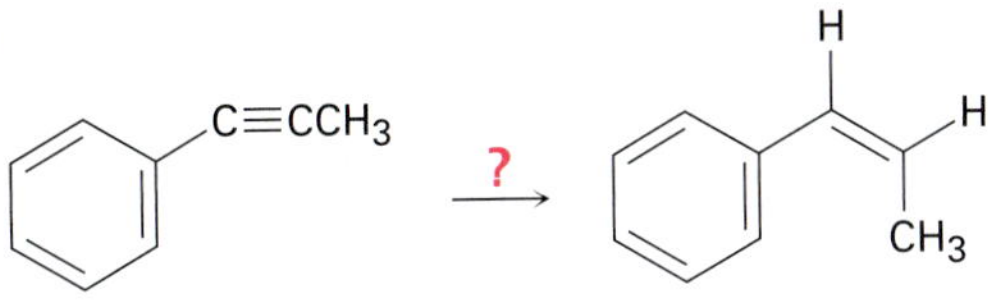

(e) $CH_3CH_2C\equiv CH \xrightarrow{?} CH_3CH_2CO_2H$

(f) $CH_3CH_2CH_2CH_2CH{=}CH_2 \xrightarrow[\text{(2 steps)}]{?} CH_3CH_2CH_2CH_2C\equiv CH$

8.29 Occasionally, chemists need to *invert* the stereochemistry of an alkene—that is, to convert a cis alkene to a trans alkene, or vice versa. There is no one-step method for doing an alkene inversion, but the transformation can be carried out by combining several reactions in the proper sequence. How would you carry out the following reactions?

(a) *trans*-5-Decene $\xrightarrow{?}$ *cis*-5-Decene

(b) *cis*-5-Decene $\xrightarrow{?}$ *trans*-5-Decene

8.30 ■ Propose structures for hydrocarbons that give the following products on oxidative cleavage by $KMnO_4$ or O_3:

(a) CO_2 + $CH_3(CH_2)_5CO_2H$

(b) CH_3CO_2H +

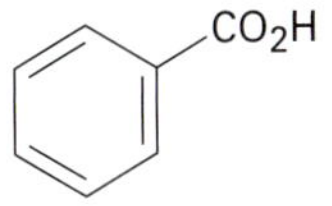

(c) $HO_2C(CH_2)_8CO_2H$

(d) CH_3CHO + $CH_3C(=O)CH_2CH_2CO_2H$ + CO_2

(e) $HC(=O)CH_2CH_2CH_2CH_2C(=O)CO_2H$ + CO_2

8.31 ■ Each of the following syntheses requires more than one step. How would you carry them out?

(a) $CH_3CH_2CH_2C\equiv CH \xrightarrow{?} CH_3CH_2CH_2CHO$

(b) $(CH_3)_2CHCH_2C\equiv CH \xrightarrow{?}$

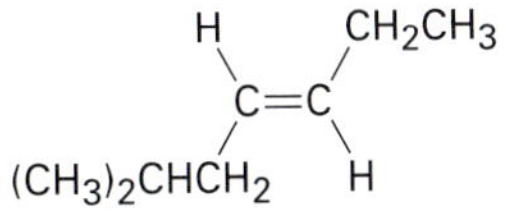

8.32 How would you carry out the following transformation? More than one step needed.

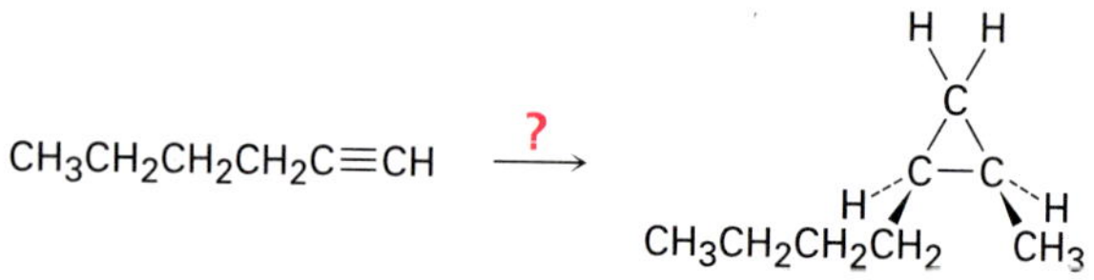

8.33 How would you carry out the following conversions? More than one step is needed in each case.

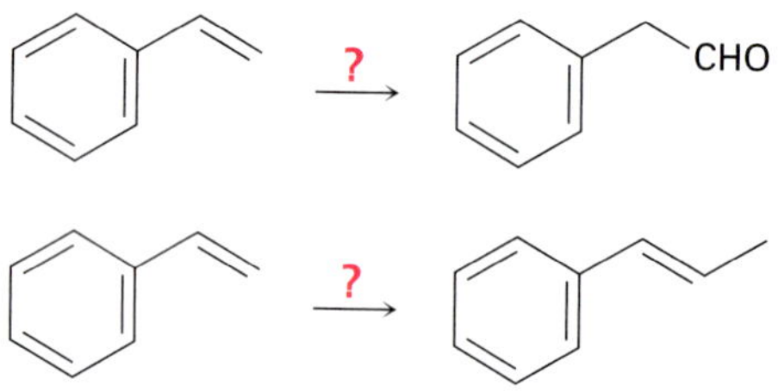

8.34 Synthesize the following compounds using 1-butyne as the only source of carbon, along with any inorganic reagents you need. More than one step may be needed.

(a) 1,1,2,2-Tetrachlorobutane (b) 1,1-Dichloro-2-ethylcyclopropane

8.35 ■ How would you synthesize the following compounds from acetylene and any alkyl halides with four or fewer carbons? More than one step may be required.

(a) $CH_3CH_2CH_2C{\equiv}CH$ **(b)** $CH_3CH_2C{\equiv}CCH_2CH_3$

(c) $CH_3CH(CH_3)CH_2CH{=}CH_2$ **(d)** $CH_3CH_2CH_2C(=O)CH_2CH_2CH_2CH_3$

(e) $CH_3CH_2CH_2CH_2CH_2CHO$

8.36 How would you carry out the following reactions to introduce deuterium into organic molecules?

(a)

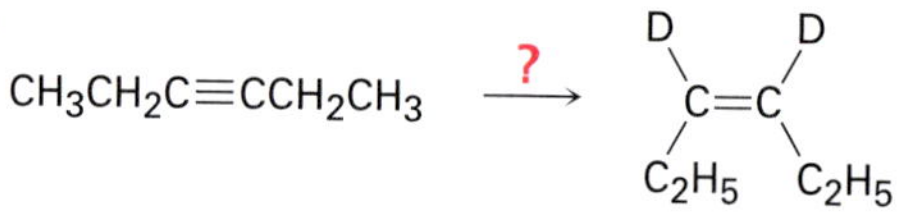

(b)

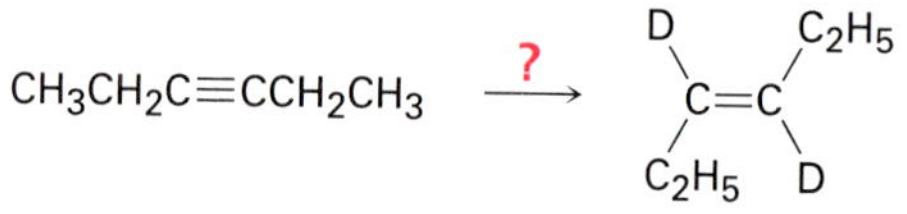

(c) $CH_3CH_2CH_2C{\equiv}CH \xrightarrow{?} CH_3CH_2CH_2C{\equiv}CD$

(d)

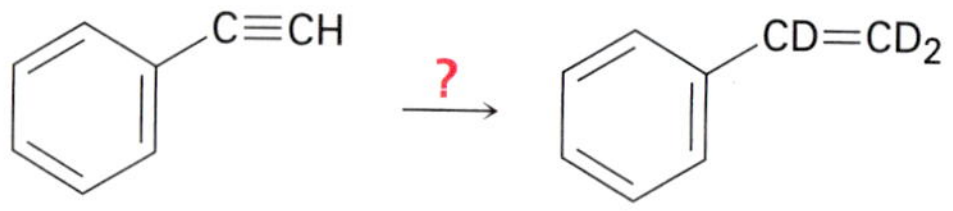

■ Assignable in OWL

8.37 How would you prepare cyclodecyne starting from acetylene and any alkyl halide needed?

8.38 The sex attractant given off by the common housefly is an alkene named *muscalure.* Propose a synthesis of muscalure starting from acetylene and any alkyl halides needed. What is the IUPAC name for muscalure?

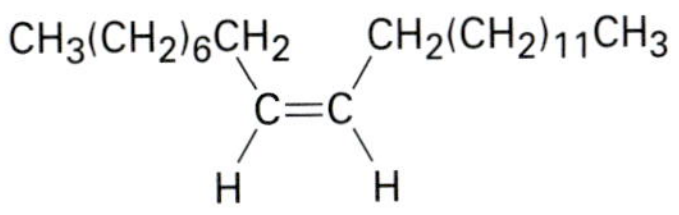

Muscalure

8.39 Compound A (C_9H_{12}) absorbed 3 equivalents of H_2 on catalytic reduction over a palladium catalyst to give B (C_9H_{18}). On ozonolysis, compound A gave, among other things, a ketone that was identified as cyclohexanone. On treatment with $NaNH_2$ in NH_3, followed by addition of iodomethane, compound A gave a new hydrocarbon, C ($C_{10}H_{14}$). What are the structures of A, B, and C?

8.40 Hydrocarbon A has the formula $C_{12}H_8$. It absorbs 8 equivalents of H_2 on catalytic reduction over a palladium catalyst. On ozonolysis, only two products are formed: oxalic acid (HO_2CCO_2H) and succinic acid ($HO_2CCH_2CH_2CO_2H$). Write the reactions, and propose a structure for A.

8.41 ■ Identify the reagents a–c in the following scheme:

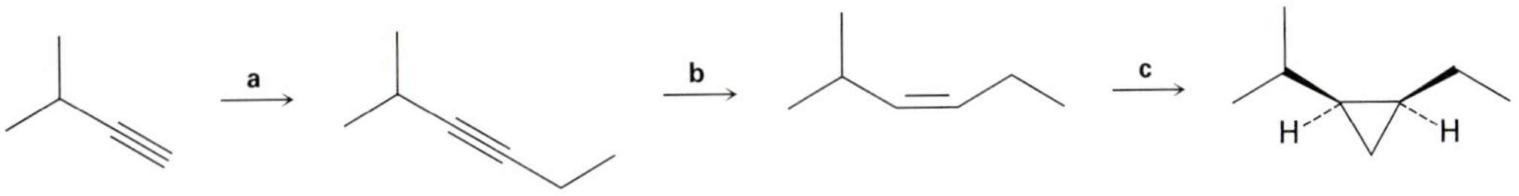

8.42 Organometallic reagents such as sodium acetylide undergo an addition reaction with ketones, giving alcohols:

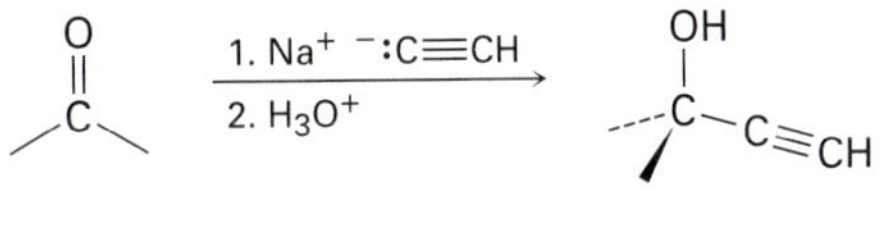

How might you use this reaction to prepare 2-methyl-1,3-butadiene, the starting material used in the manufacture of synthetic rubber?

8.43 The oral contraceptive agent Mestranol is synthesized using a carbonyl addition reaction like that shown in Problem 8.42. Draw the structure of the ketone needed.

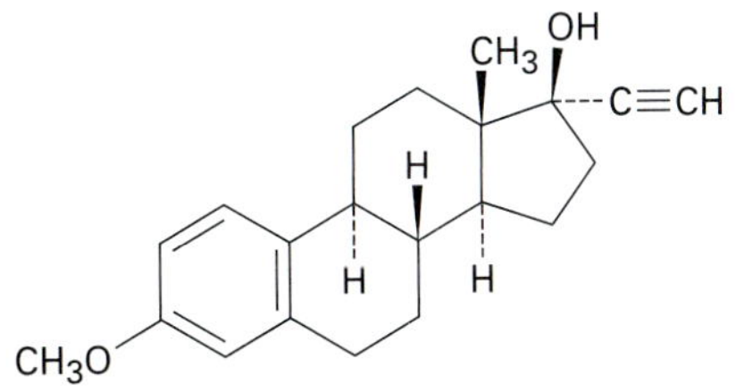

Mestranol

8.44 Erythrogenic acid, $C_{18}H_{26}O_2$, is an acetylenic fatty acid that turns a vivid r on exposure to light. On catalytic hydrogenation over a palladium cataly 5 equivalents of H_2 is absorbed, and stearic acid, $CH_3(CH_2)_{16}CO_2H$, is produce Ozonolysis of erythrogenic acid gives four products: formaldehyde, CH_2 oxalic acid, HO_2CCO_2H; azelaic acid, $HO_2C(CH_2)_7CO_2H$; and the aldehy acid $OHC(CH_2)_4CO_2H$. Draw two possible structures for erythrogenic aci and suggest a way to tell them apart by carrying out some simple reactions.

8.45 Terminal alkynes react with Br_2 and water to yield bromo ketones. For exampl

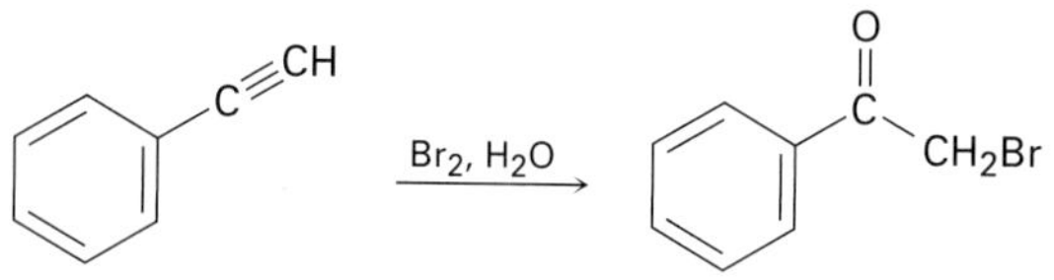

Propose a mechanism for the reaction. To what reaction of alkenes is th process analogous?

8.46 A *cumulene* is a compound with three adjacent double bonds. Draw an orbit picture of a cumulene. What kind of hybridization do the two central carbo atoms have? What is the geometric relationship of the substituents on one en to the substituents on the other end? What kind of isomerism is possibl Make a model to help see the answer.

$$R_2C{=}C{=}C{=}CR_2$$

A cumulene

8.47 Reaction of acetone with D_3O^+ yields hexadeuterioacetone. That is, all th hydrogens in acetone are exchanged for deuterium. Review the mechanism c mercuric ion–catalyzed alkyne hydration, and then propose a mechanism fo this deuterium incorporation.

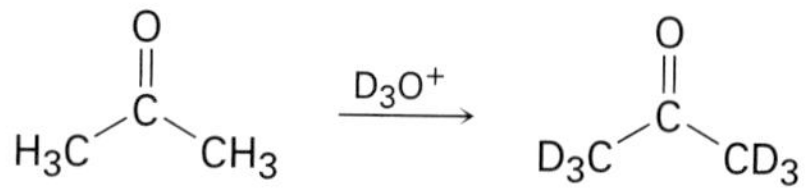

Acetone **Hexadeuterioacetone**

9 Stereochemistry

rganic KNOWLEDGE TOOLS

omsonNOW™ Throughout
is chapter, sign in at
ww.thomsonedu.com for
nline self-study and interactive
torials based on your level of
nderstanding.

Online homework for this chapter may be assigned in Organic OWL.

Are you right-handed or left-handed? You may not spend much time thinking about it, but handedness plays a surprisingly large role in your daily activities. Many musical instruments, such as oboes and clarinets, have a handedness to them; the last available softball glove always fits the wrong hand; left-handed people write in a "funny" way. The fundamental reason for these difficulties is that our hands aren't identical; rather, they're *mirror images*. When you hold a *left* hand up to a mirror, the image you see looks like a *right* hand. Try it.

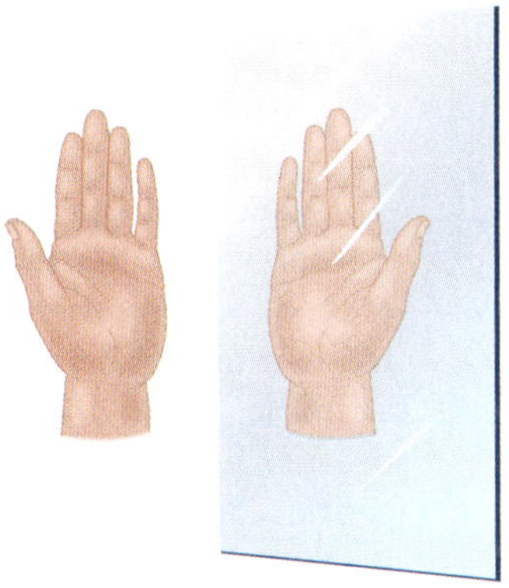

WHY THIS CHAPTER?

Handedness is also important in organic and biological chemistry, where it arises primarily as a consequence of the tetrahedral stereochemistry of sp^3-hybridized carbon atoms. Many drugs and almost all the molecules in our bodies, for instance, are handed. Furthermore, it is molecular handedness that makes possible the specific interactions between enzymes and their substrates that are so crucial to enzyme function. We'll look at handedness and its consequences in this chapter.

9.1 Enantiomers and the Tetrahedral Carbon

What causes molecular handedness? Look at generalized molecules of the ty CH_3X, CH_2XY, and CHXYZ shown in Figure 9.1. On the left are three molecul and on the right are their images reflected in a mirror. The CH_3X and CH_2 molecules are identical to their mirror images and thus are not handed. If y make molecular models of each molecule and its mirror image, you find th you can superimpose one on the other. By contrast, the CHXYZ molecule is identical to its mirror image. You can't superimpose a model of the molecule a model of its mirror image for the same reason that you can't superimpose a l hand on a right hand. They simply aren't the same.

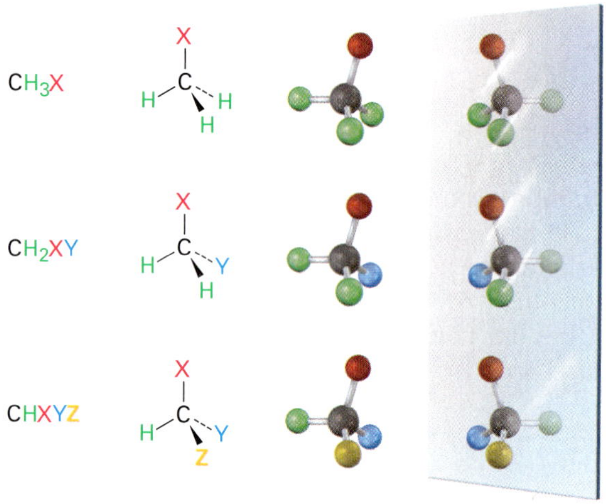

Figure 9.1 Tetrahedral carbon atoms and their mirror images. Molecules of the type CH_3X and CH_2XY are identical to their mirror images, but a molecule of the type CHXYZ is not. A CHXYZ molecule is related to its mirror image in the same way that a right hand is related to a left hand.

Molecules that are not identical to their mirror images are kinds of stere isomers called **enantiomers** (Greek *enantio*, meaning "opposite"). Enantiomers a related to each other as a right hand is related to a left hand and result whenever tetrahedral carbon is bonded to four different substituents (one need not be H). F example, lactic acid (2-hydroxypropanoic acid) exists as a pair of enantiome because there are four different groups (–H, –OH, $-CH_3$, $-CO_2H$) bonded to th central carbon atom. The enantiomers are called (+)-lactic acid and (−)-lactic aci Both are found in sour milk, but only the (+) enantiomer occurs in muscle tissu

Lactic acid: a molecule of general formula CHXYZ

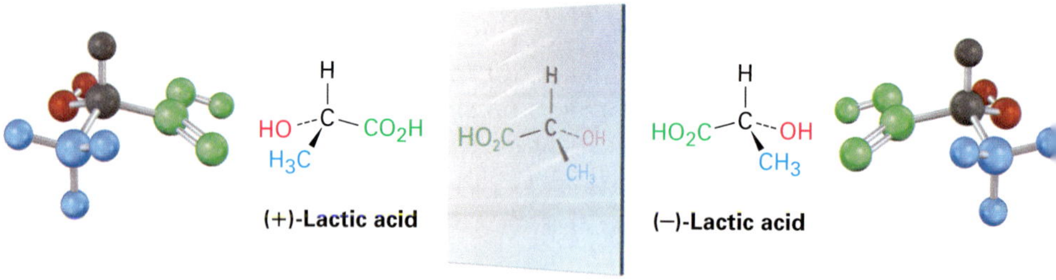

No matter how hard you try, you can't superimpose a molecule of (+)-lactic acid on a molecule of (−)-lactic acid; the two simply aren't identical. If any two groups match up, say $-H$ and $-CO_2H$, the remaining two groups don't match (Figure 9.2).

Figure 9.2 Attempts at superimposing the mirror-image forms of lactic acid. **(a)** When the $-H$ and $-OH$ substituents match up, the $-CO_2H$ and $-CH_3$ substituents don't; **(b)** when $-CO_2H$ and $-CH_3$ match up, $-H$ and $-OH$ don't. Regardless of how the molecules are oriented, they aren't identical.

9.2 The Reason for Handedness in Molecules: Chirality

ey IDEAS

est your knowledge of Key leas by using resources in homsonNOW or by answering nd-of-chapter problems marked vith ▲.

Molecules that are not identical to their mirror images, and thus exist in two enantiomeric forms, are said to be **chiral** (**ky**-ral, from the Greek *cheir,* meaning "hand"). You can't take a chiral molecule and its enantiomer and place one on the other so that all atoms coincide.

How can you predict whether a given molecule is or is not chiral? *A molecule is not chiral if it contains a plane of symmetry.* A plane of symmetry is a plane that cuts through the middle of an object (or molecule) in such a way that one half of the object is a mirror image of the other half. A laboratory flask, for example, has a plane of symmetry. If you were to cut the flask in half, one half would be a mirror image of the other half. A hand, however, does not have a plane of symmetry. One "half" of a hand is not a mirror image of the other half (Figure 9.3).

igure 9.3 The meaning of *ymmetry plane.* An object like he flask **(a)** has a symmetry lane cutting through it, making ight and left halves mirror mages. An object like a hand **(b)** as no symmetry plane; the right half" of a hand is not a mirror mage of the left half.

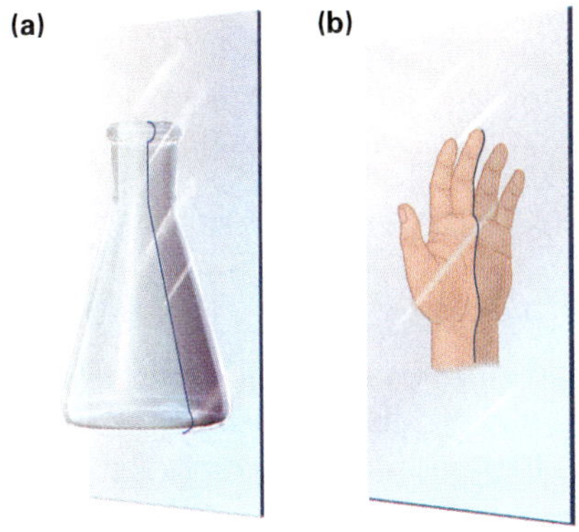

A molecule that has a plane of symmetry in any of its possible conformations must be identical to its mirror image and hence must be nonchiral, or **achiral**. Thus, propanoic acid, $CH_3CH_2CO_2H$, has a plane of symmetry when lined up as shown in Figure 9.4 and is achiral, while lactic acid, $CH_3CH(OH)CO_2H$, has no plane of symmetry in any conformation and is chiral.

Figure 9.4 The achiral propanoic acid molecule versus the chiral lactic acid molecule. Propanoic acid has a plane of symmetry that makes one side of the molecule a mirror image of the other side. Lactic acid has no such symmetry plane.

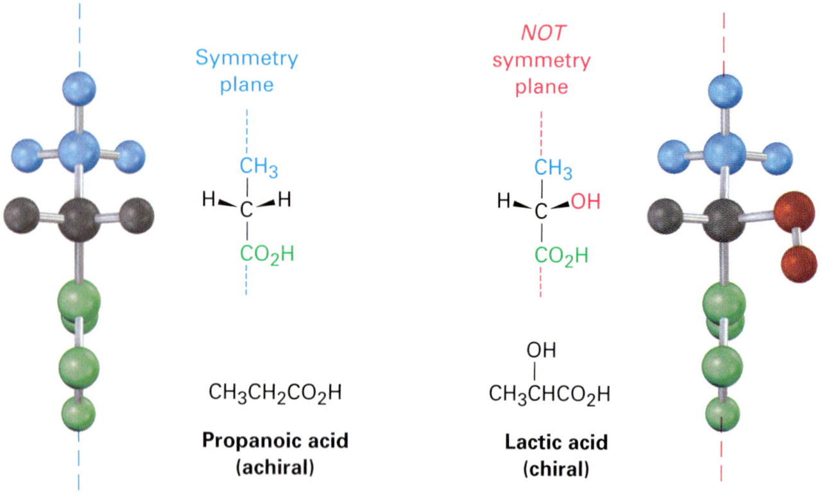

ThomsonNOW™ Click *Organic Interactive* to **practice identifying chirality centers in organic molecules.**

The most common, although not the only, cause of chirality in an organi molecule is the presence of a carbon atom bonded to four different groups—fc example, the central carbon atom in lactic acid. Such carbons are now referred t as **chirality centers**, although other terms such as *stereocenter, asymmetric cente* and *stereogenic center* have also been used formerly. Note that *chirality* is a prop erty of the entire molecule, whereas a chirality *center* is the *cause* of chirality.

Detecting chirality centers in a complex molecule takes practice because it not always immediately apparent that four different groups are bonded to given carbon. The differences don't necessarily appear right next to the chira ity center. For example, 5-bromodecane is a chiral molecule because four differ ent groups are bonded to C5, the chirality center (marked with an asterisk). butyl substituent is similar to a pentyl substituent but it isn't identical. The di ference isn't apparent until four carbon atoms away from the chirality cente but there's still a difference.

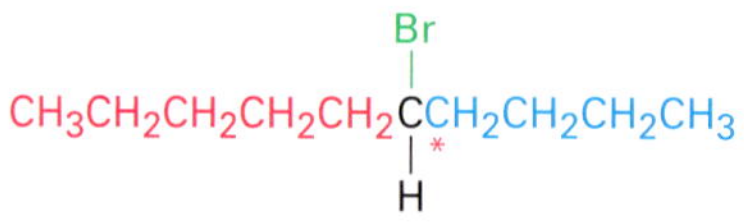

5-Bromodecane (chiral)

Substituents on carbon 5

—H

—Br

$—CH_2CH_2CH_2CH_3$ (butyl)

$—CH_2CH_2CH_2CH_2CH_3$ (pentyl)

As other possible examples, look at methylcyclohexane and 2-methylcyclo hexanone. Methylcyclohexane is achiral because no carbon atom in the mole cule is bonded to four different groups. You can immediately eliminate all $-CH_2-$ carbons and the $-CH_3$ carbon from consideration, but what about C1 on th ring? The C1 carbon atom is bonded to a $-CH_3$ group, to an $-H$ atom, and tc C2 and C6 of the ring. Carbons 2 and 6 are equivalent, however, as are carbon 3 and 5. Thus, the C6–C5–C4 "substituent" is equivalent to the C2–C3–C4 sub stituent, and methylcyclohexane is achiral. Another way of reaching the sam conclusion is to realize that methylcyclohexane has a symmetry plane, which passes through the methyl group and through C1 and C4 of the ring.

The situation is different for 2-methylcyclohexanone. 2-Methylcyclohexanone has no symmetry plane and is chiral because C2 is bonded to four different groups: a $-CH_3$ group, an $-H$ atom, a $-COCH_2-$ ring bond (C1), and a $-CH_2CH_2-$ ring bond (C3).

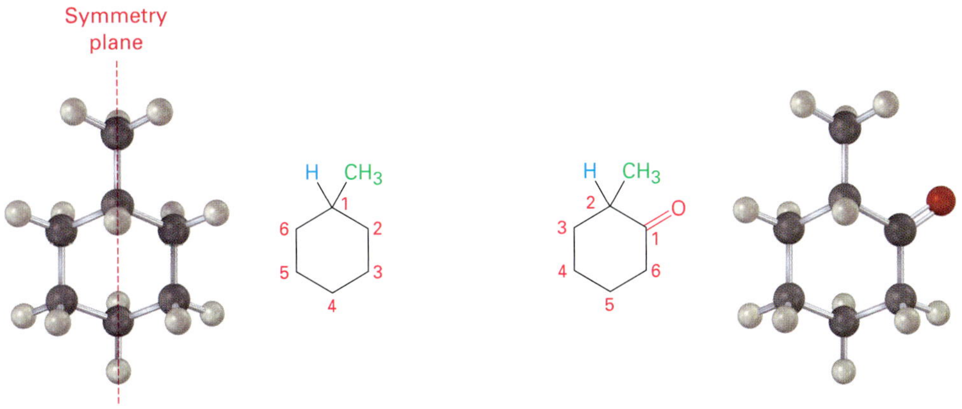

Methylcyclohexane (achiral)

2-Methylcyclohexanone (chiral)

Several more examples of chiral molecules are shown below. Check for yourself that the labeled carbons are chirality centers. You might note that carbons in $-CH_2-$, $-CH_3$, C=O, C=C, and C≡C groups can't be chirality centers. (Why?)

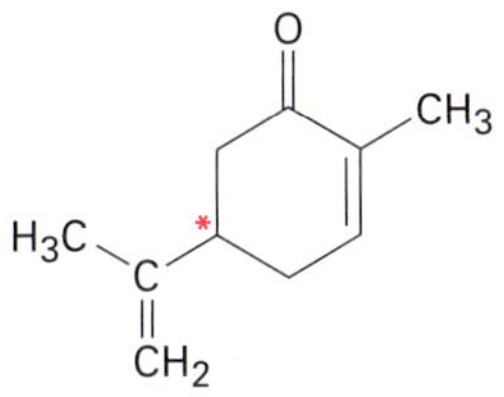

Carvone (spearmint oil)

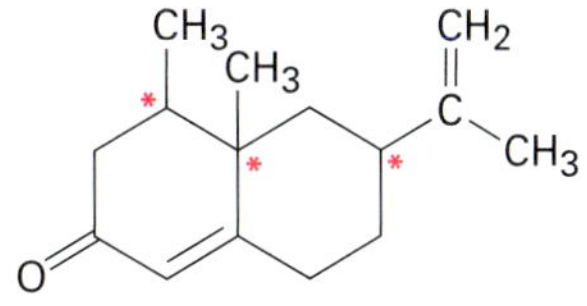

Nootkatone (grapefruit oil)

WORKED EXAMPLE 9.1 *Drawing the Three-Dimensional Structure of a Chiral Molecule*

Draw the structure of a chiral alcohol.

Strategy An alcohol is a compound that contains the $-OH$ functional group. To make an alcohol chiral, we need to have four different groups bonded to a single carbon atom, say $-H$, $-OH$, $-CH_3$, and $-CH_2CH_3$.

Solution

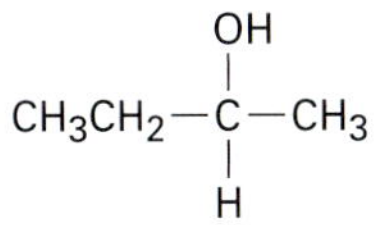

2-Butanol (chiral)

Problem 9.1 Which of the following objects are chiral?

(a) Screwdriver (b) Screw (c) Beanstalk (d) Shoe

Problem 9.2 Identify the chirality centers in the following molecules. Build molecular model you need help.

(a)

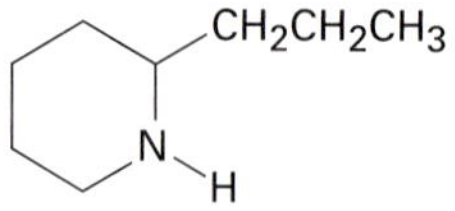

Coniine (poison hemlock)

(b)

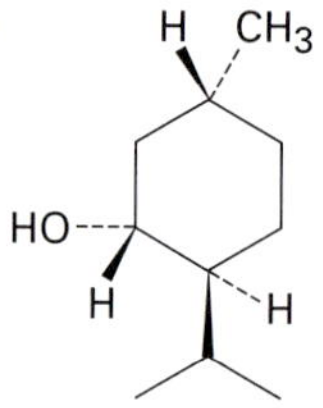

Menthol (flavoring agent)

(c)

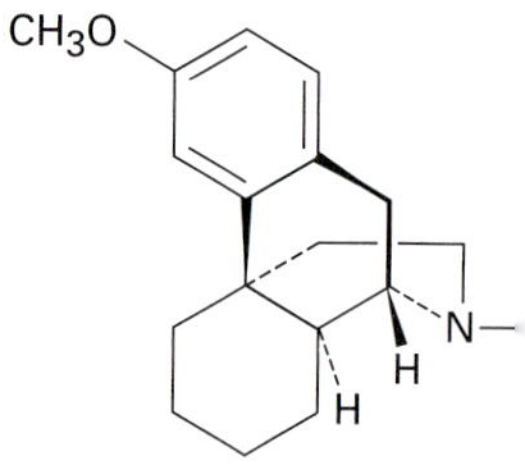

Dextromethorphan (cough suppressant)

Problem 9.3 Alanine, an amino acid found in proteins, is chiral. Draw the two enantiomers of alanine using the standard convention of solid, wedged, and dashed lines.

$$\begin{array}{c} NH_2 \\ | \\ CH_3CHCO_2H \end{array}$$ **Alanine**

Problem 9.4 Identify the chirality centers in the following molecules (yellow-green = Cl, pa yellow = F):

(a)

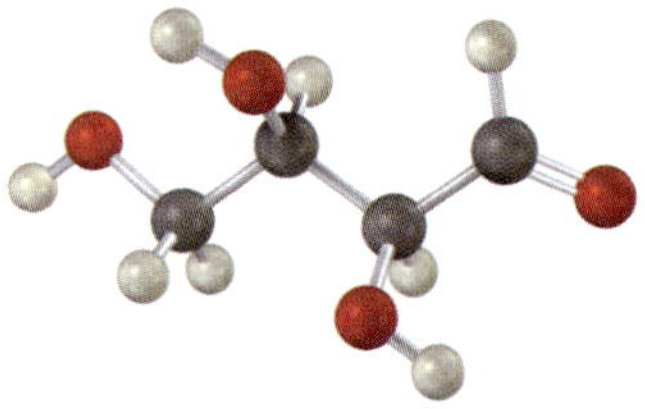

Threose (a sugar)

(b)

Enflurane (an anesthetic)

9.3 Optical Activity

The study of stereochemistry originated in the early 19th century during investigations by the French physicist Jean-Baptiste Biot into the nature of *plane-polarized light.* A beam of ordinary light consists of electromagnetic waves that oscillate in an infinite number of planes at right angles to the direction of light travel. When a beam of ordinary light is passed through a device called a *polarizer,* however, only the light waves oscillating in a single plane pass through and the light is said to be plane-polarized. Light waves in all other planes are blocked out.

Biot made the remarkable observation that when a beam of plane-polarized light passes through a solution of certain organic molecules, such as sugar

Jean-Baptiste Biot

Jean-Baptiste Biot (1774–1862) was born in Paris, France, and was educated there at the École Polytechnique. In 1800, he was appointed professor of mathematical physics at the College de France. His work on determining the optical rotation of naturally occurring molecules included an experiment on turpentine, which caught fire and nearly burned down the church building he was using for his experiments.

camphor, the plane of polarization is rotated. Not all organic substances exhibit this property, but those that do are said to be **optically active**.

The amount of rotation can be measured with an instrument called a *polarimeter,* represented in Figure 9.5. A solution of optically active organic molecules is placed in a sample tube, plane-polarized light is passed through the tube, and rotation of the polarization plane occurs. The light then goes through a second polarizer called the *analyzer.* By rotating the analyzer until the light passes through it, we can find the new plane of polarization and can tell to what extent rotation has occurred.

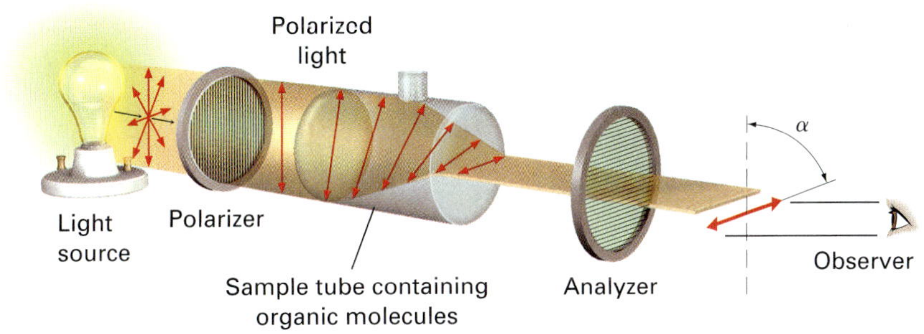

Figure 9.5 Schematic representation of a polarimeter. Plane-polarized light passes through a solution of optically active molecules, which rotate the plane of polarization.

In addition to determining the extent of rotation, we can also find the direction. From the vantage point of the observer looking directly at the analyzer, some optically active molecules rotate polarized light to the left (counterclockwise) and are said to be **levorotatory**, whereas others rotate polarized light to the right (clockwise) and are said to be **dextrorotatory**. By convention, rotation to the left is given a minus sign (−), and rotation to the right is given a plus sign (+). (−)-Morphine, for example, is levorotatory, and (+)-sucrose is dextrorotatory.

ThomsonNOW™ Click *Organic Interactive* to **learn the relationship between observed optical rotation and concentration for optically active compounds.**

The amount of rotation observed in a polarimetry experiment depends on the number of optically active molecules encountered by the light beam. This number, in turn, depends on sample concentration and sample pathlength. If the concentration of sample is doubled, the observed rotation doubles. If the concentration is kept constant but the length of the sample tube is doubled, the observed rotation is doubled. It also happens that the amount of rotation depends on the wavelength of the light used.

To express optical rotations in a meaningful way so that comparisons can be made, we have to choose standard conditions. The **specific rotation,** $[\alpha]_D$, of a compound is defined as the observed rotation when light of 589.6 nanometer (nm; 1 nm = 10^{-9} m) wavelength is used with a sample pathlength l of 1 decimeter (dm; 1 dm = 10 cm) and a sample concentration C of 1 g/mL. (Light of 589.6 nm, the so-called sodium D line, is the yellow light emitted from common sodium lamps.)

$$[\alpha]_D = \frac{\text{Observed rotation (degrees)}}{\text{Pathlength, } l \text{ (dm)} \times \text{Concentration, } C \text{ (g/mL)}} = \frac{\alpha}{l \times C}$$

When optical rotation data are expressed in this standard way, the specific rotation, $[\alpha]_D$, is a physical constant characteristic of a given optically active compound. For example, (+)-lactic acid has $[\alpha]_D = +3.82$, and (−)-lactic ac has $[\alpha]_D = -3.82$. That is, the two enantiomers rotate plane-polarized light exactly the same extent but in opposite directions. Note that specific rotation generally expressed as a unitless number. Some additional examples are listed Table 9.1.

Table 9.1 Specific Rotation of Some Organic Molecules

Compound	$[\alpha]_D$	Compound	$[\alpha]_D$
Penicillin V	+233	Cholesterol	−31.5
Sucrose	+66.47	Morphine	−132
Camphor	+44.26	Cocaine	−16
Chloroform	0	Acetic acid	0

WORKED EXAMPLE 9.2

Calculating an Optical Rotation

A 1.20 g sample of cocaine, $[\alpha]_D = -16$, was dissolved in 7.50 mL of chloroform an placed in a sample tube having a pathlength of 5.00 cm. What was the observe rotation?

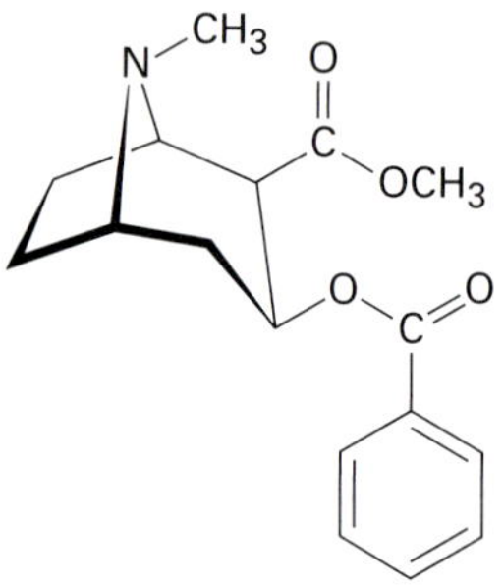

Cocaine

Strategy Observed rotation, α, is equal to specific rotation $[\alpha]_D$ times sample concentratior C, times pathlength, l: $\alpha = [\alpha]_D \times C \times l$, where $[\alpha]_D = -16$, $l = 5.00$ cm $= 0.500$ dm and $C = 1.20$ g/7.50 mL $= 0.160$ g/mL.

Solution $\alpha = -16 \times 0.500 \times 0.160 = -1.3°$.

Problem 9.5 Is cocaine (Worked Example 9.2) dextrorotatory or levorotatory?

Problem 9.6 A 1.50 g sample of coniine, the toxic extract of poison hemlock, was dissolved ir 10.0 mL of ethanol and placed in a sample cell with a 5.00 cm pathlength. Th observed rotation at the sodium D line was +1.21°. Calculate $[\alpha]_D$ for coniine.

9.4 Pasteur's Discovery of Enantiomers

Little was done after Biot's discovery of optical activity until 1848, when Loui Pasteur began work on a study of crystalline tartaric acid salts derived from wine On crystallizing a concentrated solution of sodium ammonium tartrate below

28 °C, Pasteur made the surprising observation that two distinct kinds of crystals precipitated. Furthermore, the two kinds of crystals were mirror images and were related in the same way that a right hand is related to a left hand.

Working carefully with tweezers, Pasteur was able to separate the crystals into two piles, one of "right-handed" crystals and one of "left-handed" crystals like those shown in Figure 9.6. Although the original sample, a 50:50 mixture of right and left, was optically inactive, solutions of the crystals from each of the sorted piles were optically active, and their specific rotations were equal in amount but opposite in sign.

Louis Pasteur

Louis Pasteur (1822–1895) was born at Dôle in the Jura region of France, the son of leather tanners. After receiving his doctorate from the École Normale Supérieure at age 25, his landmark discovery of tartaric acid enantiomers was made only 1 year later. Pasteur is best known for his studies in bacteriology and for his discovery of vaccines for anthrax and rabies.

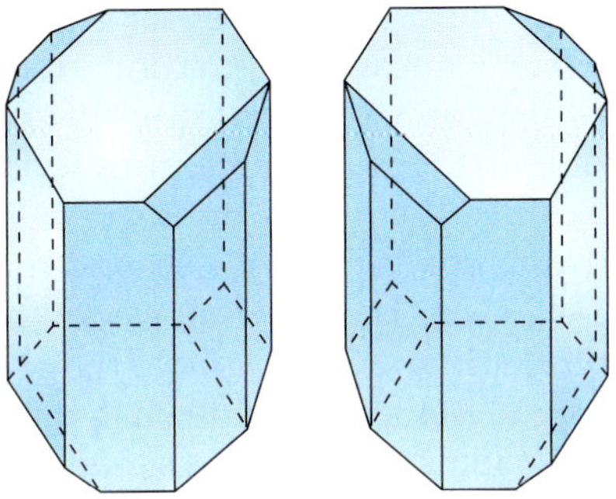

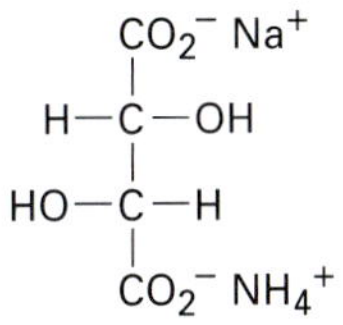

Sodium ammonium tartrate

gure 9.6 Drawings of sodium ımonium tartrate crystals ken from Pasteur's original etches. One of the crystals is ght-handed" and one is "left-nded."

Pasteur was far ahead of his time. Although the structural theory of Kekulé had not yet been proposed, Pasteur explained his results by speaking of the molecules themselves, saying, "There is no doubt that [in the *dextro* tartaric acid] there exists an asymmetric arrangement having a nonsuperimposable image. It is no less certain that the atoms of the *levo* acid have precisely the inverse asymmetric arrangement." Pasteur's vision was extraordinary, for it was not until 25 years later that his ideas regarding the asymmetric carbon atom were confirmed.

Today, we would describe Pasteur's work by saying that he had discovered enantiomers. Enantiomers, also called *optical isomers,* have identical physical properties, such as melting point and boiling point, but differ in the direction in which their solutions rotate plane-polarized light.

9.5 Sequence Rules for Specifying Configuration

ey **IDEAS**

est your knowledge of Key eas by using resources in ıomsonNOW or by answering ıd-of-chapter problems marked ith ▲.

Drawings provide a visual representation of stereochemistry, but a verbal method for indicating the three-dimensional arrangement, or **configuration**, of substituents at a chirality center is also needed. The method used employs the same sequence rules given in Section 6.5 for specifying *E* and *Z* alkene stereochemistry. Let's briefly review the sequence rules and see how they're used to specify the configuration of a chirality center. For a more thorough review, you should reread Section 6.5.

Rule 1 Look at the four atoms directly attached to the chirality center, and assign priorities in order of decreasing atomic number. The atom with the highest atomic number is ranked first; the atom with the lowest atomic number (usually hydrogen) is ranked fourth.

Rule 2 If a decision can't be reached by ranking the first atoms in the substituents, lo at the second, third, or fourth atoms outward until a difference is found.

Rule 3 Multiple-bonded atoms are equivalent to the same number of single-bond atoms. For example:

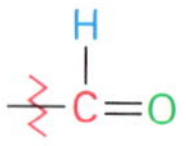

is equivalent to

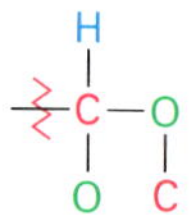

ThomsonNOW™ Click *Organic Interactive* to **assign absolute configurations using the Cahn–Ingold–Prelog rules.**

Having assigned priorities to the four groups attached to a chiral carbon, describe the stereochemical configuration around the carbon by orienting t molecule so that the group of lowest priority (4) points directly back, away fro us. We then look at the three remaining substituents, which now appear radiate toward us like the spokes on a steering wheel (Figure 9.7). If a curv arrow drawn from the highest to second-highest to third-highest priori substituent (1 → 2 → 3) is clockwise, we say that the chirality center has t ***R* configuration** (Latin *rectus,* meaning "right"). If an arrow from 1 → 2 → 3 counterclockwise, the chirality center has the ***S* configuration** (Latin *sinist* meaning "left"). To remember these assignments, think of a car's steering whe when making a *R*ight (clockwise) turn.

ThomsonNOW™ Click *Organic Interactive* to **manipulate three-dimensional models and assign *R,S* designations.**

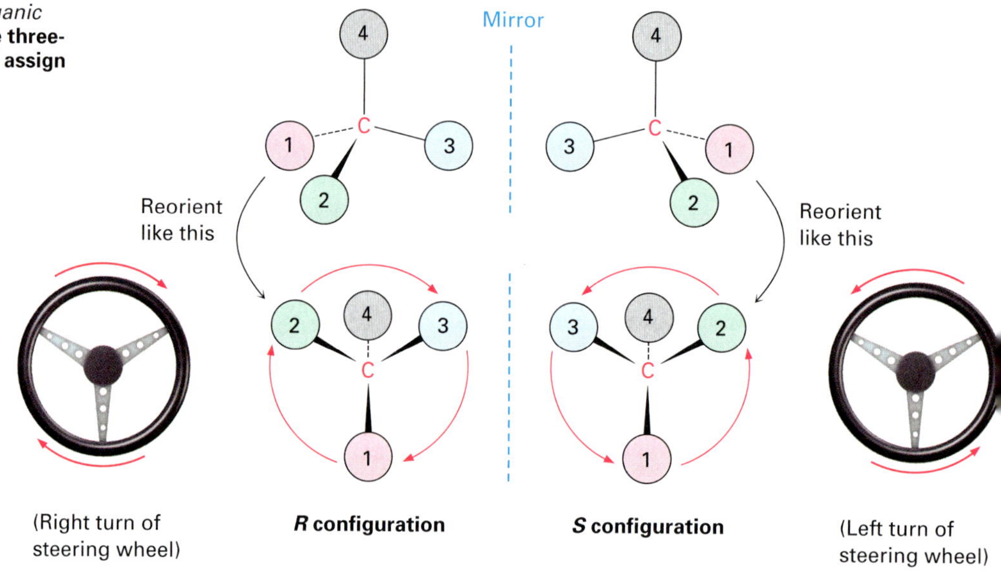

Figure 9.7 Assigning configuration to a chirality center. When the molecule is oriented so that the group of lowest priority (4) is toward the rear, the remaining three groups radiate toward the viewer like the spokes of a steering wheel. If the direction of travel 1 → 2 → 3 is clockwise (right turn), the center has the *R* configuration. If the direction of travel 1 → 2 → 3 is counterclockwise (left turn), the center is *S*.

Look at (−)-lactic acid in Figure 9.8 for an example of how to assign cor figuration. Sequence rule 1 says that −OH has priority 1 and −H has priority but it doesn't allow us to distinguish between $-CH_3$ and $-CO_2H$ becaus both groups have carbon as their first atom. Sequence rule 2, however, says that $-CO_2H$ is higher priority than $-CH_3$ because O (the second atom in $-CO_2H$) outranks H (the second atom in $-CH_3$). Now, turn the molecule so that the fourth-priority group (−H) is oriented toward the rear, away from the observer. Since a curved arrow from 1 (−OH) to 2 ($-CO_2H$) to 3 ($-CH_3$) is clockwise (right turn of the steering wheel), (−)-lactic acid has the *R* configuration. Applying the same procedure to (+)-lactic acid leads to the opposite assignment.

gure 9.8 Assignment of con-uration to **(a)** (*R*)-(−)-lactic acid d **(b)** (*S*)-(+)-lactic acid.

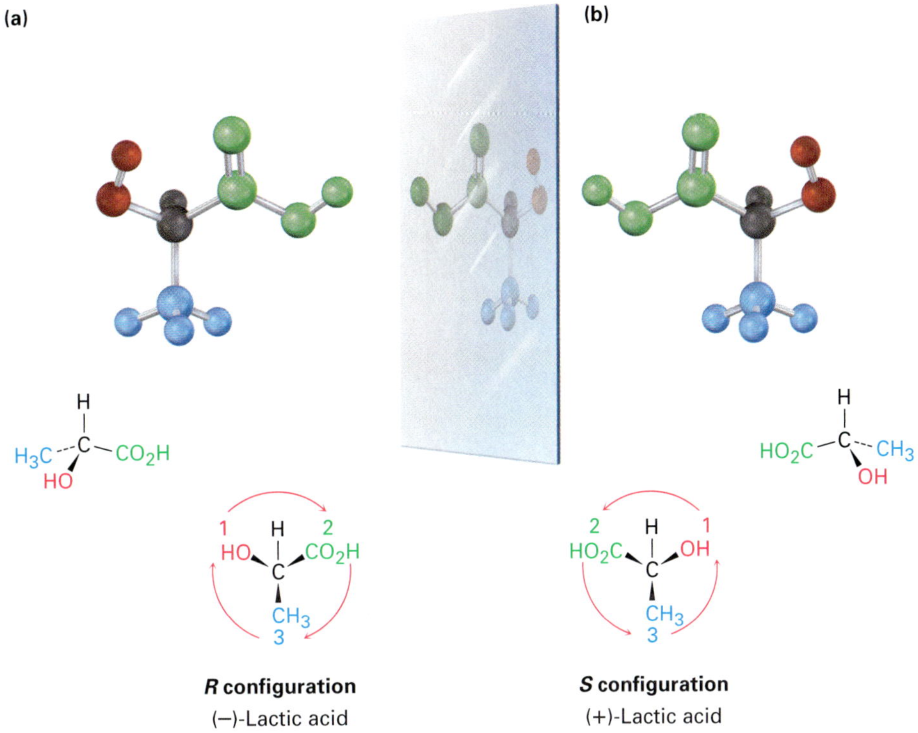

Further examples are provided by naturally occurring (−)-glyceraldehyde and (+)-alanine, which both have the *S* configuration as shown in Figure 9.9. Note that the sign of optical rotation, (+) or (−), is not related to the *R,S* designation. (*S*)-Glyceraldehyde happens to be levorotatory (−), and (*S*)-alanine happens to be dextrorotatory (+). There is no simple correlation between *R,S* configuration and direction or magnitude of optical rotation.

One further point needs to be mentioned—the matter of **absolute configuration**. How do we know that our assignments of *R,S* configuration are correct in an absolute, rather than a relative, sense? Since we can't see the molecules themselves, how do we know that the *R* configuration belongs to the dextrorotatory enantiomer of lactic acid? This difficult question was finally solved in 1951, when J. M. Bijvoet of the University of Utrecht reported an X-ray spectroscopic method for determining the absolute spatial arrangement of atoms in a molecule. Based on his results, we can say with certainty that the *R,S* conventions are correct.

Figure 9.9 Assigning configuration to **(a)** (−)-glyceraldehyde and **(b)** (+)-alanine. Both happen to have the *S* configuration, although one is levorotatory and the other is dextrorotatory.

(a)

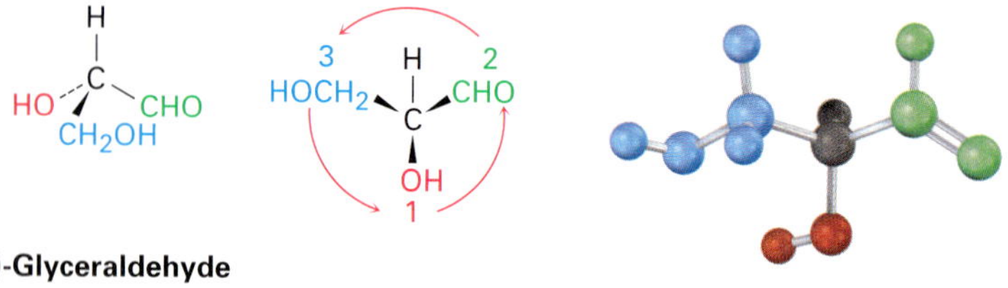

(*S*)-Glyceraldehyde
[(*S*)-(−)-2,3-Dihydroxypropanal]
$[\alpha]_D = -8.7$

(b)

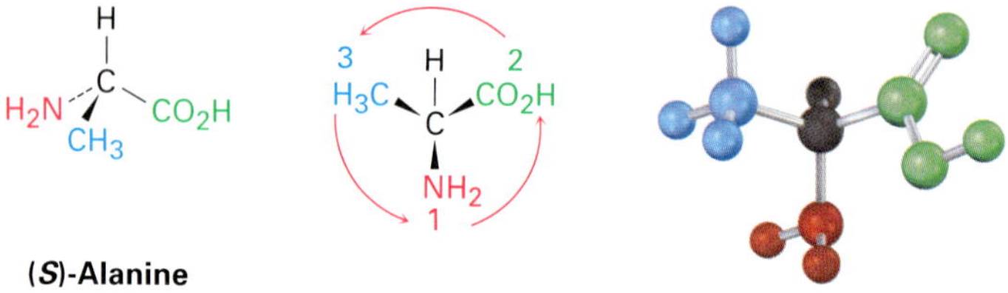

(*S*)-Alanine
[(*S*)-(+)-2-Aminopropanoic acid]
$[\alpha]_D = +8.5$

WORKED EXAMPLE 9.3 *Assigning R or S Configuration to Chirality Centers in Molecules*

Orient each of the following drawings so that the lowest-priority group is toward th rear, and then assign *R* or *S* configuration:

(a)

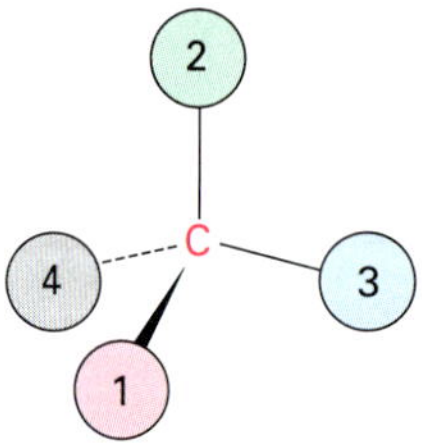

(b)

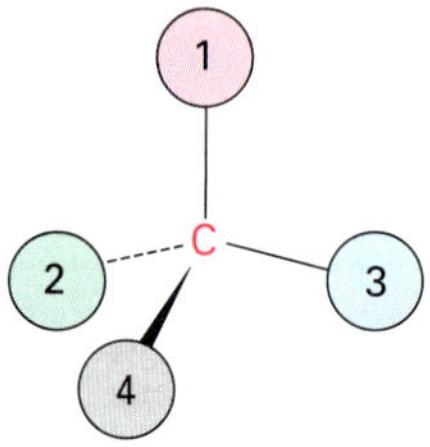

Strategy It takes practice to be able to visualize and orient a molecule in three dimension You might start by indicating where the observer must be located—180° opposite th lowest-priority group. Then imagine yourself in the position of the observer, an redraw what you would see.

Solution In (a), you would be located in front of the page toward the top right of the mole cule, and you would see group 2 to your left, group 3 to your right, and group below you. This corresponds to an *R* configuration.

(a)

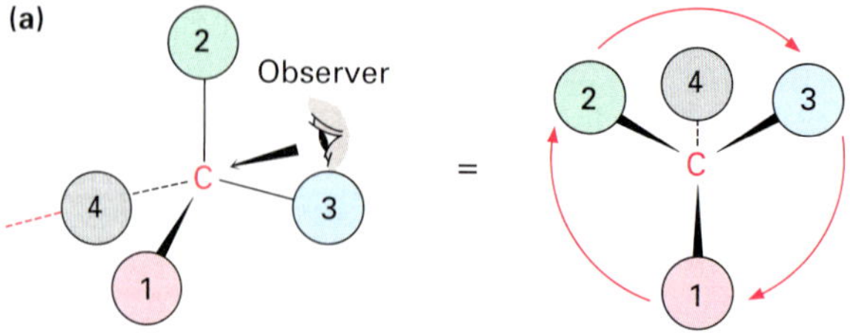

***R* configuration**

In **(b)**, you would be located behind the page toward the top left of the molecule from your point of view, and you would see group 3 to your left, group 1 to your right, and group 2 below you. This also corresponds to an *R* configuration.

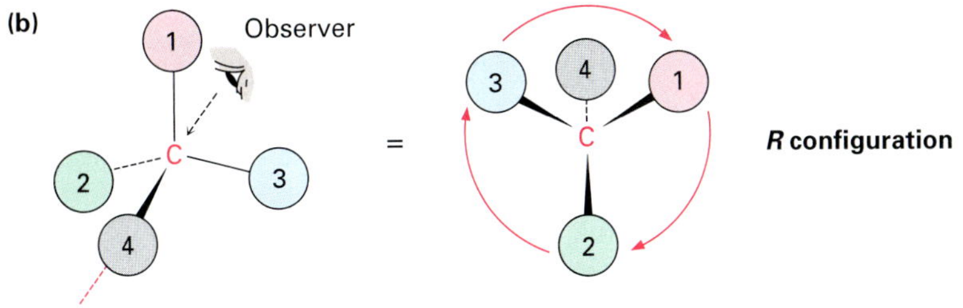

WORKED EXAMPLE 9.4 ***Drawing the Three-Dimensional Structure of a Specific Enantiomer***

Draw a tetrahedral representation of (*R*)-2-chlorobutane.

Strategy Begin by assigning priorities to the four substituents bonded to the chirality center: (1) −Cl, (2) $-CH_2CH_3$, (3) $-CH_3$, (4) −H. To draw a tetrahedral representation of the molecule, orient the lowest-priority −H group away from you and imagine that the other three groups are coming out of the page toward you. Then place the remaining three substituents such that the direction of travel 1 → 2 → 3 is clockwise (right turn), and tilt the molecule toward you to bring the rear hydrogen into view. Using molecular models is a great help in working problems of this sort.

Solution

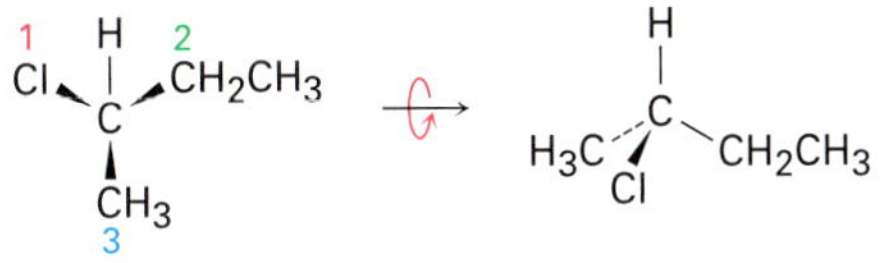

(*R*)-2-Chlorobutane

Problem 9.7 Assign priorities to the following sets of substituents:

(a) −H, −OH, $-CH_2CH_3$, $-CH_2CH_2OH$
(b) $-CO_2H$, $-CO_2CH_3$, $-CH_2OH$, −OH
(c) −CN, $-CH_2NH_2$, $-CH_2NHCH_3$, $-NH_2$
(d) −SH, $-CH_2SCH_3$, $-CH_3$, $-SSCH_3$

Problem 9.8 Orient each of the following drawings so that the lowest-priority group is toward the rear, and then assign *R* or *S* configuration:

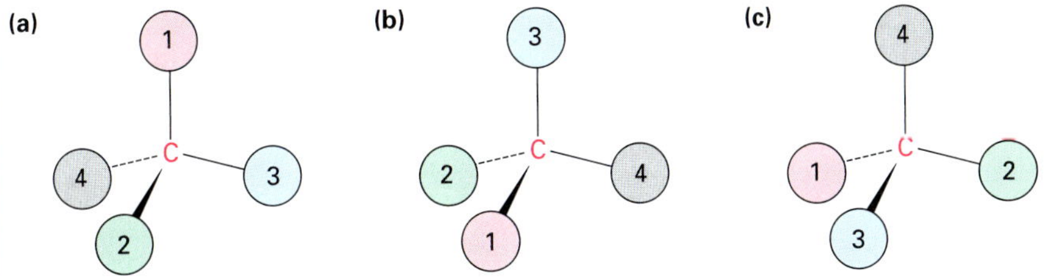

Problem 9.9 Assign *R* or *S* configuration to the chirality center in each of the following molecul

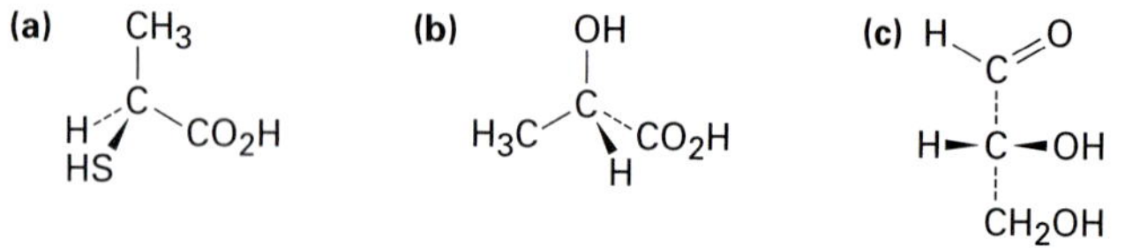

Problem 9.10 Draw a tetrahedral representation of (*S*)-2-pentanol (2-hydroxypentane).

Problem 9.11 Assign *R* or *S* configuration to the chirality center in the following molecular moc of the amino acid methionine (blue = N, yellow = S):

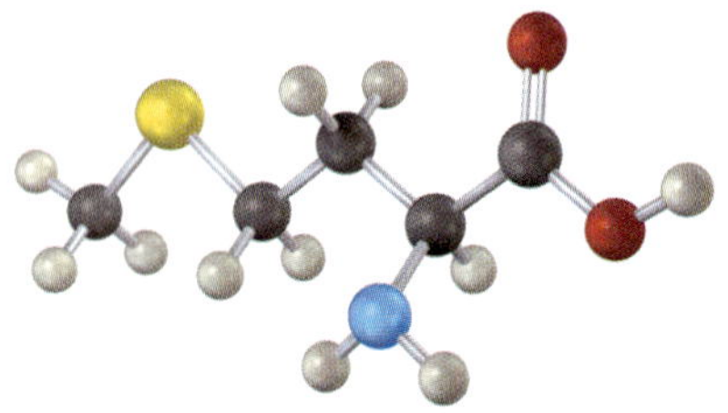

9.6 Diastereomers

Molecules like lactic acid, alanine, and glyceraldehyde are relatively simp because each has only one chirality center and only two stereoisomers. The si uation becomes more complex, however, with molecules that have more tha one chirality center. As a general rule, a molecule with *n* chirality centers ca have up to 2^n stereoisomers (although it may have fewer, as we'll see shortly Take the amino acid threonine (2-amino-3-hydroxybutanoic acid), for exampl Since threonine has two chirality centers (C2 and C3), there are four possibi stereoisomers, as shown in Figure 9.10. Check for yourself that the *R,S* configu rations are correct.

ThomsonNOW™ Click *Organic Interactive* to **use a web-based palette to draw stereoisomers.**

The four stereoisomers of 2-amino-3-hydroxybutanoic acid can be groupe into two pairs of enantiomers. The 2*R*,3*R* stereoisomer is the mirror image c 2*S*,3*S*, and the 2*R*,3*S* stereoisomer is the mirror image of 2*S*,3*R*. But what is th relationship between any two molecules that are not mirror images? What, fc example, is the relationship between the 2*R*,3*R* isomer and the 2*R*,3*S* isomer They are stereoisomers, yet they aren't enantiomers. To describe such a rela tionship, we need a new term—*diastereomer.*

Diastereomers are stereoisomers that are not mirror images. Since we use the right-hand/left-hand analogy to describe the relationship between tw enantiomers, we might extend the analogy by saying that the relationshi between diastereomers is like that of hands from different people. Your han and your friend's hand look *similar,* but they aren't identical and they aren't mi ror images. The same is true of diastereomers: they're similar, but they aren' identical and they aren't mirror images.

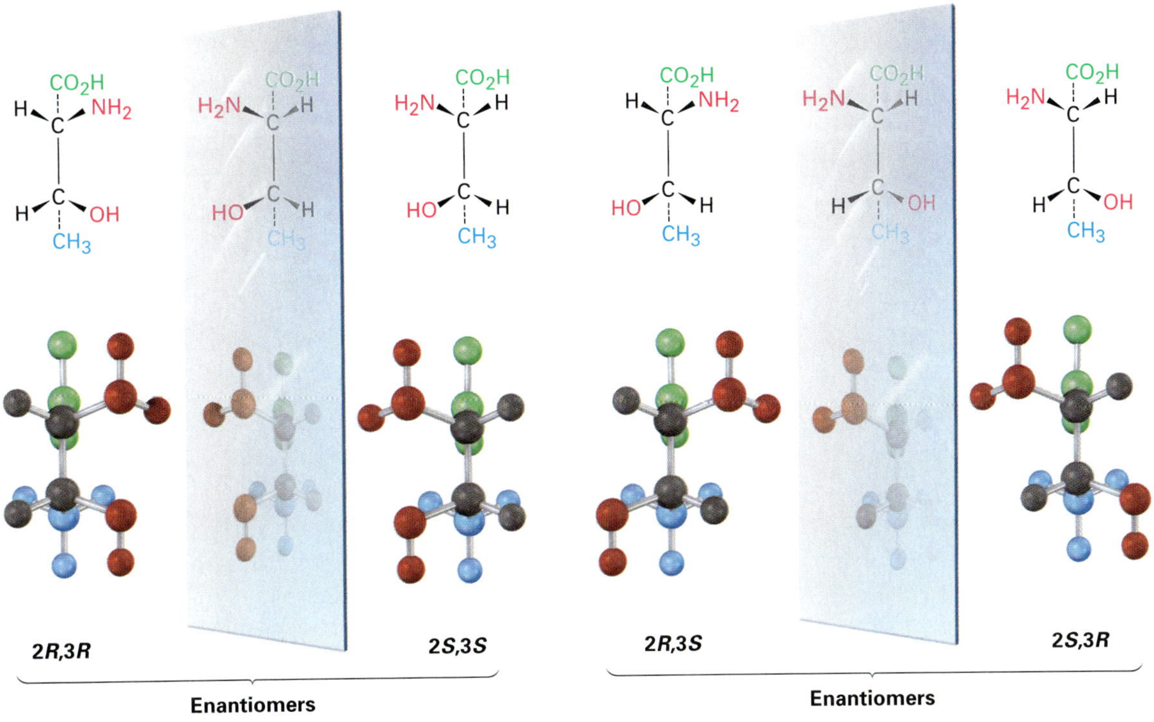

Figure 9.10 The four stereoisomers of 2-amino-3-hydroxybutanoic acid.

Note carefully the difference between enantiomers and diastereomers. Enantiomers have opposite configurations at *all* chirality centers, whereas diastereomers have opposite configurations at *some* (one or more) chirality centers but the same configuration at others. A full description of the four stereoisomers of threonine is given in Table 9.2. Of the four, only the 2*S*,3*R* isomer, $[\alpha]_D = -28.3$, occurs naturally in plants and animals and is an essential human nutrient. This result is typical: most biological molecules are chiral, and usually only one stereoisomer is found in nature.

Table 9.2 Relationships among the Four Stereoisomers of Threonine

Stereoisomer	Enantiomer	Diastereomer
2*R*,3*R*	2*S*,3*S*	2*R*,3*S* and 2*S*,3*R*
2*S*,3*S*	2*R*,3*R*	2*R*,3*S* and 2*S*,3*R*
2*R*,3*S*	2*S*,3*R*	2*R*,3*R* and 2*S*,3*S*
2*S*,3*R*	2*R*,3*S*	2*R*,3*R* and 2*S*,3*S*

In the special case where two diastereomers differ at only one chirality center but are the same at all others, we say that the compounds are **epimers**. Cholestanol and coprostanol, for instance, are both found in human feces and both have nine chirality centers. Eight of the nine are identical, but the one C5 is different. Thus, cholestanol and coprostanol are *epimeric* at C5.

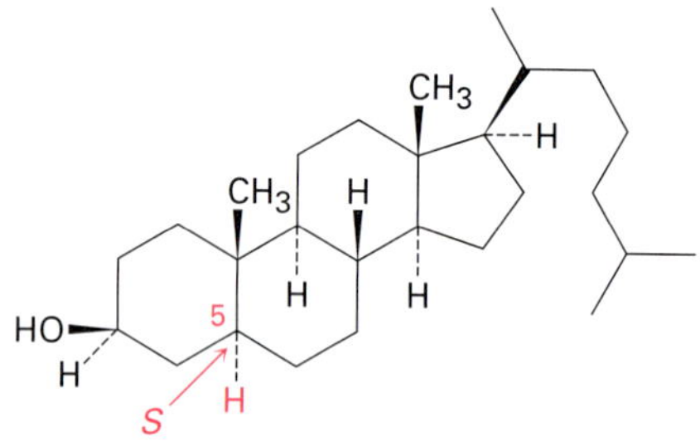

Cholestanol

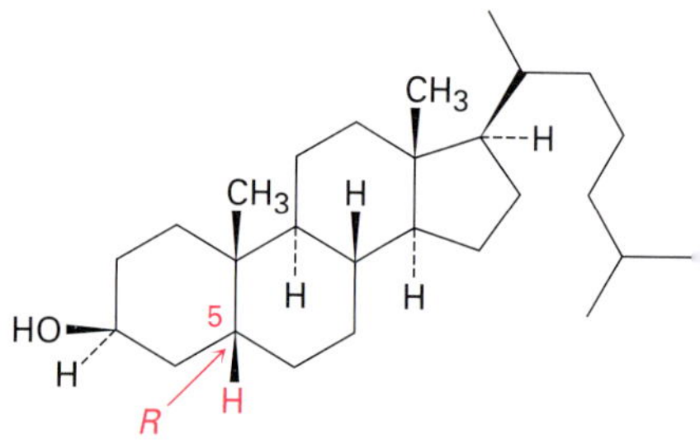

Coprostanol

Epimers

Problem 9.12 One of the following molecules **(a)**–**(d)** is D-erythrose 4-phosphate, an intermediat in the Calvin photosynthetic cycle by which plants incorporate CO_2 into carbc hydrates. If D-erythrose 4-phosphate has *R* stereochemistry at both chirality center which of the structures is it? Which of the remaining three structures is the enar tiomer of D-erythrose 4-phosphate, and which are diastereomers?

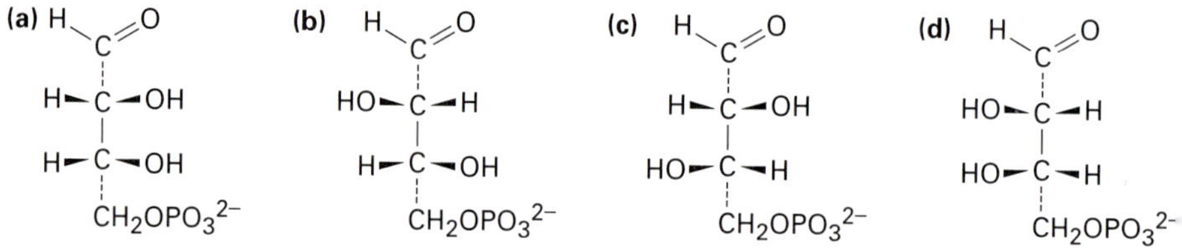

Problem 9.13 Chloramphenicol, a powerful antibiotic isolated in 1949 from the *Streptomyce venezuelae* bacterium, is active against a broad spectrum of bacterial infections and i particularly valuable against typhoid fever. Assign *R,S* configurations to the chirality centers in chloramphenicol.

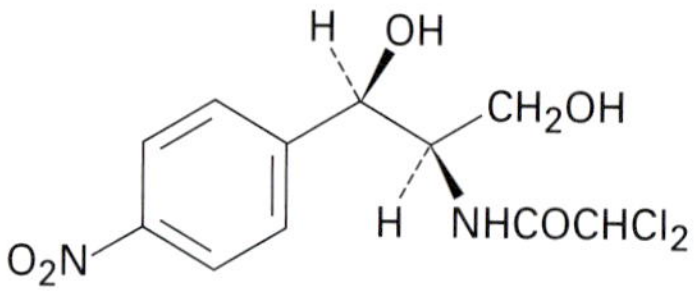

Chloramphenicol

Problem 9.14 Assign *R,S* configuration to each chirality center in the following molecular mode of the amino acid isoleucine (blue = N):

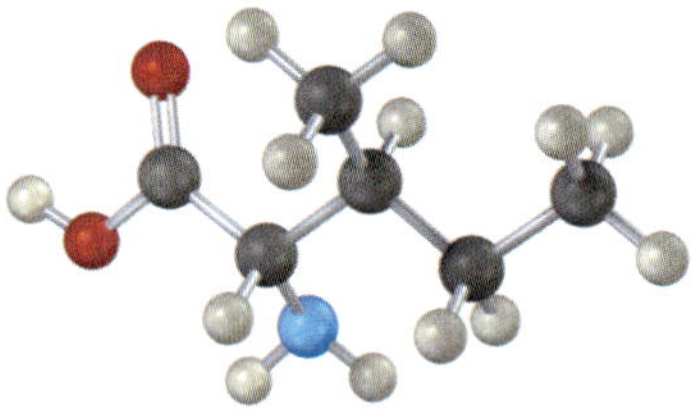

9.7 Meso Compounds

Let's look at one more example of a compound with more than one chirality center, the tartaric acid used by Pasteur. The four stereoisomers can be drawn as follows:

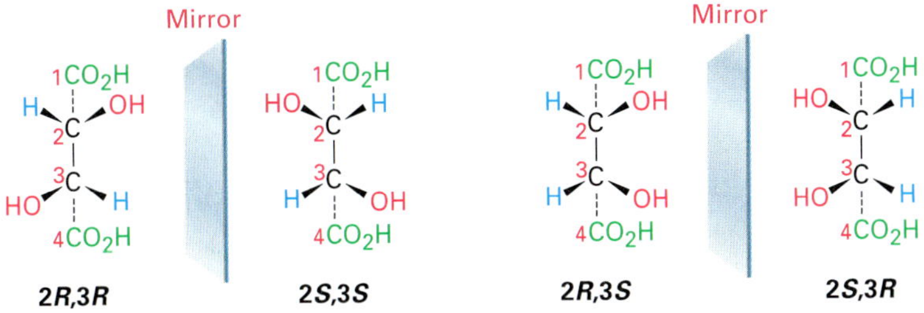

The mirror-image 2*R*,3*R* and 2*S*,3*S* structures are not identical and therefore represent a pair of enantiomers. A close look, however, shows that the 2*R*,3*S* and 2*S*,3*R* structures *are* identical, as can be seen by rotating one structure 180°.

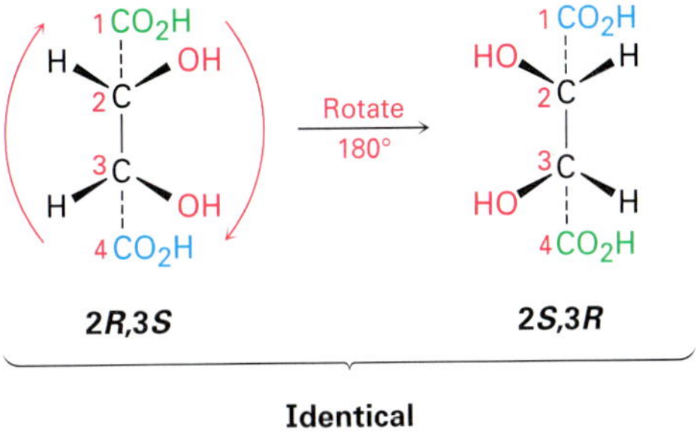

The 2*R*,3*S* and 2*S*,3*R* structures are identical because the molecule has a plane of symmetry and is therefore achiral. The symmetry plane cuts through the C2–C3 bond, making one half of the molecule a mirror image of the other half (Figure 9.11). Because of the plane of symmetry, the molecule is achiral, despite the fact that it has two chirality centers. Compounds that are achiral, yet contain chirality centers, are called **meso** (**me**-zo) **compounds**. Thus, tartaric acid exists in three stereoisomeric forms: two enantiomers and one meso form.

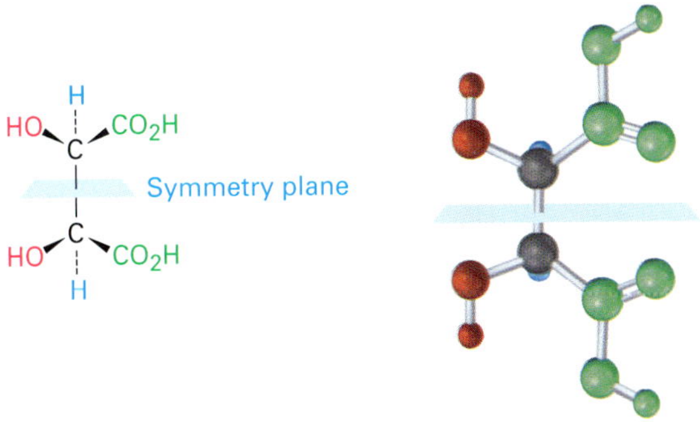

Figure 9.11 A symmetry plane through the C2–C3 bond of *meso*-tartaric acid makes the molecule achiral.

Some physical properties of the three stereoisomers are listed in Table 9. The (+)- and (−)-tartaric acids have identical melting points, solubilities, ar densities but differ in the sign of their rotation of plane-polarized light. Th meso isomer, by contrast, is diastereomeric with the (+) and (−) forms. As suc it has no mirror-image relationship to (+)- and (−)-tartaric acids, is a differe compound altogether, and has different physical properties.

Table 9.3 Some Properties of the Stereoisomers of Tartaric Acid

Stereoisomer	Melting point (°C)	$[\alpha]_D$	Density (g/cm^3)	Solubility at 20 °C ($g/100\ mL\ H_2O$)
(+)	168–170	+12	1.7598	139.0
(−)	168–170	−12	1.7598	139.0
Meso	146–148	0	1.6660	125.0

WORKED EXAMPLE 9.5

Distinguishing Chiral Compounds from Meso Compounds

Does *cis*-1,2-dimethylcyclobutane have any chirality centers? Is it chiral?

Strategy To see whether a chirality center is present, look for a carbon atom bonded to fou different groups. To see whether the molecule is chiral, look for the presence c absence of a symmetry plane. Not all molecules with chirality centers are chira overall—meso compounds are an exception.

Solution A look at the structure of *cis*-1,2-dimethylcyclobutane shows that both methyl bearing ring carbons (C1 and C2) are chirality centers. Overall, though, the com pound is achiral because there is a symmetry plane bisecting the ring betwee C1 and C2. Thus, the molecule is a meso compound.

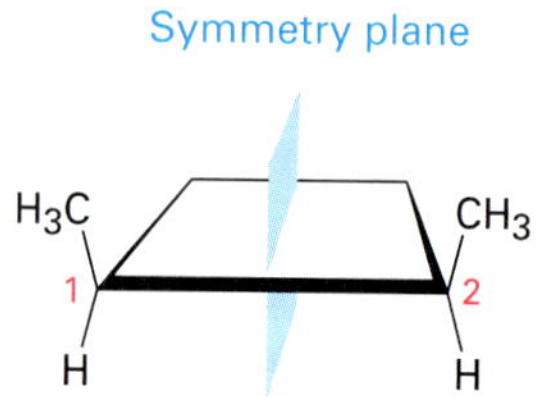

Problem 9.15 Which of the following structures represent meso compounds?

(a)
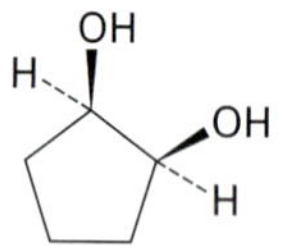

(b)
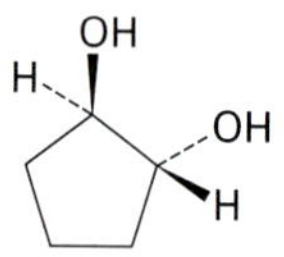

(c)
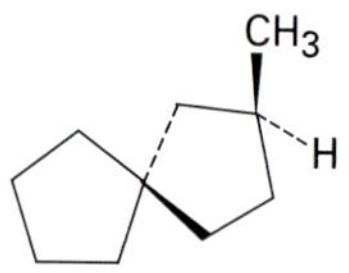

(d)
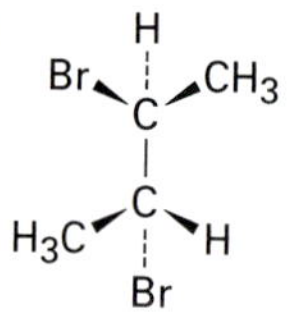

Problem 9.16 Which of the following have a meso form? (Recall that the *-ol* suffix refers to an alco- hol, ROH.)

(a) 2,3-Butanediol **(b)** 2,3-Pentanediol **(c)** 2,4-Pentanediol

Problem 9.17 Does the following structure represent a meso compound? If so, indicate the symmetry plane.

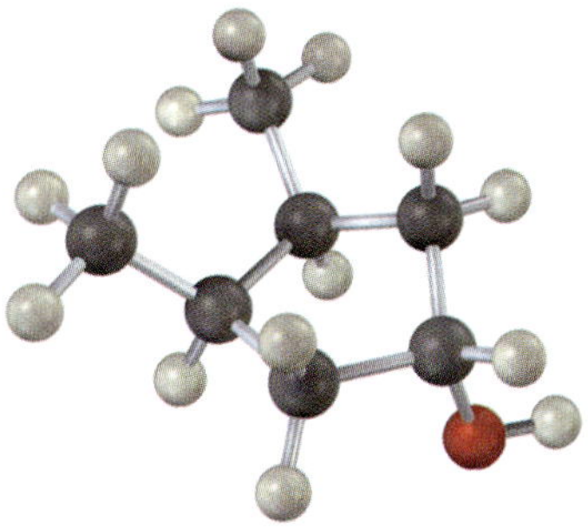

9.8 Racemic Mixtures and the Resolution of Enantiomers

Let's return for a last look at Pasteur's pioneering work. Pasteur took an optically inactive tartaric acid salt and found that he could crystallize from it two optically active forms having what we would now call the 2*R*,3*R* and 2*S*,3*S* configurations. But what was the optically inactive form he started with? It couldn't have been *meso*-tartaric acid, because *meso*-tartaric acid is a different chemical compound and can't interconvert with the two chiral enantiomers without breaking and re-forming chemical bonds.

The answer is that Pasteur started with a 50 : 50 mixture of the two chiral tartaric acid enantiomers. Such a mixture is called a **racemic** (ray-**see**-mic) **mixture**, or *racemate*, and is denoted either by the symbol (±) or the prefix *d,l* to indicate an equal mixture of dextrorotatory and levorotatory forms. Racemic mixtures show no optical rotation because the (+) rotation from one enantiomer exactly cancels the (−) rotation from the other. Through luck, Pasteur was able to separate, or **resolve**, racemic tartaric acid into its (+) and (−) enantiomers. Unfortunately, the fractional crystallization technique he used doesn't work for most racemic mixtures, so other methods are needed.

The most common method of resolution uses an acid–base reaction between a racemic mixture of chiral carboxylic acids (RCO_2H) and an amine base (RNH_2) to yield an ammonium salt.

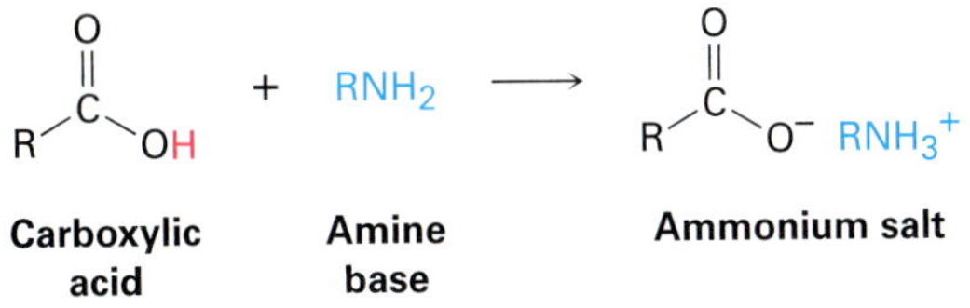

To understand how this method of resolution works, let's see what happens when a racemic mixture of chiral acids, such as (+)- and (−)-lactic acids, reacts with an achiral amine base, such as methylamine, CH_3NH_2. Stereochemically, the situation is analogous to what happens when left and right hands (chiral) pick up a ball (achiral). Both left and right hands pick up the ball equally well, and the products—ball in right hand versus ball in left hand—are mirror images. In the same way, both (+)- and (−)-lactic acid react with methylamine equally well, and the product is a racemic mixture of methylammonium (+)-lactate a methylammonium (−)-lactate (Figure 9.12).

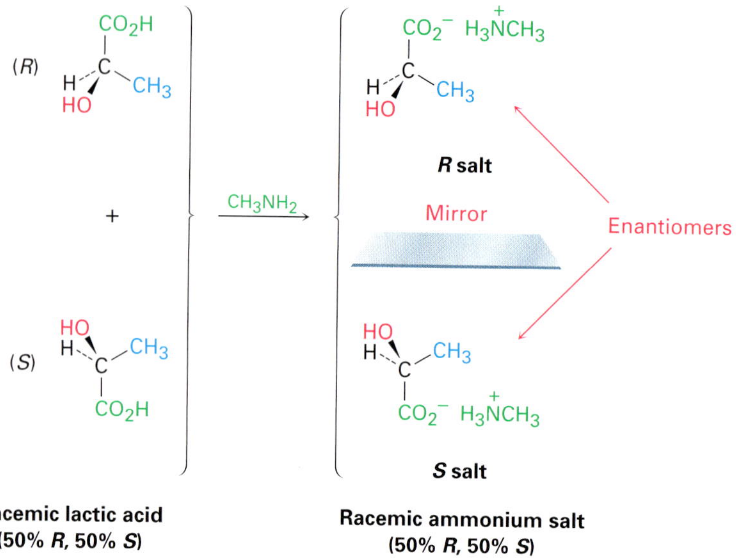

Figure 9.12 Reaction of racemic lactic acid with achiral methylamine leads to a racemic mixture of ammonium salts.

Now let's see what happens when the racemic mixture of (+)- and (−)-lact acids reacts with a single enantiomer of a chiral amine base, such as (*R*)-1-pheny ethylamine. Stereochemically, the situation is analogous to what happens whe left and right hands (chiral) put on a right-handed glove (*also chiral*). Left an right hands don't put on the same glove in the same way. The products—righ hand in right glove versus left hand in right glove—are not mirror image they're altogether different.

In the same way, (+)- and (−)-lactic acids react with (*R*)-1-phenylethylamin to give two different products (Figure 9.13). (*R*)-Lactic acid reacts wit

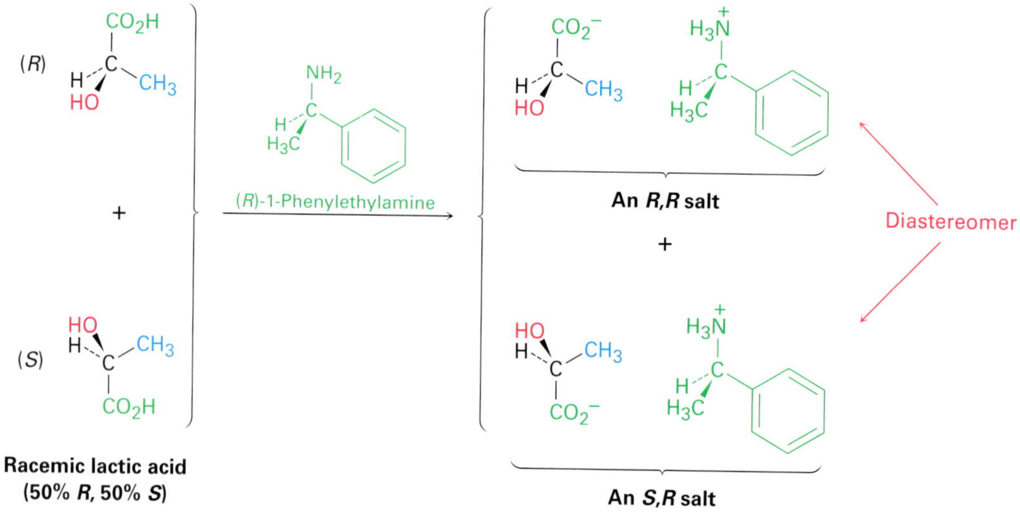

Figure 9.13 Reaction of racemic lactic acid with (*R*)-1-phenylethylamine yields a mixture of diastereomeric ammonium salts.

(*R*)-1-phenylethylamine to give the *R*,*R* salt, and (*S*)-lactic acid reacts with the *R* amine to give the *S*,*R* salt. The two salts are diastereomers; they are different compounds, with different chemical and physical properties. It may therefore be possible to separate them by crystallization or some other means. Once separated, acidification of the two diastereomeric salts with a strong acid then allows us to isolate the two pure enantiomers of lactic acid and to recover the chiral amine for reuse.

WORKED EXAMPLE 9.6

Predicting the Chirality of a Product

We'll see in Section 21.3 that carboxylic acids (RCO_2H) react with alcohols (R′OH) to form esters (RCO_2R'). Suppose that (±)-lactic acid reacts with CH_3OH to form the ester, methyl lactate. What stereochemistry would you expect the product(s) to have? What is the relationship of the products?

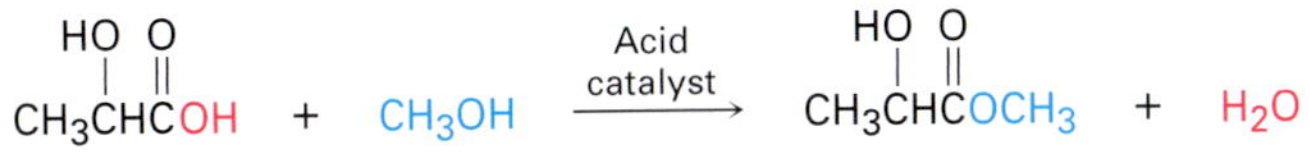

Lactic acid **Methanol** **Methyl lactate**

Solution Reaction of a racemic acid with an achiral alcohol such as methanol yields a racemic mixture of mirror-image (enantiomeric) products.

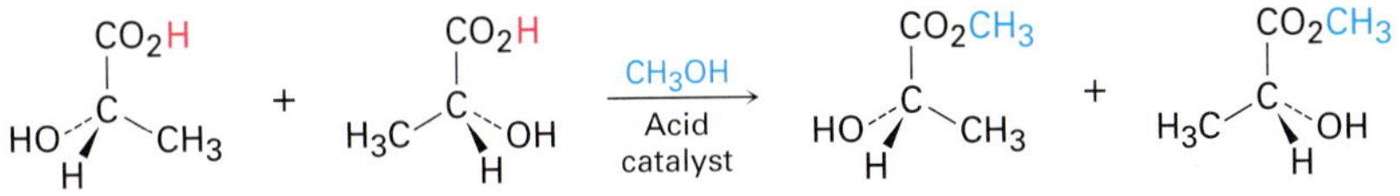

(*S*)-Lactic acid **(*R*)-Lactic acid** **Methyl (*S*)-lactate** **Methyl (*R*)-lactate**

Problem 9.18 Suppose that acetic acid (CH_3CO_2H) reacts with (*S*)-2-butanol to form an ester (see Worked Example 9.6). What stereochemistry would you expect the product(s) to have? What is the relationship of the products?

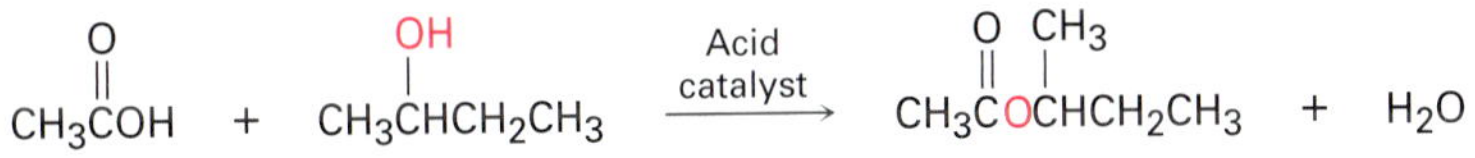

Acetic acid **2-Butanol** ***sec*-Butyl acetate**

Problem 9.19 What stereoisomers would result from reaction of (±)-lactic acid with (*S*)-1-phenylethylamine, and what is the relationship between them?

9.9 A Review of Isomerism

As noted on several previous occasions, isomers are compounds that have the same chemical formula but different structures. We've seen several kinds of isomers in the past few chapters, and it's a good idea at this point to see how they relate to one another (Figure 9.14).

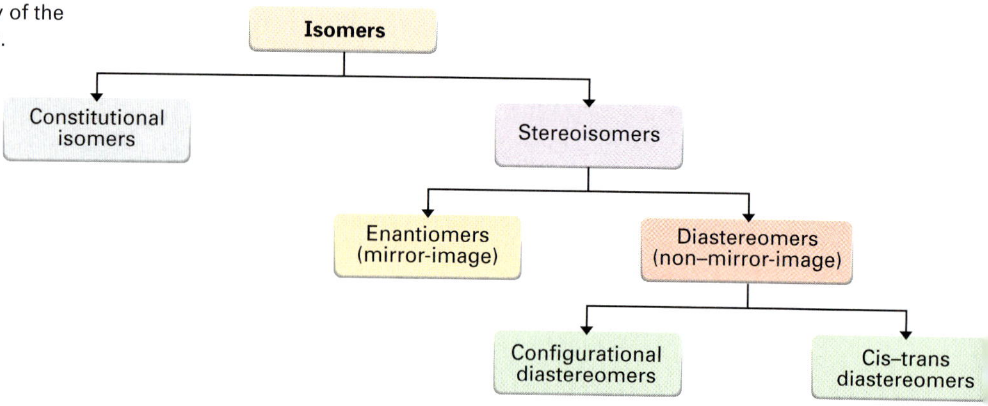

Figure 9.14 A summary of the different kinds of isomers.

There are two fundamental types of isomers, both of which we've no encountered: constitutional isomers and stereoisomers.

- **Constitutional isomers** (Section 3.2) are compounds whose atoms are cor nected differently. Among the kinds of constitutional isomers we've seen ar skeletal, functional, and positional isomers.

Different carbon skeletons

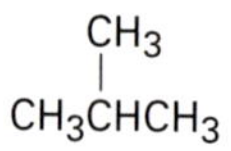

2-Methylpropane and $CH_3CH_2CH_2CH_3$ **Butane**

Different functional groups

CH_3CH_2OH **Ethyl alcohol** and CH_3OCH_3 **Dimethyl ether**

Different position of functional groups

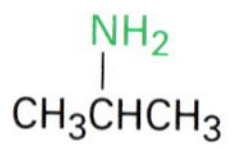

Isopropylamine and $CH_3CH_2CH_2NH_2$ **Propylamine**

- **Stereoisomers** (Section 4.2) are compounds whose atoms are connected ir the same order but with a different geometry. Among the kinds of stereoisomer we've seen are enantiomers, diastereomers, and cis–trans isomers (both ir alkenes and in cycloalkanes). Actually, cis–trans isomers are just another kind o diastereomers because they are non–mirror-image stereoisomers.

Enantiomers (nonsuperimposable mirror-image stereoisomers)

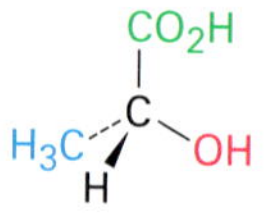

(*R*)-Lactic acid

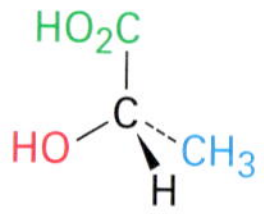

(*S*)-Lactic acid

Diastereomers (nonsuperimposable, non–mirror-image stereoisomers)

Configurational diastereomers

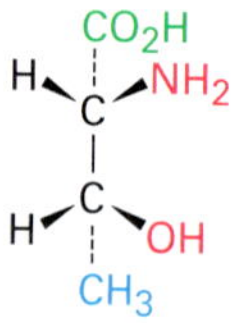

2*R*,3*R*-2-Amino-3-hydroxybutanoic acid

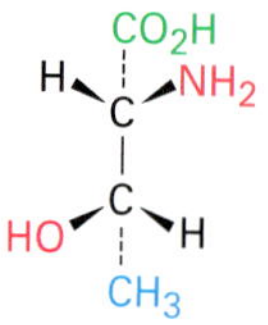

2*R*,3*S*-2-Amino-3-hydroxybutanoic acid

Cis–trans diastereomers (substituents on same side or opposite side of double bond or ring)

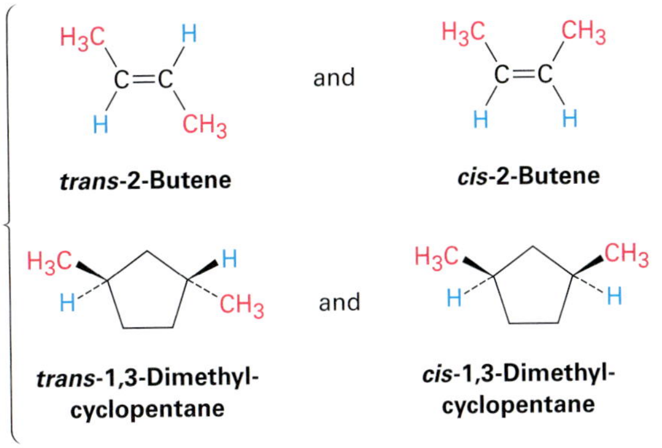

Problem 9.20 What kinds of isomers are the following pairs?

(a) (*S*)-5-Chloro-2-hexene and chlorocyclohexane

(b) (2*R*,3*R*)-Dibromopentane and (2*S*,3*R*)-dibromopentane

9.10 Stereochemistry of Reactions: Addition of H_2O to an Achiral Alkene

omsonNOW™ Click *Organic* *nteractive* to **predict the products nd stereochemistry of alkene ddition reactions.**

Most of the biochemical reactions that take place in the body, as well as many organic reactions in the laboratory, yield products with chirality centers. For example, acid-catalyzed addition of H_2O to 1-butene in the laboratory yields 2-butanol, a chiral alcohol. What is the stereochemistry of this chiral product? If a single enantiomer is formed, is it *R* or *S*? If a mixture of enantiomers is formed, how much of each? In fact, the 2-butanol produced is a racemic mixture of *R* and *S* enantiomers. Let's see why.

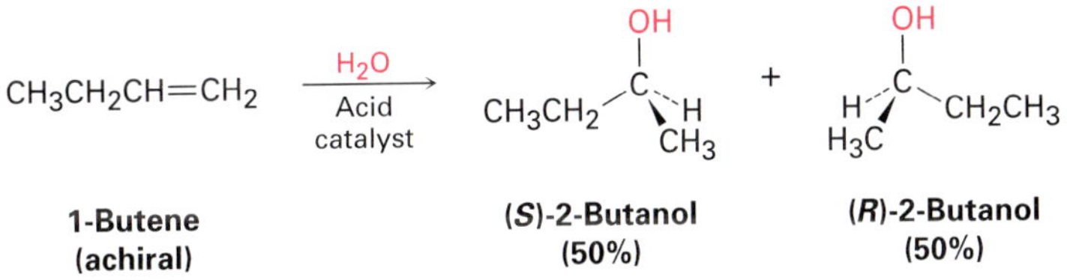

To understand why a racemic product results from the reaction of H_2O with 1-butene, think about the reaction mechanism. 1-Butene is first protonated to yield an intermediate secondary (2°) carbocation. Since the trivalent carbon is sp^2-hybridized and planar, the cation has no chirality centers, has a plane of symmetry, and is achiral. As a result, it can react with H_2O equally well from either the top or the bottom. Reaction from the top leads to (*S*)-2-butanol through transition state 1 (TS 1) in Figure 9.15, and reaction from the bottom leads to *R* product through TS 2. *The two transition states are mirror images.* They therefore have identical energies, form at identical rates, and are equally likely to occur.

As a general rule, formation of a new chirality center by reaction between two achiral reactants always leads to a racemic mixture of enantiomeric products. Put another way, optical activity can't appear from nowhere. An optically active product can only result by starting with an optically active reactant or environment.

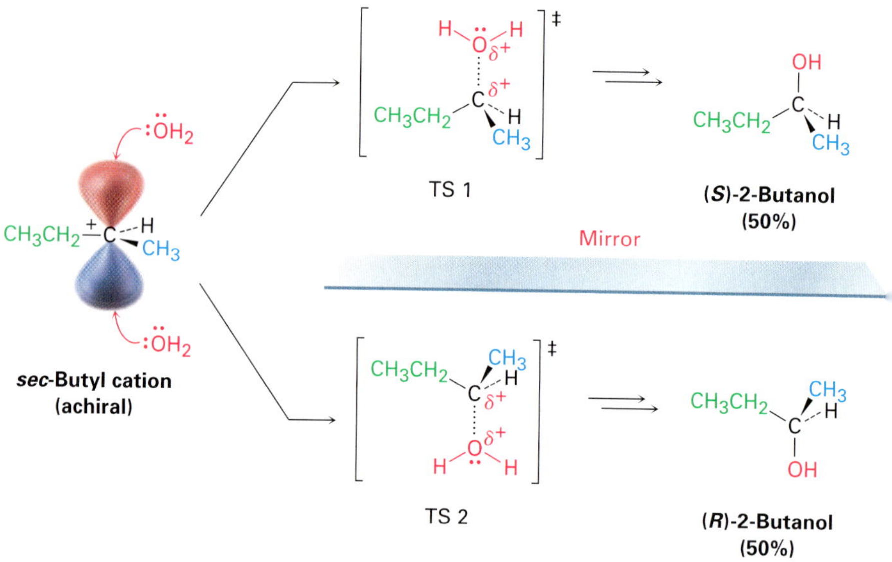

Figure 9.15 Reaction of H_2O with the *sec*-butyl carbocation. Reaction from the top leads to *S* product and is the mirror image of reaction from the bottom, which leads to *R* product. Since both are equally likely, a racemic mixture of products is formed. The dotted C···O bond in the transition state indicates partial bond formation.

In contrast to laboratory reactions, enzyme-catalyzed reactions often give a single enantiomer of a chiral product, even when the substrate is achiral. One step in the citric acid cycle of food metabolism, for instance, is the aconitase-catalyzed addition of water to (*Z*)-aconitate (usually called *cis*-aconitate) to give isocitrate.

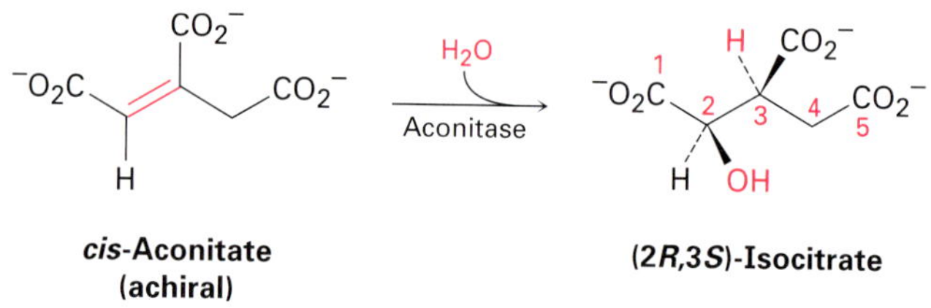

Even though the *cis*-aconitate substrate is achiral, only the (2*R*,3*S*) enantiomer of the product is formed. We'll look at the reason for this stereospecificity in Section 9.14.

9.11 Stereochemistry of Reactions: Addition of H_2O to a Chiral Alkene

The reaction discussed in the previous section involves addition to an achiral alkene and forms an optically inactive, racemic mixture of the two enantiomeric products. What would happen, though, if we were to carry out the reaction on a *single* enantiomer of a *chiral* reactant? For example, what stereochemical result would be obtained from addition of H_2O to a chiral alkene, such as

(*R*)-4-methyl-1-hexene? The product of the reaction, 4-methyl-2-hexanol, has two chirality centers and so has four possible stereoisomers.

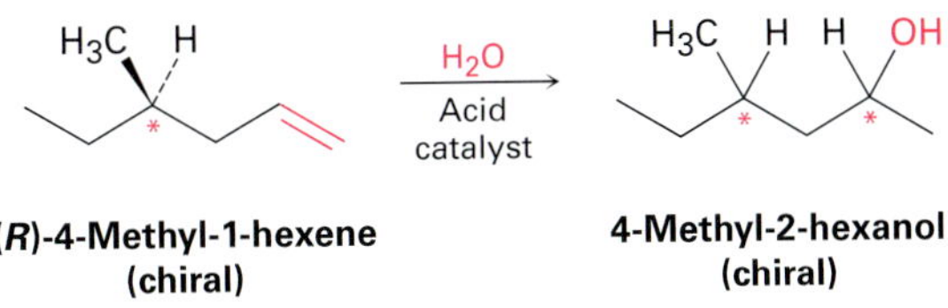

Let's think about the two chirality centers separately. What about the configuration at C4, the methyl-bearing carbon atom? Since C4 has the *R* configuration in the starting material and this chirality center is unaffected by the reaction, its configuration is unchanged. Thus, the configuration at C4 in the product remains *R* (assuming that the relative priorities of the four attached groups are not changed by the reaction).

What about the configuration at C2, the newly formed chirality center? As illustrated in Figure 9.16, the stereochemistry at C2 is established by reaction of H_2O with a carbocation intermediate in the usual manner. But this carbocation does not have a plane of symmetry; it is chiral because of the chirality center at C4. Because the carbocation has no plane of symmetry, it does not react equally well from top and bottom faces. One of the two faces is likely, for steric reasons, to be a bit more accessible than the other face, leading to a mixture of *R* and *S* products in some ratio other than 50:50. Thus, two diastereomeric products, (2*R*,4*R*)-4-methyl-2-hexanol and (2*S*,4*R*)-4-methyl-2-hexanol, are formed in unequal amounts, and the mixture is optically active.

As a general rule, the reaction of a chiral reactant with an achiral reactant leads to unequal amounts of diastereomeric products. If the chiral reactant is optically active because only one enantiomer is used rather than a racemic mixture, then the products are also optically active.

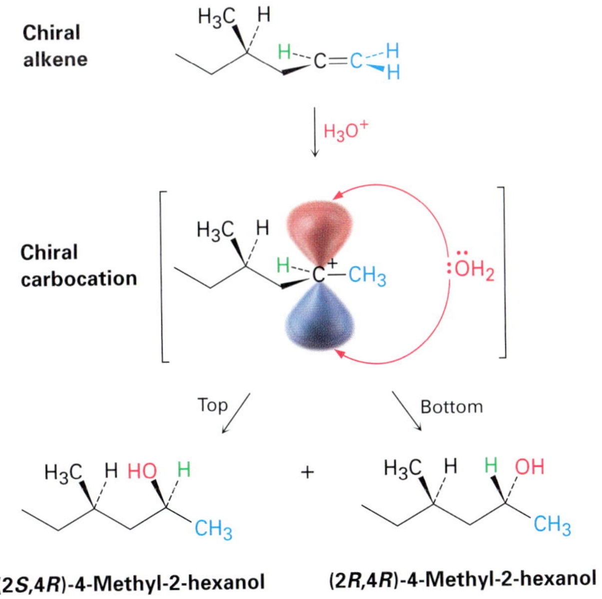

Figure 9.16 Stereochemistry of the addition of H_2O to the chiral alkene, (*R*)-4-methyl-1-hexene. A mixture of diastereomeric 2*R*,4*R* and 2*S*,4*R* products is formed in unequal amounts because reaction of the chiral carbocation intermediate is not equally likely from top and bottom. The product mixture is optically active.

Problem 9.21 What products are formed from acid-catalyzed hydration of racemic (±)-4-methy 1-hexene? What can you say about the relative amounts of the products? Is the proc uct mixture optically active?

Problem 9.22 What products are formed from hydration of 4-methylcyclopentene? What can yo say about the relative amounts of the products?

9.12 Chirality at Nitrogen, Phosphorus, and Sulfur

The most common cause of chirality is the presence of four different sub stituents bonded to a tetrahedral atom, but that atom doesn't necessaril have to be carbon. Nitrogen, phosphorus, and sulfur are all commonl encountered in organic molecules, and all can be chirality centers. We know for instance, that trivalent nitrogen is tetrahedral, with its lone pair of elec trons acting as the fourth "substituent" (Section 1.10). Is trivalent nitroger chiral? Does a compound such as ethylmethylamine exist as a pair o enantiomers?

The answer is both yes and no. Yes in principle, but no in practice. Triva lent nitrogen compounds undergo a rapid umbrella-like inversion that inter converts enantiomers. We therefore can't isolate individual enantiomers except in special cases.

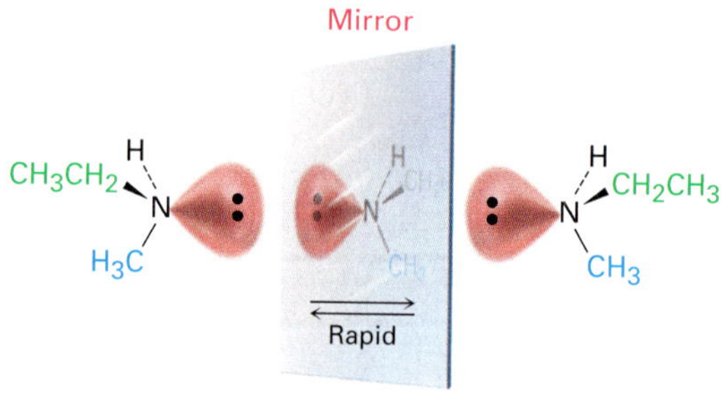

A similar situation occurs in trivalent phosphorus compounds, or *phosphines*. It turns out, though, that inversion at phosphorus is substantially slower than inversion at nitrogen, so stable chiral phosphines *can* be isolated. (*R*)- and (*S*)-methylpropylphenylphosphine, for example, are configurationally stable for several hours at 100 °C. We'll see the importance of phosphine chirality in Section 26.7 in connection with the synthesis of chiral amino acids.

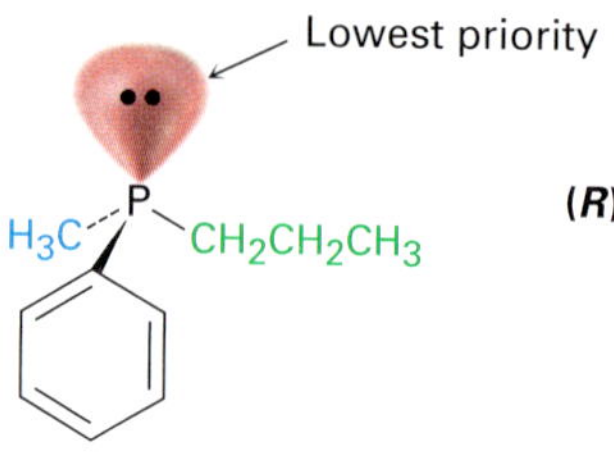

(*R*)-Methylpropylphenylphosphine (configurationally stable)

Divalent sulfur compounds are achiral, but trivalent sulfur compounds called *sulfonium salts* (R_3S^+) can be chiral. Like phosphines, sulfonium salts undergo relatively slow inversion, so chiral sulfonium salts are configurationally stable and can be isolated. The best known example is the coenzyme *S*-adenosylmethionine, the so-called biological methyl donor, which is involved in many metabolic pathways as a source of CH_3 groups. (The "*S*" in the name *S*-adenosylmethionine stands for *sulfur* and means that the adenosyl group is attached to the sulfur atom of methionine.) The molecule has *S* stereochemistry at sulfur and is configurationally stable for several days at room temperature. Its *R* enantiomer is also known but has no biological activity.

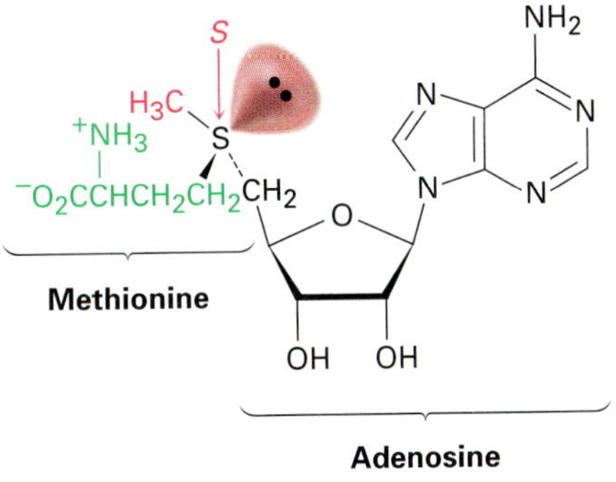

(*S*)-*S*-Adenosylmethionine

9.13 Prochirality

Closely related to the concept of chirality, and particularly important in biological chemistry, is the notion of *prochirality.* A molecule is said to be **prochiral** if can be converted from achiral to chiral in a single chemical step. For instance, an unsymmetrical ketone like 2-butanone is prochiral because it can be converted to the chiral alcohol 2-butanol by addition of hydrogen, as we'll see in Section 17.4.

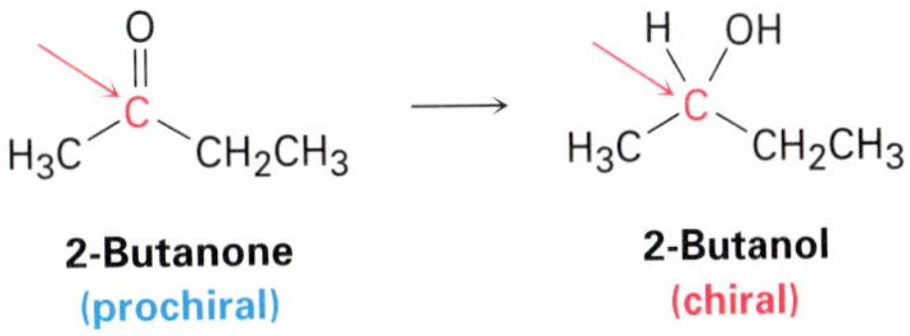

Which enantiomer of 2-butanol is produced depends on which face of the planar carbonyl carbon undergoes reaction. To distinguish between the possibilities, we use the stereochemical descriptors *Re* and *Si*. Assign priorities to the three groups attached to the trigonal, sp^2-hybridized carbon, and imagine curved arrows from the highest to second-highest to third-highest priority substituents. The face on which the arrows curve clockwise is designated ***Re*** (similar to *R*), and the face on which the arrows curve counterclockwise is designated ***Si*** (similar to *S*). In this particular example, addition of hydrogen from the *Re* faces gives (*S*)-2-butanol, and addition from the *Si* face giv(*R*)-2-butanol.

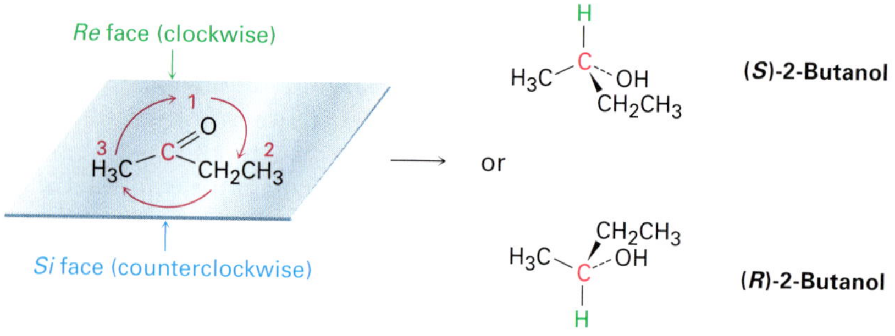

In addition to compounds with planar, sp^2-hybridized carbons, compound with tetrahedral, sp^3-hybridized atoms can also be prochiral. An sp^3-hybridize atom is said to be a **prochirality center** if, by changing one of its attache groups, it becomes a chirality center. The $-CH_2OH$ carbon atom of ethanol, fc instance, is a prochirality center because changing one of its attached $-H$ atom converts it into a chirality center.

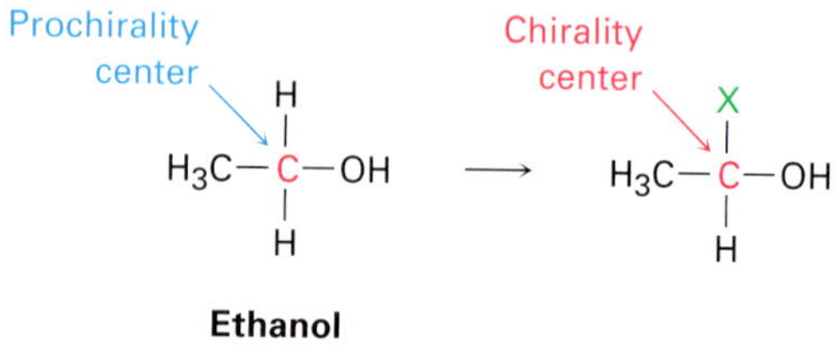

Ethanol

To distinguish between the two identical atoms (or groups of atoms) on a prochirality center, we imagine a change that will raise the priority of one atom over the other without affecting its priority with respect to other attached groups. On the $-CH_2OH$ carbon of ethanol, for instance, we might imagine replacing one of the ^1H atoms (protium) by ^2H (deuterium). The newly introduced ^2H atom is higher in priority than the remaining ^1H atom but remains lower in priority than other groups attached to the carbon. Of the two identical atoms in the original compound, that atom whose replacement leads to an *R* chirality center is said to be ***pro-R*** and that atom whose replacement leads to an *S* chirality center is ***pro-S***.

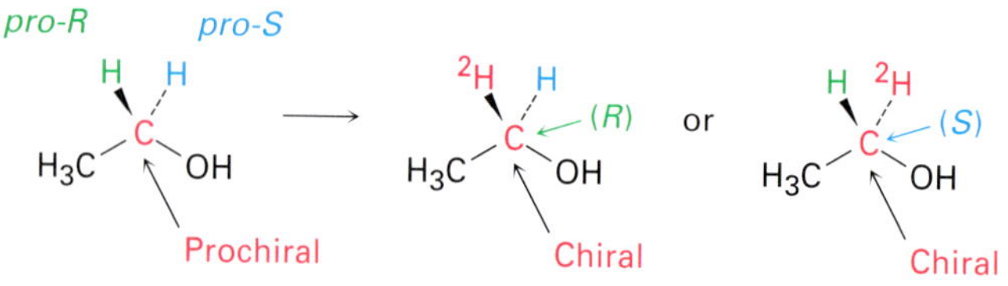

A large number of biological reactions involve prochiral compounds. One of the steps in the citric acid cycle by which food is metabolized, for instance, is the addition of H_2O to fumarate to give malate. Addition of –OH occurs on the *Si* face of a fumarate carbon and gives (*S*)-malate as product.

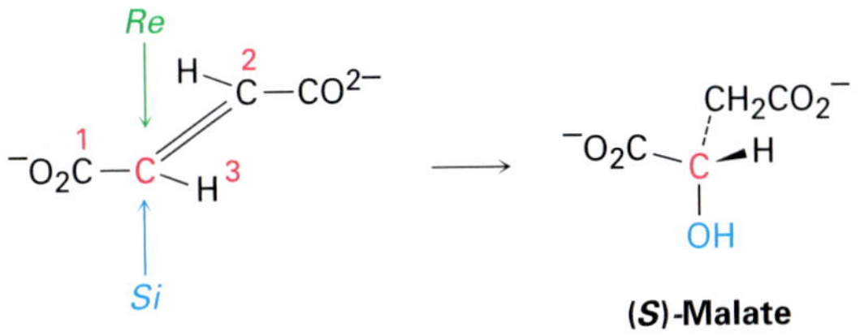

As another example, studies with deuterium-labeled substrates have shown that the reaction of ethanol with the coenzyme NAD^+ catalyzed by yeast alcohol dehydrogenase occurs with exclusive removal of the *pro-R* hydrogen from ethanol and with addition only to the *Re* face of NAD^+.

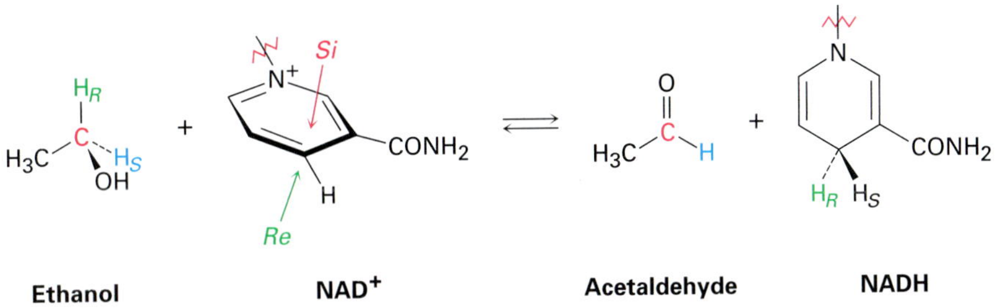

Elucidating the stereochemistry of reaction at prochirality centers is a powerful method for studying detailed mechanisms in biochemical reactions. As just one example, the conversion of citrate to (*cis*)-aconitate in the citric acid cycle has been shown to occur with loss of a *pro-R* hydrogen, implying that the reaction takes place by an anti elimination mechanism. That is, the OH and H groups leave from opposite sides of the molecule.

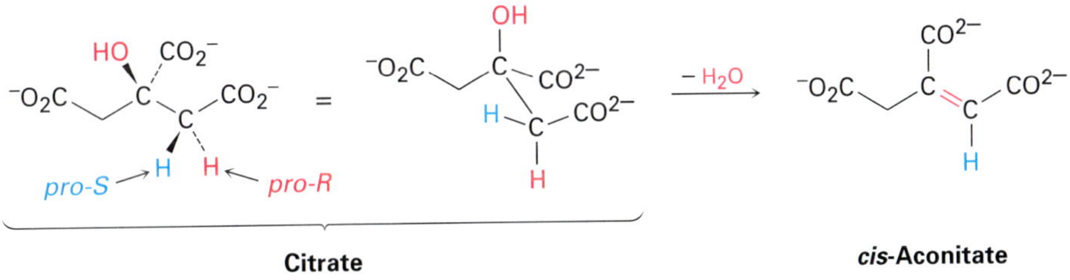

Problem 9.23 Identify the indicated hydrogens in the following molecules as *pro-R* or *pro-S*:

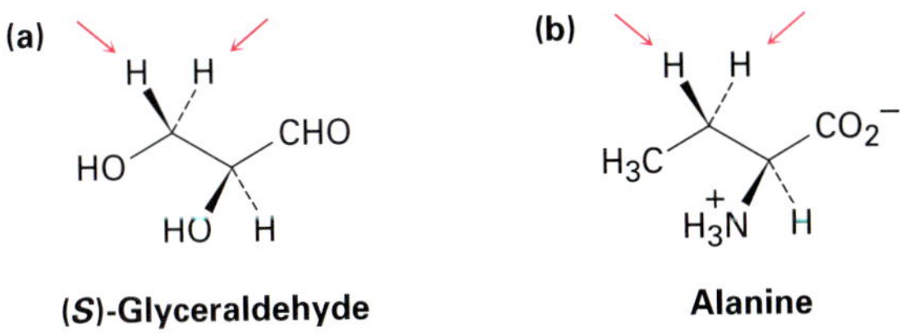

Problem 9.24 Identify the indicated faces of carbon atoms in the following molecules as *Re* or *Si*

(a)

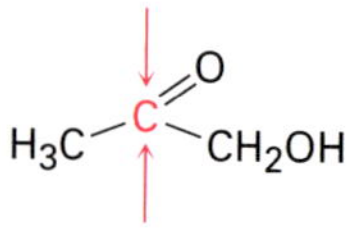

Hydroxyacetone

(b)

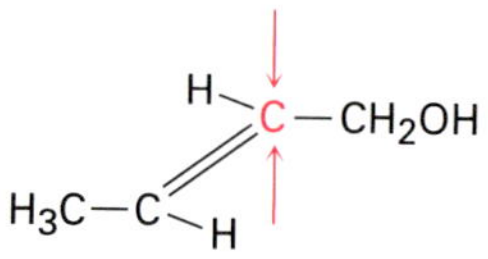

Crotyl alcohol

Problem 9.25 Lactic acid buildup in tired muscles results from reduction of pyruvate. If the reaction occurs from the *Re* face, what is the stereochemistry of the product?

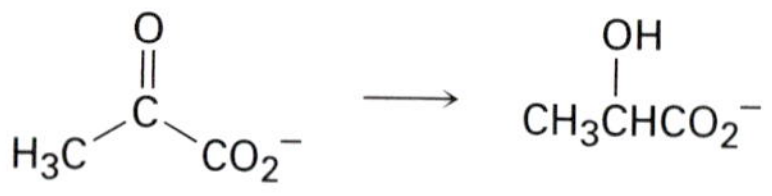

Pyruvate **Lactate**

Problem 9.26 The aconitase-catalyzed addition of water to *cis*-aconitate in the citric acid cycle occurs with the following stereochemistry. Does the addition of the OH group occur on the *Re* or the *Si* face of the substrate? What about the addition of the H? Does the reaction have syn or anti stereochemistry?

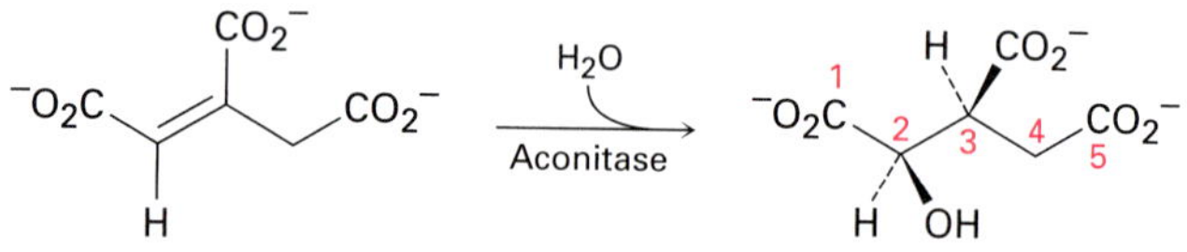

***cis*-Aconitate** **(2*R*,3*S*)-Isocitrate**

9.14 Chirality in Nature and Chiral Environments

Although the different enantiomers of a chiral molecule have the same physical properties, they usually have different biological properties. For example, the (+) enantiomer of limonene has the odor of oranges, but the (−) enantiomer has the odor of pine trees.

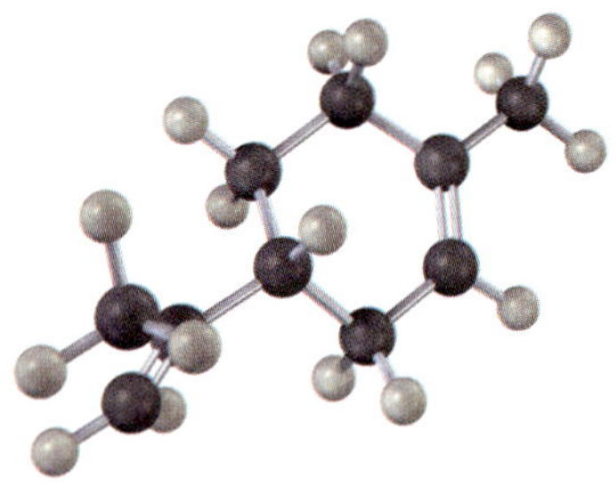

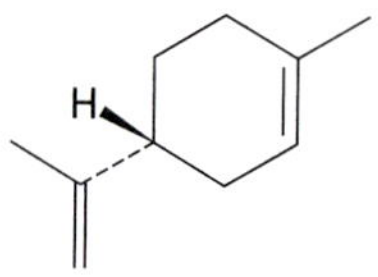

(+)-Limonene (in citrus fruits)

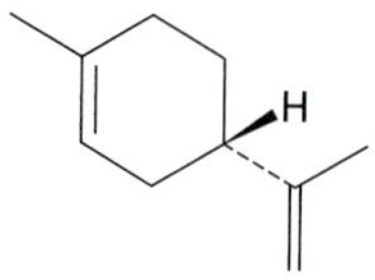

(−)-Limonene (in pine trees)

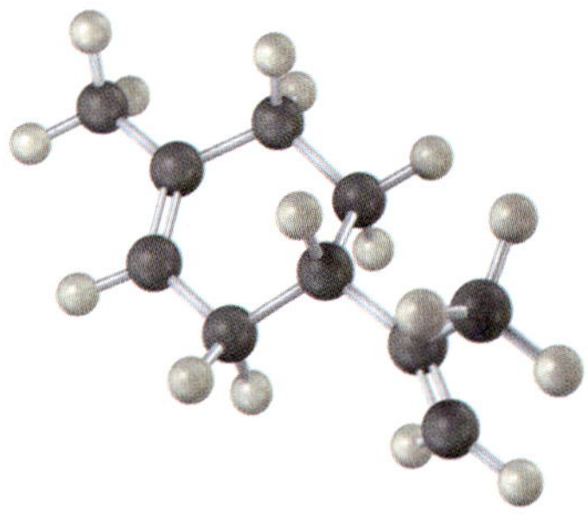

More dramatic examples of how a change in chirality can affect the biological properties of a molecule are found in many drugs, such as fluoxetine, a heavily prescribed medication sold under the trade name Prozac. Racemic fluoxetine is an extraordinarily effective antidepressant but has no activity against migraine. The pure *S* enantiomer, however, works remarkably well in preventing migraine. The *Focus On* "Chiral Drugs" at the end of this chapter gives other examples.

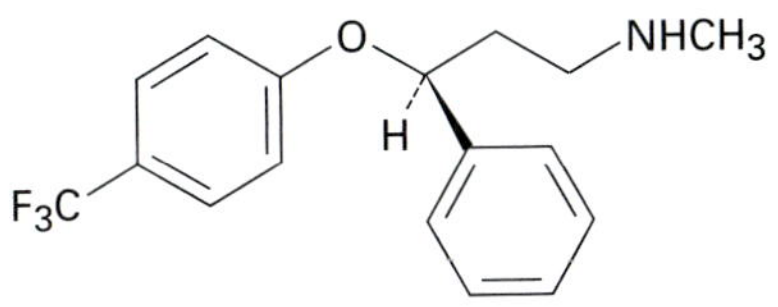

(*S*)-Fluoxetine (prevents migraine)

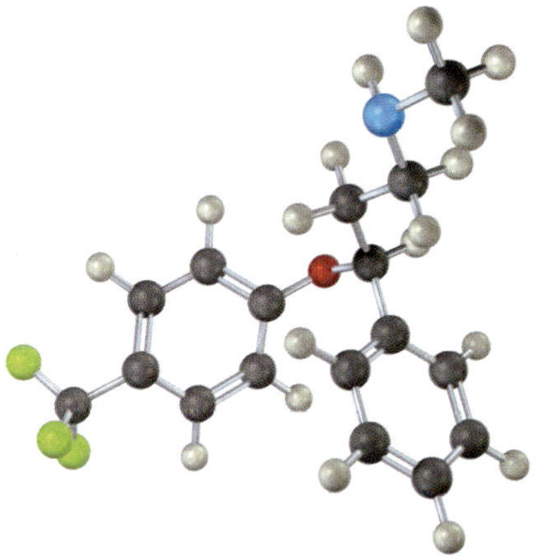

Why do different enantiomers have different biological properties? To have a biological effect, a substance typically must fit into an appropriate receptor that has an exactly complementary shape. But because biological receptors are chiral, only one enantiomer of a chiral substrate can fit in, just as only a right hand will fit into right-handed glove. The mirror-image enantiomer will be a misfit, like a left hand in a right-handed glove. A representation of the interaction between a chiral molecule and a chiral biological receptor is shown in Figure 9.17: one enantiomer fits the receptor perfectly, but the other does not.

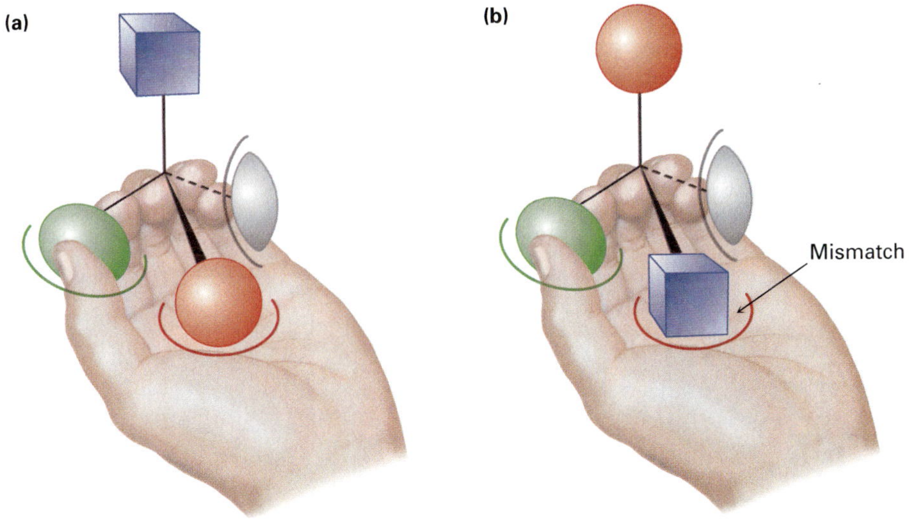

igure 9.17 Imagine that a left and interacts with a chiral bject, much as a biological eceptor interacts with a chiral nolecule. **(a)** One enantiomer fits nto the hand perfectly: green humb, red palm, and gray pinkie inger, with the blue substituent xposed. **(b)** The other enantiomer, however, can't fit into the and. When the green thumb nd gray pinkie finger interact ppropriately, the palm holds blue substituent rather than a ed one, with the red substituent xposed.

The hand-in-glove fit of a chiral substrate into a chiral receptor is relatively straightforward, but it's less obvious how a prochiral substrate can undergo a selective reaction. Take the reaction of ethanol with NAD^+ catalyzed by yeast alcohol dehydrogenase. As we saw at the end of Section 9.13, the reaction occurs with exclusive removal of the *pro-R* hydrogen from ethanol and with addition only to the *Re* face of the NAD^+ carbon.

We can understand this result by imagining that the chiral enzyme recept again has three binding sites, as was previously the case in Figure 9.17. Wh green and gray substituents of a prochiral substrate are held appropriately, ho ever, only one of the two red substituents—say, the *pro-S* one— is also held whi the other, *pro-R*, substituent is exposed for reaction.

We describe the situation by saying that the receptor provides a **chir environment** for the substrate. In the absence of a chiral environment, the tw red substituents are chemically identical, but in the presence of the chiral env ronment, they are chemically distinctive (Figure 9.18a). The situation is simil to what happens when you pick up a coffee mug. By itself, the mug has a plar of symmetry and is achiral. You could, if you wanted, drink from on either side the handle. When you pick up the mug, however, your hand provides a chir environment so one side becomes much more accessible and easier to drin from than the other (Figure 9.18b).

(a) **(b)**

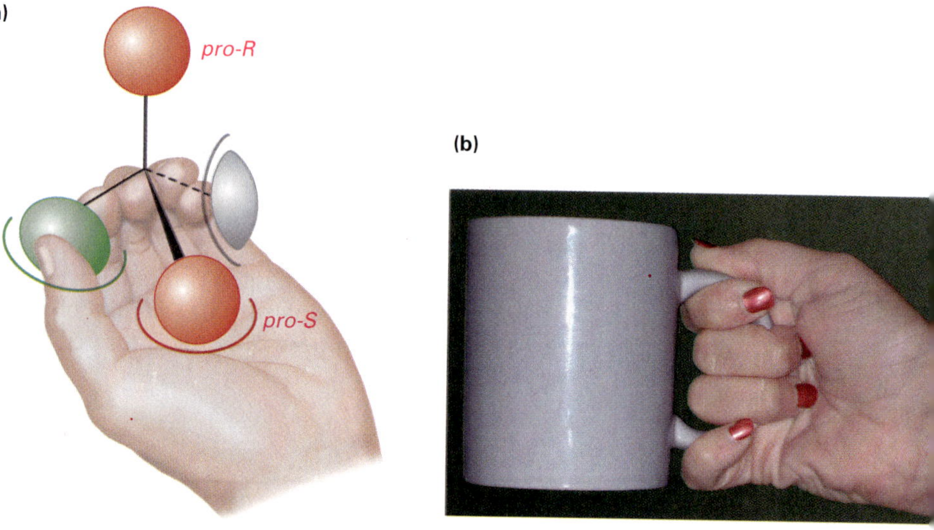

Figure 9.18 **(a)** When a prochiral molecule is held in a chiral environment, the two seemingly identical substituents (red) are distinguishable. **(b)** Similarly, when an achiral coffee mug is held in the chiral environment of your hand, it's much easier to drink from one side than the other because the two sides of the mug are now distinguishable.

Focus On . . .

Chiral Drugs

The hundreds of different pharmaceutical agents approved for use by the U.S. Food and Drug Administration come from many sources (see the Chapter 5 *Focus On*). Many drugs are isolated directly from plants or bacteria, and others are made by chemical modification of naturally occurring compounds, but an

(continued)

estimated 33% are made entirely in the laboratory and have no relatives in nature.

Those drugs that come from natural sources, either directly or after chemical modification, are usually chiral and are generally found only as a single enantiomer rather than as a racemic mixture. Penicillin V, for example, an antibiotic isolated from the *Penicillium* mold, has the 2*S*,5*R*,6*R* configuration. Its enantiomer, which does not occur naturally but can be made in the laboratory, has no antibiotic activity.

The *S* enantiomer of ibuprofen soothes the aches and pains of athletic injuries much more effectively than the *R* enantiomer.

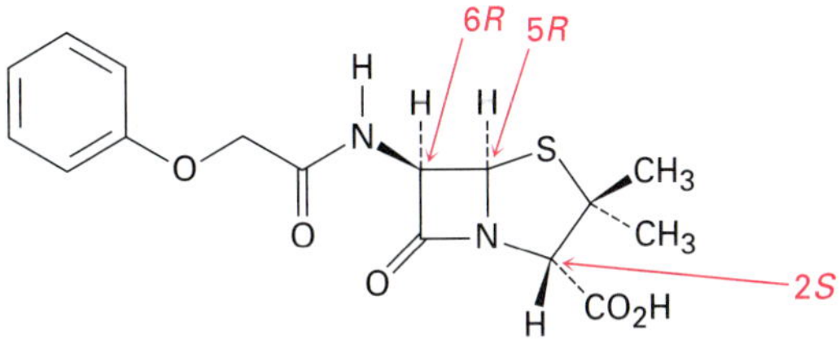

Penicillin V (2*S*,5*R*,6*R* configuration)

In contrast to drugs from natural sources, those drugs that are made entirely in the laboratory are either achiral or, if chiral, are often produced and sold as racemic mixtures. Ibuprofen, for example, has one chirality center and is sold commercially under such trade names as Advil, Nuprin, and Motrin as a racemic mixture of *R* and *S*. It turns out, however, that only the *S* enantiomer is active as an analgesic and anti-inflammatory agent. The *R* enantiomer of ibuprofen is inactive, although it is slowly converted in the body to the active *S* form.

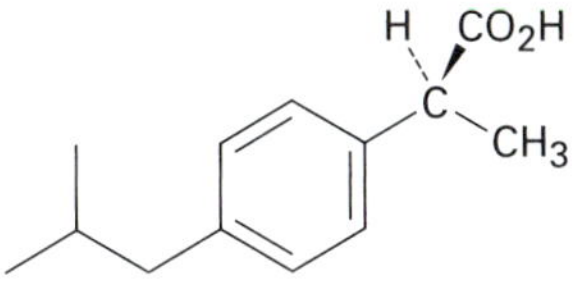

(*S*)-Ibuprofen
(an active analgesic agent)

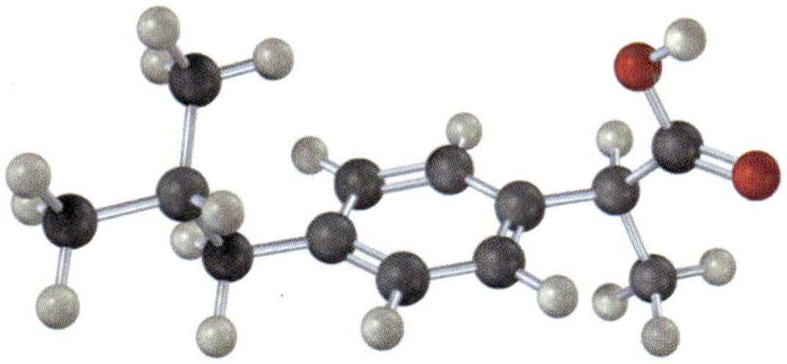

Not only is it chemically wasteful to synthesize and administer an enantiomer that doesn't serve the intended purpose, many examples are now known where the presence of the "wrong" enantiomer in a racemic mixture

(continued)

either affects the body's ability to utilize the "right" enantiomer or has unintended pharmacological effects of its own. The presence of (*R*)-ibuprofen in the racemic mixture, for instance, slows substantially the rate at which the *S* enantiomer takes effect in the body, from 12 minutes to 38 minutes.

To get around this problem, pharmaceutical companies attempt to devise methods of *enantioselective synthesis,* which allow them to prepare only a single enantiomer rather than a racemic mixture. Viable methods have already been developed for the preparation of (*S*)-ibuprofen, which is now being marketed in Europe. We'll look further into enantioselective synthesis in the Chapter 19 *Focus On.*

SUMMARY AND KEY WORDS

An object or molecule that is not superimposable on its mirror image is said t be **chiral**, meaning "handed." A chiral molecule is one that does not contain plane of symmetry cutting through it so that one half is a mirror image of th other half. The most common cause of chirality in organic molecules is the pre ence of a tetrahedral, sp^3-hybridized carbon atom bonded to four differer groups—a so-called **chirality center**. Chiral compounds can exist as a pair c nonsuperimposable, mirror-image stereoisomers called **enantiomers**. Enar tiomers are identical in all physical properties except for their **optical activity** or direction in which they rotate plane-polarized light.

The stereochemical **configuration** of a carbon atom can be specified as eithe ***R*** *(rectus)* or ***S*** *(sinister)* by using the Cahn–Ingold–Prelog sequence rules. Firs assign priorities to the four substituents on the chiral carbon atom, and then or ent the molecule so that the lowest-priority group points directly back. If a curve arrow drawn in the direction of decreasing priority ($1 \rightarrow 2 \rightarrow 3$) for the remainin three groups is clockwise, the chirality center has the *R* configuration. If the direc tion is counterclockwise, the chirality center has the *S* configuration.

Some molecules have more than one chirality center. Enantiomers have oppo site configuration at all chirality centers, whereas **diastereomers** have the sam configuration in at least one center but opposite configurations at the others **Epimers** are diastereomers that differ in configuration at only one chirality cente A compound with *n* chirality centers can have a maximum of 2^n stereoisomers.

Meso compounds contain chirality centers but are achiral overall because they have a plane of symmetry. **Racemic mixtures**, or *racemates,* are 50:50 mixtures o (+) and (−) enantiomers. Racemic mixtures and individual diastereomers differ ir their physical properties, such as solubility, melting point, and boiling point.

Many reactions give chiral products. If the reactants are optically inactive the products are also optically inactive. If one or both of the reactants is opti cally active, the product can also be optically active.

A molecule is **prochiral** if can be converted from achiral to chiral in a sin gle chemical step. A prochiral sp^2-hybridized atom has two faces, described a either ***Re*** or ***Si***. An sp^3-hybridized atom is a **prochirality center** if, by changing one of its attached atoms, a chirality center results. The atom whose replace ment leads to an *R* chirality center is ***pro-R***, and the atom whose replacement leads to an *S* chirality center is ***pro-S***.

EXERCISES

Organic KNOWLEDGE TOOLS

ThomsonNOW™ Sign in at **www.thomsonedu.com** to assess your knowledge of this chapter's topics by taking a pre-test. The pre-test will link you to interactive organic chemistry resources based on your score in each concept area.

Online homework for this chapter may be assigned in Organic OWL.

■ indicates problems assignable in Organic OWL.

▲ denotes problems linked to Key Ideas of this chapter and testable in ThomsonNOW.

VISUALIZING CHEMISTRY

(Problems 9.1–9.26 appear within the chapter.)

9.27 Which of the following structures are identical? (Yellow-green = Cl.)

(a)

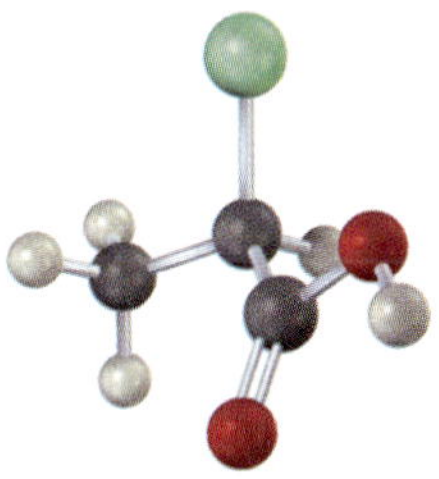

(b)

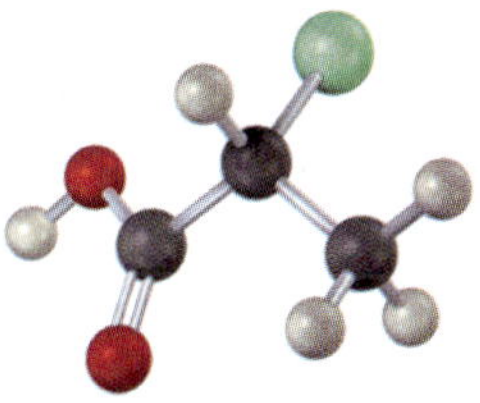

(c)

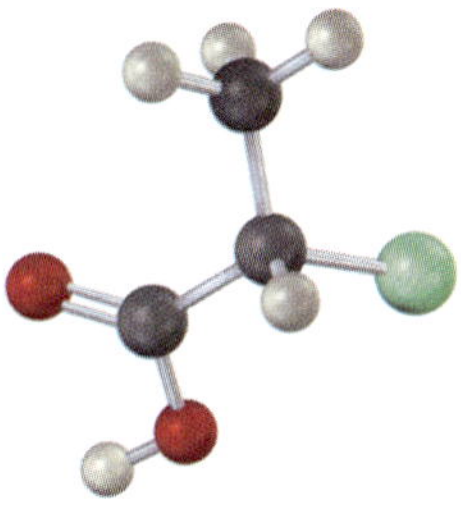

(d)

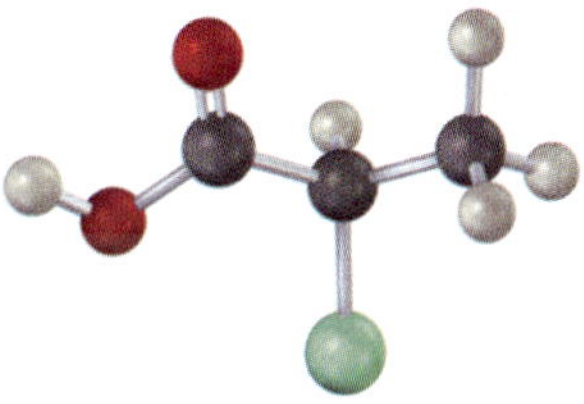

9.28 ■ ▲ Assign *R* or *S* configuration to the chirality centers in the following molecules (blue = N):

(a)

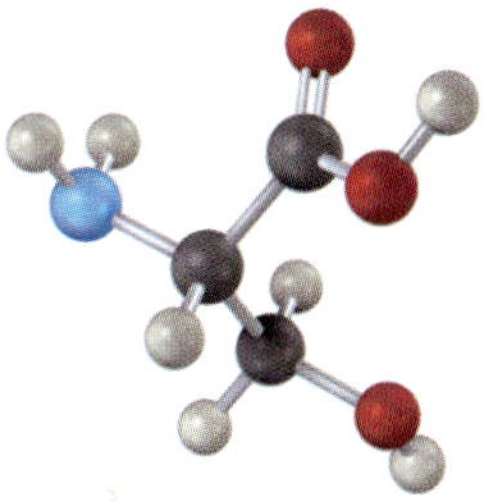

Serine

(b)

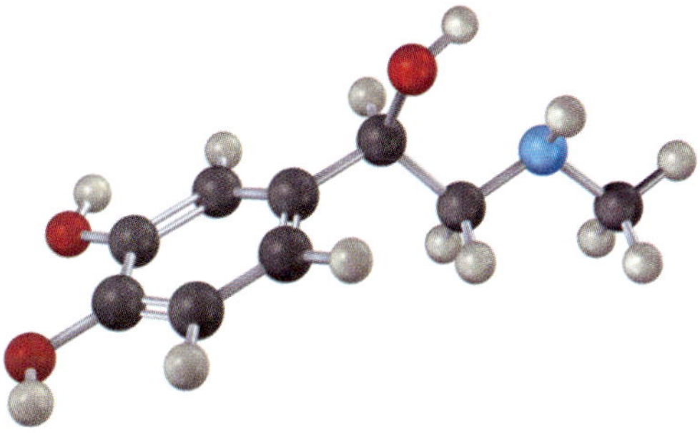

Adrenaline

9.29 Which, if any, of the following structures represent meso compounds? (Blue = N, yellow-green = Cl.)

(a)

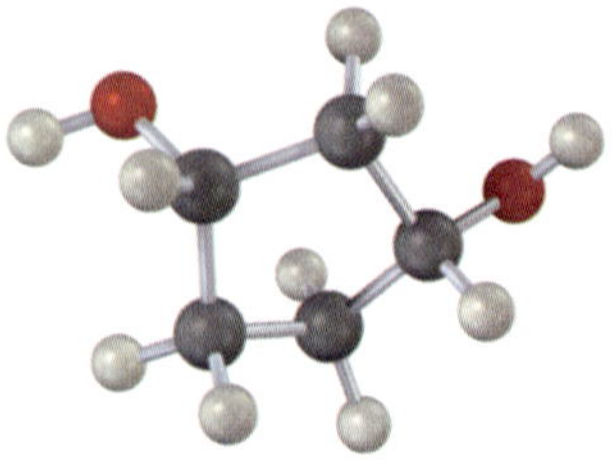

(b)

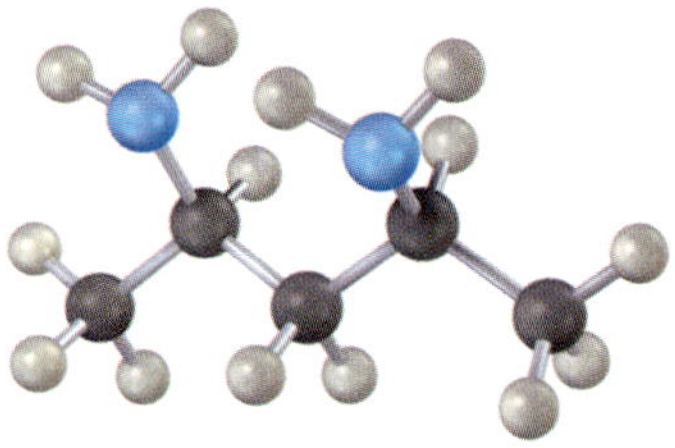

(c)

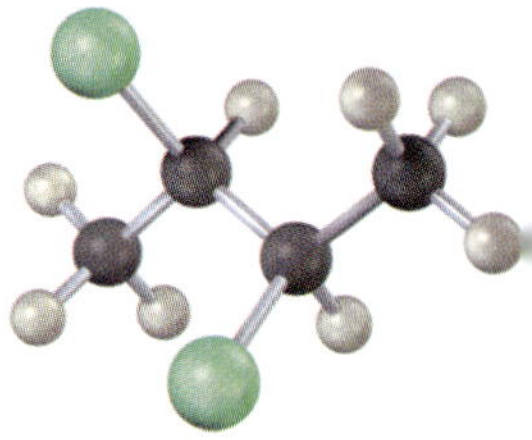

9.30 ■ ▲ Assign *R* or *S* configuration to each chirality center in pseudoephedrine, an over-the-counter decongestant found in cold remedies (blue = N).

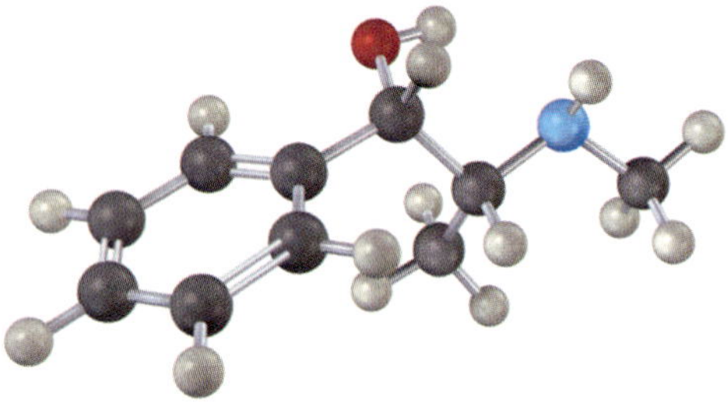

ADDITIONAL PROBLEMS

9.31 ■ ▲ Which of the following compounds are chiral? Draw them, and label the chirality centers.

(a) 2,4-Dimethylheptane **(b)** 5-Ethyl-3,3-dimethylheptane
(c) *cis*-1,4-Dichlorocyclohexane **(d)** 4,5-Dimethyl-2,6-octadiyne

9.32 ▲ Draw chiral molecules that meet the following descriptions:

(a) A chloroalkane, $C_5H_{11}Cl$ **(b)** An alcohol, $C_6H_{14}O$
(c) An alkene, C_6H_{12} **(d)** An alkane, C_8H_{18}

9.33 ▲ Eight alcohols have the formula $C_5H_{12}O$. Draw them. Which are chiral?

9.34 Draw the nine chiral molecules that have the formula $C_6H_{13}Br$.

9.35 ■ Draw compounds that fit the following descriptions:

(a) A chiral alcohol with four carbons
(b) A chiral carboxylic acid with the formula $C_5H_{10}O_2$
(c) A compound with two chirality centers
(d) A chiral aldehyde with the formula C_3H_5BrO

9.36 Which of the following objects are chiral?

(a) A basketball **(b)** A fork **(c)** A wine glass
(d) A golf club **(e)** A monkey wrench **(f)** A snowflake

9.37 Erythronolide B is the biological precursor of erythromycin, a broad-spectrum antibiotic. How many chirality centers does erythronolide B have?

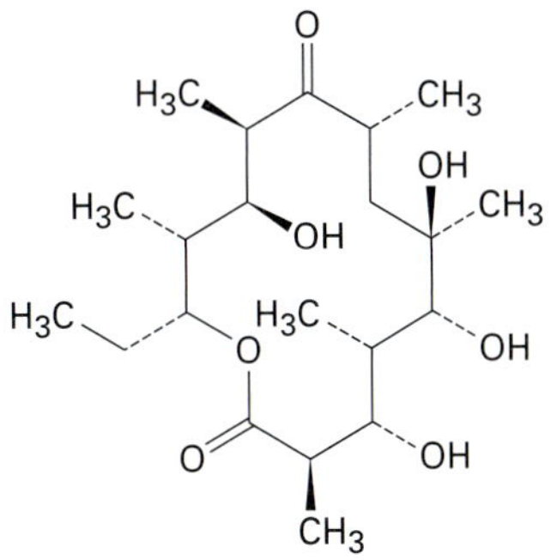

Erythronolide B

9.38 Draw examples of the following:
(a) A meso compound with the formula C_8H_{18}
(b) A meso compound with the formula C_9H_{20}
(c) A compound with two chirality centers, one *R* and the other *S*

9.39 What is the relationship between the specific rotations of (2*R*,3*R*)-dichloropentane and (2*S*,3*S*)-dichloropentane? Between (2*R*,3*S*)-dichloropentane and (2*R*,3*R*)-dichloropentane?

9.40 What is the stereochemical configuration of the enantiomer of (2*S*,4*R*)-2,4-octanediol?

9.41 What are the stereochemical configurations of the two diastereomers of (2*S*,4*R*)-2,4-octanediol?

9.42 Orient each of the following drawings so that the lowest-priority group is toward the rear, and then assign *R* or *S* configuration:

(a)
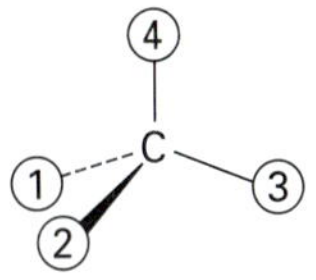

(b)
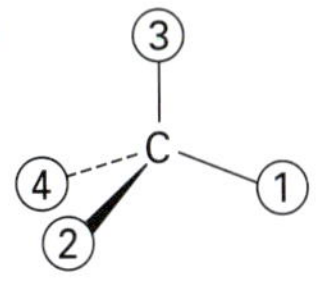

(c)
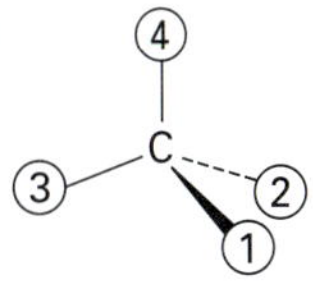

9.43 ■ Assign Cahn–Ingold–Prelog priorities to the following sets of substituents:

(a) —CH=CH$_2$, —CH(CH$_3$)$_2$, —C(CH$_3$)$_3$, —CH$_2$CH$_3$

(b) —C≡CH, —CH=CH$_2$, —C(CH$_3$)$_3$, 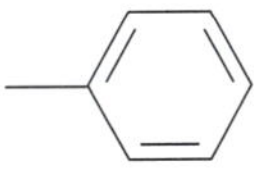

(c) —CO$_2$CH$_3$, —COCH$_3$, —CH$_2$OCH$_3$, —CH$_2$CH$_3$

(d) —C≡N, —CH$_2$Br, —CH$_2$CH$_2$Br, —Br

9.44 ■ Assign *R* or *S* configurations to the chirality centers in the following molecules:

(a)
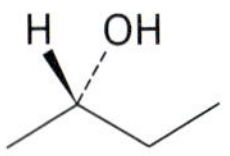

(b)
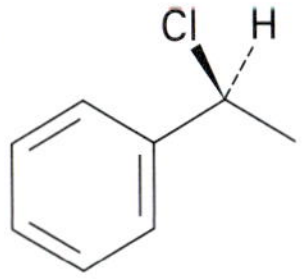

(c)
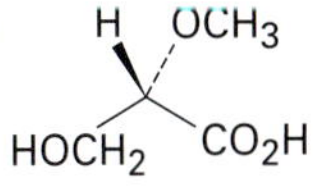

9.45 ■ Assign *R* or *S* configuration to each chirality center in the following molecul

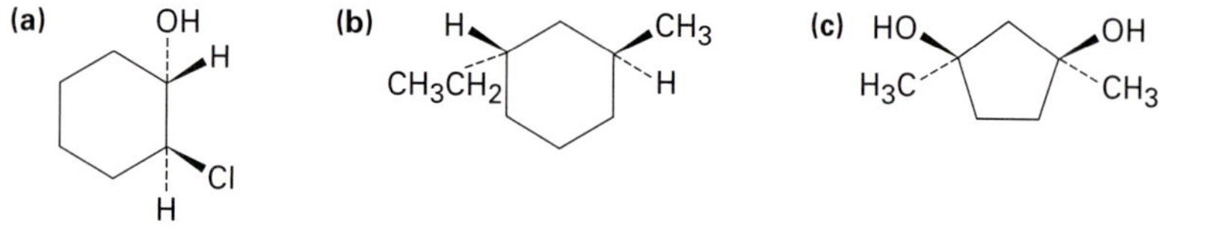

9.46 Assign *R* or *S* configuration to each chirality center in the following biologic molecules:

(a)

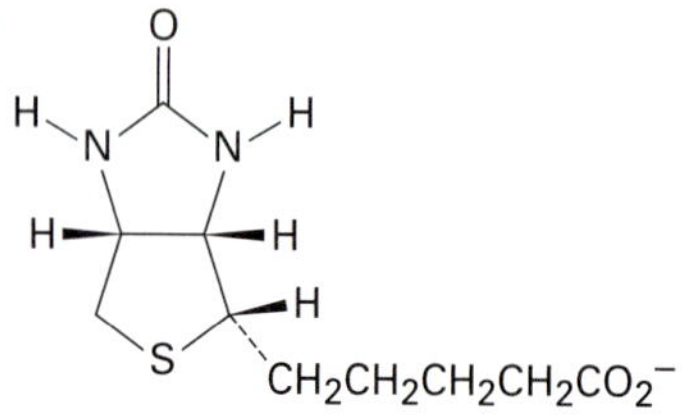

Biotin

(b)

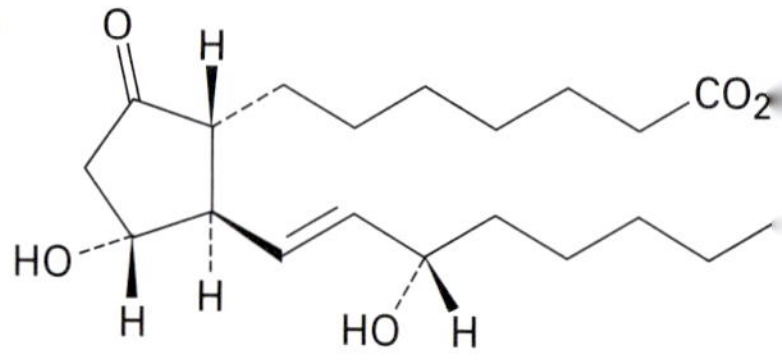

Prostaglandin E_1

9.47 ■ Draw tetrahedral representations of the following molecules:
(a) (*S*)-2-Chlorobutane **(b)** (*R*)-3-Chloro-1-pentene

9.48 Draw tetrahedral representations of the two enantiomers of the amino aci cysteine, $HSCH_2CH(NH_2)CO_2H$, and identify each as *R* or *S*.

9.49 The naturally occurring form of the amino acid cysteine (Problem 9.48) has th *S* configuration at its chirality center. On treatment with a mild oxidizin agent, two cysteines join to give cystine, a disulfide. Assuming that the chira ity center is not affected by the reaction, is cystine optically active?

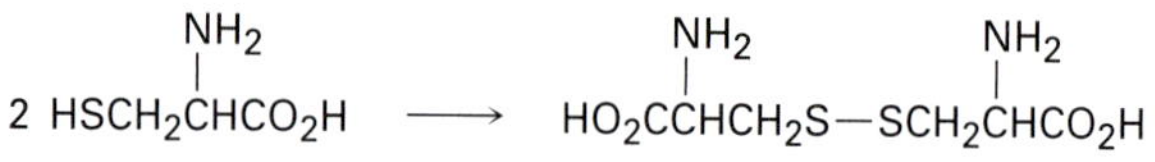

Cysteine **Cystine**

9.50 ■ Which of the following pairs of structures represent the same enantiome and which represent different enantiomers?

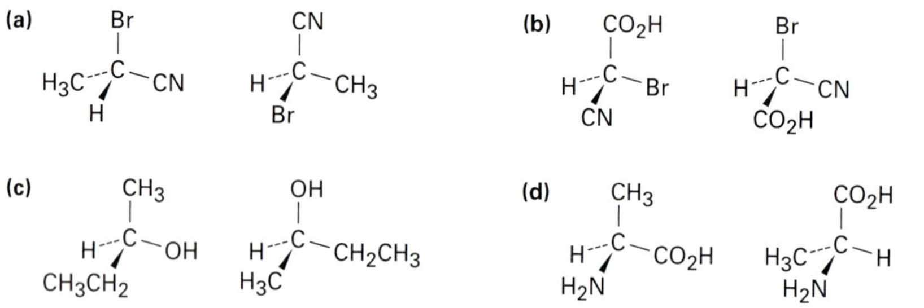

9.51 Assign *R* or *S* configuration to each chirality center in the following molecules:

(a)

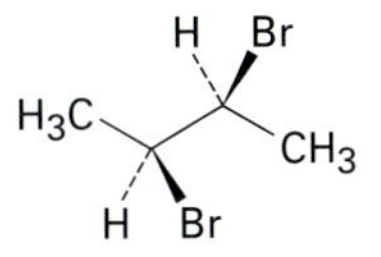

(b)

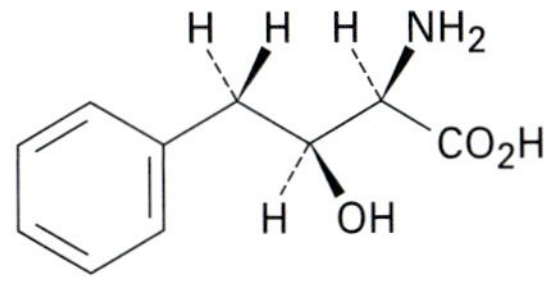

9.52 ■ Draw tetrahedral representations of the following molecules:

(a) The 2*S*,3*R* enantiomer of 2,3-dibromopentane

(b) The meso form of 3,5-heptanediol

9.53 Draw the meso form of each of the following molecules, and indicate the plane of symmetry in each:

(a) $CH_3CH(OH)CH_2CH_2CH(OH)CH_3$ (OH above each CH)

(b)

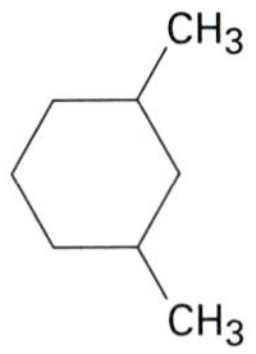

(c)

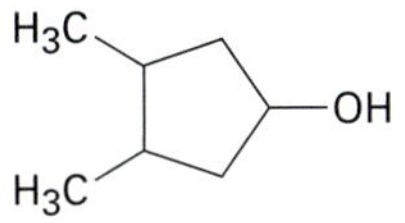

9.54 Assign *R* or *S* configurations to the chirality centers in ascorbic acid (vitamin C).

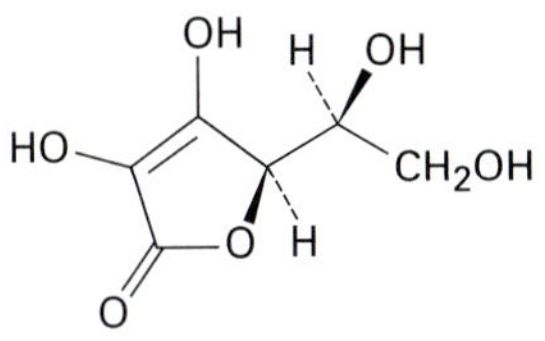

Ascorbic acid

9.55 Assign *R* or *S* stereochemistry to the chirality centers in the following Newman projections:

(a)

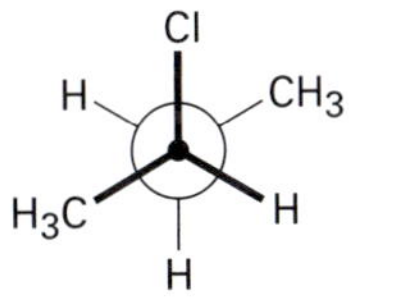

(b)

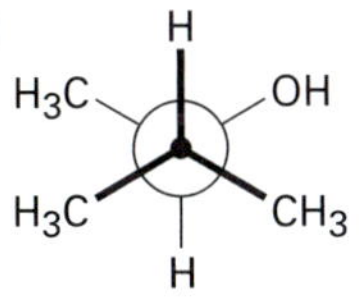

9.56 Xylose is a common sugar found in many types of wood, including maple and cherry. Because it is much less prone to cause tooth decay than sucrose, xylose has been used in candy and chewing gum. Assign *R* or *S* configurations to the chirality centers in xylose.

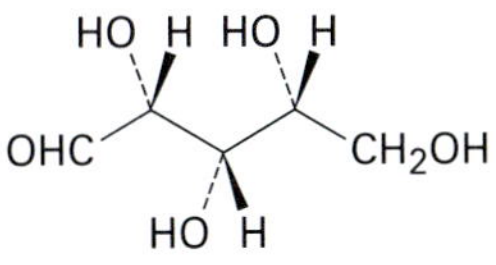

(+)-Xylose

9.57 Ribose, an essential part of ribonucleic acid (RNA), has the following structur

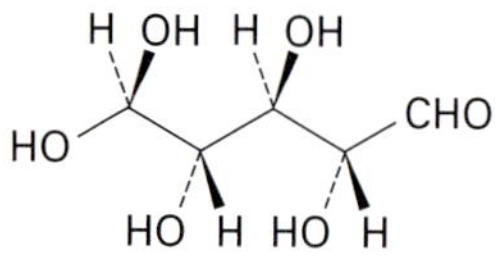

Ribose

(a) How many chirality centers does ribose have? Identify them.
(b) How many stereoisomers of ribose are there?
(c) Draw the structure of the enantiomer of ribose.
(d) Draw the structure of a diastereomer of ribose.

9.58 On catalytic hydrogenation over a platinum catalyst, ribose (Problem 9.57) converted into ribitol. Is ribitol optically active or inactive? Explain.

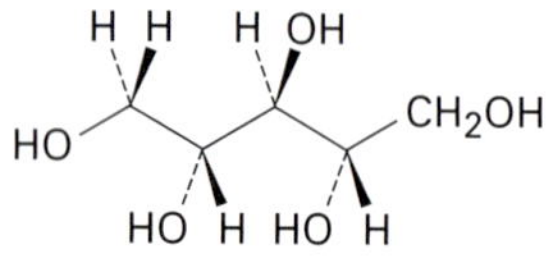

Ribitol

9.59 ■ Hydroxylation of *cis*-2-butene with OsO_4 yields 2,3-butanediol. Wha stereochemistry do you expect for the product? (Review Section 7.8.)

9.60 Hydroxylation of *trans*-2-butene with OsO_4 also yields 2,3-butanediol. Wha stereochemistry do you expect for the product?

9.61 *cis*-4-Octene reacts with a peroxyacid to yield 4,5-epoxyoctane. Is the produc chiral? How many chirality centers does it have? How would you describe stereochemically? (Review Section 7.8.)

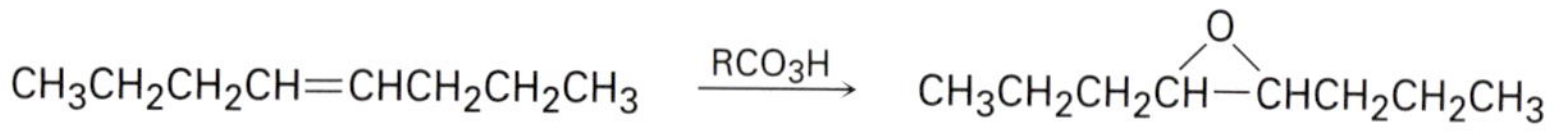

4-Octene **4,5-Epoxyoctane**

9.62 Answer Problem 9.61 for the epoxidation of *trans*-4-octene.

9.63 ■ Identify the indicated hydrogens in the following molecules as *pro-R* or *pro-S*

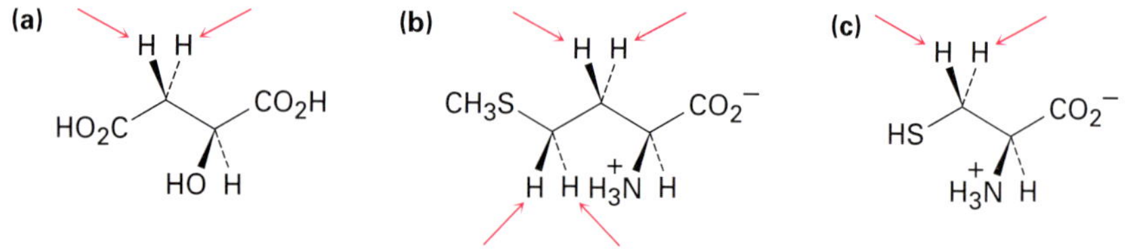

Malic acid **Methionine** **Cysteine**

9.64 ■ Identify the indicated faces in the following molecules as *Re* or *Si*:

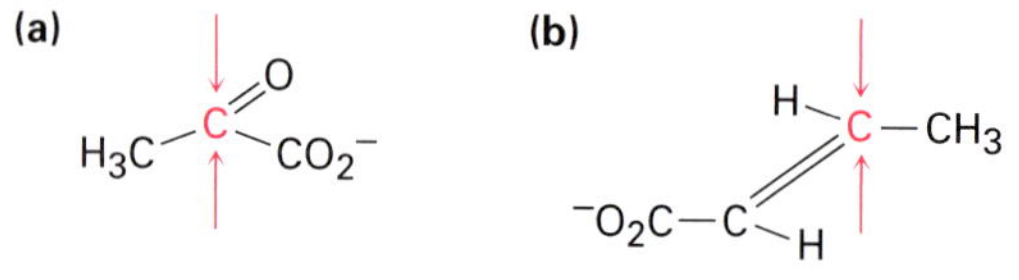

Pyruvate **Crotonate**

9.65 Draw all possible stereoisomers of 1,2-cyclobutanedicarboxylic acid, and indicate the interrelationships. Which, if any, are optically active? Do the same for 1,3-cyclobutanedicarboxylic acid.

9.66 Compound A, C_7H_{12}, was found to be optically active. On catalytic reduction over a palladium catalyst, 2 equivalents of hydrogen were absorbed, yielding compound B, C_7H_{16}. On ozonolysis of A, two fragments were obtained. One fragment was identified as acetic acid. The other fragment, compound C, was an optically active carboxylic acid, $C_5H_{10}O_2$. Write the reactions, and draw structures for A, B, and C.

9.67 Compound A, $C_{11}H_{16}O$, was found to be an optically active alcohol. Despite its apparent unsaturation, no hydrogen was absorbed on catalytic reduction over a palladium catalyst. On treatment of A with dilute sulfuric acid, dehydration occurred and an optically inactive alkene B, $C_{11}H_{14}$, was produced as the major product. Alkene B, on ozonolysis, gave two products. One product was identified as propanal, CH_3CH_2CHO. Compound C, the other product, was shown to be a ketone, C_8H_8O. How many degrees of unsaturation does A have? Write the reactions, and identify A, B, and C.

9.68 One of the steps in fat metabolism is the hydration of crotonate to yield 3-hydroxybutyrate. The reaction occurs by addition of $-OH$ to the *Si* face at C3, followed by protonation at C2, also from the *Si* face. Draw the product of the reaction, showing the stereochemistry of each step.

Crotonate **3-Hydroxybutyrate**

9.69 The dehydration of citrate to yield *cis*-aconitate, a step in the citric acid cycle, involves the *pro-R* "arm" of citrate rather than the *pro-S* arm. Which of the following two products is formed?

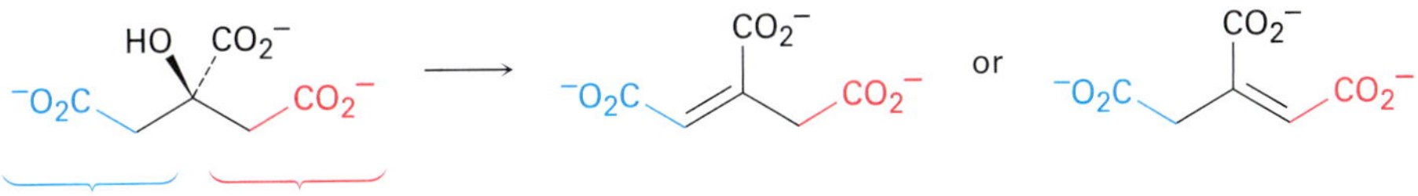

Citrate ***cis*-Aconitate**

9.70 The first step in the metabolism of glycerol formed by digestion of fats is phosphorylation of the *pro-R* $-CH_2OH$ group by reaction with ATP to give the corresponding glycerol phosphate. Show the stereochemistry of the product.

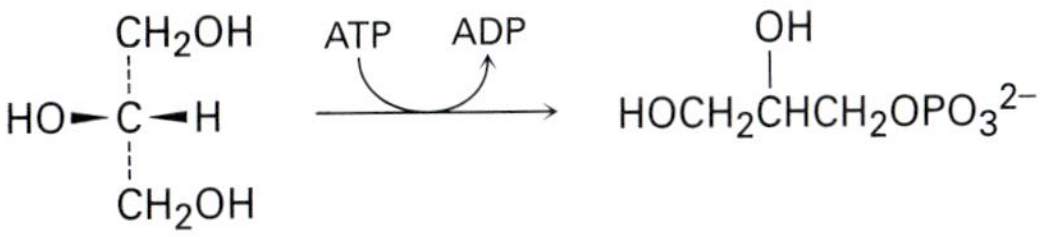

Glycerol **Glycerol phosphate**

9.71 One of the steps in fatty-acid biosynthesis is the dehydration of (*R*)-3-hydrox butyryl ACP to give *trans*-crotonyl ACP. Does the reaction remove the *pro-R* the *pro-S* hydrogen from C2?

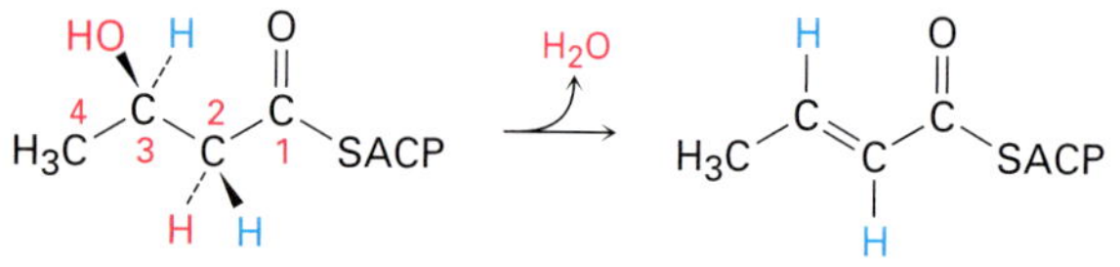

(*R*)-3-Hydroxybutyryl ACP **_trans_-Crotonyl ACP**

9.72 *Allenes* are compounds with adjacent carbon–carbon double bonds. Mar allenes are chiral, even though they don't contain chirality cente Mycomycin, for example, a naturally occurring antibiotic isolated from t bacterium *Nocardia acidophilus*, is chiral and has $[\alpha]_D = -130$. Explain wh mycomycin is chiral. Making a molecular model should be helpful.

$$HC{\equiv}C{-}C{\equiv}C{-}CH{=}C{=}CH{-}CH{=}CH{-}CH{=}CH{-}CH_2CO_2H$$

Mycomycin

9.73 Long before chiral allenes were known (Problem 9.72), the resolution 4-methylcyclohexylideneacetic acid into two enantiomers had been carrie out. Why is it chiral? What geometric similarity does it have to allenes?

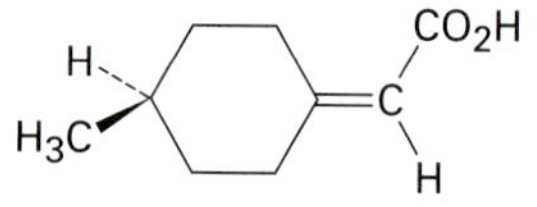

4-Methylcyclohexylideneacetic acid

9.74 (*S*)-1-Chloro-2-methylbutane undergoes light-induced reaction with Cl_2 b a radical mechanism to yield a mixture of products, among which a 1,4-dichloro-2-methylbutane and 1,2-dichloro-2-methylbutane.

(a) Write the reaction, showing the correct stereochemistry of the reactant.

(b) One of the two products is optically active, but the other is optically inac tive. Which is which?

(c) What can you conclude about the stereochemistry of radical chlorinatio reactions?

9.75 Draw the structure of a meso compound that has five carbons and thre chirality centers.

9.76 How many stereoisomers of 2,4-dibromo-3-chloropentane are there? Draw them, and indicate which are optically active.

9.77 Draw both *cis*- and *trans*-1,4-dimethylcyclohexane in their most stable chai conformations.

(a) How many stereoisomers are there of *cis*-1,4-dimethylcyclohexane, and how many of *trans*-1,4-dimethylcyclohexane?

(b) Are any of the structures chiral?

(c) What are the stereochemical relationships among the various stereo isomers of 1,4-dimethylcyclohexane?

9.78 Draw both *cis*- and *trans*-1,3-dimethylcyclohexane in their most stable chair conformations.

(a) How many stereoisomers are there of *cis*-1,3-dimethylcyclohexane, and how many of *trans*-1,3-dimethylcyclohexane?

(b) Are any of the structures chiral?

(c) What are the stereochemical relationships among the various stereoisomers of 1,3-dimethylcyclohexane?

9.79 *cis*-1,2-Dimethylcyclohexane is optically inactive even though it has two chirality centers. Explain.

9.80 We'll see in the next chapter that alkyl halides react with nucleophiles to give substitution products by a mechanism that involves *inversion* of stereochemistry at carbon:

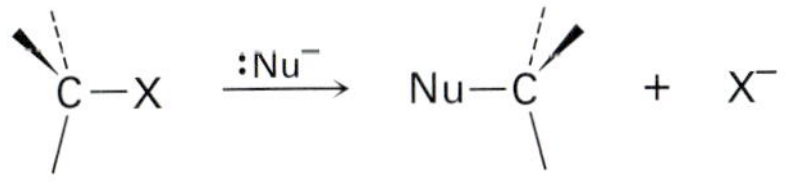

Draw the reaction of (*S*)-2-bromobutane with HS^- ion to yield 2-butanethiol, $CH_3CH_2CH(SH)CH_3$. What is the stereochemistry of the product?

9.81 ■ Ketones react with acetylide ion (Section 8.7) to give alcohols. For example, the reaction of sodium acetylide with 2-butanone yields 3-methyl-1-pentyn-3-ol:

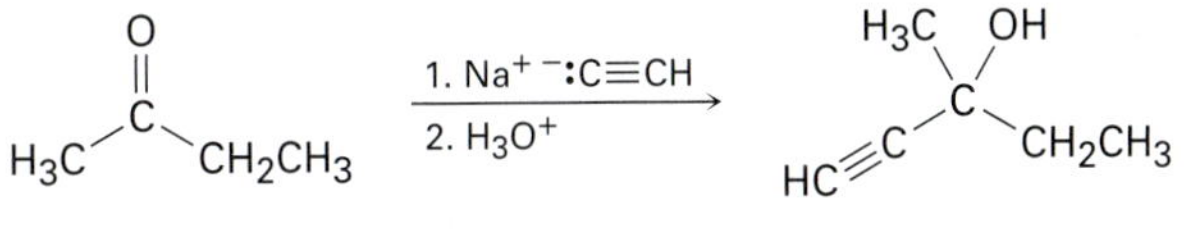

2-Butanone **3-Methyl-1-pentyne-3-ol**

(a) Is the product chiral? Is it optically active?

(b) How many stereoisomers of the product are formed, what are their stereochemical relationships, and what are their relative amounts?

9.82 Imagine that another reaction similar to that in Problem 9.81 is carried out between sodium acetylide and (*R*)-2-phenylpropanal to yield 1-phenyl-3-butyn-2-ol:

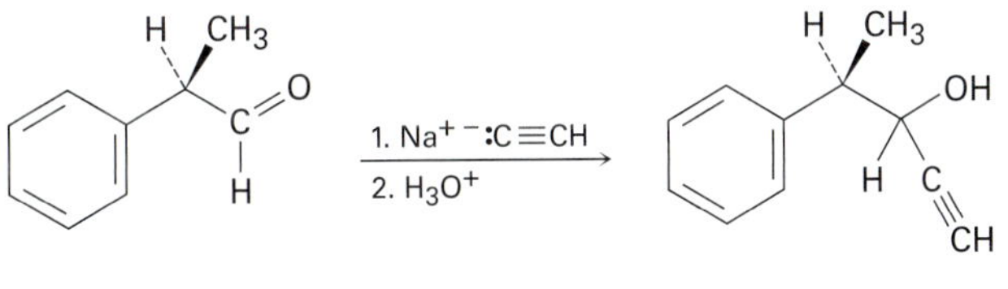

(*R*)-2-Phenylpropanal **1-Phenyl-3-butyn-2-ol**

(a) Is the product chiral? Is it optically active?

(b) How many stereoisomers of 1-phenyl-3-butyn-2-ol are formed, what are their stereochemical relationships, and what are their relative amounts?

10 Organohalides

Organic **KNOWLEDGE TOOLS**

ThomsonNOW™ Throughout this chapter, sign in at **www.thomsonedu.com** for online self-study and interactive tutorials based on your level of understanding.

Online homework for this chapter may be assigned in Organic OWL.

Now that we've covered the chemistry of hydrocarbons, it's time to start look ing at more complex substances that contain elements in addition to C and H We'll begin by discussing the chemistry of **organohalides**, compounds tha contain one or more halogen atoms.

Halogen-substituted organic compounds are widespread throughout nature and approximately 5000 organohalides have been found in algae and variou other marine organisms. Chloromethane, for example, is released in large amount by oceanic kelp, as well as by forest fires and volcanoes. Halogen-containing com pounds also have a vast array of industrial applications, including their use as sol vents, inhaled anesthetics in medicine, refrigerants, and pesticides.

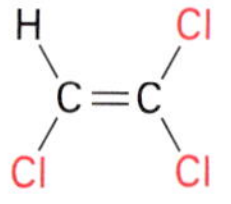

Trichloroethylene
(a solvent)

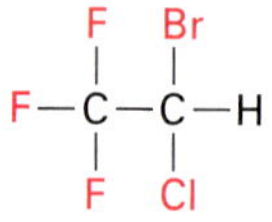

Halothane
(an inhaled anesthetic)

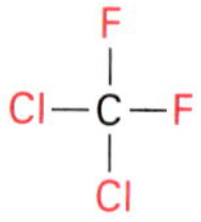

Dichlorodifluoromethane
(a refrigerant)

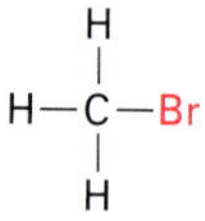

Bromomethane
(a fumigant)

Still other halo-substituted compounds are providing important leads tc new medicines. The compound epibatidine, for instance, has been isolated from the skin of Ecuadorian frogs and found to be more than 200 times as potent as morphine at blocking pain in animals.

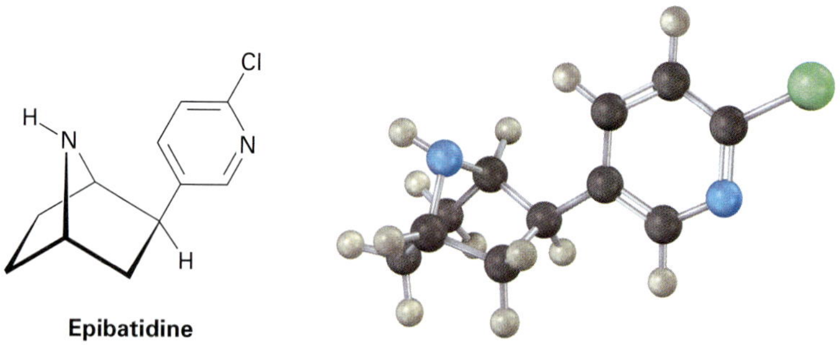

Epibatidine
(from the Ecuadorian frog *Epipedobates tricolor*)

A large variety of organohalides are known. The halogen might be bonded to an alkynyl group (C≡C–X), a vinylic group (C=C–X), an aromatic ring (Ar–X), or an alkyl group. We'll be concerned in this chapter, however, primarily with **alkyl halides**, compounds with a halogen atom bonded to a saturated, sp^3-hybridized carbon atom.

WHY THIS CHAPTER?

Alkyl halides are encountered less frequently than their oxygen-containing relatives alcohols and ethers, but some of the *kinds* of reactions they undergo—nucleophilic substitutions and eliminations—*are* encountered frequently. Thus, alkyl halide chemistry acts as a relatively simple model for many mechanistically similar but structurally more complex reactions found in biomolecules. We'll begin in this chapter with a look at how to name and prepare alkyl halides, and we'll see several of their reactions. Then in the following chapter, we'll make a detailed study of the substitution and elimination reactions of alkyl halides—two of the most important and well-studied reaction types in organic chemistry.

10.1 Naming Alkyl Halides

homsonNOW™ Click *Organic nteractive* to **practice assigning JPAC names to organic halides.**

Although members of the class are commonly called *alkyl halides,* they are named systematically as *haloalkanes* (Section 3.4), treating the halogen as a substituent on a parent alkane chain. There are three steps:

Step 1 **Find the longest chain, and name it as the parent.** If a double or triple bond is present, the parent chain must contain it.

Step 2 **Number the carbons of the parent chain beginning at the end nearer the first substituent, whether alkyl or halo.** Assign each substituent a number according to its position on the chain.

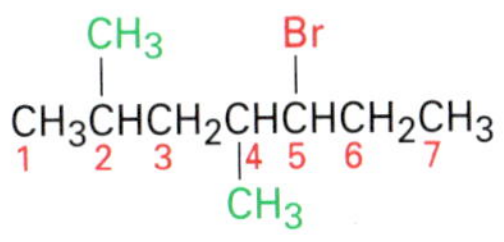

5-Bromo-2,4-dimethylheptane

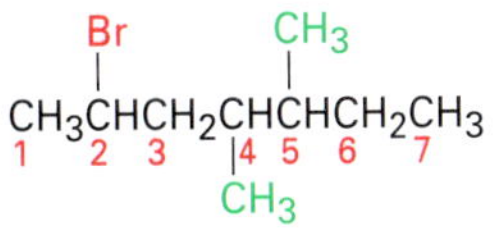

2-Bromo-4,5-dimethylheptane

If different halogens are present, number all and list them in alphabetical order when writing the name.

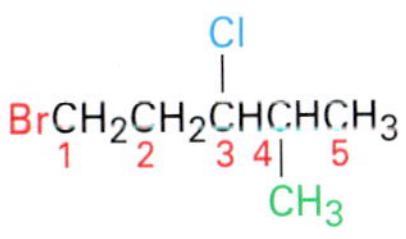

1-Bromo-3-chloro-4-methylpentane

Step 3 **If the parent chain can be properly numbered from either end by step begin at the end nearer the substituent that has alphabetical precedence.**

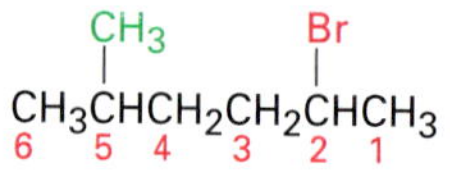

2-Bromo-5-methylhexane
(*NOT* 5-bromo-2-methylhexane)

ThomsonNOW™ Click *Organic Interactive* to **use a web-based palette to draw structures for alkyl halides, based on their IUPAC names.**

In addition to their systematic names, many simple alkyl halides are al named by identifying first the alkyl group and then the halogen. For exampl CH_3I can be called either iodomethane or methyl iodide. Such names are we entrenched in the chemical literature and in daily usage, but they won't be use in this book.

CH_3I

Iodomethane
(or methyl iodide)

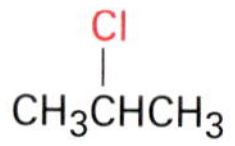

2-Chloropropane
(or isopropyl chloride)

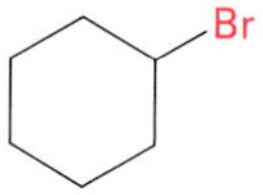

Bromocyclohexane
(or cyclohexyl bromide)

Problem 10.1 Give IUPAC names for the following alkyl halides:

(a) $CH_3CH_2CH_2CH_2I$

(b) 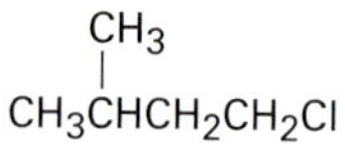

(c) 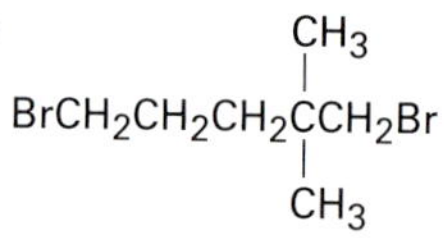

(d) 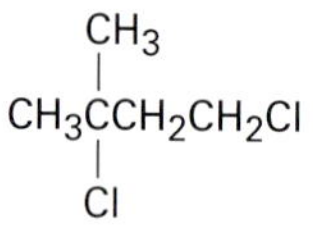

(e) 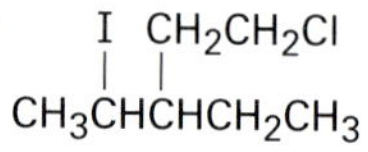

(d) 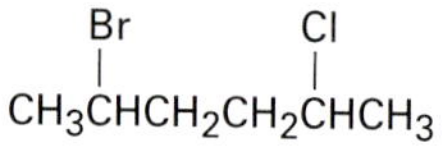

Problem 10.2 Draw structures corresponding to the following IUPAC names:

(a) 2-Chloro-3,3-dimethylhexane
(b) 3,3-Dichloro-2-methylhexane
(c) 3-Bromo-3-ethylpentane
(d) 1,1-Dibromo-4-isopropylcyclohexane
(e) 4-*sec*-Butyl-2-chlorononane
(f) 1,1-Dibromo-4-*tert*-butylcyclohexane

10.2 Structure of Alkyl Halides

Halogens increase in size going down the periodic table, so the lengths of the corresponding carbon–halogen bonds increase accordingly (Table 10.1). In addition, C–X bond strengths decrease going down the periodic table. As we've been doing consistently thus far, we'll continue to use the abbreviation X to represent any of the halogens F, Cl, Br, or I.

Table 10.1 A Comparison of the Halomethanes

Halomethane	Bond length (pm)	Bond strength (kJ/mol)	Bond strength (kcal/mol)	Dipole moment (*D*)
CH_3F	139	452	108	1.85
CH_3Cl	178	351	84	1.87
CH_3Br	193	293	70	1.81
CH_3I	214	234	56	1.62

In an earlier discussion of bond polarity in functional groups (Section 5.4), we noted that halogens are more electronegative than carbon. The C–X bond is therefore polar, with the carbon atom bearing a slight positive charge ($\delta+$) and the halogen a slight negative charge ($\delta-$). This polarity results in a substantial dipole moment for all the halomethanes (Table 10.1) and implies that the alkyl halide C–X carbon atom should behave as an electrophile in polar reactions. We'll see in the next chapter that much of the chemistry of alkyl halides is indeed dominated by their electrophilic behavior.

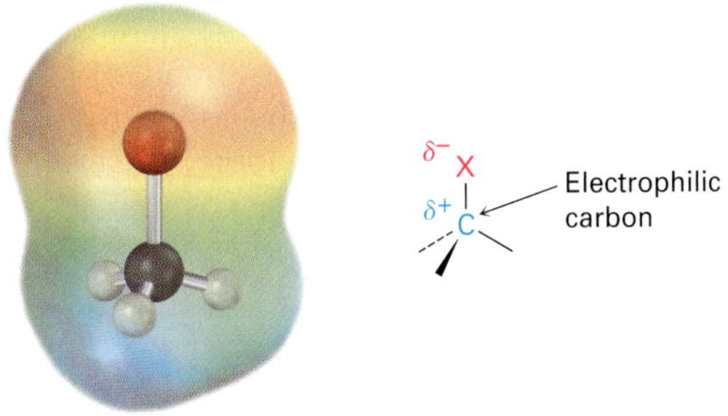

10.3 Preparing Alkyl Halides from Alkanes: Radical Halogenation

Structurally simple alkyl halides can sometimes be prepared by reaction of an alkane with Cl_2 or Br_2 through a radical chain-reaction pathway (Section 5.3). Although inert to most reagents, alkanes react readily with Cl_2 or Br_2 in the presence of light to give alkyl halide substitution products. The reaction occurs by the radical mechanism shown in Figure 10.1 for chlorination.

Recall from Section 5.3 that radical substitution reactions require three kinds of steps: *initiation, propagation,* and *termination.* Once an initiation step has started the process by producing radicals, the reaction continues in a self-sustaining cycle. The cycle requires two repeating propagation steps in which a radical, the halogen, and the alkane yield alkyl halide product plus more radical to carry on the chain. The chain is occasionally terminated by the combination of two radicals.

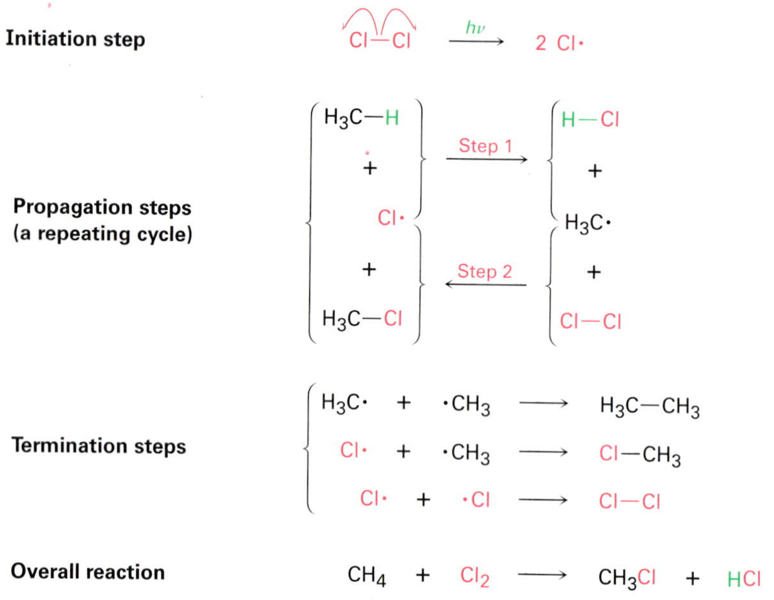

Figure 10.1 Mechanism of the radical chlorination of methane. Three kinds of steps are required: initiation, propagation, and termination. The propagation steps are a repeating cycle, with Cl· a reactant in step 1 and a product in step 2, and with ·CH_3 a product in step 1 and a reactant in step 2. (The symbol *hν* shown in the initiation step is the standard way of indicating irradiation with light.)

Although interesting from a mechanistic point of view, alkane halogenation is a poor synthetic method for preparing alkyl halides because mixtures of products invariably result. For example, chlorination of methane does not stop cleanly at the monochlorinated stage but continues to give a mixture of dichloro, trichloro, and even tetrachloro products.

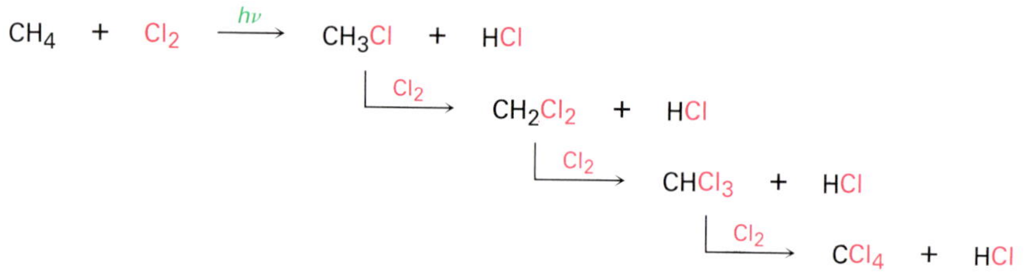

The situation is even worse for chlorination of alkanes that have more than one sort of hydrogen. For example, chlorination of butane gives two monochlorinated products in addition to dichlorobutane, trichlorobutane, and so on. Thirty percent of the monochloro product is 1-chlorobutane, and seventy percent is 2-chlorobutane.

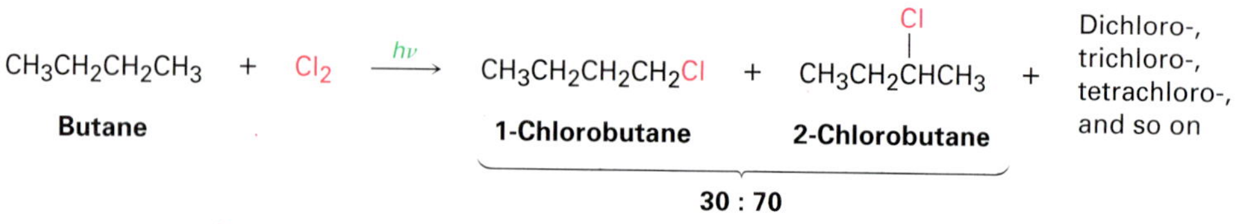

As another example, 2-methylpropane yields 2-chloro-2-methylpropane and 1-chloro-2-methylpropane in the ratio 35:65, along with more highly chlorinated products.

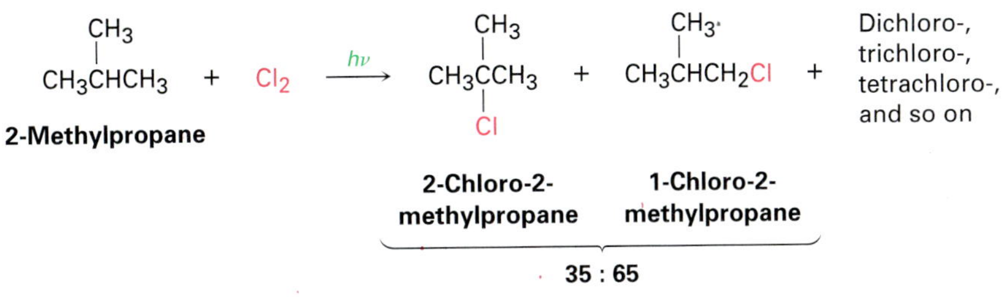

From these and similar reactions, it's possible to calculate a reactivity order toward chlorination for different sorts of hydrogen atoms in a molecule. Take the butane chlorination, for instance. Butane has six equivalent primary hydrogens ($-CH_3$) and four equivalent secondary hydrogens ($-CH_2-$). The fact that butane yields 30% of 1-chlorobutane product means that *each one* of the six primary hydrogens is responsible for 30% ÷ 6 = 5% of the product. Similarly, the fact that 70% of 2-chlorobutane is formed means that each of the four secondary hydrogens is responsible for 70% ÷ 4 = 17.5% of the product. Thus, reaction of a secondary hydrogen happens 17.5% ÷ 5% = 3.5 times as often as reaction of a primary hydrogen.

A similar calculation for the chlorination of 2-methylpropane indicates that each of the nine primary hydrogens accounts for 65% ÷ 9 = 7.2% of the product, while the single tertiary hydrogen (R_3CH) accounts for 35% of the product. Thus, a tertiary hydrogen is 35 ÷ 7.2 = 5 times as reactive as a primary hydrogen toward chlorination.

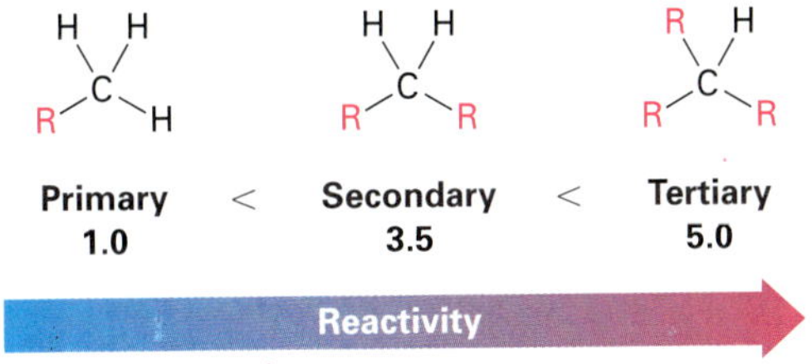

What are the reasons for the observed reactivity order of alkane hydrogens toward radical chlorination? A look at the bond dissociation energies given previously in Table 5.3 on page 156 hints at the answer. The data in Table 5.3 indicate that a tertiary C–H bond (390 kJ/mol; 93 kcal/mol) is weaker than a secondary C–H bond (401 kJ/mol; 96 kcal/mol), which is in turn weaker than a primary C–H bond (420 kJ/mol; 100 kcal/mol). Since less energy is needed to break a tertiary C–H bond than to break a primary or secondary C–H bond, the resultant tertiary radical is more stable than a primary or secondary radical.

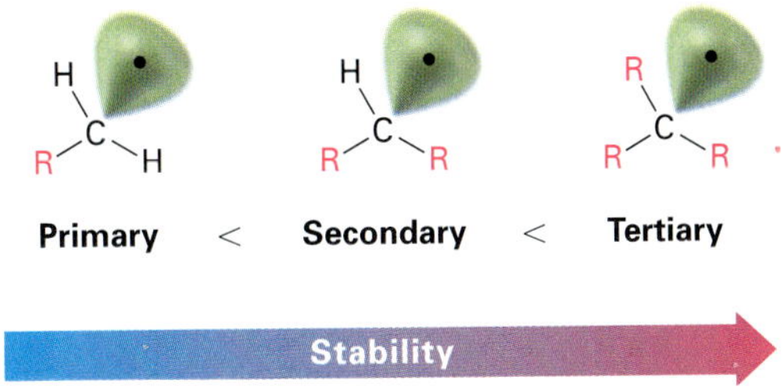

An explanation of the relationship between reactivity and bond strength in radical chlorination reactions relies on the Hammond postulate, discussed in Section 6.10 to explain why more stable carbocations form faster than less stable ones in alkene electrophilic addition reactions. An energy diagram for the formation of an alkyl radical during alkane chlorination is shown in Figure 10.2. Although the hydrogen abstraction step is slightly exergonic, there is nevertheless a certain amount of developing radical character in the transition state. Since the increasing alkyl substitution that stabilizes the radical intermediate also stabilizes the transition state leading to that intermediate, the more stable radical forms faster than the less stable one.

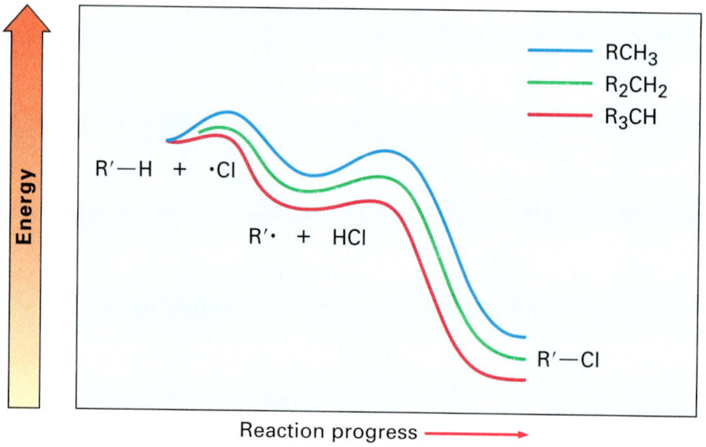

Figure 10.2 Energy diagram for alkane chlorination. The relative rates of formation of tertiary, secondary, and primary radicals are the same as their stability order.

In contrast with alkane chlorination, alkane bromination is usually much more selective. In its reaction with 2-methylpropane, for example, bromine abstracts the tertiary hydrogen with greater than 99% selectivity, as opposed to the 35:65 mixture observed in the corresponding chlorination.

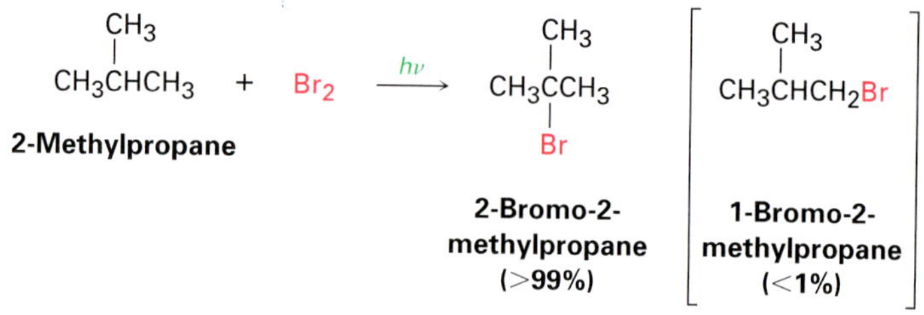

The enhanced selectivity of alkane bromination over chlorination can be explained by turning once again to the Hammond postulate. In comparing the abstractions of an alkane hydrogen by Cl· and Br· radicals, reaction with Br· is less exergonic. As a result, the transition state for bromination resembles the alkyl radical more closely than does the transition state for chlorination, and the stability of that radical is therefore more important for bromination than for chlorination.

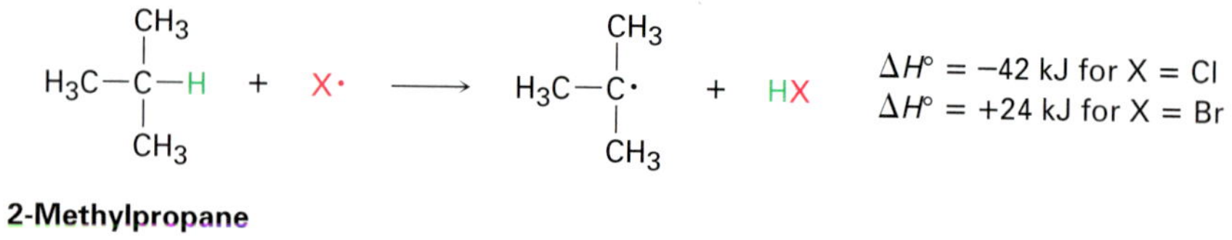

Problem 10.3 Draw and name all monochloro products you would expect to obtain from radical chlorination of 2-methylpentane. Which, if any, are chiral?

Problem 10.4 Taking the relative reactivities of 1°, 2°, and 3° hydrogen atoms into account, what product(s) would you expect to obtain from monochlorination of 2-methylbutane? What would the approximate percentage of each product be? (Don't forget to take into account the number of each sort of hydrogen.)

10.4 Preparing Alkyl Halides from Alkenes: Allylic Bromination

We've already seen several methods for preparing alkyl halides from alkenes, including the reactions of HX and X_2 with alkenes in electrophilic addition reactions (Sections 6.7 and 7.2). The hydrogen halides HCl, HBr, and HI react with alkenes by a polar mechanism to give the product of Markovnikov addition. Bromine and chlorine undergo anti addition through halonium ion intermediates to give 1,2-dihalogenated products.

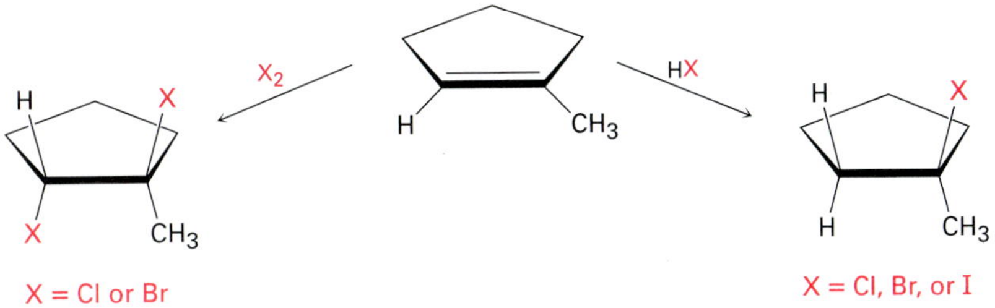

Another method for preparing alkyl halides from alkenes is by reaction with *N*-bromosuccinimide (abbreviated NBS) in the presence of light to give products resulting from substitution of hydrogen by bromine at the **allylic** position—the position *next to* the double bond. Cyclohexene, for example, gives 3-bromocyclohexene.

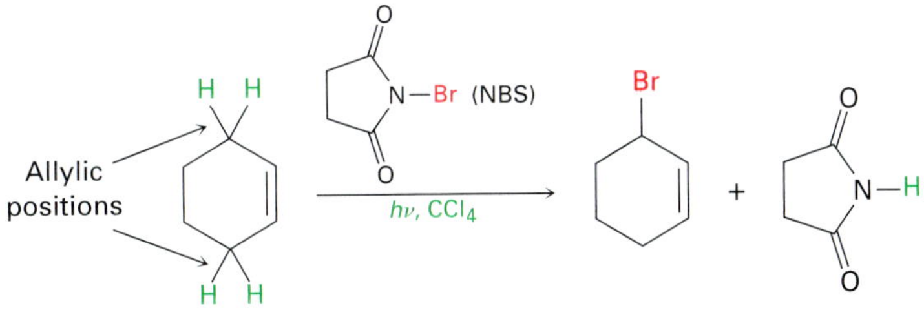

This allylic bromination with NBS is analogous to the alkane halogenation reaction discussed in the previous section and occurs by a radical chain reaction pathway. As in alkane halogenation, Br· radical abstracts an allylic hydrogen atom of the alkene, thereby forming an allylic radical plus HBr. This allylic radical then reacts with Br_2 to yield the product and a Br· radical, which cycles back into the first step and carries on the chain. The Br_2 results from reaction of NBS with the HBr formed in the first step.

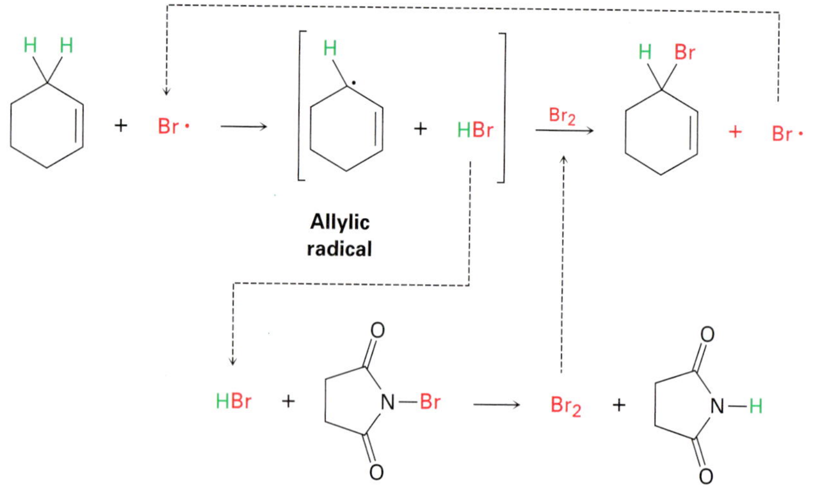

Why does bromination with NBS occur exclusively at an allylic position rather than elsewhere in the molecule? The answer, once again, is found by looking at bond dissociation energies to see the relative stabilities of various kinds of radicals.

There are three sorts of C–H bonds in cyclohexene, and Table 5.3 gives an estimate of their relative strengths. Although a typical secondary alkyl C–H bond has a strength of about 400 kJ/mol (96 kcal/mol) and a typical vinylic C–H bond has a strength of 445 kJ/mol (106 kcal/mol), an *allylic* C–H bond has a strength of only about 360 kJ/mol (87 kcal/mol). An allylic radical is therefore more stable than a typical alkyl radical with the same substitution by about 40 kJ/mol (9 kcal/mol).

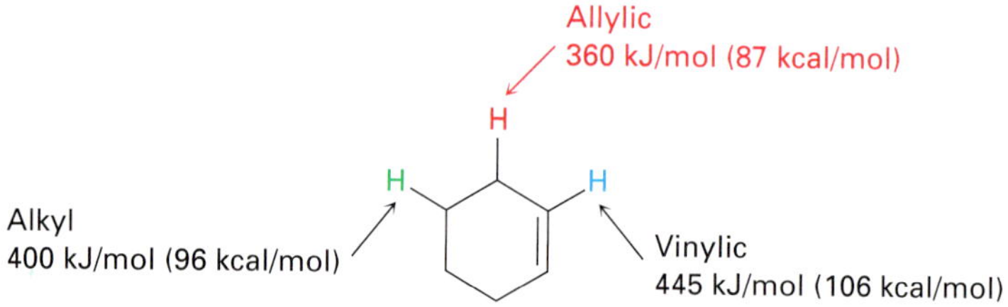

We can thus expand the stability ordering to include vinylic and allylic radicals.

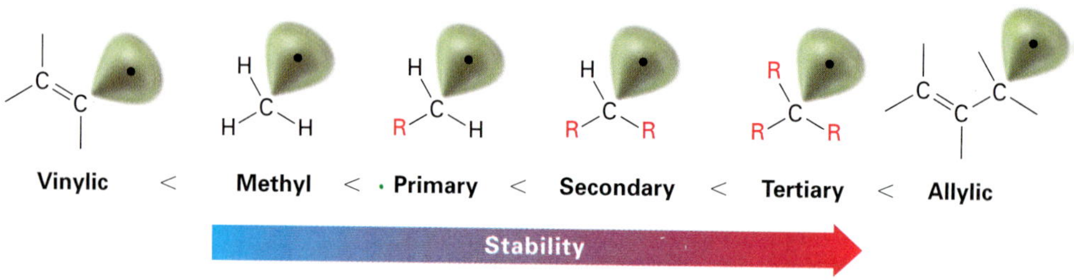

10.5 Stability of the Allyl Radical: Resonance Revisited

To see why allylic radicals are so stable, look at the orbital picture in Figure 10.3. The radical carbon atom with an unpaired electron can adopt sp^2 hybridization, placing the unpaired electron in a *p* orbital and giving a structure that is electronically symmetrical. The *p* orbital on the central carbon can therefore overlap equally well with a *p* orbital on *either* of the two neighboring carbons.

Because the allyl radical is electronically symmetrical, it can be drawn in either of two resonance forms—with the unpaired electron on the left and the double bond on the right or with the unpaired electron on the right and the double bond on the left. Neither structure is correct by itself; the true structure of the allyl radical is a resonance hybrid of the two. (You might want to review Sections 2.4–2.6 to brush up on resonance.) As noted in Section 2.5, the greater the number of resonance forms, the greater the stability of a compound because bonding electrons are attracted to more nuclei. An allyl radical, with two resonance forms, is therefore more stable than a typical alkyl radical, which has only a single structure.

Active Figure 10.3 An orbital view of the allyl radical. The *p* orbital on the central carbon can overlap equally well with a *p* orbital on either neighboring carbon, giving rise to two equivalent resonance structures. *Sign in at* **www.thomsonedu.com** *to see a simulation based on this figure and to take a short quiz.*

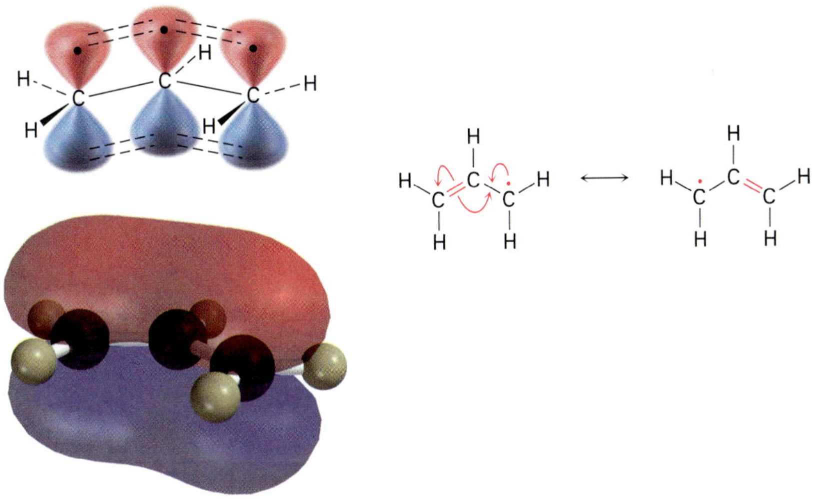

In molecular orbital terms, the stability of the allyl radical is due to the fact that the unpaired electron is **delocalized**, or spread out, over an extended π orbital network rather than localized at only one site, as shown by the computer-generated MO in Fig 10.3. This delocalization is particularly apparent in the so-called spin density surface in Figure 10.4, which shows the calculated location of the unpaired electron. The two terminal carbons share the unpaired electron equally.

In addition to its effect on stability, delocalization of the unpaired electron in the allyl radical has other chemical consequences. Because the unpaired electron is delocalized over both ends of the π orbital system, reaction with Br_2 can occur at either end. As a result, allylic bromination of an unsymmetrical alkene often leads to a mixture of products. For example, bromination of 1-octene gives a mixture of 3-bromo-1-octene and 1-bromo-2-octene. The two products are not formed in equal amounts, however, because the intermediate allylic radical is

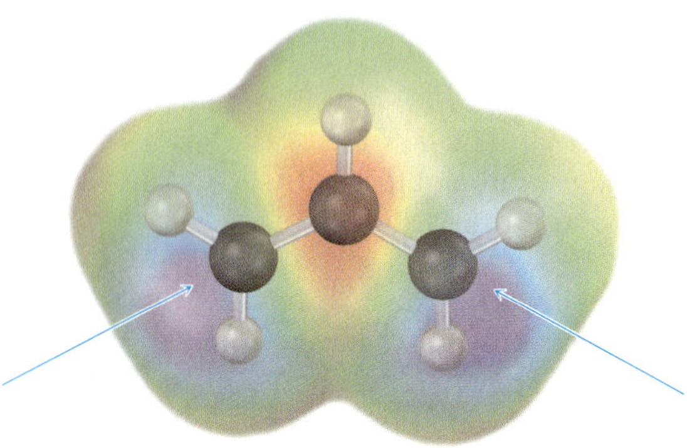

Active Figure 10.4 The spin density surface of the allyl radical locates the position of the unpaired electron (blue) and shows that it is equally shared between the two terminal carbons. *Sign in at* **www .thomsonedu.com** *to see a simulation based on this figure and to take a short quiz.*

not symmetrical and reaction at the two ends is not equally likely. Reaction at the less hindered, primary end is favored.

$CH_3CH_2CH_2CH_2CH_2CH_2CH{=}CH_2$

1-Octene

↓ NBS, CCl_4

$CH_3CH_2CH_2CH_2CH_2\dot{C}HCH{=}CH_2 \longleftrightarrow CH_3CH_2CH_2CH_2CH_2CH{=}CH\dot{C}H_2$

↓

$CH_3CH_2CH_2CH_2CH_2CH(Br)CH{=}CH_2 \; + \; CH_3CH_2CH_2CH_2CH_2CH{=}CHCH_2Br$

3-Bromo-1-octene (17%) **1-Bromo-2-octene (83%)** **(53 : 47 trans : cis)**

The products of allylic bromination reactions are useful for conversion into dienes by dehydrohalogenation with base. Cyclohexene can be converted into 1,3-cyclohexadiene, for example.

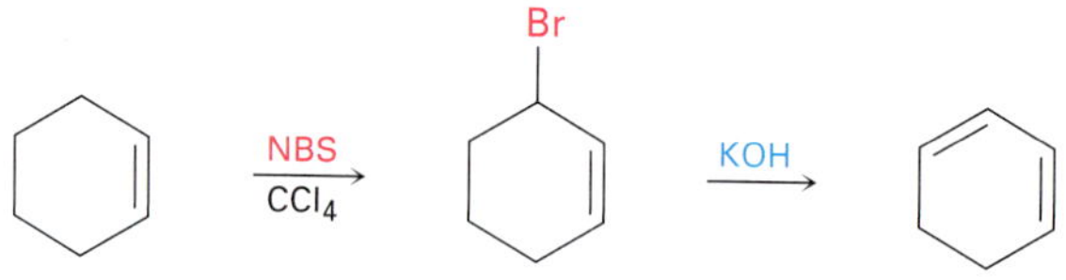

Cyclohexene **3-Bromocyclohexene** **1,3-Cyclohexadiene**

WORKED EXAMPLE 10.1 *Predicting the Product of an Allylic Bromination Reaction*

What products would you expect from reaction of 4,4-dimethylcyclohexene with NBS?

Strategy Draw the alkene reactant, and identify the allylic positions. In this case, there are two different allylic positions; we'll label them **A** and **B**. Now abstract an allylic hydrogen from each position to generate the two corresponding allylic radicals. Each of the two allylic radicals can add a Br atom at either end (**A** or **a**; **B** or **b**) to give a mixture of up to four products. Draw and name the products. In the present instance, the "two" products from reaction at position **B** are identical, so a total of only three products are formed in this reaction.

Solution

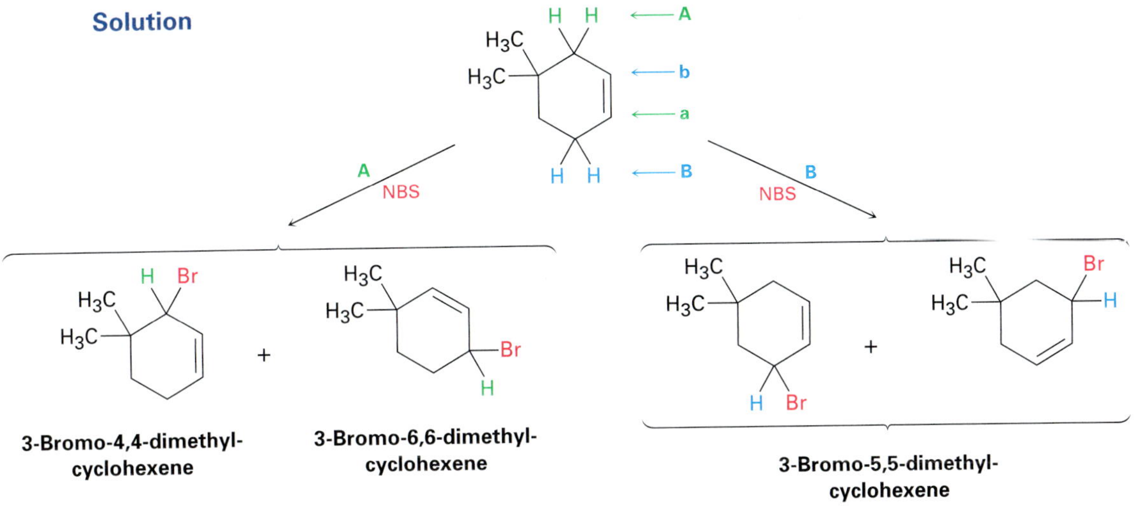

Problem 10.5 Draw three resonance forms for the cyclohexadienyl radical.

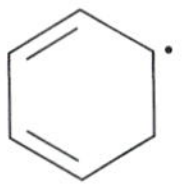

Cyclohexadienyl radical

Problem 10.6 The major product of the reaction of methylenecyclohexane with *N*-bromosuccinimide is 1-(bromomethyl)cyclohexene. Explain.

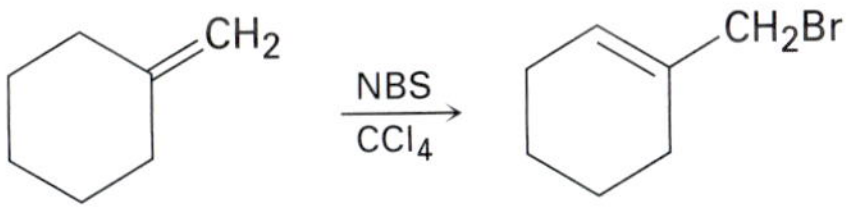

Major product

Problem 10.7 What products would you expect from reaction of the following alkenes with NBS? If more than one product is formed, show the structures of all.

(a)

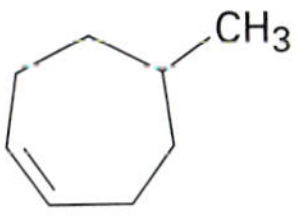

(b)

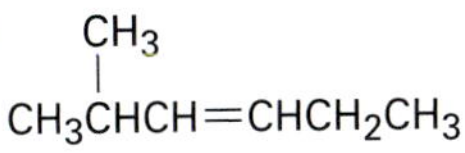

10.6 Preparing Alkyl Halides from Alcohols

ThomsonNOW™ Click *Organic Interactive* to **use a web-based palette to design a synthesis of alkyl halides, beginning with alcohols.**

The most generally useful method for preparing alkyl halides is to make them from alcohols, which themselves can be obtained from carbonyl compounds, as we'll see in Sections 17.4 and 17.5. Because of the importance of the process, many different methods have been developed to transform alcohols into alkyl halides. The simplest method is to treat the alcohol with HCl, HBr, or HI. For reasons that will be discussed in Section 11.5, the reaction works best with tertiary alcohols, R_3COH. Primary and secondary alcohols react much more slowly and at higher temperatures.

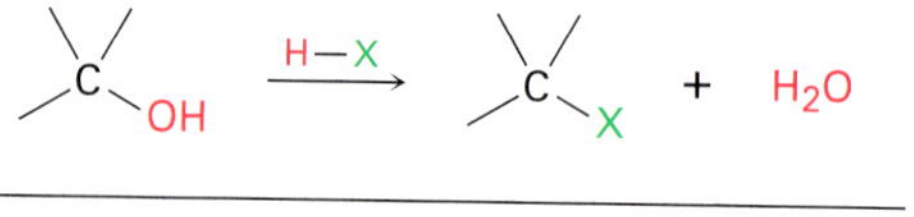

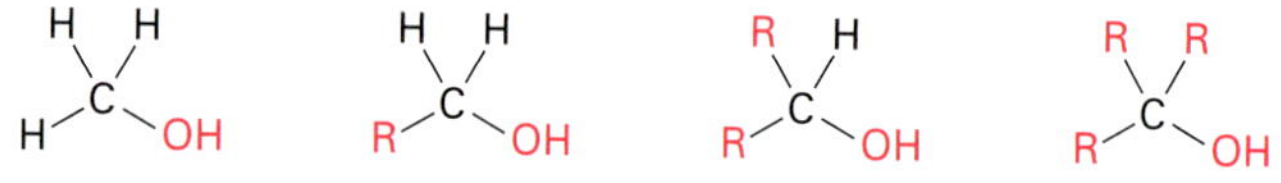

Methyl < **Primary** < **Secondary** < **Tertiary**

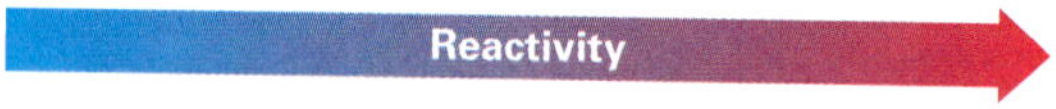

The reaction of HX with a tertiary alcohol is so rapid that it's often carried out simply by bubbling the pure HCl or HBr gas into a cold ether solution of the alcohol. 1-Methylcyclohexanol, for example, is converted into 1-chloro-1-methylcyclohexane by treating with HCl.

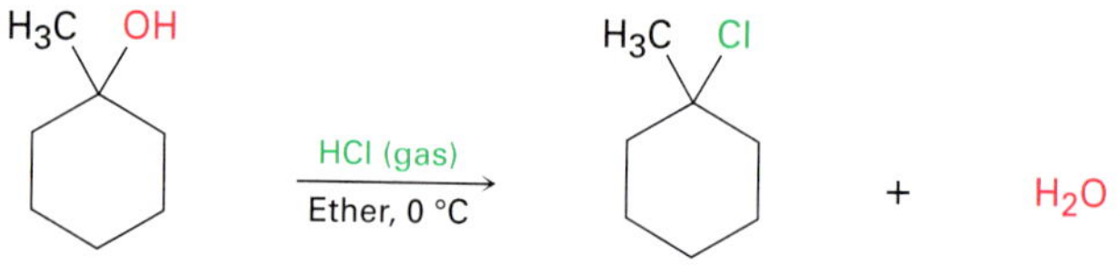

1-Methylcyclohexanol **1-Chloro-1-methylcyclohexane (90%)**

Primary and secondary alcohols are best converted into alkyl halides by treatment with either thionyl chloride ($SOCl_2$) or phosphorus tribromide (PBr_3). These reactions, which normally take place readily under mild conditions, are less acidic and less likely to cause acid-catalyzed rearrangements than the HX method.

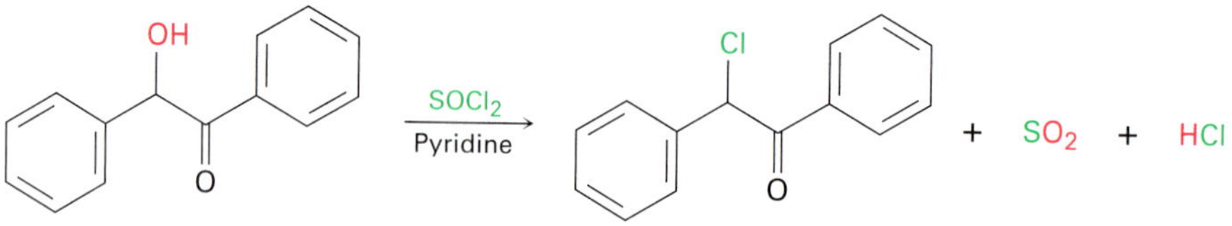

Benzoin **(86%)**

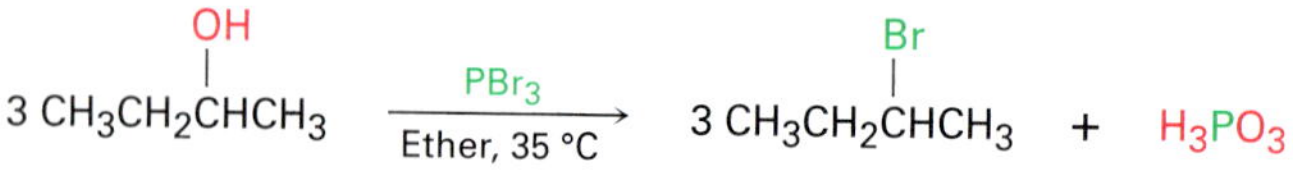

2-Butanol **2-Bromobutane (86%)**

As the preceding examples indicate, the yields of these $SOCl_2$ and PBr_3 reactions are generally high, and other functional groups such as ethers, carbonyls, and aromatic rings don't usually interfere. We'll look at the mechanisms of these substitution reactions in the next chapter.

Problem 10.8 How would you prepare the following alkyl halides from the corresponding alcohols?

(a) 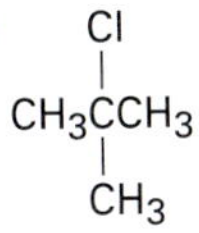

(b) 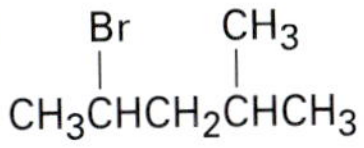

(c) 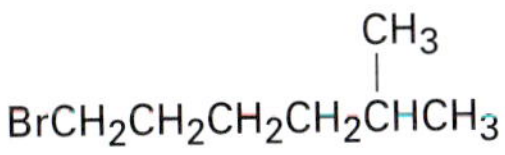

(d) 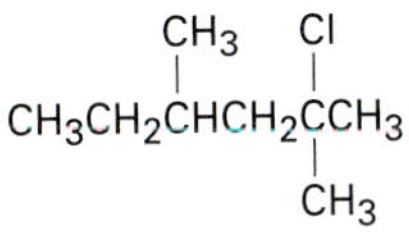

10.7 Reactions of Alkyl Halides: Grignard Reagents

François Auguste Victor Grignard

François Auguste Victor Grignard (1871–1935) was born in Cherbourg, France, and received his Ph.D. at the University of Lyon in 1901. During his doctoral work under Philippe Barbier, Grignard discovered the preparation and usefulness of organomagnesium reagents. He became professor of chemistry at Nancy and at Lyon, and he won the Nobel Prize in chemistry in 1912. During World War I, he was drafted into the French army as a Corporal (a Nobel Prize–winning Corporal!), where he developed a method for detecting German war gases.

Alkyl halides, RX, react with magnesium metal in ether or tetrahydrofuran (THF) solvent to yield alkylmagnesium halides, RMgX. The products, called **Grignard reagents** after their discoverer, Victor Grignard, are examples of *organometallic* compounds because they contain a carbon–metal bond. In addition to alkyl halides, Grignard reagents can also be made from alkenyl (vinylic) and aryl (aromatic) halides. The halogen can be Cl, Br, or I, although chlorides are less reactive than bromides and iodides. Organofluorides rarely react with magnesium.

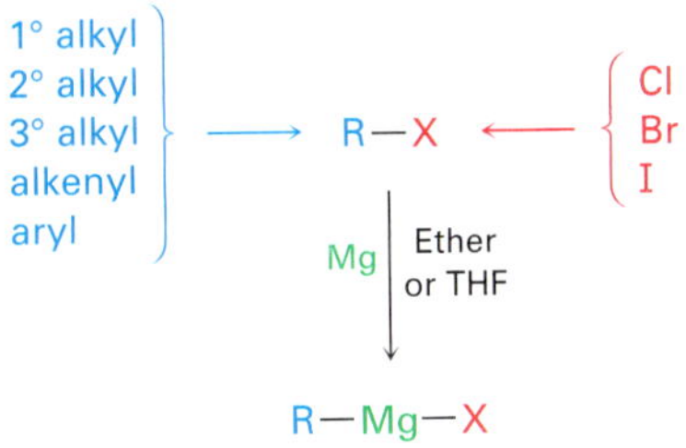

As you might expect from the discussion of electronegativity and bond polarity in Section 5.4, the carbon–magnesium bond is polarized, making the carbon atom of Grignard reagents both nucleophilic and basic. An electrostatic potential map of methylmagnesium iodide, for instance, indicates the electron-rich (red) character of the carbon bonded to magnesium.

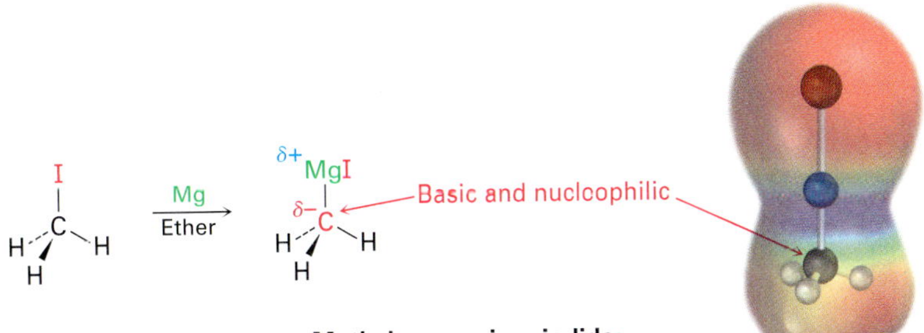

Iodomethane **Methylmagnesium iodide**

In a formal sense, a Grignard reagent is the magnesium salt, $R_3C^- \ {}^+MgX$, of a carbon acid, $R_3C{-}H$. But because hydrocarbons are such weak acids, with pK_a's in the range of 44 to 60 (Section 8.7), carbon anions are very strong bases. Grignard reagents therefore react with such weak acids as H_2O, ROH, RCO_2H, and RNH_2 to abstract a proton and yield hydrocarbons. Thus, an organic halide can be reduced to a hydrocarbon by converting it to a Grignard reagent followed by protonation, $R{-}X \rightarrow R{-}MgX \rightarrow R{-}H$.

$$CH_3CH_2CH_2CH_2CH_2CH_2Br \xrightarrow[\text{Ether}]{\text{Mg}} CH_3CH_2CH_2CH_2CH_2CH_2MgBr \xrightarrow{H_2O} CH_3CH_2CH_2CH_2CH_2CH_3$$

1-Bromohexane **1-Hexylmagnesium bromide** **Hexane (85%)**

We'll see many more uses of Grignard reagents as sources for carbon nucleophiles in later chapters.

Problem 10.9 How strong a base would you expect a Grignard reagent to be? Look at Table 8.1 on page 271, and then predict whether the following reactions will occur as written. (The pK_a of NH_3 is 35.)

(a) $CH_3MgBr + H{-}C{\equiv}C{-}H \rightarrow CH_4 + H{-}C{\equiv}C{-}MgBr$

(b) $CH_3MgBr + NH_3 \rightarrow CH_4 + H_2N{-}MgBr$

Problem 10.10 How might you replace a halogen substituent by a deuterium atom if you wanted to prepare a deuterated compound?

10.8 Organometallic Coupling Reactions

Many other kinds of organometallic compounds can be prepared in a manner similar to that of Grignard reagents. For instance, alkyllithium reagents, RLi, can be prepared by the reaction of an alkyl halide with lithium metal. Alkyllithiums are both nucleophiles and strong bases, and their chemistry is similar in many respects to that of alkylmagnesium halides.

$$CH_3CH_2CH_2CH_2Br \xrightarrow[\text{Pentane}]{\text{2 Li}} CH_3CH_2CH_2CH_2Li + LiBr$$

1-Bromobutane **Butyllithium**

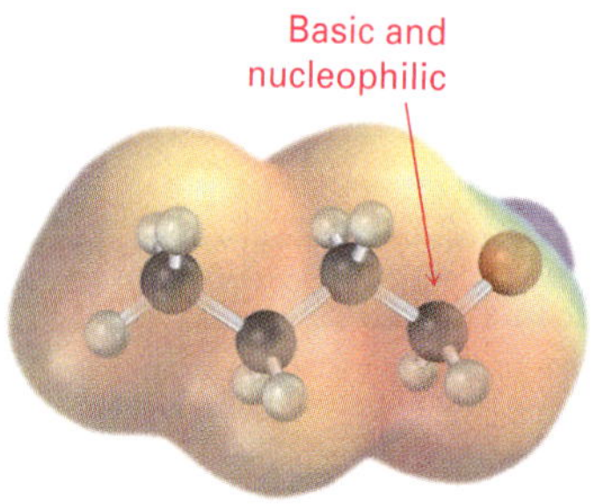

One particularly valuable reaction of alkyllithiums is in making lithium diorganocopper compounds, LiR_2Cu, by reaction with copper(I) iodide in

Henry Gilman

Henry Gilman (1893–1986) was born in Boston, Massachusetts, and received his Ph.D. in 1918 at Harvard. He then became professor of chemistry at Iowa State University (1919–1962), where he remained active until his death at age 93. An extremely prolific researcher, Gilman published more than 1000 scientific papers during his career. Remarkably, he lost much of his eyesight at age 53 but still went on to accomplish some of his finest work in later years.

diethyl ether as solvent. Called **Gilman reagents**, lithium diorganocopper compounds are useful because they undergo a *coupling* reaction with organochlorides, organobromides, and organoiodides (but not fluorides). One of the alkyl groups from the Gilman reagent replaces the halogen of the organohalide, forming a new carbon–carbon bond and yielding a hydrocarbon product. Lithium dimethylcopper, for example, reacts with 1-iododecane to give undecane in 90% yield.

$$2\ CH_3Li + CuI \xrightarrow{\text{Ether}} (CH_3)_2Cu^-\ Li^+ + LiI$$

Methyllithium → **Lithium dimethylcopper (a Gilman reagent)**

$$(CH_3)_2CuLi + CH_3(CH_2)_8CH_2I \xrightarrow[0\ °C]{\text{Ether}} CH_3(CH_2)_8CH_2CH_3 + LiI + CH_3Cu$$

Lithium dimethylcopper + **1-Iododecane** → **Undecane (90%)**

This organometallic coupling reaction is useful in organic synthesis because it forms carbon–carbon bonds, thereby making possible the preparation of larger molecules from smaller ones. As the following examples indicate, the coupling reaction can be carried out on aryl and vinylic halides as well as on alkyl halides.

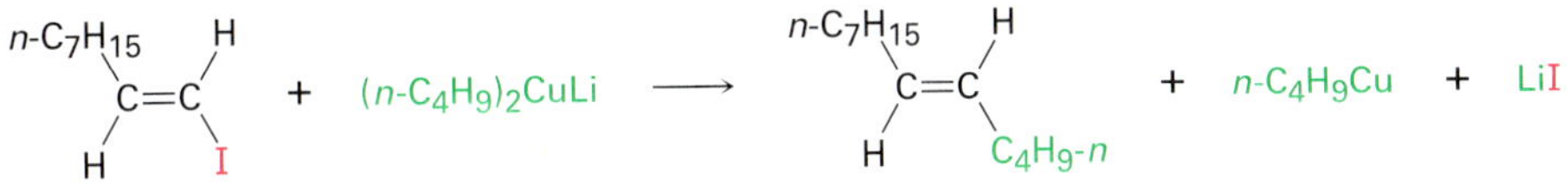

***trans*-1-Iodo-1-nonene** → ***trans*-5-Tridecene (71%)**

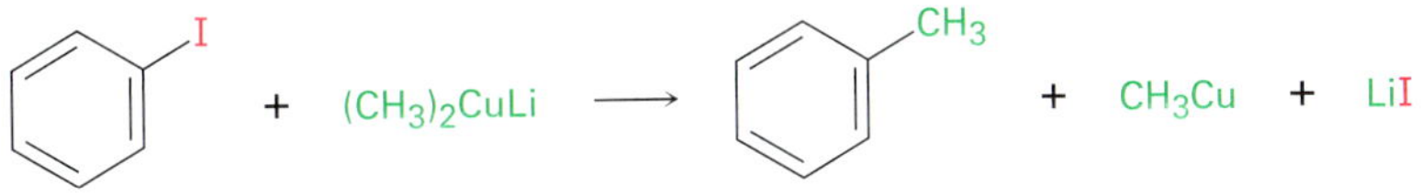

Iodobenzene → **Toluene (91%)**

The mechanism of the reaction involves initial formation of a triorganocopper intermediate, followed by coupling and loss of RCu. The coupling is not a typical polar nucleophilic substitution reaction of the sort considered in the next chapter.

$$R{-}X + [R'{-}Cu{-}R']^-\ Li^+ \longrightarrow \left[\begin{array}{c} R \\ | \\ R'{-}Cu{-}R' \end{array}\right] \longrightarrow R{-}R' + R'{-}Cu$$

ThomsonNOW™ Click *Organic Interactive* to **learn more about the preparation of organometallics and their use in coupling reactions.**

In addition to the coupling reaction of diorganocopper reagents with organohalides, related processes also occur with other organometallics, particularly organopalladium compounds. One of the more commonly used procedures is the palladium-catalyzed reaction of an aryl or vinyl substituted organotin reagent with an organohalide. The organotin is itself usually formed by reaction of an organolithium such as vinyllithium with tributyltin chloric Bu_3SnCl. For example:

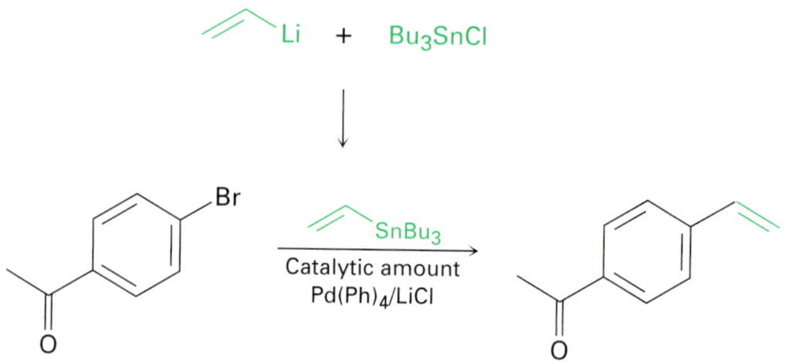

para-**Bromoacetophenone** *para*-**Vinylacetophenone**

Problem 10.11 How would you carry out the following transformations using an organocopper co pling reaction? More than one step is required in each case.

(a)

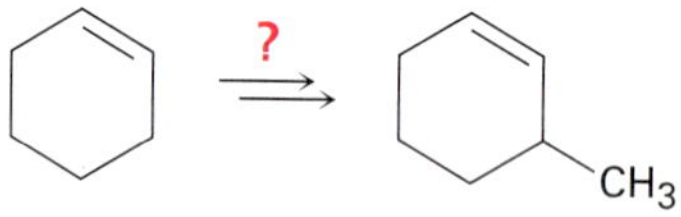

(b) $CH_3CH_2CH_2CH_2Br \xrightarrow{?} CH_3CH_2CH_2CH_2CH_2CH_2CH_2CH_3$

(c) $CH_3CH_2CH_2CH{=}CH_2 \xrightarrow{?} CH_3CH_2CH_2CH_2CH_2CH_2CH_2CH_2CH_2CH_3$

10.9 Oxidation and Reduction in Organic Chemistry

We've pointed out on several occasions that some of the reactions discussed ir this and earlier chapters are either *oxidations* or *reductions.* As noted in Section 7.7 and 7.8, an organic oxidation results in a loss of electron density by carbon caused either by bond formation between carbon and a more electronegativ atom (usually O, N, or a halogen) or by bond-breaking between carbon and a les electronegative atom (usually H). Conversely, an organic reduction results in a gain of electron density by carbon, caused either by bond formation betweer carbon and a less electronegative atom or by bond-breaking between carbon anc a more electronegative atom.

Oxidation Decreases electron density on carbon by:
- forming one of these: C–O C–N C–X
- or breaking this: C–H

Reduction Increases electron density on carbon by:
- forming this: C–H
- or breaking one of these: C–O C–N C–X

Based on these definitions, the chlorination reaction of methane to yield chloromethane is an oxidation because a C–H bond is broken and a C–Cl bond is formed. The conversion of an alkyl chloride to an alkane via a Grignard reagent followed by protonation is a reduction, however, because a C–Cl bond is broken and a C–H bond is formed.

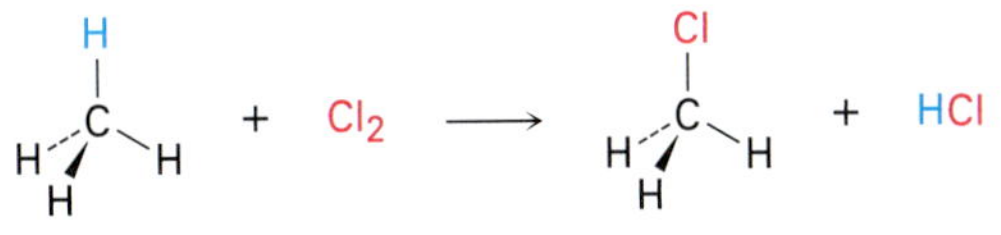

Methane Chloromethane

Oxidation: C–H bond broken and C–Cl bond formed

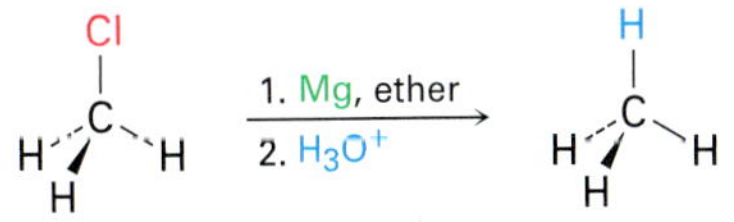

Chloromethane Methane

Reduction: C–Cl bond broken and C–H bond formed

As other examples, the reaction of an alkene with Br_2 to yield a 1,2-dibromide is an oxidation because two C–Br bonds are formed, but the reaction of an alkene with HBr to yield an alkyl bromide is neither an oxidation nor a reduction because both a C–H and a C–Br bond are formed.

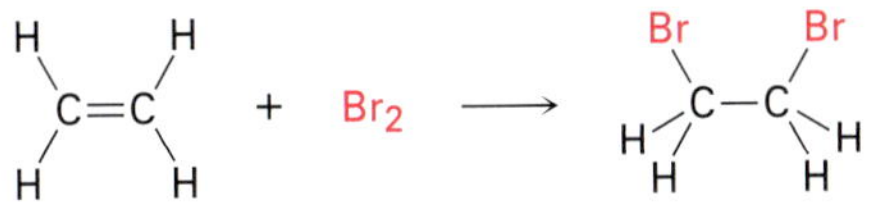

Ethylene 1,2-Dibromoethane

Oxidation: Two new bonds formed between carbon and a more electronegative element

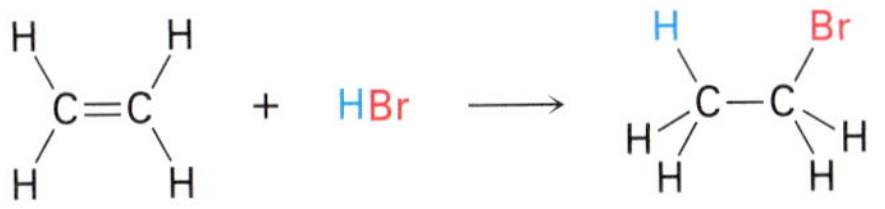

Ethylene Bromoethane

Neither oxidation nor reduction: One new C–H bond and one new C–Br bond formed

A list of compounds of increasing oxidation level is shown in Figure 10.5. Alkanes are at the lowest oxidation level because they have the maximum possible number of C–H bonds per carbon, and CO_2 is at the highest level because it has the maximum possible number of C–O bonds per carbon. Any reaction that converts a compound from a lower level to a higher level is an oxidation, any reaction that converts a compound from a higher level to a lower level is a reduction, and any reaction that doesn't change the level is neither an oxidation nor a reduction.

gure 10.5 Oxidation levels f some common types of ompounds.

CH_3CH_3	$H_2C{=}CH_2$	$HC{\equiv}CH$		
	CH_3OH	$H_2C{=}O$	HCO_2H	CO_2
	CH_3Cl	CH_2Cl_2	$CHCl_3$	CCl_4
	CH_3NH_2	$H_2C{=}NH$	$HC{\equiv}N$	

Low oxidation level → High oxidation level

Worked Example 10.2 shows how to compare the oxidation levels of diffe ent compounds with the same number of carbon atoms.

WORKED EXAMPLE 10.2 Comparing Oxidation Levels of Compounds

Rank the following compounds in order of increasing oxidation level:

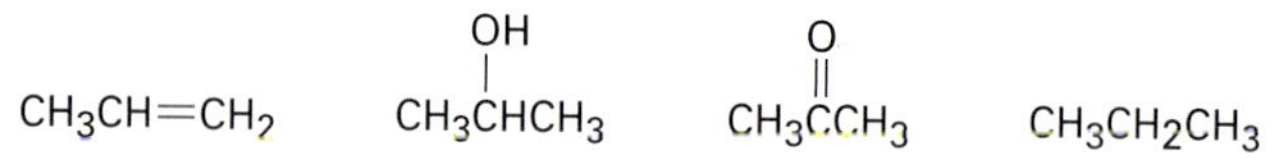

Strategy Compounds that have the same number of carbon atoms can be compared b adding the number of C–O, C–N, and C–X bonds in each and then subtracting th number of C–H bonds. The larger the resultant value, the higher the oxidation leve

Solution The first compound (propene) has six C–H bonds, giving an oxidation level of –6 the second (2-propanol) has one C–O bond and seven C–H bonds, giving an oxi dation level of –6; the third (acetone) has two C–O bonds and six C–H bonds, giv ing an oxidation level of –4; and the fourth (propane) has eight C–H bonds, givin an oxidation level of –8. Thus, the order of increasing oxidation level is

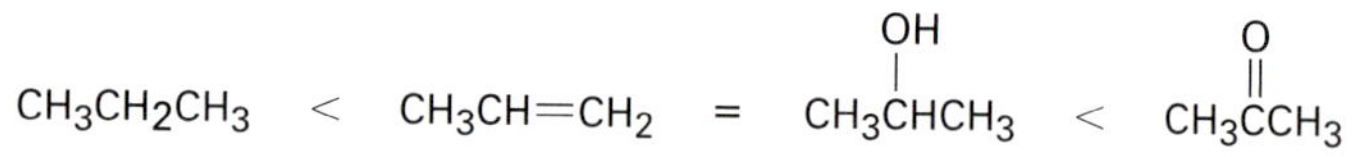

Problem 10.12 Rank each of the following series of compounds in order of increasing oxidation level:

(a)

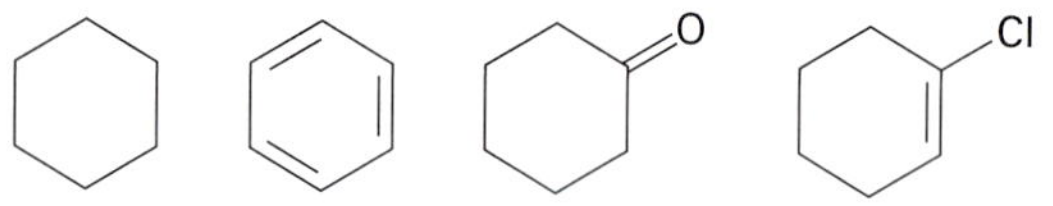

(b) CH_3CN $CH_3CH_2NH_2$ $H_2NCH_2CH_2NH_2$

Problem 10.13 Tell whether each of the following reactions is an oxidation, a reduction, or neither.

(a)

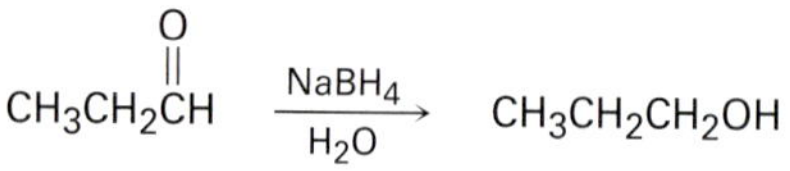

(b)

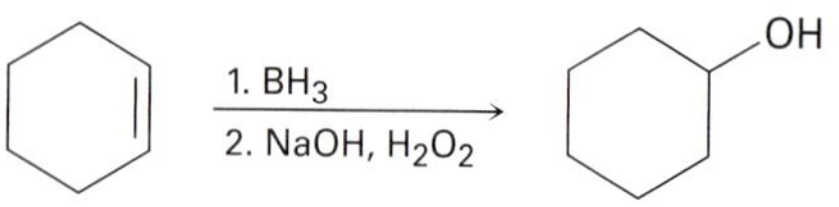

Focus On . . .

Naturally Occurring Organohalides

Marine corals secrete organohalogen compounds that act as a feeding deterrent to starfish.

As recently as 1970, only about 30 naturally occurring organohalogen compounds were known. It was simply assumed that chloroform, halogenated phenols, chlorinated aromatic compounds called PCBs, and other such substances found in the environment were industrial pollutants. Now, only a third of a century later, the situation is quite different. More than 5000 organohalogen compounds have been found to occur naturally, and tens of thousands more surely exist. From a simple compound like chloromethane to an extremely complex one like vancomycin, a remarkably diverse range of organohalogen compounds exists in plants, bacteria, and animals. Many even have valuable physiological activity. Vancomycin, for instance, is a powerful antibiotic produced by the bacterium *Amycolatopsis orientalis* and used clinically to treat methicillin-resistant *Staphylococcus aureus* (MRSA).

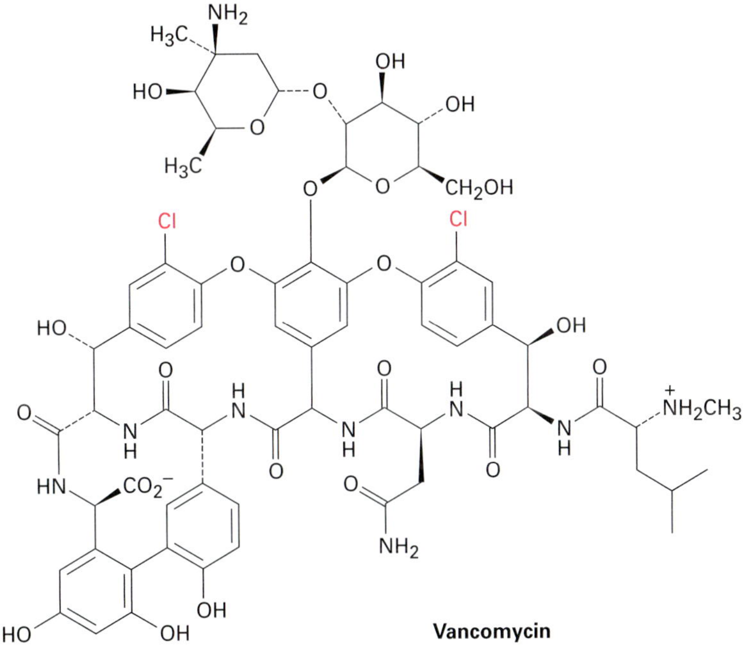

Vancomycin

Some naturally occurring organohalogen compounds are produced in massive quantities. Forest fires, volcanoes, and marine kelp release up to *5 million tons* of CH_3Cl per year, for example, while annual industrial emissions

(continued)

total about 26,000 tons. Termites are thought to release as much as 10^8 kg of chloroform per year. A detailed examination of the Okinawan acorn worm *Ptychodera flava* found that the 64 million worms living in a 1 km^2 study area excreted nearly 8000 pounds per year of bromophenols and bromoindoles, compounds previously thought to be nonnatural pollutants.

Why do organisms produce organohalogen compounds, many of which are undoubtedly toxic? The answer seems to be that many organisms use organohalogen compounds for self-defense, either as feeding deterrents, as irritants to predators, or as natural pesticides. Marine sponges, coral, and sea hares, for example, release foul-tasting organohalogen compounds that deter fish, starfish, and other predators from eating them. Even humans appear to produce halogenated compounds as part of their defense against infection. The human immune system contains a peroxidase enzyme capable of carrying out halogenation reactions on fungi and bacteria, thereby killing the pathogen. And most remarkable of all, even free chlorine—Cl_2—has been found to be present in humans.

Much remains to be learned—only a few hundred of the more than 500,000 known species of marine organisms have been examined—but it is clear that organohalogen compounds are an integral part of the world around us.

SUMMARY AND KEY WORDS

alkyl halide, 333
allylic, 339
delocalized, 341
Gilman reagent (LiR_2Cu), 347
Grignard reagent (RMgX), 345
organohalide, 332

Alkyl halides contain a halogen bonded to a saturated, sp^3-hybridized carbon atom. The C–X bond is polar, and alkyl halides can therefore behave as electrophiles.

Simple alkyl halides can be prepared by radical halogenation of alkanes, but mixtures of products usually result. The reactivity order of alkanes toward halogenation is identical to the stability order of radicals: $R_3C\cdot > R_2CH\cdot > RCH_2\cdot$. Alkyl halides can also be prepared from alkenes by reaction with *N*-bromosuccinimide (NBS) to give the product of **allylic** bromination. The NBS bromination of alkenes takes place through an intermediate allylic radical, which is stabilized by resonance.

Alcohols react with HX to form alkyl halides, but the reaction works well only for tertiary alcohols, R_3COH. Primary and secondary alkyl halides are normally prepared from alcohols using either $SOCl_2$ or PBr_3. Alkyl halides react with magnesium in ether solution to form organomagnesium halides, called **Grignard reagents (RMgX)**. Because Grignard reagents are both nucleophilic and basic, they react with acids to yield hydrocarbons. The overall result of Grignard formation and protonation is the conversion of an alkyl halide into an alkane (RX → RMgX → RH).

Alkyl halides also react with lithium metal to form organolithium reagents, RLi. In the presence of CuI, these form diorganocoppers, or **Gilman reagents (LiR_2Cu)**. Gilman reagents react with alkyl halides to yield coupled hydrocarbon products.

In organic chemistry, an *oxidation* is a reaction that causes a decrease in electron density on carbon, either by bond formation between carbon and a more electronegative atom (usually oxygen, nitrogen, or a halogen) or by bond-breaking between carbon and a less electronegative atom (usually hydrogen). Conversely, a *reduction* causes an increase of electron density on carbon, either by bond-breaking between carbon and a more electronegative atom or by bond formation between carbon and a less electronegative atom. Thus, the halogenation of an alkane to yield an alkyl halide is an oxidation, while the conversion of an alkyl halide to an alkane by protonation of a Grignard reagent is a reduction.

SUMMARY OF REACTIONS

1. Preparation of alkyl halides
   (a) From alkenes by allylic bromination (Section 10.4)

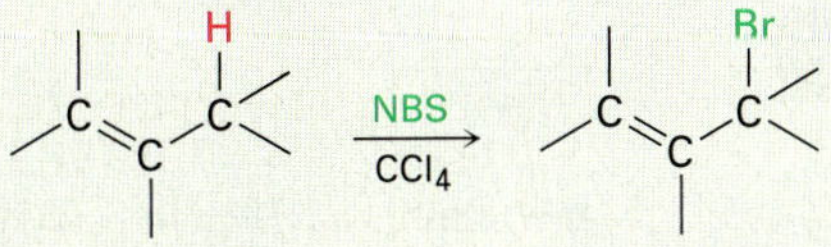

   (b) From alcohols (Section 10.6)
   (1) Reaction with HX

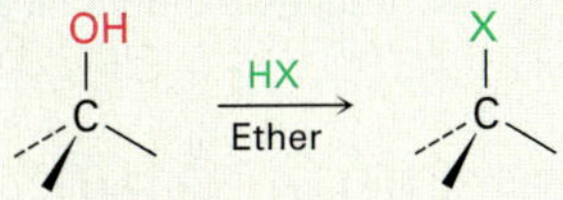

Reactivity order: 3° > 2° > 1°

   (2) Reaction of 1° and 2° alcohols with $SOCl_2$

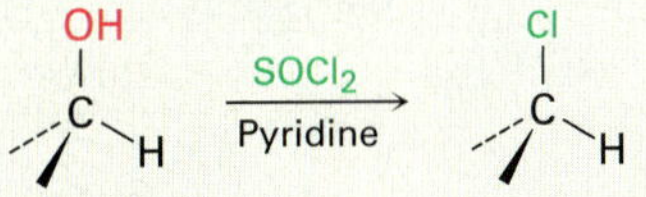

   (3) Reaction of 1° and 2° alcohols with PBr_3

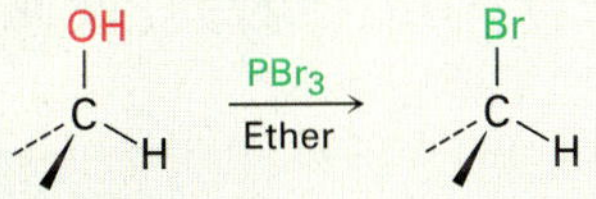

2. Reactions of alkyl halides
   (a) Formation of Grignard (organomagnesium) reagents (Section 10.7)

$$\mathrm{R{-}X} \xrightarrow[\text{Ether}]{\text{Mg}} \mathrm{R{-}Mg{-}X}$$

   (b) Formation of Gilman (diorganocopper) reagents (Section 10.8)

$$\mathrm{R{-}X} \xrightarrow[\text{Pentane}]{\text{2 Li}} \mathrm{R{-}Li} + \mathrm{LiX}$$

$$2\ \mathrm{R{-}Li} + \mathrm{CuI} \xrightarrow{\text{In ether}} [\mathrm{R{-}Cu{-}R}]^{-}\ \mathrm{Li}^{+} + \mathrm{LiI}$$

(c) Organometallic coupling (Section 10.8)

$$R_2CuLi + R'{-}X \xrightarrow{\text{In ether}} R{-}R' + RCu + LiX$$

(d) Reduction of alkyl halides to alkanes (Section 10.7)

$$R{-}X \xrightarrow[\text{Ether}]{\text{Mg}} R{-}Mg{-}X \xrightarrow{H_3O^+} R{-}H + HOMgX$$

EXERCISES

Organic KNOWLEDGE TOOLS

ThomsonNOW™ Sign in at **www.thomsonedu.com** to assess your knowledge of this chapter's topics by taking a pre-test. The pre-test will link you to interactive organic chemistry resources based on your score in each concept area.

Online homework for this chapter may be assigned in Organic OWL.

■ indicates problems assignable in Organic OWL.

VISUALIZING CHEMISTRY

(Problems 10.1–10.13 appear within the chapter.)

10.14 ■ Give a IUPAC name for each of the following alkyl halides (yellow-green = Cl):

(a)

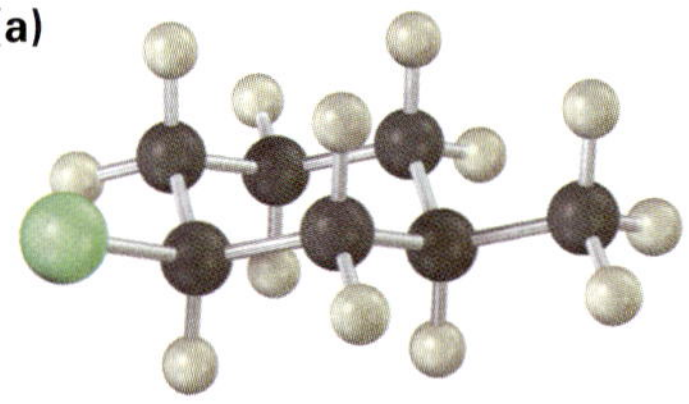

(b)

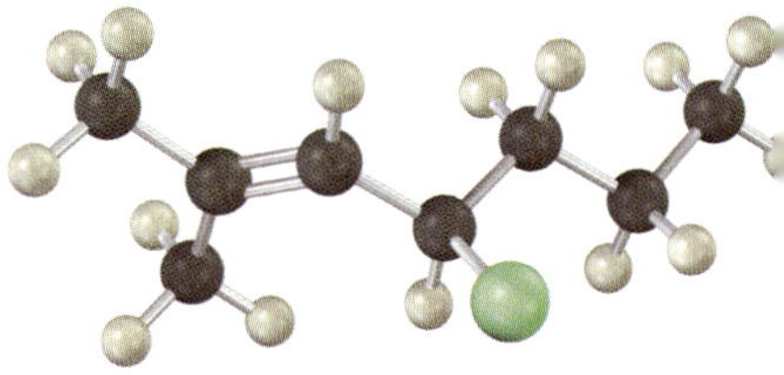

10.15 ■ Show the product(s) of reaction of the following alkenes with NBS:

(a)

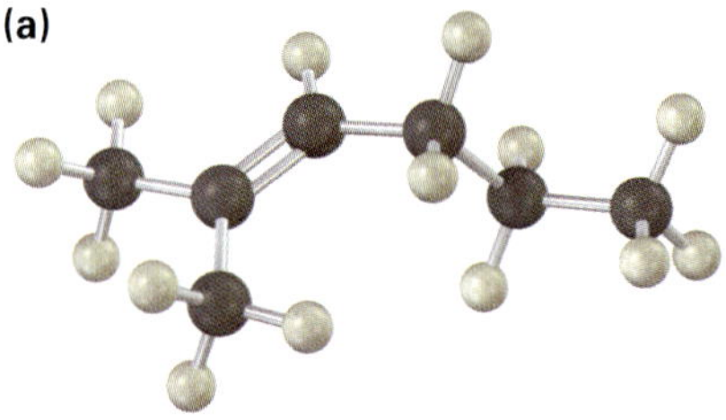

(b)

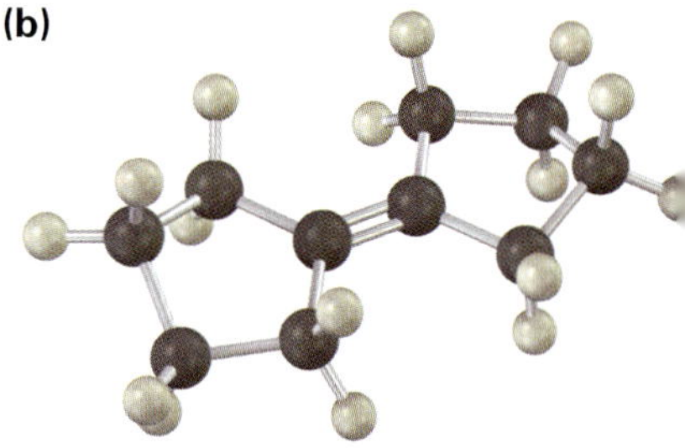

10.16 The following alkyl bromide can be prepared by reaction of the alcohol (*S*)-2-pentanol with PBr_3. Name the compound, assign (*R*) or (*S*) stereochemistry, and tell whether the reaction of the alcohol occurs with retention of the same stereochemistry or with a change in stereochemistry (reddish brown = Br).

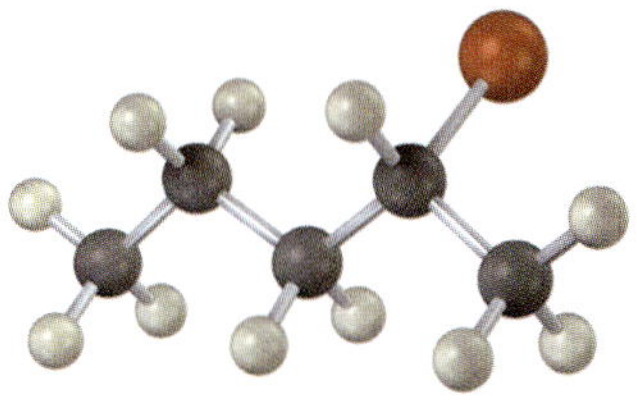

ADDITIONAL PROBLEMS

10.17 ■ Name the following alkyl halides:

(a)
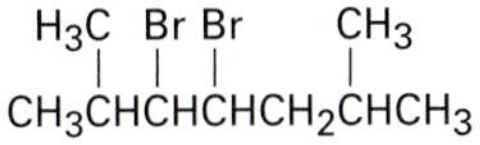

(b)
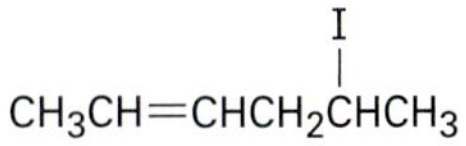

(c)
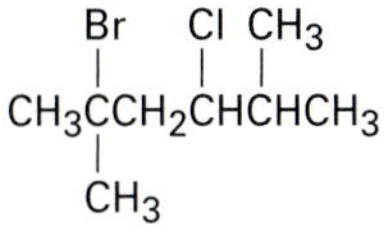

(d)
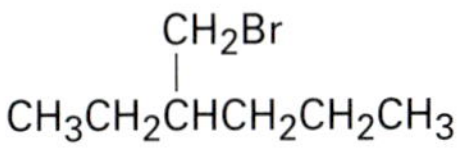

(e) $ClCH_2CH_2CH_2C{\equiv}CCH_2Br$

10.18 ■ Draw structures corresponding to the following IUPAC names:
(a) 2,3-Dichloro-4-methylhexane
(b) 4-Bromo-4-ethyl-2-methylhexane
(c) 3-Iodo-2,2,4,4-tetramethylpentane
(d) *cis*-1-Bromo-2-ethylcyclopentane

10.19 ■ Draw and name the monochlorination products you might obtain by radical chlorination of 2-methylbutane. Which of the products are chiral? Are any of the products optically active?

10.20 A chemist requires a large amount of 1-bromo-2-pentene as starting material for a synthesis and decides to carry out an NBS allylic bromination reaction. What is wrong with the following synthesis plan? What side products would form in addition to the desired product?

$$CH_3CH_2CH{=}CHCH_3 \xrightarrow[CCl_4]{NBS} CH_3CH_2CH{=}CHCH_2Br$$

10.21 What product(s) would you expect from the reaction of 1-methylcyclohexene with NBS? Would you use this reaction as part of a synthesis?

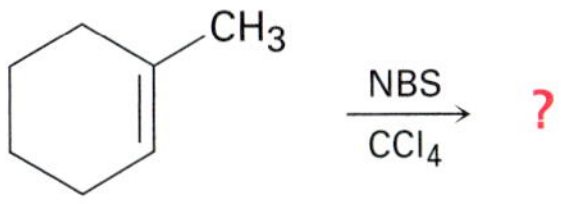

10.22 ■ How would you prepare the following compounds, starting with cyclopentene and any other reagents needed?
(a) Chlorocyclopentane
(b) Methylcyclopentane
(c) 3-Bromocyclopentene
(d) Cyclopentanol
(e) Cyclopentylcyclopentane
(f) 1,3-Cyclopentadiene

10.23 ■ Predict the product(s) of the following reactions:

(a) 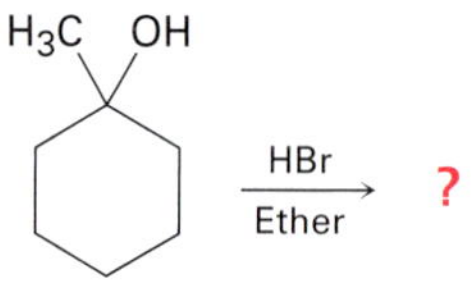

(b) $CH_3CH_2CH_2CH_2OH \xrightarrow{SOCl_4}$?

(c) 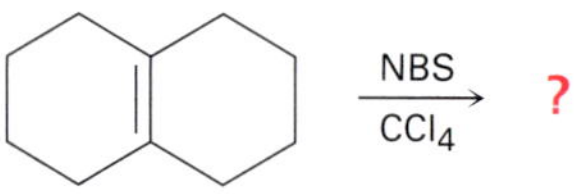

(d) 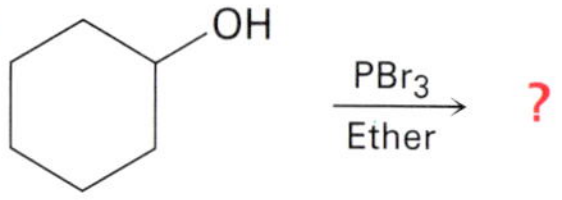

(e) $CH_3CH_2CHBrCH_3 \xrightarrow[\text{Ether}]{Mg}$ A? $\xrightarrow{H_2O}$ B?

(f) $CH_3CH_2CH_2CH_2Br \xrightarrow[\text{Pentane}]{Li}$ A? $\xrightarrow{CuI}$ B?

(g) $CH_3CH_2CH_2CH_2Br + (CH_3)_2CuLi \xrightarrow{\text{Ether}}$?

10.24 (*S*)-3-Methylhexane undergoes radical bromination to yield optically inacti 3-bromo-3-methylhexane as the major product. Is the product chiral? Wh conclusions can you draw about the radical intermediate?

10.25 Assume that you have carried out a radical chlorination reaction on (*R*)-2-chlor pentane and have isolated (in low yield) 2,4-dichloropentane. How man stereoisomers of the product are formed and in what ratio? Are any of th isomers optically active? (See Problem 10.24.)

10.26 What product(s) would you expect from the reaction of 1,4-hexadiene wit NBS? What is the structure of the most stable radical intermediate?

10.27 Alkylbenzenes such as toluene (methylbenzene) react with NBS to give pro ucts in which bromine substitution has occurred at the position next to th aromatic ring (the *benzylic* position). Explain, based on the bond dissociatio energies in Table 5.3 on page 156.

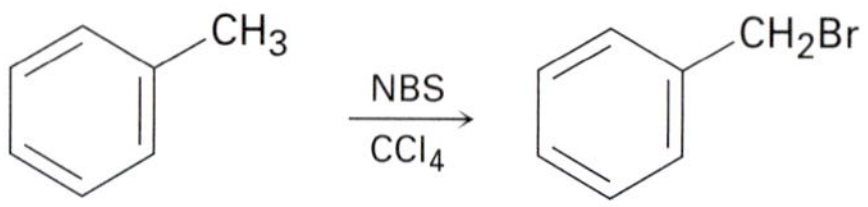

10.28 Draw resonance structures for the benzyl radical, $C_6H_5CH_2\cdot$, the intermediat produced in the NBS bromination reaction of toluene (Problem 10.27).

10.29 What product would you expect from the reaction of 1-phenyl-2-butene wit NBS? Explain.

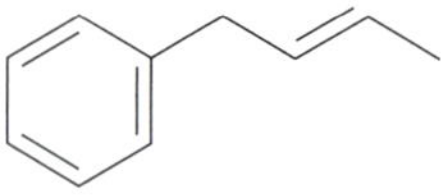

1-Phenyl-2-butene

10.30 ■ Draw resonance structures for the following species:

(a) $CH_3CH{=}CHCH{=}CHCH{=}CH\overset{+}{C}H_2$ (b) 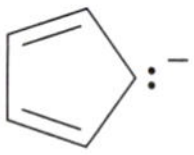 (c) $CH_3C{\equiv}\overset{+}{N}{-}\ddot{\underset{\cdot\cdot}{O}}:^-$

■ Assignable in OWL

10.31 ■ Rank the compounds in each of the following series in order of increasing oxidation level:

(a)

$CH_3CH{=}CHCH_3 \quad CH_3CH_2CH{=}CH_2 \quad CH_3CH_2CH_2\overset{O}{\overset{\|}{C}}H \quad CH_3CH_2CH_2\overset{O}{\overset{\|}{C}}OH$

(b)

$CH_3CH_2CH_2NH_2 \quad CH_3CH_2CH_2Br \quad CH_3\overset{O}{\overset{\|}{C}}CH_2Cl \quad BrCH_2CH_2CH_2Cl$

10.32 Which of the following compounds have the same oxidation level, and which have different levels?

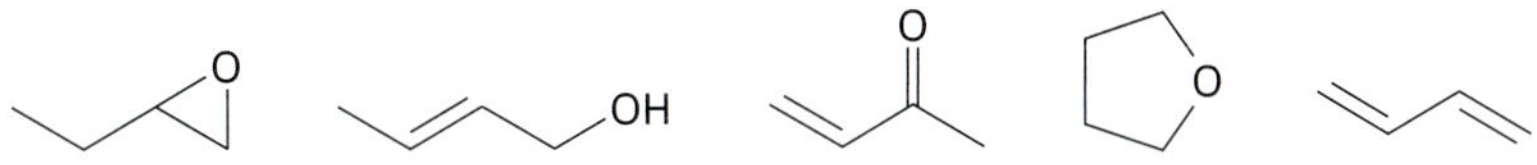

1 2 3 4 5

10.33 ■ Tell whether each of the following reactions is an oxidation, a reduction, or neither:

(a)

$$CH_3CH_2OH \xrightarrow{CrO_3} CH_3\overset{O}{\overset{\|}{C}}H$$

(b)

$$H_2C{=}CH\overset{O}{\overset{\|}{C}}CH_3 + NH_3 \longrightarrow H_2NCH_2CH_2\overset{O}{\overset{\|}{C}}CH_3$$

(c)

$$CH_3CH_2\overset{Br}{\overset{|}{C}}HCH_3 \xrightarrow[2.\ H_2O]{1.\ Mg} CH_3CH_2CH_2CH_3$$

10.34 ■ How would you carry out the following syntheses?

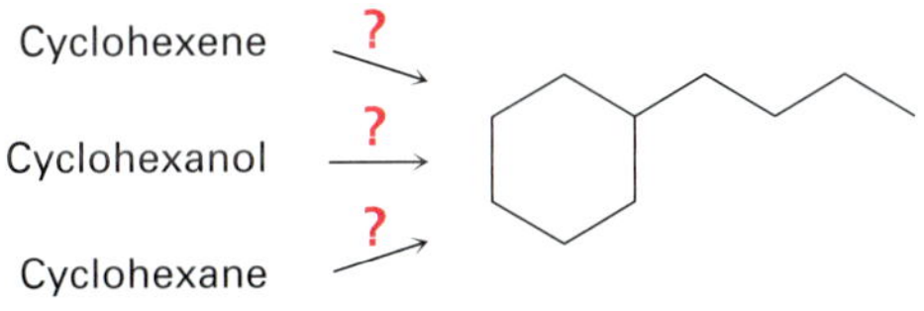

10.35 The syntheses shown here are unlikely to occur as written. What is wrong with each?

(a) $CH_3CH_2CH_2F \xrightarrow[2.\ H_3O^+]{1.\ Mg} CH_3CH_2CH_3$

(b)

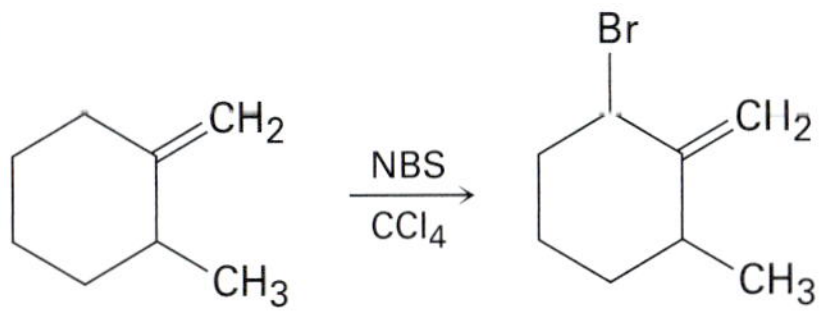

(c)

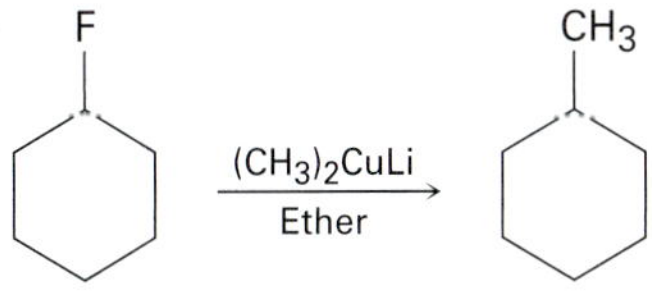

10.36 Why do you suppose it's not possible to prepare a Grignard reagent fron bromo alcohol such as 4-bromo-1-pentanol? Give another example of a m ecule that is unlikely to form a Grignard reagent.

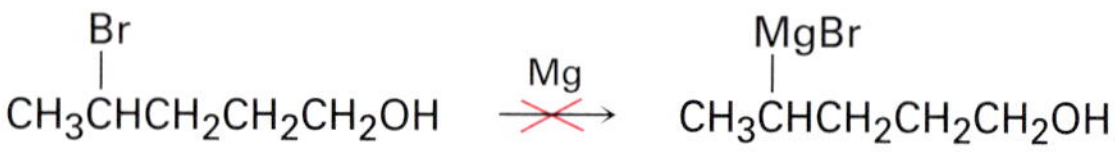

10.37 ■ Addition of HBr to a double bond with an ether (−OR) substituent occ regiospecifically to give a product in which the −Br and −OR are bonded the same carbon. Draw the two possible carbocation intermediates in t electrophilic addition reaction, and explain using resonance why t observed product is formed.

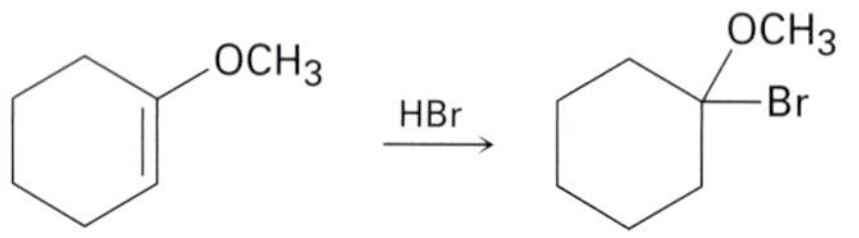

10.38 Alkyl halides can be reduced to alkanes by a radical reaction with tributyl hydride, $(C_4H_9)_3SnH$, in the presence of light ($h\nu$). Propose a radical cha mechanism by which the reaction might occur. The initiation step is the lig induced homolytic cleavage of the Sn−H bond to yield a tributyltin radica

$$R{-}X \quad + \quad (C_4H_9)_3SnH \xrightarrow{h\nu} R{-}H \quad + \quad (C_4H_9)_3SnX$$

10.39 Identify the reagents a–c in the following scheme:

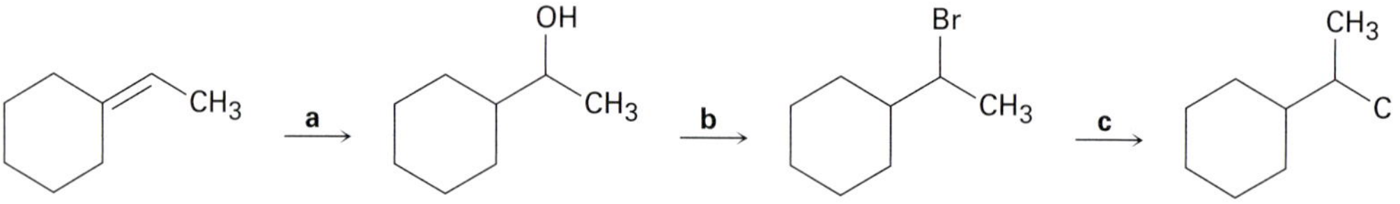

10.40 Tertiary alkyl halides, R_3CX, undergo spontaneous dissociation to yield carbocation, R_3C^+, plus halide ion. Which do you think reacts fast $(CH_3)_3CBr$ or $H_2C{=}CHC(CH_3)_2Br$? Explain.

10.41 In light of the fact that tertiary alkyl halides undergo spontaneous dissoc tion to yield a carbocation plus halide ion (Problem 10.40), propose a mec anism for the following reaction:

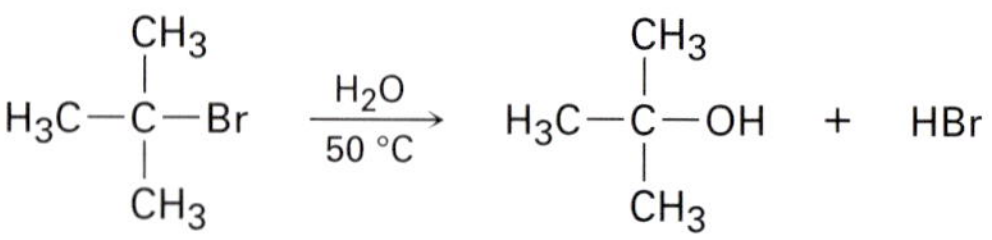

10.42 ■ Carboxylic acids (RCO_2H; $pK_a \approx 5$) are approximately 10^{11} times mo acidic than alcohols (ROH; $pK_a \approx 16$). In other words, a carboxylate ic (RCO_2^-) is more stable than an alkoxide ion (RO^-). Explain, using resonanc

■ Assignable in OWL

11 Reactions of Alkyl Halides: Nucleophilic Substitutions and Eliminations

)rganic KNOWLEDGE TOOLS

ıomsonNOW™ Throughout ıis chapter, sign in at ,www.thomsonedu.com for nline self-study and interactive ıtorials based on your level of nderstanding.

Online homework for this chapter may be assigned in Organic OWL.

We saw in the preceding chapter that the carbon–halogen bond in an alkyl halide is polar and that the carbon atom is electron-poor. Thus, alkyl halides are electrophiles, and much of their chemistry involves polar reactions with nucleophiles and bases. Alkyl halides do one of two things when they react with a nucleophile/base, such as hydroxide ion: either they undergo *substitution* of the X group by the nucleophile, or they undergo *elimination* of HX to yield an alkene.

Substitution

Elimination

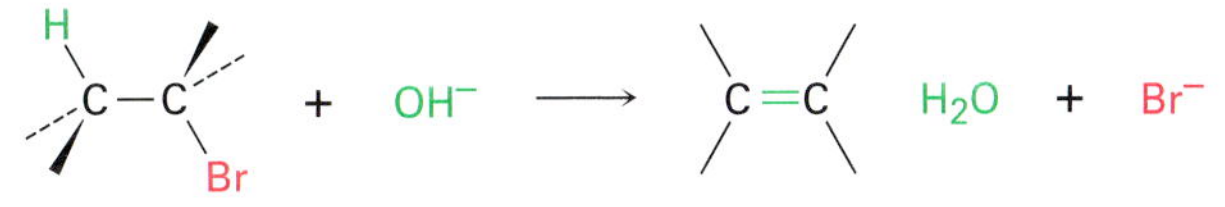

WHY THIS CHAPTER?

Nucleophilic substitution and base-induced elimination are two of the most widely occurring and versatile reaction types in organic chemistry, both in the laboratory and in biological pathways. We'll look at them closely in this chapter to see how they occur, what their characteristics are, and how they can be used.

11.1 The Discovery of Nucleophilic Substitution Reactions

In 1896, the German chemist Paul Walden made a remarkable discovery. He found that the pure enantiomeric (+)- and (−)-malic acids could be interconverted through a series of simple substitution reactions. When Walden treated (−)-malic acid with PCl_5, he isolated (+)-chlorosuccinic acid. This, on treatment with wet Ag_2O, gave (+)-malic acid. Similarly, reaction of (+)-malic acid with

Paul Walden

Paul Walden (1863–1957) was born in Cesis, Latvia, to German parents who died while he was still a child. He received his Ph.D. in Leipzig, Germany, and returned to Russia as professor of chemistry at Riga Polytechnic (1882–1919). Following the Russian Revolution, he went back to Germany as professor at the University of Rostock (1919–1934) and later at the University of Tübingen.

PCl_5 gave (−)-chlorosuccinic acid, which was converted into (−)-malic ac when treated with wet Ag_2O. The full cycle of reactions reported by Walden shown in Figure 11.1.

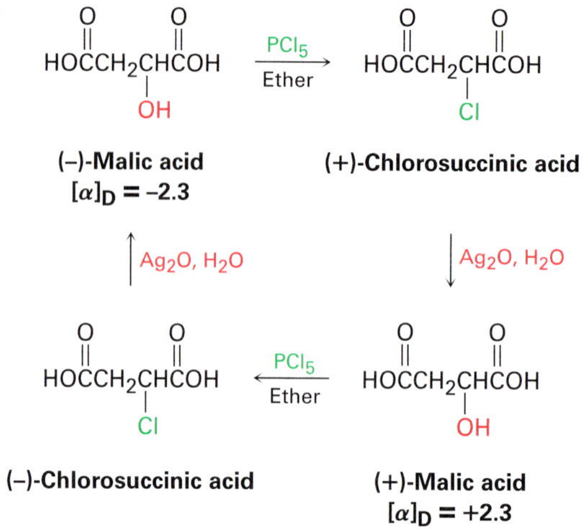

Figure 11.1 Walden's cycle of reactions interconverting (+)- and (−)-malic acids.

At the time, the results were astonishing. The eminent chemist Emil Fisch called Walden's discovery "the most remarkable observation made in the fie of optical activity since the fundamental observations of Pasteur." Becaus (−)-malic acid was converted into (+)-malic acid, *some reactions in the cycle mu have occurred with a change, or inversion, in configuration at the chirality center.* Bu which ones, and how? (Remember from Section 9.5 that the direction of ligh rotation and the configuration of a chirality center aren't directly related. Yo can't tell by looking at the sign of rotation whether a change in configuratio has occurred during a reaction.)

Today, we refer to the transformations taking place in Walden's cycle a **nucleophilic substitution reactions** because each step involves the substitu tion of one nucleophile (chloride ion, Cl^-, or hydroxide ion, HO^-) by anothe Nucleophilic substitution reactions are one of the most common and versati reaction types in organic chemistry.

$$R{-}X \;+\; Nu{:}^- \longrightarrow R{-}Nu \;+\; X{:}^-$$

Following the work of Walden, a further series of investigations was unde taken during the 1920s and 1930s to clarify the mechanism of nucleophilic sub stitution reactions and to find out how inversions of configuration occu Among the first series studied was one that interconverted the two enantiomer of 1-phenyl-2-propanol (Figure 11.2).

Although this particular series of reactions involves nucleophilic substitu tion of an alkyl *p*-toluenesulfonate (called a *tosylate*) rather than an alkyl halide exactly the same type of reaction is involved as that studied by Walden. For al practical purposes, the *entire* tosylate group acts as if it were simply a haloge substituent. In fact, when you see a tosylate substituent in a molecule, do a men tal substitution and tell yourself that you're dealing with an alkyl halide.

ure 11.2 A Walden cycle ·rconverting (+) and (−) enan- ·ners of 1-phenyl-2-propanol. ·rality centers are marked by ·erisks, and the bonds broken ·ach reaction are indicated by wavy lines.

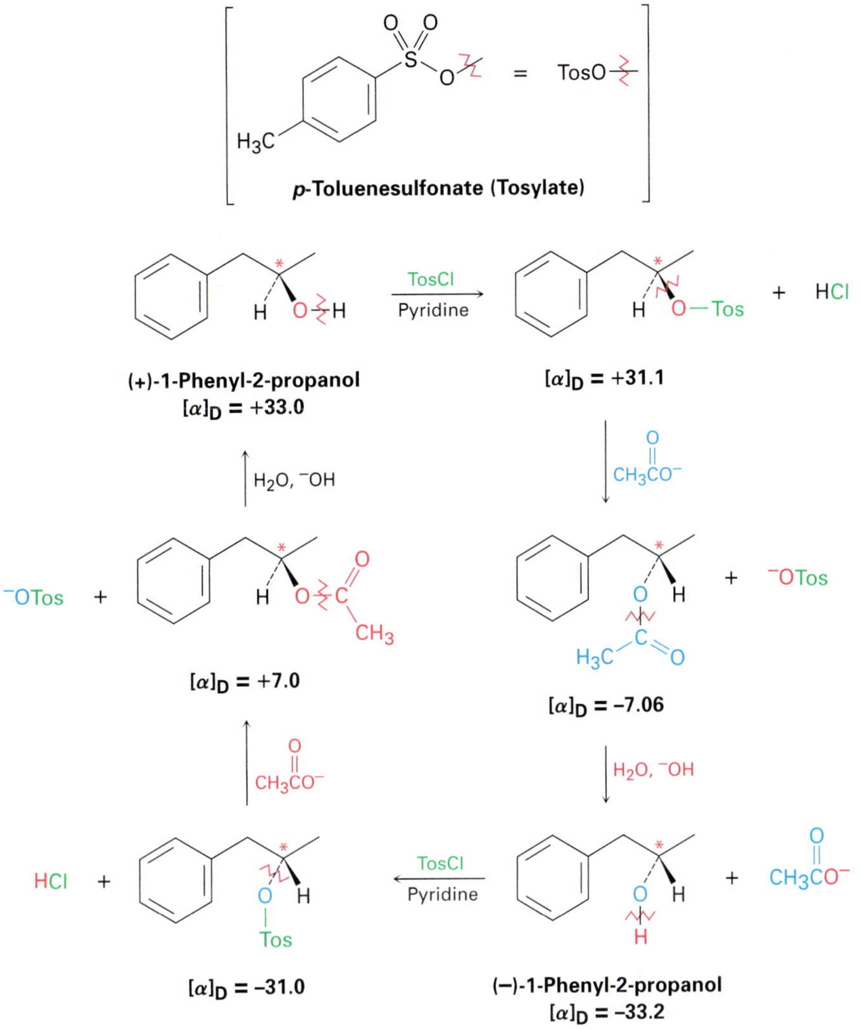

In the three-step reaction sequence shown in Figure 11.2, (+)-1-phenyl-2-propanol is interconverted with its (−) enantiomer, so at least one of the three steps must involve an inversion of configuration at the chirality center. The first step, formation of a toluenesulfonate, occurs by breaking the O−H bond of the alcohol rather than the C−O bond to the chiral carbon, so the configuration around carbon is unchanged. Similarly, the third step, hydroxide ion cleavage of the acetate, takes place without breaking the C−O bond at the chirality center. *The inversion of stereochemical configuration must therefore take place in the second step, the nucleophilic substitution of tosylate ion by acetate ion.*

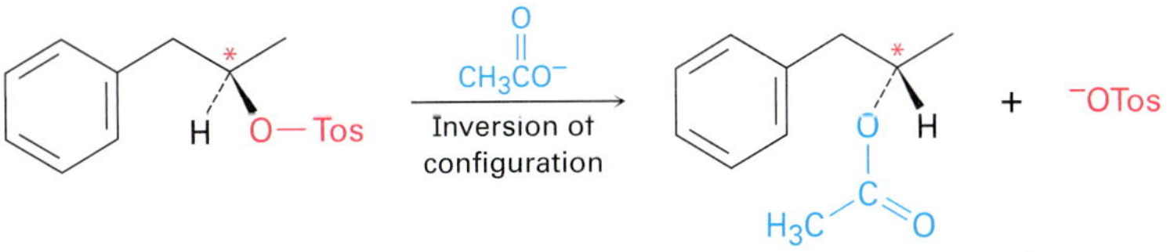

From this and nearly a dozen other series of similar reactions, workers cc cluded that the nucleophilic substitution reaction of a primary or seconda alkyl halide or tosylate always proceeds with inversion of configuratic (Tertiary alkyl halides and tosylates, as we'll see shortly, give different stere chemical results and react by a different mechanism.)

WORKED EXAMPLE 11.1

Predicting the Stereochemistry of a Nucleophilic Substitution Reaction

What product would you expect from a nucleophilic substitution reaction (*R*)-1-bromo-1-phenylethane with cyanide ion, $^-C{\equiv}N$, as nucleophile? Show t stereochemistry of both reactant and product, assuming that inversion of config ration occurs.

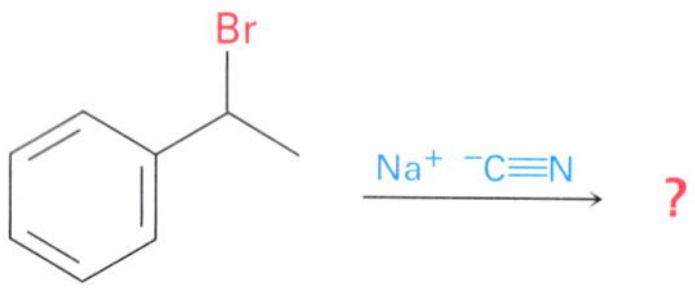

Strategy Draw the *R* enantiomer of the reactant, and then change the configuration of t chirality center while replacing the ^-Br with a ^-CN.

Solution

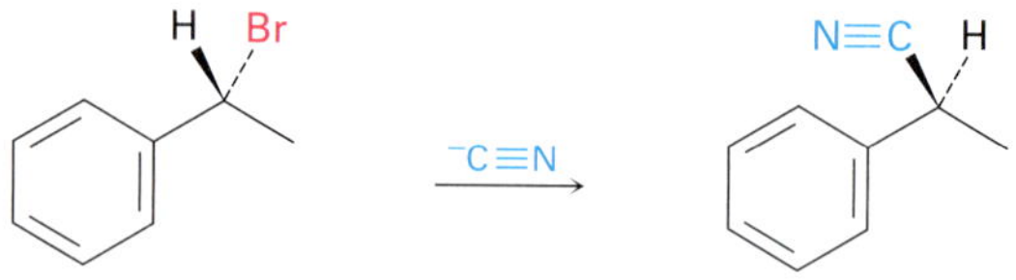

(*R*)-1-Bromo-1-phenylethane **(*S*)-2-Phenylpropanenitrile**

Problem 11.1 What product would you expect to obtain from a nucleophilic substitution reactic of (*S*)-2-bromohexane with acetate ion, $CH_3CO_2^-$? Assume that inversion of confi uration occurs, and show the stereochemistry of both reactant and product.

11.2 The S_N2 Reaction

In every chemical reaction, there is a direct relationship between the rate which the reaction occurs and the concentrations of the reactants. When w measure this relationship, we measure the **kinetics** of the reaction. For exampl let's look at the kinetics of a simple nucleophilic substitution—the reaction c CH_3Br with OH^- to yield CH_3OH plus Br^-—to see what can be learned.

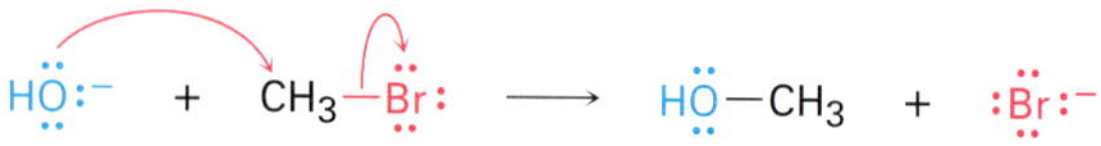

At a given temperature and concentration of reactants, the substitutio occurs at a certain rate. If we double the concentration of OH^-, the frequenc of encounter between the reaction partners doubles and we find that the reac tion rate also doubles. Similarly, if we double the concentration of CH_3Br, th

ThomsonNOW™ Click *Organic Process* to **view an animation showing the stereochemistry of the S_N2 reaction.**

reaction rate again doubles. We call such a reaction, in which the rate is linearly dependent on the concentrations of two species, a **second-order reaction.** Mathematically, we can express this second-order dependence of the nucleophilic substitution reaction by setting up a *rate equation.* As either [RX] or [$^-$OH] changes, the rate of the reaction changes proportionately.

$$\text{Reaction rate} = \text{Rate of disappearance of reactant}$$
$$= k \times [\text{RX}] \times [^-\text{OH}]$$

where [RX] = CH_3Br concentration in molarity
[$^-$OH] = $^-$OH concentration in molarity
k = A constant value (the rate constant)

Edward Davies Hughes

Edward Davies Hughes (1906–1963) was born in Criccieth, North Wales, and earned two doctoral degrees: a Ph.D. from Wales and a D.Sc. from the University of London, working with Christopher Ingold. From 1930 to 1963, he was professor of chemistry at University College, London.

A mechanism that accounts for both the inversion of configuration and the second-order kinetics that are observed with nucleophilic substitution reactions was suggested in 1937 by E. D. Hughes and Christopher Ingold, who formulated what they called the **S_N2 reaction**—short for *substitution, nucleophilic, bimolecular.* (*Bimolecular* means that two molecules, nucleophile and alkyl halide, take part in the step whose kinetics are measured.)

The essential feature of the S_N2 mechanism is that it takes place in a single step without intermediates when the incoming nucleophile reacts with the alkyl halide or tosylate (the *substrate*) from a direction opposite the group that is displaced (the *leaving group*). As the nucleophile comes in on one side of the substrate and bonds to the carbon, the halide or tosylate departs from the other side, thereby inverting the stereochemical configuration. The process is shown in Figure 11.3 for the reaction of (*S*)-2-bromobutane with HO$^-$ to give (*R*)-2-butanol.

igure 11.3 MECHANISM: he mechanism of the S_N2 reacon. The reaction takes place in single step when the incoming ucleophile approaches from a rection 180° away from the aving halide ion, thereby verting the stereochemistry t carbon.

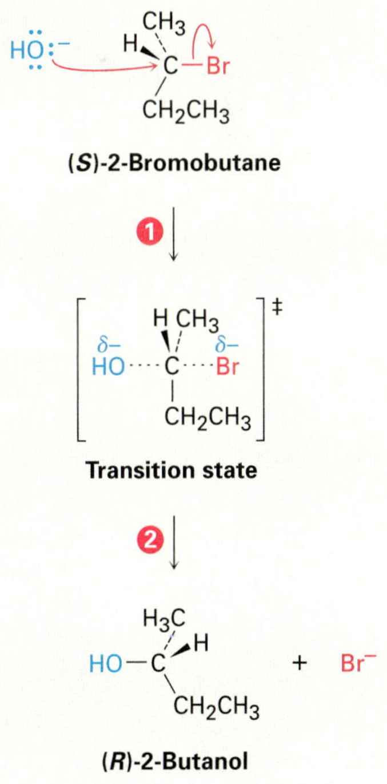

1. The nucleophile $^-$OH uses its lone-pair electrons to attack the alkyl halide carbon 180° away from the departing halogen. This leads to a transition state with a partially formed C–OH bond and a partially broken C–Br bond.

2. The stereochemistry at carbon is inverted as the C–OH bond forms fully and the bromide ion departs with the electron pair from the former C–Br bond.

As shown in Figure 11.3, the S_N2 reaction occurs when an electron pair c the nucleophile Nu^- forces out the group $X:^-$, which takes with it the electrc pair from the former C–X bond. This occurs through a transition state in whic the new Nu–C bond is partially forming at the same time that the old C– bond is partially breaking and in which the negative charge is shared by bot the incoming nucleophile and the outgoing halide ion. The transition state fc this inversion has the remaining three bonds to carbon in a planar arrangemer (Figure 11.4).

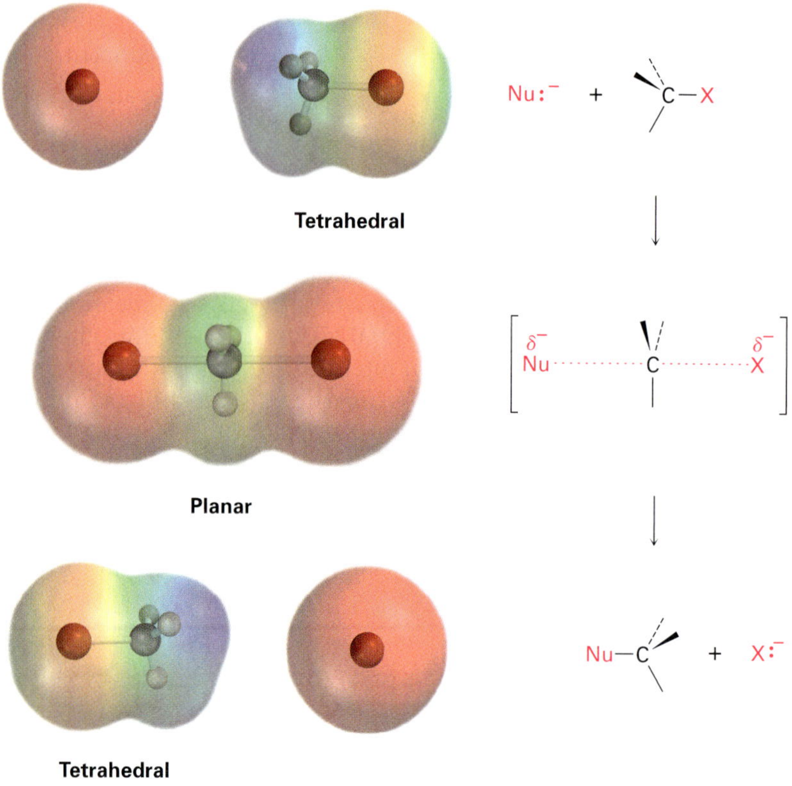

Figure 11.4 The transition state of an S_N2 reaction has a planar arrangement of the carbon atom and the remaining three groups. Electrostatic potential maps show that negative charge (red) is delocalized in the transition state.

The mechanism proposed by Hughes and Ingold is fully consistent witl experimental results, explaining both stereochemical and kinetic data. Thus the requirement for backside approach of the entering nucleophile from direction 180° away from the departing X group causes the stereochemistry o the substrate to invert, much like an umbrella turning inside out in th wind. The Hughes–Ingold mechanism also explains why second-order kinet ics are found: the S_N2 reaction occurs in a single step that involves both alky halide and nucleophile. Two molecules are involved in the step whose rate i measured.

Problem 11.2 What product would you expect to obtain from S_N2 reaction of OH^- with (*R*)-2-bromo butane? Show the stereochemistry of both reactant and product.

Problem 11.3 Assign configuration to the following substance, and draw the structure of the product that would result on nucleophilic substitution reaction with HS^- (reddish brown = Br):

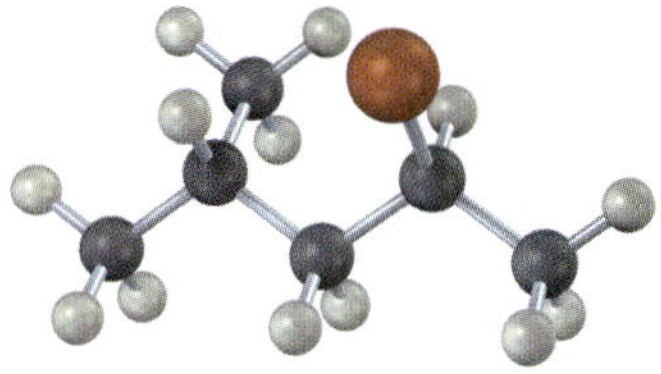

11.3 Characteristics of the S_N2 Reaction

ey IDEAS

st your knowledge of Key eas by using resources in omsonNOW or by answering d-of-chapter problems marked th ▲.

Now that we have a good picture of how S_N2 reactions occur, we need to see how they can be used and what variables affect them. Some S_N2 reactions are fast, and some are slow; some take place in high yield and others, in low yield. Understanding the factors involved can be of tremendous value. Let's begin by recalling a few things about reaction rates in general.

The rate of a chemical reaction is determined by $\Delta G^\ddagger$, the energy difference between reactant ground state and transition state. A change in reaction conditions can affect $\Delta G^\ddagger$ either by changing the reactant energy level or by changing the transition-state energy level. Lowering the reactant energy or raising the transition-state energy increases $\Delta G^\ddagger$ and decreases the reaction rate; raising the reactant energy or decreasing the transition-state energy decreases $\Delta G^\ddagger$ and increases the reaction rate (Figure 11.5). We'll see examples of all these effects as we look at S_N2 reaction variables.

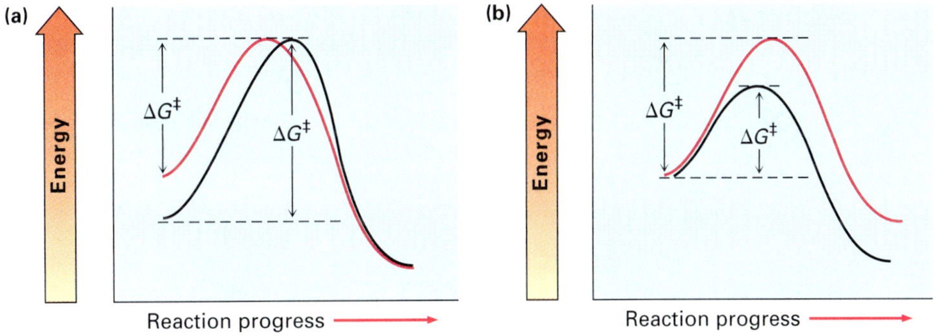

gure 11.5 The effects of anges in reactant and ansition-state energy levels reaction rate. **(a)** A higher actant energy level (red rve) corresponds to a ster reaction (smaller $\Delta G^\ddagger$).) A higher transition-state ergy level (red curve) corre- onds to a slower reaction rger $\Delta G^\ddagger$).

The Substrate: Steric Effects in the S_N2 Reaction

The first S_N2 reaction variable to look at is the structure of the substrate. Because the S_N2 transition state involves partial bond formation between the incoming nucleophile and the alkyl halide carbon atom, it seems reasonable that a hindered, bulky substrate should prevent easy approach of the nucleophile, making bond formation difficult. In other words, the transition state for reaction of a sterically hindered alkyl halide, whose carbon atom is "shielded" from approach of the incoming nucleophile, is higher in energy and forms more slowly than the corresponding transition state for a less hi dered alkyl halide (Figure 11.6).

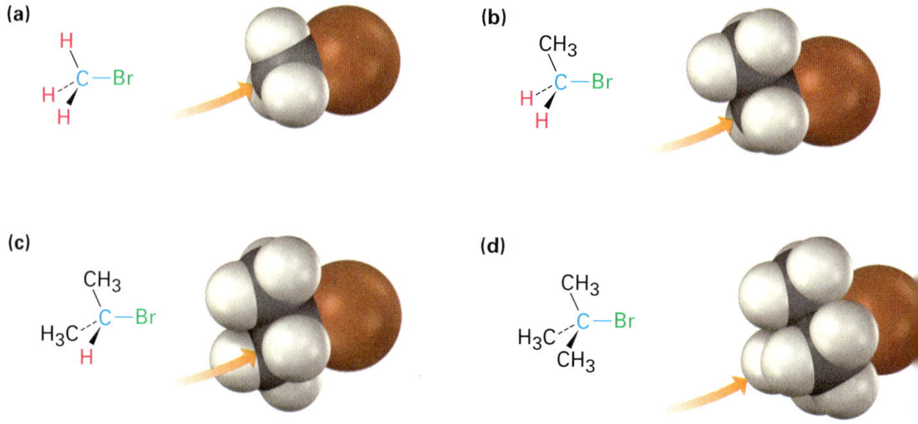

Figure 11.6 Steric hindrance to the S_N2 reaction. As the computer-generated models indicate, the carbon atom in **(a)** bromomethane is readily accessible, resulting in a fast S_N2 reaction. The carbon atoms in **(b)** bromoethane (primary), **(c)** 2-bromopropane (secondary), and **(d)** 2-bromo-2-methylpropane (tertiary) are successively more hindered, resulting in successively slower S_N2 reactions.

As Figure 11.6 shows, the difficulty of nucleophilic approach increases a the three substituents bonded to the halo-substituted carbon atom increase i size. Methyl halides are by far the most reactive substrates in S_N2 reactions, fo lowed by primary alkyl halides such as ethyl and propyl. Alkyl branching at th reacting center, as in isopropyl halides (2°), slows the reaction greatly, and fu ther branching, as in *tert*-butyl halides (3°), effectively halts the reaction. Eve branching one carbon removed from the reacting center, as in 2,2-dimethy propyl *(neopentyl)* halides, greatly slows nucleophilic displacement. As a resul *S_N2 reactions occur only at relatively unhindered sites* and are normally useful onl with methyl halides, primary halides, and a few simple secondary halides. Relative reactivities for some different substrates are as follows:

$$R{-}Br + Cl^- \longrightarrow R{-}Cl + Br^-$$

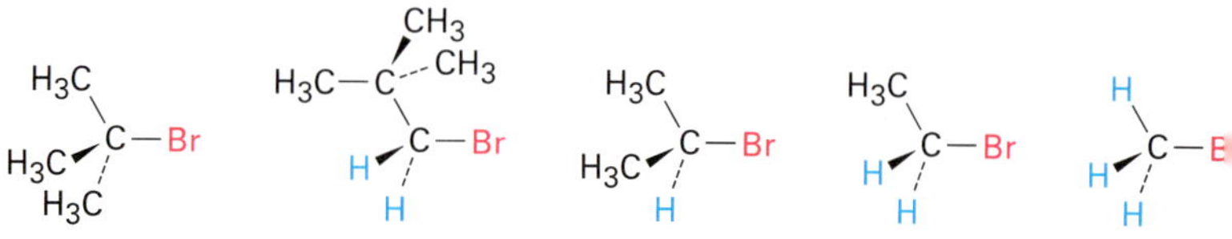

	Tertiary	Neopentyl	Secondary	Primary	Methyl
Relative reactivity	< 1	1	500	40,000	2,000,00

Although not shown in the preceding reactivity order, vinylic halide ($R_2C{=}CRX$) and aryl halides are unreactive toward S_N2 reaction. This lack o reactivity is probably due to steric factors, because the incoming nucleophil would have to approach in the plane of the carbon–carbon double bond to carry out a backside displacement.

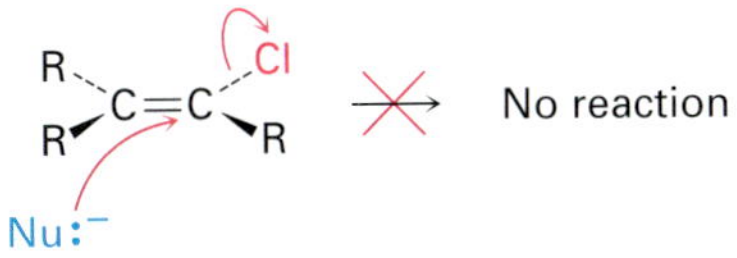

Vinylic halide

Aryl halide

The Nucleophile

Another variable that has a major effect on the S_N2 reaction is the nature of the nucleophile. Any species, either neutral or negatively charged, can act as a nucleophile as long as it has an unshared pair of electrons, that is, as long as it is a Lewis base. If the nucleophile is negatively charged, the product is neutral; if the nucleophile is neutral, the product is positively charged.

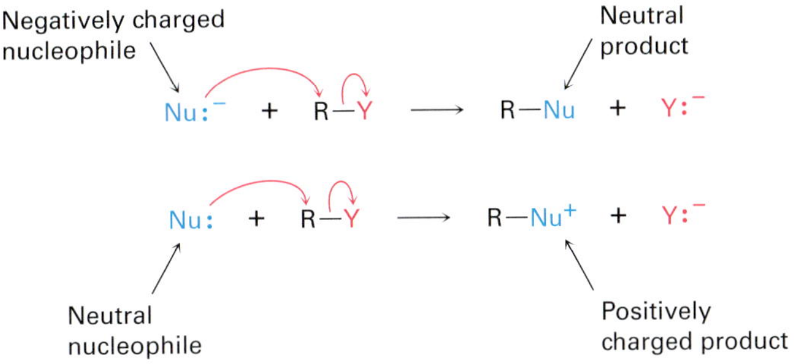

A wide array of substances can be prepared using nucleophilic substitution reactions. In fact, we've already seen examples in previous chapters. The reaction of an acetylide anion with an alkyl halide (Section 8.8), for instance, is an S_N2 reaction in which the acetylide nucleophile replaces halide.

$$R{-}C{\equiv}C{:}^- \; + \; CH_3Br \xrightarrow[\text{reaction}]{S_N2} R{-}C{\equiv}C{-}CH_3 \; + \; Br^-$$

An acetylide anion

Table 11.1 lists some nucleophiles in the order of their reactivity, shows the products of their reactions with bromomethane, and gives the relative rates of their reactions. Clearly, there are large differences in the rates at which various nucleophiles react.

What are the reasons for the reactivity differences observed in Table 11.1? Why do some reactants appear to be much more "nucleophilic" than others? The answers to these questions aren't straightforward. Part of the problem is that the term *nucleophilicity* is imprecise. The term is usually taken to be a measure of the affinity of a nucleophile for a carbon atom in the S_N2 reaction, but the reactivity of a given nucleophile can change from one reaction to the next. The exact nucleophilicity of a species in a given reaction depends on the substrate, the solvent, and even the reactant concentrations. Detailed

Table 11.1 Some S_N2 Reactions with Bromomethane

$$Nu{:}^- + CH_3Br \rightarrow CH_3Nu + Br^-$$

Nucleophile		Product		
Formula	**Name**	**Formula**	**Name**	**Relative rate of reaction**
H_2O	Water	$CH_3OH_2^+$	Methylhydronium ion	1
$CH_3CO_2^-$	Acetate	$CH_3CO_2CH_3$	Methyl acetate	500
NH_3	Ammonia	$CH_3NH_3^+$	Methylammonium ion	700
Cl^-	Chloride	CH_3Cl	Chloromethane	1,000
HO^-	Hydroxide	CH_3OH	Methanol	10,000
CH_3O^-	Methoxide	CH_3OCH_3	Dimethyl ether	25,000
I^-	Iodide	CH_3I	Iodomethane	100,000
^-CN	Cyanide	CH_3CN	Acetonitrile	125,000
HS^-	Hydrosulfide	CH_3SH	Methanethiol	125,000

explanations for the observed nucleophilicities aren't always simple, but som trends can be detected in the data of Table 11.1.

- **Nucleophilicity roughly parallels basicity** when comparing nucleophil that have the same reacting atom. For example, OH^- is both more basic an more nucleophilic than acetate ion, $CH_3CO_2^-$, which in turn is more bas and more nucleophilic than H_2O. Since "nucleophilicity" is usually taken a the affinity of a Lewis base for a carbon atom in the S_N2 reaction and "basi ity" is the affinity of a base for a proton, it's easy to see why there might be correlation between the two kinds of behavior.
- **Nucleophilicity usually increases going down a column of the periodi table.** Thus, HS^- is more nucleophilic than HO^-, and the halide reactivit order is $I^- > Br^- > Cl^-$. Going down the periodic table, elements have the valence electrons in successively larger shells where they are successively fa ther from the nucleus, less tightly held, and consequently more reactive. Th matter is complex, though, and the nucleophilicity order can change depend ing on the solvent.
- **Negatively charged nucleophiles are usually more reactive than neutra ones.** As a result, S_N2 reactions are often carried out under basic condition rather than neutral or acidic conditions.

Problem 11.4 What product would you expect from S_N2 reaction of 1-bromobutane with each c the following?

(a) NaI **(b)** KOH **(c)** $H{-}C{\equiv}C{-}Li$ **(d)** NH_3

Problem 11.5 Which substance in each of the following pairs is more reactive as a nucleophile Explain.

(a) $(CH_3)_2N^-$ or $(CH_3)_2NH$ **(b)** $(CH_3)_3B$ or $(CH_3)_3N$ **(c)** H_2O or H_2S

The Leaving Group

Still another variable that can affect the S_N2 reaction is the nature of the group displaced by the incoming nucleophile. Because the leaving group is expelled with a negative charge in most S_N2 reactions, the best leaving groups are those that best stabilize the negative charge in the transition state. The greater the extent of charge stabilization by the leaving group, the lower the energy of the transition state and the more rapid the reaction. But as we saw in Section 2.8, those groups that best stabilize a negative charge are also the weakest bases. Thus, weak bases such as Cl^-, Br^-, and tosylate ion make good leaving groups, while strong bases such as OH^- and NH_2^-make poor leaving groups.

	OH^-, NH_2^-, OR^-	F^-	Cl^-	Br^-	I^-	$TosO^-$
Relative reactivity	<<1	1	200	10,000	30,000	60,000

Leaving group reactivity →

It's just as important to know which are poor leaving groups as to know which are good, and the preceding data clearly indicate that F^-, HO^-, RO^-, and H_2N^- are not displaced by nucleophiles. In other words, alkyl fluorides, alcohols, ethers, and amines do not typically undergo S_N2 reactions. To carry out an S_N2 reaction with an alcohol, it's necessary to convert the ^-OH into a better leaving group. This, in fact, is just what happens when a primary or secondary alcohol is converted into either an alkyl chloride by reaction with $SOCl_2$ or an alkyl bromide by reaction with PBr_3 (Section 10.6).

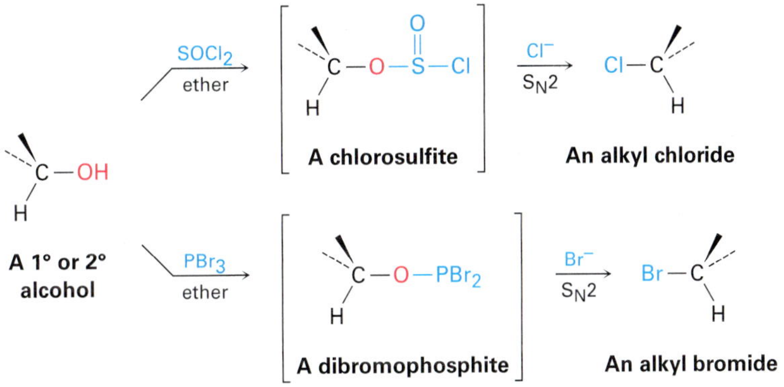

Alternatively, an alcohol can be made more reactive toward nucleophilic substitution by treating it with *para*-toluenesulfonyl chloride to form a tosylate. As noted on several previous occasions, tosylates are even more reactive than halides in nucleophilic substitutions. Note that tosylate formation does not change the configuration of the oxygen-bearing carbon because the C–O bond is not broken.

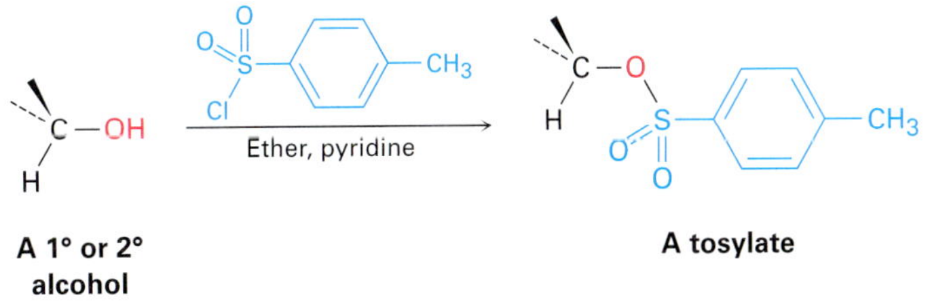

The one general exception to the rule that ethers don't typically under S_N2 reactions occurs with epoxides, the three-membered cyclic ethers th we saw in Section 7.8. Epoxides, because of the angle strain in the thre membered ring, are much more reactive than other ethers. They react wi aqueous acid to give 1,2-diols, as we saw in Section 7.8, and they react rea ily with many other nucleophiles as well. Propene oxide, for instance, rea with HCl to give 1-chloro-2-propanol by S_N2 backside attack on the less hi dered primary carbon atom. We'll look at the process in more detail Section 18.6.

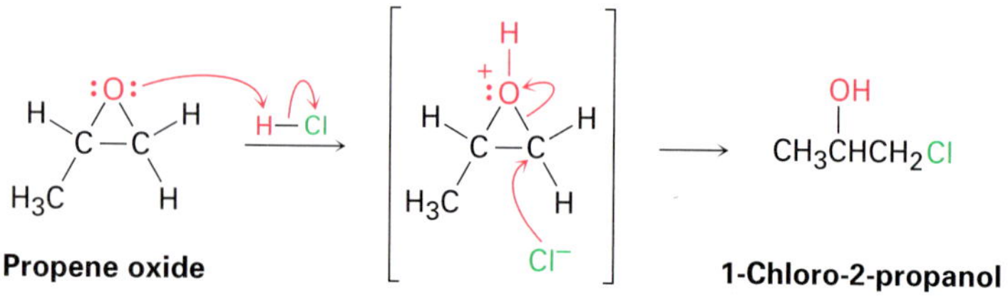

Propene oxide

1-Chloro-2-propanol

Problem 11.6 Rank the following compounds in order of their expected reactivity towa S_N2 reaction:

CH_3Br, CH_3OTos, $(CH_3)_3CCl$, $(CH_3)_2CHCl$

The Solvent

The rates of S_N2 reactions are strongly affected by the solvent. *Protic solvents*—those that contain an –OH or –NH group—are generally the worst for S_N2 rea tions, while *polar aprotic solvents,* which are polar but don't have an –OH o –NH group, are the best.

Protic solvents, such as methanol and ethanol, slow down S_N2 reactions b **solvation** of the reactant nucleophile. The solvent molecules hydrogen bond t the nucleophile and form a "cage" around it, thereby lowering its energy an reactivity.

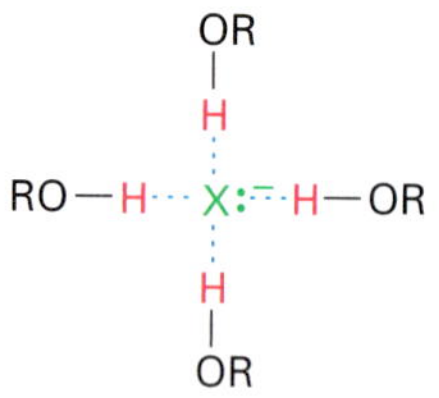

A solvated anion (reduced nucleophilicity due to enhanced ground-state stability)

In contrast with protic solvents, which *decrease* the rates of S_N2 reactions b *lowering* the ground-state energy of the nucleophile, polar aprotic solvent *increase* the rates of S_N2 reactions by *raising* the ground-state energy o the nucleophile. Acetonitrile (CH_3CN), dimethylformamide [$(CH_3)_2NCHO$

abbreviated DMF], dimethyl sulfoxide [$(CH_3)_2SO$, abbreviated DMSO], and hexamethylphosphoramide {[$(CH_3)_2N]_3PO$, abbreviated HMPA} are particularly useful. These solvents can dissolve many salts because of their high polarity, but they tend to solvate metal cations rather than nucleophilic anions. As a result, the bare unsolvated anions have a greater nucleophilicity, and S_N2 reactions take place at correspondingly faster rates. For instance, a rate increase of 200,000 has been observed on changing from methanol to HMPA for the reaction of azide ion with 1-bromobutane.

$$CH_3CH_2CH_2CH_2{-}Br + N_3^- \longrightarrow CH_3CH_2CH_2CH_2{-}N_3 + Br^-$$

Solvent	CH_3OH	H_2O	DMSO	DMF	CH_3CN	HMPA
Relative reactivity	1	7	1300	2800	5000	200,000

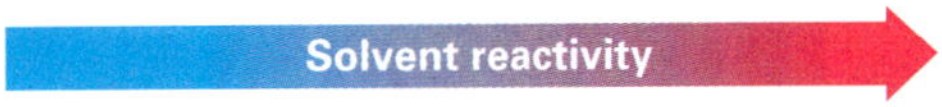

Problem 11.7 Organic solvents such as benzene, ether, and chloroform are neither protic nor strongly polar. What effect would you expect these solvents to have on the reactivity of a nucleophile in S_N2 reactions?

A Summary of S_N2 Reaction Characteristics

The effects on S_N2 reactions of the four variables—substrate structure, nucleophile, leaving group, and solvent—are summarized in the following statements and in the energy diagrams of Figure 11.7:

Substrate Steric hindrance raises the energy of the S_N2 transition state, increasing $\Delta G^{\ddagger}$ and decreasing the reaction rate (Figure 11.7a). As a result, S_N2 reactions are best for methyl and primary substrates. Secondary substrates react slowly, and tertiary substrates do not react by an S_N2 mechanism.

Nucleophile Basic, negatively charged nucleophiles are less stable and have a higher ground-state energy than neutral ones, decreasing $\Delta G^{\ddagger}$ and increasing the S_N2 reaction rate (Figure 11.7b).

Leaving group Good leaving groups (more stable anions) lower the energy of the transition state, decreasing $\Delta G^{\ddagger}$ and increasing the S_N2 reaction rate (Figure 11.7c).

Solvent Protic solvents solvate the nucleophile, thereby lowering its ground-state energy, increasing $\Delta G^{\ddagger}$, and decreasing the S_N2 reaction rate. Polar aprotic solvents surround the accompanying cation but not the nucleophilic anion, thereby raising the ground-state energy of the nucleophile, decreasing $\Delta G^{\ddagger}$, and increasing the reaction rate (Figure 11.7d).

homsonNOW™ Click *Organic nteractive* to **use a web-based palette to predict products from simple S_N2 reactions.**

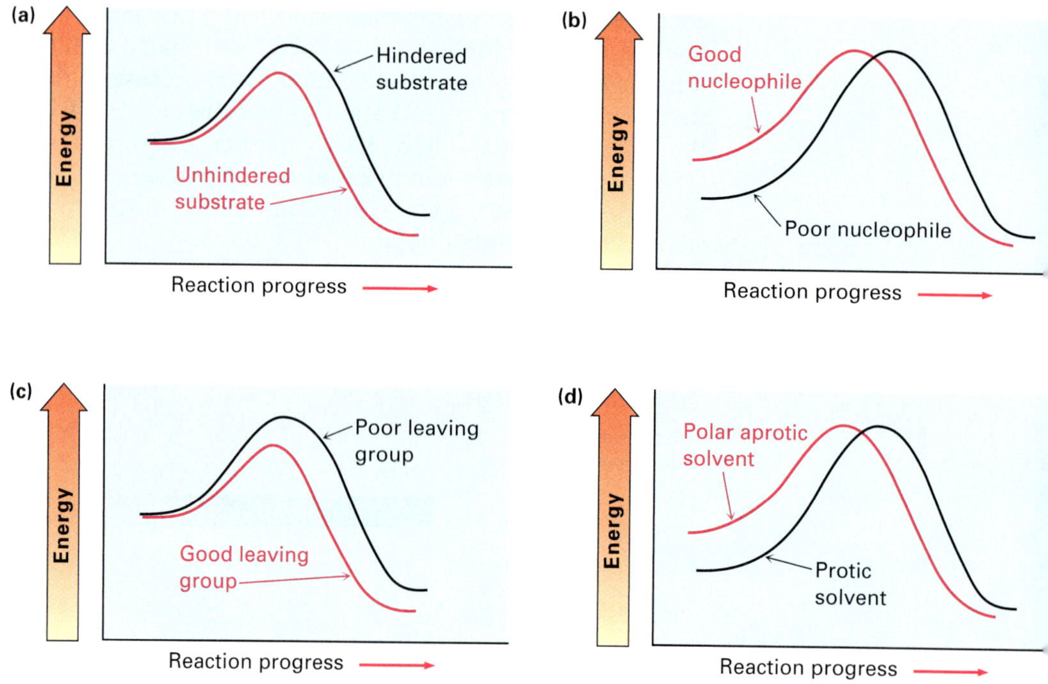

Figure 11.7 Energy diagrams showing the effects of **(a)** substrate, **(b)** nucleophile, **(c)** leavin group, and **(d)** solvent on S_N2 reaction rates. Substrate and leaving group effects are felt prim rily in the transition state. Nucleophile and solvent effects are felt primarily in the reactant ground state.

11.4 The S_N1 Reaction

As we've seen, the S_N2 reaction is best when carried out with an unhindered sub strate and a negatively charged nucleophile in a polar aprotic solvent, but it worst when carried out with a hindered substrate and a neutral nucleophile in protic solvent. You might therefore expect the reaction of a tertiary substrate (hir dered) with water (neutral, protic) to be among the slowest of substitution reac tions. Remarkably, however, the opposite is true. The reaction of the tertiar halide $(CH_3)_3CBr$ with H_2O to give the alcohol 2-methyl-2-propanol is more tha *1 million times* as fast as the corresponding reaction of CH_3Br to give methanol.

$$R{-}Br + H_2O \longrightarrow R{-}OH + HBr$$

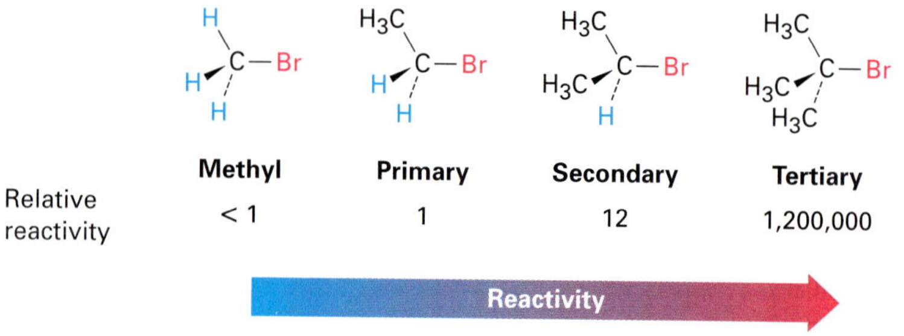

What's going on here? Clearly, a nucleophilic substitution reaction is occur ring, yet the reactivity order seems backward. These reactions can't be taking plac by the S_N2 mechanism we've been discussing, and we must therefore conclude that they are occurring by *an alternative substitution mechanism.* This alternative mechanism is called the **S_N1 reaction** (for *substitution, nucleophilic, unimolecular*).

In contrast to the S_N2 reaction of CH_3Br with OH^-, the S_N1 reaction of $(CH_3)_3CBr$ with H_2O has a rate that depends only on the alkyl halide concentration and is independent of the H_2O concentration. In other words, the reaction is a **first-order process**; the concentration of the nucleophile does not appear in the rate equation.

$$\text{Reaction rate} = \text{Rate of disappearance of alkyl halide}$$
$$= k \times [\text{RX}]$$

To explain this result, we need to learn more about kinetics measurements. Many organic reactions occur in several steps, one of which is usually slower than the others. We call this slow step the *rate-limiting step,* or *rate-determining step.* No reaction can proceed faster than its rate-limiting step, which acts as a kind of traffic jam, or bottleneck. In the S_N1 reaction of $(CH_3)_3CBr$ with H_2O, the fact that the nucleophile does not appear in the first-order rate equation means that the alkyl halide is involved in a *unimolecular* rate-limiting step. But if the nucleophile is not involved in the rate-limiting step, then it must be involved in some other, non–rate-limiting step. The mechanism shown in Figure 11.8 accounts for these observations.

Figure 11.8 MECHANISM: The mechanism of the S_N1 reaction of 2-bromo-2-methylpropane with H_2O involves three steps. The first step—spontaneous, unimolecular dissociation of the alkyl bromide to yield a carbocation—is rate-limiting.

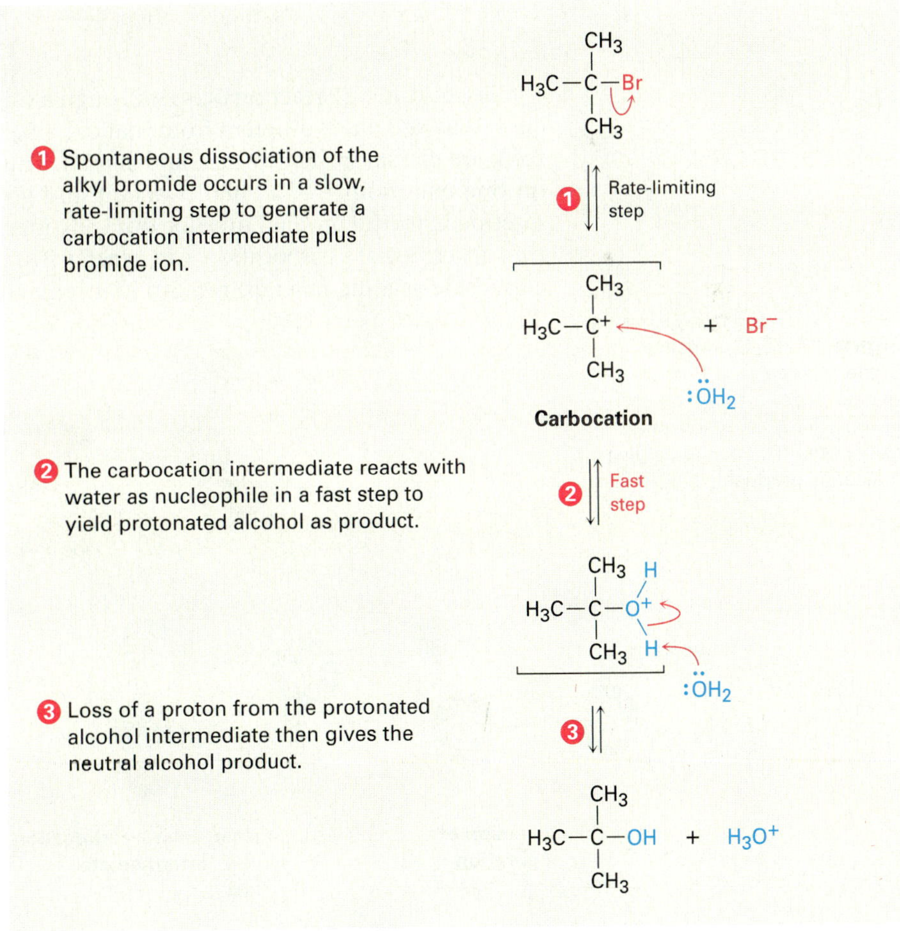

ThomsonNOW™ Click *Organic Process* to **view animations showing the S_N1 reaction of 2-methyl-2-propanol with HCl and the S_N1 solvolysis of 2-chloro-2-methylpropane.**

Unlike what happens in an S_N2 reaction, where the leaving group is displac at the same time the incoming nucleophile approaches, an S_N1 reaction tak place by loss of the leaving group *before* the nucleophile approaches. 2-Brom 2-methylpropane spontaneously dissociates to the *tert*-butyl carbocation pl Br^- in a slow, rate-limiting step, and the intermediate carbocation is then imm diately trapped by the nucleophile water in a faster second step. *Water is not a rea tant in the step whose rate is measured.* The energy diagram is shown in Figure 11.

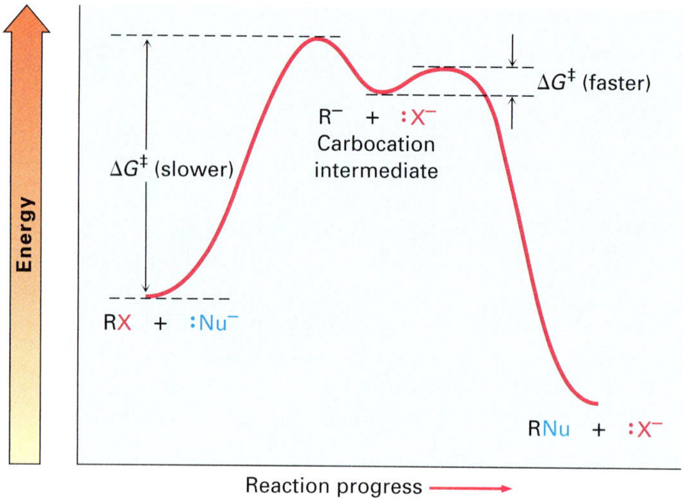

Figure 11.9 An energy diagram for an S_N1 reaction. The slower, rate-limiting step is the spontaneous dissociation of the alkyl halide to give a carbocation intermediate. Reaction of the carbocation with a nucleophile then occurs in a second, faster step.

Because an S_N1 reaction occurs through a carbocation intermediate, its stereo chemical outcome is different from that of an S_N2 reaction. Carbocations, as we'v seen, are planar, sp^2-hybridized, and achiral. Thus, if we carry out an S_N1 reactio on one enantiomer of a chiral reactant and go through an achiral carbocatio intermediate, the product must be optically inactive (Section 9.10). The symme rical intermediate carbocation can react with a nucleophile equally well fron either side, leading to a racemic, 50:50 mixture of enantiomers (Figure 11.10).

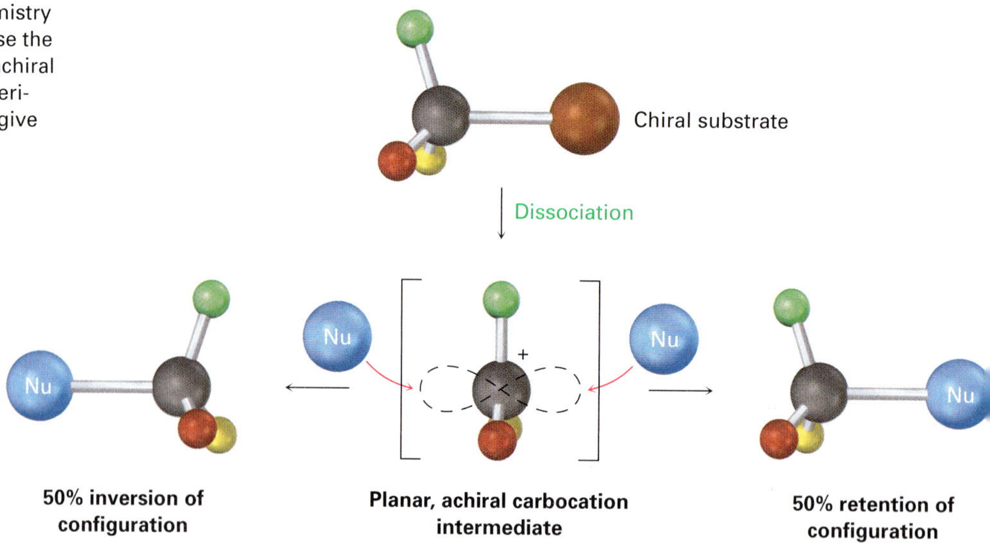

Figure 11.10 Stereochemistry of the S_N1 reaction. Because the reaction goes through an achiral intermediate, an enantiomerically pure reactant should give a racemic product.

The conclusion that S_N1 reactions on enantiomerically pure substrates should give racemic products is nearly, but not exactly, what is found. In fact, few S_N1 displacements occur with complete racemization. Most give a minor (0%–20%) excess of inversion. The reaction of (*R*)-6-chloro-2,6-dimethyloctane with H_2O, for example, leads to an alcohol product that is approximately 80% racemized and 20% inverted (80% *R,S* + 20% *S* is equivalent to 40% *R* + 60% *S*).

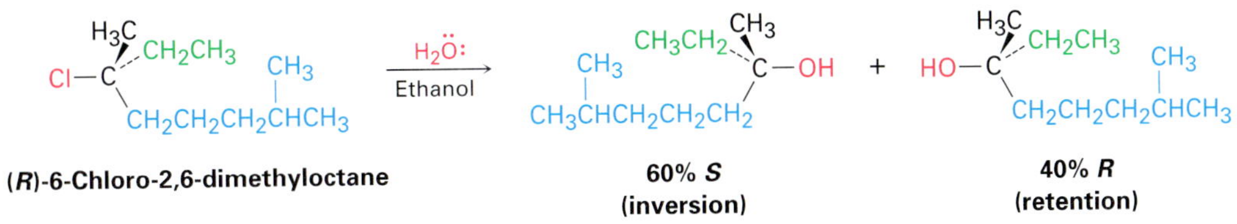

Saul Winstein

Saul Winstein (1912–1969) was born in Montreal, Canada, and received his Ph.D. in 1938 at the California Institute of Technology. From 1942 to 1969, he was professor of chemistry at the University of California, Los Angeles, where he devoted his scientific career to the study of organic reaction mechanisms, particularly those involving carbocations.

This lack of complete racemization in most S_N1 reactions is due to the fact that *ion pairs* are involved. According to this explanation, first proposed by Saul Winstein, dissociation of the substrate occurs to give a structure in which the two ions are still loosely associated and in which the carbocation is effectively shielded from reaction on one side by the departing anion. If a certain amount of substitution occurs before the two ions fully diffuse apart, then a net inversion of configuration will be observed (Figure 11.11).

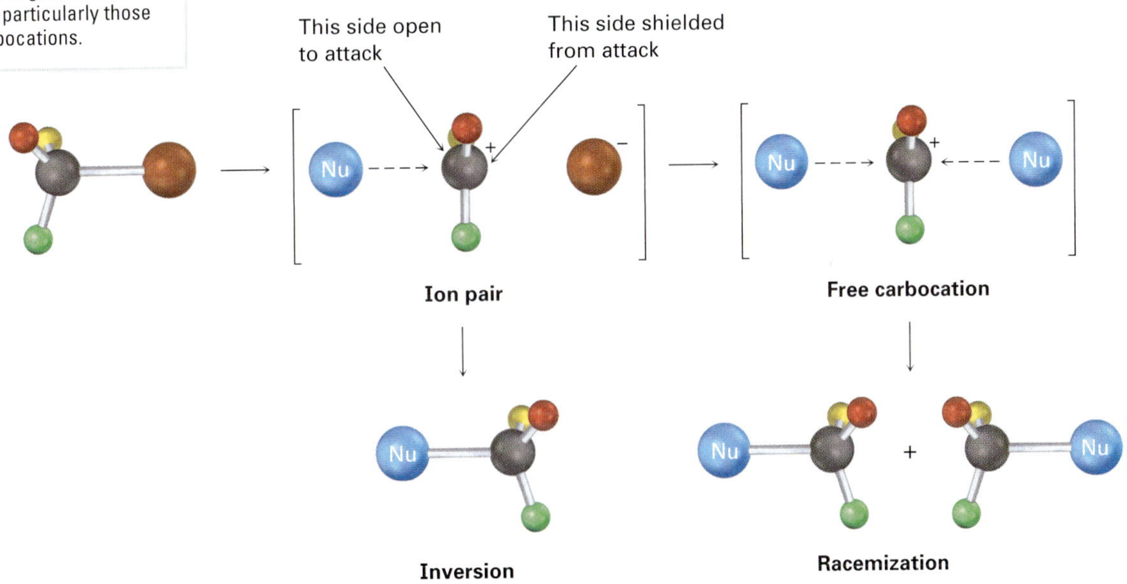

Figure 11.11 Ion pairs in an S_N1 reaction. The leaving group shields one side of the carbocation intermediate from reaction with the nucleophile, thereby leading to some inversion of configuration rather than complete racemization.

Problem 11.8 What product(s) would you expect from reaction of (*S*)-3-chloro-3-methyloctane with acetic acid? Show the stereochemistry of both reactant and product.

Problem 11.9 Among the numerous examples of S_N1 reactions that occur with incomplete racemization is one reported by Winstein in 1952. The optically pure tosylate of

2,2-dimethyl-1-phenyl-1-propanol ($[\alpha]_D = -30.3°$) was heated in acetic acid to yi
the corresponding acetate ($[\alpha]_D = +5.3°$). If complete inversion had occurred, t
optically pure acetate would have had $[\alpha]_D = +53.6°$. What percentage racemizati
and what percentage inversion occurred in this reaction?

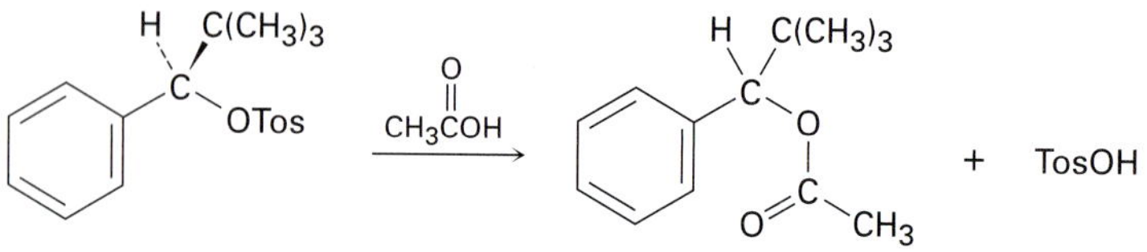

$[\alpha]_D = -30.3$

Observed $[\alpha]_D = +5.3$
(optically pure $[\alpha]_D = +53.6$)

Problem 11.10 Assign configuration to the following substrate, and show the stereochemistry ar identity of the product you would obtain by S_N1 reaction with water (reddi: brown = Br):

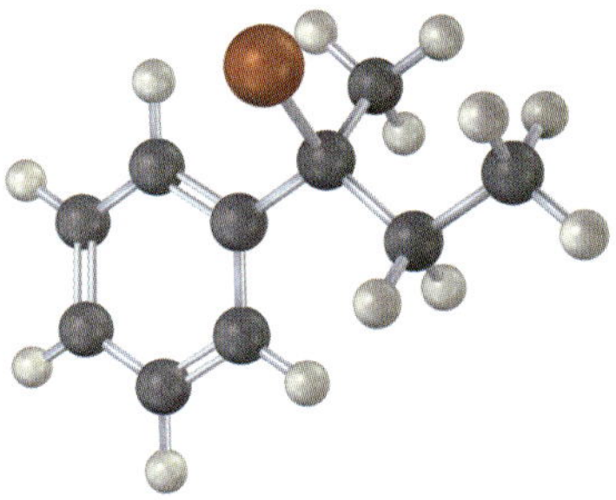

11.5 Characteristics of the S_N1 Reaction

Key **IDEAS**

Test your knowledge of Key Ideas by using resources in ThomsonNOW or by answering end-of-chapter problems marked with ▲.

Just as the S_N2 reaction is strongly influenced by the structure of the substrat the leaving group, the nucleophile, and the solvent, the S_N1 reaction is similarl influenced. Factors that lower $\Delta G^‡$, either by lowering the energy level of th transition state or by raising the energy level of the ground state, favor faste S_N1 reactions. Conversely, factors that raise $\Delta G^‡$, either by raising the energ level of the transition state or by lowering the energy level of the reactant, slo down the S_N1 reaction.

The Substrate

According to the Hammond postulate (Section 6.10), any factor that stabilizes high-energy intermediate also stabilizes the transition state leading to tha intermediate. Since the rate-limiting step in an S_N1 reaction is the spontaneou unimolecular dissociation of the substrate to yield a carbocation, the reaction i favored whenever a stabilized carbocation intermediate is formed. The more sta ble the carbocation intermediate, the faster the S_N1 reaction.

We saw in Section 6.9 that the stability order of alkyl carbocations i $3° > 2° > 1° > -CH_3$. To this list we must also add the resonance-stabilize allyl and benzyl cations. Just as allylic *radicals* are unusually stable because th unpaired electron can be delocalized over an extended π orbital system (Section 10.5), so allylic and benzylic *carbocations* are unusually stable. (The word **benzylic** means "next to an aromatic ring.") As Figure 11.12 indicates, an allylic cation has two resonance forms. In one form the double bond is on the "left"; in the other form it's on the "right." A benzylic cation has five resonance forms, all of which make substantial contributions to the overall resonance hybrid.

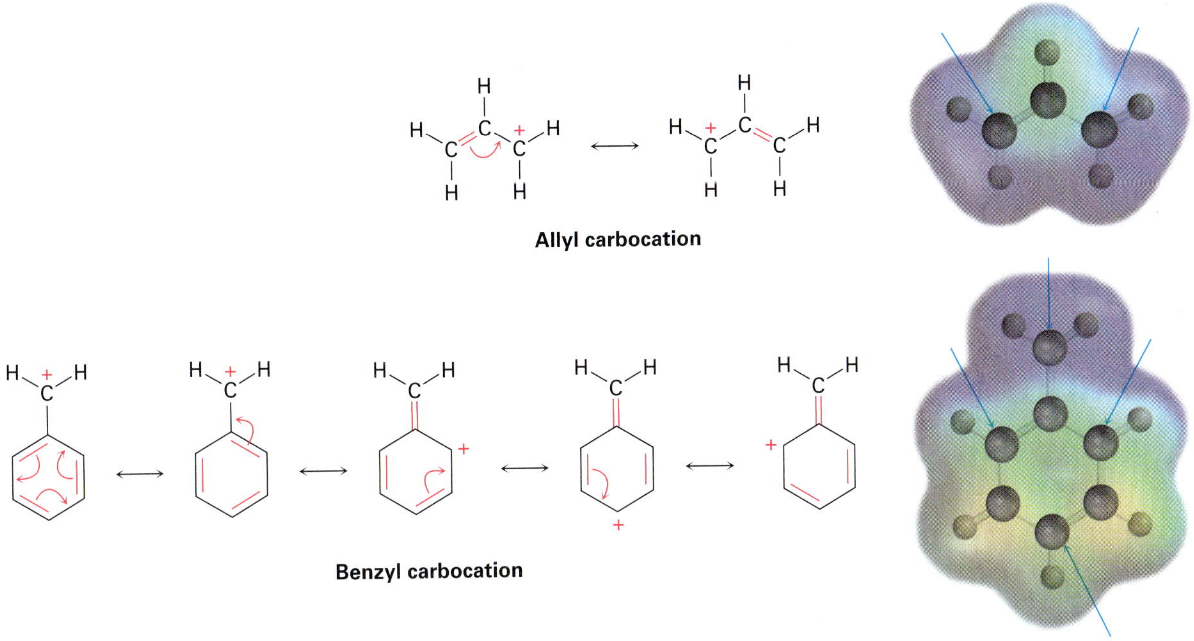

Figure 11.12 Resonance forms of the allyl and benzyl carbocations. Electrostatic potential maps show that the positive charge (blue) is delocalized over the π system in both. Electron-poor atoms are indicated by blue arrows.

Because of resonance stabilization, a *primary* allylic or benzylic carbocation is about as stable as a *secondary* alkyl carbocation and a *secondary* allylic or benzylic carbocation is about as stable as a *tertiary* alkyl carbocation. This stability order of carbocations is the same as the order of S_N1 reactivity for alkyl halides and tosylates.

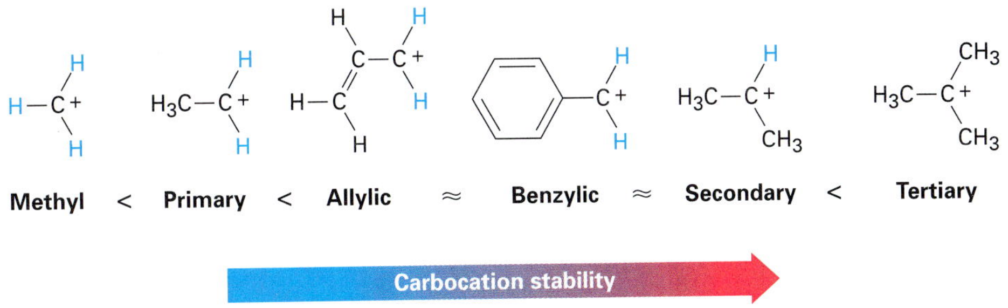

Parenthetically, we might also note that primary allylic and benzylic substrates are particularly reactive in S_N2 reactions as well as in S_N1 reactions.

Allylic and benzylic C–X bonds are about 50 kJ/mol (12 kcal/mol) weaker tha the corresponding saturated bonds and are therefore more easily broken.

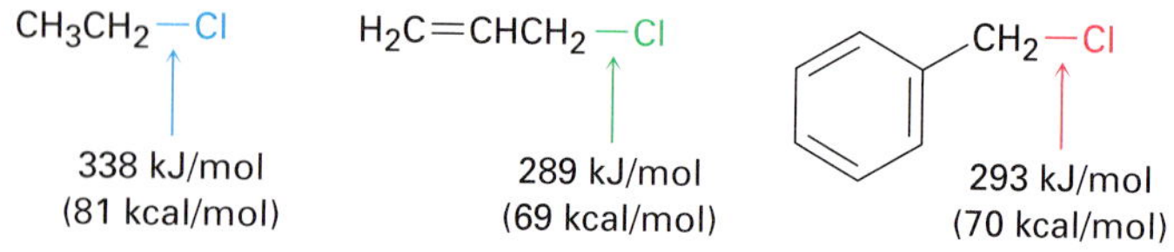

Problem 11.11 Rank the following substances in order of their expected S_N1 reactivity:

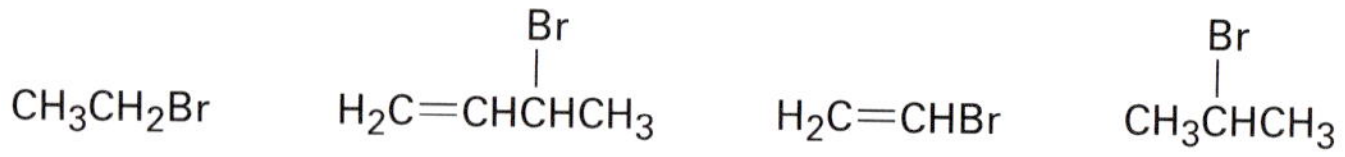

Problem 11.12 3-Bromo-1-butene and 1-bromo-2-butene undergo S_N1 reaction at nearly the sam rate even though one is a secondary halide and the other is primary. Explain.

The Leaving Group

We said during the discussion of S_N2 reactivity that the best leaving groups are thos that are most stable, that is, those that are the conjugate bases of strong acids. Ar identical reactivity order is found for the S_N1 reaction because the leaving group i directly involved in the rate-limiting step. Thus, the S_N1 reactivity order is

$$HO^- < Cl^- < Br^- < I^- \approx TosO^- \quad H_2O$$

Note that in the S_N1 reaction, which is often carried out under acidic con- ditions, neutral water can act as a leaving group. This occurs, for example, wher an alkyl halide is prepared from a tertiary alcohol by reaction with HBr or HC (Section 10.6). The alcohol is first protonated and then spontaneously loses H_2C to generate a carbocation, which reacts with halide ion to give the alkyl halide (Figure 11.13). Knowing that an S_N1 reaction is involved in the conversion of alcohols to alkyl halides explains why the reaction works well only for tertiary alcohols. Tertiary alcohols react fastest because they give the most stable carbocation intermediates.

The Nucleophile

The nature of the nucleophile plays a major role in the S_N2 reaction but does not affect an S_N1 reaction. Because the S_N1 reaction occurs through a rate-limiting step in which the added nucleophile has no part, the nucleophile can't affect the reaction rate. The reaction of 2-methyl-2-propanol with HX, for instance, occurs at the same rate regardless of whether X is Cl, Br, or I. Furthermore, neutral nucleophiles are just as effective as negatively charged ones, so S_N1 reactions frequently occur under neutral or acidic conditions.

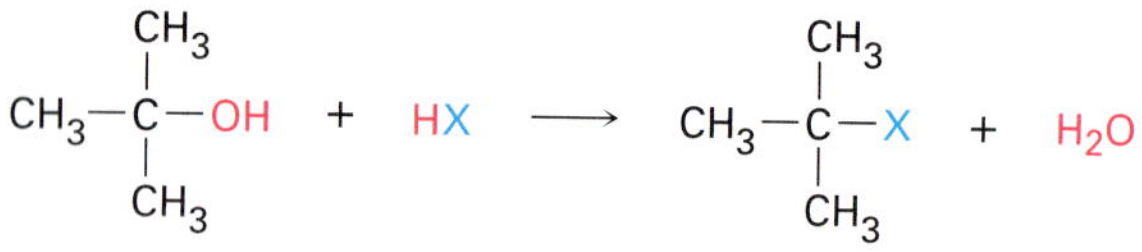

2-Methyl-2-propanol (Same rate for X = Cl, Br, I)

gure 11.13 MECHANISM: e mechanism of the S_N1 reac- n of a tertiary alcohol with HBr yield an alkyl halide. Neutral ater is the leaving group.

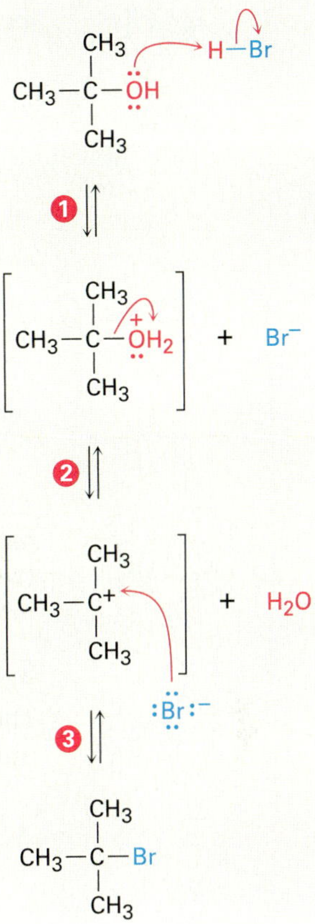

1 The –OH group is first protonated by HBr.

2 Spontaneous dissociation of the protonated alcohol occurs in a slow, rate-limiting step to yield a carbocation intermediate plus water.

3 The carbocation intermediate reacts with bromide ion in a fast step to yield the neutral substitution product.

The Solvent

What about solvent? Do solvents have the same effect in S_N1 reactions that they have in S_N2 reactions? The answer is both yes and no. Yes, solvents have a large effect on S_N1 reactions, but no, the reasons for the effects on S_N1 and S_N2 reactions are not the same. Solvent effects in the S_N2 reaction are due largely to stabilization or destabilization of the nucleophile *reactant.* Solvent effects in the S_N1 reaction, however, are due largely to stabilization or destabilization of the *transition state.*

The Hammond postulate says that any factor stabilizing the intermediate carbocation should increase the rate of an S_N1 reaction. Solvation of the carbocation—the interaction of the ion with solvent molecules—has just such an effect. Solvent molecules orient around the carbocation so that the electron-rich ends of the solvent dipoles face the positive charge (Figure 11.14), thereby lowering the energy of the ion and favoring its formation.

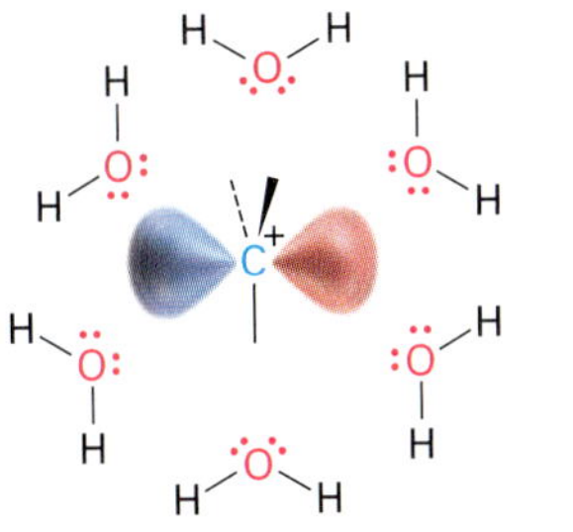

Figure 11.14 Solvation of a carbocation by water. The electron-rich oxygen atoms of solvent molecules orient around the positively charged carbocation and thereby stabilize it.

The properties of a solvent that contribute to its ability to stabilize ions by solvation are related to the solvent's polarity. S_N1 reactions take place much more rapidly in strongly polar solvents, such as water and methanol, than in less polar solvents, such as ether and chloroform. In the reaction of 2-chloro-2-methylpropane, for example, a rate increase of 100,000 is observed on going from ethanol (less polar) to water (more polar). The rate increases on going from a hydrocarbon solvent to water are so large they car be measured accurately.

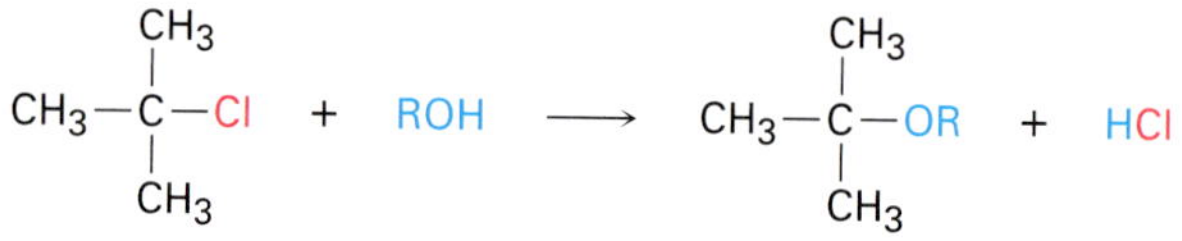

	Ethanol	40% Water/ 60% Ethanol	80% Water/ 20% Ethanol	Water
Relative reactivity	1	100	14,000	100,000

Solvent reactivity →

It should be emphasized again that both the S_N1 and the S_N2 reaction sho solvent effects but that they do so for different reasons. S_N2 reactions are *di favored* in protic solvents because the *ground-state energy* of the nucleophile is low ered by solvation. S_N1 reactions are *favored* in protic solvents because th *transition-state energy* leading to carbocation intermediate is lowered by solvatior

S_N1 Reaction Characteristics: A Summary

The effects on S_N1 reactions of the four variables—substrate, leaving grouŗ nucleophile, and solvent—are summarized in the following statements:

Substrate The best substrates yield the most stable carbocations. A a result, S_N1 reactions are best for tertiary, allylic, an benzylic halides.

Leaving group Good leaving groups increase the reaction rate by lower ing the energy level of the transition state for carbocatio formation.

Nucleophile The nucleophile must be nonbasic to prevent a competi tive elimination of HX (Section 11.7), but otherwise doe not affect the reaction rate. Neutral nucleophiles worl well.

Solvent Polar solvents stabilize the carbocation intermediate by solvation, thereby increasing the reaction rate.

WORKED EXAMPLE 11.2 ***Predicting the Mechanism of a Nucleophilic Substitution Reaction***

Predict whether each of the following substitution reactions is likely to be S_N1 or S_N2

(a)

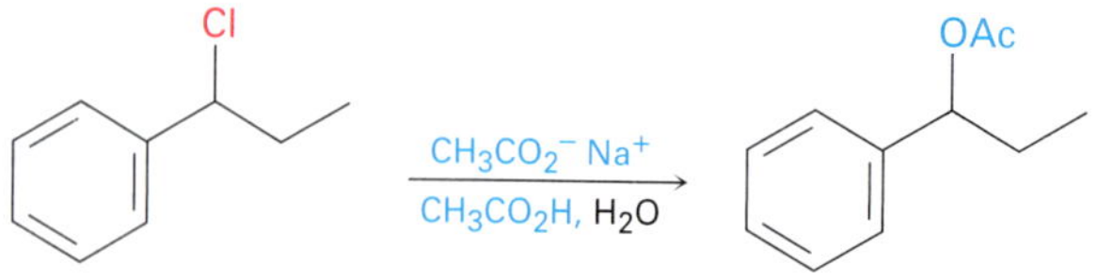

(b)

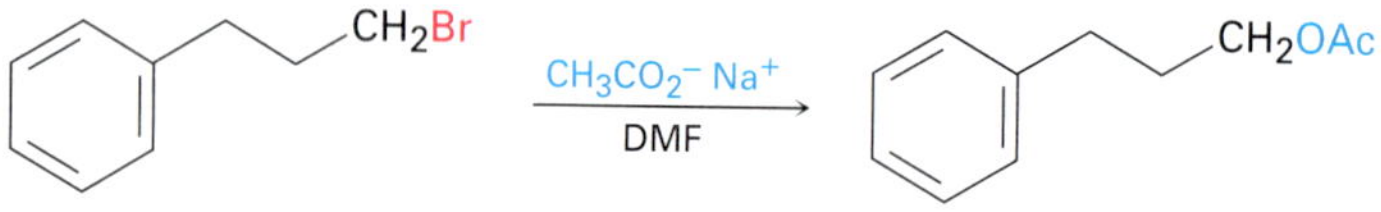

Strategy Look at the substrate, leaving group, nucleophile, and solvent. Then decide from the summaries at the ends of Sections 11.3 and 11.5 whether an S_N1 or an S_N2 reaction is favored. S_N1 reactions are favored by tertiary, allylic, or benzylic substrates, by good leaving groups, by nonbasic nucleophiles, and by protic solvents. S_N2 reactions are favored by primary substrates, by good leaving groups, by good nucleophiles, and by polar aprotic solvents.

Solution
(a) This is likely to be an S_N1 reaction because the substrate is secondary and benzylic, the nucleophile is weakly basic, and the solvent is protic.
(b) This is likely to be an S_N2 reaction because the substrate is primary, the nucleophile is a reasonably good one, and the solvent is polar aprotic.

Problem 11.13 Predict whether each of the following substitution reactions is likely to be S_N1 or S_N2:

(a)

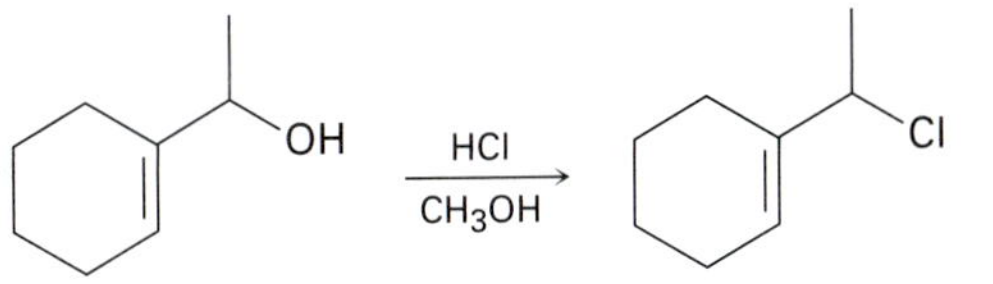

(b)

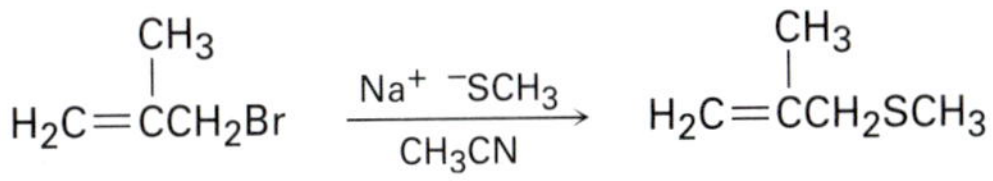

11.6 Biological Substitution Reactions

Both S_N1 and S_N2 reactions are well known in biological chemistry, particularly in the pathways for biosynthesis of the many thousands of terpenes (Chapter 6 *Focus On*). Unlike what typically happens in the laboratory, however, the substrate in a biological substitution reaction is often an organodiphosphate rather than an alkyl halide. Thus, the leaving group is the diphosphate ion, abbreviated PP_i, rather than a halide ion. In fact, it's useful to think of the diphosphate group as the "biological equivalent" of a halogen. The dissociation of an organodiphosphate in a biological reaction is typically assisted by complexation to a divalent metal cation such as Mg^{2+} to help neutralize charge.

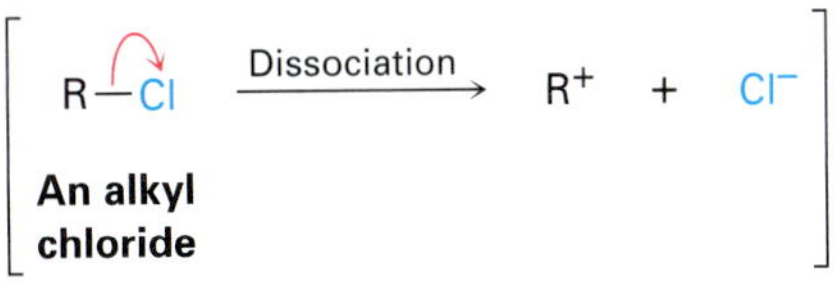

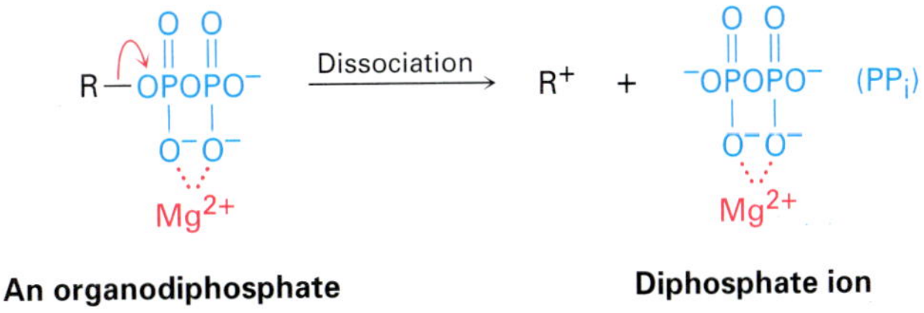

Two S_N1 reactions occur during the biosynthesis of geraniol, a fragrant alc hol found in roses and used in perfumery. Geraniol biosynthesis begins with di sociation of dimethylallyl diphosphate to give an allylic carbocation, which reac with isopentenyl diphosphate (Figure 11.15). From the viewpoint of isopenten diphosphate, the reaction is an electrophilic alkene addition, but from the view point of dimethylallyl diphosphate, the process in an S_N1 reaction in which th carbocation intermediate reacts with a double bond as the nucleophile.

Following this initial S_N1 reaction, loss of the *pro-R* hydrogen gives gerany diphosphate, itself an allylic diphosphate that dissociates a second time. Rea tion of the geranyl carbocation with water in a second S_N1 reaction, followed b loss of a proton, then yields geraniol.

Figure 11.15 Biosynthesis of geraniol from dimethylallyl diphosphate. Two S_N1 reactions occur, both with diphosphate ion as the leaving group.

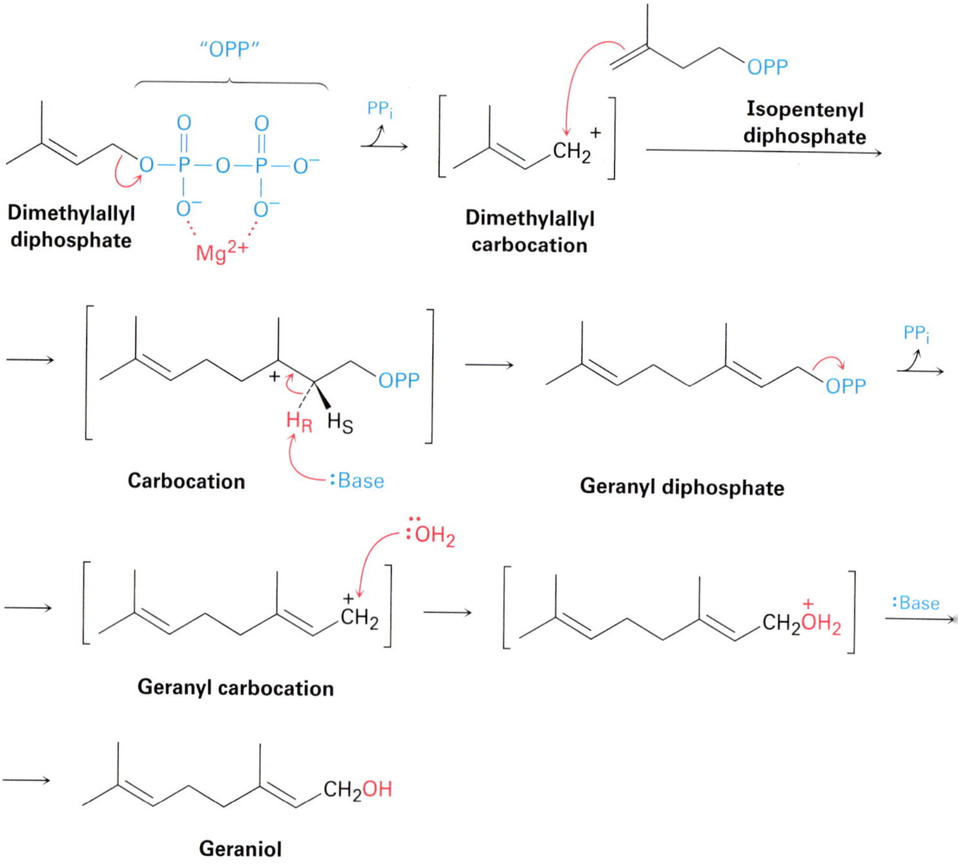

S_N2 reactions are involved in almost all biological methylations, which transfer a $-CH_3$ group from an electrophilic donor to a nucleophile. The donor is *S*-adenosylmethionine (abbreviated SAM), which contains a positively charged sulfur (a sulfonium ion; Section 9.12), and the leaving group is the neutral *S*-adenosylhomocysteine molecule. In the biosynthesis of epinephrine (adrenaline) from norepinephrine, for instance, the nucleophilic nitrogen atom of norepinephrine attacks the electrophilic methyl carbon atom of *S*-adenosylmethionine in an S_N2 reaction, displacing *S*-adenosylhomocysteine (Figure 11.16). In effect, *S*-adenosylmethionine is simply a biological equivalent of CH_3Cl.

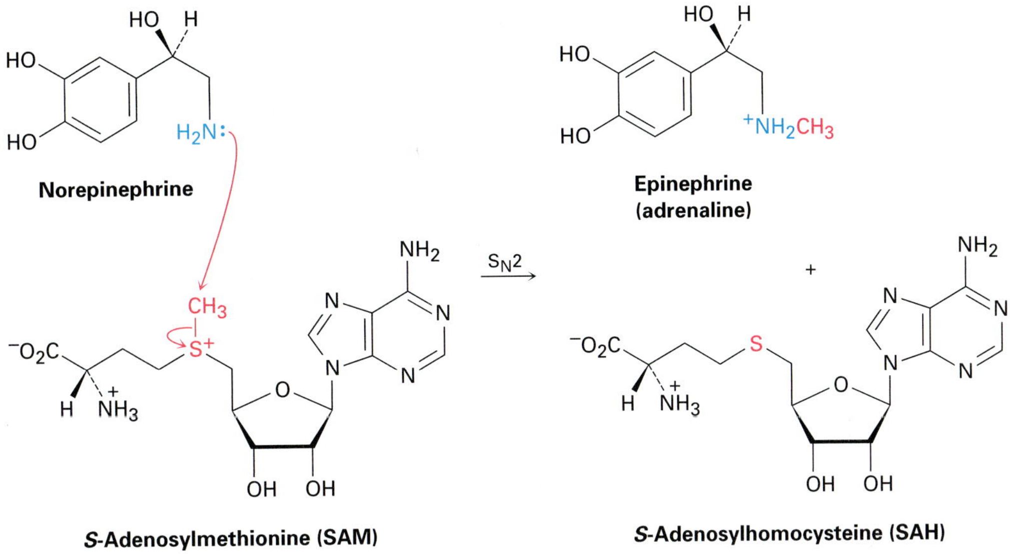

Figure 11.16 The biosynthesis of epinephrine from norepinephrine occurs by an S_N2 reaction with *S*-adenosylmethionine.

Problem 11.14 Review the mechanism of geraniol biosynthesis shown in Figure 11.15, and then propose a mechanism for the biosynthesis of limonene from linalyl diphosphate.

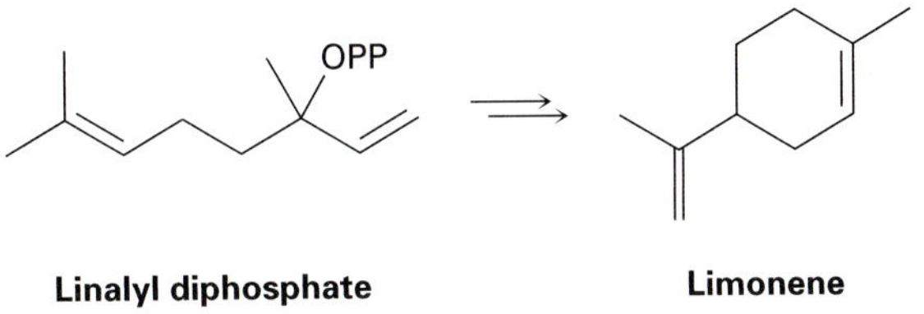

11.7 Elimination Reactions of Alkyl Halides: Zaitsev's Rule

Key IDEAS

Test your knowledge of Key Ideas by using resources in ThomsonNOW or by answering end-of-chapter problems marked with ▲.

We said at the beginning of this chapter that two kinds of reactions can happen when a nucleophile/Lewis base reacts with an alkyl halide. The nucleophile can either substitute for the halide by reaction at carbon or cause elimination of HX by reaction at a neighboring hydrogen:

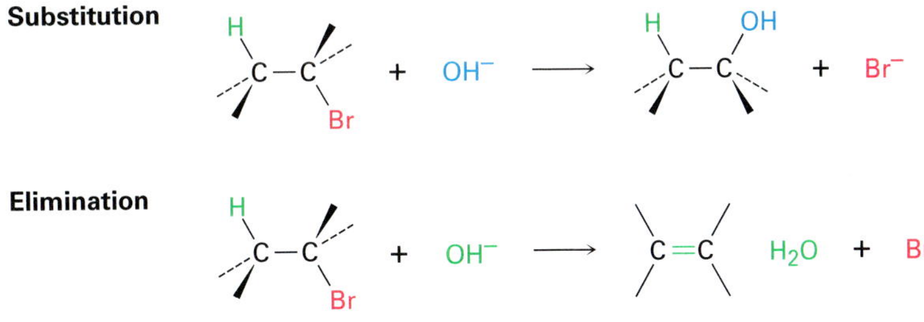

Elimination reactions are more complex than substitution reactions for several reasons. There is, for example, the problem of regiochemistry. What products result by loss of HX from an unsymmetrical halide? In fact, elimina tion reactions almost always give mixtures of alkene products, and the best w can usually do is to predict which will be the major product.

According to **Zaitsev's rule**, formulated in 1875 by the Russian chemis Alexander Zaitsev, base-induced elimination reactions generally (although no always) give the more stable alkene product—that is, the alkene with more alky substituents on the double-bond carbons. In the following two cases, for exam ple, the more highly substituted alkene product predominates.

Zaitsev's rule In the elimination of HX from an alkyl halide, the more highly substitute alkene product predominates.

Alexander M. Zaitsev

Alexander M. Zaitsev (1841–1910) was born in Kazan, Russia, and received his Ph.D. from the University of Leipzig in 1866. He was professor at the University of Kazan (1870–1903) and at Kiev University, and many of his students went on to assume faculty positions throughout Russia.

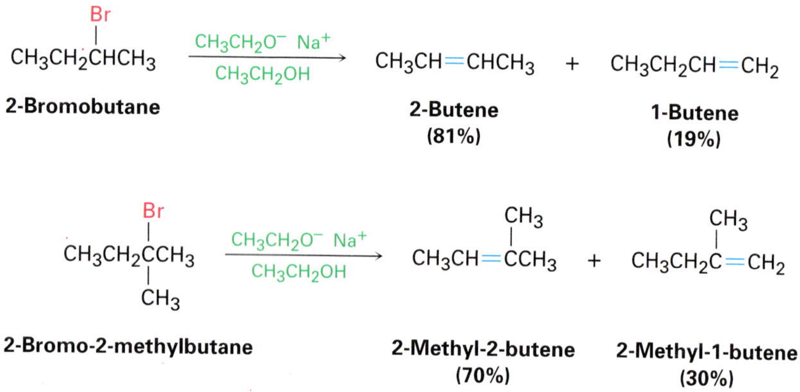

A second factor that complicates a study of elimination reactions is tha they can take place by different mechanisms, just as substitutions can. We'l consider three of the most common mechanisms—the E1, E2, and E1cB reactions—which differ in the timing of C–H and C–X bond-breaking. In the E1 reaction, the C–X bond breaks first to give a carbocation intermediate that undergoes subsequent base abstraction of H^+ to yield the alkene. In the E2 reaction, base-induced C–H bond cleavage is simultaneous with C–X bond cleavage, giving the alkene in a single step. In the E1cB reaction (cB for "conjugate base"), base abstraction of the proton occurs first, giving a carbon anion, or *carbanion* intermediate. This anion, the conjugate base of the reactant "acid," then undergoes loss of X^- in a subsequent step to give the alkene. All three mechanisms occur frequently in the laboratory, but the E1cB mechanism predominates in biological pathways.

E1 Reaction: C–X bond breaks first to give a carbocation intermediate, followed by base removal of a proton to yield the alkene.

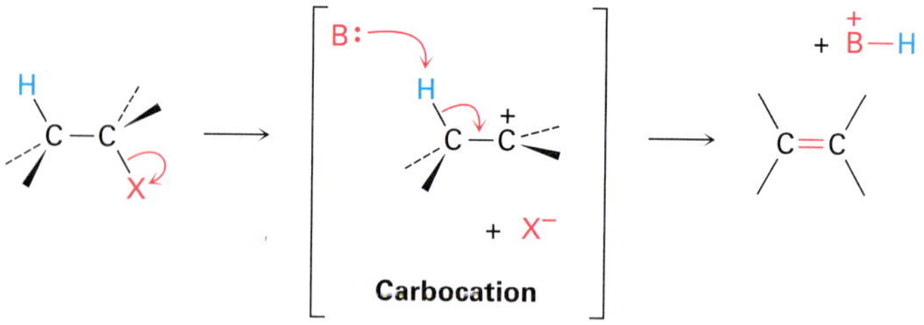

E2 Reaction: C–H and C–X bonds break simultaneously, giving the alkene in a single step without intermediates.

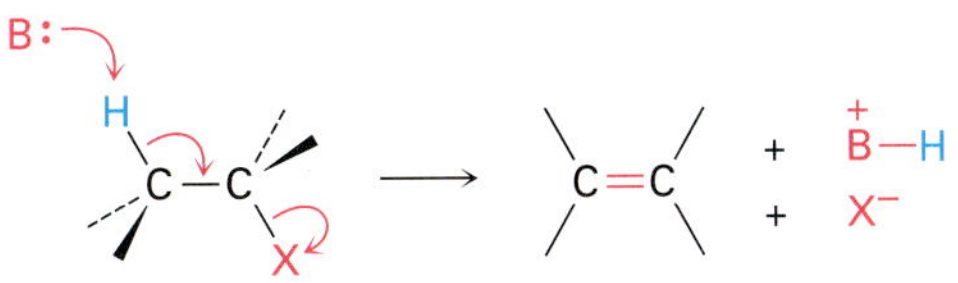

E1cB Reaction: C–H bond breaks first, giving a carbanion intermediate that loses X^- to form the alkene.

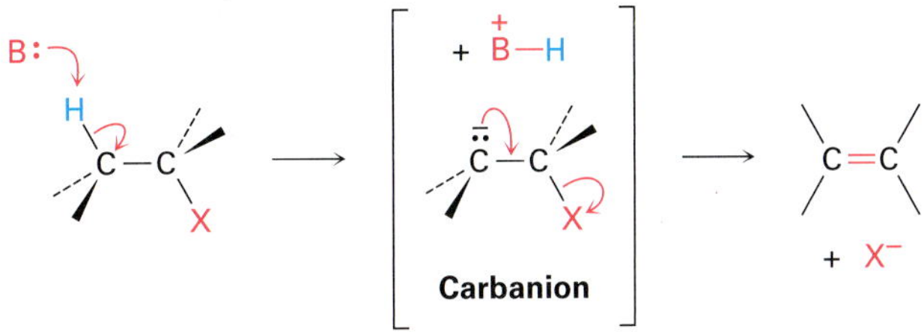

WORKED EXAMPLE 11.3

Predicting the Product of an Elimination Reaction

What product would you expect from reaction of 1-chloro-1-methylcyclohexane with KOH in ethanol?

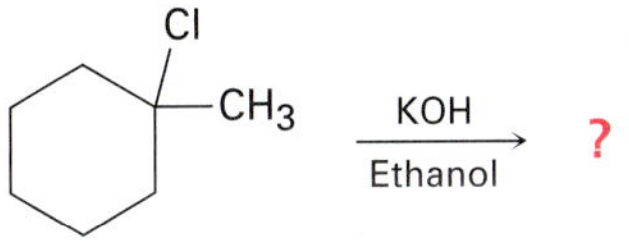

Strategy Treatment of an alkyl halide with a strong base such as KOH yields an alkene. To find the products in a specific case, locate the hydrogen atoms on each carbon next to the leaving group. Then generate the potential alkene products by removing HX in as many ways as possible. The major product will be the one that has the most highly substituted double bond—in this case, 1-methylcyclohexene.

Solution

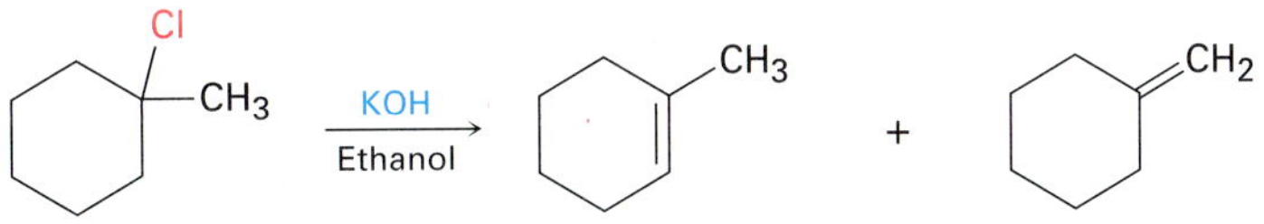

1-Chloro-1-methyl-cyclohexane **1-Methylcyclohexene (major)** **Methylenecyclohexane (minor)**

Problem 11.15 Ignoring double-bond stereochemistry, what products would you expect from elimination reactions of the following alkyl halides? Which will be the major product in each case?

(a) 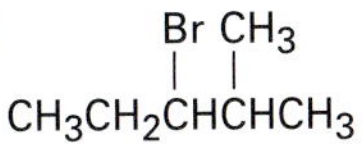 **(b)** 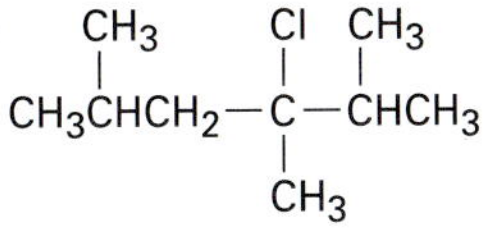 **(c)** 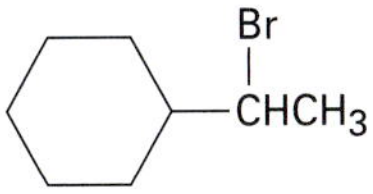

Problem 11.16 What alkyl halides might the following alkenes have been made from?

(a)
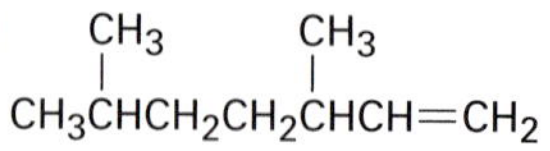

(b)
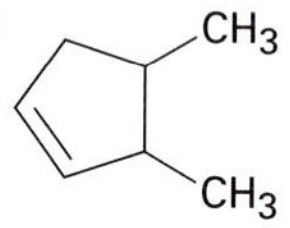

11.8 The E2 Reaction and the Deuterium Isotope Effect

The **E2 reaction** (for *elimination, bimolecular*) occurs when an alkyl halide is treated with a strong base, such as hydroxide ion or alkoxide ion (RO^-). It is the most commonly occurring pathway for elimination and can be formulated as shown in Figure 11.17.

Figure 11.17 MECHANISM: Mechanism of the E2 reaction of an alkyl halide. The reaction takes place in a single step through a transition state in which the double bond begins to form at the same time the H and X groups are leaving.

1 Base (B:) attacks a neighboring hydrogen and begins to remove the H at the same time as the alkene double bond starts to form and the X group starts to leave.

2 Neutral alkene is produced when the C–H bond is fully broken and the X group has departed with the C–X bond electron pair.

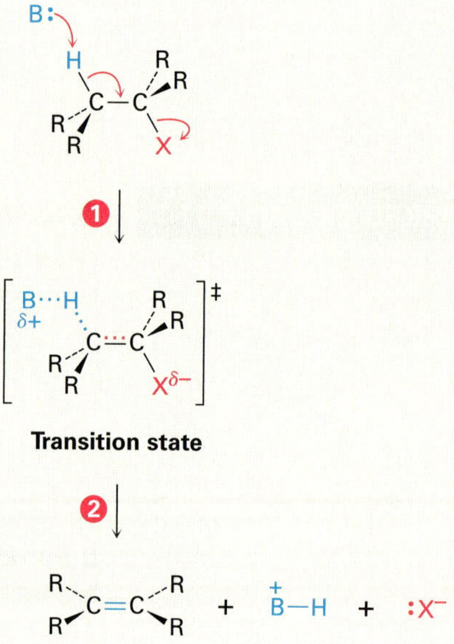

ThomsonNOW™ Click *Organic Process* to **view an animation showing the mechanism of an E2 elimination reaction.**

Like the S_N2 reaction, the E2 reaction takes place in one step without intermediates. As the base begins to abstract H^+ from a carbon next to the leaving group, the C–H bond begins to break, a C=C bond begins to form, and the leaving group begins to depart, taking with it the electron pair from the C–X bond. Among the pieces of evidence supporting this mechanism is that E2 reactions show second-order kinetics and follow the rate law: rate = $k \times [RX] \times [Base]$. That is, both base and alkyl halide take part in the rate-limiting step.

A second piece of evidence in support of the E2 mechanism is provided by a phenomenon known as the **deuterium isotope effect**. For reasons that we won't go into, a carbon–*hydrogen* bond is weaker by about 5 kJ/mol (1.2 kcal/mol) than the corresponding carbon–*deuterium* bond. Thus, a C–H bond is more easily broken than an equivalent C–D bond, and the rate of C–H bond cleavage is faster. For instance, the base-induced elimination of HBr from 1-bromo-2-phenylethane proceeds 7.11 times as fast as the corresponding omsonNOW™ Click *Organic teractive* to **use a web-based alette to predict products from mple elimination reactions.**

elimination of DBr from 1-bromo-2,2-dideuterio-2-phenylethane. This result tells us that the C–H (or C–D) bond is broken *in the rate-limiting step,* consistent with our picture of the E2 reaction as a one-step process. If it were otherwise, we couldn't measure a rate difference.

(H)—Faster reaction
(D)—Slower reaction

Yet a third piece of mechanistic evidence involves the stereochemistry of E2 eliminations. As shown by a large number of experiments, E2 reactions occur with *periplanar* geometry, meaning that all four reacting atoms—the hydrogen, the two carbons, and the leaving group—lie in the same plane. Two such geometries are possible: **syn periplanar** geometry, in which the H and the X are on the same side of the molecule, and **anti periplanar** geometry, in which the H and the X are on opposite sides of the molecule. Of the two, anti periplanar geometry is energetically preferred because it allows the substituents on the two carbons to adopt a staggered relationship, whereas syn geometry requires that the substituents be eclipsed.

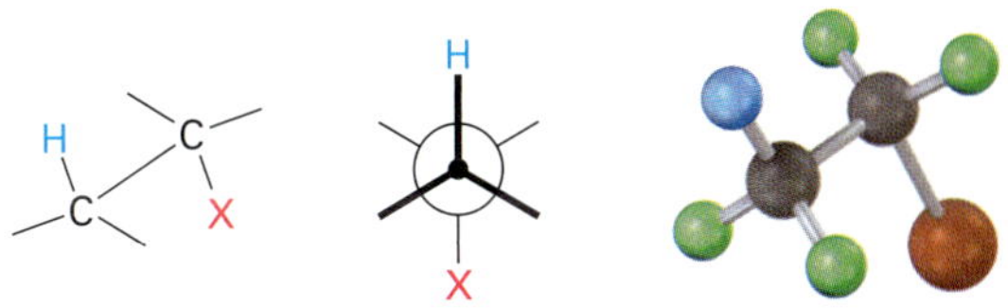

Anti periplanar geometry
(staggered, lower energy)

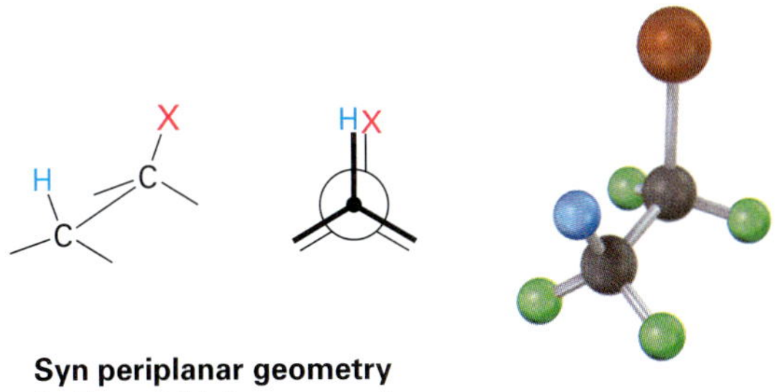

Syn periplanar geometry
(eclipsed, higher energy)

What's so special about periplanar geometry? Because the sp^3 σ orbitals in the reactant C–H and C–X bonds must overlap and become p π orbitals in the alkene product, there must also be some overlap in the transition state. This can occur most easily if all the orbitals are in the same plane to begin with—that is, if they're periplanar (Figure 11.18).

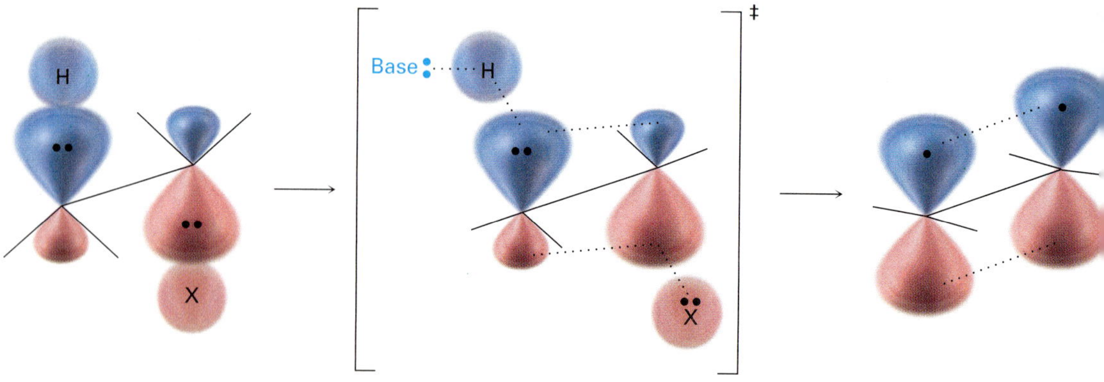

Anti periplanar reactant **Anti transition state** **Alkene product**

Figure 11.18 The transition state for the E2 reaction of an alkyl halide with base. Overlap c the developing *p* orbitals in the transition state requires periplanar geometry of the reactant.

It might help to think of E2 elimination reactions with periplanar geometry a being similar to S_N2 reactions with 180° geometry. In an S_N2 reaction, an electro pair from the incoming nucleophile pushes out the leaving group on the opposit side of the molecule. In an E2 reaction, an electron pair from a neighborin C–H bond pushes out the leaving group on the opposite side of the molecule.

S_N2 reaction (backside attack) **E2 reaction (anti periplanar)**

Anti periplanar geometry for E2 eliminations has specific stereochemica consequences that provide strong evidence for the proposed mechanism. To take just one example, *meso*-1,2-dibromo-1,2-diphenylethane undergoe E2 elimination on treatment with base to give only the *E* alkene. None of the isomeric *Z* alkene is formed because the transition state leading to the *Z* alken would have to have syn periplanar geometry and would thus be higher in energy

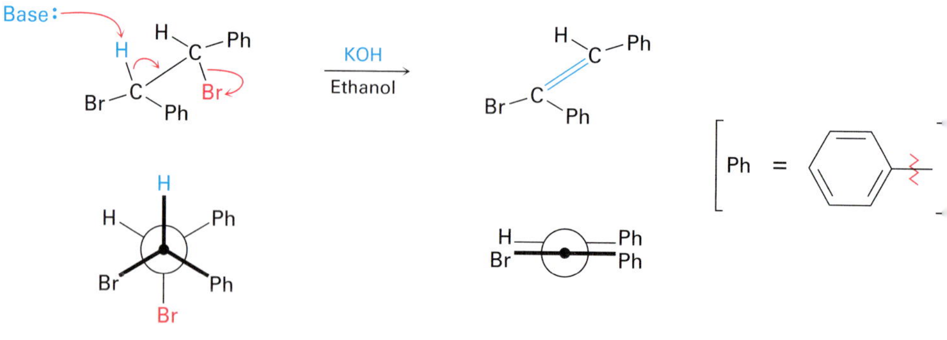

***meso*-1,2-Dibromo-1,2-diphenylethane (anti periplanar geometry)**

(*E*)-1-Bromo-1,2-diphenylethylene

WORKED EXAMPLE 11.4

Predicting the Double-Bond Stereochemistry of the Product in an E2 Reaction

What stereochemistry do you expect for the alkene obtained by E2 elimination of (1*S*,2*S*)-1,2-dibromo-1,2-diphenylethane?

Strategy Draw (1*S*,2*S*)-1,2-dibromo-1,2-diphenylethane so that you can see its stereochemistry and so that the −H and −Br groups to be eliminated are anti periplanar. Then carry out the elimination while keeping all substituents in approximately their same positions, and see what alkene results.

Solution Anti periplanar elimination of HBr gives (*Z*)-1-bromo-1,2-diphenylethylene.

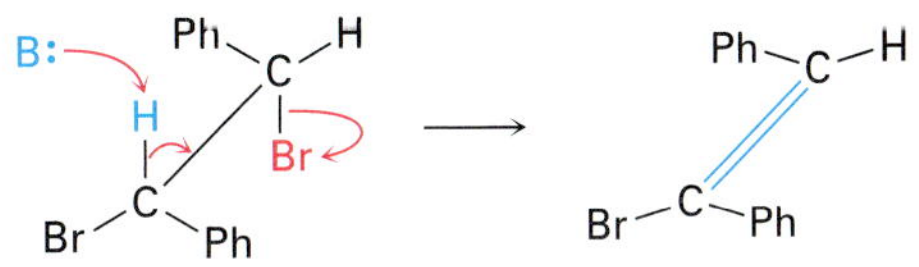

Problem 11.17 What stereochemistry do you expect for the alkene obtained by E2 elimination of (1*R*,2*R*)-1,2-dibromo-1,2-diphenylethane? Draw a Newman projection of the reacting conformation.

Problem 11.18 What stereochemistry do you expect for the trisubstituted alkene obtained by E2 elimination of the following alkyl halide on treatment with KOH? (Reddish brown = Br.)

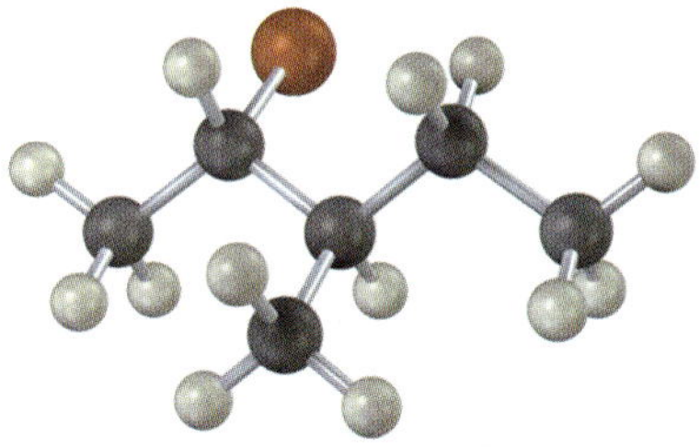

11.9 The E2 Reaction and Cyclohexane Conformation

Derek H. R. Barton

Derek H. R. Barton (1918–1998) was born in Gravesend, England, and received both Ph.D. and D.Sc. degrees from Imperial College, London. Among his numerous positions were those as professor at Imperial College, the University of London, Glasgow, Institut de Chimie des Substances Naturelles, and Texas A&M University. Barton received the Nobel Prize in chemistry in 1969 and was knighted by Queen Elizabeth in 1972.

Anti periplanar geometry for E2 reactions is particularly important in cyclohexane rings, where chair geometry forces a rigid relationship between the substituents on neighboring carbon atoms (Section 4.8). As pointed out by Derek Barton in a landmark 1950 paper, much of the chemical reactivity of substituted cyclohexanes is controlled by their conformation. Let's look at the E2 dehydrohalogenation of chlorocyclohexanes to see an example.

The anti periplanar requirement for E2 reactions overrides Zaitsev's rule and can be met in cyclohexanes only if the hydrogen and the leaving group are trans diaxial (Figure 11.19). If either the leaving group or the hydrogen is equatorial, E2 elimination can't occur.

Figure 11.19 The geometric requirement for E2 reaction in a substituted cyclohexane. The leaving group and the hydrogen must both be axial for anti periplanar elimination to occur.

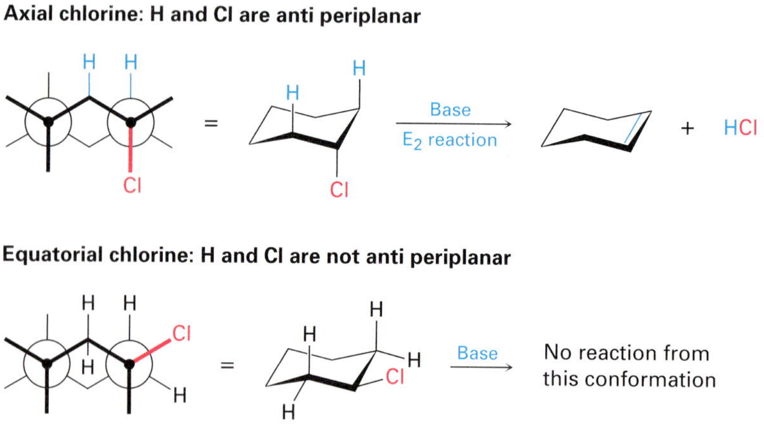

The elimination of HCl from the isomeric menthyl and neomenthyl chlo rides shown in Figure 11.20 gives a good illustration of this trans-diaxial requir ment. Neomenthyl chloride undergoes elimination of HCl on reaction wit ethoxide ion 200 times as fast as menthyl chloride. Furthermore, neomenth chloride yields 3-menthene as the major alkene product, whereas menthyl chlo ride yields 2-menthene.

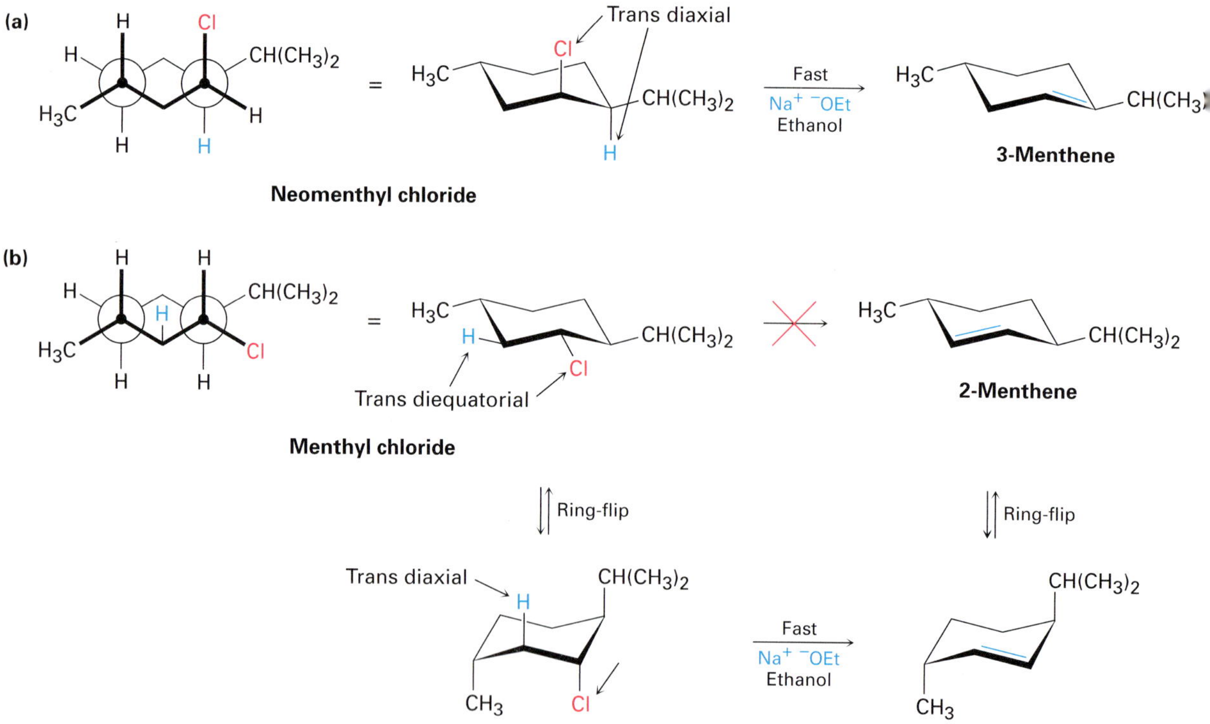

Active Figure 11.20 Dehydrochlorination of menthyl and neomenthyl chlorides. **(a)** Neomenthyl chloride loses HCl directly from its more stable conformation, but **(b)** menthyl chloride must first ring-flip before HCl loss can occur. The abbreviation "Et" represents an ethyl group. *Sign in at* **www.thomsonedu.com** *to see a simulation based on this figure and to take a short quiz.*

The difference in reactivity between the isomeric menthyl chlorides is due to the difference in their conformations. Neomenthyl chloride has the conformation shown in Figure 11.20a, with the methyl and isopropyl groups equatorial and the chlorine axial—a perfect geometry for E2 elimination. Loss of the hydrogen atom at C4 occurs easily to yield the more substituted alkene product, 3-menthene, as predicted by Zaitsev's rule.

Menthyl chloride, by contrast, has a conformation in which all three substituents are equatorial (Figure 11.20b). To achieve the necessary geometry for elimination, menthyl chloride must first ring-flip to a higher-energy chair conformation, in which all three substituents are axial. E2 elimination then occurs with loss of the only trans-diaxial hydrogen available, leading to the non-Zaitsev product 2-menthene. The net effect of the simple change in chlorine stereochemistry is a 200-fold change in reaction rate and a complete change of product. The chemistry of the molecule is controlled by its conformation.

Problem 11.19 Which isomer would you expect to undergo E2 elimination faster, *trans*-1-bromo-4-*tert*-butylcyclohexane or *cis*-1-bromo-4-*tert*-butylcyclohexane? Draw each molecule in its more stable chair conformation, and explain your answer.

11.10 The E1 and E1cB Reactions

The E1 Reaction

Just as the E2 reaction is analogous to the S_N2 reaction, the S_N1 reaction has a close analog called the **E1 reaction** (for *elimination, unimolecular*). The E1 reaction can be formulated as shown in Figure 11.21 for the elimination of HCl from 2-chloro-2-methylpropane.

igure 11.21 MECHANISM: lechanism of the E1 reaction. wo steps are involved, the first f which is rate-limiting, and a arbocation intermediate is resent.

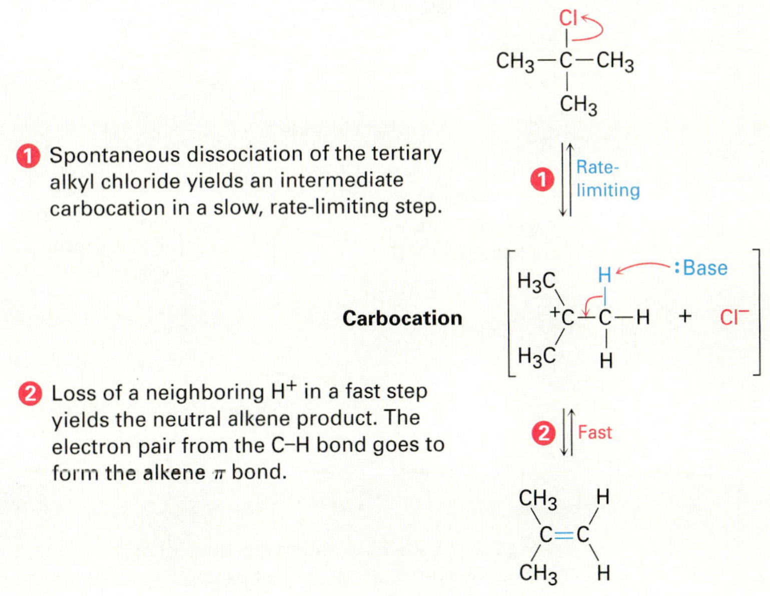

hornsonNOW Click *Organic Process* to **view an animation showing the mechanism of an E1 elimination reaction.**

E1 eliminations begin with the same unimolecular dissociation we saw in th S_N1 reaction, but the dissociation is followed by loss of H^+ from the adjacer carbon rather than by substitution. In fact, the E1 and S_N1 reactions normall occur together whenever an alkyl halide is treated in a protic solvent with a nor basic nucleophile. Thus, the best E1 substrates are also the best S_N1 substrates, an mixtures of substitution and elimination products are usually obtained. For exan ple, when 2-chloro-2-methylpropane is warmed to 65 °C in 80% aqueous ethano a 64:36 mixture of 2-methyl-2-propanol (S_N1) and 2-methylpropene (E1) result

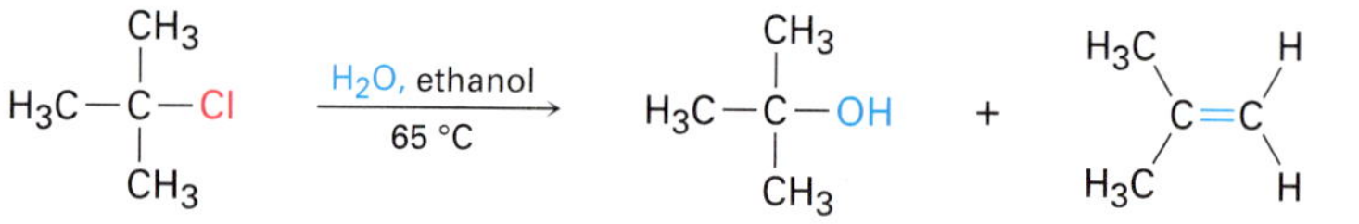

2-Chloro-2-methylpropane **2-Methyl-2-propanol (64%)** **2-Methylpropene (36%)**

Much evidence has been obtained in support of the E1 mechanism. Fo example, E1 reactions show first-order kinetics, consistent with a rate-limitin spontaneous dissociation process. Furthermore, E1 reactions show no deu terium isotope effect because rupture of the C–H (or C–D) bond occurs *after* th rate-limiting step rather than during it. Thus, we can't measure a rate differenc between a deuterated and nondeuterated substrate.

A final piece of evidence involves the stereochemistry of elimination Unlike the E2 reaction, where anti periplanar geometry is required, there is n geometric requirement on the E1 reaction because the halide and the hydroger are lost in separate steps. We might therefore expect to obtain the more stabl (Zaitsev's rule) product from E1 reaction, which is just what we find. To returr to a familiar example, menthyl chloride loses HCl under E1 conditions in a pola solvent to give a mixture of alkenes in which the Zaitsev product, 3-menthene predominates (Figure 11.22).

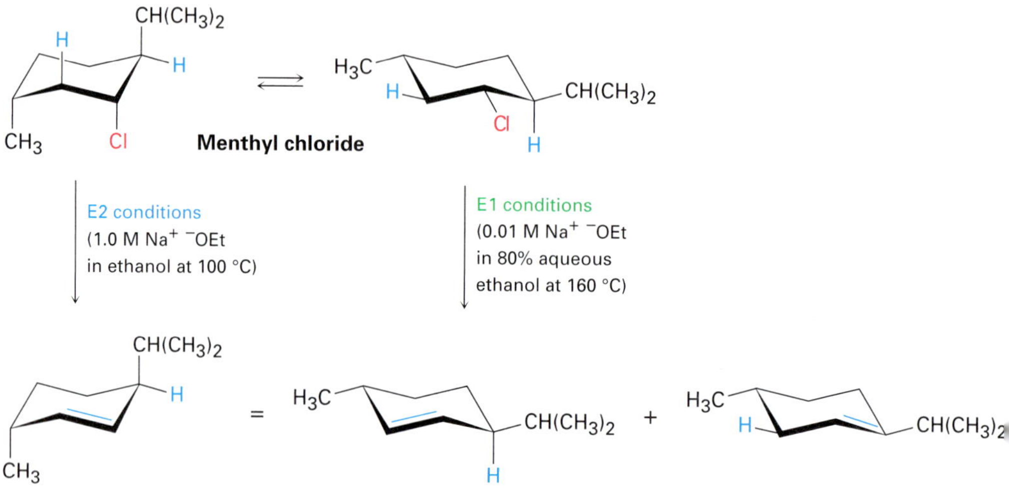

2-Menthene (100%) **2-Menthene (32%)** **3-Menthene (68%)**

Figure 11.22 Elimination reactions of menthyl chloride. E2 conditions (strong base in 100% ethanol) lead to 2-menthene through an anti periplanar elimination, whereas E1 conditions (dilute base in 80% aqueous ethanol) lead to a mixture of 2-menthene and 3-menthene.

The E1cB Reaction

In contrast to the E1 reaction, which involves a carbocation intermediate, the **E1cB reaction** takes place through a *carbanion* intermediate. Base-induced abstraction of a proton in a slow, rate-limiting step gives an anion, which expels a leaving group on the adjacent carbon. The reaction is particularly common in substrates that have a poor leaving group, such as $^-$OH, two carbons removed from a carbonyl group, HO—C—CH—C=O. The poor leaving group disfavors the alternative E1 and E2 possibilities, and the carbonyl group makes the adjacent hydrogen unusually acidic by resonance stabilization of the anion intermediate. We'll look at this acidifying effect of a carbonyl group in Section 22.5.

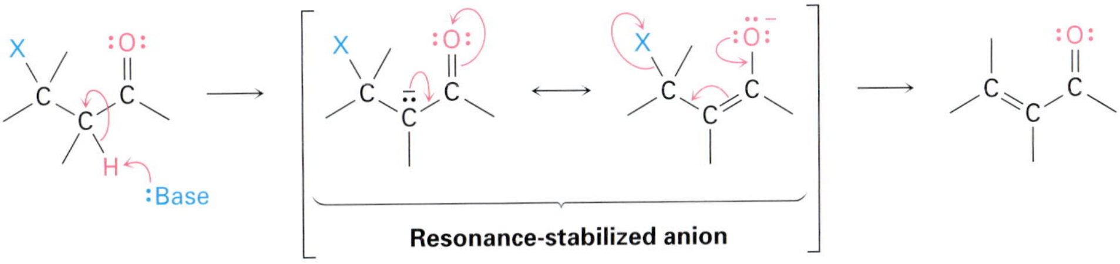

11.11 Biological Elimination Reactions

All three elimination reactions—E2, E1, and E1cB—occur in biological pathways, but the E1cB mechanism is particularly common. The substrate is usually an alcohol, and the H atom removed is usually adjacent to a carbonyl group, just as in laboratory reactions. Thus, 3-hydroxy carbonyl compounds are frequently converted to unsaturated carbonyl compounds by elimination reactions. A typical example occurs during the biosynthesis of fats when a 3-hydroxybutyryl thioester is dehydrated to the corresponding unsaturated (crotonyl) thioester. The base in this reaction is a histidine amino acid in the enzyme, and loss of the $^-$OH group is assisted by simultaneous protonation.

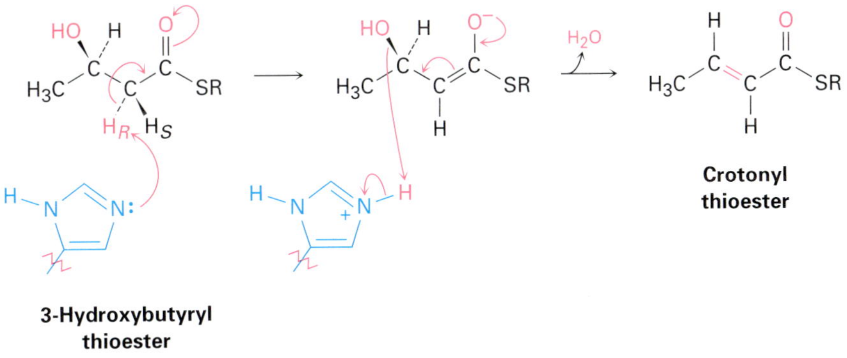

11.12 A Summary of Reactivity: S_N1, S_N2, E1, E1cB, and E2

S_N1, S_N2, E1, E1cB, E2—how can you keep it all straight and predict what will happen in any given case? Will substitution or elimination occur? Will the reaction be bimolecular or unimolecular? There are no rigid answers to

ThomsonNOW™ Click *Organic Interactive* to **use a web-based palette to design syntheses using substitution and elimination reactions.**

these questions, but it's possible to recognize some trends and make son generalizations.

- **Primary alkyl halides** S_N2 substitution occurs if a good nucleophile is use E2 elimination occurs if a strong base is used, and E1cB elimination occurs the leaving group is two carbons away from a carbonyl group.
- **Secondary alkyl halides** S_N2 substitution occurs if a weakly basic nucle phile is used in a polar aprotic solvent, E2 elimination predominates if strong base is used, and E1cB elimination takes place if the leaving group two carbons away from a carbonyl group. Secondary allylic and benzylic alk halides can also undergo S_N1 and E1 reactions if a weakly basic nucleophile used in a protic solvent.
- **Tertiary alkyl halides** E2 elimination occurs when a base is used, b S_N1 substitution and E1 elimination occur together under neutral condition such as in pure ethanol or water. E1cB elimination takes place if the leavin group is two carbons away from a carbonyl group.

WORKED EXAMPLE 11.5

Predicting the Product and Mechanism of Reactions

Tell whether each of the following reactions is likely to be S_N1, S_N2, E1, E1cB, c E2, and predict the product of each:

(a)

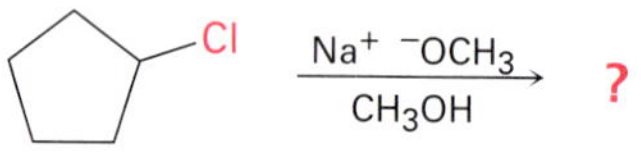

(b)

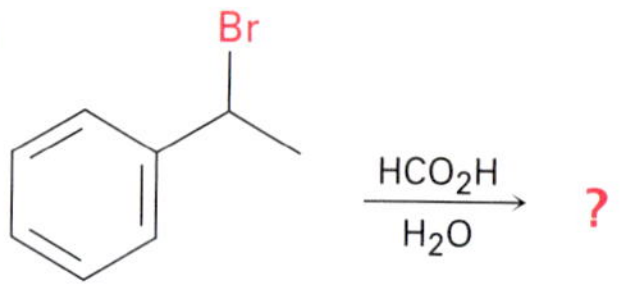

Strategy Look carefully in each reaction at the structure of the substrate, the leaving grou the nucleophile, and the solvent. Then decide from the preceding summary whic kind of reaction is likely to be favored.

Solution (a) A secondary, nonallylic substrate can undergo an S_N2 reaction with a good nucle phile in a polar aprotic solvent but will undergo an E2 reaction on treatment wit a strong base in a protic solvent. In this case, E2 reaction is likely to predominate

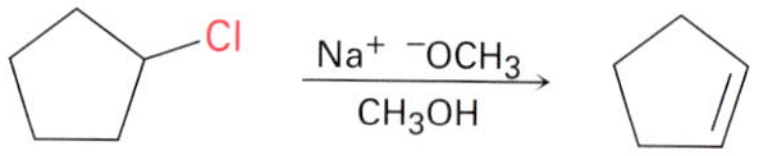

E2 reaction

(b) A secondary benzylic substrate can undergo an S_N2 reaction on treatment wit a nonbasic nucleophile in a polar aprotic solvent and will undergo an E2 reac tion on treatment with a base. Under protic conditions, such as aqueous formi acid (HCO_2H), an S_N1 reaction is likely, along with some E1 reaction.

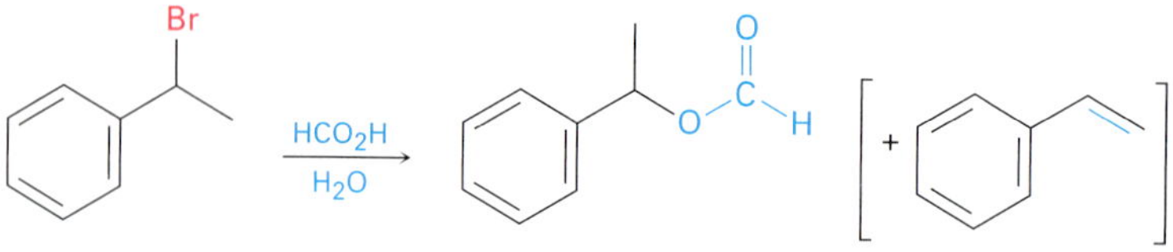

S_N1 **E1**

Problem 11.20 Tell whether each of the following reactions is likely to be S_N1, S_N2, E1, E1cB, or E2:

(a) $CH_3CH_2CH_2CH_2Br \xrightarrow[THF]{NaN_3} CH_3CH_2CH_2CH_2N{=}N{=}N$

(b) $CH_3CH_2\overset{Cl}{\overset{|}{C}}HCH_2CH_3 \xrightarrow[Ethanol]{KOH} CH_3CH_2CH{=}CHCH_3$

(c)

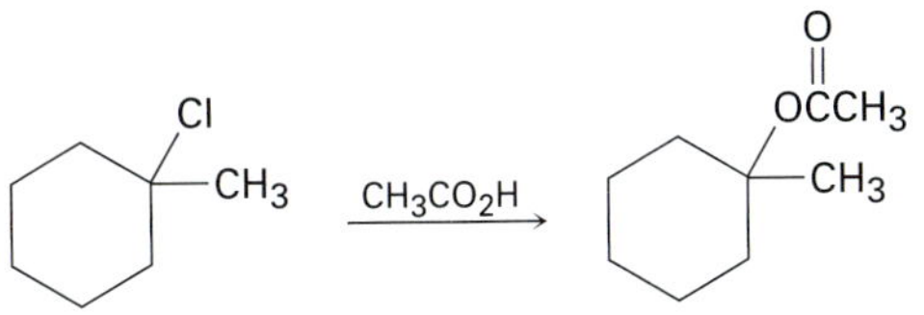

(d)

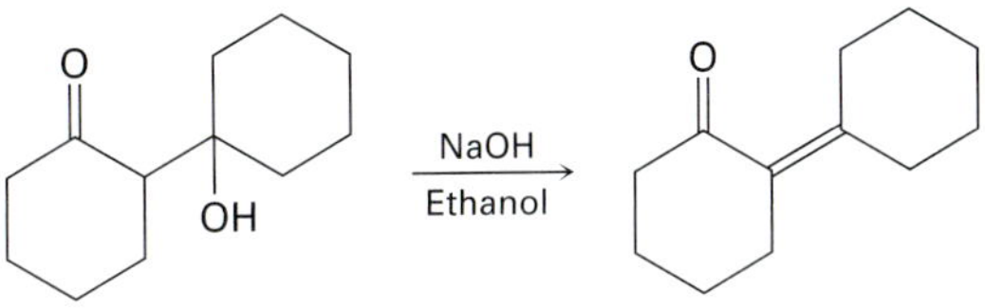

Focus On . . .

Green Chemistry

© Rachel Dulson

Let's hope disasters like this are never repeated.

Organic chemistry in the 20th century changed the world, giving us new medicines, insecticides, adhesives, textiles, dyes, building materials, composites, and all manner of polymers. But these advances did not come without a cost: every chemical process produces wastes that must be dealt with, including reaction solvents and toxic by-products that might evaporate into the air or be leached into ground water if not disposed of properly. Even apparently harmless by-products must be safely buried or otherwise sequestered. As always, there's no such thing as a free lunch; with the good also comes the bad.

It may never be possible to make organic chemistry completely benign, but awareness of the environmental problems caused by many chemical processes has grown dramatically in recent years, giving rise to a movement called *green chemistry.* Green chemistry is the design and implementation of chemical products and processes that reduce waste and attempt to eliminate the generation of hazardous substances. There are 12 principles of green chemistry:

(continued)

Prevent waste. Waste should be prevented rather than treated or cleaned up after it has been created.

Maximize atom economy. Synthetic methods should maximize the incorporation of all materials used in a process into the final product so that waste is minimized.

Use less hazardous processes. Synthetic methods should use reactants and generate wastes with minimal toxicity to health and the environment.

Design safer chemicals. Chemical products should be designed to have minimal toxicity.

Use safer solvents. Minimal use should be made of solvents, separation agents, and other auxiliary substances in a reaction.

Design for energy efficiency. Energy requirements for chemical processes should be minimized, with reactions carried out at room temperature if possible.

Use renewable feedstocks. Raw materials should come from renewable sources when feasible.

Minimize derivatives. Syntheses should be designed with minimal use of protecting groups to avoid extra steps and reduce waste.

Use catalysis. Reactions should be catalytic rather than stoichiometric.

Design for degradation. Products should be designed to be biodegradable at the end of their useful lifetimes.

Monitor pollution in real time. Processes should be monitored in real time for the formation of hazardous substances.

Prevent accidents. Chemical substances and processes should minimize the potential for fires, explosions, or other accidents.

The 12 principles won't all be met in most real-world applications, but they provide a worthy goal to aim for and they can make chemists think more carefully about the environmental implications of their work. Success stories are already occurring, and more are in progress. Approximately 7 million pounds per year of ibuprofen (6 billion tablets!) is now made by a "green" process that produces approximately 99% less waste than the process it replaces. Only three steps are needed, the anhydrous HF solvent used in the first step is recovered and reused, and the second and third steps are catalytic.

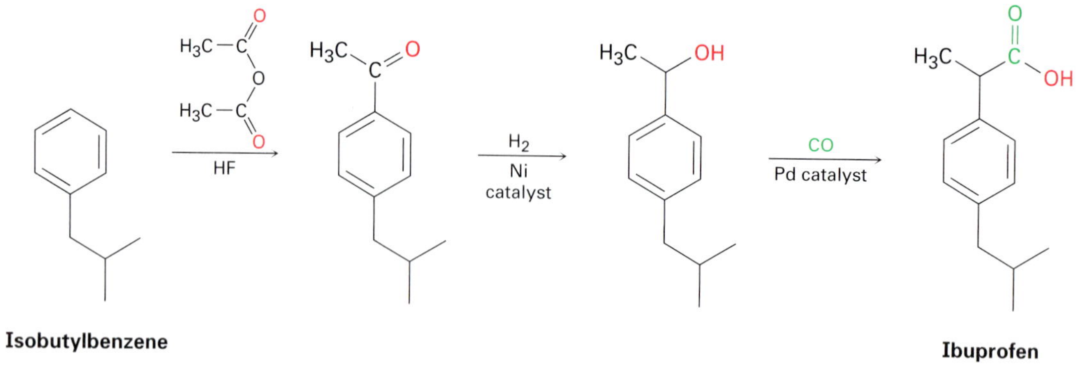

SUMMARY AND KEY WORDS

anti periplanar, 387
benzylic, 377
deuterium isotope effect, 386
E1 reaction, 391
E1cB reaction, 393
E2 reaction, 386
first-order reaction, 373
kinetics, 362
nucleophilic substitution reaction, 360
second-order reaction, 363
S_N1 reaction, 373
S_N2 reaction, 363
solvation, 370
syn periplanar, 387
Zaitsev's rule, 384

The reaction of an alkyl halide or tosylate with a nucleophile/base results either in *substitution* or in *elimination*. Nucleophilic substitutions are of two types: **S_N2 reactions** and **S_N1 reactions**. In the S_N2 reaction, the entering nucleophile approaches the halide from a direction 180° away from the leaving group, resulting in an umbrella-like inversion of configuration at the carbon atom. The reaction is kinetically **second-order** and is strongly inhibited by increasing steric bulk of the reactants. Thus, S_N2 reactions are favored for primary and secondary substrates.

The S_N1 reaction occurs when the substrate spontaneously dissociates to a carbocation in a slow rate-limiting step, followed by a rapid reaction with the nucleophile. As a result, S_N1 reactions are kinetically **first-order** and take place with racemization of configuration at the carbon atom. They are most favored for tertiary substrates. Both S_N1 and S_N2 reactions occur in biological pathways, although the leaving group is typically a diphosphate ion rather than a halide.

Eliminations of alkyl halides to yield alkenes occur by three mechanisms: **E2 reactions**, **E1 reactions**, and **E1cB reactions**, which differ in the timing of C–H and C–X bond-breaking. In the E2 reaction, C–H and C–X bond-breaking occur simultaneously when a base abstracts H^+ from one carbon at the same time the leaving group departs from the neighboring carbon. The reaction takes place preferentially through an **anti periplanar** transition state in which the four reacting atoms—hydrogen, two carbons, and leaving group—are in the same plane. The reaction shows second-order kinetics and a **deuterium isotope effect**, and it occurs when a secondary or tertiary substrate is treated with a strong base. These elimination reactions usually give a mixture of alkene products in which the more highly substituted alkene predominates (**Zaitsev's rule**).

In the E1 reaction, C–X bond-breaking occurs first. The substrate dissociates to yield a carbocation in the slow rate-limiting step before losing H^+ from an adjacent carbon in a second step. The reaction shows first-order kinetics and no deuterium isotope effect and occurs when a tertiary substrate reacts in polar, nonbasic solution.

In the E1cB reaction, C–H bond-breaking occurs first. A base abstracts a proton to give an anion, followed by loss of the leaving group from the adjacent carbon in a second step. The reaction is favored when the leaving group is two carbons removed from a carbonyl, which stabilizes the intermediate anion by resonance. Biological elimination reactions typically occur by this E1cB mechanism.

In general, substrates react in the following way:

RCH_2X (primary) ⟶ Mostly S_N2 substitution

R_2CHX (secondary) ⟶ S_N2 substitution with nonbasic nucleophiles; E2 elimination with strong bases

R_3CX (tertiary) ⟶ Mostly E2 elimination (S_N1 substitution and E1 elimination in nonbasic solvents)

SUMMARY OF REACTIONS

1. Nucleophilic substitutions
   (a) S_N1 reaction of 3°, allylic, and benzylic halides (Sections 11.4 and 11.5)

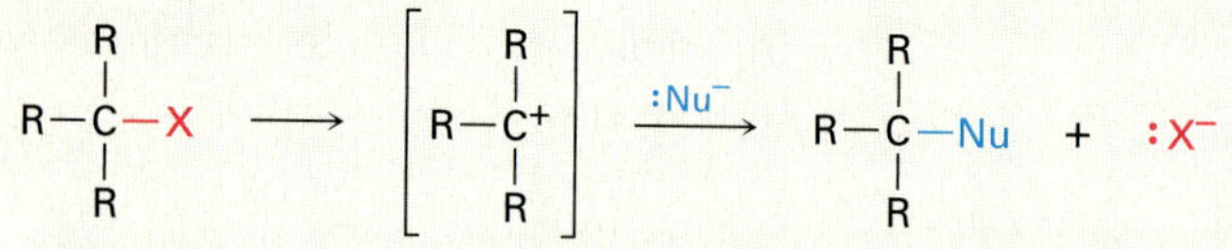

   (b) S_N2 reaction of 1° and simple 2° halides (Sections 11.2 and 11.3)

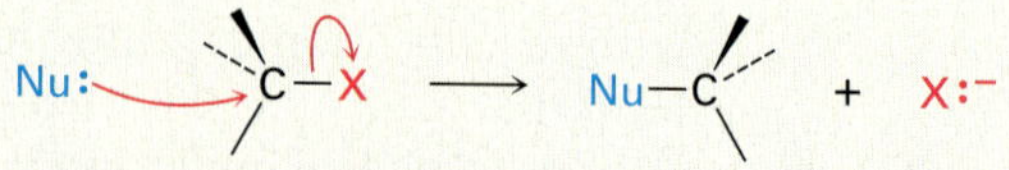

2. Eliminations
   (a) E1 reaction (Section 11.10)

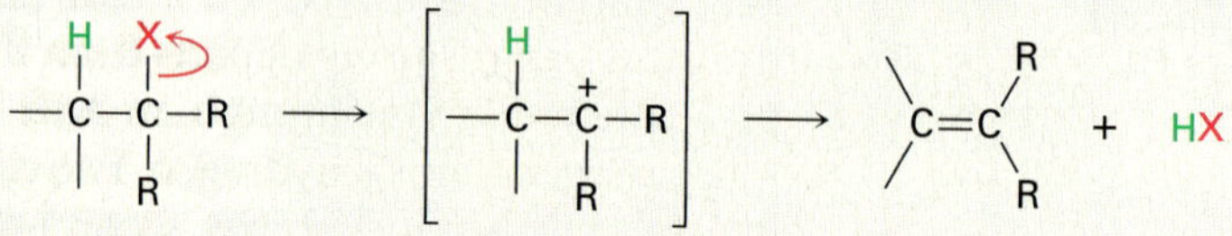

   (b) E1cB reaction (Section 11.10)

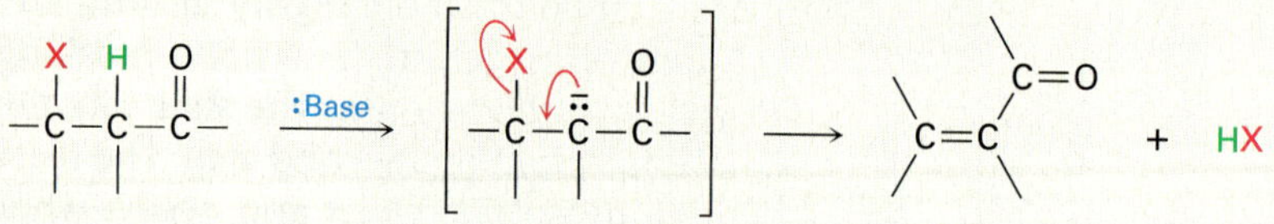

   (c) E2 reaction (Section 11.8)

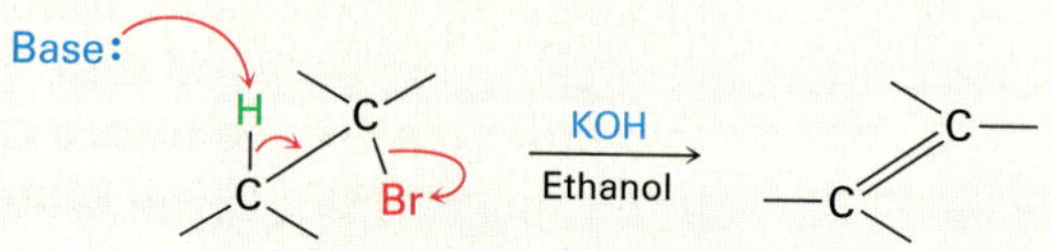

EXERCISES

Organic KNOWLEDGE TOOLS

ThomsonNOW™ Sign in at **www.thomsonedu.com** to assess your knowledge of this chapter's topics by taking a pre-test. The pre-test will link you to interactive organic chemistry resources based on your score in each concept area.

Online homework for this chapter may be assigned in Organic OWL.

■ indicates problems assignable in Organic OWL.

▲ denotes problems linked to Key Ideas of this chapter and testable in ThomsonNOW.

VISUALIZING CHEMISTRY

(Problems 11.1–11.20 appear within the chapter.)

11.21 ■ Write the product you would expect from reaction of each of the following alkyl halides with (i) Na^+ $^-SCH_3$ and (ii) Na^+ ^-OH (yellow-green = Cl):

(a)

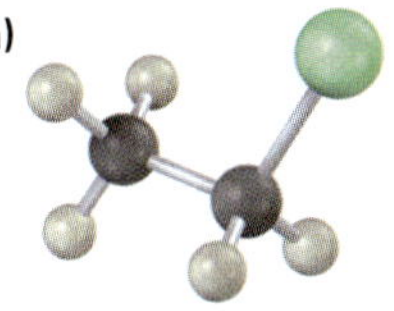

(b)

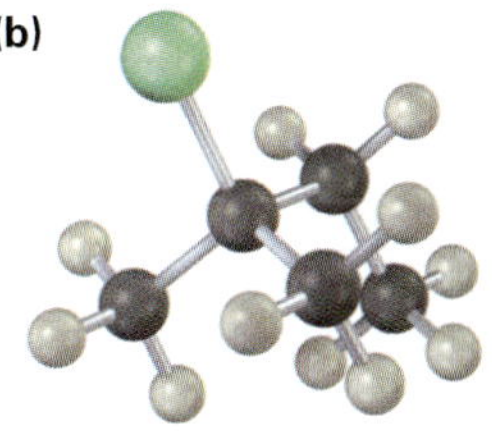

(c)

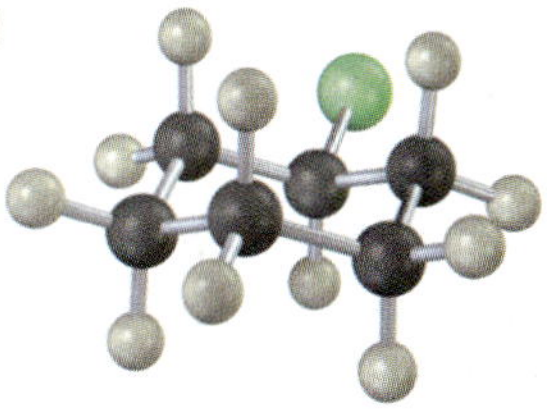

11.22 ■ From what alkyl bromide was the following alkyl acetate made by S_N2 reaction? Write the reaction, showing all stereochemistry.

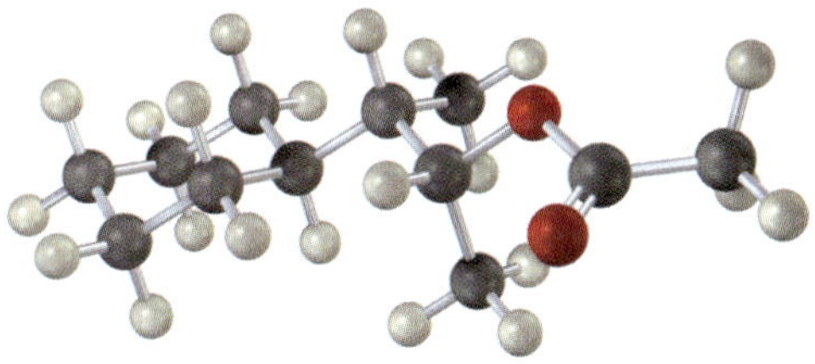

11.23 ■ Assign *R* or *S* configuration to the following molecule, write the product you would expect from S_N2 reaction with NaCN, and assign *R* or *S* configuration to the product (yellow-green = Cl):

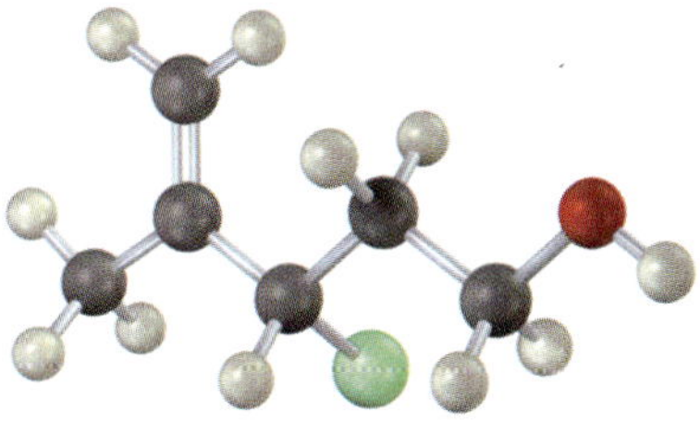

11.24 ■ Draw the structure and assign *Z* or *E* stereochemistry to the product y expect from E2 reaction of the following molecule with NaOH (yello green = Cl):

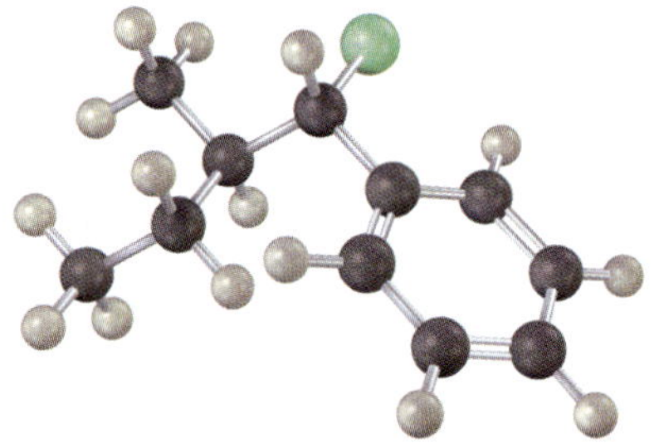

ADDITIONAL PROBLEMS

11.25 ■ Which compound in each of the following pairs will react faster in a S_N2 reaction with OH^-?

(a) CH_3Br or CH_3I (b) CH_3CH_2I in ethanol or in dimethyl sulfoxi
(c) $(CH_3)_3CCl$ or CH_3Cl (d) $H_2C{=}CHBr$ or $H_2C{=}CHCH_2Br$

11.26 ■ What effect would you expect the following changes to have on the rate the S_N2 reaction of 1-iodo-2-methylbutane with cyanide ion?

(a) The CN^- concentration is halved, and the 1-iodo-2-methylbutane co centration is doubled.
(b) Both the CN^- and the 1-iodo-2-methylbutane concentrations are triple

11.27 ■ What effect would you expect the following changes to have on the rate the reaction of ethanol with 2-iodo-2-methylbutane?

(a) The concentration of the halide is tripled.
(b) The concentration of the ethanol is halved by adding diethyl ether as a inert solvent.

11.28 How might you prepare each of the following molecules using a nucleophil substitution reaction at some step?

(a)
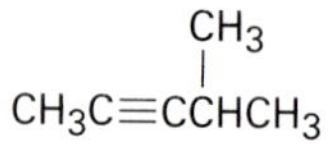

(b)
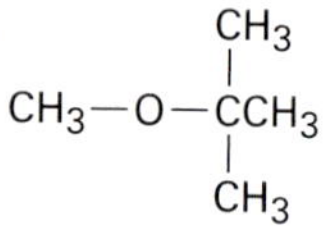

(c) $CH_3CH_2CH_2CH_2CN$ (d) $CH_3CH_2CH_2NH$

11.29 ■ ▲ Which reaction in each of the following pairs would you expect to be faster

(a) The SN2 displacement by I^- on CH_3Cl or on CH_3OTos
(b) The S_N2 displacement by $CH_3CO_2^-$ on bromoethane or on bromocyclo hexane
(c) The S_N2 displacement on 2-bromopropane by $CH_3CH_2O^-$ or by CN^-
(d) The S_N2 displacement by $HC{\equiv}C^-$ on bromomethane in benzene or in acetonitrile

11.30 ■ What products would you expect from the reaction of 1-bromopropan with each of the following?

(a) $NaNH_2$ (b) $KOC(CH_3)_3$ (c) NaI
(d) NaCN (e) $NaC{\equiv}CH$ (f) Mg, then H_2O

11.31 ■ Which reactant in each of the following pairs is more nucleophilic? Explain

(a) $^-NH_2$ or NH_3 (b) H_2O or $CH_3CO_2^-$ (c) BF_3 or F^-
(d) $(CH_3)_3P$ or $(CH_3)_3N$ (e) I^- or Cl^- (f) $^-C{\equiv}N$ or ^-OCH

11.32 Propose structures for compounds that fit the following descriptions:

(a) An alkyl halide that gives a mixture of three alkenes on E2 reaction
(b) An organohalide that will not undergo nucleophilic substitution
(c) An alkyl halide that gives the non-Zaitsev product on E2 reaction
(d) An alcohol that reacts rapidly with HCl at 0 °C

11.33 Draw all isomers of C_4H_9Br, name them, and arrange them in order of decreasing reactivity in the S_N2 reaction.

11.34 The following Walden cycle has been carried out. Explain the results, and indicate where Walden inversion is occurring.

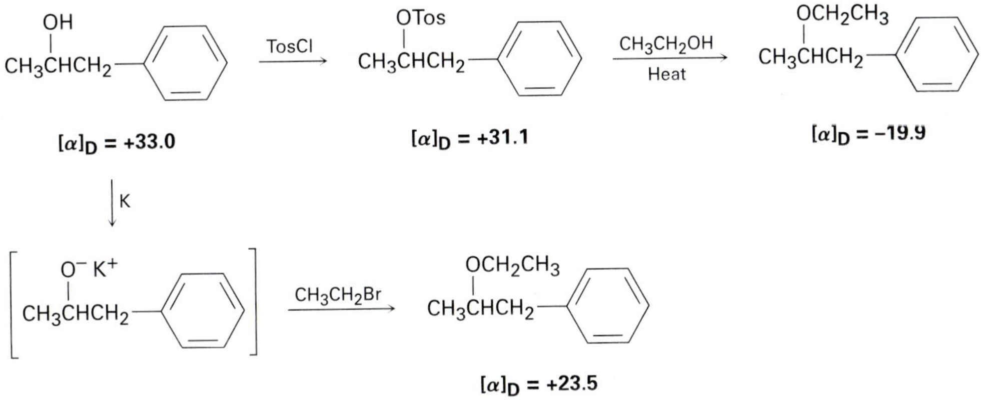

11.35 ■ The reactions shown below are unlikely to occur as written. Tell what is wrong with each, and predict the actual product.

(a)

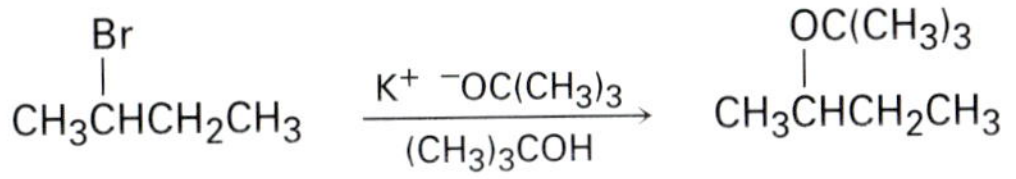

(b)

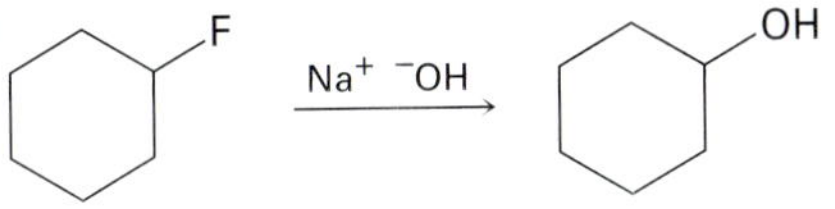

(c)

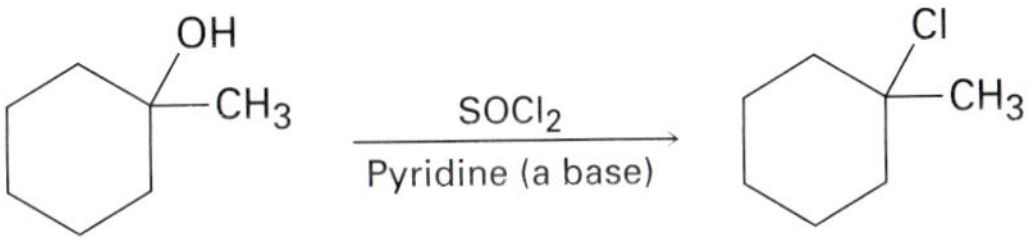

11.36 ■ Order each of the following sets of compounds with respect to S_N1 reactivity:

(a)

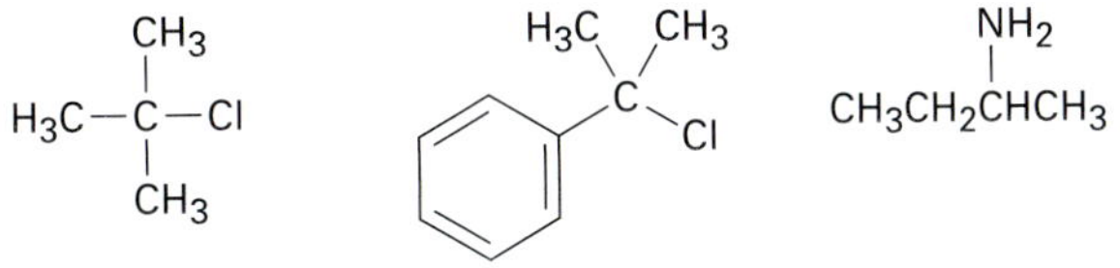

(b) $(CH_3)_3CCl$ $(CH_3)_3CBr$ $(CH_3)_3COH$

(c)

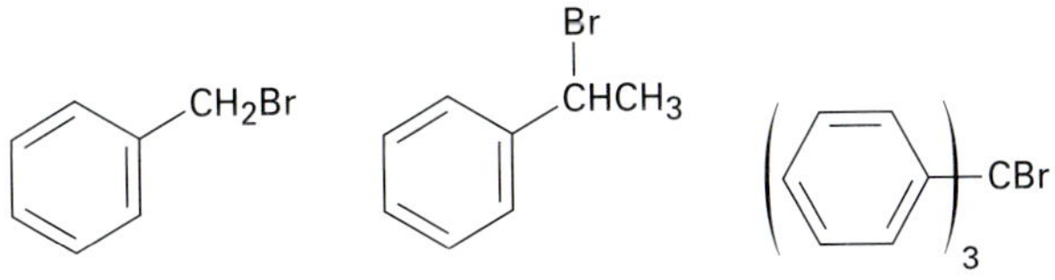

11.37 ■ Order each of the following sets of compounds with respect to S_N2 reactivi

(a) 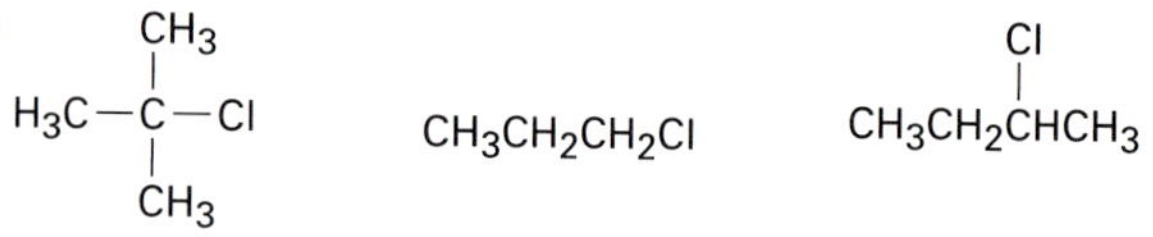

(b) 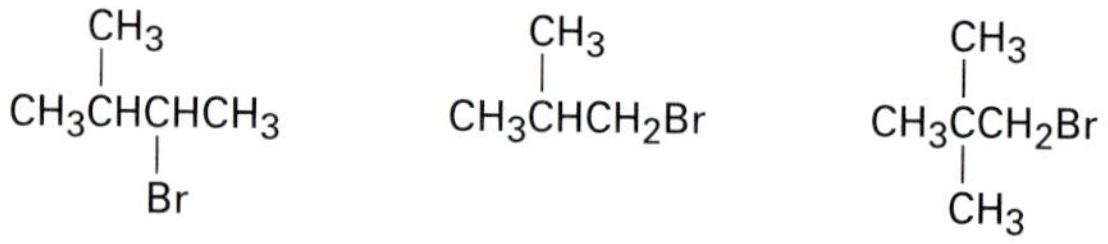

(c) $CH_3CH_2CH_2OCH_3$ $CH_3CH_2CH_2OTos$ $CH_3CH_2CH_2Br$

11.38 ■ Predict the product and give the stereochemistry resulting from reaction each of the following nucleophiles with (*R*)-2-bromooctane:
(a) ^{-}CN (b) $CH_3CO_2^{-}$ (c) CH_3S^{-}

11.39 (*R*)-2-Bromooctane undergoes racemization to give (±)-2-bromooctane whe treated with NaBr in dimethyl sulfoxide. Explain.

11.40 Reaction of the following *S* tosylate with cyanide ion yields a nitrile produ that also has *S* stereochemistry. Explain.

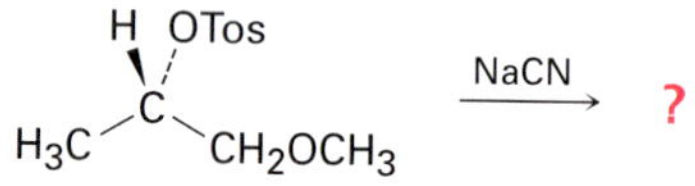

(*S* stereochemistry)

11.41 ■ Ethers can often be prepared by S_N2 reaction of alkoxide ions, RO^-, wit alkyl halides. Suppose you wanted to prepare cyclohexyl methyl ether. Whic of the two possible routes shown below would you choose? Explain.

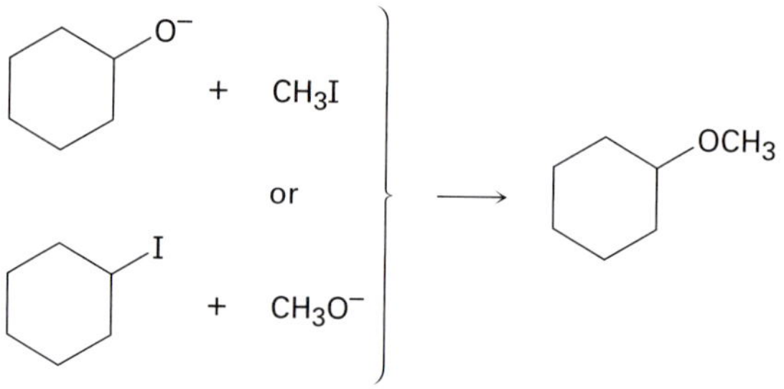

11.42 We saw in Section 7.8 that bromohydrins are converted into epoxides wher treated with base. Propose a mechanism, using curved arrows to show the electron flow.

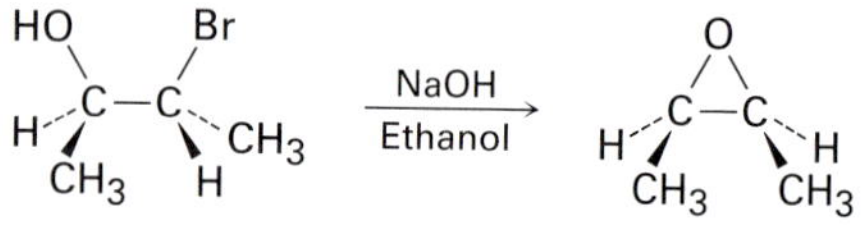

11.43 Show the stereochemistry of the epoxide (see Problem 11.42) you would obtain by formation of a bromohydrin from *trans*-2-butene, followed by treatment with base.

11.44 In light of your answer to Problem 11.42, what product might you expect from treatment of 4-bromo-1-butanol with base?

$$BrCH_2CH_2CH_2CH_2OH \xrightarrow{\text{Base}} ?$$

11.45 ▲ The following tertiary alkyl bromide does not undergo a nucleophilic substitution reaction by either S_N1 or S_N2 mechanisms. Explain.

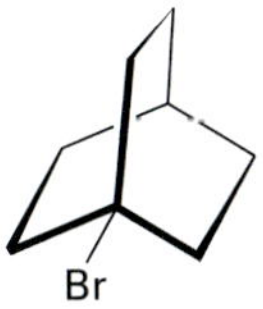

11.46 In addition to not undergoing substitution reactions, the alkyl bromide shown in Problem 11.45 also fails to undergo an elimination reaction when treated with base. Explain.

11.47 1-Chloro-1,2-diphenylethane can undergo E2 elimination to give either *cis*- or *trans*-1,2-diphenylethylene (stilbene). Draw Newman projections of the reactive conformations leading to both possible products, and suggest a reason why the trans alkene is the major product.

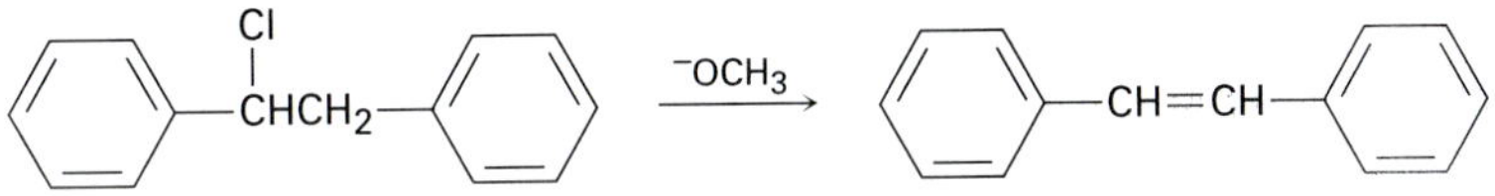

1-Chloro-1,2-diphenylethane ***trans*-1,2-Diphenylethylene**

11.48 ■ Predict the major alkene product of the following E1 reaction:

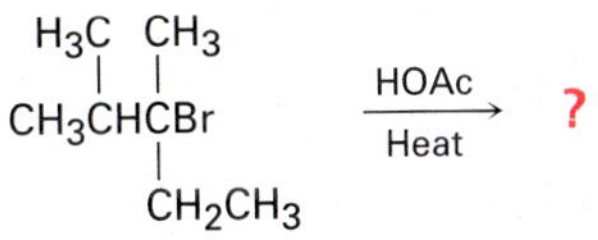

11.49 The tosylate of (2*R*,3*S*)-3-phenyl-2-butanol undergoes E2 elimination on treatment with sodium ethoxide to yield (*Z*)-2-phenyl-2-butene. Explain, using Newman projections.

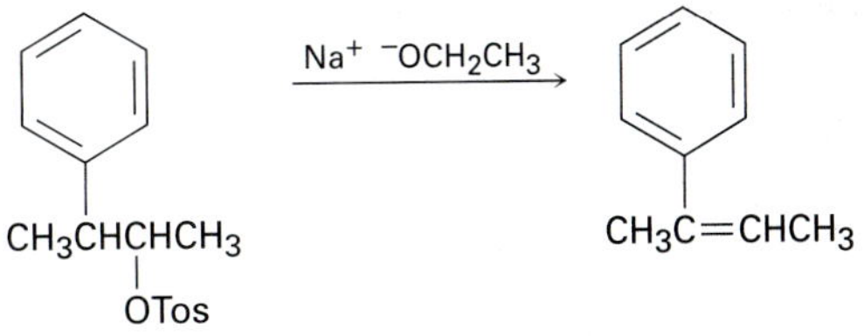

11.50 In light of your answer to Problem 11.49, which alkene, *E* or *Z*, would yo expect from an E2 reaction on the tosylate of (2*R*,3*R*)-3-phenyl-2-butanol Which alkene would result from E2 reaction on the (2*S*,3*R*) and (2*S*,3*S*) tos lates? Explain.

11.51 How can you explain the fact that *trans*-1-bromo-2-methylcyclohexane yields th non-Zaitsev elimination product 3-methylcyclohexene on treatment with base

***trans*-1-Bromo-2-methylcyclohexane** **3-Methylcyclohexene**

11.52 ■ Predict the product(s) of the following reaction, indicating stereochemistry where necessary:

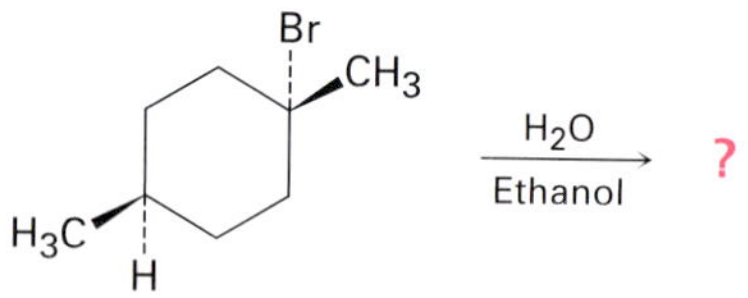

11.53 Metabolism of *S*-Adenosylhomocysteine (Section 11.6) involves the following sequence. Propose a mechanism for the second step.

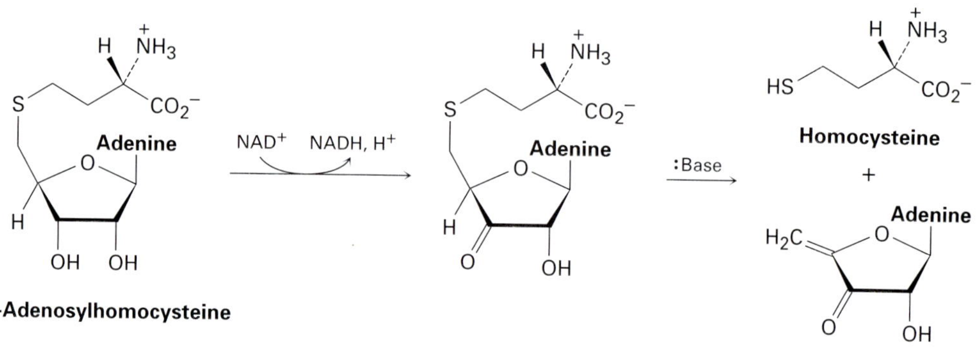

***S*-Adenosylhomocysteine**

11.54 Reaction of iodoethane with CN^- yields a small amount of *isonitrile*, $CH_3CH_2N{\equiv}C$, along with the nitrile $CH_3CH_2C{\equiv}N$ as the major product. Write electron-dot structures for both products, assign formal charges as necessary, and propose mechanisms to account for their formation.

11.55 ▲ Alkynes can be made by dehydrohalogenation of vinylic halides in a reaction that is essentially an E2 process. In studying the stereochemistry of this elimination, it was found that (*Z*)-2-chloro-2-butenedioic acid reacts 50 times as fast as the corresponding *E* isomer. What conclusion can you draw about the stereochemistry of eliminations in vinylic halides? How does this result compare with eliminations of alkyl halides?

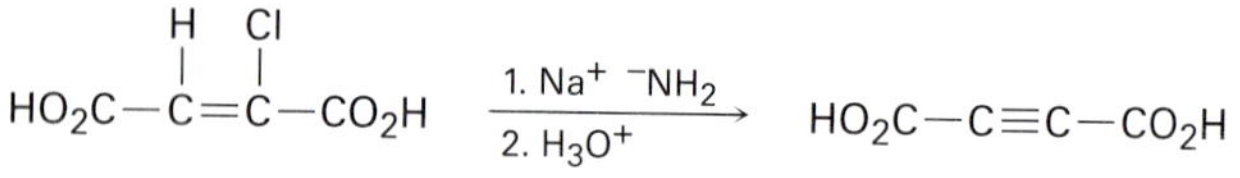

11.56 (*S*)-2-Butanol slowly racemizes on standing in dilute sulfuric acid. Explain.

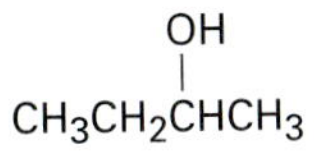

2-Butanol

11.57 Reaction of HBr with (*R*)-3-methyl-3-hexanol leads to racemic 3-bromo-3-methylhexane. Explain.

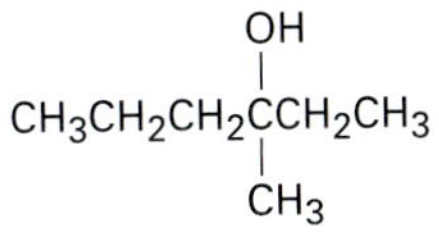

3-Methyl-3-hexanol

11.58 Treatment of 1-bromo-2-deuterio-2-phenylethane with strong base leads to a mixture of deuterated and nondeuterated phenylethylenes in an approximately 7:1 ratio. Explain.

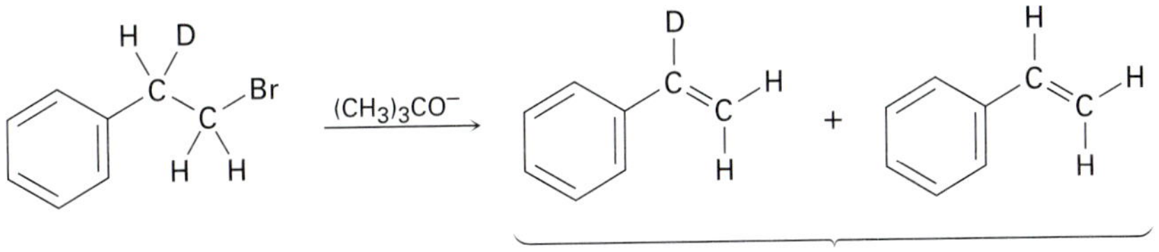

7 : 1 ratio

11.59 ▲ Propose a structure for an alkyl halide that gives only (*E*)-3-methyl-2-phenyl-2-pentene on E2 elimination. Make sure you indicate the stereochemistry.

11.60 One step in the urea cycle for ridding the body of ammonia is the conversion of argininosuccinate to the amino acid arginine plus fumarate. Propose a mechanism for the reaction, and show the structure of arginine.

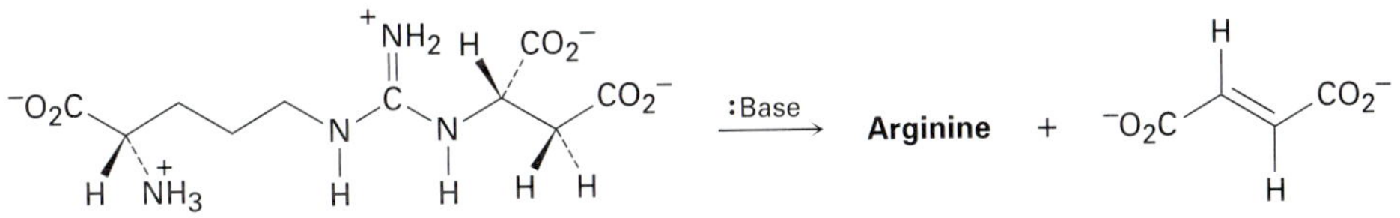

Argininosuccinate

Fumarate

11.61 Although anti periplanar geometry is preferred for E2 reactions, it isn't absolutely necessary. The deuterated bromo compound shown here reacts with strong base to yield an undeuterated alkene. Clearly, a syn elimination has occurred. Make a molecular model of the reactant, and explain the result.

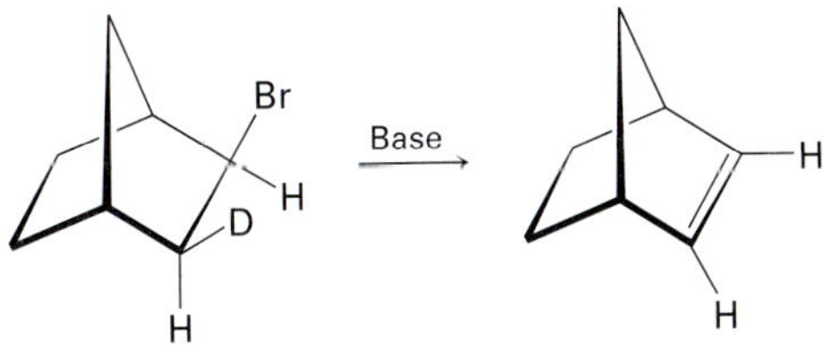

11.62 In light of your answer to Problem 11.61, explain why one of the followi isomers undergoes E2 reaction approximately 100 times as fast as the oth Which isomer is more reactive, and why?

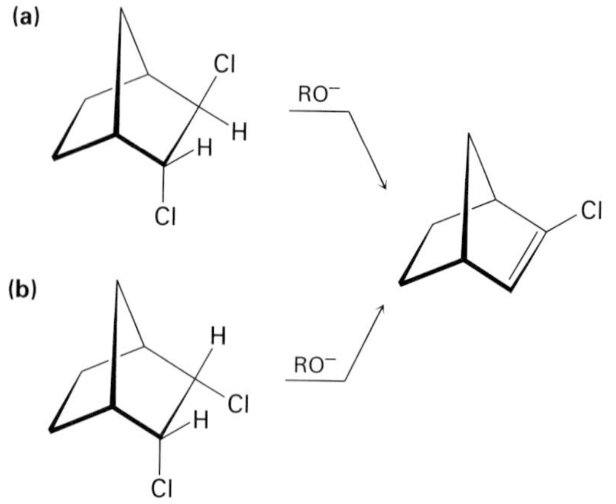

11.63 There are eight diastereomers of 1,2,3,4,5,6-hexachlorocyclohexane. Dra each in its more stable chair conformation. One isomer loses HCl in a E2 reaction nearly 1000 times more slowly than the others. Which isome reacts so slowly, and why?

11.64 Methyl esters (RCO_2CH_3) undergo a cleavage reaction to yield carboxylat ions plus iodomethane on heating with LiI in dimethylformamide:

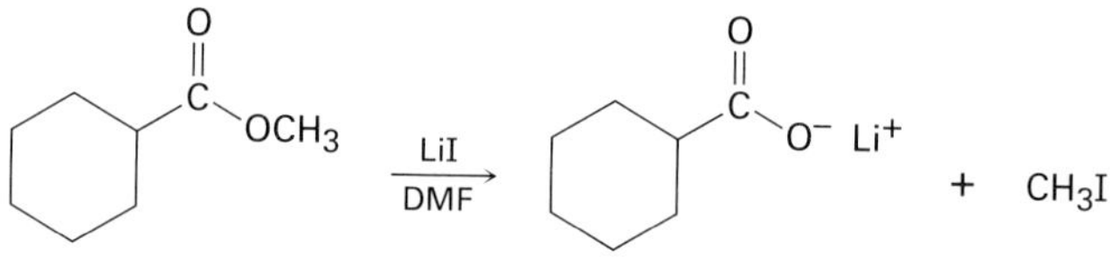

The following evidence has been obtained: (1) The reaction occurs mucl faster in DMF than in ethanol. (2) The corresponding ethyl este ($RCO_2CH_2CH_3$) cleaves approximately 10 times more slowly than the methy ester. Propose a mechanism for the reaction. What other kinds of experimen tal evidence could you gather to support your hypothesis?

11.65 The reaction of 1-chlorooctane with $CH_3CO_2^-$ to give octyl acetate is greatly accelerated by adding a small quantity of iodide ion. Explain.

11.66 Compound X is optically inactive and has the formula $C_{16}H_{16}Br_2$. On treatment with strong base, X gives hydrocarbon Y, $C_{16}H_{14}$. Compound Y absorbs 2 equivalents of hydrogen when reduced over a palladium catalyst and reacts with ozone to give two fragments. One fragment, Z, is an aldehyde with formula C_7H_6O. The other fragment is glyoxal, $(CHO)_2$. Write the reactions involved, and suggest structures for X, Y, and Z. What is the stereochemistry of X?

11.67 When a primary alcohol is treated with *p*-toluenesulfonyl chloride at room temperature in the presence of an organic base such as pyridine, a tosylate is formed. When the same reaction is carried out at higher temperature, an alkyl chloride is often formed. Explain.

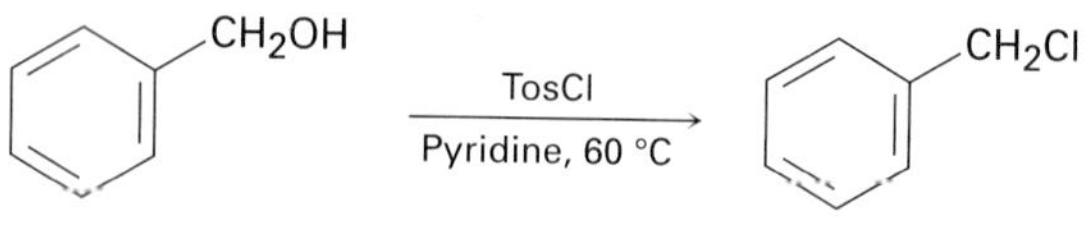

11.68 S_N2 reactions take place with inversion of configuration, and S_N1 reactions take place with racemization. The following substitution reaction, however, occurs with complete *retention* of configuration. Propose a mechanism.

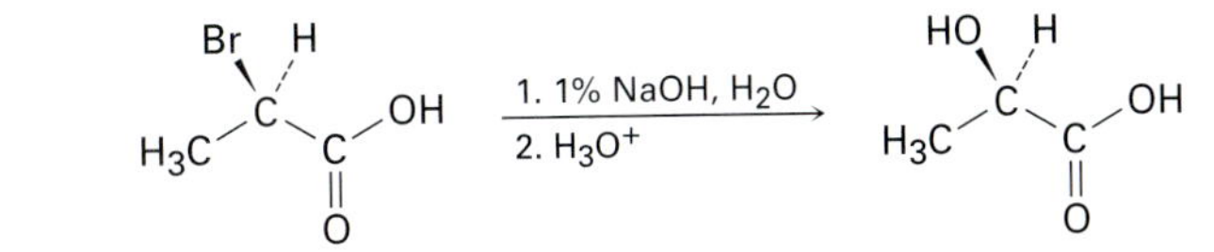

11.69 Propose a mechanism for the following reaction, an important step in the laboratory synthesis of proteins:

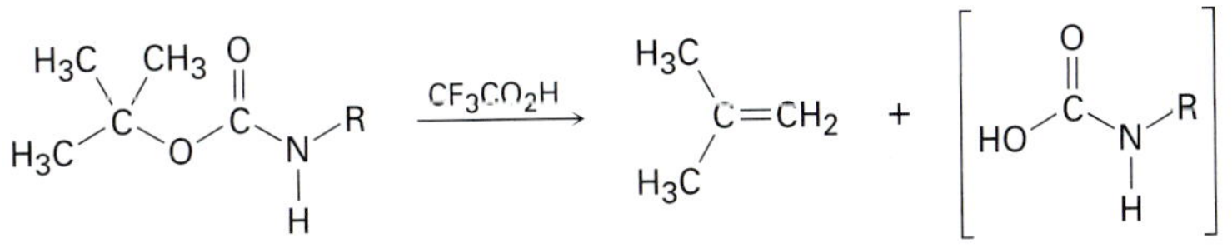

11.70 The amino acid methionine is formed by a methylation reaction of homocysteine with *N*-methyltetrahydrofolate. The stereochemistry of the reaction has been probed by carrying out the transformation using a donor with a "chiral methyl group" that contains protium (H), deuterium (D), and tritium (T) isotopes of hydrogen. Does the methylation reaction occur with inversion or retention of configuration?

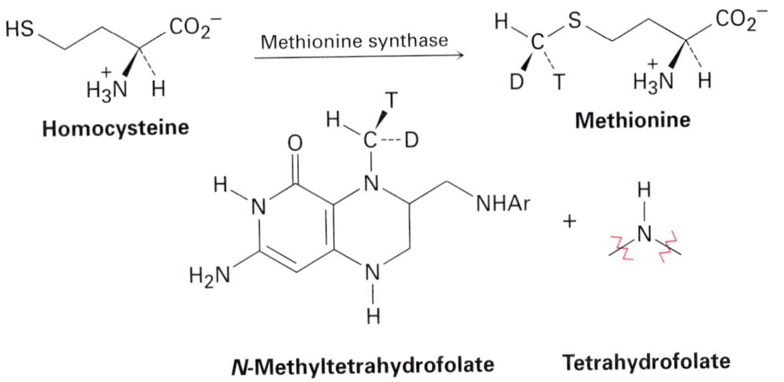

11.71 Amines are converted into alkenes by a two-step process called the *Hofmann elimination*. S_N2 reaction of the amine with an excess of CH_3I in the first step yields an intermediate that undergoes E2 reaction when treated with silver oxide as base. Pentylamine, for example, yields 1-pentene. Propose a structure for the intermediate, and explain why it undergoes ready elimination.

$$CH_3CH_2CH_2CH_2CH_2NH_2 \xrightarrow[\text{2. } Ag_2O,\ H_2O]{\text{1. Excess } CH_3I} CH_3CH_2CH_2CH{=}CH_2$$

12 Structure Determination: Mass Spectrometry and Infrared Spectroscopy

Organic **KNOWLEDGE TOOLS**

ThomsonNOW™ Throughout this chapter, sign in at **www.thomsonedu.com** for online self-study and interactive tutorials based on your level of understanding.

Online homework for this chapter may be assigned in Organic OWL.

Practically everything we've said in previous chapters has been stated without any proof. We said in Section 6.8, for instance, that Markovnikov's rule is followed in alkene electrophilic addition reactions and that treatment of 1-butene with HCl yields 2-chlorobutane rather than 1-chlorobutane. Similarly, we said in Section 11.7 that Zaitsev's rule is followed in elimination reactions and that treatment of 2-chlorobutane with NaOH yields 2-butene rather than 1-butene. But how do we know that these statements are correct? The answer to these and many thousands of similar questions is that the structures of the reaction products have been determined experimentally.

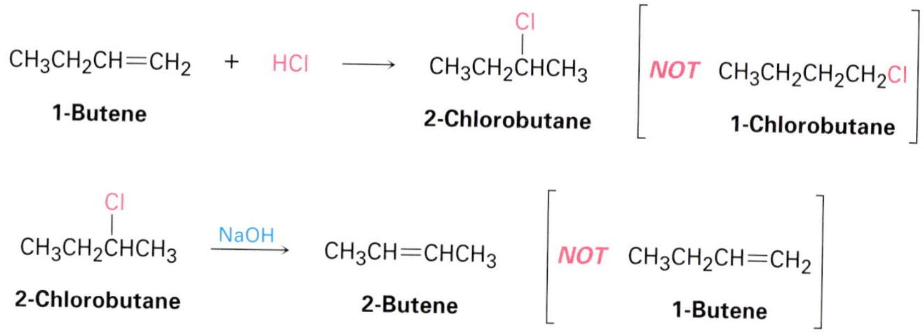

Determining the structure of an organic compound was a difficult and time-consuming process in the 19th and early 20th centuries, but powerful techniques are now available that greatly simplify the problem. In this and the next chapter, we'll look at four such techniques—mass spectrometry (MS), infrared (IR) spectroscopy, ultraviolet spectroscopy (UV), and nuclear magnetic resonance spectroscopy (NMR)—and we'll see the kind of information that can be obtained from each.

Mass spectrometry	What is the size and formula?
Infrared spectroscopy	What functional groups are present?

Ultraviolet spectroscopy	Is a conjugated π electron system present?
Nuclear magnetic resonance spectroscopy	What is the carbon–hydrogen framework?

WHY THIS CHAPTER?

Finding the structures of new molecules, whether small ones synthesized in the laboratory or large proteins and nucleic acids found in living organisms, is central to progress in chemistry and biochemistry. We can only scratch the surface of structure determination in this book, but after reading this and the following chapter, you should have a good idea of the range of structural techniques available and of how and when each is used.

12.1 Mass Spectrometry of Small Molecules: Magnetic-Sector Instruments

At its simplest, **mass spectrometry (MS)** is a technique for measuring the mass, and therefore the molecular weight (MW), of a molecule. In addition, it's often possible to gain structural information about a molecule by measuring the masses of the fragments produced when molecules are broken apart.

More than 20 different kinds of commercial mass spectrometers are available depending on the intended application, but all have three basic parts: an *ionization source* in which sample molecules are given an electrical charge, a *mass analyzer* in which ions are separated by their mass-to-charge ratio, and a *detector* in which the separated ions are observed and counted.

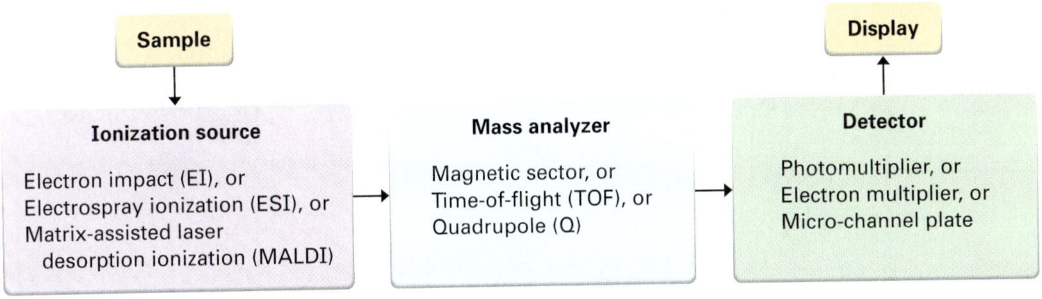

Perhaps the most common mass spectrometer used for routine purposes in the laboratory is the electron-impact, magnetic-sector instrument shown schematically in Figure 12.1. A small amount of sample is vaporized into the ionization source, where it is bombarded by a stream of high-energy electrons. The energy of the electron beam can be varied but is commonly around 70 electron volts (eV), or 6700 kJ/mol. When a high-energy electron strikes an organic molecule, it dislodges a valence electron from the molecule, producing a *cation radical*—*cation* because the molecule has lost an electron and now has a positive charge; *radical* because the molecule now has an odd number electrons.

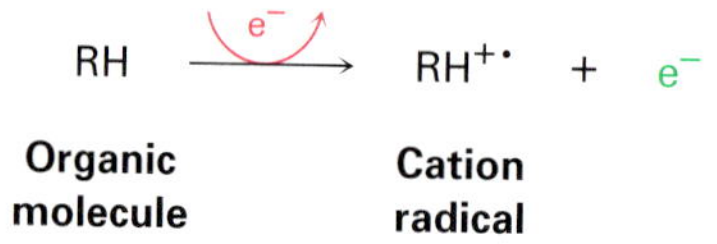

Electron bombardment transfers so much energy that most of the catio radicals *fragment* after formation. They fly apart into smaller pieces, some c which retain the positive charge, and some of which are neutral. The fragmen then flow through a curved pipe in a strong magnetic field, which deflects ther into different paths according to their mass-to-charge ratio (m/z). Neutral frag ments are not deflected by the magnetic field and are lost on the walls of th pipe, but positively charged fragments are sorted by the mass spectrometer ont a detector, which records them as peaks at the various m/z ratios. Since the num ber of charges z on each ion is usually 1, the value of m/z for each ion is simpl its mass m. Masses up to approximately 2500 atomic mass units (amu) can b analyzed.

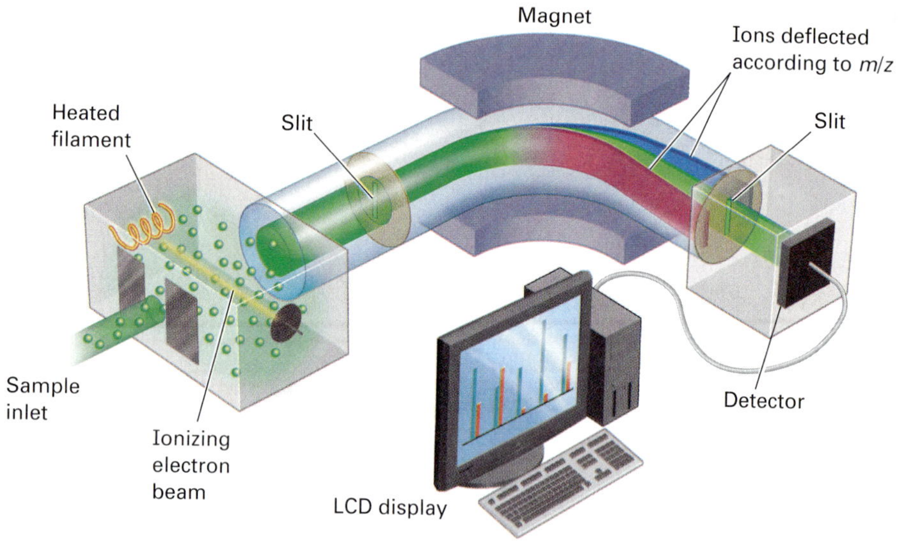

Figure 12.1 A representation of an electron-ionization, magnetic-sector mass spectrometer. Molecules are ionized by collision with high-energy electrons, causing some of the molecules to fragment. Passage of the charged fragments through a magnetic field then sorts them according to their mass.

The **mass spectrum** of a compound is typically presented as a bar graph with masses (m/z values) on the x axis and intensity, or relative abundance of ions of a given m/z striking the detector, on the y axis. The tallest peak, assigned an intensity of 100%, is called the **base peak**, and the peak that corresponds to the unfragmented cation radical is called the **parent peak** or the *molecular ion* (M^+). Figure 12.2 shows the mass spectrum of propane.

Mass spectral fragmentation patterns are usually complex, and the molecular ion is often not the base peak. The mass spectrum of propane in Figure 12.2, for instance, shows a molecular ion at $m/z = 44$ that is only about 30% as high as the base peak at $m/z = 29$. In addition, many other fragment ions are present.

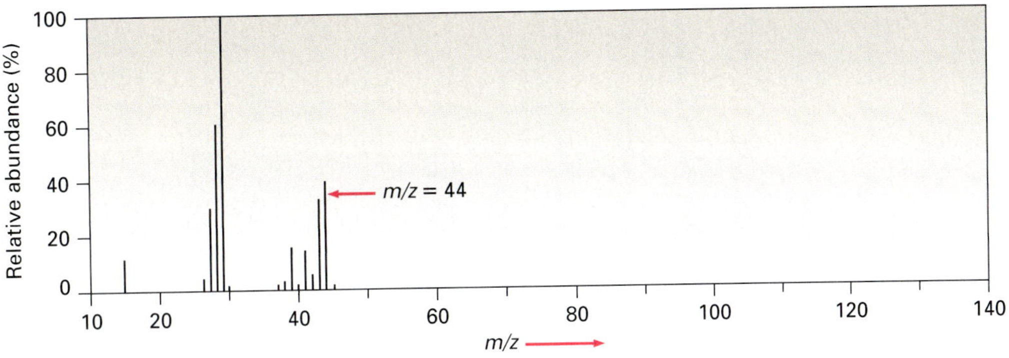

Figure 12.2 Mass spectrum of propane (C_3H_8; MW = 44).

12.2 Interpreting Mass Spectra

ThomsonNOW Click *Organic Interactive* to **learn to utilize mass spectrometry to deduce molecular structures.**

What kinds of information can we get from a mass spectrum? Certainly the most obvious information is the molecular weight, which in itself can be invaluable. For example, if we were given samples of hexane (MW = 86), 1-hexene (MW = 84), and 1-hexyne (MW = 82), mass spectrometry would easily distinguish them.

Some instruments, called *double-focusing mass spectrometers,* have such high resolution that they provide exact mass measurements accurate to 5 ppm, or about 0.0005 amu, making it possible to distinguish between two formulas with the same nominal mass. For example, both C_5H_{12} and C_4H_8O have MW = 72, but they differ slightly beyond the decimal point: C_5H_{12} has an exact mass of 72.0939 amu, whereas C_4H_8O has an exact mass of 72.0575 amu. A high-resolution instrument can easily distinguish between them. Note, however, that exact mass measurements refer to molecules with specific isotopic compositions. Thus, the sum of the exact atomic masses of the specific isotopes in a molecule is measured—1.00783 amu for ^{1}H, 12.00000 amu for ^{12}C, 14.00307 amu for ^{14}N, 15.99491 amu for ^{16}O, and so forth—rather than the sum of the average atomic masses as found on a periodic table.

Unfortunately, not every compound shows a molecular ion in its mass spectrum. Although M^+ is usually easy to identify if it's abundant, some compounds, such as 2,2-dimethylpropane, fragment so easily that no molecular ion is observed (Figure 12.3). In such cases, alternative "soft" ionization methods that do not use electron bombardment can prevent or minimize fragmentation.

Knowing the molecular weight makes it possible to narrow greatly the choices of molecular formula. For example, if the mass spectrum of an unknown compound shows a molecular ion at m/z = 110, the molecular formula is likely to be C_8H_{14}, $C_7H_{10}O$, $C_6H_6O_2$, or $C_6H_{10}N_2$. There are always a number of molecular formulas possible for all but the lowest molecular weights, and computer programs can easily generate a list of choices.

A further point about mass spectrometry, noticeable in the spectrum of propane (Figure 12.2), is that the peak for the molecular ion is not at the highest m/z value. There is also a small peak at M+1 because of the presence of different isotopes in the molecules. Although ^{12}C is the most abundant carbon isotope, a small amount (1.10% natural abundance) of ^{13}C is also present. Thus, a certain

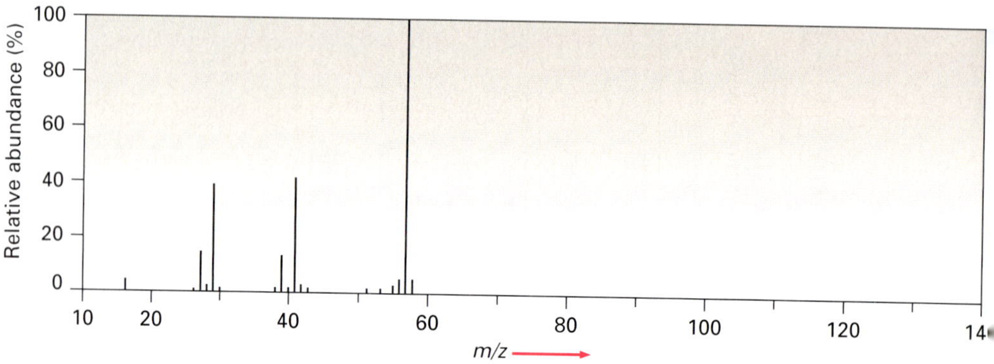

Figure 12.3 Mass spectrum of 2,2-dimethylpropane (C_5H_{12}; MW = 72). No molecular ion i observed when electron-impact ionization is used. (What do you think is the structure of the M^+ peak at $m/z = 57$?)

percentage of the molecules analyzed in the mass spectrometer are likely to con tain a ^{13}C atom, giving rise to the observed M+1 peak. In addition, a small amoun of ^2H (deuterium; 0.015% natural abundance) is present, making a further contri bution to the M+1 peak.

Mass spectrometry would be useful even if molecular weight and formula were the only information that could be obtained, but in fact we can get much more. For one thing, the mass spectrum of a compound serves as a kind of "molecular fingerprint." Each organic compound fragments in a unique way depending on its structure, and the likelihood of two compounds having identical mass spectra is small. Thus, it's sometimes possible to identify an unknown by computer-based matching of its mass spectrum to one of the more than 390,000 mass spectra recorded in a database called the *Registry of Mass Spectral Data.*

It's also possible to derive structural information about a molecule by interpreting its fragmentation pattern. Fragmentation occurs when the high-energy cation radical flies apart by spontaneous cleavage of a chemical bond. One of the two fragments retains the positive charge and is a carbocation, while the other fragment is a neutral radical.

Not surprisingly, the positive charge often remains with the fragment that is best able to stabilize it. In other words, a relatively stable carbocation is often formed during fragmentation. For example, 2,2-dimethylpropane tends to fragment in such a way that the positive charge remains with the *tert*-butyl group. 2,2-Dimethylpropane therefore has a base peak at $m/z = 57$, corresponding to $C_4H_9^+$ (Figure 12.3).

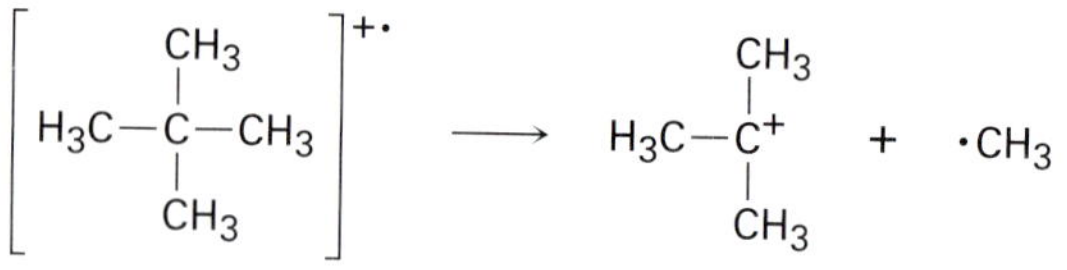

$m/z = 57$

Because mass-spectral fragmentation patterns are usually complex, it's often difficult to assign structures to fragment ions. Most hydrocarbons fragment in many ways, as the mass spectrum of hexane shown in Figure 12.4 demonstrates. The hexane spectrum shows a moderately abundant molecular ion at $m/z = 86$

and fragment ions at $m/z = 71$, 57, 43, and 29. Since all the carbon–carbon bonds of hexane are electronically similar, all break to a similar extent, giving rise to the observed mixture of ions.

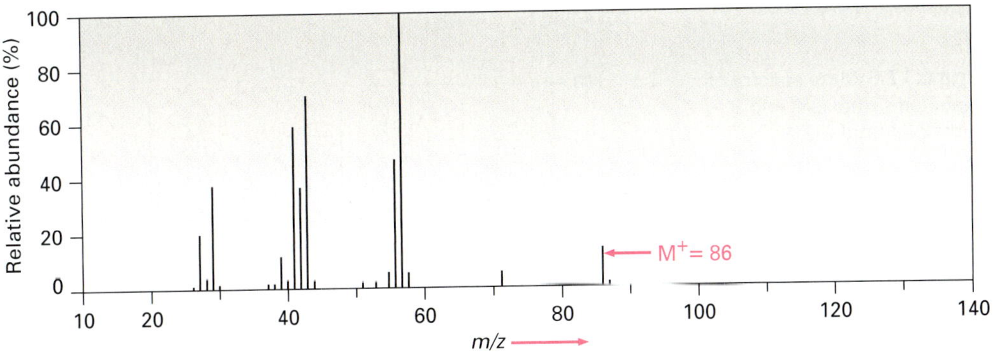

Figure 12.4 Mass spectrum of hexane (C_6H_{14}; MW = 86). The base peak is at $m/z = 57$, and numerous other ions are present.

Figure 12.5 shows how the hexane fragments might arise. The loss of a methyl radical from the hexane cation radical ($M^+ = 86$) gives rise to a fragment of mass 71; the loss of an ethyl radical accounts for a fragment of mass 57; the loss of a propyl radical accounts for a fragment of mass 43; and the loss of a butyl radical accounts for a fragment of mass 29. With skill and practice, it's sometimes possible to analyze the fragmentation pattern of an unknown compound and work backward to a structure that is compatible with the data.

ctive **Figure 12.5** Fragmen-tion of hexane in a mass ɔectrometer. *Sign in at* **ww.thomsonedu.com** *to ee a simulation based on this gure and to take a short quiz.*

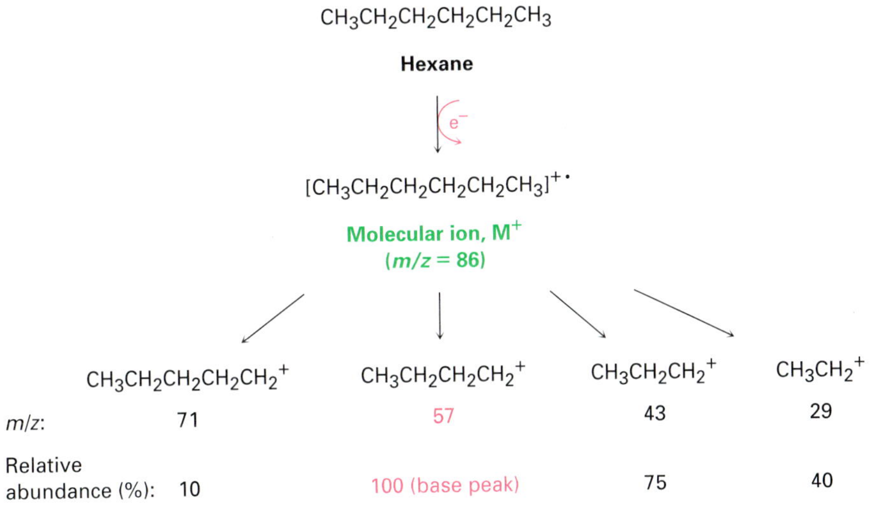

An example of how information from fragmentation patterns can be used to solve structural problems is given in Worked Example 12.1. This example is a simple one, but the principles used are broadly applicable for organic structure determination by mass spectrometry. We'll see in the next section and in later chapters that specific functional groups, such as alcohols, ketones, aldehydes, and amines, show specific kinds of mass spectral fragmentations that can be interpreted to provide structural information.

WORKED EXAMPLE 12.1 *Using Mass Spectra to Identify Compounds*

Assume that you have two unlabeled samples, one of methylcyclohexane and t other of ethylcyclopentane. How could you use mass spectrometry to tell the apart? The mass spectra of both are shown in Figure 12.6.

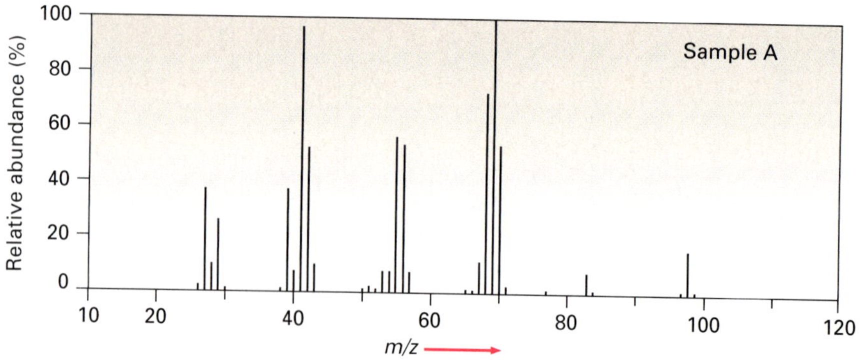

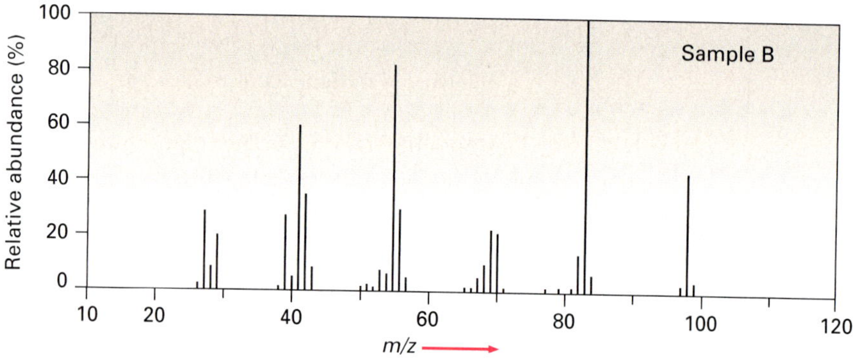

Figure 12.6 Mass spectra of unlabeled samples A and B for Worked Example 12.1.

Strategy Look at the possible structures and decide on how they differ. Then think about hov any of these differences in structure might give rise to differences in mass spectra Methylcyclohexane, for instance, has a $-CH_3$ group, and ethylcyclopentane has a $-CH_2CH_3$ group, which should affect the fragmentation patterns.

Solution Both mass spectra show molecular ions at $M^+ = 98$, corresponding to C_7H_{14}, bu they differ in their fragmentation patterns. Sample A has its base peak at $m/z = 69$ corresponding to the loss of a CH_2CH_3 group (29 mass units), but B has a rather smal peak at $m/z = 69$. Sample B shows a base peak at $m/z = 83$, corresponding to the loss of a CH_3 group (15 mass units), but sample A has only a small peak at $m/z = 83$. We can therefore be reasonably certain that A is ethylcyclopentane and B is methyl-cyclohexane.

Problem 12.1 The male sex hormone testosterone contains C, H, and O and has a mass of 288.2089 amu as determined by high-resolution mass spectrometry. What is the likely molecular formula of testosterone?

Problem 12.2 Two mass spectra are shown in Figure 12.7. One spectrum is that of 2-methyl-2-pentene; the other is of 2-hexene. Which is which? Explain.

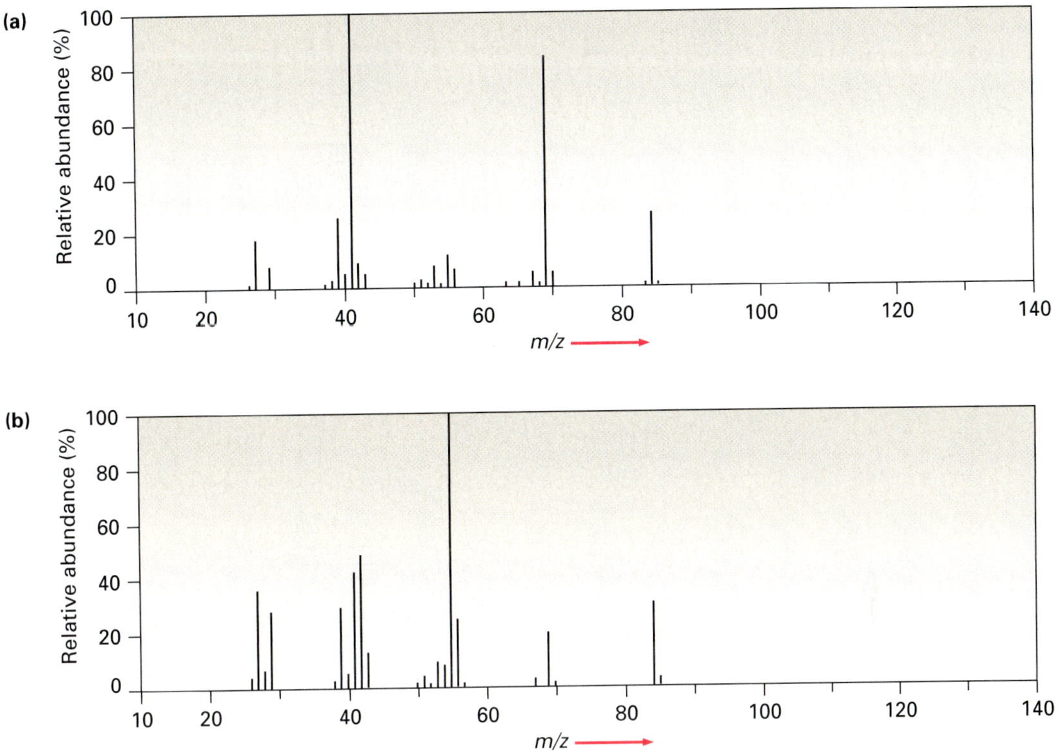

Figure 12.7 Mass spectra for Problem 12.2.

12.3 Mass Spectrometry of Some Common Functional Groups

As each functional group is discussed in future chapters, mass-spectral fragmentations characteristic of that group will be described. As a preview, though, we'll point out some distinguishing features of several common functional groups.

Alcohols

Alcohols undergo fragmentation in the mass spectrometer by two pathways: *alpha cleavage* and *dehydration.* In the α-cleavage pathway, a C–C bond nearest the hydroxyl group is broken, yielding a neutral radical plus a resonance-stabilized, oxygen-containing cation.

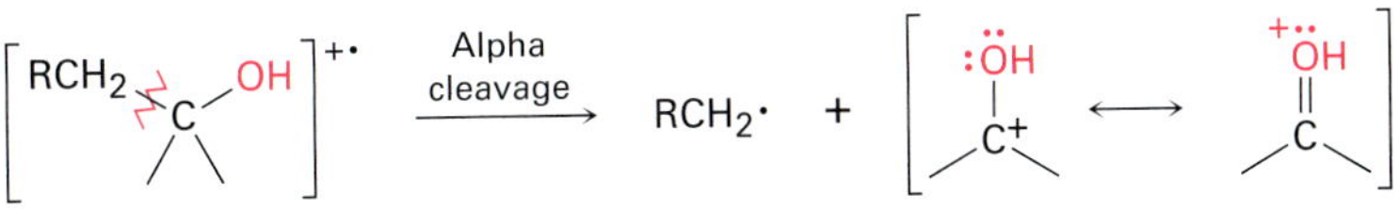

In the dehydration pathway, water is eliminated, yielding an alkene radical cation with a mass 18 units less than M^+.

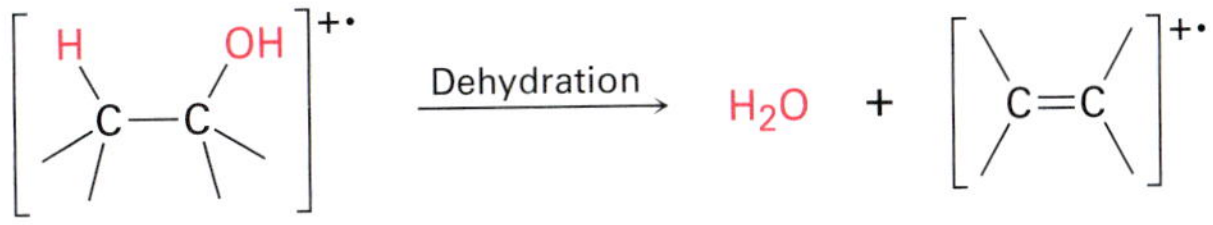

Amines

Aliphatic amines undergo a characteristic α cleavage in the mass spectromete similar to that observed for alcohols. A C–C bond nearest the nitrogen atom broken, yielding an alkyl radical and a resonance-stabilized, nitrogen-containin cation.

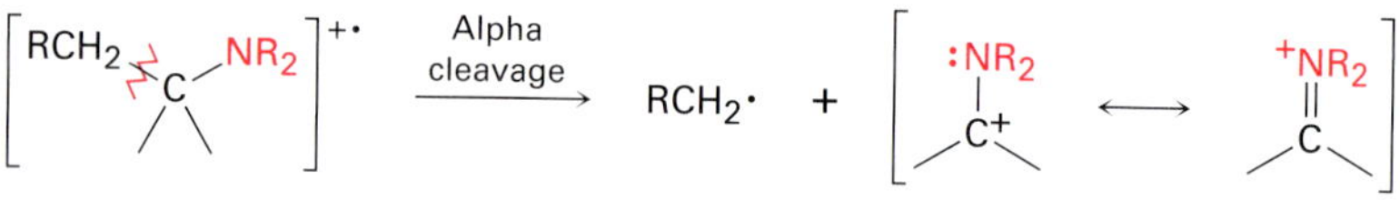

Carbonyl Compounds

Ketones and aldehydes that have a hydrogen on a carbon three atoms awa from the carbonyl group undergo a characteristic mass-spectral cleavage calle the *McLafferty rearrangement.* The hydrogen atom is transferred to the carbony oxygen, a C–C bond is broken, and a neutral alkene fragment is produced. Th charge remains with the oxygen-containing fragment.

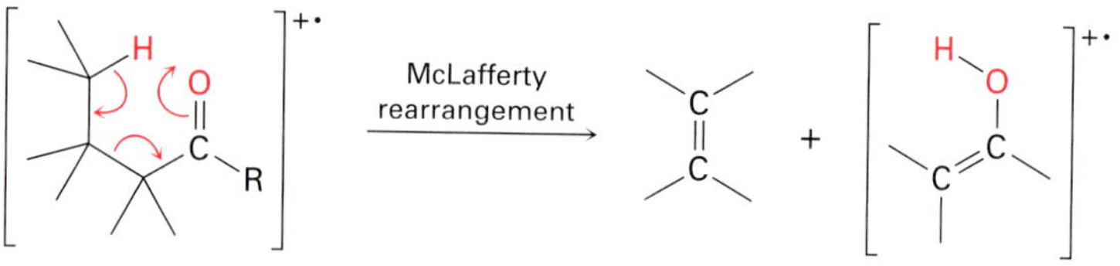

In addition, ketones and aldehydes frequently undergo α cleavage of the bond between the carbonyl group and the neighboring carbon. Alpha cleavag yields a neutral radical and a resonance-stabilized acyl cation.

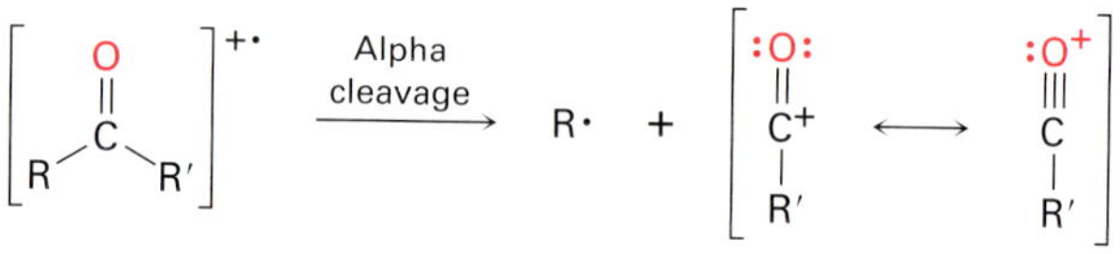

WORKED EXAMPLE 12.2 ***Identifying Fragmentation Patterns in a Mass Spectrum***

The mass spectrum of 2-methyl-3-pentanol is shown in Figure 12.8. What fragments can you identify?

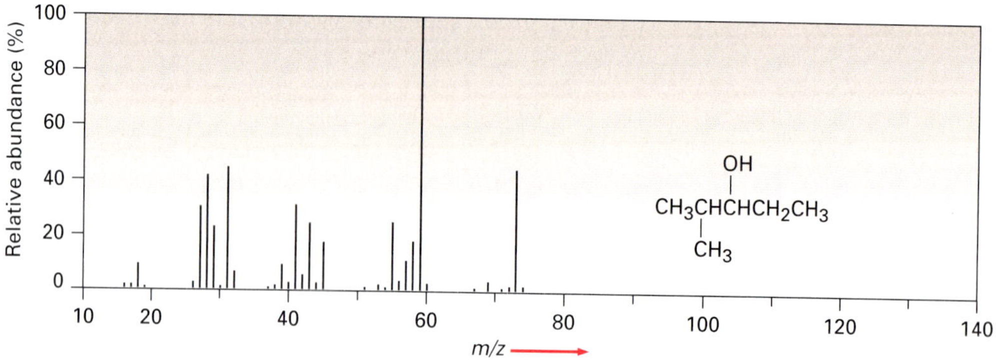

Figure 12.8 Mass spectrum of 2-methyl-3-pentanol, Worked Example 12.2.

Strategy Calculate the mass of the molecular ion, and identify the functional groups in the molecule. Then write the fragmentation processes you might expect, and compare the masses of the resultant fragments with those peaks present in the spectrum.

Solution 2-Methyl-3-pentanol, an open-chain alcohol, has $M^+ = 102$ and might be expected to fragment by α cleavage and by dehydration. These processes would lead to fragment ions of $m/z = 84$, 73, and 59. Of the three expected fragments, dehydration is not observed (no $m/z = 84$ peak), but both α cleavages take place ($m/z = 73, 59$).

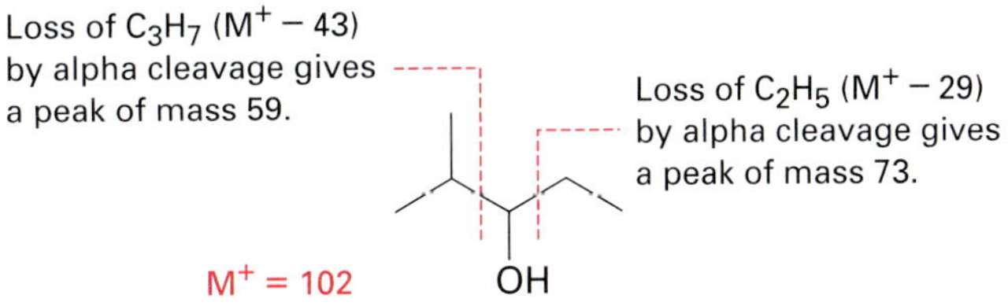

Problem 12.3 What are the masses of the charged fragments produced in the following cleavage pathways?

(a) Alpha cleavage of 2-pentanone ($CH_3COCH_2CH_2CH_3$)

(b) Dehydration of cyclohexanol (hydroxycyclohexane)

(c) McLafferty rearrangement of 4-methyl-2-pentanone [$CH_3COCH_2CH(CH_3)_2$]

(d) Alpha cleavage of triethylamine [$(CH_3CH_2)_3N$]

Problem 12.4 List the masses of the parent ion and of several fragments you might expect to find in the mass spectrum of the following molecule:

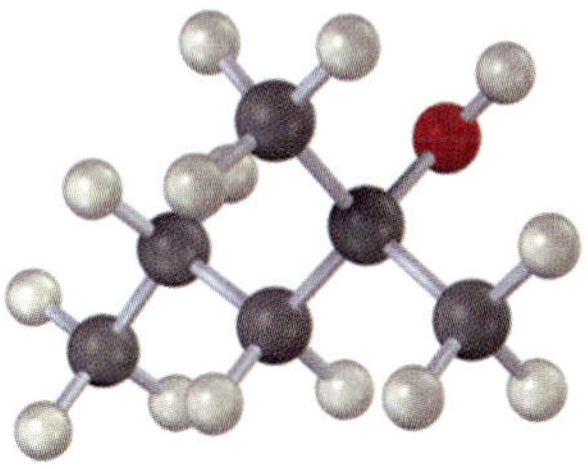

12.4 Mass Spectrometry in Biological Chemistry: Time-of-Flight (TOF) Instruments

Most biochemical analyses by MS use either electrospray ionization (ESI) or matrix-assisted laser desorption ionization (MALDI), typically linked to a time-of-flight (TOF) mass analyzer. Both ESI and MALDI are "soft" ionization methods that produce charged molecules with little fragmentation, even with biological samples of very high molecular weight.

In an ESI source, the sample M is dissolved in a polar solvent and sprayed through a steel capillary tube. As it exits the tube, it is subjected to a high voltage that causes it to become protonated by removing H^+ ions from the solvent. The volatile solvent is then evaporated, giving variably protonated sample molecules ($M+H_n^{n+}$). In a MALDI source, the sample is adsorbed onto a suitab matrix compound, such as 2,5-dihydroxybenzoic acid, which is ionized by short burst of laser light. The matrix compound then transfers the energy to th sample and protonates it, forming $M+H_n^{n+}$ ions.

Following ion formation, the variably protonated sample molecules a electrically focused into a small packet with a narrow spatial distribution, an the packet is given a sudden kick of energy by an accelerator electrode. Sinc each molecule in the packet is given the same energy, $E = mv^2/2$, it begins mov ing with a velocity that depends on the square root of its mass, $v = \sqrt{2E/m}$ Lighter molecules move faster, and heavier molecules move slower. The analyze itself, called the *drift tube,* is simply an electrically grounded metal tube insid which the different charged molecules become separated as they move along a different velocities and take different amounts of time to complete their pa sage. The TOF technique is considerably more sensitive than the magnetic se tor alternative, and protein samples of up to 100 kilodaltons (100,000 amu) ca be separated with a mass accuracy of 3 ppm. Figure 12.9 shows a MALDI–TO spectrum of chicken egg-white lysozyme, MW = 14,306.7578 daltons. (Bio chemists generally use the unit *dalton,* abbreviated Da, instead of amu.)

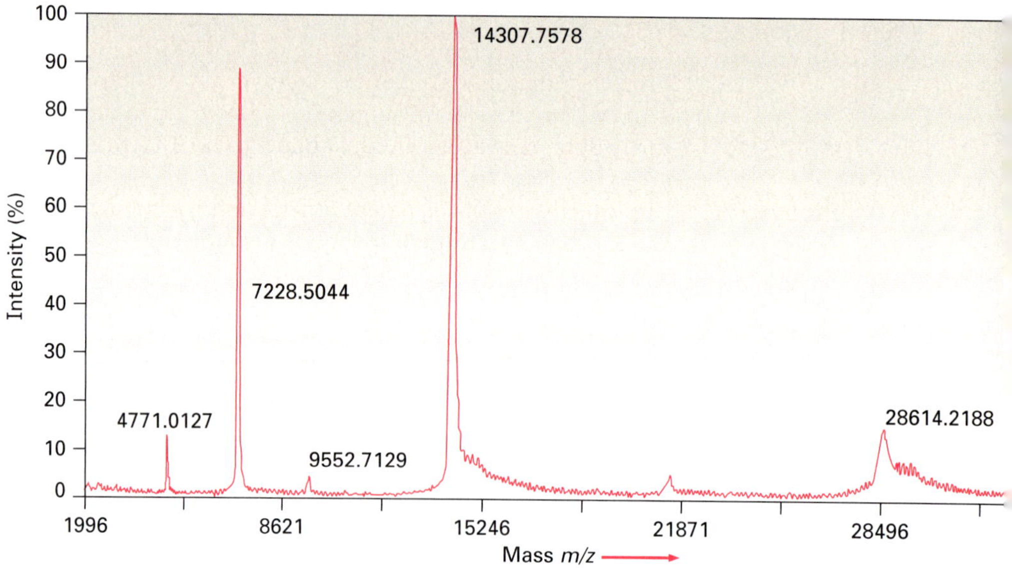

Figure 12.9 MALDI–TOF mass spectrum of chicken egg-white lysozyme. The peak at 14,307.7578 daltons (amu) is due to the monoprotonated protein, $M+H^+$, and that at 28,614.2188 daltons is due to an impurity formed by dimerization of the protein. Other peaks are various protonated species, $M+H_n^{n+}$.

12.5 Spectroscopy and the Electromagnetic Spectrum

Infrared, ultraviolet, and nuclear magnetic resonance spectroscopies differ from mass spectrometry in that they are nondestructive and involve the interaction of molecules with electromagnetic energy rather than with an ionizing source. Before beginning a study of these techniques, however, let's briefly review the nature of radiant energy and the electromagnetic spectrum.

Visible light, X rays, microwaves, radio waves, and so forth, are all different kinds of *electromagnetic radiation.* Collectively, they make up the **electromagnetic**

spectrum, shown in Figure 12.10. The electromagnetic spectrum is arbitrarily divided into regions, with the familiar visible region accounting for only a small portion, from 3.8×10^{-7} m to 7.8×10^{-7} m in wavelength. The visible region is flanked by the infrared and ultraviolet regions.

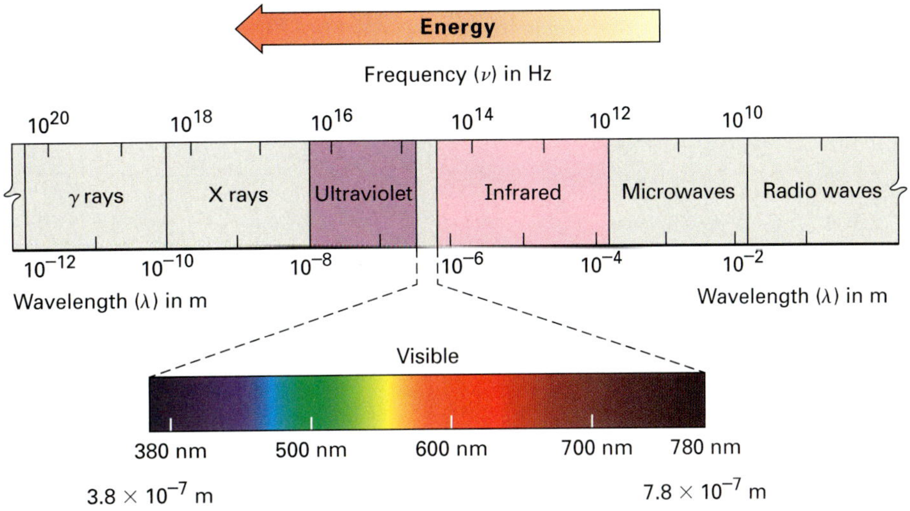

gure 12.10 The electromag-tic spectrum covers a continu-s range of wavelengths and quencies, from radio waves at e low-frequency end to gamma rays at the high-frequency d. The familiar visible region counts for only a small portion ar the middle of the spectrum.

Electromagnetic radiation is often said to have dual behavior. In some respects, it has the properties of a particle (called a *photon*), yet in other respects it behaves as an energy wave. Like all waves, electromagnetic radiation is characterized by a *wavelength*, a *frequency*, and an *amplitude* (Figure 12.11). The **wavelength**, λ (Greek lambda), is the distance from one wave maximum to the next. The **frequency**, ν (Greek nu), is the number of waves that pass by a fixed point per unit time, usually given in reciprocal seconds (s^{-1}), or **hertz, Hz** (1 Hz = 1 s^{-1}). The **amplitude** is the height of a wave, measured from midpoint to peak. The intensity of radiant energy, whether a feeble glow or a blinding glare, is proportional to the square of the wave's amplitude.

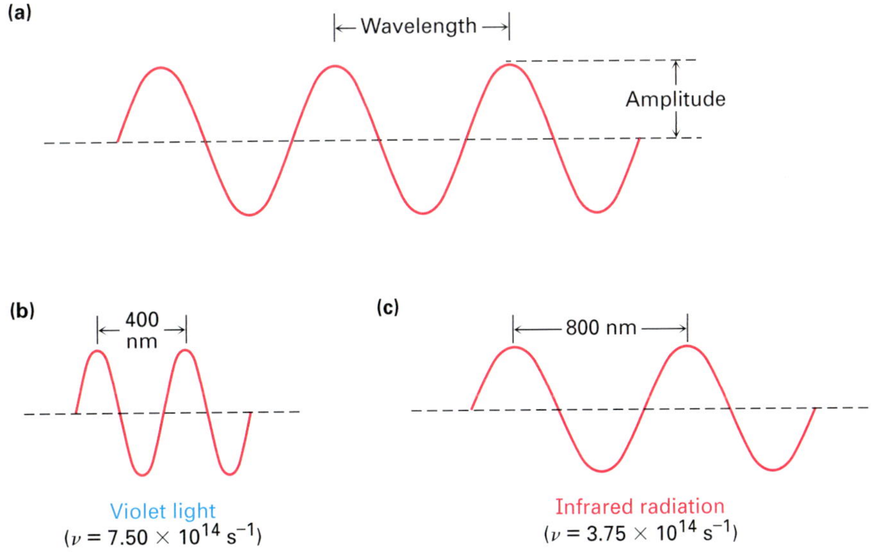

igure 12.11 Electromagnetic aves are characterized by a avelength, a frequency, and an mplitude. **(a)** Wavelength (λ) is e distance between two successive wave maxima. Amplitude is the height of the wave easured from the center. **)–(c)** What we perceive as different kinds of electromagnetic adiation are simply waves ith different wavelengths nd frequencies.

Multiplying the wavelength of a wave in meters (m) by its frequency reciprocal seconds (s^{-1}) gives the speed of the wave in meters per second (m/ The rate of travel of all electromagnetic radiation in a vacuum is a co stant value, commonly called the "speed of light" and abbreviated c. numerical value is defined as exactly $2.997\,924\,58 \times 10^8$ m/s, usua rounded off to 3.00×10^8 m/s.

$$\text{Wavelength} \times \text{Frequency} = \text{Speed}$$

$$\lambda\ (\text{m}) \times \nu\ (\text{s}^{-1}) = c\ (\text{m/s})$$

$$\lambda = \frac{c}{\nu} \quad \text{or} \quad \nu = \frac{c}{\lambda}$$

Just as matter comes only in discrete units called atoms, electromagne energy is transmitted only in discrete amounts called *quanta*. The amou of energy, ϵ, corresponding to 1 quantum of energy (1 photon) of a given fr quency, ν, is expressed by the Planck equation

$$\varepsilon = h\nu = \frac{hc}{\lambda}$$

where h = Planck's constant (6.62×10^{-34} J · s = 1.58×10^{-34} cal · s).

The Planck equation says that the energy of a given photon varies *direct* with its frequency ν but *inversely* with its wavelength λ. High frequencies ar short wavelengths correspond to high-energy radiation such as gamma rays; lo frequencies and long wavelengths correspond to low-energy radiation such radio waves. Multiplying ϵ by Avogadro's number N_A gives the same equation more familiar units, where E represents the energy of Avogadro's number (or "mole") of photons of wavelength λ:

$$E = \frac{N_A hc}{\lambda} = \frac{1.20 \times 10^{-4}\ \text{kJ/mol}}{\lambda\ (\text{m})} \quad \text{or} \quad \frac{2.86 \times 10^{-5}\ \text{kcal/mol}}{\lambda\ (\text{m})}$$

When an organic compound is exposed to a beam of electromagnetic rad ation, it absorbs energy of some wavelengths but passes, or transmits, energy o other wavelengths. If we irradiate the sample with energy of many differen wavelengths and determine which are absorbed and which are transmitted, w can measure the **absorption spectrum** of the compound.

An example of an absorption spectrum—that of ethanol exposed to infrare radiation—is shown in Figure 12.12. The horizontal axis records the wave length, and the vertical axis records the intensity of the various energy absorp tions in percent transmittance. The baseline corresponding to 0% absorption (o 100% transmittance) runs along the top of the chart, so a downward spik means that energy absorption has occurred at that wavelength.

The energy a molecule gains when it absorbs radiation must be distrib uted over the molecule in some way. With infrared radiation, the absorbe energy causes bonds to stretch and bend more vigorously. With ultraviole radiation, the energy causes an electron to jump from a lower-energy orbita to a higher-energy one. Different radiation frequencies affect molecules i

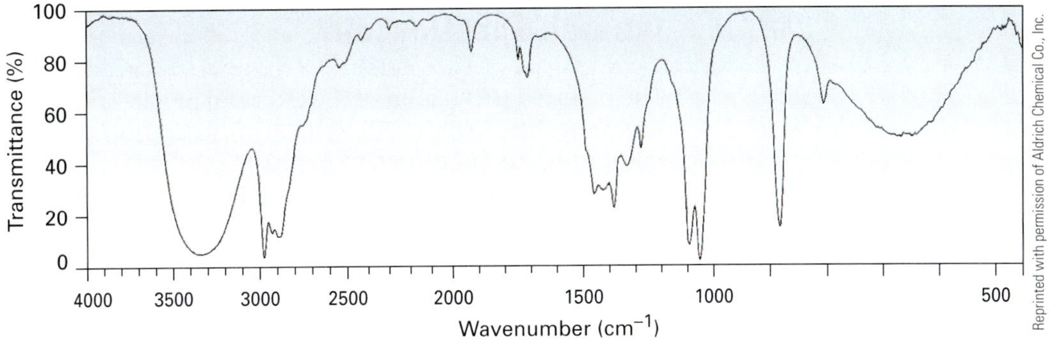

Figure 12.12 An infrared absorption spectrum of ethyl alcohol, CH_3CH_2OH. A transmittance of 100% means that all the energy is passing through the sample, whereas a lower transmittance means that some energy is being absorbed. Thus, each downward spike corresponds to an energy absorption.

different ways, but each provides structural information when the results are interpreted.

There are many kinds of spectroscopies, which differ according to the region of the electromagnetic spectrum that is used. We'll look at three—infrared spectroscopy, ultraviolet spectroscopy, and nuclear magnetic resonance spectroscopy. Let's begin by seeing what happens when an organic sample absorbs infrared energy.

WORKED EXAMPLE 12.3

Correlating Energy and Frequency of Radiation

Which is higher in energy, FM radio waves with a frequency of 1.015×10^8 Hz (101.5 MHz) or visible green light with a frequency of 5×10^{14} Hz?

Strategy Remember the equations $\epsilon = h\nu$ and $\epsilon = hc/\lambda$, which say that energy increases as frequency increases and as wavelength decreases.

Solution Since visible light has a higher frequency than radio waves, it is higher in energy.

Problem 12.5 Which has higher energy, infrared radiation with $\lambda = 1.0 \times 10^{-6}$ m or an X ray with $\lambda = 3.0 \times 10^{-9}$ m? Radiation with $\nu = 4.0 \times 10^9$ Hz or with $\lambda = 9.0 \times 10^{-6}$ m?

Problem 12.6 It's useful to develop a feeling for the amounts of energy that correspond to different parts of the electromagnetic spectrum. Calculate the energies of each of the following kinds of radiation

(a) A gamma ray with $\lambda = 5.0 \times 10^{-11}$ m
(b) An X ray with $\lambda = 3.0 \times 10^{-9}$ m
(c) Ultraviolet light with $\nu = 6.0 \times 10^{15}$ Hz
(d) Visible light with $\nu = 7.0 \times 10^{14}$ Hz
(e) Infrared radiation with $\lambda = 2.0 \times 10^{-5}$ m
(f) Microwave radiation with $\nu = 1.0 \times 10^{11}$ Hz

12.6 Infrared Spectroscopy

The **infrared (IR)** region of the electromagnetic spectrum covers the range fro just above the visible (7.8×10^{-7} m) to approximately 10^{-4} m, but only th midportion from 2.5×10^{-6} m to 2.5×10^{-5} m is used by organic chemists (Fi ure 12.13). Wavelengths within the IR region are usually given in micromete (1 μm = 10^{-6} m), and frequencies are given in wavenumbers rather than i hertz. The **wavenumber** ($\tilde{\nu}$) is the reciprocal of the wavelength in centimeter and is therefore expressed in units of cm^{-1}.

$$\text{Wavenumber:} \quad \tilde{\nu}\ (\text{cm}^{-1}) = \frac{1}{\lambda\ (\text{cm})}$$

Thus, the useful IR region is from 4000 to 400 cm^{-1}, corresponding to energie of 48.0 kJ/mol to 4.80 kJ/mol (11.5–1.15 kcal/mol).

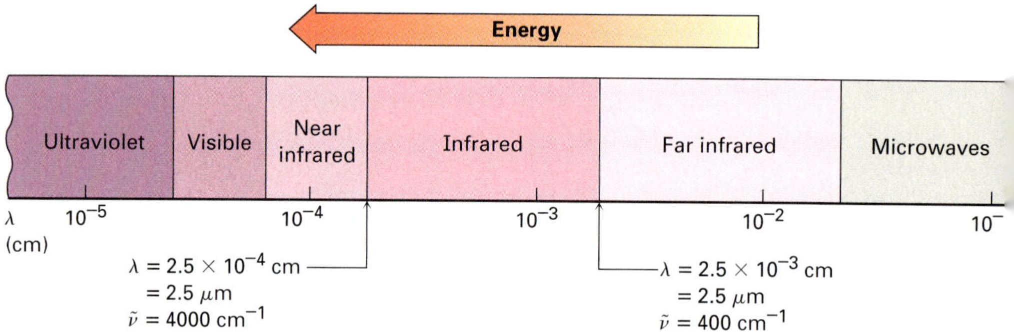

Figure 12.13 The infrared region of the electromagnetic spectrum.

Why does an organic molecule absorb some wavelengths of IR radiation bu not others? All molecules have a certain amount of energy and are in constan motion. Their bonds stretch and contract, atoms wag back and forth, and othe molecular vibrations occur. Following are some of the kinds of allowed vibrations

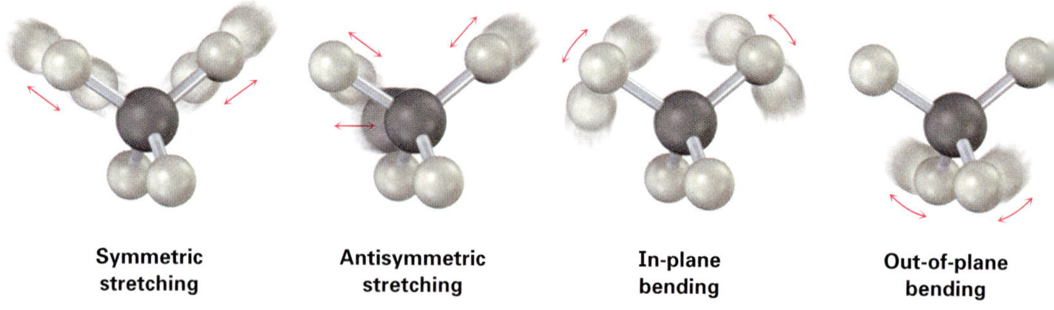

The amount of energy a molecule contains is not continuously variable but is *quantized*. That is, a molecule can stretch or bend only at specific frequencies corresponding to specific energy levels. Take bond-stretching, for example Although we usually speak of bond lengths as if they were fixed, the numbers given are really averages. In fact, a typical C–H bond with an average bond length of 110 pm is actually vibrating at a specific frequency, alternately stretching and contracting as if there were a spring connecting the two atoms.

When a molecule is irradiated with electromagnetic radiation, energy is absorbed if the frequency of the radiation matches the frequency of the vibration. The result of this energy absorption is an increased amplitude for the vibration; in other words, the "spring" connecting the two atoms stretches and compresses a bit further. Since each frequency absorbed by a molecule corresponds to a specific molecular motion, we can find what kinds of motions a molecule has by measuring its IR spectrum. By then interpreting those motions, we can find out what kinds of bonds (functional groups) are present in the molecule.

IR spectrum → What molecular motions? → What functional groups?

12.7 Interpreting Infrared Spectra

omsonNOW™ Click *Organic teractive* to **learn to utilize frared spectrometry to deduce olecular structures.**

Complete interpretation of an IR spectrum is difficult because most organic molecules have dozens of different bond stretching and bending motions, and thus have dozens of absorptions. On the one hand, this complexity is a problem because it generally limits the laboratory use of IR spectroscopy to pure samples of fairly small molecules—little can be learned from IR spectroscopy of large, complex biomolecules. On the other hand, the complexity is useful because an IR spectrum serves as a unique fingerprint of a compound. In fact, the complex region of the IR spectrum from 1500 cm^{-1} to around 400 cm^{-1} is called the *fingerprint region*. If two samples have identical IR spectra, they are almost certainly identical compounds.

Fortunately, we don't need to interpret an IR spectrum fully to get useful structural information. Most functional groups have characteristic IR absorption bands that don't change from one compound to another. The C=O absorption of a ketone is almost always in the range 1680 to 1750 cm^{-1}; the O–H absorption of an alcohol is almost always in the range 3400 to 3650 cm^{-1}; the C=C absorption of an alkene is almost always in the range 1640 to 1680 cm^{-1}; and so forth. By learning where characteristic functional-group absorptions occur, it's possible to get structural information from IR spectra. Table 12.1 lists the characteristic IR bands of some common functional groups.

Look at the IR spectra of hexane, 1-hexene, and 1-hexyne in Figure 12.14 to see an example of how IR spectroscopy can be used. Although all three IR spectra contain many peaks, there are characteristic absorptions of the C=C and C≡C functional groups that allow the three compounds to be distinguished. Thus, 1-hexene shows a characteristic C=C absorption at 1660 cm^{-1} and a vinylic =C–H absorption at 3100 cm^{-1}, whereas 1-hexyne has a C≡C absorption at 2100 cm^{-1} and a terminal alkyne ≡C–H absorption at 3300 cm^{-1}.

It helps in remembering the position of specific IR absorptions to divide the IR region from 4000 to 400 cm^{-1} into four parts, as shown in Figure 12.15.

- The region from 4000 to 2500 cm^{-1} corresponds to absorptions caused by N–H, C–H, and O–H single-bond stretching motions. N–H and O–H bonds absorb in the 3300 to 3600 cm^{-1} range; C–H bond-stretching occurs near 3000 cm^{-1}.
- The region from 2500 to 2000 cm^{-1} is where triple-bond stretching occurs. Both C≡N and C≡C bonds absorb here.

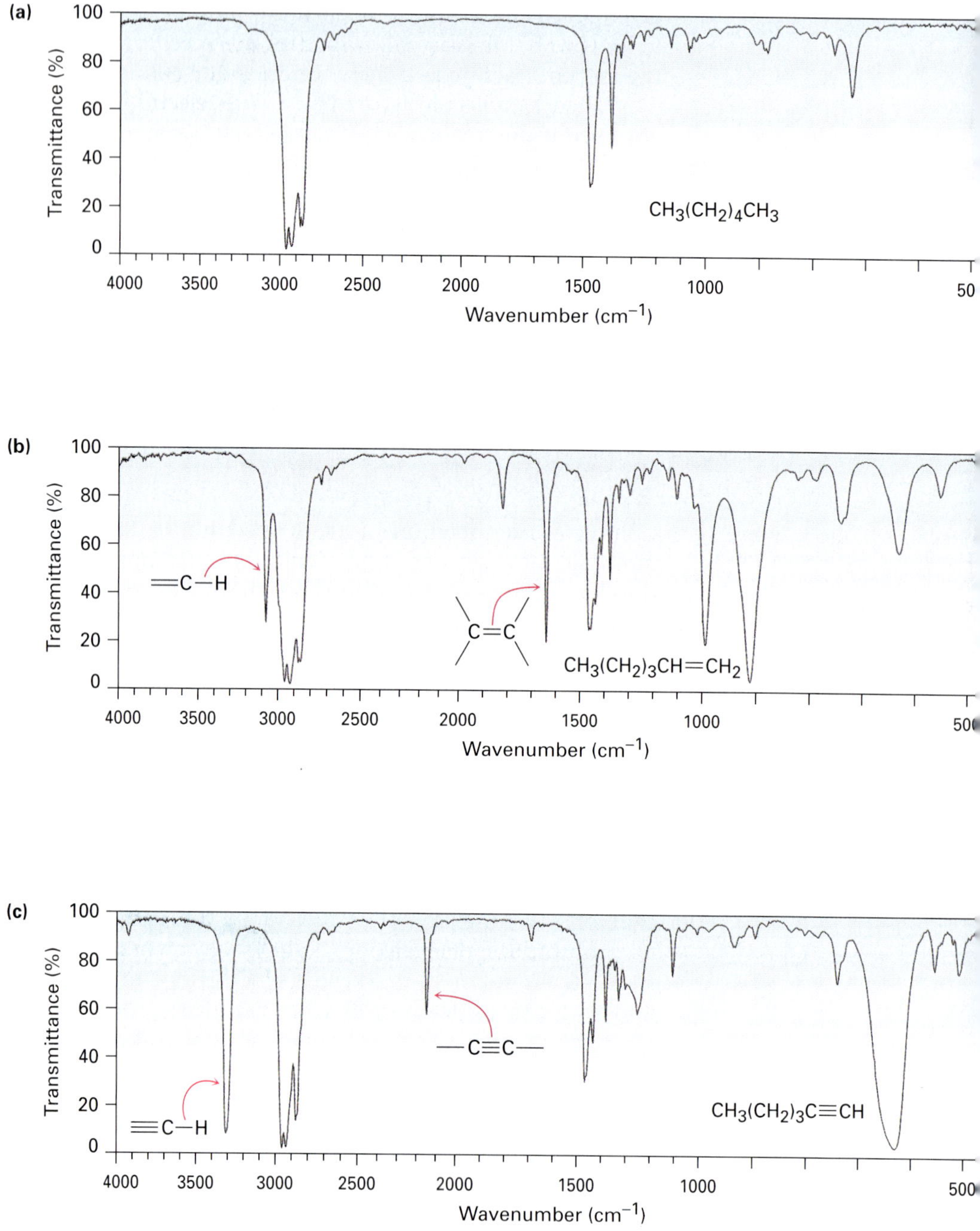

Figure 12.14 IR spectra of **(a)** hexane, **(b)** 1-hexene, and **(c)** 1-hexyne. Spectra like these are easily obtained on milligram amounts of material in a few minutes using commercially available instruments.

able 12.1 Characteristic IR Absorptions of Some Functional Groups

ınctional Group	Absorption (cm^{-1})	Intensity	Functional Group	Absorption (cm^{-1})	Intensity
lkane			Amine		
C–H	2850–2960	Medium	N–H	3300–3500	Medium
lkene			C–N	1030–1230	Medium
=C–H	3020–3100	Medium	Carbonyl compound		
C=C	1640–1680	Medium	C=O	1670–1780	Strong
lkyne			Carboxylic acid		
≡C–H	3300	Strong	O–H	2500–3100	Strong, broad
C≡C	2100–2260	Medium	Nitrile		
lkyl halide			C≡N	2210–2260	Medium
C–Cl	600–800	Strong	Nitro		
C–Br	500–600	Strong	NO_2	1540	Strong
lcohol					
O–H	3400–3650	Strong, broad			
C–O	1050–1150	Strong			
rene					
C–H	3030	Weak			
Aromatic ring	1660–2000	Weak			
	1450–1600	Medium			

- The region from 2000 to 1500 cm^{-1} is where double bonds (C=O, C=N, and C=C) absorb. Carbonyl groups generally absorb in the range 1680 to 1750 cm^{-1}, and alkene stretching normally occurs in the narrow range 1640 to 1680 cm^{-1}.
- The region below 1500 cm^{-1} is the fingerprint portion of the IR spectrum. A large number of absorptions due to a variety of C–C, C–O, C–N, and C–X single-bond vibrations occur here.

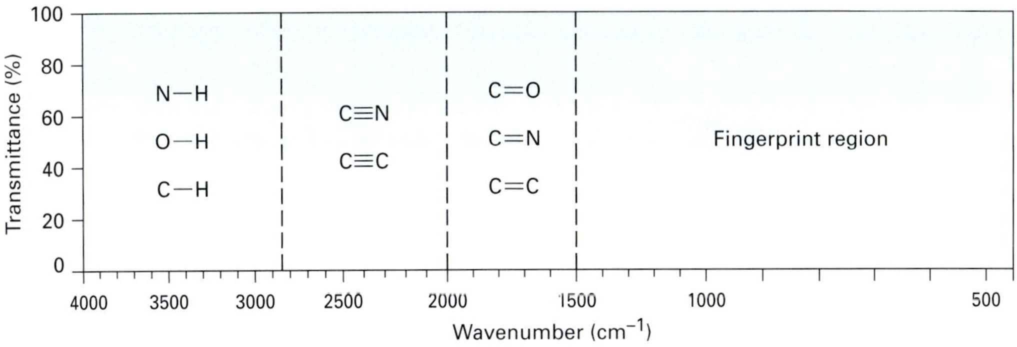

Figure 12.15 The four regions of the infrared spectrum: single bonds to hydrogen, triple bonds, double bonds, and fingerprint.

Why do different functional groups absorb where they do? As noted pre ously, a good analogy is that of two weights (atoms) connected by a spri (a bond). Short, strong bonds vibrate at a higher energy and higher frequen than do long, weak bonds, just as a short, strong spring vibrates faster thar long, weak spring. Thus, triple bonds absorb at a higher frequency than doul bonds, which in turn absorb at a higher frequency than single bonds. In ad tion, springs connecting small weights vibrate faster than springs connecti large weights. Thus, C–H, O–H, and N–H bonds vibrate at a higher frequen than bonds between heavier C, O, and N atoms.

WORKED EXAMPLE 12.4 ***Distinguishing Isomeric Compounds by IR Spectroscopy***

Acetone (CH_3COCH_3) and 2-propen-1-ol ($H_2C{=}CHCH_2OH$) are isomers. Hc could you distinguish them by IR spectroscopy?

Strategy Identify the functional groups in each molecule, and refer to Table 12.1.

Solution Acetone has a strong C=O absorption at 1715 cm^{-1}, while 2-propen-1-ol has –OH absorption at 3500 cm^{-1} and a C=C absorption at 1660 cm^{-1}.

Problem 12.7 What functional groups might the following molecules contain?

(a) A compound with a strong absorption at 1710 cm^{-1}

(b) A compound with a strong absorption at 1540 cm^{-1}

(c) A compound with strong absorptions at 1720 cm^{-1} and at 2500 to 3100 cm^{-}

Problem 12.8 How might you use IR spectroscopy to distinguish between the following pairs isomers?

(a) CH_3CH_2OH and CH_3OCH_3

(b) Cyclohexane and 1-hexene

(c) $CH_3CH_2CO_2H$ and $HOCH_2CH_2CHO$

12.8 Infrared Spectra of Some Common Functional Groups

As each functional group is discussed in future chapters, the spectroscopic pro erties of that group will be described. For the present, we'll point out some di tinguishing features of the hydrocarbon functional groups already studied an briefly preview some other common functional groups. We should also poir out, however, that in addition to interpreting absorptions that *are* present in a IR spectrum, it's also possible to get structural information by noticing whic absorptions are *not* present. If the spectrum of a compound has no absorptior at 3300 and 2150 cm^{-1}, the compound is not a terminal alkyne; if the spectrur has no absorption near 3400 cm^{-1}, the compound is not an alcohol; and so or

Alkanes

The IR spectrum of an alkane is fairly uninformative because no function groups are present and all absorptions are due to C–H and C–C bond Alkane C–H bonds show a strong absorption from 2850 to 2960 cm^{-1}, an saturated C–C bonds show a number of bands in the 800 to 1300 cm^{-1} rang

Since most organic compounds contain saturated alkane-like portions, most organic compounds have these characteristic IR absorptions. The C–H and C–C bands are clearly visible in the three spectra shown in Figure 12.14.

Alkanes	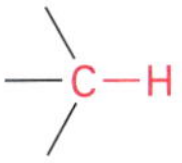	2850–2960 cm^{-1}
	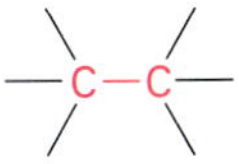	800–1300 cm^{-1}

Alkenes

Alkenes show several characteristic stretching absorptions. Vinylic =C–H bonds absorb from 3020 to 3100 cm^{-1}, and alkene C=C bonds usually absorb near 1650 cm^{-1}, although in some cases the peaks can be rather small and difficult to see clearly. Both absorptions are visible in the 1-hexene spectrum in Figure 12.14b.

Monosubstituted and disubstituted alkenes have characteristic =C–H out-of-plane bending absorptions in the 700 to 1000 cm^{-1} range, thereby allowing the substitution pattern on a double bond to be determined. Monosubstituted alkenes such as 1-hexene show strong characteristic bands at 910 and 990 cm^{-1}, and 2,2-disubstituted alkenes ($R_2C{=}CH_2$) have an intense band at 890 cm^{-1}.

Alkenes	=C–H	3020–3100 cm^{-1}
	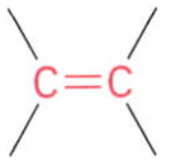	1640–1680 cm^{-1}
	$RCH{=}CH_2$	910 and 990 cm^{-1}
	$R_2C{=}CH_2$	890 cm^{-1}

Alkynes

Alkynes show a C≡C stretching absorption at 2100 to 2260 cm^{-1}, an absorption that is much more intense for terminal alkynes than for internal alkynes. In fact, symmetrically substituted triple bonds like that in 3-hexyne show no absorption at all, for reasons we won't go into. Terminal alkynes such as 1-hexyne also have a characteristic ≡C–H stretch at 3300 cm^{-1} (Figure 12.14c). This band is diagnostic for terminal alkynes because it is fairly intense and quite sharp.

Alkynes	–C≡C–	2100–2260 cm^{-1}
	≡C–H	3300 cm^{-1}

Aromatic Compounds

Aromatic compounds such as benzene have a weak C–H stretching absorption at 3030 cm^{-1}, a series of weak absorptions in the 1660 to 2000 cm^{-1} range, and a second series of medium-intensity absorptions in the 1450 to 1600 cm^{-1} region. These latter absorptions are due to complex molecular motions of the entire ring. The IR spectrum of phenylacetylene, shown in Figure 12.17 at t end of this section, gives an example.

Aromatic compounds 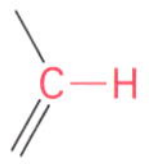 3030 cm^{-1} (weak)

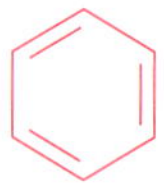 1660–2000 cm^{-1} (weak)
1450–1600 cm^{-1} (medium)

Alcohols

The O–H functional group of alcohols is easy to spot. Alcohols have a chara teristic band in the range 3400 to 3650 cm^{-1} that is usually broad and intens If present, it's hard to miss this band or to confuse it with anything else.

Alcohols —O—H 3400–3650 cm^{-1} (broad, intense)

Amines

The N–H functional group of amines is also easy to spot in the IR, with a cha acteristic absorption in the 3300 to 3500 cm^{-1} range. Although alcohols absor in the same range, an N–H absorption is much sharper and less intense than a O–H band.

Amines 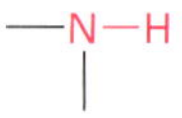 3300–3500 cm^{-1} (sharp, medium intensity)

Carbonyl Compounds

Carbonyl functional groups are the easiest to identify of all IR absorption because of their sharp, intense peak in the range 1670 to 1780 cm^{-1}. Mos important, the exact position of absorption within the range can often be use to identify the exact kind of carbonyl functional group—aldehyde, keton ester, and so forth.

Aldehydes Saturated aldehydes absorb at 1730 cm^{-1}; aldehydes next to eithe a double bond or an aromatic ring absorb at 1705 cm^{-1}.

Aldehydes

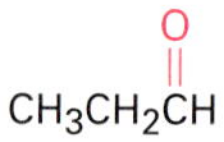

1730 cm^{-1}

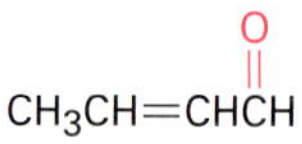

1705 cm^{-1}

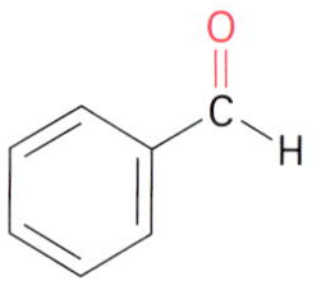

1705 cm^{-1}

Ketones Saturated open-chain ketones and six-membered cyclic ketones absorb at 1715 cm^{-1}, five-membered cyclic ketones absorb at 1750 cm^{-1}, and ketones next to a double bond or an aromatic ring absorb at 1690 cm^{-1}.

Ketones

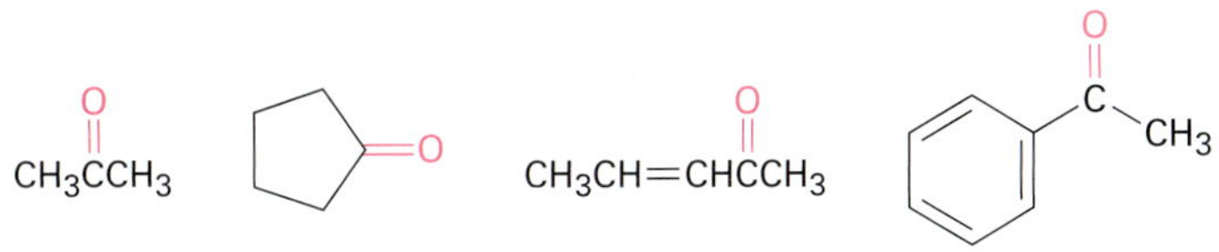

1715 cm^{-1} 1750 cm^{-1} 1690 cm^{-1} 1690 cm^{-1}

Esters Saturated esters absorb at 1735 cm^{-1}; esters next to either an aromatic ring or a double bond absorb at 1715 cm^{-1}.

Esters

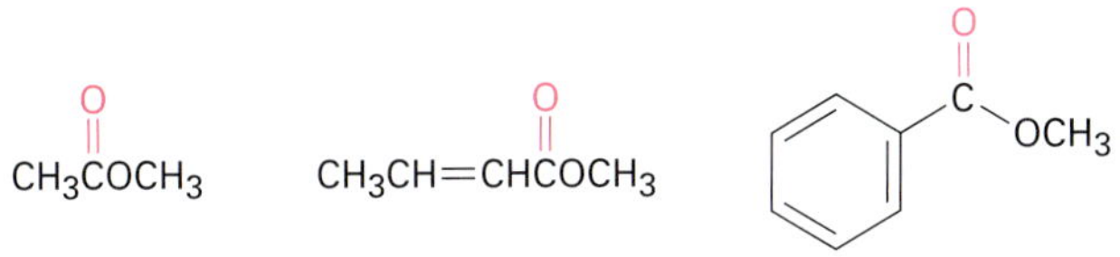

1735 cm^{-1} 1715 cm^{-1} 1715 cm^{-1}

WORKED EXAMPLE 12.5

Predicting IR Absorptions of Compounds

Where might the following compounds have IR absorptions?

(a) 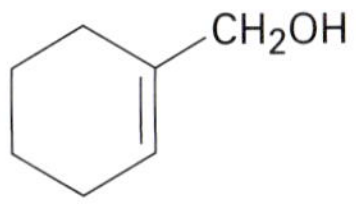 **(b)** 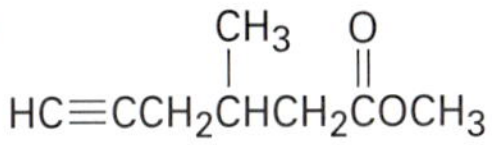

Strategy Identify the functional groups in each molecule, and then check Table 12.1 to see where those groups absorb.

Solution (a) *Absorptions:* 3400–3650 cm^{-1} (O–H), 3020–3100 cm^{-1} (=C–H), 1640–1680 cm^{-1} (C=C). This molecule has an alcohol O–H group and an alkene double bond.

(b) *Absorptions:* 3300 cm^{-1} (≡C–H), 2100–2260 cm^{-1} (C≡C), 1735 cm^{-1} (C=O). This molecule has a terminal alkyne triple bond and a saturated ester carbonyl group.

WORKED EXAMPLE 12.6

Identifying Functional Groups from an IR Spectrum

The IR spectrum of an unknown compound is shown in Figure 12.16. What functional groups does the compound contain?

Strategy All IR spectra have many absorptions, but those useful for identifying specific functional groups are usually found in the region from 1500 cm^{-1} to 3300 cm^{-1}. Pay particular attention to the carbonyl region (1670–1780 cm^{-1}), the aromatic region

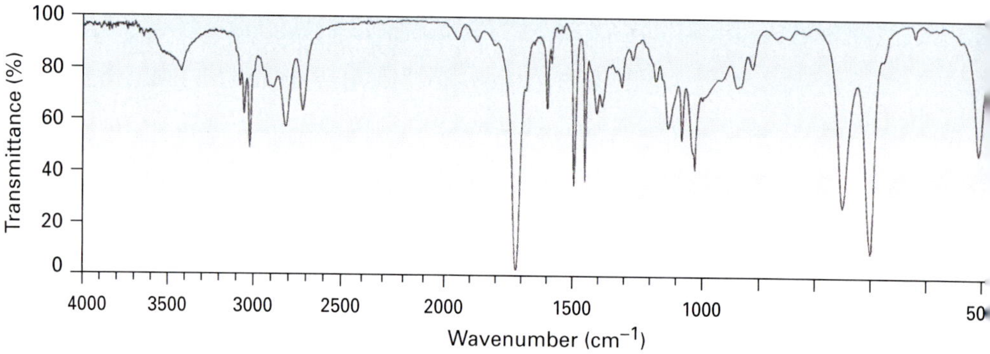

Figure 12.16 The IR spectrum for Worked Example 12.6.

(1660–2000 cm^{-1}), the triple-bond region (2000–2500 cm^{-1}), and the C–H regic (2500–3500 cm^{-1}).

Solution The spectrum shows an intense absorption at 1725 cm^{-1} due to a carbonyl group (pe haps an aldehyde, –CHO), a series of weak absorptions from 1800 to 2000 cm^{-1}, cha acteristic of aromatic compounds, and a C–H absorption near 3030 cm^{-1}, al characteristic of aromatic compounds. In fact, the compound is phenylacetaldehyd

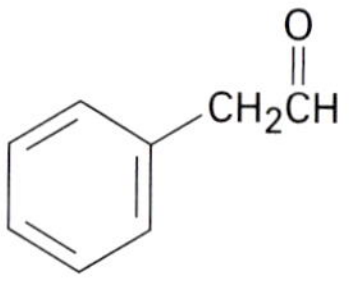

Phenylacetaldehyde

Problem 12.9 The IR spectrum of phenylacetylene is shown in Figure 12.17. What absorptio bands can you identify?

Problem 12.10 Where might the following compounds have IR absorptions?

(a)

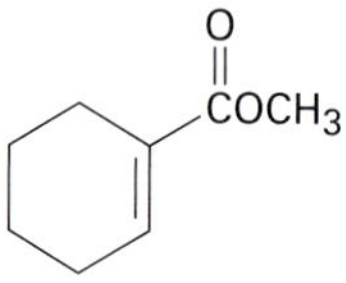

(b) $HC\equiv CCH_2CH_2\overset{O}{\overset{\|}{C}}H$

(c)

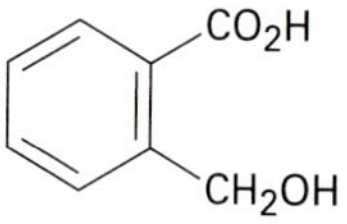

Problem 12.11 Where might the following compound have IR absorptions?

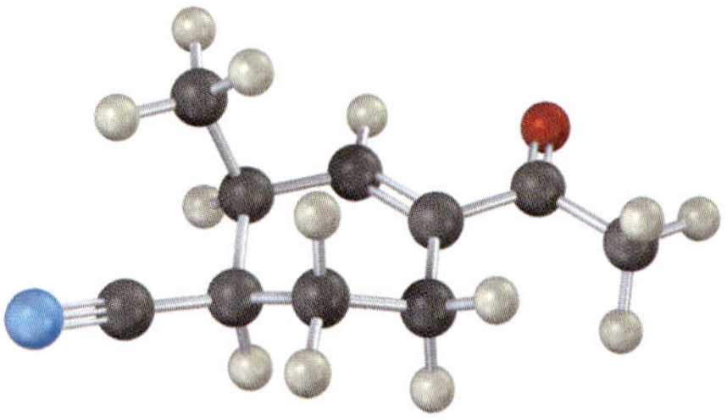

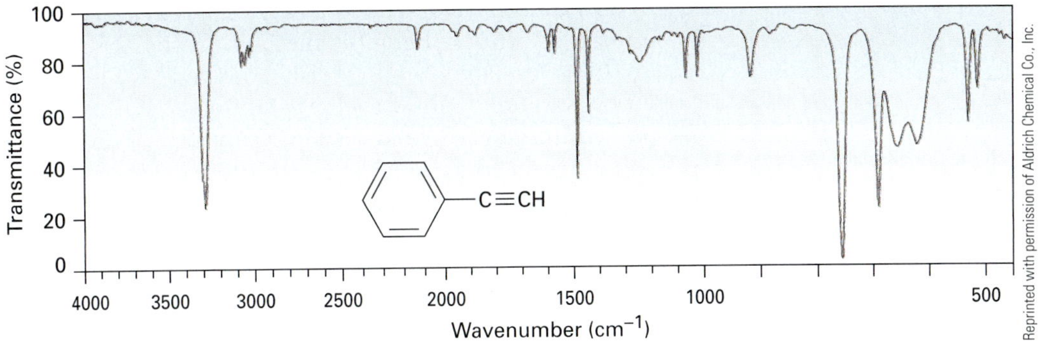

Figure 12.17 The IR spectrum of phenylacetylene, Problem 12.9.

Focus On . . .

Chromatography: Purifying Organic Compounds

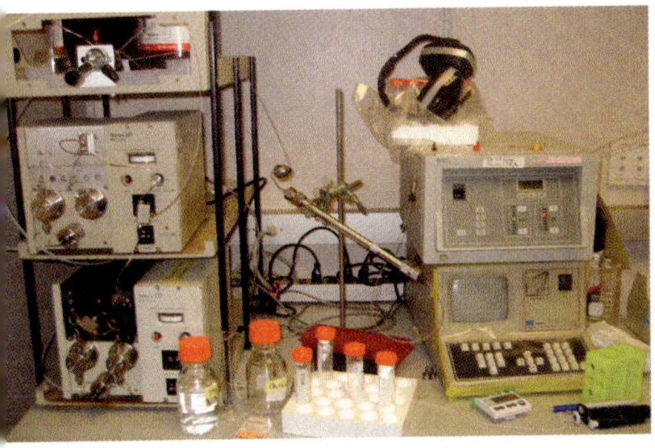

High-pressure liquid chromatography (HPLC) is used to separate and purify the products of laboratory reactions.

Even before a new organic substance has its structure determined, it must be purified by separating it from solvents and all contaminants. Purification was an enormously time-consuming, hit-or-miss proposition in the 19th and early 20th centuries, but powerful instruments developed in the last few decades now simplify the problem.

Most organic purification is done by *chromatography* (literally, "color writing"), a separation technique that dates from the work of the Russian chemist Mikhail Tswett in 1903. Tswett accomplished the separation of the pigments in green leaves by dissolving the leaf extract in an organic solvent and allowing the solution to run down through a vertical glass tube packed with chalk powder. Different pigments passed down the column at different rates, leaving a series of colored bands on the white chalk column.

A variety of chromatographic techniques are now in common use, all of which work on a similar principle. The mixture to be separated is dissolved in a solvent, called the *mobile phase,* and passed over an adsorbent material, called the *stationary phase.* Because different compounds adsorb to the stationary phase to different extents, they migrate along the phase at different rates and are separated as they emerge *(elute)* from the end of the chromatography column.

(continued)

Liquid chromatography, or *column chromatography,* is perhaps the most often used chromatographic method. As in Tswett's original experiments, a mixture of organic compounds is dissolved in a suitable solvent and adsorbed onto a stationary phase such as alumina (Al_2O_3) or silica gel (hydrated SiO_2) packed into a glass column. More solvent is then passed down the column, and different compounds elute at different times.

The time at which a compound is eluted is strongly influenced by its polarity. Molecules with polar functional groups are generally adsorbed more strongly and therefore migrate through the stationary phase more slowly than nonpolar molecules. A mixture of an alcohol and an alkene, for example, can be easily separated with liquid chromatography because the nonpolar alkene passes through the column much faster than the more polar alcohol.

High-pressure liquid chromatography (HPLC) is a variant of the simple column technique, based on the discovery that chromatographic separations are vastly improved if the stationary phase is made up of very small, uniformly sized spherical particles. Small particle size ensures a large surface area for better adsorption, and a uniform spherical shape allows a tight, uniform packing of particles. In practice, coated SiO_2 microspheres of 3.5 to 5 μm diameter are often used.

High-pressure pumps operating at up to 6000 psi are required to force solvent through a tightly packed HPLC column, and electronic detectors are used to monitor the appearance of material eluting from the column. Alternatively, the column can be interfaced to a mass spectrometer to determine the mass spectrum of every substance as it elutes. Figure 12.18 shows the results of HPLC analysis of a mixture of 10 fat-soluble vitamins on 5 μm silica spheres with acetonitrile as solvent.

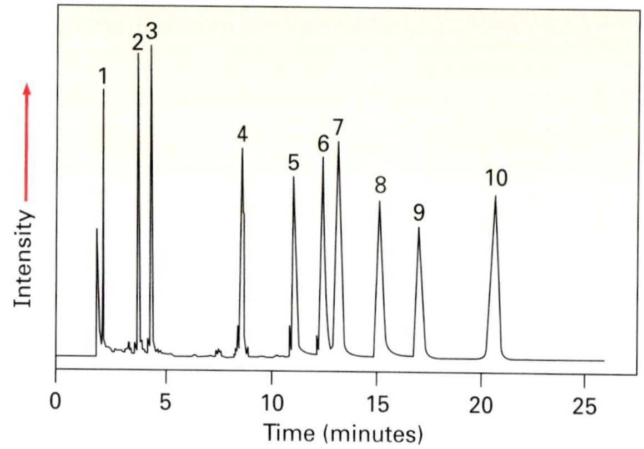

Figure 12.18 Results of an HPLC analysis of a mixture of ten fat-soluble vitamins.

1. Menadione (vitamin K_3)
2. Retinol (vitamin A)
3. Retinol acetate
4. Menaquinone (vitamin K_2)
5. δ-Tocopherol
6. Ergocalciferol (vitamin D_2)
7. Cholecalciferol (vitamin D_3)
8. α-Tocopherol (vitamin E)
9. α-Tocopherol acetate
10. Phylloquinone (vitamin K_1)

SUMMARY AND KEY WORDS

The structure of an organic molecule is usually determined using spectroscopic methods such as mass spectrometry and infrared spectroscopy. **Mass spectrometry (MS)** tells the molecular weight and formula of a molecule; **infrared (IR) spectroscopy** identifies the functional groups present in the molecule.

In small-molecule mass spectrometry, molecules are first ionized by collision with a high-energy electron beam. The ions then fragment into smaller pieces, which are magnetically sorted according to their mass-to-charge ratio (m/z). The ionized sample molecule is called the *molecular ion*, M^+, and measurement of its mass gives the molecular weight of the sample. Structural clues about unknown samples can be obtained by interpreting the fragmentation pattern of the molecular ion. Mass-spectral fragmentations are usually complex, however, and interpretation is often difficult. In biological mass spectrometry, molecules are protonated using either electrospray ionization (ESI) or matrix-assisted laser desorption ionization (MALDI), and the protonated molecules are separated by time-of-flight (TOF).

Infrared spectroscopy involves the interaction of a molecule with **electromagnetic radiation**. When an organic molecule is irradiated with infrared energy, certain **frequencies** are absorbed by the molecule. The frequencies absorbed correspond to the amounts of energy needed to increase the amplitude of specific molecular vibrations such as bond-stretchings and bond-bendings. Since every functional group has a characteristic combination of bonds, every functional group has a characteristic set of infrared absorptions. For example, the terminal alkyne ≡C–H bond absorbs IR radiation of 3300 cm^{-1} frequency, and the alkene C=C bond absorbs in the range 1640 to 1680 cm^{-1}. By observing which frequencies of infrared radiation are absorbed by a molecule and which are not, it's possible to determine the functional groups a molecule contains.

absorption spectrum, 420
amplitude, 419
base peak, 410
electromagnetic spectrum, 418
frequency (ν), 419
hertz (Hz), 419
infrared spectroscopy (IR), 422
mass spectrometry (MS), 409
mass spectrum, 410
parent peak, 410
wavelength (λ), 419
wavenumber ($\tilde{\nu}$), 422

EXERCISES

Organic KNOWLEDGE TOOLS

ThomsonNOW™ Sign in at **www.thomsonedu.com** to assess your knowledge of this chapter's topics by taking a pre-test. The pre-test will link you to interactive organic chemistry resources based on your score in each concept area.

Online homework for this chapter may be assigned in Organic OWL.

■ indicates problems assignable in Organic OWL.

VISUALIZING CHEMISTRY

(Problems 12.1–12.11 appear within the chapter.)

12.12 ■ Where in the IR spectrum would you expect each of the following mol cules to absorb?

(a)

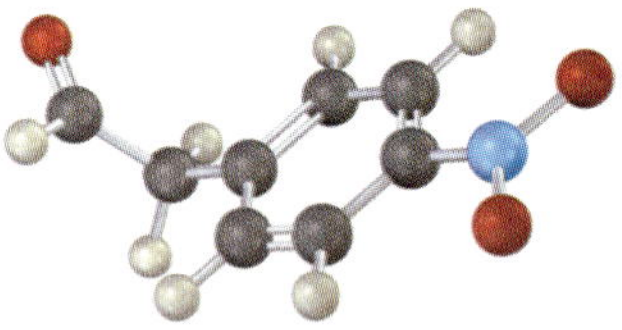

(b)

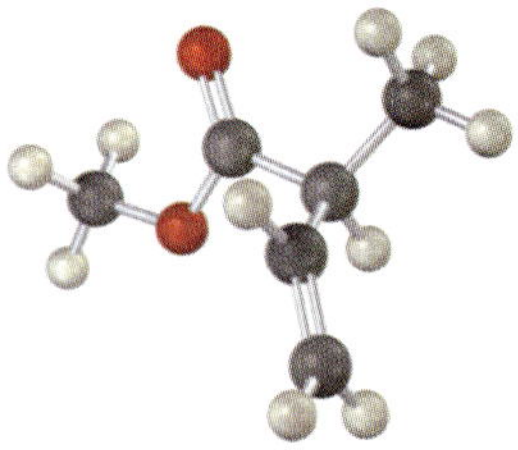

(c)

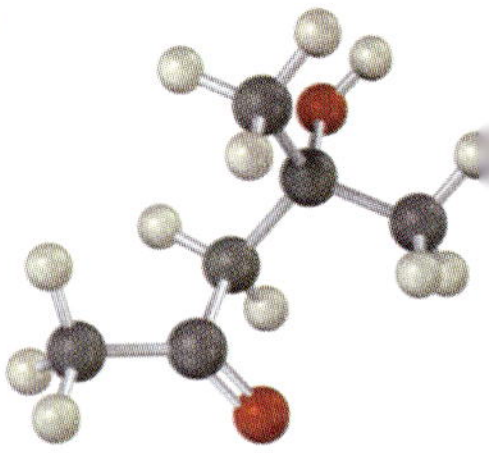

12.13 ■ Show the structures of the likely fragments you would expect in the mas spectra of the following molecules:

(a)

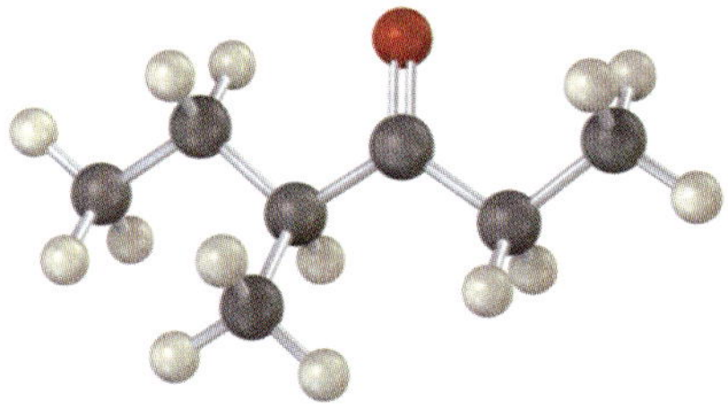

(b)

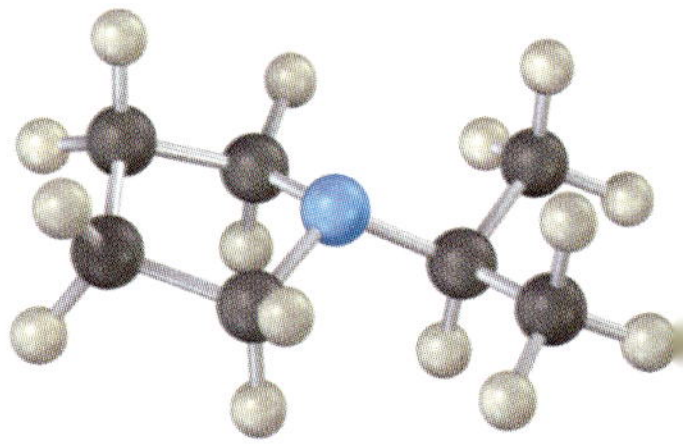

ADDITIONAL PROBLEMS

12.14 ■ Propose structures for compounds that fit the following mass-spectral data
(a) A hydrocarbon with $M^+ = 132$ **(b)** A hydrocarbon with $M^+ = 166$
(c) A hydrocarbon with $M^+ = 84$

12.15 ■ Write molecular formulas for compounds that show the following molecular ions in their high-resolution mass spectra. Assume that C, H, N, and O might be present, and use the exact atomic masses given in Section 12.2.
(a) $M^+ = 98.0844$ **(b)** $M^+ = 123.0320$

12.16 ■ Camphor, a saturated monoketone from the Asian camphor tree, is used among other things as a moth repellent and as a constituent of embalming fluid. If camphor has $M^+ = 152.1201$ by high-resolution mass spectrometry, what is its molecular formula? How many rings does camphor have?

12.17 The *nitrogen rule* of mass spectrometry says that a compound containing an odd number of nitrogens has an odd-numbered molecular ion. Conversely, a compound containing an even number of nitrogens has an even-numbered M^+ peak. Explain.

12.18 ■ In light of the nitrogen rule mentioned in Problem 12.17, what is the molecular formula of pyridine, $M^+ = 79$?

12.19 ■ Nicotine is a diamino compound isolated from dried tobacco leaves. Nicotine has two rings and $M^+ = 162.1157$ by high-resolution mass spectrometry. Give a molecular formula for nicotine, and calculate the number of double bonds.

12.20 ■ The hormone cortisone contains C, H, and O, and shows a molecular ion at $M^+ = 360.1937$ by high-resolution mass spectrometry. What is the molecular formula of cortisone? (The degree of unsaturation of cortisone is 8.)

12.21 ■ Halogenated compounds are particularly easy to identify by their mass spectra because both chlorine and bromine occur naturally as mixtures of two abundant isotopes. Chlorine occurs as ^{35}Cl (75.8%) and ^{37}Cl (24.2%); bromine occurs as ^{79}Br (50.7%) and ^{81}Br (49.3%). At what masses do the molecular ions occur for the following formulas? What are the relative percentages of each molecular ion?

(a) Bromomethane, CH_3Br **(b)** 1-Chlorohexane, $C_6H_{13}Cl$

12.22 ■ By knowing the natural abundances of minor isotopes, it's possible to calculate the relative heights of M^+ and $M+1$ peaks. If ^{13}C has a natural abundance of 1.10%, what are the relative heights of the M^+ and $M+1$ peaks in the mass spectrum of benzene, C_6H_6?

12.23 ■ Propose structures for compounds that fit the following data:

(a) A ketone with $M^+ = 86$ and fragments at $m/z = 71$ and $m/z = 43$

(b) An alcohol with $M^+ = 88$ and fragments at $m/z = 73$, $m/z = 70$, and $m/z = 59$

12.24 2-Methylpentane (C_6H_{14}) has the mass spectrum shown. Which peak represents M^+? Which is the base peak? Propose structures for fragment ions of $m/z = 71$, 57, 43, and 29. Why does the base peak have the mass it does?

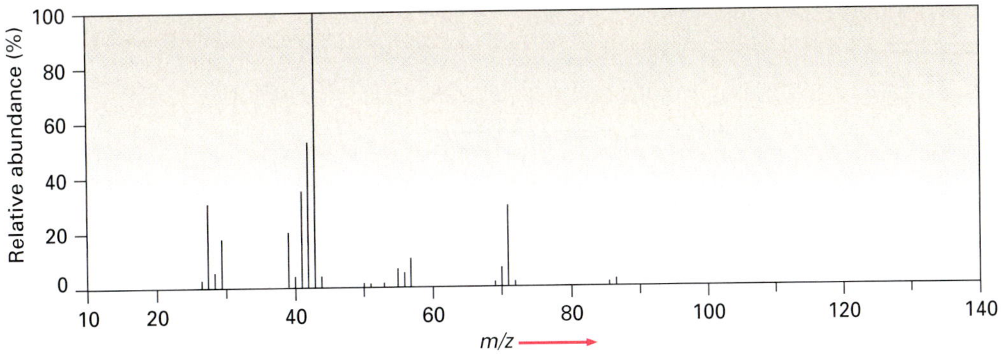

12.25 Assume that you are in a laboratory carrying out the catalytic hydrogenation of cyclohexene to cyclohexane. How could you use a mass spectrometer to determine when the reaction is finished?

12.26 What fragments might you expect in the mass spectra of the followi compounds?

(a)
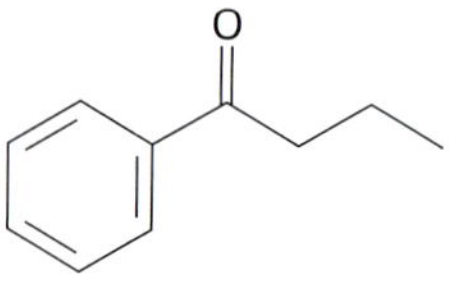

(b)
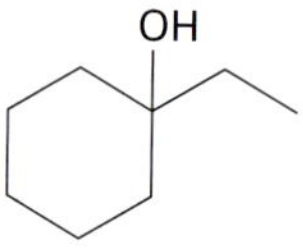

(c)
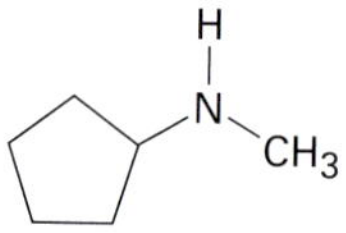

12.27 ■ How might you use IR spectroscopy to distinguish among the three isome 1-butyne, 1,3-butadiene, and 2-butyne?

12.28 Would you expect two enantiomers such as (*R*)-2-bromobutane ar (*S*)-2-bromobutane to have identical or different IR spectra? Explain.

12.29 Would you expect two diastereomers such as *meso*-2,3-dibromobutane ar (2*R*,3*R*)-dibromobutane to have identical or different IR spectra? Explain.

12.30 ■ Propose structures for compounds that meet the following descriptions:

(a) C_5H_8, with IR absorptions at 3300 and 2150 cm^{-1}

(b) C_4H_8O, with a strong IR absorption at 3400 cm^{-1}

(c) C_4H_8O, with a strong IR absorption at 1715 cm^{-1}

(d) C_8H_{10}, with IR absorptions at 1600 and 1500 cm^{-1}

12.31 ■ How could you use infrared spectroscopy to distinguish between the fo lowing pairs of isomers?

(a) $HC{\equiv}CCH_2NH_2$ and $CH_3CH_2C{\equiv}N$

(b) CH_3COCH_3 and CH_3CH_2CHO

12.32 Two infrared spectra are shown. One is the spectrum of cyclohexane, and th other is the spectrum of cyclohexene. Identify them, and explain your answe

(a)
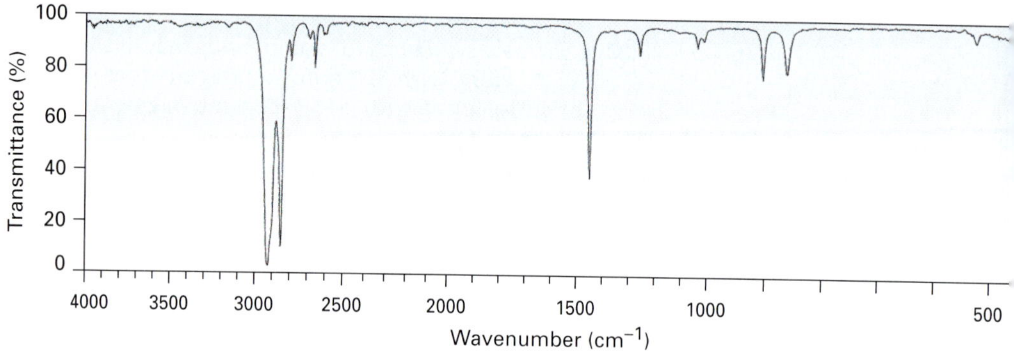

(b)
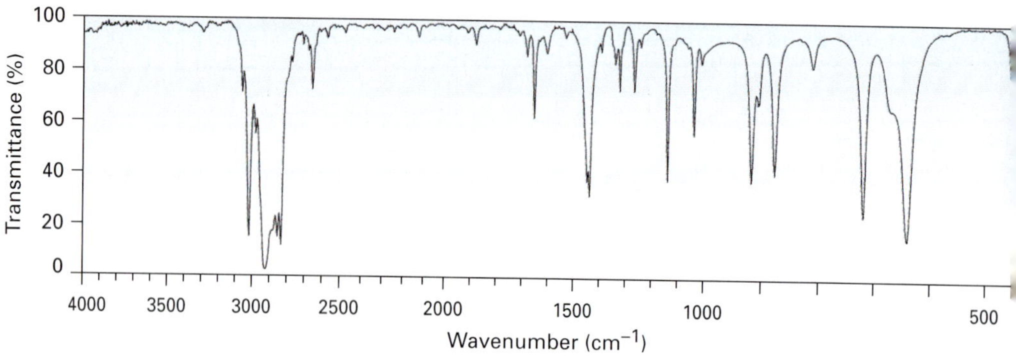

12.33 At what approximate positions might the following compounds show IR absorptions?

(a) 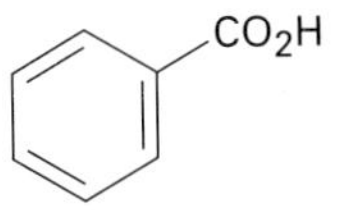

(b) 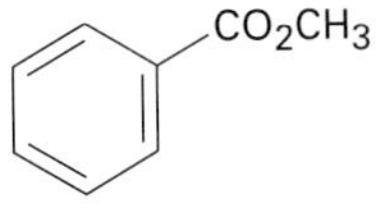

(c) 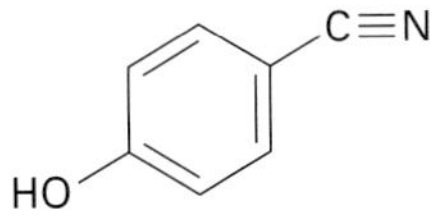

(d) 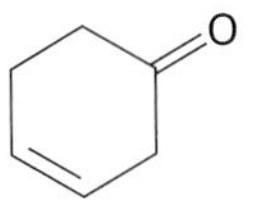

(e) $CH_3\overset{O}{\overset{\|}{C}}CH_2CH_2\overset{O}{\overset{\|}{C}}OCH_3$

12.34 How would you use infrared spectroscopy to distinguish between the following pairs of constitutional isomers?

(a) $CH_3C{\equiv}CCH_3$ and $CH_3CH_2C{\equiv}CH$

(b) $CH_3\overset{O}{\overset{\|}{C}}CH{=}CHCH_3$ and $CH_3\overset{O}{\overset{\|}{C}}CH_2CH{=}CH_2$

(c) $H_2C{=}CHOCH_3$ and CH_3CH_2CHO

12.35 At what approximate positions might the following compounds show IR absorptions?

(a) $CH_3CH_2\overset{O}{\overset{\|}{C}}CH_3$

(b) $CH_3\underset{}{\overset{CH_3}{\overset{|}{C}}}HCH_2C{\equiv}CH$

(c) $CH_3\overset{CH_3}{\overset{|}{C}}HCH_2CH{=}CH_2$

(d) $CH_3CH_2CH_2\overset{O}{\overset{\|}{C}}OCH_3$

(e) 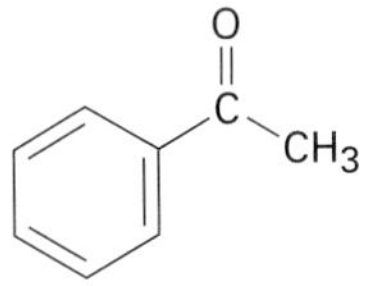

(f) 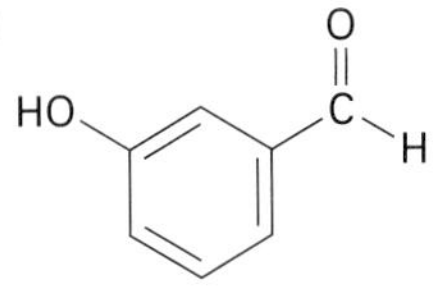

12.36 Assume you are carrying out the dehydration of 1-methylcyclohexanol to yield 1-methylcyclohexene. How could you use infrared spectroscopy to determine when the reaction is complete?

12.37 Assume that you are carrying out the base-induced dehydrobromination of 3-bromo-3-methylpentane (Section 11.7) to yield an alkene. How could you use IR spectroscopy to tell which of two possible elimination products is formed?

12.38 Which is stronger, the C=O bond in an ester (1735 cm^{-1}) or the C=O bond in a saturated ketone (1715 cm^{-1})? Explain.

12.39 Carvone is an unsaturated ketone responsible for the odor of spearmint. If carvone has $M^+ = 150$ in its mass spectrum and contains three double bonds and one ring, what is its molecular formula?

12.40 Carvone (Problem 12.39) has an intense infrared absorption at 1690 cm^{-1}. What kind of ketone does carvone contain?

12.41 The **(a)** mass spectrum and the **(b)** infrared spectrum of an unknown hydr carbon are shown. Propose as many structures as you can.

(a)

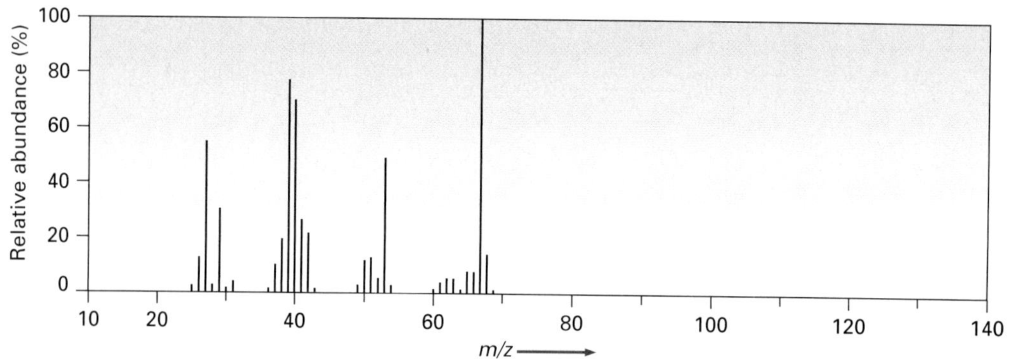

(b)

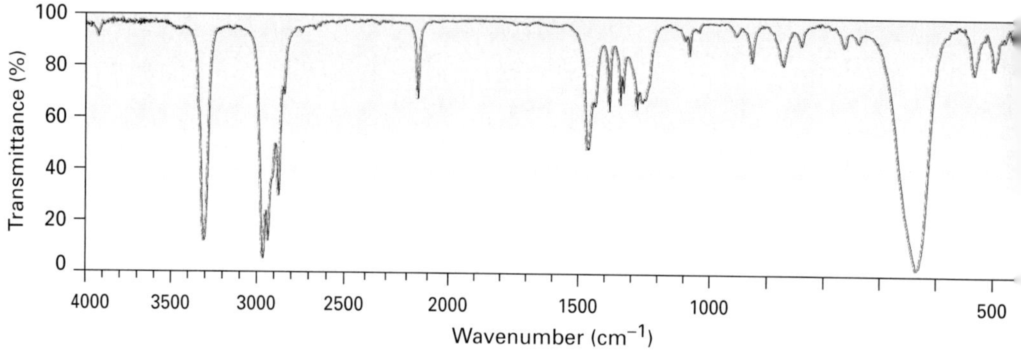

12.42 The **(a)** mass spectrum and the **(b)** infrared spectrum of another unknown hydrocarbon are shown. Propose as many structures as you can.

(a)

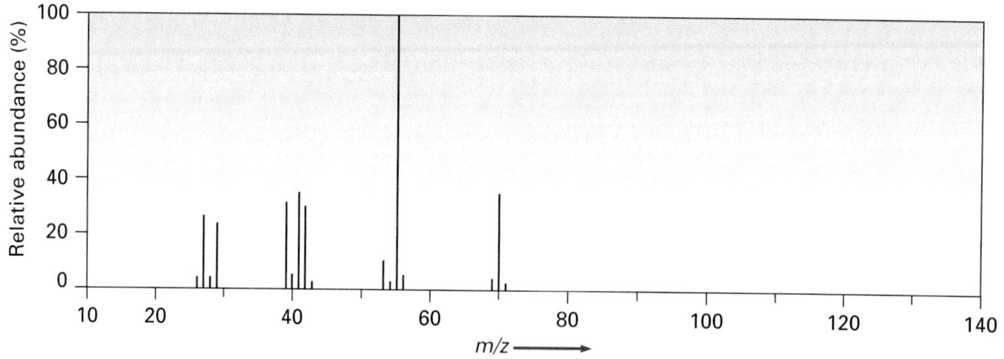

(b)

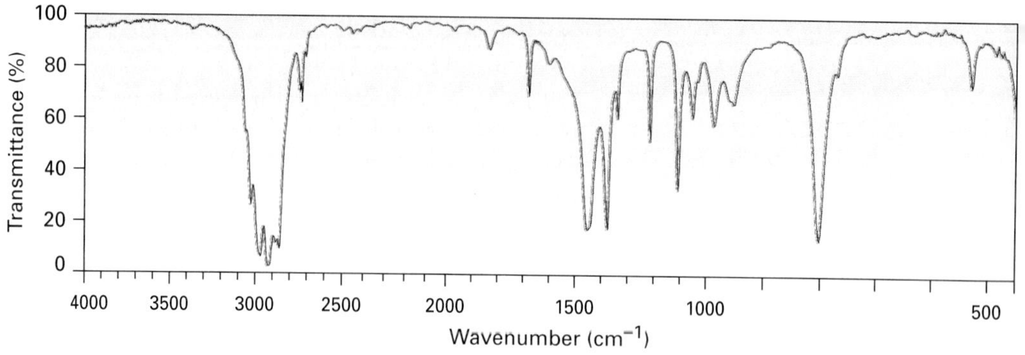

■ Assignable in OWL

12.43 ■ Propose structures for compounds that meet the following descriptions:

(a) An optically active compound $C_5H_{10}O$ with an IR absorption at 1730 cm^{-1}

(b) A non–optically active compound C_5H_9N with an IR absorption at 2215 cm^{-1}

12.44 4-Methyl-2-pentanone and 3-methylpentanal are isomers. Explain how you could tell them apart, both by mass spectrometry and by infrared spectroscopy.

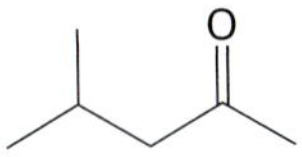

4-Methyl-2-pentanone

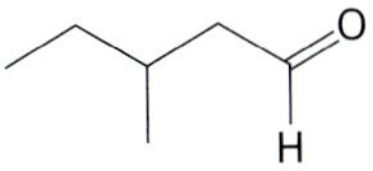

3-Methylpentanal

12.45 Grignard reagents undergo a general and very useful reaction with ketones. Methylmagnesium bromide, for example, reacts with cyclohexanone to yield a product with the formula $C_7H_{14}O$. What is the structure of this product if it has an IR absorption at 3400 cm^{-1}?

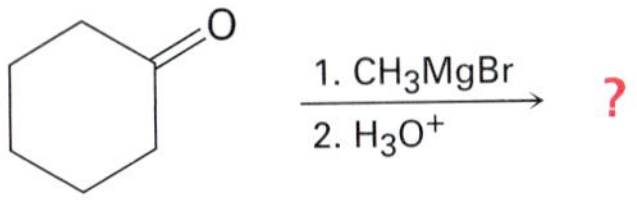

Cyclohexanone

12.46 Ketones undergo a reduction when treated with sodium borohydride, $NaBH_4$. What is the structure of the compound produced by reaction of 2-butanone with $NaBH_4$ if it has an IR absorption at 3400 cm^{-1} and $M^+ = 74$ in the mass spectrum?

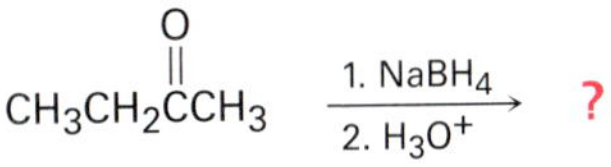

2-Butanone

12.47 Nitriles, R—C≡N, undergo a hydrolysis reaction when heated with aqueous acid. What is the structure of the compound produced by hydrolysis of propanenitrile, $CH_3CH_2C{\equiv}N$, if it has IR absorptions at 2500 to 3100 cm^{-1} and 1710 cm^{-1} and has $M^+ = 74$?

13 Structure Determination: Nuclear Magnetic Resonance Spectroscopy

Organic **KNOWLEDGE TOOLS**

ThomsonNOW™ Throughout this chapter, sign in at **www.thomsonedu.com** for online self-study and interactive tutorials based on your level of understanding.

Online homework for this chapter may be assigned in Organic OWL.

Nuclear magnetic resonance (NMR) spectroscopy is the most valuable spec troscopic technique available to organic chemists. It's the method of structur determination that organic chemists turn to first.

We saw in Chapter 12 that mass spectrometry gives a molecule's formula an infrared spectroscopy identifies a molecule's functional groups. Nuclear mag netic resonance spectroscopy does not replace either of these techniques; rathe it complements them by "mapping" a molecule's carbon–hydrogen framewor Taken together, mass spectrometry, IR, and NMR make it possible to determin the structures of even very complex molecules.

Mass spectrometry	Molecular size and formula
Infrared spectroscopy	Functional groups
NMR spectroscopy	Map of carbon–hydrogen framework

WHY THIS CHAPTER?

The opening sentence above says it all. NMR is by far the most valuable spec troscopic technique for structure determination. Although we'll just give ar overview of the subject in this chapter, focusing on NMR applications to smal molecules, more advanced NMR techniques are also used in biological chem istry to study protein structure and folding.

13.1 Nuclear Magnetic Resonance Spectroscopy

Many kinds of atomic nuclei behave as if they were spinning about an axis, much as the earth spins daily. Because they're positively charged, these spinning nuclei act like tiny bar magnets and interact with an external magnetic field, denoted $\boldsymbol{B}_0$. Not all nuclei act this way, but fortunately for organic chemists, both the proton (^1H) and the ^{13}C nucleus do have spins. (In speaking about NMR, the words *proton* and *hydrogen* are often used interchangeably.) Let's see what the consequences of nuclear spin are and how we can use the results.

In the absence of an external magnetic field, the spins of magnetic nuclei are oriented randomly. When a sample containing these nuclei is placed between the poles of a strong magnet, however, the nuclei adopt specific orientations, much as a compass needle orients in the earth's magnetic field. A spinning ^{1}H or ^{13}C nucleus can orient so that its own tiny magnetic field is aligned either with (parallel to) or against (antiparallel to) the external field. The two orientations don't have the same energy, however, and aren't equally likely. The parallel orientation is slightly lower in energy by an amount that depends on the strength of the external field, making this spin state very slightly favored over the antiparallel orientation (Figure 13.1).

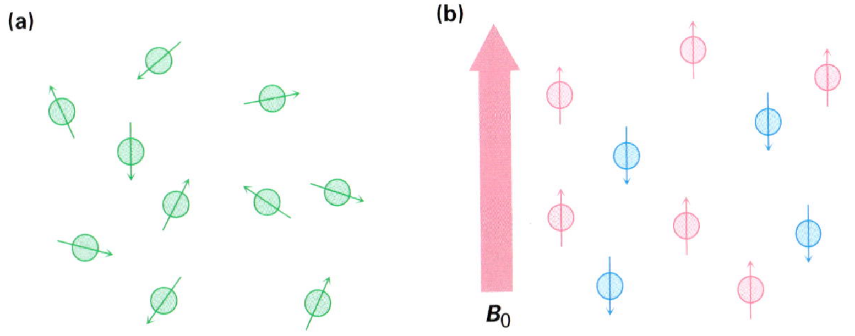

igure 13.1 **(a)** Nuclear spins re oriented randomly in the bsence of an external magnetic eld but **(b)** have a specific orienation in the presence of an xternal field, B_0. Some of the pins (red) are aligned parallel to he external field while others blue) are antiparallel. The parallel spin state is slightly lower in nergy and therefore favored.

If the oriented nuclei are now irradiated with electromagnetic radiation of the proper frequency, energy absorption occurs and the lower-energy state "spin-flips" to the higher-energy state. When this spin-flip occurs, the magnetic nuclei are said to be in resonance with the applied radiation—hence the name *nuclear magnetic resonance.*

The exact frequency necessary for resonance depends both on the strength of the external magnetic field and on the identity of the nuclei. If a very strong magnetic field is applied, the energy difference between the two spin states is larger and higher-frequency (higher-energy) radiation is required for a spin-flip. If a weaker magnetic field is applied, less energy is required to effect the transition between nuclear spin states (Figure 13.2).

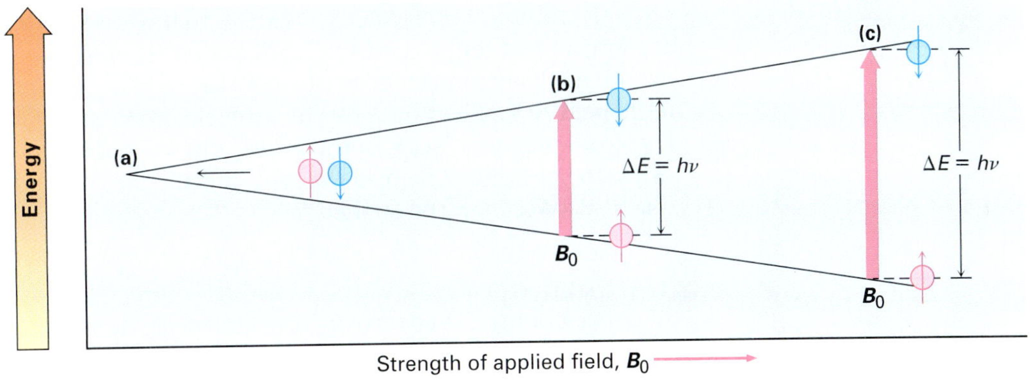

Figure 13.2 The energy difference ΔE between nuclear spin states depends on the strength of the applied magnetic field. Absorption of energy with frequency ν converts a nucleus from a lower spin state to a higher spin state. Spin states **(a)** have equal energies in the absence of an applied magnetic field but **(b)** have unequal energies in the presence of a magnetic field. At $\nu = 200$ MHz, $\Delta E = 8.0 \times 10^{-5}$ kJ/mol (1.9×10^{-5} kcal/mol). **(c)** The energy difference between spin states is greater at larger applied fields. At $\nu = 500$ MHz, $\Delta E = 2.0 \times 10^{-4}$ kJ/mol.

Table 13.1 The NMR Behavior of Some Common Nuclei

Magnetic nuclei	Nonmagnetic nuclei
^{1}H	^{12}C
^{13}C	^{16}C
^{2}H	^{32}S
^{14}N	
^{19}F	
^{31}P	

In practice, superconducting magnets that produce enormously powerfu fields up to 21.2 tesla (T) are sometimes used, but field strengths in the range c 4.7 to 7.0 T are more common. At a magnetic field strength of 4.7 T, so-calle radiofrequency (rf) energy in the 200 MHz range (1 MHz = 10^6 Hz) brings ^{1}H nucleus into resonance, and rf energy of 50 MHz brings a ^{13}C nucleus int resonance. At the highest field strength currently available in commercia instruments (21.2 T), 900 MHz energy is required for ^{1}H spectroscopy. Thes energies needed for NMR are much smaller than those required for IR spec troscopy; 200 MHz rf energy corresponds to only 8.0×10^{-5} kJ/mol versus th 4.8 to 48 kJ/mol needed for IR spectroscopy.

^{1}H and ^{13}C nuclei are not unique in their ability to exhibit the NMR phe nomenon. All nuclei with an odd number of protons (^{1}H, ^{2}H, ^{14}N, ^{19}F, ^{31}P, fo example) and all nuclei with an odd number of neutrons (^{13}C, for example show magnetic properties. Only nuclei with even numbers of both protons anc neutrons (^{12}C, ^{16}O) do not give rise to magnetic phenomena (Table 13.1).

Problem 13.1 The amount of energy required to spin-flip a nucleus depends both on the strengtl of the external magnetic field and on the nucleus. At a field strength of 4.7 T, r energy of 200 MHz is required to bring a ^{1}H nucleus into resonance, but energy of only 187 MHz will bring a ^{19}F nucleus into resonance. Calculate the amount o energy required to spin-flip a ^{19}F nucleus. Is this amount greater or less than that required to spin-flip a ^{1}H nucleus?

Problem 13.2 Calculate the amount of energy required to spin-flip a proton in a spectrometer oper- ating at 300 MHz. Does increasing the spectrometer frequency from 200 to 300 MHz increase or decrease the amount of energy necessary for resonance?

13.2 The Nature of NMR Absorptions

From the description thus far, you might expect all ^{1}H nuclei in a molecule to absorb energy at the same frequency and all ^{13}C nuclei to absorb at the same frequency. If so, we would observe only a single NMR absorption band in the ^{1}H or ^{13}C spectrum of a molecule, a situation that would be of little use. In fact, the absorption frequency is not the same for all ^{1}H or all ^{13}C nuclei.

All nuclei in molecules are surrounded by electrons. When an external magnetic field is applied to a molecule, the electrons moving around nuclei set up tiny local magnetic fields of their own. These local magnetic fields act in opposition to the applied field so that the effective field actually felt by the nucleus is a bit weaker than the applied field.

$$\boldsymbol{B}_{\text{effective}} = \boldsymbol{B}_{\text{applied}} - \boldsymbol{B}_{\text{local}}$$

In describing this effect of local fields, we say that nuclei are **shielded** from the full effect of the applied field by the surrounding electrons. Because each specific nucleus in a molecule is in a slightly different electronic environment, each nucleus is shielded to a slightly different extent and the effective magnetic field felt by each is slightly different. These tiny differences in the effective magnetic fields experienced by different nuclei can be detected, and we thus see a distinct NMR signal for each chemically distinct ^{13}C or ^{1}H nucleus in a molecule. As a result, an NMR spectrum effectively maps the carbon–hydrogen framework of an organic molecule. With practice, it's possible to read the map and derive structural information.

Figure 13.3 shows both the ^{1}H and the ^{13}C NMR spectra of methyl acetate, $CH_3CO_2CH_3$. The horizontal axis shows the effective field strength felt by the nuclei, and the vertical axis indicates the intensity of absorption of rf energy. Each peak in the NMR spectrum corresponds to a chemically distinct ^{1}H or ^{13}C nucleus in the molecule. (Note that NMR spectra are formatted with the zero absorption line at the *bottom,* whereas IR spectra are formatted with the zero absorption line at the *top;* Section 12.5.) Note also that ^{1}H and ^{13}C spectra can't be observed simultaneously on the same spectrometer because different amounts of energy are required to spin-flip the different kinds of nuclei. The two spectra must be recorded separately.

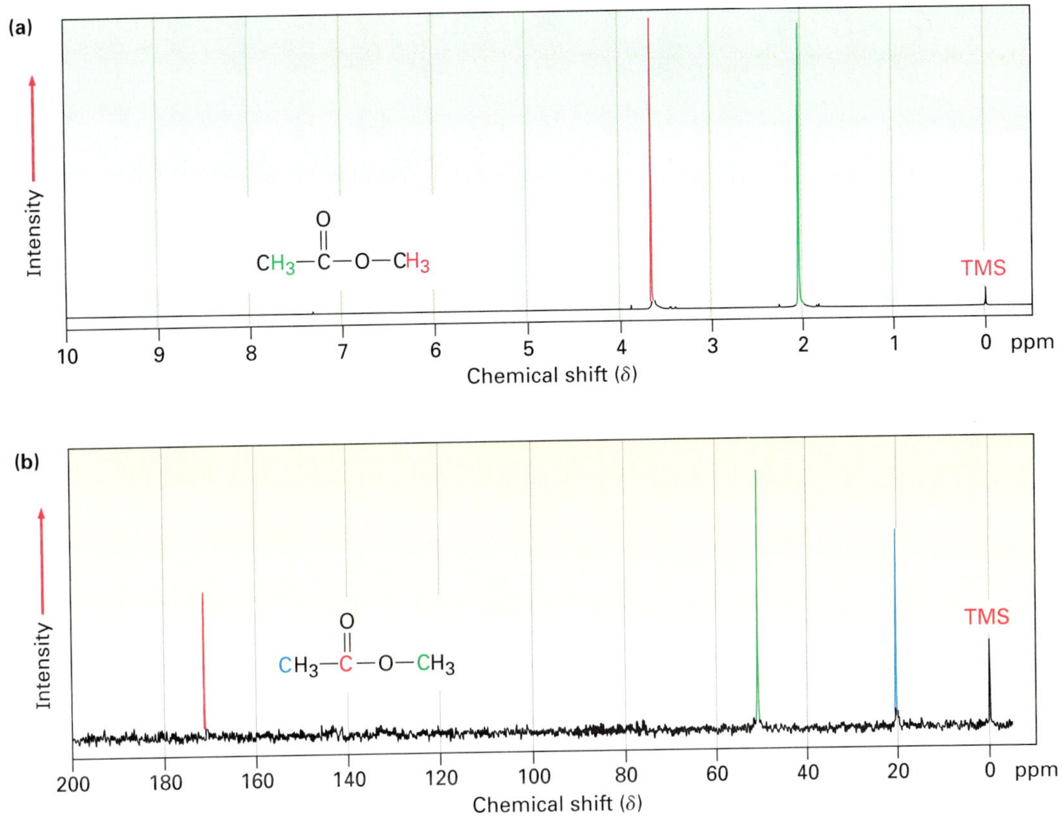

Active Figure 13.3 **(a)** The ^{1}H NMR spectrum and **(b)** the ^{13}C NMR spectrum of methyl acetate, $CH_3CO_2CH_3$. The small peak labeled "TMS" at the far right of each spectrum is a calibration peak, as explained in Section 13.3. *Sign in at* **www.thomsonedu.com** *to see a simulation based on this figure and to take a short quiz.*

The ^{13}C spectrum of methyl acetate in Figure 13.3b shows three peaks, one for each of the three chemically distinct carbon atoms in the molecule. The ^{1}H NMR spectrum in Figure 13.3a shows only two peaks, however, even though methyl acetate has six hydrogens. One peak is due to the $CH_3C{=}O$ hydrogens, and the other to the $-OCH_3$ hydrogens. Because the three hydrogens in each methyl group have the same electronic environment, they are shielded to the same extent and are said to be *equivalent. Chemically equivalent nuclei always show a single absorption.* The two methyl groups themselves, however, are nonequivalent, so the two sets of hydrogens absorb at different positions.

The operation of a basic NMR spectrometer is illustrated in Figure 13.4. A organic sample is dissolved in a suitable solvent (usually deuteriochlorofor $CDCl_3$, which has no hydrogens) and placed in a thin glass tube between the pol of a magnet. The strong magnetic field causes the ^{1}H and ^{13}C nuclei in the mol cule to align in one of the two possible orientations, and the sample is irradiat with rf energy. If the frequency of the rf irradiation is held constant and t strength of the applied magnetic field is varied, each nucleus comes into resonan at a slightly different field strength. A sensitive detector monitors the absorptio of rf energy, and the electronic signal is then amplified and displayed as a peak.

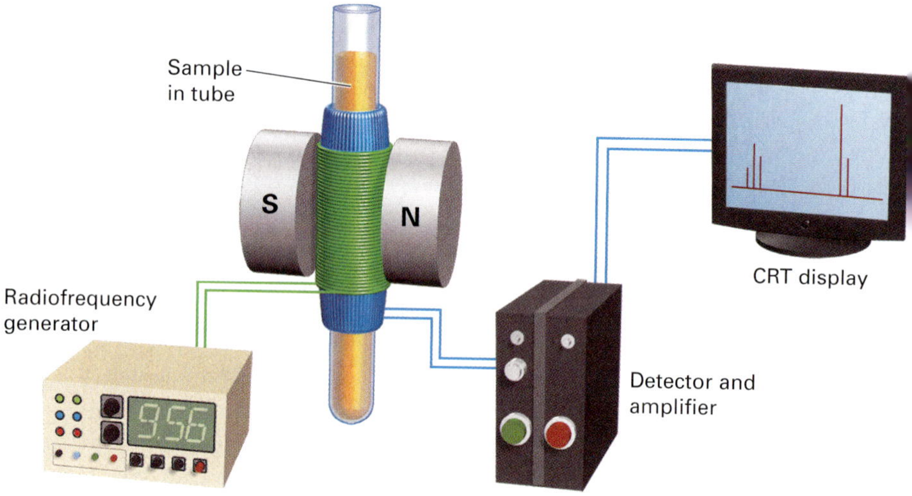

Figure 13.4 Schematic operation of an NMR spectrometer. A thin glass tube containing the sample solution is placed between the poles of a strong magnet and irradiated with rf energy.

NMR spectroscopy differs from IR spectroscopy (Sections 12.6–12.8) in tha the timescales of the two techniques are quite different. The absorption o infrared energy by a molecule giving rise to a change in vibrational amplitude i an essentially instantaneous process (about 10^{-13} s), but the NMR proces is much slower (about 10^{-3} s). This difference in timescales between IR and NMR spectroscopy is analogous to the difference between cameras operating at very fast and very slow shutter speeds. The fast camera (IR) takes an instantaneous picture and "freezes" the action. If two rapidly interconverting species are present, IR spectroscopy records the spectrum of both. The slow camera (NMR), however, takes a blurred, time-averaged picture. If two species interconverting faster than 10^3 times per second are present in a sample, NMR records only a single, averaged spectrum, rather than separate spectra of the two discrete species.

Because of this blurring effect, NMR spectroscopy can be used to measure the rates and activation energies of very fast processes. In cyclohexane, for example, a ring-flip (Section 4.6) occurs so rapidly at room temperature that axial and equatorial hydrogens can't be distinguished by NMR; only a single, averaged ^{1}H NMR absorption is seen for cyclohexane at 25 °C. At −90 °C, however, the ring-flip is slowed down enough that two absorption peaks are seen, one for the six axial hydrogens and one for the six equatorial hydrogens. Knowing the temperature and the rate at which signal blurring begins to occur, it's possible to calculate that the activation energy for the cyclohexane ring-flip is 45 kJ/mol (10.8 kcal/mol).

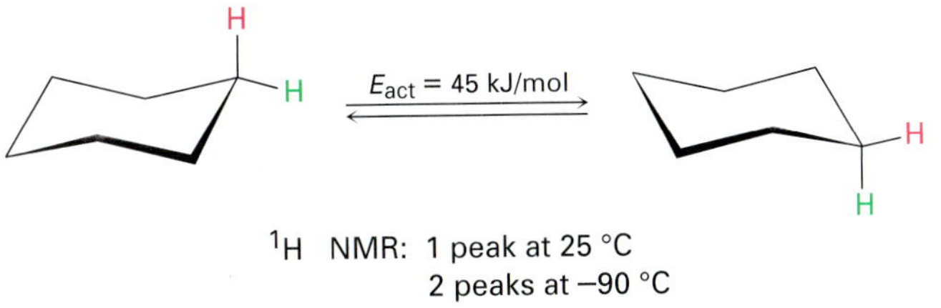

Problem 13.3 2-Chloropropene shows signals for three kinds of protons in its ^{1}H NMR spectrum. Explain.

13.3 Chemical Shifts

NMR spectra are displayed on charts that show the applied field strength increasing from left to right (Figure 13.5). Thus, the left part of the chart is the low-field, or **downfield**, side, and the right part is the high-field, or **upfield**, side. Nuclei that absorb on the downfield side of the chart require a lower field strength for resonance, implying that they have relatively less shielding. Nuclei that absorb on the upfield side require a higher field strength for resonance, implying that they have relatively more shielding.

To define the position of an absorption, the NMR chart is calibrated and a reference point is used. In practice, a small amount of tetramethylsilane [TMS; $(CH_3)_4Si$] is added to the sample so that a reference absorption peak is produced when the spectrum is run. TMS is used as reference for both ^{1}H and ^{13}C measurements because it produces in both a single peak that occurs upfield of other absorptions normally found in organic compounds. The ^{1}H and ^{13}C spectra of methyl acetate in Figure 13.3 have the TMS reference peak indicated.

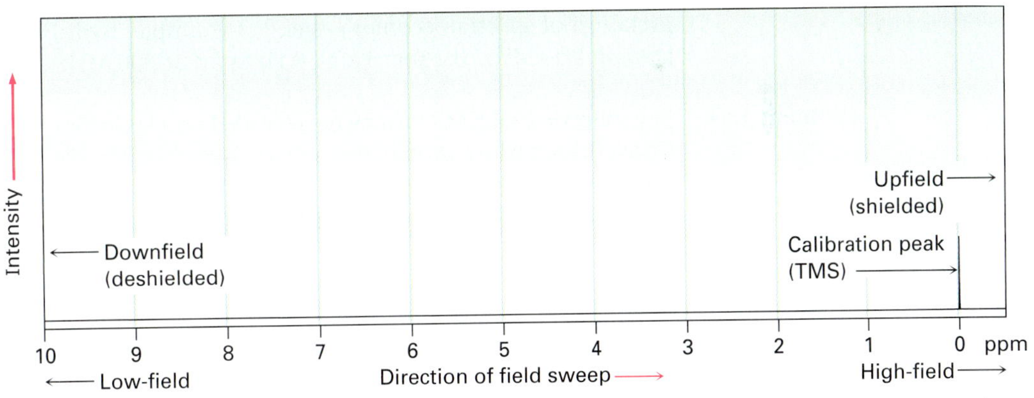

Figure 13.5 The NMR chart. The downfield, deshielded side is on the left, and the upfield, shielded side is on the right. The tetramethylsilane (TMS) absorption is used as reference point.

The position on the chart at which a nucleus absorbs is called its **chemical shift**. The chemical shift of TMS is set as the zero point, and other absorptions normally occur downfield, to the left on the chart. NMR charts are calibrated using an arbitrary scale called the **delta (δ) scale**, where 1 δ equals 1 part per million (1 ppm) of the spectrometer operating frequency. For example, if we were measuring the ^{1}H NMR spectrum of a sample using an instrument operating at

200 MHz, 1 δ would be 1 millionth of 200,000,000 Hz, or 200 Hz. If we were measuring the spectrum using a 500 MHz instrument, 1 δ = 500 Hz. The following equation can be used for any absorption:

$$\delta = \frac{\text{Observed chemical shift (number of Hz away from TMS)}}{\text{Spectrometer frequency in MHz}}$$

Although this method of calibrating NMR charts may seem complex, there is a good reason for it. As we saw earlier, the rf frequency required to bring a given nucleus into resonance depends on the spectrometer's magnetic field strength. But because there are many different kinds of spectrometers with many different magnetic field strengths available, chemical shifts given in frequency units (Hz) vary from one instrument to another. Thus, a resonance that occurs at 120 Hz downfield from TMS on one spectrometer might occur at 600 Hz downfield from TMS on another spectrometer with a more powerful magnet.

By using a system of measurement in which NMR absorptions are expressed in relative terms (parts per million relative to spectrometer frequency) rather than absolute terms (Hz), it's possible to compare spectra obtained on different instruments. *The chemical shift of an NMR absorption in δ units is constant, regardless of the operating frequency of the spectrometer.* A ^{1}H nucleus that absorbs at 2.0 δ on a 200 MHz instrument also absorbs at 2.0 δ on a 500 MHz instrument.

The range in which most NMR absorptions occur is quite narrow. Almost all ^{1}H NMR absorptions occur 0 to 10 δ downfield from the proton absorption of TMS, and almost all ^{13}C absorptions occur 1 to 220 δ downfield from the carbon absorption of TMS. Thus, there is a considerable likelihood that accidental overlap of nonequivalent signals will occur. The advantage of using an instrument with higher field strength (say, 500 MHz) rather than lower field strength (200 MHz) is that different NMR absorptions are more widely separated at the higher field strength. The chances that two signals will accidentally overlap are therefore lessened, and interpretation of spectra becomes easier. For example, two signals that are only 20 Hz apart at 200 MHz (0.1 ppm) are 50 Hz apart at 500 MHz (still 0.1 ppm).

Problem 13.4 The following ^{1}H NMR peaks were recorded on a spectrometer operating at 200 MHz. Convert each into δ units.

(a) $CHCl_3$; 1454 Hz **(b)** CH_3Cl; 610 Hz
(c) CH_3OH; 693 Hz **(d)** CH_2Cl_2; 1060 Hz

Problem 13.5 When the ^{1}H NMR spectrum of acetone, CH_3COCH_3, is recorded on an instrument operating at 200 MHz, a single sharp resonance at 2.1 δ is seen.

(a) How many Hz downfield from TMS does the acetone resonance correspond to?
(b) If the ^{1}H NMR spectrum of acetone were recorded at 500 MHz, what would the position of the absorption be in δ units?
(c) How many Hz downfield from TMS does this 500 MHz resonance correspond to?

13.4 ^{13}C NMR Spectroscopy: Signal Averaging and FT–NMR

Everything we've said thus far about NMR spectroscopy applies to both ^{1}H and ^{13}C spectra. Now, though, let's focus only on ^{13}C spectroscopy because it's much easier to interpret. What we learn now about interpreting ^{13}C spectra will simplify the subsequent discussion of ^{1}H spectra.

particularly distinct in ^{13}C NMR and are always found at the low-field end of the spectrum, from 160 to 220 δ. Figure 13.8 shows the ^{13}C NMR spectra of 2-butanone and *para*-bromoacetophenone and indicates the peak assignments. Note that the C=O carbons are at the left edge of the spectrum in each case.

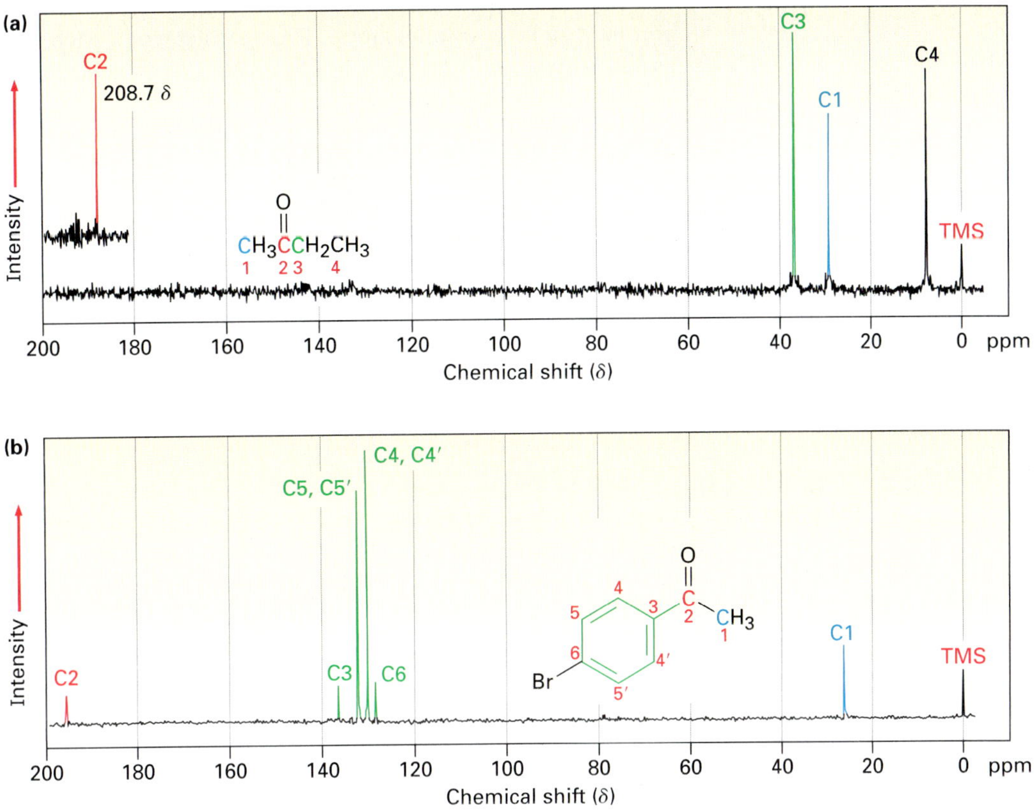

Figure 13.8 Carbon-13 NMR spectra of **(a)** 2-butanone and **(b)** *para*-bromoacetophenone.

The ^{13}C NMR spectrum of *para*-bromoacetophenone is interesting in several ways. Note particularly that only six carbon absorptions are observed, even though the molecule contains eight carbons. *para*-Bromoacetophenone has a symmetry plane that makes ring carbons 4 and 4′, and ring carbons 5 and 5′ equivalent. (Remember from Section 2.4 that aromatic rings have two resonance forms.) Thus, the six ring carbons show only four absorptions in the 128 to 137 δ range.

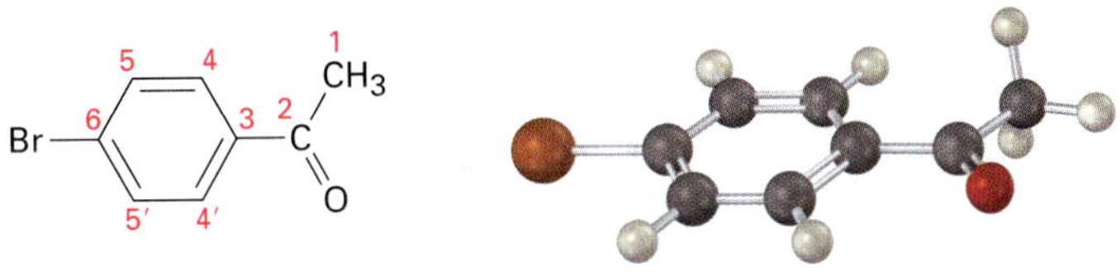

para-Bromoacetophenone

A second interesting point about both spectra in Figure 13.8 is that the peaks aren't uniform in size. Some peaks are larger than others even though they are one-carbon resonances (except for the two 2-carbon peaks of *para*-bromoacetophenone). This difference in peak size is a general feature of ^{13}C NMR spectra.

WORKED EXAMPLE 13.1 *Predicting Chemical Shifts in ^{13}C NMR Spectra*

At what approximate positions would you expect ethyl acrylat $H_2C{=}CHCO_2CH_2CH_3$, to show ^{13}C NMR absorptions?

Strategy Identify the distinct carbons in the molecule, and note whether each is alkyl, vinyli aromatic, or in a carbonyl group. Then predict where each absorbs, using Figure 13 as necessary.

Solution Ethyl acrylate has five distinct carbons: two different C=C, one C=O, one O–C, ar one alkyl C. From Figure 13.7, the likely absorptions are

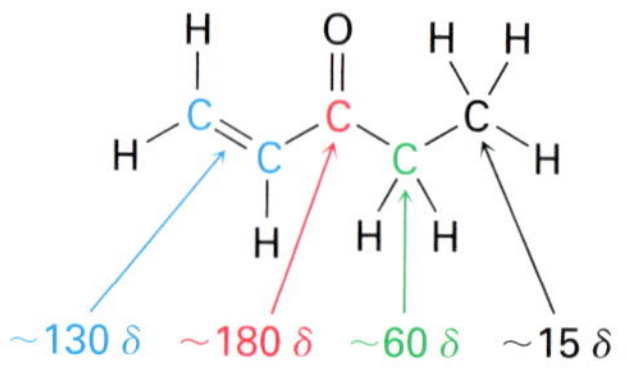

The actual absorptions are at 14.1, 60.5, 128.5, 130.3, and 166.0 δ.

Problem 13.6 Predict the number of carbon resonance lines you would expect in the ^{13}C NM spectra of the following compounds:

(a) Methylcyclopentane **(b)** 1-Methylcyclohexene

(c) 1,2-Dimethylbenzene **(d)** 2-Methyl-2-butene

(e)

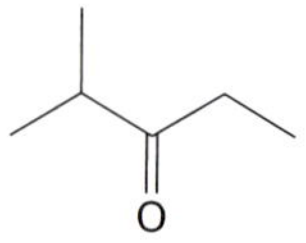

(f)

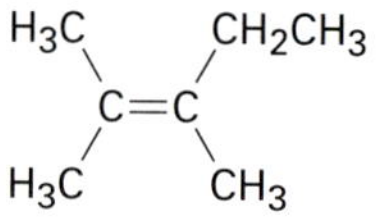

Problem 13.7 Propose structures for compounds that fit the following descriptions:

(a) A hydrocarbon with seven lines in its ^{13}C NMR spectrum

(b) A six-carbon compound with only five lines in its ^{13}C NMR spectrum

(c) A four-carbon compound with three lines in its ^{13}C NMR spectrum

Problem 13.8 Assign the resonances in the ^{13}C NMR spectrum of methyl propanoate $CH_3CH_2CO_2CH_3$ (Figure 13.9).

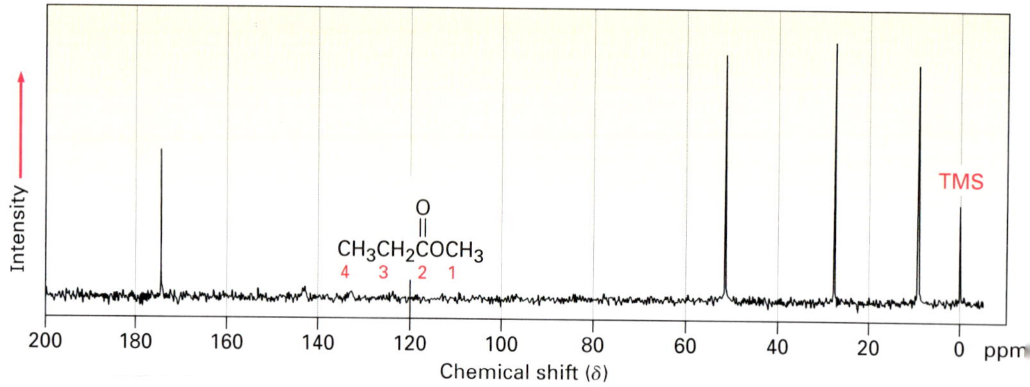

Figure 13.9 ^{13}C NMR spectrum of methyl propanoate, Problem 13.8.

13.6 DEPT ^{13}C NMR Spectroscopy

Techniques developed in recent years make it possible to obtain large amounts of information from ^{13}C NMR spectra. For example, *DEPT–NMR,* for *distortionless enhancement by polarization transfer,* allows us to determine the number of hydrogens attached to each carbon in a molecule.

A DEPT experiment is usually done in three stages, as shown in Figure 13.10 for 6-methyl-5-hepten-2-ol. The first stage is to run an ordinary spectrum (called

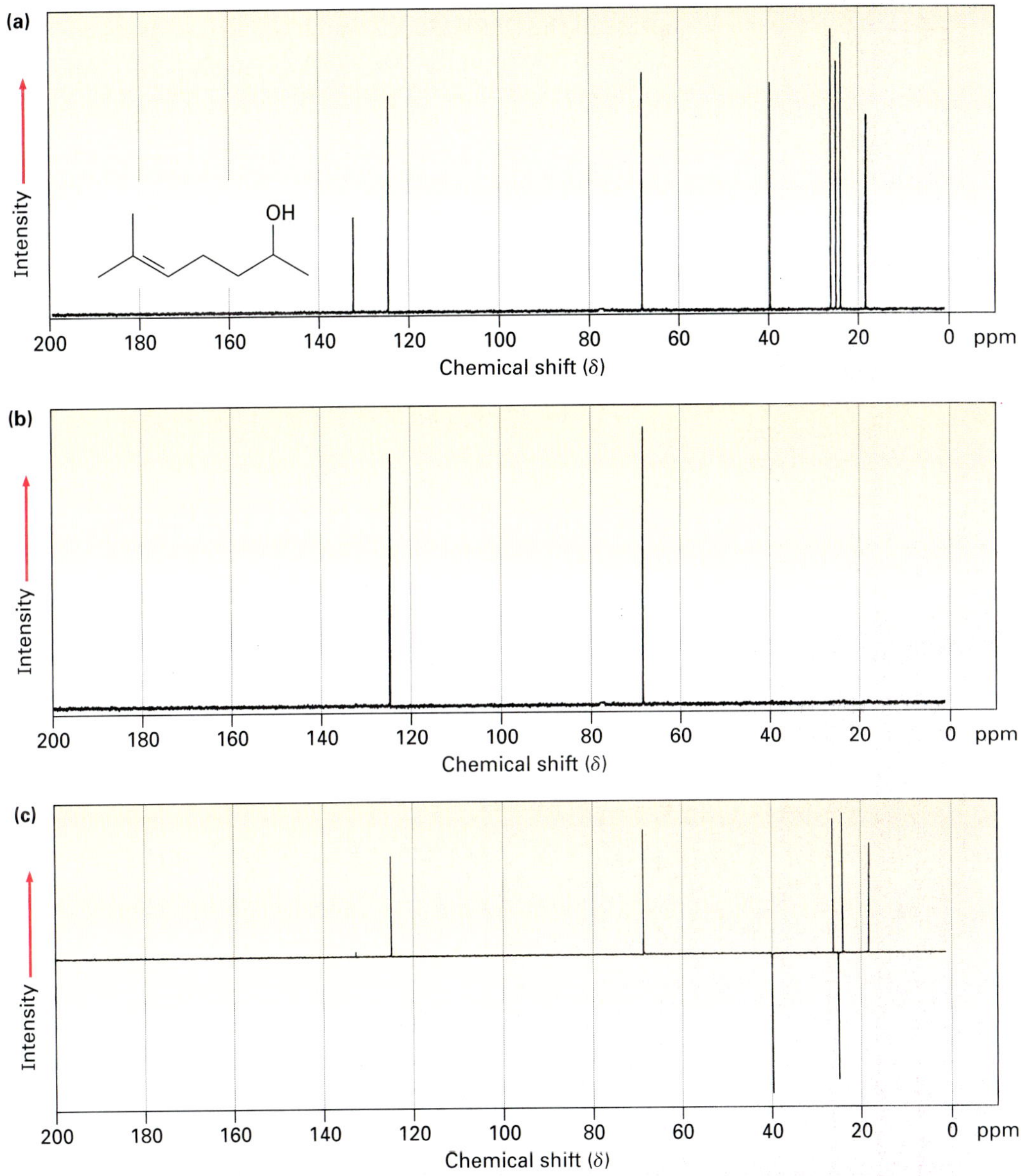

Figure 13.10 DEPT–NMR spectra for 6-methyl-5-hepten-2-ol. Part **(a)** is an ordinary broadband-decoupled spectrum, which shows signals for all eight carbons. Part **(b)** is a DEPT-90 spectrum, which shows only signals for the two CH carbons. Part **(c)** is a DEPT-135 spectrum, which shows positive signals for the two CH and three CH_3 carbons and negative signals for the two CH_2 carbons.

a *broadband-decoupled spectrum*) to locate the chemical shifts of all carbons. Next, a second spectrum called a DEPT-90 is run, using special conditions under which only signals due to CH carbons appear. Signals due to CH_3, CH_2, and quaternary carbons are absent. Finally, a third spectrum called a DEPT-135 is run, using conditions under which CH_3 and CH resonances appear as positive signals, CH_2 resonances appear as *negative* signals—that is, as peaks below the baseline—and quaternary carbons are again absent.

Putting together the information from all three spectra makes it possible to tell the number of hydrogens attached to each carbon. The CH carbons are identified in the DEPT-90 spectrum, the CH_2 carbons are identified as the negative peaks in the DEPT-135 spectrum, the CH_3 carbons are identified by subtracting the CH peaks from the positive peaks in the DEPT-135 spectrum, and quaternary carbons are identified by subtracting all peaks in the DEPT-135 spectrum from the peaks in the broadband-decoupled spectrum.

Broadband-decoupled	**DEPT-90**	**DEPT-135**
C, CH, CH_2, CH_3	CH	CH_3, CH are positive CH_2 is negative

C	Subtract DEPT-135 from broadband-decoupled spectrum
CH	DEPT-90
CH_2	Negative DEPT-135
CH_3	Subtract DEPT-90 from positive DEPT-135

WORKED EXAMPLE 13.2 *Assigning a Chemical Structure from a ^{13}C NMR Spectrum*

Propose a structure for an alcohol, $C_4H_{10}O$, that has the following ^{13}C NMR spectral data:

Broadband-decoupled ^{13}C NMR: 19.0, 31.7, 69.5 δ
DEPT-90: 31.7 δ
DEPT-135: positive peak at 19.0 δ, negative peak at 69.5 δ

Strategy As noted in Section 6.2, it usually helps with compounds of known formula but unknown structure to calculate the compound's degree of unsaturation. In the present instance, a formula of $C_4H_{10}O$ corresponds to a saturated, open-chain molecule.

To gain information from the ^{13}C data, let's begin by noting that the unknown alcohol has *four* carbon atoms, yet has only *three* NMR absorptions, which implies that two of the carbons must be equivalent. Looking at chemical shifts, two of the absorptions are in the typical alkane region (19.0 and 31.7 δ), while one is in the region of a carbon bonded to an electronegative atom (69.5 δ)—oxygen in this instance. The DEPT-90 spectrum tells us that the alkyl carbon at 31.7 δ is tertiary (CH); the DEPT-135 spectrum tells us that the alkyl carbon at 19.0 δ is a methyl (CH_3) and that the carbon bonded to oxygen (69.5 δ) is secondary (CH_2). The two equivalent carbons are probably both methyls bonded to the same tertiary carbon, $(CH_3)_2CH-$. We can now put the pieces together to propose a structure: 2-methyl-1-propanol.

Solution

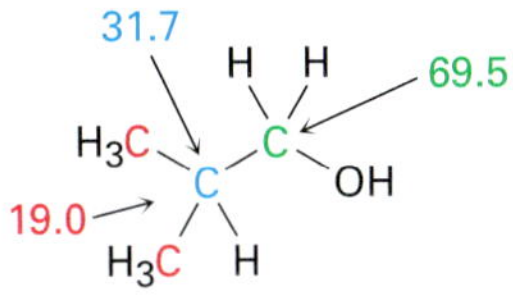

2-Methyl-1-propanol

Problem 13.9 Assign a chemical shift to each carbon in 6-methyl-5-hepten-2-ol (Figure 13.10).

Problem 13.10 Estimate the chemical shift of each carbon in the following molecule. Predict which carbons will appear in the DEPT-90 spectrum, which will give positive peaks in the DEPT-135 spectrum, and which will give negative peaks in the DEPT-135 spectrum.

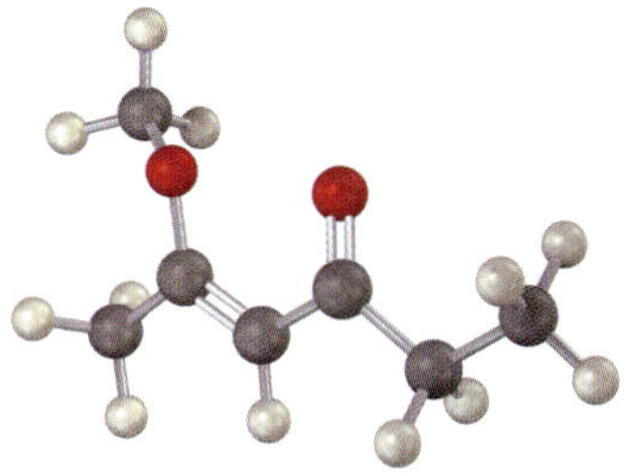

Problem 13.11 Propose a structure for an aromatic hydrocarbon, $C_{11}H_{16}$, that has the following ^{13}C NMR spectral data:

Broadband-decoupled ^{13}C NMR: 29.5, 31.8, 50.2, 125.5, 127.5, 130.3, 139.8 δ
DEPT-90: 125.5, 127.5, 130.3 δ
DEPT-135: positive peaks at 29.5, 125.5, 127.5, 130.3 δ; negative peak at 50.2 δ

13.7 Uses of ^{13}C NMR Spectroscopy

The information derived from ^{13}C NMR spectroscopy is extraordinarily useful for structure determination. Not only can we count the number of nonequivalent carbon atoms in a molecule, we can also get information about the electronic environment of each carbon and can even find how many protons each is attached to. As a result, we can answer many structural questions that go unanswered by IR spectroscopy or mass spectrometry.

Here's an example: how might we prove that E2 elimination of an alkyl halide gives the more highly substituted alkene (Zaitsev's rule, Section 11.7)? Does reaction of 1-chloro-1-methylcyclohexane with strong base lead predominantly to 1-methylcyclohexene or to methylenecyclohexane?

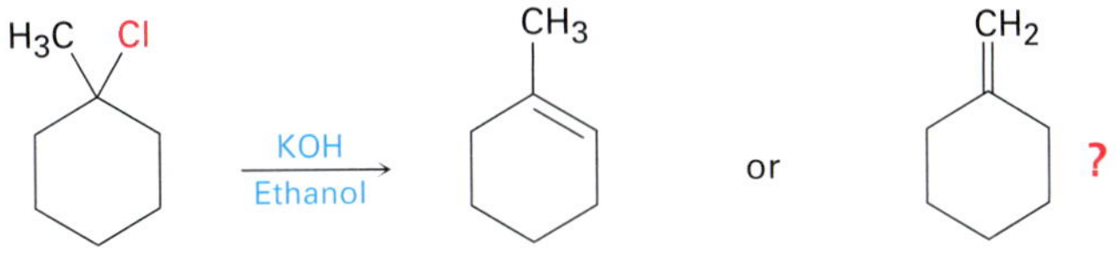

1-Chloro-1-methylcyclohexane **1-Methylcyclohexene** **Methylenecyclohexane**

1-Methylcyclohexene will have five sp^3-carbon resonances in the 20 to 50 δ range and two sp^2-carbon resonances in the 100 to 150 δ range. Methylenecyclohexane, however, because of its symmetry, will have only three sp^3-carbon resonance peaks and two sp^2-carbon peaks. The spectrum of the actual reactic product, shown in Figure 13.11, clearly identifies 1-methylcyclohexene the product of this E2 reaction.

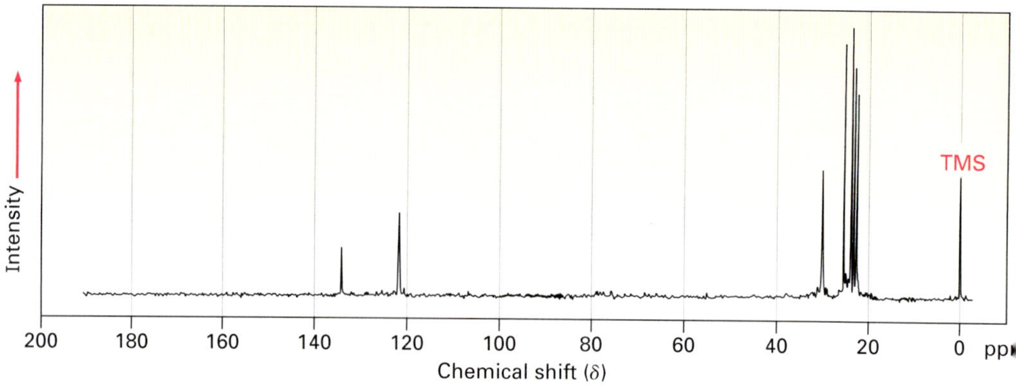

Figure 13.11 The ^{13}C NMR spectrum of 1-methylcyclohexene, the E2 reaction product from treatment of 1-chloro-1-methylcyclohexane with base.

Problem 13.12 We saw in Section 8.3 that addition of HBr to a terminal alkyne leads to th Markovnikov addition product, with the Br bonding to the more highly substitute carbon. How could you use ^{13}C NMR to identify the product of the addition c 1 equivalent of HBr to 1-hexyne?

13.8 ^1H NMR Spectroscopy and Proton Equivalence

Having looked at ^{13}C spectra, let's now focus on ^1H NMR spectroscopy. Becaus each electronically distinct hydrogen in a molecule has its own unique absorp tion, one use of ^1H NMR is to find out how many kinds of electronically non equivalent hydrogens are present. In the ^1H NMR spectrum of methyl acetat shown previously in Figure 13.3a, for instance, there are two signals, corre sponding to the two kinds of nonequivalent protons present, $CH_3C{=}O$ proton and $-OCH_3$ protons.

For relatively small molecules, a quick look at a structure is often enough to decide how many kinds of protons are present and thus how many NMR absorptions might appear. If in doubt, though, the equivalence or nonequiva lence of two protons can be determined by comparing the structures tha would be formed if each hydrogen were replaced by an X group. There are fou possibilities.

- One possibility is that the protons are chemically unrelated and thus non equivalent. If so, the products formed on replacement of H by X would be dif ferent constitutional isomers. In butane, for instance, the $-CH_3$ protons are different from the $-CH_2-$ protons, would give different products on replace ment by X, and would likely show different NMR absorptions.

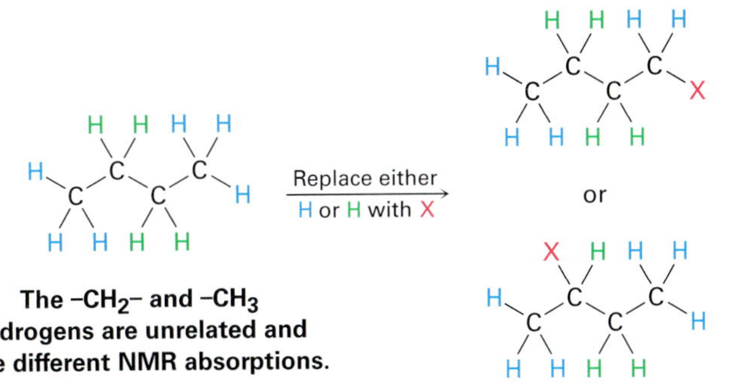

The –CH$_2$– and –CH$_3$ hydrogens are unrelated and have different NMR absorptions.

The two replacement products are constitutional isomers.

- A second possibility is that the protons are chemically identical and thus electronically equivalent. If so, the same product would be formed regardless of which H is replaced by X. In butane, for instance, the six $-CH_3$ hydrogens on C1 and C4 are identical, would give the identical structure on replacement by X, and would show the identical NMR absorption. Such protons are said to be **homotopic**.

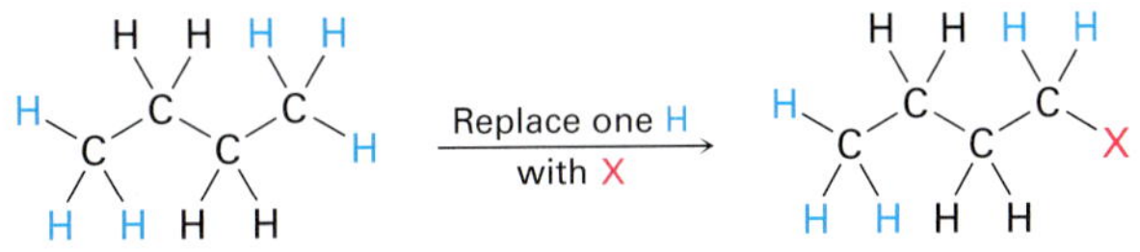

The 6 –CH$_3$ hydrogens are *homotopic* and have the same NMR absorptions.

Only one replacement product is possible.

- The third possibility is a bit subtler. Although they might at first seem homotopic, the two $-CH_2-$ hydrogens on C2 in butane (and the two $-CH_2-$ hydrogens on C3) are in fact *not* identical. Replacement of a hydrogen at C2 (or C3) would form a new chirality center, so different enantiomers (Section 9.1) would result depending on whether the *pro-R* or *pro-S* hydrogen were replaced (Section 9.13). Such hydrogens, whose replacement by X would lead to different enantiomers, are said to be **enantiotopic**. Enantiotopic hydrogens, even though not identical, are nevertheless electronically equivalent and thus have the same NMR absorption.

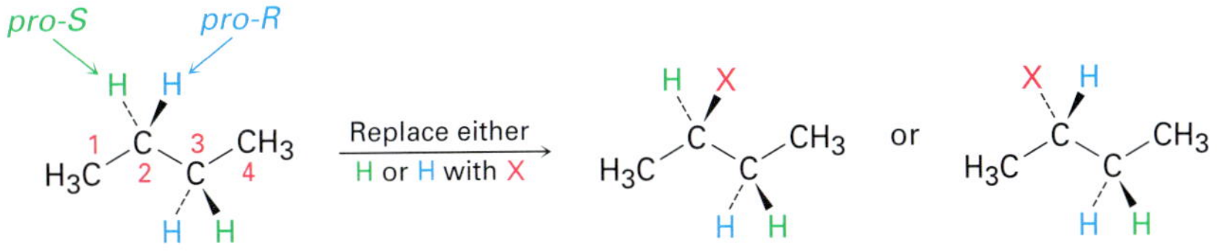

The two hydrogens on C2 (and the two hydrogens on C3) are *enantiotopic* and have the same NMR absorption.

The two possible replacement products are enantiomers.

- The fourth possibility arises in chiral molecules, such as (*R*)-2-butanol. The two $-CH_2-$ hydrogens at C3 are neither homotopic nor enantiotopic. Since replacement of a hydrogen at C3 would form a *second* chirality center, different *diastereomers* (Section 9.6) would result depending on whether the *pro-R* or *pro-S* hydrogen were replaced. Such hydrogens, whose replacement by X leads to different diastereomers, are said to be **diastereotopic**. Diastereotopic hydrogens are neither chemically nor electronically equivalent. They are completely different and would likely show different NMR absorptions.

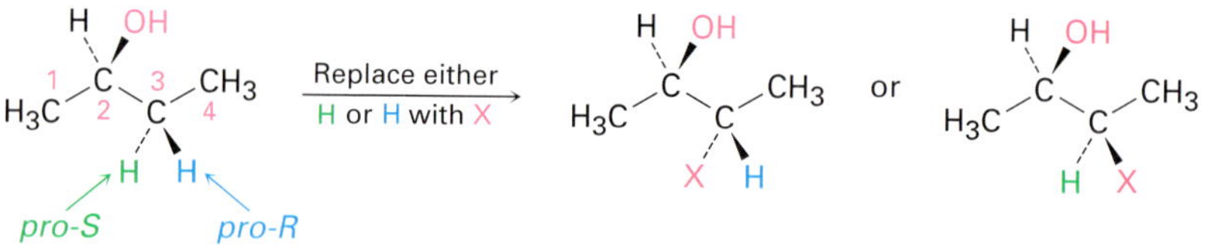

The two hydrogens on C3 are *diastereotopic* and have different NMR absorptions.

The two possible replacement products are diastereomers.

Problem 13.13 Identify the indicated sets of protons as unrelated, homotopic, enantiotopic, or diastereotopic:

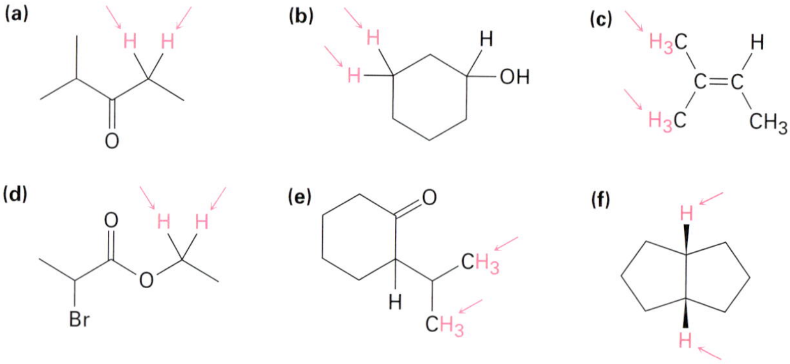

Problem 13.14 How many kinds of electronically nonequivalent protons are present in each of the following compounds, and thus how many NMR absorptions might you expect in each?

(a) CH_3CH_2Br **(b)** $CH_3OCH_2CH(CH_3)_2$ **(c)** $CH_3CH_2CH_2NO_2$
(d) Methylbenzene **(e)** 2-Methyl-1-butene **(f)** *cis*-3-Hexene

Problem 13.15 How many absorptions would you expect (*S*)-malate, an intermediate in carbohydrate metabolism, to have in its ^1H NMR spectrum? Explain.

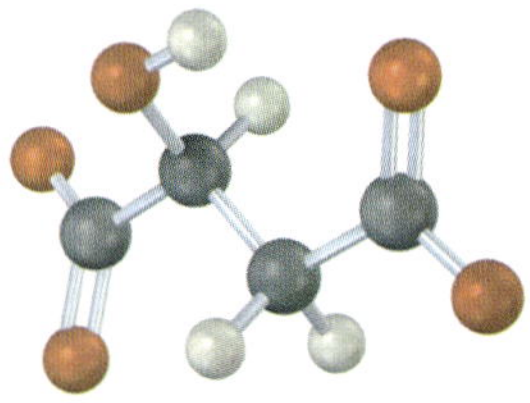

(*S*)-Malate

13.9 Chemical Shifts in ^{1}H NMR Spectroscopy

We said previously that differences in chemical shifts are caused by the small local magnetic fields of electrons surrounding the different nuclei. Nuclei that are more strongly shielded by electrons require a higher applied field to bring them into resonance and therefore absorb on the right side of the NMR chart. Nuclei that are less strongly shielded need a lower applied field for resonance and therefore absorb on the left of the NMR chart.

Most ^{1}H chemical shifts fall within the range of 0 to 10 δ, which can be divided into the five regions shown in Table 13.2. By remembering the positions of these regions, it's often possible to tell at a glance what kinds of protons a molecule contains.

Table 13.2 **Regions of the ^{1}H NMR Spectrum**

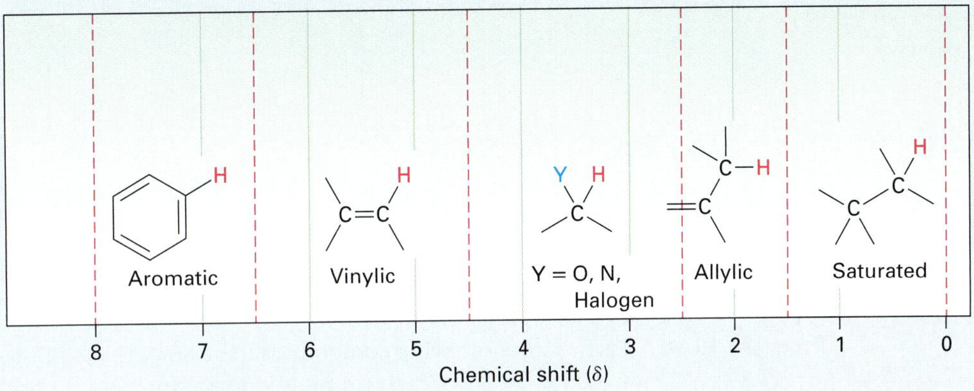

Table 13.3 shows the correlation of ^{1}H chemical shift with electronic environment in more detail. In general, protons bonded to saturated, *sp*^{3}-hybridized carbons absorb at higher fields, whereas protons bonded to *sp*^{2}-hybridized carbons absorb at lower fields. Protons on carbons that are bonded to electronegative atoms, such as N, O, or halogen, also absorb at lower fields.

WORKED EXAMPLE 13.3 ***Predicting Chemical Shifts in ^{1}H NMR Spectra***

Methyl 2,2-dimethylpropanoate $(CH_3)_3CCO_2CH_3$ has two peaks in its ^{1}H NMR spectrum. What are their approximate chemical shifts?

Strategy Identify the types of hydrogens in the molecule, and note whether each is alkyl, vinylic, or next to an electronegative atom. Then predict where each absorbs, using Table 13.3 if necessary.

Solution The $-OCH_3$ protons absorb around 3.5 to 4.0 δ because they are on carbon bonded to oxygen. The $(CH_3)_3C-$ protons absorb near 1.0 δ because they are typical alkane-like protons.

Table 13.3 Correlation of ^{1}H Chemical Shift with Environment

Type of hydrogen		Chemical shift (δ)	Type of hydrogen		Chemical shift (δ)
Reference	$Si(CH_3)_4$	0			
Alkyl (primary)	$—CH_3$	0.7–1.3	Alcohol	—C—O—H	2.5–5.0
Alkyl (secondary)	$—CH_2—$	1.2–1.6			
Alkyl (tertiary)	—CH—	1.4–1.8	Alcohol, ether	H—C—O—	3.3–4.5
Allylic	C=C—C(H)—	1.6–2.2	Vinylic	C=C—H	4.5–6.5
			Aryl	Ar—H	
Methyl ketone	$—C(=O)—CH_3$	2.0–2.4			
Aromatic methyl	$Ar—CH_3$	2.4–2.7	Aldehyde	—C(=O)—H	9.7–10.0
Alkynyl	—C≡C—H	2.5–3.0			
Alkyl halide	H—C—Hal	2.5–4.0	Carboxylic acid	—C(=O)—O—H	11.0–12.0

Problem 13.16 Each of the following compounds has a single ^{1}H NMR peak. Approximately where would you expect each compound to absorb?

(a)

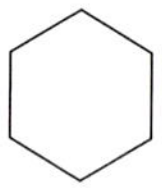

(b)

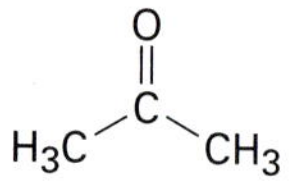

(c)

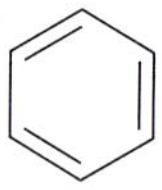

(d) CH_2Cl_2

(e)

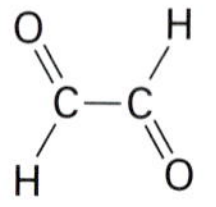

(f)

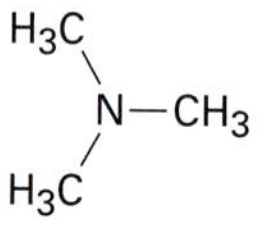

Problem 13.17 Identify the different kinds of nonequivalent protons in the following molecule, and tell where you would expect each to absorb:

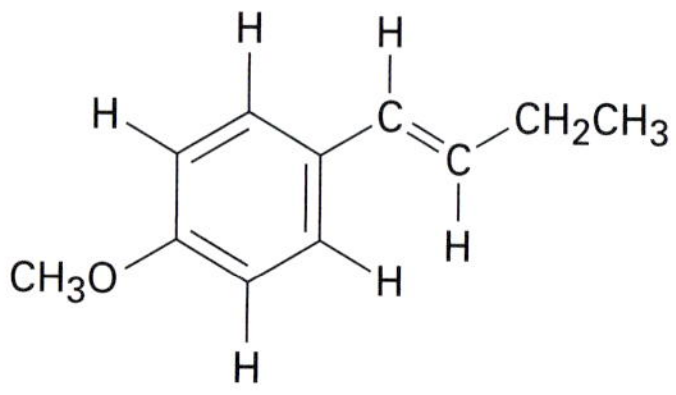

13.10 Integration of ^{1}H NMR Absorptions: Proton Counting

Look at the ^{1}H NMR spectrum of methyl 2,2-dimethylpropanoate in Figure 13.12. There are two peaks, corresponding to the two kinds of protons, but the peaks aren't the same size. The peak at 1.2 δ, due to the $(CH_3)_3C-$ protons, is larger than the peak at 3.7 δ, due to the $-OCH_3$ protons.

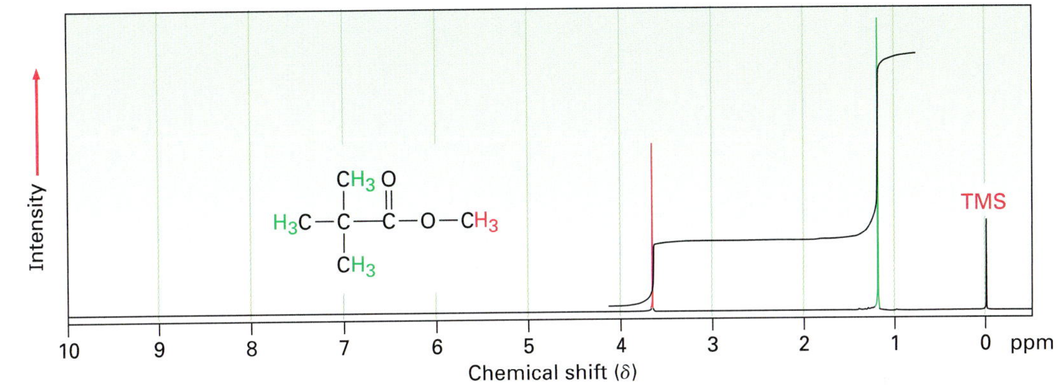

Figure 13.12 The ^{1}H NMR spectrum of methyl 2,2-dimethylpropanoate. Integrating the peaks in a "stair-step" manner shows that they have a 1:3 ratio, corresponding to the ratio of the numbers of protons (3:9) responsible for each peak.

The area under each peak is proportional to the number of protons causing that peak. By electronically measuring, or **integrating**, the area under each peak, it's possible to measure the relative numbers of the different kinds of protons in a molecule. If desired, the integrated peak area can be superimposed over the spectrum as a "stair-step" line, with the height of each step proportional to the area under the peak, and therefore proportional to the relative number of protons causing the peak. To compare the size of one peak against another, simply take a ruler and measure the heights of the various steps. For example, the two steps for the peaks in methyl 2,2-dimethylpropanoate are found to have a 1:3 (or 3:9) height ratio when integrated—exactly what we expect since the three $-OCH_3$ protons are equivalent and the nine $(CH_3)_3C-$ protons are equivalent.

Problem 13.18 How many peaks would you expect in the ^{1}H NMR spectrum of 1,4-dimethylbenzene (*para*-xylene, or *p*-xylene)? What ratio of peak areas would you expect on integration of the spectrum? Refer to Table 13.3 for approximate chemical shifts, and sketch what the spectrum would look like. (Remember from Section 2.4 that aromatic rings have two resonance forms.)

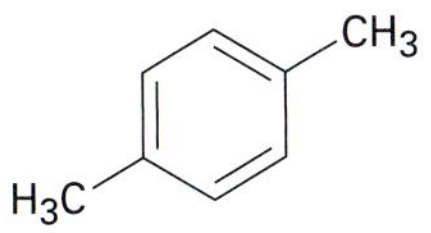

p-Xylene

13.11 Spin–Spin Splitting in ^1H NMR Spectra

In the ^1H NMR spectra we've seen thus far, each different kind of proton i molecule has given rise to a single peak. It often happens, though, that t absorption of a proton splits into multiple peaks, called a **multiplet**. For exa ple, in the ^1H NMR spectrum of bromoethane shown in Figure 13.13, t $-CH_2Br$ protons appear as four peaks (a *quartet*) centered at 3.42 δ a the $-CH_3$ protons appear as three peaks (a *triplet*) centered at 1.68 δ.

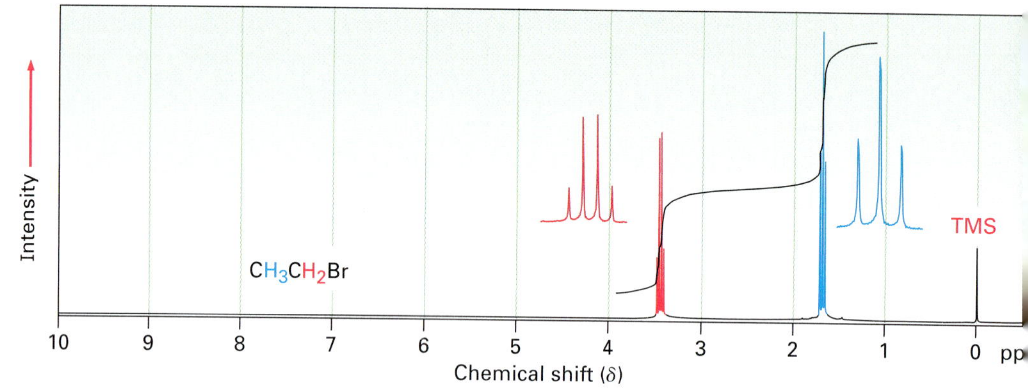

Figure 13.13 The ^1H NMR spectrum of bromoethane, CH_3CH_2Br. The $-CH_2Br$ protons appear as a quartet at 3.42 δ, and the $-CH_3$ protons appear as a triplet at 1.68 δ.

Called **spin–spin splitting**, multiple absorptions of a nucleus are caused b the interaction, or **coupling**, of the spins of nearby nuclei. In other words, th tiny magnetic field produced by one nucleus affects the magnetic field felt b neighboring nuclei. Look at the $-CH_3$ protons in bromoethane, for exampl The three equivalent $-CH_3$ protons are neighbored by two other magnet nuclei—the two protons on the adjacent $-CH_2Br$ group. Each of the neighbo ing $-CH_2Br$ protons has its own nuclear spin, which can align either with against the applied field, producing a tiny effect that is felt by the $-CH_3$ proton

There are three ways in which the spins of the two $-CH_2Br$ protons ca align, as shown in Figure 13.14. If both proton spins align with the applied fiel the total effective field felt by the neighboring $-CH_3$ protons is slightly larg than it would otherwise be. Consequently, the applied field necessary to caus resonance is slightly reduced. Alternatively, if one of the $-CH_2Br$ proton spir aligns with the field and one aligns against the field, there is no effect on th neighboring $-CH_3$ protons. (There are two ways this arrangement can occu depending on which of the two proton spins aligns which way.) Finally, if bot $-CH_2Br$ proton spins align against the applied field, the effective field felt by th $-CH_3$ protons is slightly smaller than it would otherwise be and the applied fiel needed for resonance is slightly increased.

Any given molecule has only one of the three possible alignments of $-CH_2B$ spins, but in a large collection of molecules, all three spin states are represented i a 1:2:1 statistical ratio. We therefore find that the neighboring $-CH_3$ proton come into resonance at three slightly different values of the applied field, and w see a 1:2:1 triplet in the NMR spectrum. One resonance is a little above where i would be without coupling, one is at the same place it would be without coupling, and the third resonance is a little below where it would be without coupling.

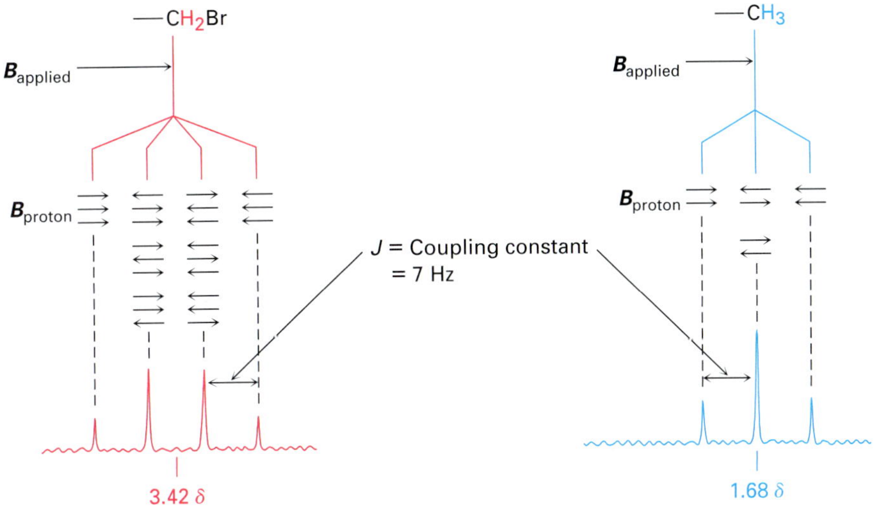

igure 13.14 The origin of oin–spin splitting in bromo-:hane. The nuclear spins of eighboring protons, indicated y horizontal arrows, align either ith or against the applied field, ausing the splitting of absorp-ons into multiplets.

In the same way that the $-CH_3$ absorption of bromoethane is split into a triplet, the $-CH_2Br$ absorption is split into a quartet. The three spins of the neighboring $-CH_3$ protons can align in four possible combinations: all three with the applied field, two with and one against (three ways), one with and two against (three ways), or all three against. Thus, four peaks are produced for the $-CH_2Br$ protons in a 1:3:3:1 ratio.

As a general rule, called the ***n* + 1 rule**, protons that have n equivalent neighboring protons show $n + 1$ peaks in their NMR spectrum. For example, the spectrum of 2-bromopropane in Figure 13.15 shows a doublet at 1.71 δ and a seven-line multiplet, or *septet,* at 4.28 δ. The septet is caused by splitting of the $-CHBr-$ proton signal by six equivalent neighboring protons on the two methyl groups ($n = 6$ leads to $6 + 1 = 7$ peaks). The doublet is due to signal splitting of the six equivalent methyl protons by the single $-CHBr-$ proton ($n = 1$ leads to 2 peaks). Integration confirms the expected 6:1 ratio.

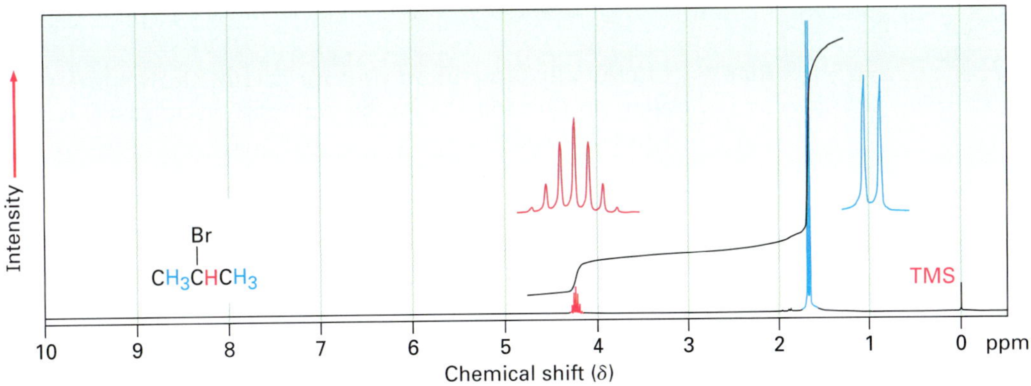

Figure 13.15 The ¹H NMR spectrum of 2-bromopropane. The $-CH_3$ proton signal at 1.71 δ is split into a doublet, and the $-CHBr-$ proton signal at 4.28 δ is split into a septet. Note that the distance between peaks—the *coupling constant*—is the same in both multiplets. Note also that the outer two peaks of the septet are so small as to be nearly lost.

The distance between peaks in a multiplet is called the **coupling constan** denoted *J*. Coupling constants are measured in hertz and generally fall in th range 0 to 18 Hz. The exact value of the coupling constant between two neig boring protons depends on the geometry of the molecule, but a typical value fo an open-chain alkane is $J = 6$ to 8 Hz. The same coupling constant is shared b both groups of hydrogens whose spins are coupled and is independent o spectrometer field strength. In bromoethane, for instance, the $-CH_2Br$ protor are coupled to the $-CH_3$ protons and appear as a quartet with $J = 7$ Hz. Th $-CH_3$ protons appear as a triplet with the same $J = 7$ Hz coupling constant.

Because coupling is a reciprocal interaction between two adjacent groups o protons, it's sometimes possible to tell which multiplets in a complex NMR spe trum are related to each other. If two multiplets have the same coupling cor stant, they are probably related, and the protons causing those multiplets ar therefore adjacent in the molecule.

The most commonly observed coupling patterns and the relative intensiti of lines in their multiplets are listed in Table 13.4. Note that it's not possible fo a given proton to have *five* equivalent neighboring protons. (Why not?) A six line multiplet, or sextet, is therefore found only when a proton has five *non equivalent* neighboring protons that coincidentally happen to be coupled wit an identical coupling constant *J*.

Table 13.4 Some Common Spin Multiplicities

Number of equivalent adjacent protons	Multiplet	Ratio of intensities
0	Singlet	1
1	Doublet	1:1
2	Triplet	1:2:1
3	Quartet	1:3:3:1
4	Quintet	1:4:6:4:1
6	Septet	1:6:15:20:15:6:1

Spin–spin splitting in ^{1}H NMR can be summarized in three rules.

Rule 1 **Chemically equivalent protons do not show spin–spin splitting.** The equiv- alent protons may be on the same carbon or on different carbons, but their sig- nals don't split.

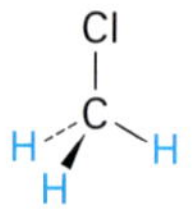

Three C–H protons are chemically equivalent; no splitting occurs.

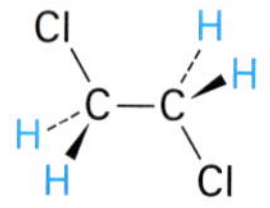

Four C–H protons are chemically equivalent; no splitting occurs.

Rule 2 **The signal of a proton that has *n* equivalent neighboring protons is split into a multiplet of *n* + 1 peaks with coupling constant *J*.** Protons that are farther than two carbon atoms apart don't usually couple, although they sometimes show small coupling when they are separated by a π bond.

Rule 3 **Two groups of protons coupled to each other have the same coupling constant, *J*.**

The spectrum of *para*-methoxypropiophenone in Figure 13.16 further illustrates the three rules. The downfield absorptions at 6.91 and 7.93 δ are due to the four aromatic ring protons. There are two kinds of aromatic protons, each of which gives a signal that is split into a doublet by its neighbor. The $-OCH_3$ signal is unsplit and appears as a sharp singlet at 3.84 δ. The $-CH_2-$ protons next to the carbonyl group appear at 2.93 δ in the region expected for protons on carbon next to an unsaturated center, and their signal is split into a quartet by coupling with the protons of the neighboring methyl group. The methyl protons appear as a triplet at 1.20 δ in the usual upfield region.

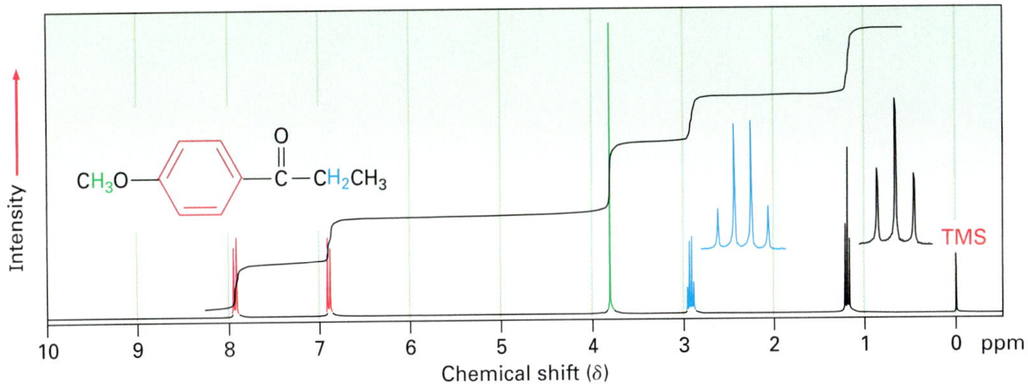

Figure 13.16 The ^{1}H NMR spectrum of *para*-methoxypropiophenone.

One further question needs to be answered before leaving the topic of spin–spin splitting. Why is spin–spin splitting seen only for ^{1}H NMR? Why is there no splitting of *carbon* signals into multiplets in ^{13}C NMR? After all, you might expect that the spin of a given ^{13}C nucleus would couple with the spin of an adjacent magnetic nucleus, either ^{13}C or ^{1}H.

No coupling of a ^{13}C nucleus with nearby *carbons* is seen because the low natural abundance makes it unlikely that two ^{13}C nuclei will be adjacent. No coupling of a ^{13}C nucleus with nearby *hydrogens* is seen because ^{13}C spectra, as previously noted (Section 13.6), are normally recorded using broadband decoupling. At the same time that the sample is irradiated with a pulse of rf energy to cover the *carbon* resonance frequencies, it is also irradiated by a second band of rf energy covering all the *hydrogen* resonance frequencies. This second irradiation makes the hydrogens spin-flip so rapidly that their local magnetic fields average to zero and no coupling with carbon spins occurs.

WORKED EXAMPLE 13.4 *Assigning a Chemical Structure from a ^{1}H NMR Spectrum*

Propose a structure for a compound, $C_5H_{12}O$, that fits the following ^{1}H NMR dat 0.92 δ (3 H, triplet, J = 7 Hz), 1.20 δ (6 H, singlet), 1.50 δ (2 H, quartet, J = 7 Hz 1.64 δ (1 H, broad singlet).

Strategy As noted in Worked Example 13.2, it's best to begin solving structural problems b calculating a molecule's degree of unsaturation. In the present instance, a formula c $C_5H_{12}O$ corresponds to a saturated, open-chain molecule, either an alcohol or a ether.

To interpret the NMR information, let's look at each absorption individuall The three-proton absorption at 0.92 δ is due to a methyl group in an alkane-like env ronment, and the triplet splitting pattern implies that the CH_3 is next to a CH_2 Thus, our molecule contains an ethyl group, CH_3CH_2–. The six-proton singlet a 1.20 δ is due to two equivalent alkane-like methyl groups attached to a carbon wit no hydrogens, $(CH_3)_2C$, and the two-proton quartet at 1.50 δ is due to the CH_2 c the ethyl group. All 5 carbons and 11 of the 12 hydrogens in the molecule are nov accounted for. The remaining hydrogen, which appears as a broad one-proton sin glet at 1.64 δ, is probably due to an OH group, since there is no other way to accoun for it. Putting the pieces together gives the structure.

Solution

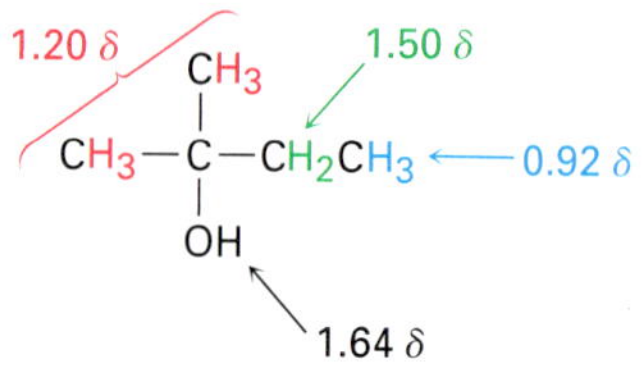

2-Methyl-2-butanol

Problem 13.19 Predict the splitting patterns you would expect for each proton in the following molecules:

(a) $CHBr_2CH_3$ **(b)** $CH_3OCH_2CH_2Br$ **(c)** $ClCH_2CH_2CH_2Cl$

(d)

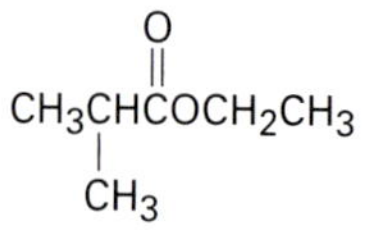

(e)

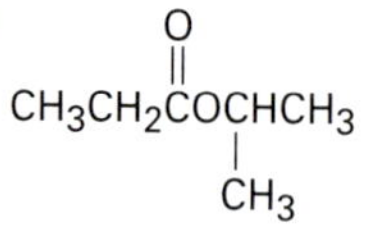

(f)

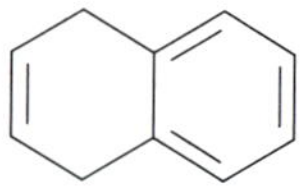

Problem 13.20 Draw structures for compounds that meet the following descriptions:

(a) C_2H_6O; one singlet **(b)** C_3H_7Cl; one doublet and one septet

(c) $C_4H_8Cl_2O$; two triplets **(d)** $C_4H_8O_2$; one singlet, one triplet, and one quartet

Problem 13.21 The integrated ^{1}H NMR spectrum of a compound of formula $C_4H_{10}O$ is shown in Figure 13.17. Propose a structure.

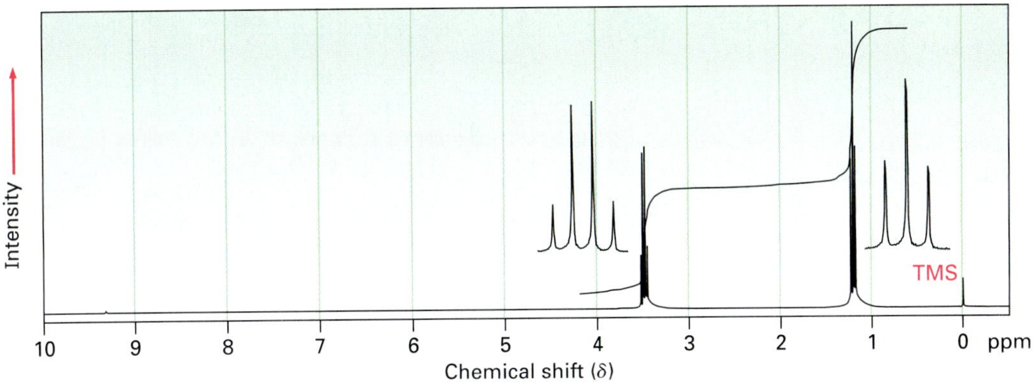

Figure 13.17 An integrated ^{1}H NMR spectrum for Problem 13.21.

13.12 More Complex Spin–Spin Splitting Patterns

In the ^{1}H NMR spectra we've seen so far, the chemical shifts of different protons have been distinct and the spin–spin splitting patterns have been straightforward. It often happens, however, that different kinds of hydrogens in a molecule have accidentally *overlapping* signals. The spectrum of toluene (methylbenzene) in Figure 13.18, for example, shows that the five aromatic ring protons give a complex, overlapping pattern, even though they aren't all equivalent.

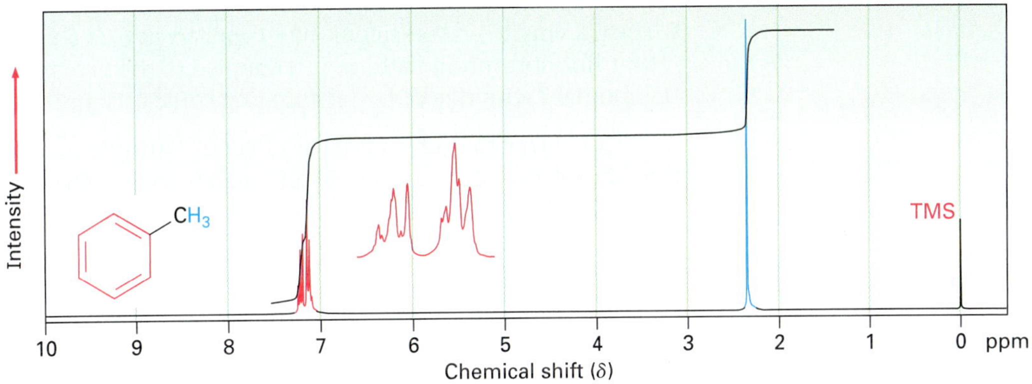

Figure 13.18 The ^{1}H NMR spectrum of toluene, showing the accidental overlap of the five nonequivalent aromatic ring protons.

Yet another complication in ^{1}H NMR spectroscopy arises when a signal is split by two or more *nonequivalent* kinds of protons, as is the case with *trans*-cinnamaldehyde, isolated from oil of cinnamon (Figure 13.19). Although the $n + 1$ rule predicts splitting caused by equivalent protons, splittings caused by nonequivalent protons are more complex.

To understand the ^{1}H NMR spectrum of *trans*-cinnamaldehyde, we have to isolate the different parts and look at the signal of each proton individually.

- The five aromatic proton signals (black in Figure 13.19) overlap into a complex pattern with a large peak at 7.42 δ and a broad absorption at 7.57 δ.

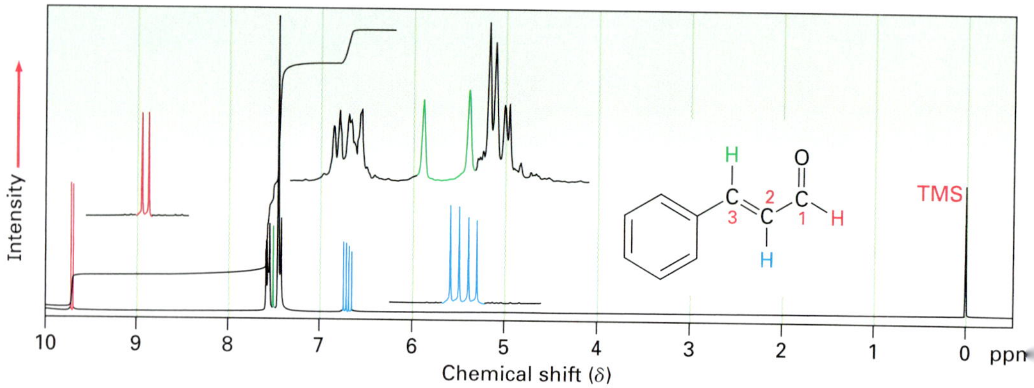

Active Figure 13.19 The ^{1}H NMR spectrum of *trans*-cinnamaldehyde. The signal of the proton at C2 (blue) is split into four peaks—a doublet of doublets—by the two nonequivalent neighboring protons. *Sign in at* **www.thomsonedu.com** *to see a simulation based on this figure and to take a short quiz.*

- The aldehyde proton signal at C1 (red) appears in the normal downfield position at 9.69 δ and is split into a doublet with $J = 6$ Hz by the adjacent proton at C2.
- The vinylic proton at C3 (green) is next to the aromatic ring and is therefore shifted downfield from the normal vinylic region. This C3 proton signal appears as a doublet centered at 7.49 δ. Because it has one neighbor proton at C2, its signal is split into a doublet, with $J = 12$ Hz.
- The C2 vinylic proton signal (blue) appears at 6.73 δ and shows an interesting four-line absorption pattern. It is coupled to the two nonequivalent protons at C1 and C3 with two different coupling constants: $J_{1\text{-}2} = 6$ Hz and $J_{2\text{-}3} = 12$ Hz.

A good way to understand the effect of multiple coupling such as occurs for the C2 proton of *trans*-cinnamaldehyde is to draw a *tree diagram,* like that in Figure 13.20. The diagram shows the individual effect of each coupling constant on the overall pattern. Coupling with the C3 proton splits the signal of the C2 proton in *trans*-cinnamaldehyde into a doublet with $J = 12$ Hz. Further coupling with the aldehyde proton then splits each peak of the doublet into new doublets, and we therefore observe a four-line spectrum for the C2 proton.

Active Figure 13.20 A tree diagram for the C2 proton of *trans*-cinnamaldehyde shows how it is coupled to the C1 and C3 protons with different coupling constants. *Sign in at* **www.thomsonedu.com** *to see a simulation based on this figure and to take a short quiz.*

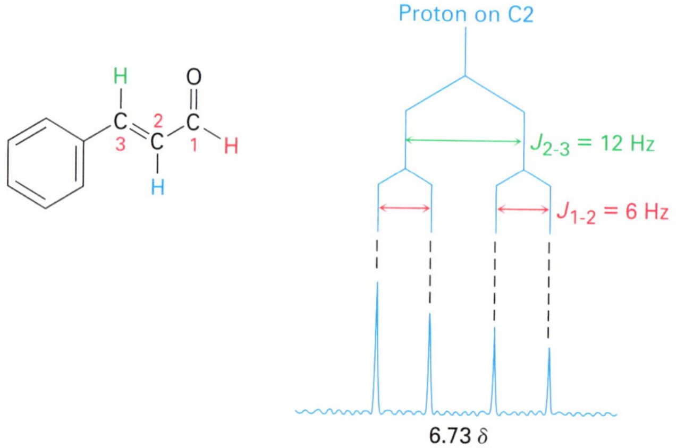

One further point evident in the cinnamaldehyde spectrum is that the four peaks of the C2 proton signal are not all the same size. The two left-hand peaks are somewhat larger than the two right-hand peaks. Such a size difference occurs whenever coupled nuclei have similar chemical shifts—in this case, 7.49 δ for the C3 proton and 6.73 δ for the C2 proton. The peaks nearer the signal of the coupled partner are always larger, and the peaks farther from the signal of the coupled partner are always smaller. Thus, the left-hand peaks of the C2 proton multiplet at 6.73 δ are closer to the C3 proton absorption at 7.49 δ and are larger than the right-hand peaks. At the same time, the *right-hand* peak of the C3 proton doublet at 7.49 δ is larger than the left-hand peak because it is closer to the C2 proton multiplet at 6.73 δ. This skewing effect on multiplets can often be useful because it tells where to look in the spectrum to find the coupled partner: look toward the direction of the larger peaks.

Problem 13.22 3-Bromo-1-phenyl-1-propene shows a complex NMR spectrum in which the vinylic proton at C2 is coupled with both the C1 vinylic proton ($J = 16$ Hz) and the C3 methylene protons ($J = 8$ Hz). Draw a tree diagram for the C2 proton signal, and account for the fact that a five-line multiplet is observed.

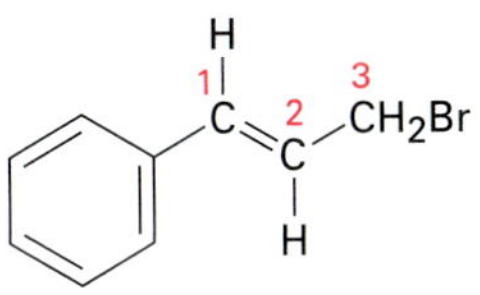

3-Bromo-1-phenyl-1-propene

13.13 Uses of ^1H NMR Spectroscopy

homsonNOW Click *Organic nteractive* to **learn to utilize H NMR spectroscopy to deduce nolecular structures.**

NMR can be used to help identify the product of nearly every reaction run in the laboratory. For example, we said in Section 7.5 that hydroboration/oxidation of alkenes occurs with non-Markovnikov regiochemistry to yield the less highly substituted alcohol. With the help of NMR, we can now prove this statement.

Does hydroboration/oxidation of methylenecyclohexane yield cyclohexylmethanol or 1-methylcyclohexanol?

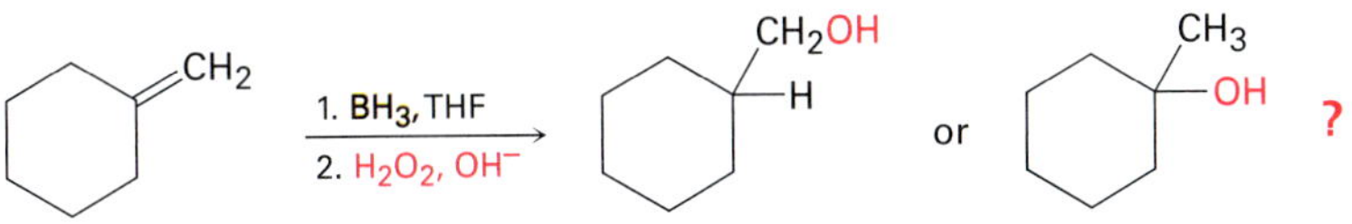

Methylenecyclohexane **Cyclohexylmethanol** **1-Methylcyclohexanol**

The ^1H NMR spectrum of the reaction product is shown in Figure 13.21a. The spectrum shows a two-proton peak at 3.40 δ, indicating that the product has a $-CH_2-$ group bonded to an electronegative oxygen atom ($-CH_2OH$). Furthermore, the spectrum shows *no* large three-proton singlet absorption near 1 δ, where we would expect the signal of a quaternary $-CH_3$ group to appear. (Figure 13.21b gives the spectrum of 1-methylcyclohexanol, the alternative product.) Thus, it's clear that cyclohexylmethanol is the reaction product.

(a)

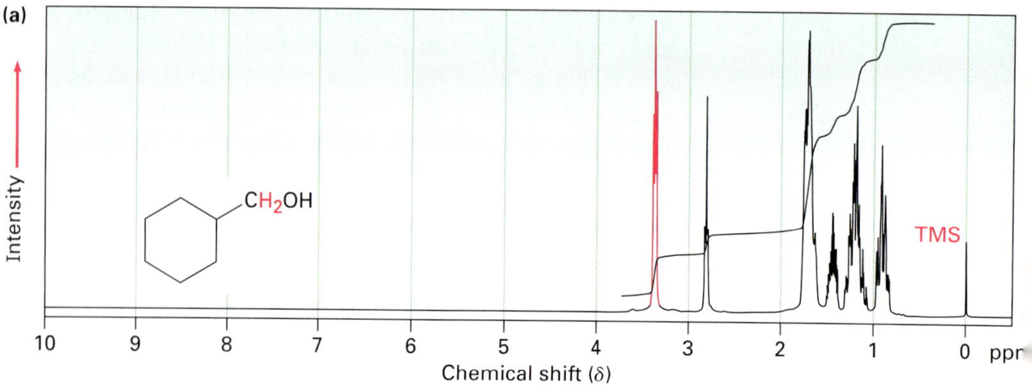

(b)

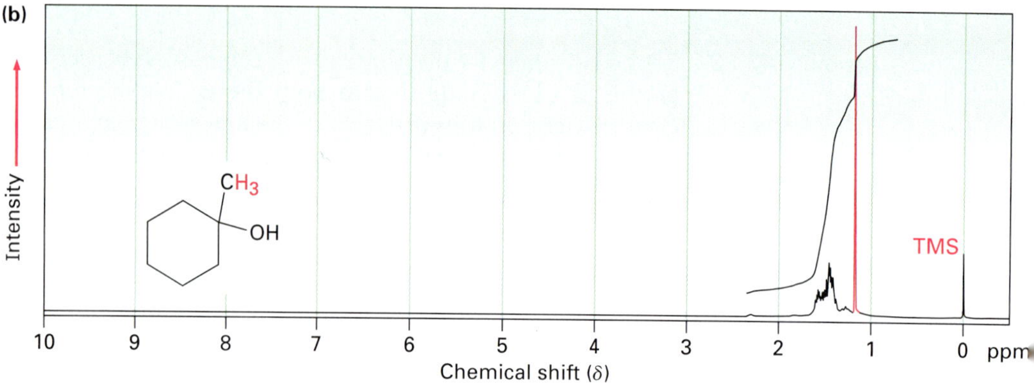

Figure 13.21 **(a)** The ^{1}H NMR spectrum of cyclohexylmethanol, the product from hydroboration/oxidation of methylenecyclohexane, and **(b)** the ^{1}H NMR spectrum of 1-methylcyclohexanol, the possible alternative reaction product.

Problem 13.23 How could you use ^{1}H NMR to determine the regiochemistry of electrophilic addition to alkenes? For example, does addition of HCl to 1-methylcyclohexene yield 1-chloro-1-methylcyclohexane or 1-chloro-2-methylcyclohexane?

Focus On . . .

Magnetic Resonance Imaging (MRI)

As practiced by organic chemists, NMR spectroscopy is a powerful method of structure determination. A small amount of sample, typically a few milligrams or less, is dissolved in a small amount of solvent, the solution is placed in a thin glass tube, and the tube is placed into the narrow (1–2 cm) gap between the poles of a strong magnet. Imagine, though, that a much larger NMR instrument were available. Instead of a few milligrams, the sample size could be tens of kilograms; instead of a narrow gap between magnet poles, the gap

(continued)

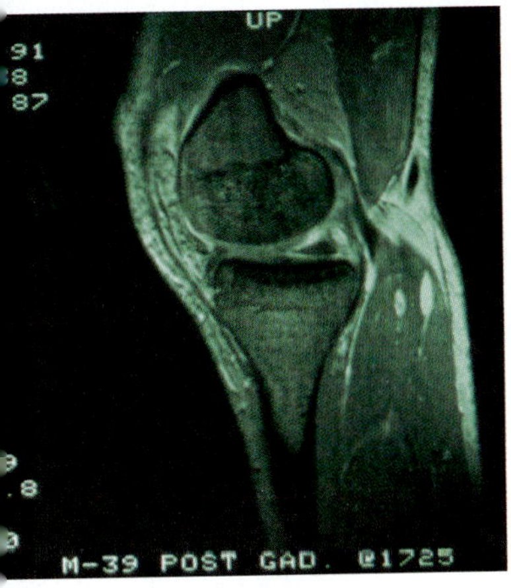

If you're a runner, you really don't want this to happen to you. The MRI of this left knee shows the presence of a ganglion cyst.

could be large enough for a whole person to climb into so that an NMR spectrum of body parts could be obtained. That large instrument is exactly what's used for *magnetic resonance imaging (MRI),* a diagnostic technique of enormous value to the medical community.

Like NMR spectroscopy, MRI takes advantage of the magnetic properties of certain nuclei, typically hydrogen, and of the signals emitted when those nuclei are stimulated by radiofrequency energy. Unlike what happens in NMR spectroscopy, though, MRI instruments use data manipulation techniques to look at the three-dimensional *location* of magnetic nuclei in the body rather than at the chemical nature of the nuclei. As noted, most MRI instruments currently look at hydrogen, present in abundance wherever there is water or fat in the body.

The signals detected by MRI vary with the density of hydrogen atoms and with the nature of their surroundings, allowing identification of different types of tissue and even allowing the visualization of motion. For example, the volume of blood leaving the heart in a single stroke can be measured, and heart motion can be observed. Soft tissues that don't show up well on X rays can be seen clearly, allowing diagnosis of brain tumors, strokes, and other conditions. The technique is also valuable in diagnosing damage to knees or other joints and is a noninvasive alternative to surgical explorations.

Several types of atoms in addition to hydrogen can be detected by MRI, and the applications of images based on ^{31}P atoms are being explored. The technique holds great promise for studies of metabolism.

SUMMARY AND KEY WORDS

chemical shift, 445
coupling, 460
coupling constant (*J*), 462
delta (δ) scale, 445
diastereotopic, 456
downfield, 445
enantiotopic, 455
FT–NMR, 447
homotopic, 455
integration, 459
multiplet, 460
n + 1 rule, 461
nuclear magnetic resonance (NMR) spectroscopy, 440
shielding, 442
spin–spin splitting, 460
upfield, 445

When magnetic nuclei such as ^{1}H and ^{13}C are placed in a strong magnetic field, their spins orient either with or against the field. On irradiation with radiofrequency (rf) waves, energy is absorbed and the nuclei "spin-flip" from the lower-energy state to the higher-energy state. This absorption of rf energy is detected, amplified, and displayed as a **nuclear magnetic resonance (NMR) spectrum**.

Each electronically distinct ^{1}H or ^{13}C nucleus in a molecule comes into resonance at a slightly different value of the applied field, thereby producing a unique absorption signal. The exact position of each peak is called the **chemical shift**. Chemical shifts are caused by electrons setting up tiny local magnetic fields that **shield** a nearby nucleus from the applied field.

The NMR chart is calibrated in **delta units** (δ), where 1 δ = 1 ppm of spectrometer frequency. Tetramethylsilane (TMS) is used as a reference point because it shows both ^{1}H and ^{13}C absorptions at unusually high values of the applied magnetic field. The TMS absorption occurs at the right-hand (**upfield**) side of the chart and is arbitrarily assigned a value of 0 δ.

Most ^{13}C spectra are run on Fourier-transform NMR (**FT–NMR**) spectrometers using broadband decoupling of proton spins so that each chemically distinct carbon shows a single unsplit resonance line. As with ^{1}H NMR, the chemical shift of each ^{13}C signal provides information about a carbon's chemical environment in the sample. In addition, the number of protons attached to each carbon can be determined using the DEPT–NMR technique.

In ^{1}H NMR spectra, the area under each absorption peak can be electro cally **integrated** to determine the relative number of hydrogens responsible f each peak. In addition, neighboring nuclear spins can **couple**, causing t **spin–spin splitting** of NMR peaks into **multiplets**. The NMR signal of a hydr gen neighbored by n equivalent adjacent hydrogens splits into $n + 1$ peaks (t **n + 1 rule**) with **coupling constant** J.

EXERCISES

Organic KNOWLEDGE TOOLS

ThomsonNOW™ Sign in at **www.thomsonedu.com** to assess your knowledge of this chapter's topics by taking a pre-test. The pre-test will link you to interactive organic chemistry resources based on your score in each concept area.

Online homework for this chapter may be assigned in Organic OWL.

■ indicates problems assignable in Organic OWL.

VISUALIZING CHEMISTRY

(Problems 13.1–13.23 appear within the chapter.)

13.24 ■ Into how many peaks would you expect the ^{1}H NMR signals of the indicate protons to be split? (Yellow-green = Cl.)

(a)

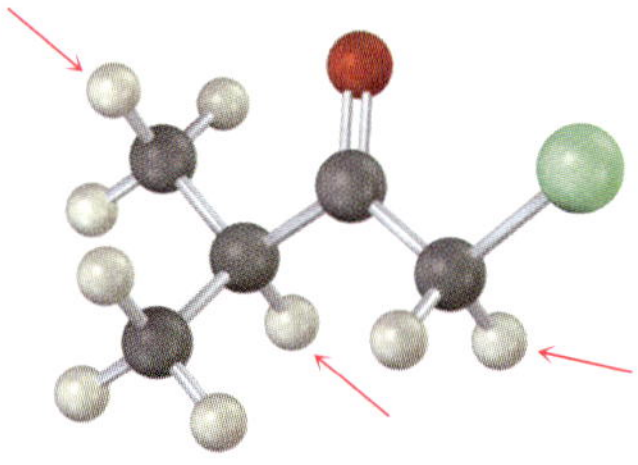

(b)

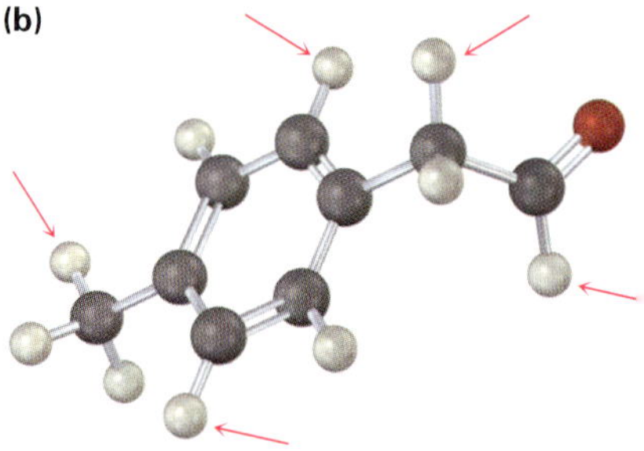

13.25 ■ How many absorptions would you expect the following compound to hav in its ^{1}H and ^{13}C NMR spectra?

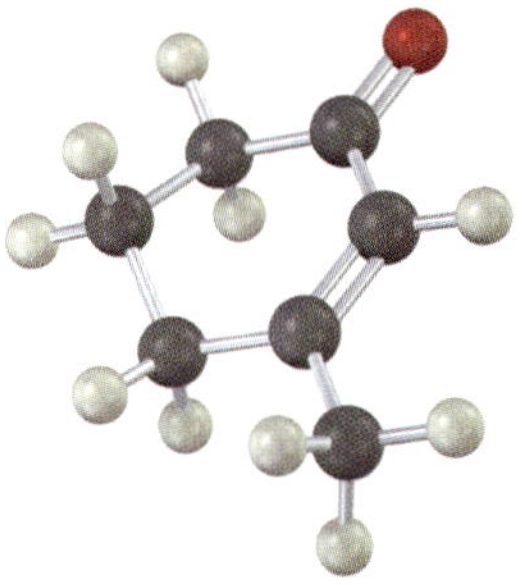

13.26 Sketch what you might expect the ^1H and ^{13}C NMR spectra of the following compound to look like (yellow-green = Cl):

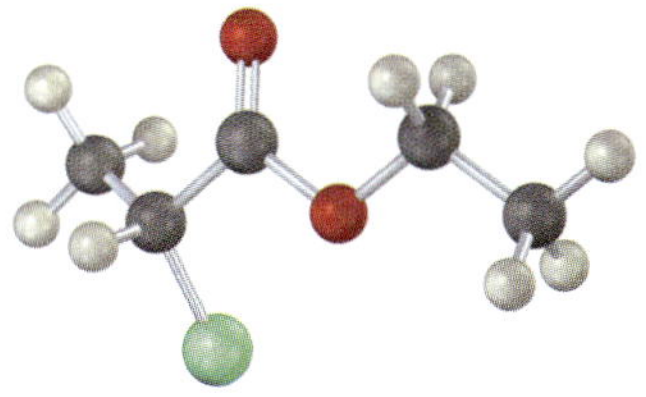

13.27 ■ How many electronically nonequivalent kinds of protons and how many kinds of carbons are present in the following compound? Don't forget that cyclohexane rings can ring-flip.

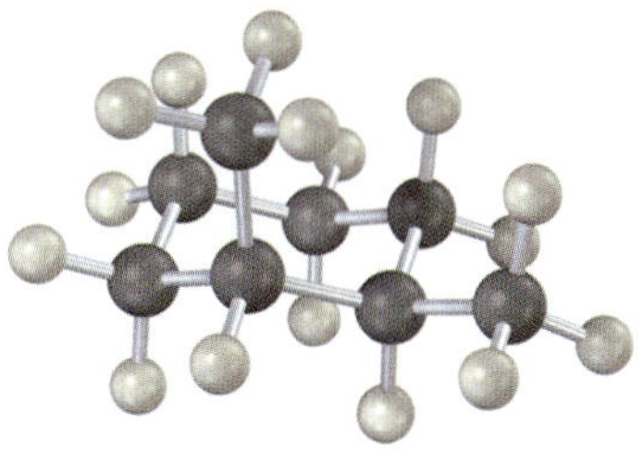

13.28 ■ Identify the indicated protons in the following molecules as unrelated, homotopic, enantiotopic, or diastereotopic:

(a)

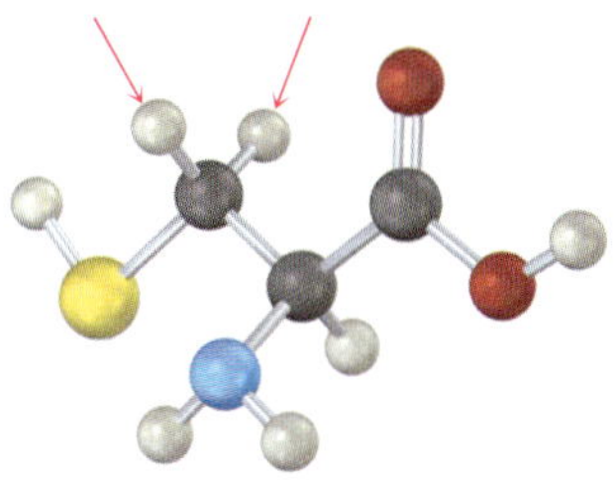

Cysteine

(b)

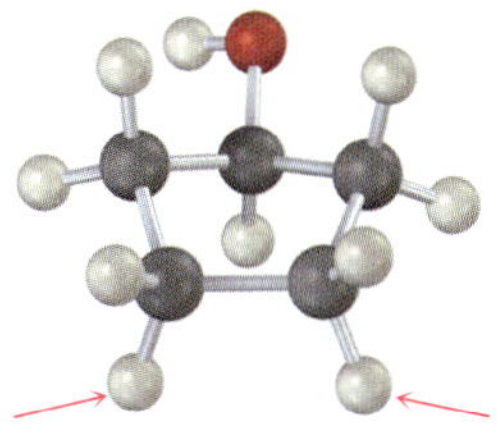

ADDITIONAL PROBLEMS

ThomsonNOW™ Click *Organic Interactive* to **learn to use ^{13}C NMR, ^{1}H NMR, infrared, and mass spectrometry together to deduce molecular structures.**

13.29 ■ The following ^1H NMR absorptions were obtained on a spectrometer operating at 200 MHz and are given in hertz downfield from the TMS standard. Convert the absorptions to δ units.

(a) 436 Hz (b) 956 Hz (c) 1504 Hz

13.30 ■ The following ^1H NMR absorptions were obtained on a spectrometer operating at 300 MHz. Convert the chemical shifts from δ units to hertz downfield from TMS.

(a) 2.1 δ (b) 3.45 δ (c) 6.30 δ (d) 7.70 δ

13.31 When measured on a spectrometer operating at 200 MHz, chloroform ($CHCl_3$) shows a single sharp absorption at 7.3 δ.

(a) How many parts per million downfield from TMS does chloroform absorb?

(b) How many hertz downfield from TMS would chloroform absorb if the measurement were carried out on a spectrometer operating at 360 MHz?

(c) What would be the position of the chloroform absorption in δ units when measured on a 360 MHz spectrometer?

13.32 How many signals would you expect each of the following molecules to have in its ^1H and ^{13}C spectra?

(a) 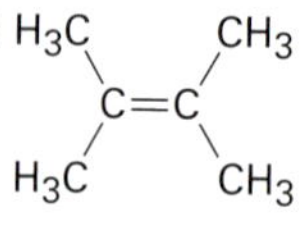

(b) 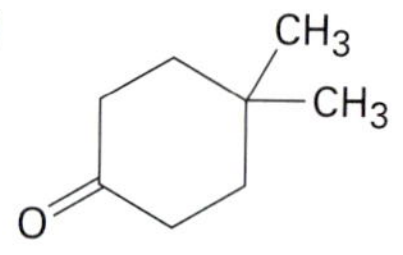

(c) 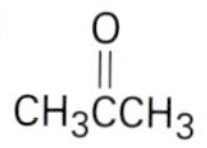

(d) 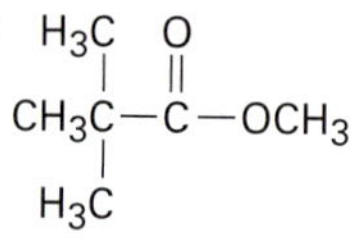

(e) 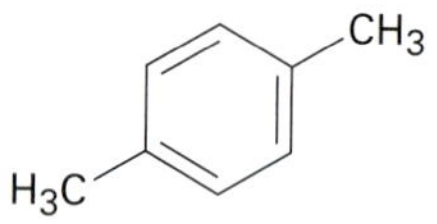

(f) 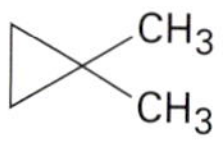

13.33 ■ How many absorptions would you expect to observe in the ^{13}C NMR spectra of the following compounds?

(a) 1,1-Dimethylcyclohexane
(b) $CH_3CH_2OCH_3$
(c) *tert*-Butylcyclohexane
(d) 3-Methyl-1-pentyne
(e) *cis*-1,2-Dimethylcyclohexane
(f) Cyclohexanone

13.34 ■ Suppose you ran a DEPT-135 spectrum for each substance in Problem 13.33. Which carbon atoms in each molecule would show positive peaks and which would show negative peaks?

13.35 Why do you suppose accidental overlap of signals is much more common in ^1H NMR than in ^{13}C NMR?

13.36 ■ Is a nucleus that absorbs at 6.50 δ more shielded or less shielded than a nucleus that absorbs at 3.20 δ? Does the nucleus that absorbs at 6.50 δ require a stronger applied field or a weaker applied field to come into resonance than the nucleus that absorbs at 3.20 δ?

13.37 ■ Identify the indicated sets of protons as unrelated, homotopic, enantiotopic, or diastereotopic:

(a) 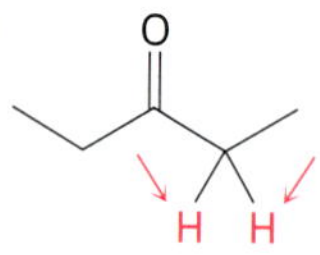

(b) 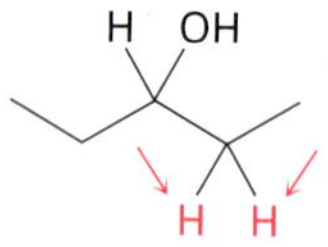

(c) 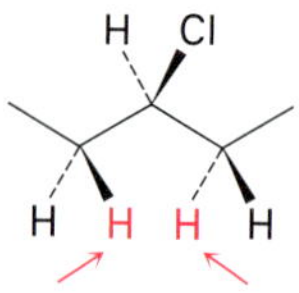

13.38 How many types of nonequivalent protons are present in each of the following molecules?

(a) 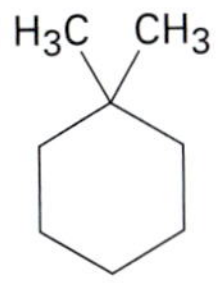

(b) $CH_3CH_2CH_2OCH_3$

(c) 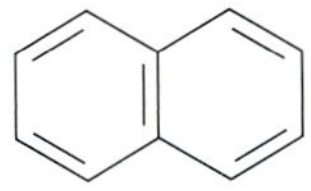

Naphthalene

(d) 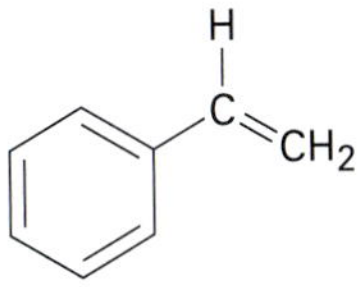

Styrene

(e) 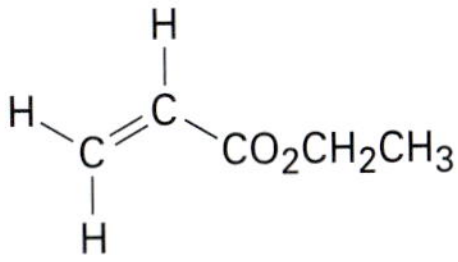

Ethyl acrylate

13.39 ■ Identify the indicated sets of protons as unrelated, homotopic, enantiotopic, or diastereotopic:

(a) 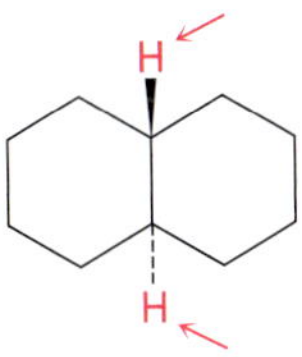

(b) 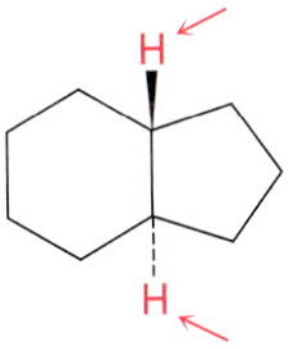

(c) 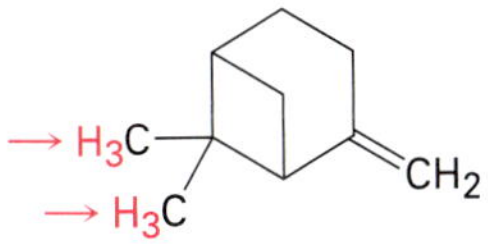

13.40 ■ The following compounds all show a single line in their ^1H NMR spectra. List them in expected order of increasing chemical shift:

CH_4, CH_2Cl_2, cyclohexane, CH_3COCH_3, $H_2C{=}CH_2$, benzene

13.41 ■ Predict the splitting pattern for each kind of hydrogen in the following molecules:
(a) $(CH_3)_3CH$ (b) $CH_3CH_2CO_2CH_3$ (c) *trans*-2-Butene

13.42 Predict the splitting pattern for each kind of hydrogen in isopropyl propanoate, $CH_3CH_2CO_2CH(CH_3)_2$.

13.43 The acid-catalyzed dehydration of 1-methylcyclohexanol yields a mixture of two alkenes. How could you use ^1H NMR to help you decide which was which?

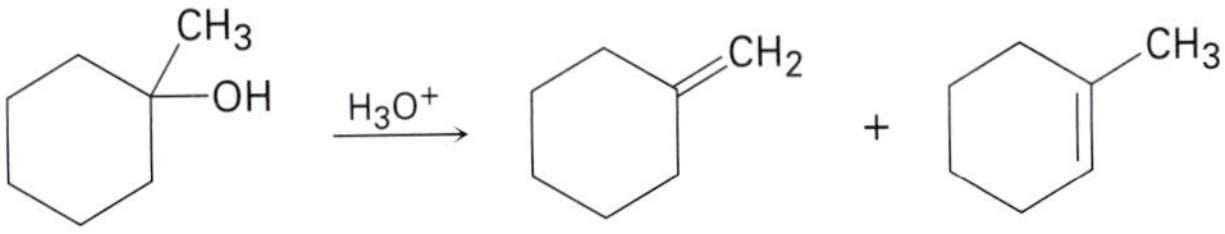

13.44 How could you use ^1H NMR to distinguish between the following pairs of isomers?

(a) $CH_3CH{=}CHCH_2CH_3$ and 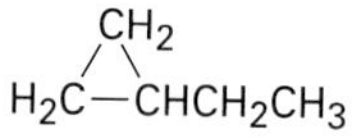

(b) $CH_3CH_2OCH_2CH_3$ and $CH_3OCH_2CH_2CH_3$

(c) $CH_3COCH_2CH_3$ (C=O) and $CH_3CH_2CCH_3$ (C=O)

(d) $H_2C{=}C(CH_3)CCH_3$ (C=O) and $CH_3CH{=}CHCCH_3$ (C=O)

13.45 Propose structures for compounds with the following formulas that show only one peak in their ^1H NMR spectra:
(a) C_5H_{12} (b) C_5H_{10} (c) $C_4H_8O_2$

13.46 How many ^{13}C NMR absorptions would you expect for *cis*-1,3-dimethylcyclohexane? For *trans*-1,3-dimethylcyclohexane? Explain.

13.47 Assume that you have a compound with formula C_3H_6O.
(a) How many double bonds and/or rings does your compound contain?
(b) Propose as many structures as you can that fit the molecular formula.
(c) If your compound shows an infrared absorption peak at 1715 cm^{-1}, what functional group does it have?
(d) If your compound shows a single ^1H NMR absorption peak at 2.1 δ, what is its structure?

13.48 How could you use ^1H and ^{13}C NMR to help you distinguish among the following isomeric compounds of formula C_4H_8?

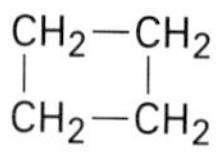 $H_2C{=}CHCH_2CH_3$ $CH_3CH{=}CHCH_3$ $CH_3C(CH_3){=}CH_2$

13.49 How could you use ^1H NMR, ^{13}C NMR, and IR spectroscopy to help you distinguish between the following structures?

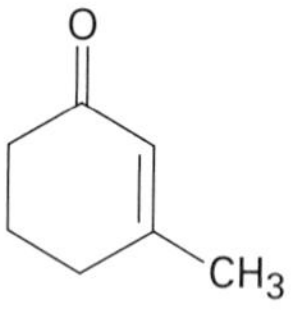

3-Methyl-2-cyclohexenone

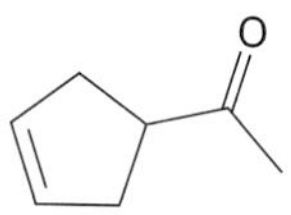

3-Cyclopentenyl methyl ketone

13.50 The compound whose ^{1}H NMR spectrum is shown has the molecular formula $C_3H_6Br_2$. Propose a structure.

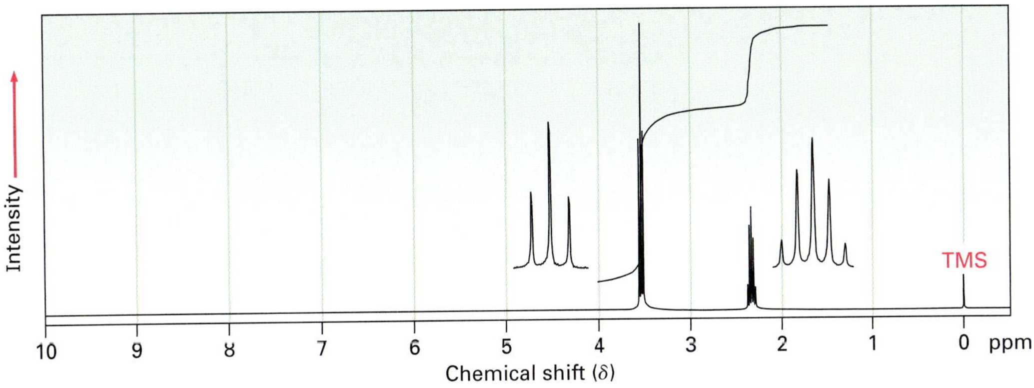

13.51 ■ Propose structures for compounds that fit the following ^{1}H NMR data:

(a) $C_5H_{10}O$
0.95 δ (6 H, doublet, J = 7 Hz)
2.10 δ (3 H, singlet)
2.43 δ (1 H, multiplet)

(b) C_3H_5Br
2.32 δ (3 H, singlet)
5.35 δ (1 H, broad singlet)
5.54 δ (1 H, broad singlet)

13.52 The compound whose ^{1}H NMR spectrum is shown has the molecular formula $C_4H_7O_2Cl$ and has an infrared absorption peak at 1740 cm^{-1}. Propose a structure.

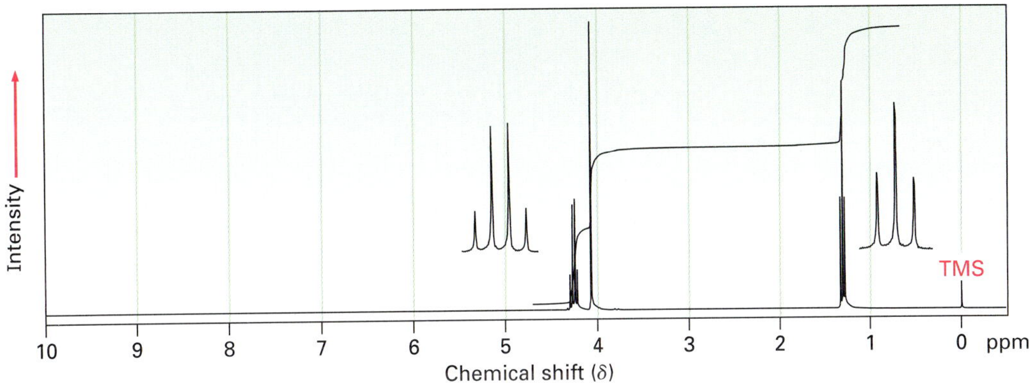

13.53 Propose structures for compounds that fit the following ^{1}H NMR data:

(a) $C_4H_6Cl_2$
2.18 δ (3 H, singlet)
4.16 δ (2 H, doublet, J = 7 Hz)
5.71 δ (1 H, triplet, J = 7 Hz)

(b) $C_{10}H_{14}$
1.30 δ (9 H, singlet)
7.30 δ (5 H, singlet)

(c) C_4H_7BrO
2.11 δ (3 H, singlet)
3.52 δ (2 H, triplet, J = 6 Hz)
4.40 δ (2 H, triplet, J = 6 Hz)

(d) $C_9H_{11}Br$
2.15 δ (2 H, quintet, J = 7 Hz)
2.75 δ (2 H, triplet, J = 7 Hz)
3.38 δ (2 H, triplet, J = 7 Hz)
7.22 δ (5 H, singlet)

■ Assignable in OWL

13.54 Propose structures for the two compounds whose ^{1}H NMR spectra are shown.
(a) C_4H_9Br

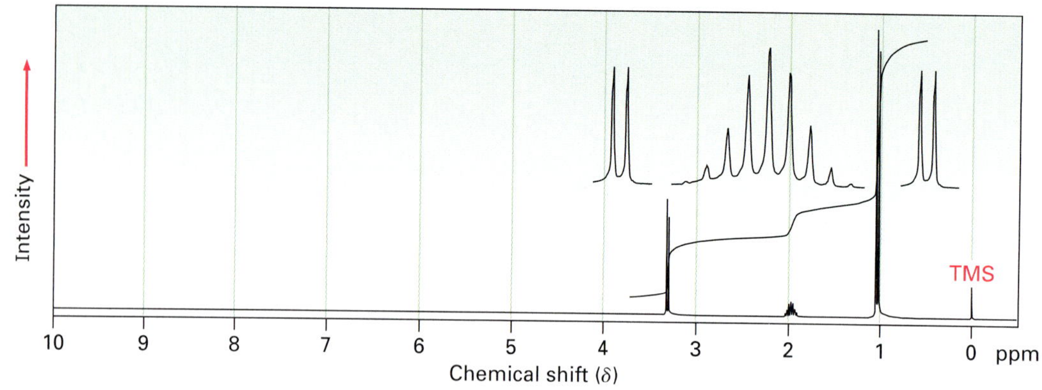

(b) $C_4H_8Cl_2$

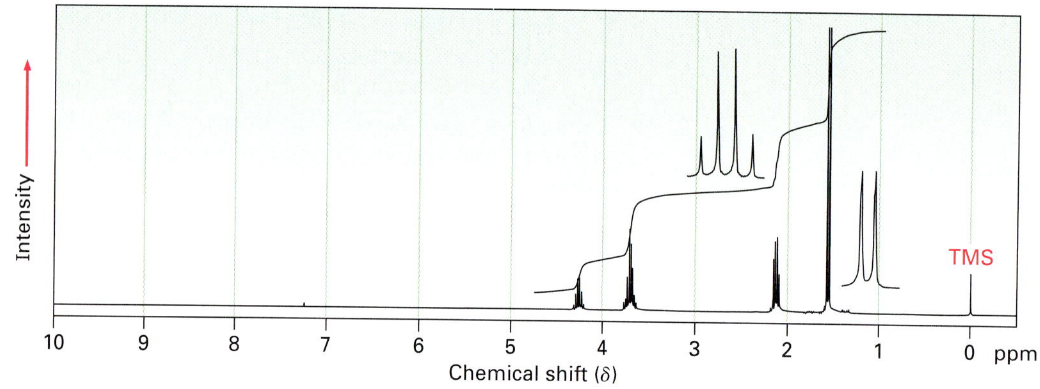

13.55 Long-range coupling between protons more than two carbon atoms apart is sometimes observed when π bonds intervene. An example is found in 1-methoxy-1-buten-3-yne. Not only does the acetylenic proton, H_a, couple with the vinylic proton H_b, it also couples with the vinylic proton H_c, *four* carbon atoms away. The data are:

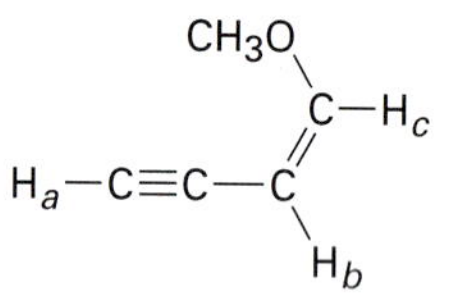

1-Methoxy-1-buten-3-yne

H_a (3.08 δ) H_b (4.52 δ) H_c (6.35 δ)

$J_{a\text{-}b}$ = 3 Hz $J_{a\text{-}c}$ = 1 Hz $J_{b\text{-}c}$ = 7 Hz

Construct tree diagrams that account for the observed splitting patterns of H_a, H_b, and H_c.

13.56 Assign as many of the resonances as you can to specific carbon atoms in the ^{13}C NMR spectrum of ethyl benzoate.

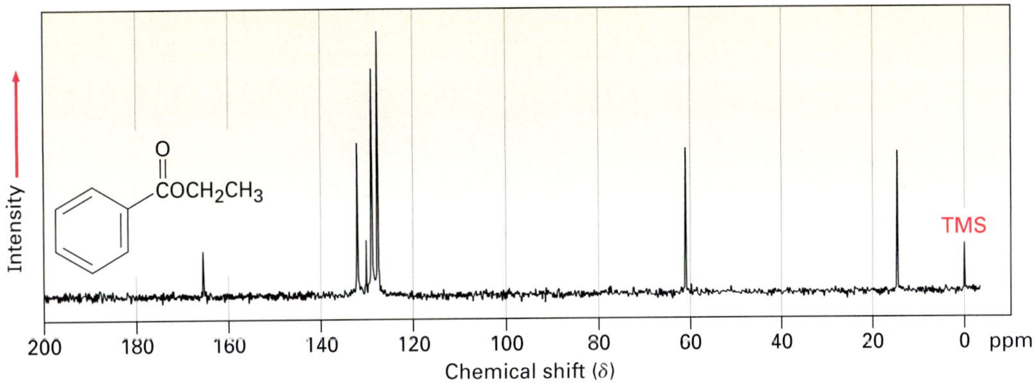

13.57 The ^{1}H and ^{13}C NMR spectra of compound **A**, C_8H_9Br, are shown. Propose a structure for **A**, and assign peaks in the spectra to your structure.

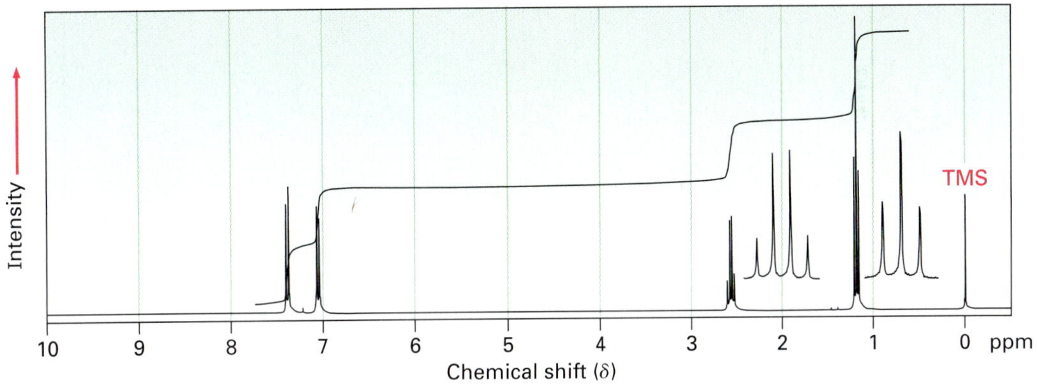

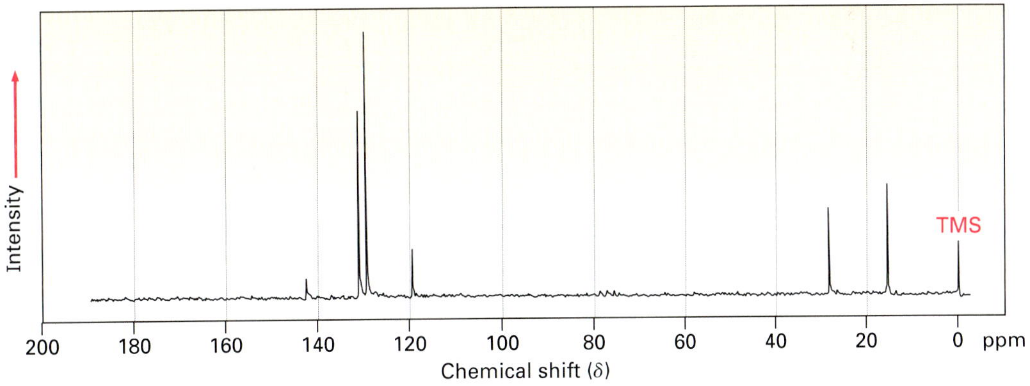

13.58 ■ Propose structures for the three compounds whose ^{1}H NMR spectra are shown.

(a) $C_5H_{10}O$

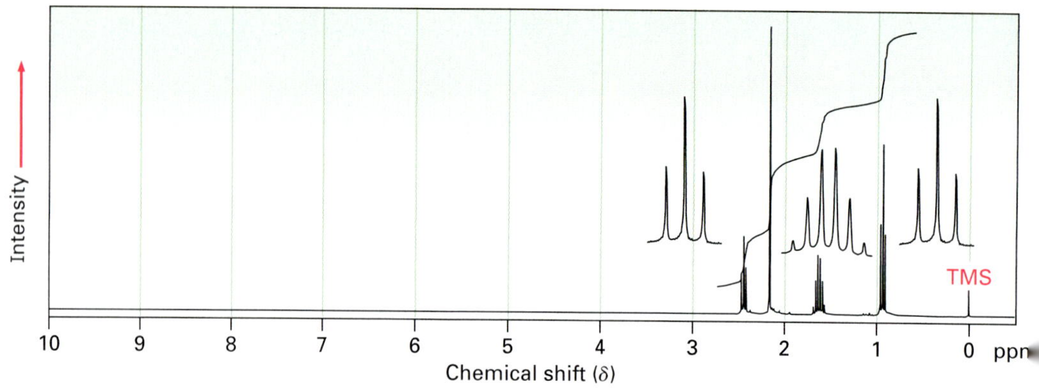

(b) C_7H_7Br

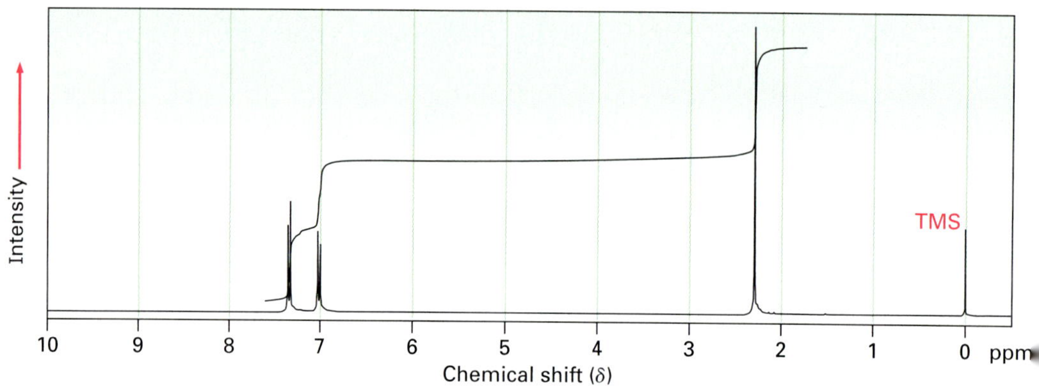

(c) C_8H_9Br

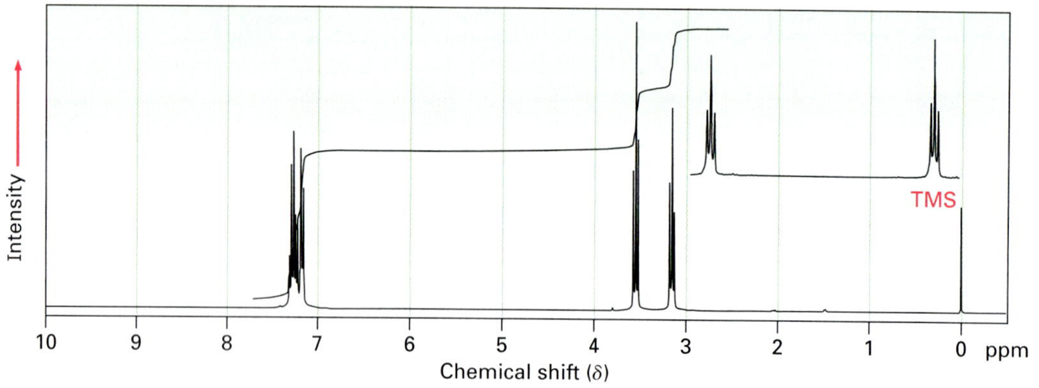

13.59 The mass spectrum and ^{13}C NMR spectrum of a hydrocarbon are shown. Propose a structure for this hydrocarbon, and explain the spectral data.

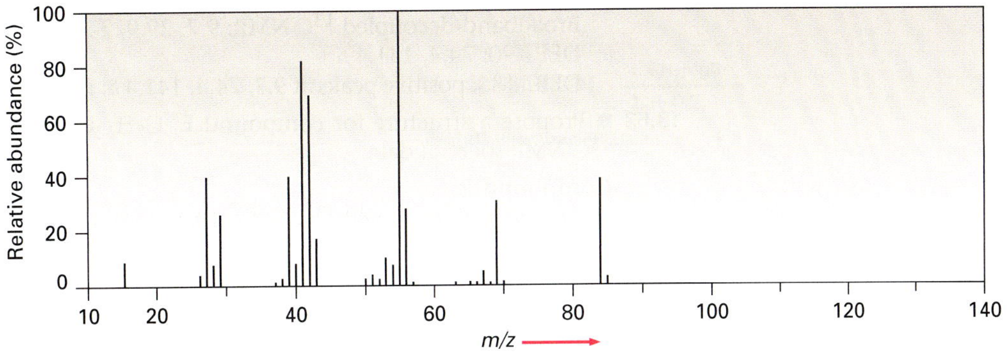

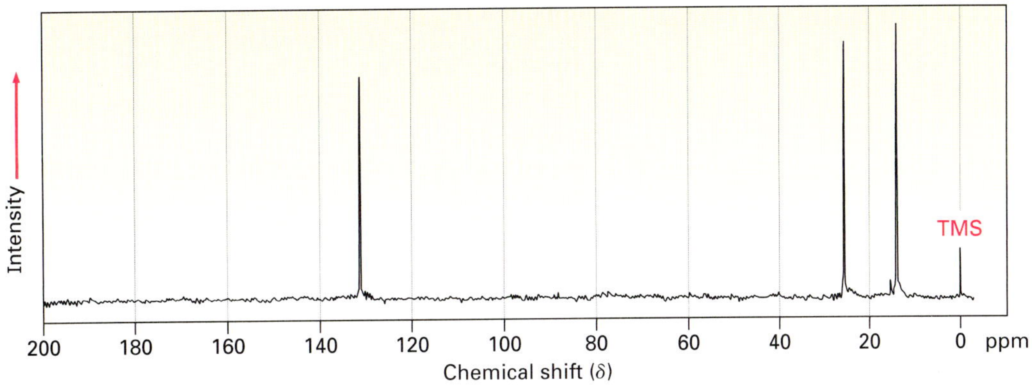

13.60 ■ Compound **A**, a hydrocarbon with $M^+ = 96$ in its mass spectrum, has the ^{13}C spectral data that follow. On reaction with BH_3 followed by treatment with basic H_2O_2, **A** is converted into **B**, whose ^{13}C spectral data are also given. Propose structures for **A** and **B**.

Compound A

Broadband-decoupled ^{13}C NMR: 26.8, 28.7, 35.7, 106.9, 149.7 δ
DEPT-90: no peaks
DEPT-135: no positive peaks; negative peaks at 26.8, 28.7, 35.7, 106.9 δ

Compound B

Broadband-decoupled ^{13}C NMR: 26.1, 26.9, 29.9, 40.5, 68.2 δ
DEPT-90: 40.5 δ
DEPT-135: positive peak at 40.5 δ; negative peaks at 26.1, 26.9, 29.9, 68.2 δ

13.61 ■ Propose a structure for compound **C**, which has $M^+ = 86$ in its mass spectrum, an IR absorption at 3400 cm^{-1}, and the following ^{13}C NMR spectral data:

Compound C

Broadband-decoupled ^{13}C NMR: 30.2, 31.9, 61.8, 114.7, 138.4 δ
DEPT-90: 138.4 δ
DEPT-135: positive peak at 138.4 δ; negative peaks at 30.2, 31.9, 61.8, 114.7 δ

APPENDIX

A Nomenclature of Polyfunctional Organic Compounds

With more than 30 million organic compounds now known and thousands more being created daily, naming them all is a real problem. Part of the problem is due to the sheer complexity of organic structures, but part is also due to the fact that chemical names have more than one purpose. For Chemical Abstracts Service (CAS), which catalogs and indexes the worldwide chemical literature, each compound must have only one correct name. It would be chaos if half the entries for CH_3Br were indexed under "M" for methyl bromide and half under "B" for bromomethane. Furthermore, a CAS name must be strictly systematic so that it can be assigned and interpreted by computers; common names are not allowed.

People, however, have different requirements than computers. For people—which is to say chemists in their spoken and written communications—it's best that a chemical name be pronounceable and that it be as easy as possible to assign and interpret. Furthermore, it's convenient if names follow historical precedents, even if that means a particularly well-known compound might have more than one name. People can readily understand that bromomethane and methyl bromide both refer to CH_3Br.

As noted in the text, chemists overwhelmingly use the nomenclature system devised and maintained by the International Union of Pure and Applied Chemistry, or IUPAC. Rules for naming monofunctional compounds were given throughout the text as each new functional group was introduced, and a list of where these rules can be found is given in Table A.1.

Table A.1 Nomenclature Rules for Functional Groups

Functional group	Text section	Functional group	Text section
Acid anhydrides	21.1	Aromatic compounds	15.1
Acid halides	21.1	Carboxylic acids	20.1
Acyl phosphates	21.1	Cycloalkanes	4.1
Alcohols	17.1	Esters	21.1
Aldehydes	19.1	Ethers	18.1
Alkanes	3.4	Ketones	19.1
Alkenes	6.3	Nitriles	20.1
Alkyl halides	10.1	Phenols	17.1
Alkynes	8.1	Sulfides	18.8
Amides	21.1	Thioesters	21.1
Amines	24.1	Thiols	18.8

Naming a monofunctional compound is reasonably straightforward, but even experienced chemists often encounter problems when faced with naming a complex polyfunctional compound. Take the following compound, for instance. It has three functional groups, ester, ketone, and C=C, but how should it be named? As an ester with an *-oate* ending, a ketone with an *-one* ending, or an alkene with an *-ene* ending? It's actually named methyl 3-(2-oxo-6-cyclohexenyl)propanoate.

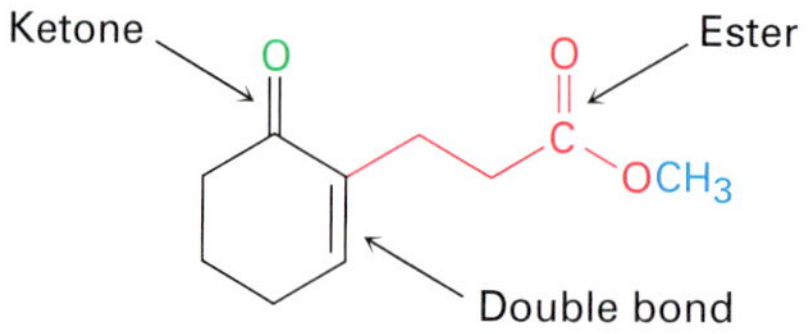

Methyl 3-(2-oxo-6-cylohexenyl)propanoate

The name of a polyfunctional organic molecule has four parts—suffix, parent, prefixes, and locants—which must be identified and expressed in the proper order and format. Let's look at each of the four.

Name Part 1. The Suffix: Functional-Group Precedence

Although a polyfunctional organic molecule might contain several different functional groups, we must choose just one suffix for nomenclature purposes. It's not correct to use two suffixes. Thus, keto ester **1** must be named either as a ketone with an *-one* suffix or as an ester with an *-oate* suffix but can't be named as an *-onoate*. Similarly, amino alcohol **2** must be named either as an alcohol (*-ol*) or as an amine (*-amine*) but can't be named as an *-olamine* or *-aminol*.

1. 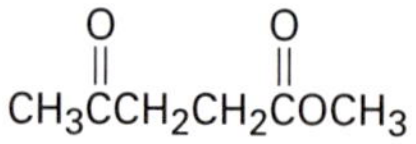

2. 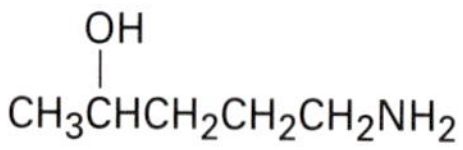

The only exception to the rule requiring a single suffix is when naming compounds that have double or triple bonds. Thus, the unsaturated acid $H_2C{=}CHCH_2CO_2H$ is 3-butenoic acid, and the acetylenic alcohol $HC{\equiv}CCH_2CH_2CH_2OH$ is 5-pentyn-1-ol.

How do we choose which suffix to use? Functional groups are divided into two classes, **principal groups** and **subordinate groups**, as shown in Table A.2. Principal groups can be cited either as prefixes or as suffixes, while subordinate groups are cited only as prefixes. Within the principal groups, an order of priority has been established, with the proper suffix for a given compound determined by choosing the principal group of highest priority. For example, Table A.2 indicates that keto ester **1** should be named as an ester rather than as a ketone because an ester functional group is higher in priority than a ketone. Similarly, amino alcohol **2** should be named as an alcohol rather than as an amine.

Table A.2 Classification of Functional Groups[a]

Functional group	Name as suffix	Name as prefix
Principal groups		
Carboxylic acids	-oic acid -carboxylic acid	carboxy
Acid anhydrides	-oic anhydride -carboxylic anhydride	—
Esters	-oate -carboxylate	alkoxycarbonyl
Thioesters	-thioate -carbothioate	alkylthiocarbonyl
Acid halides	-oyl halide -carbonyl halide	halocarbonyl
Amides	-amide -carboxamide	carbamoyl
Nitriles	-nitrile -carbonitrile	cyano
Aldehydes	-al -carbaldehyde	oxo
Ketones	-one	oxo
Alcohols	-ol	hydroxy
Phenols	-ol	hydroxy
Thiols	-thiol	mercapto
Amines	-amine	amino
Imines	-imine	imino
Ethers	ether	alkoxy
Sulfides	sulfide	alkylthio
Disulfides	disulfide	—
Alkenes	-ene	—
Alkynes	-yne	—
Alkanes	-ane	—
Subordinate groups		
Azides	—	azido
Halides	—	halo
Nitro compounds	—	nitro

[a]Principal groups are listed in order of decreasing priority; subordinate groups have no priority order.

Thus, the name of **1** is methyl 4-oxopentanoate, and the name of **2** is 5-amino-2-pentanol. Further examples are shown:

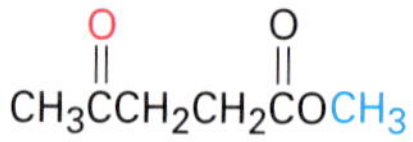

1. Methyl 4-oxopentanoate
(an ester with a ketone group)

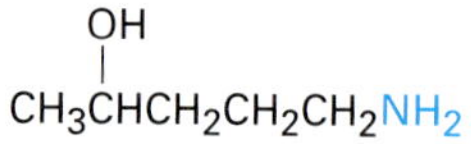

2. 5-Amino-2-pentanol
(an alcohol with an amine group)

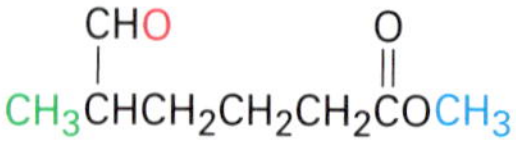

3. Methyl 5-methyl-6-oxohexanoate
(an ester with an aldehyde group)

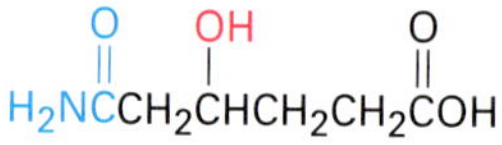

4. 5-Carbamoyl-4-hydroxypentanoic acid
(a carboxylic acid with amide and alcohol groups)

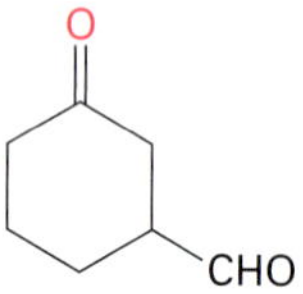

5. 3-Oxocyclohexanecarbaldehyde
(an aldehyde with a ketone group)

Name Part 2. The Parent: Selecting the Main Chain or Ring

The parent, or base, name of a polyfunctional organic compound is usually easy to identify. If the principal group of highest priority is part of an open chain, the parent name is that of the longest chain containing the largest number of principal groups. For example, compounds **6** and **7** are isomeric aldehydo amides, which must be named as amides rather than as aldehydes according to Table A.2. The longest chain in compound **6** has six carbons, and the substance is therefore named 5-methyl-6-oxohexanamide. Compound **7** also has a chain of six carbons, but the longest chain that contains both principal functional groups has only four carbons. The correct name of **7** is 4-oxo-3-propylbutanamide.

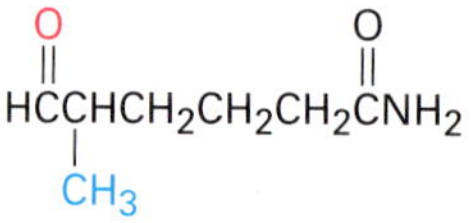

6. 5-Methyl-6-oxohexanamide

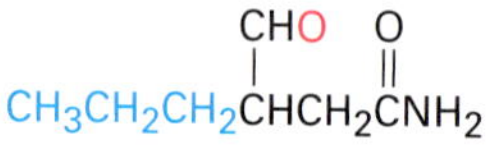

7. 4-Oxo-3-propylbutanamide

If the highest-priority principal group is attached to a ring, the parent name is that of the ring system. Compounds **8** and **9**, for instance, are isomeric keto nitriles and must both be named as nitriles according to Table A.2. Substance **8** is named as a benzonitrile because the −CN functional group is a substituent on the aromatic ring, but substance **9** is named as an acetonitrile because the −CN functional group is on an open chain. The correct names are 2-acetyl-(4-bromomethyl)benzonitrile (**8**) and (2-acetyl-4-bromophenyl)acetonitrile (**9**).

As further examples, compounds **10** and **11** are both keto acids and must be named as acids, but the parent name in **10** is that of a ring system (cyclohexanecarboxylic acid) and the parent name in **11** is that of an open chain (propanoic acid). The full names are *trans*-2-(3-oxopropyl)cyclohexanecarboxylic acid (**10**) and 3-(2-oxocyclohexyl)propanoic acid (**11**).

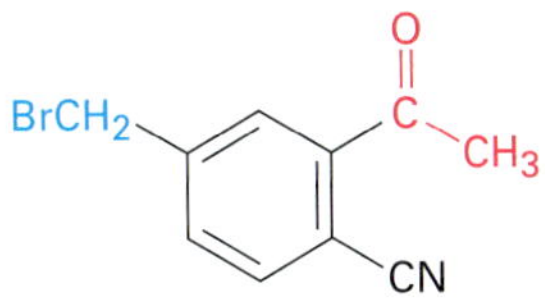

8. 2-Acetyl-(4-bromomethyl)benzonitrile

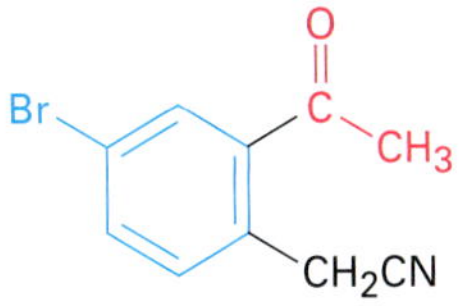

9. (2-Acetyl-4-bromophenyl)acetonitrile

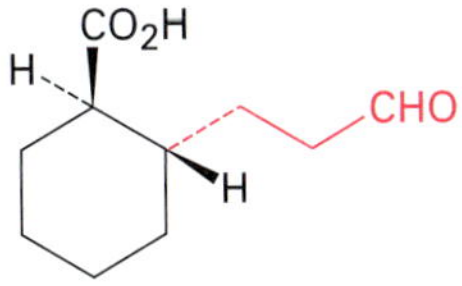

10. *trans*-2-(3-oxopropyl)cyclohexanecarboxylic acid

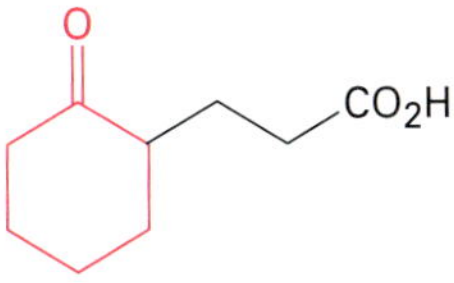

11. 3-(2-Oxocyclohexyl)propanoic acid

Name Parts 3 and 4. The Prefixes and Locants

With parent name and suffix established, the next step is to identify and give numbers, or *locants,* to all substituents on the parent chain or ring. These substituents include all alkyl groups and all functional groups other than the one cited in the suffix. For example, compound **12** contains three different functional groups (carboxyl, keto, and double bond). Because the carboxyl group is highest in priority and because the longest chain containing the functional groups has seven carbons, compound **12** is a heptenoic acid. In addition, the main chain has a keto (oxo) substituent and three methyl groups. Numbering from the end nearer the highest-priority functional group, compound **12** is named (*E*)-2,5,5-trimethyl-4-oxo-2-heptenoic acid. Look back at some of the other compounds we've named to see other examples of how prefixes and locants are assigned.

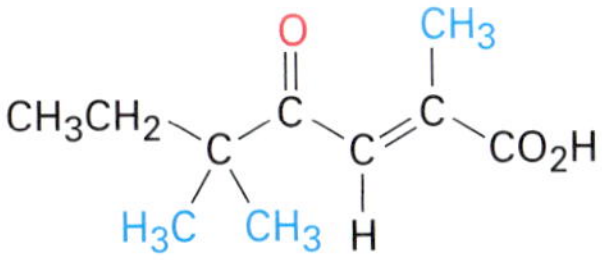

12. (*E*)-2,5,5-Trimethyl-4-oxo-2-heptenoic acid

Writing the Name

With the name parts established, the entire name is then written out. Several additional rules apply:

1. **Order of prefixes.** When the substituents have been identified, the main chain has been numbered, and the proper multipliers such as *di-* and *tri-* have been assigned, the name is written with the substituents listed in alphabetical,

rather than numerical, order. Multipliers such as *di-* and *tri-* are not used for alphabetization purposes, but the prefix *iso-* is used.

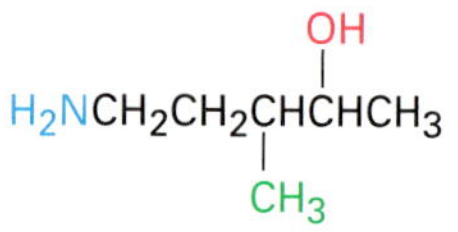

13. 5-Amino-3-methyl-2-pentanol

2. **Use of hyphens; single- and multiple-word names.** The general rule is to determine whether the parent is itself an element or compound. If so, then the name is written as a single word; if not, then the name is written as multiple words. Methylbenzene is written as one word, for instance, because the parent—benzene—is itself a compound. Diethyl ether, however, is written as two words because the parent—ether—is a class name rather than a compound name. Some further examples follow:

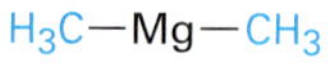

14. Dimethylmagnesium
(one word, because magnesium is an element)

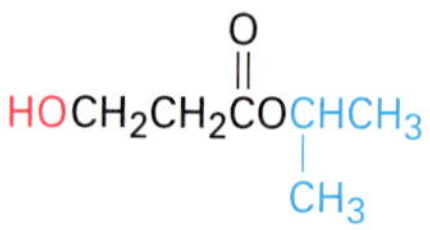

15. Isopropyl 3-hydroxypropanoate
(two words, because "propanoate" is not a compound)

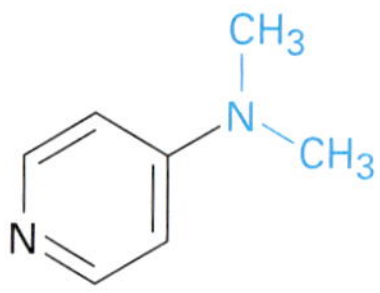

16. 4-(Dimethylamino)pyridine
(one word, because pyridine is a compound)

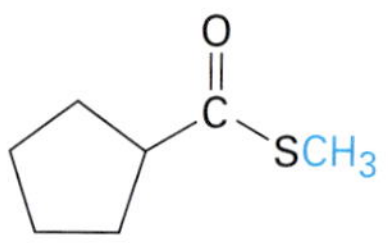

17. Methyl cyclopentanecarbothioate
(two words, because "cyclopentanecarbothioate" is not a compound)

3. **Parentheses.** Parentheses are used to denote complex substituents when ambiguity would otherwise arise. For example, chloromethylbenzene has two substituents on a benzene ring, but (chloromethyl)benzene has only one complex substituent. Note that the expression in parentheses is not set off by hyphens from the rest of the name.

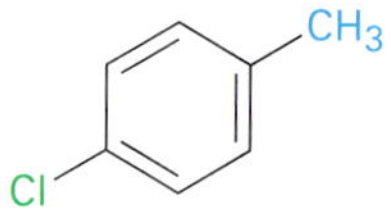

18. *p*-Chloromethylbenzene

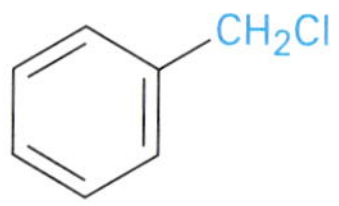

19. (Chloromethyl)benzene

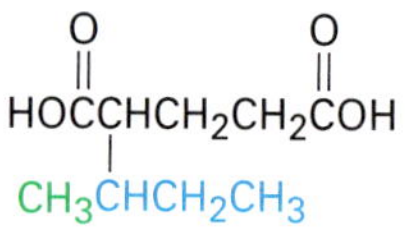

20. 2-(1-Methylpropyl)pentanedioic acid

Additional Reading

Further explanations of the rules of organic nomenclature can be found online at http://www.acdlabs.com/iupac/nomenclature/ and in the following references:

1. "A Guide to IUPAC Nomenclature of Organic Compounds," CRC Press, Boca Raton, FL, 1993.
2. "Nomenclature of Organic Chemistry, Sections A, B, C, D, E, F, and H," International Union of Pure and Applied Chemistry, Pergamon Press, Oxford, 1979.

APPENDIX

B Acidity Constants for Some Organic Compounds

Compound	pK_a
CH_3SO_3H	−1.8
$CH(NO_2)_3$	0.1
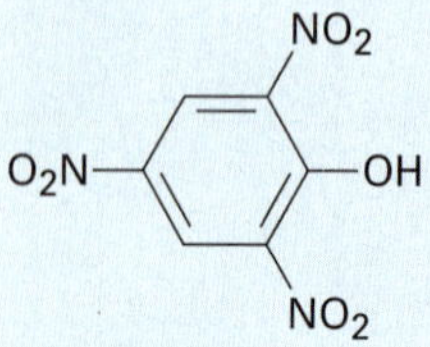	0.3
CCl_3CO_2H	0.5
CF_3CO_2H	0.5
CBr_3CO_2H	0.7
$HO_2CC{\equiv}CCO_2H$	1.2; 2.5
HO_2CCO_2H	1.2; 3.7
$CHCl_2CO_2H$	1.3
$CH_2(NO_2)CO_2H$	1.3
$HC{\equiv}CCO_2H$	1.9
Z $HO_2CCH{=}CHCO_2H$	1.9; 6.3
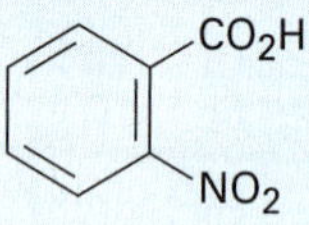	2.4
CH_3COCO_2H	2.4
$NCCH_2CO_2H$	2.5
$CH_3C{\equiv}CCO_2H$	2.6
CH_2FCO_2H	2.7
CH_2ClCO_2H	2.8
$HO_2CCH_2CO_2H$	2.8; 5.6
CH_2BrCO_2H	2.9
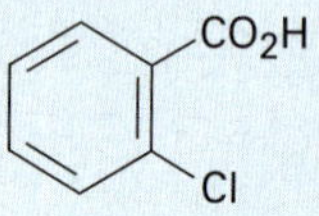	3.0
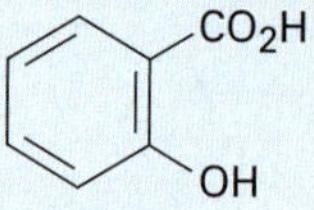	3.0
CH_2ICO_2H	3.2
$CHOCO_2H$	3.2
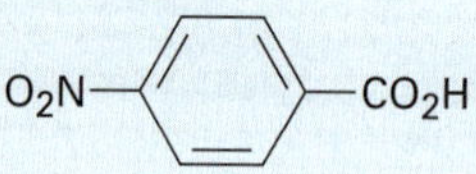	3.4
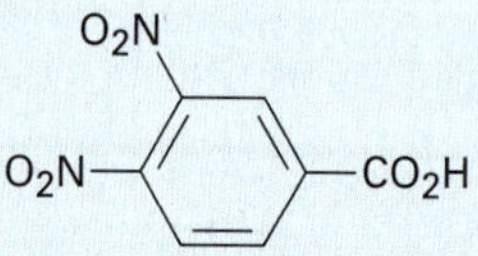	3.5
$HSCH_2CO_2H$	3.5; 10.2
$CH_2(NO_2)_2$	3.6
$CH_3OCH_2CO_2H$	3.6
$CH_3COCH_2CO_2H$	3.6
$HOCH_2CO_2H$	3.7
HCO_2H	3.7
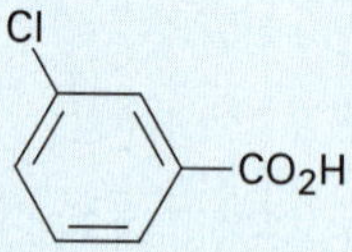	3.8
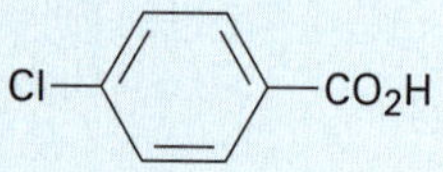	4.0
$CH_2BrCH_2CO_2H$	4.0
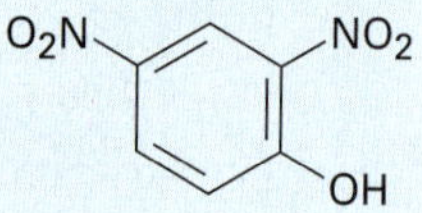	4.1
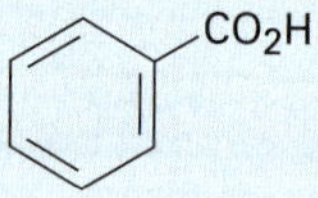	4.2
$H_2C{=}CHCO_2H$	4.2
$HO_2CCH_2CH_2CO_2H$	4.2; 5.7
$HO_2CCH_2CH_2CH_2CO_2H$	4.3; 5.4
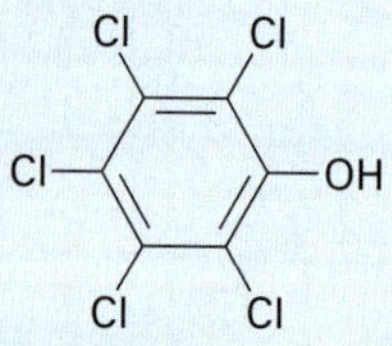	4.5
$H_2C{=}C(CH_3)CO_2H$	4.7
CH_3CO_2H	4.8

Compound	pK_a
$CH_3CH_2CO_2H$	4.8
$(CH_3)_3CCO_2H$	5.0
$CH_3COCH_2NO_2$	5.1
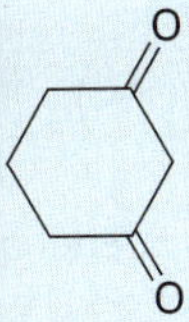	5.3
$O_2NCH_2CO_2CH_3$	5.8
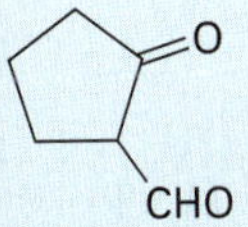	5.8
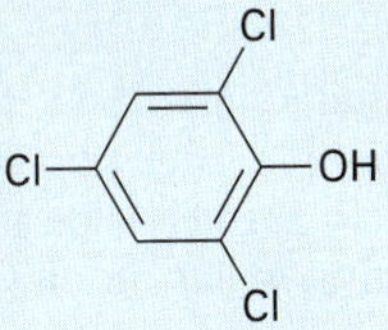	6.2
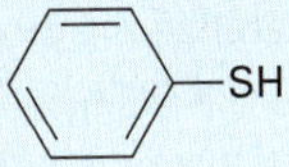	6.6
HCO_3H	7.1
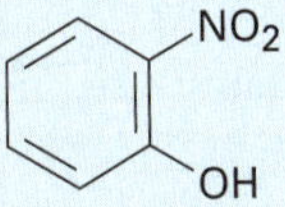	7.2
$(CH_3)_2CHNO_2$	7.7
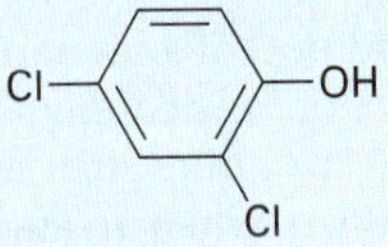	7.8
CH_3CO_3H	8.2
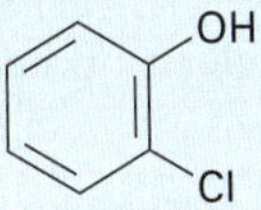	8.5
$CH_3CH_2NO_2$	8.5
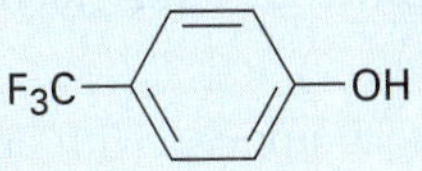	8.7

Compound	pK_a
$CH_3COCH_2COCH_3$	9.0
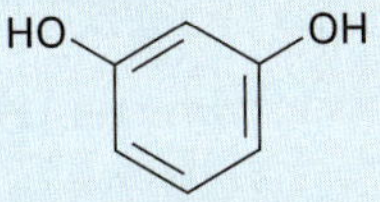	9.3; 11.1
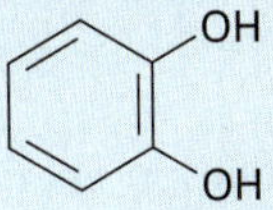	9.3; 12.6
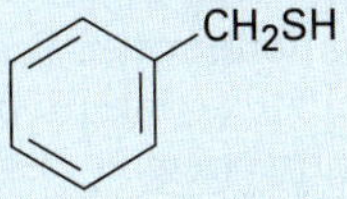	9.4
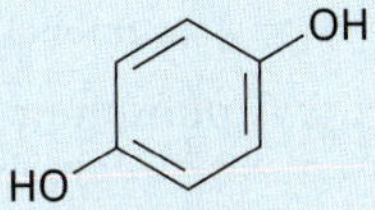	9.9; 11.5
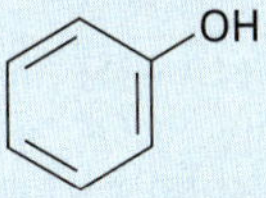	9.9
$CH_3COCH_2SOCH_3$	10.0
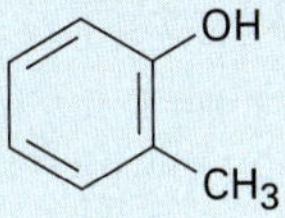	10.3
CH_3NO_2	10.3
CH_3SH	10.3
$CH_3COCH_2CO_2CH_3$	10.6
CH_3COCHO	11.0
$CH_2(CN)_2$	11.2
CCl_3CH_2OH	12.2
Glucose	12.3
$(CH_3)_2C{=}NOH$	12.4
$CH_2(CO_2CH_3)_2$	12.9
$CHCl_2CH_2OH$	12.9
$CH_2(OH)_2$	13.3
$HOCH_2CH(OH)CH_2OH$	14.1
CH_2ClCH_2OH	14.3
	15.0

Compound	pK_a
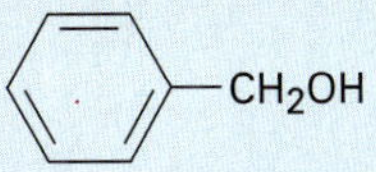	15.4
CH_3OH	15.5
$H_2C{=}CHCH_2OH$	15.5
CH_3CH_2OH	16.0
$CH_3CH_2CH_2OH$	16.1
CH_3COCH_2Br	16.1
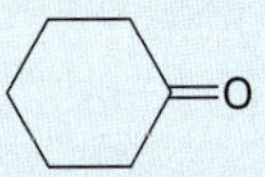	16.7
CH_3CHO	17
$(CH_3)_2CHCHO$	17
$(CH_3)_2CHOH$	17.1
$(CH_3)_3COH$	18.0
CH_3COCH_3	19.3
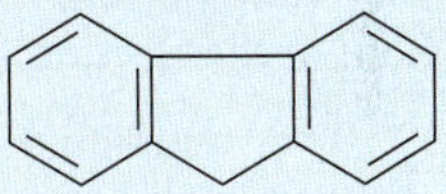	23
$CH_3CO_2CH_2CH_3$	25
$HC{\equiv}CH$	25
CH_3CN	25
$CH_3SO_2CH_3$	28
$(C_6H_5)_3CH$	32
$(C_6H_5)_2CH_2$	34
CH_3SOCH_3	35
NH_3	36
$CH_3CH_2NH_2$	36
$(CH_3CH_2)_2NH$	40
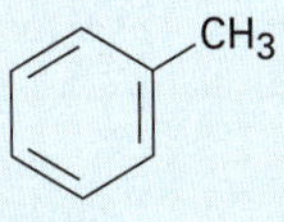	41
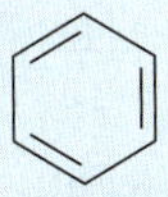	43
$H_2C{=}CH_2$	44
CH_4	~60

An acidity list covering more than 5000 organic compounds has been published: E.P. Serjeant and B. Dempsey (eds.), "Ionization Constants of Organic Acids in Aqueous Solution," IUPAC Chemical Data Series No. 23, Pergamon Press, Oxford, 1979.

APPENDIX

C Glossary

Absolute configuration (Section 9.5): The exact three-dimensional structure of a chiral molecule. Absolute configurations are specified verbally by the Cahn–Ingold–Prelog *R,S* convention and are represented on paper by Fischer projections.

Absorbance (Section 14.7): In optical spectroscopy, the logarithm of the intensity of the incident light divided by the intensity of the light transmitted through a sample; $A = \log I_0/I$.

Absorption spectrum (Section 12.5): A plot of wavelength of incident light versus amount of light absorbed. Organic molecules show absorption spectra in both the infrared and the ultraviolet regions of the electromagnetic spectrum.

Acetal (Section 19.10): A functional group consisting of two $-OR$ groups bonded to the same carbon, $R_2C(OR')_2$. Acetals are often used as protecting groups for ketones and aldehydes.

Acetoacetic ester synthesis (Section 22.7): The synthesis of a methyl ketone by alkylation of an alkyl halide, followed by hydrolysis and decarboxylation.

Acetyl group (Section 19.1): The CH_3CO- group.

Acetylide anion (Section 8.7): The anion formed by removal of a proton from a terminal alkyne.

Achiral (Section 9.2): Having a lack of handedness. A molecule is achiral if it has a plane of symmetry and is thus superimposable on its mirror image.

Acid anhydride (Section 21.1): A functional group with two acyl groups bonded to a common oxygen atom, RCO_2COR'.

Acid halide (Section 21.1): A functional group with an acyl group bonded to a halogen atom, RCOX.

Acidity constant, K_a (Section 2.8): A measure of acid strength. For any acid HA, the acidity constant is given by the expression $K_a = K_{eq}[H_2O] = \dfrac{[H_3O^+][A^-]}{[HA]}$.

Activating group (Section 16.4): An electron-donating group such as hydroxyl ($-OH$) or amino ($-NH_2$) that increases the reactivity of an aromatic ring toward electrophilic aromatic substitution.

Activation energy (Section 5.9): The difference in energy between ground state and transition state in a reaction. The amount of activation energy determines the rate at which the reaction proceeds. Most organic reactions have activation energies of 40–100 kJ/mol.

Acyl group (Sections 16.3, 19.1): A $-COR$ group.

Acyl phosphate (Section 21.8): A functional group with an acyl group bonded to a phosphate, $RCO_2PO_3^{2-}$ or $RCO_2PO_3R'^-$.

Acylation (Sections 16.3, 21.4): The introduction of an acyl group, $-COR$, onto a molecule. For example, acylation of an alcohol yields an ester, acylation of an amine yields an amide, and acylation of an aromatic ring yields an alkyl aryl ketone.

Acylium ion (Section 16.3): A resonance-stabilized carbocation in which the positive charge is located at a carbonyl-group carbon, $R-\overset{+}{C}=O \leftrightarrow R-C\equiv O^+$. Acylium ions are strongly electrophilic and are involved as intermediates in Friedel–Crafts acylation reactions.

Adams catalyst (Section 7.7): The PtO_2 catalyst used for hydrogenations.

1,2-Addition (Sections 14.2, 19.13): The addition of a reactant to the two ends of a double bond.

1,4-Addition (Sections 14.2, 19.13): Addition of a reactant to the ends of a conjugated π system. Conjugated dienes yield 1,4 adducts when treated with electrophiles such as HCl. Conjugated enones yield 1,4 adducts when treated with nucleophiles such as cyanide ion.

Addition reaction (Section 5.1): The reaction that occurs when two reactants add together to form a single new product with no atoms "left over."

Adrenocortical hormone (Section 27.6): A steroid hormone secreted by the adrenal glands. There are two types of adrenocortical hormones: mineralocorticoids and glucocorticoids.

Alcohol (Chapter 17 introduction): A compound with an $-OH$ group bonded to a saturated, alkane-like carbon, ROH.

Aldaric acid (Section 25.6): The dicarboxylic acid resulting from oxidation of an aldose.

Aldehyde (Chapter 19 introduction): A compound containing the $-CHO$ functional group.

Alditol (Section 25.6): The polyalcohol resulting from reduction of the carbonyl group of a sugar.

Aldol reaction (Section 23.1): The carbonyl condensation reaction of an aldehyde or ketone to give a β-hydroxy carbonyl compound.

Aldonic acid (Section 25.6): The monocarboxylic acid resulting from mild oxidation of the $-CHO$ group of an aldose.

Aldose (Section 25.1): A carbohydrate with an aldehyde functional group.

Alicyclic (Section 4.1): An aliphatic cyclic hydrocarbon such as a cycloalkane or cycloalkene.

Aliphatic (Section 3.2): A nonaromatic hydrocarbon such as a simple alkane, alkene, or alkyne.

Alkaloid (Chapter 2 *Focus On*): A naturally occurring organic base, such as morphine.

Alkane (Section 3.2): A compound of carbon and hydrogen that contains only single bonds.

Alkene (Chapter 6 introduction): A hydrocarbon that contains a carbon–carbon double bond, $R_2C{=}CR_2$.

Alkoxide ion (Section 17.2): The anion RO^- formed by deprotonation of an alcohol.

Alkoxymercuration reaction (Section 18.2): A method for synthesizing ethers by mercuric-ion catalyzed addition of an alcohol to an alkene.

Alkyl group (Section 3.3): The partial structure that remains when a hydrogen atom is removed from an alkane.

Alkylamine (Section 24.1): An amino-substituted alkane.

Alkylation (Sections 8.8, 16.3, 18.2, 22.7): Introduction of an alkyl group onto a molecule. For example, aromatic rings can be alkylated to yield arenes, and enolate anions can be alkylated to yield α-substituted carbonyl compounds.

Alkyne (Chapter 8 introduction): A hydrocarbon that contains a carbon–carbon triple bond, $RC{\equiv}CR$.

Allyl group (Section 6.3): A $H_2C{=}CHCH_2-$ substituent.

Allylic (Section 10.5): The position next to a double bond. For example, $H_2C{=}CHCH_2Br$ is an allylic bromide.

α-Amino acid (Section 26.1): A difunctional compound with an amino group on the carbon atom next to a carboxyl group, $RCH(NH_2)CO_2H$.

α Anomer (Section 25.5): The cyclic hemiacetal form of a sugar that has the hemiacetal $-OH$ group on the side of the ring opposite the terminal $-CH_2OH$.

α Helix (Section 26.9): The coiled secondary structure of a protein.

α Position (Chapter 22 introduction): The position next to a carbonyl group.

α-Substitution reaction (Section 22.2): The substitution of the α hydrogen atom of a carbonyl compound by reaction with an electrophile.

Amide (Chapter 21 introduction): A compound containing the $-CONR_2$ functional group.

Amidomalonate synthesis (Section 26.3): A method for preparing α-amino acids by alkylation of diethyl amidomalonate with an alkyl halide.

Amine (Chapter 24 introduction): A compound containing one or more organic substituents bonded to a nitrogen atom, RNH_2, R_2NH, or R_3N.

Amino acid (*See* α-Amino acid; Section 26.1)

Amino sugar (Section 25.7): A sugar with one of its $-OH$ groups replaced by $-NH_2$.

Amphiprotic (Section 26.1): Capable of acting either as an acid or as a base. Amino acids are amphiprotic.

Amplitude (Section 12.5): The height of a wave measured from the midpoint to the maximum. The intensity of radiant energy is proportional to the square of the wave's amplitude.

Anabolism (Section 29.1): The group of metabolic pathways that build up larger molecules from smaller ones.

Androgen (Section 27.6): A male steroid sex hormone.

Angle strain (Section 4.3): The strain introduced into a molecule when a bond angle is deformed from its ideal value. Angle strain is particularly important in small-ring cycloalkanes, where it results from compression of bond angles to less than their ideal tetrahedral values.

Annulation (Section 23.12): The building of a new ring onto an existing molecule.

Anomers (Section 25.5): Cyclic stereoisomers of sugars that differ only in their configuration at the hemiacetal (anomeric) carbon.

Antarafacial (Section 30.6): A pericyclic reaction that takes place on opposite faces of the two ends of a π electron system.

Anti conformation (Section 3.7): The geometric arrangement around a carbon–carbon single bond in which the two largest substituents are 180° apart as viewed in a Newman projection.

Anti periplanar (Section 11.8): Describing a stereochemical relationship whereby two bonds on adjacent carbons lie in the same plane at an angle of 180°.

Anti stereochemistry (Section 7.2): The opposite of syn. An anti addition reaction is one in which the two ends of the double bond are attacked from different sides. An anti elimination reaction is one in which the two groups leave from opposite sides of the molecule.

Antiaromatic (Section 15.3): Referring to a planar, conjugated molecule with $4n$ π electrons. Delocalization of the π electrons leads to an increase in energy.

Antibonding MO (Section 1.11): A molecular orbital that is higher in energy than the atomic orbitals from which it is formed.

Anticodon (Section 28.5): A sequence of three bases on tRNA that reads the codons on mRNA and brings the correct amino acids into position for protein synthesis.

Arene (Section 15.1): An alkyl-substituted benzene.

Arenediazonium salt (Section 24.7): An aromatic compound $Ar{-}\overset{+}{N}{\equiv}N\ X^-$; used in the Sandmeyer reaction.

Aromaticity (Chapter 15 introduction): The special characteristics of cyclic conjugated molecules. These characteristics include unusual stability, the presence of a ring current in the ^1H NMR spectrum, and a tendency to undergo substitution reactions rather than addition reactions on treatment with electrophiles. Aromatic molecules are planar, cyclic, conjugated species that have $4n + 2$ π electrons.

Arylamine (Section 24.1): An amino-substituted aromatic compound, $Ar{-}NH_2$.

Atactic (Section 31.2): A chain-growth polymer in which the substituents are randomly oriented along the backbone.

Atomic mass (Section 1.1): The weighted average mass of an element's naturally occurring isotopes.

Atomic number, *Z* (Section 1.1): The number of protons in the nucleus of an atom.

ATZ Derivative (Section 26.6): An anilinothiazolinone, formed from an amino acid during Edman degradation of a peptide.

Aufbau principle (Section 1.3): The rules for determining the electron configuration of an atom.

Axial bond (Section 4.6): A bond to chair cyclohexane that lies along the ring axis perpendicular to the rough plane of the ring.

Azide synthesis (Section 24.6): A method for preparing amines by S_N2 reaction of an alkyl halide with azide ion, followed by reduction.

Azo compound (Section 24.8): A compound with the general structure $R{-}N{=}N{-}R'$.

Backbone (Section 26.4): The continuous chain of atoms running the length of a polymer.

Base peak (Section 12.1): The most intense peak in a mass spectrum.

Basicity constant, K_b (Section 24.3): A measure of base strength. For any base B, the basicity constant is given by the expression

$$B + H_2O \rightleftharpoons BH^+ + OH^-$$

$$K_b = \frac{[BH^+]\,[OH^-]}{[B]}$$

Bent bonds (Section 4.4): The bonds in small rings such as cyclopropane that bend away from the internuclear line and overlap at a slight angle, rather than head-on. Bent bonds are highly strained and highly reactive.

Benzoyl group (Section 19.1): The $C_6H_5CO{-}$ group.

Benzyl group (Section 15.1): The $C_6H_5CH_2{-}$ group.

Benzylic (Section 11.5): The position next to an aromatic ring.

Benzyne (Section 16.8): An unstable compound having a triple bond in a benzene ring.

β Anomer (Section 25.5): The cyclic hemiacetal form of a sugar that has the hemiacetal $-OH$ group on the same side of the ring as the terminal $-CH_2OH$.

β-Diketone (Section 22.5): A 1,3-diketone.

β-Keto ester (Section 22.5): A 3-oxoester.

β-Oxidation pathway (Section 29.3): The metabolic pathway for degrading fatty acids.

β-Pleated sheet (Section 26.9): A type of secondary structure of a protein.

Betaine (Section 19.11): A neutral dipolar molecule with nonadjacent positive and negative charges. For example, the adduct of a Wittig reagent with a carbonyl compound is a betaine.

Bicycloalkane (Section 4.9): A cycloalkane that contains two rings.

Bimolecular reaction (Section 11.2): A reaction whose rate-limiting step occurs between two reactants.

Block copolymer (Section 31.3): A polymer in which different blocks of identical monomer units alternate with one another.

Boat cyclohexane (Section 4.5): A conformation of cyclohexane that bears a slight resemblance to a boat. Boat cyclohexane has no angle strain but has a large number of

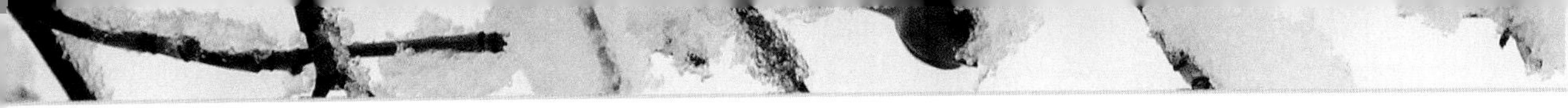

clipsing interactions that make it less stable than chair yclohexane.

Boc derivative (Section 26.7): A butyloxycarbonyl amide protected amino acid.

Bond angle (Section 1.6): The angle formed between two djacent bonds.

Bond dissociation energy, *D* (Section 5.8): The amount of energy needed to break a bond homolytically and produce wo radical fragments.

Bond length (Section 1.5): The equilibrium distance between he nuclei of two atoms that are bonded to each other.

Bond strength (Section 1.5): An alternative name for bond dissociation energy.

Bonding MO (Section 1.11): A molecular orbital that is ower in energy than the atomic orbitals from which it s formed.

Branched-chain alkane (Section 3.2): An alkane that contains a branching connection of carbons as opposed to a straight-chain alkane.

Bridgehead atom (Section 4.9): An atom that is shared by more than one ring in a polycyclic molecule.

Bromohydrin (Section 7.3): A 1,2-disubstituted bromoalcohol; obtained by addition of HOBr to an alkene.

Bromonium ion (Section 7.2): A species with a divalent, positively charged bromine, R_2Br^+.

Brønsted–Lowry acid (Section 2.7): A substance that donates a hydrogen ion (proton; H^+) to a base.

Brønsted–Lowry base (Section 2.7): A substance that accepts H^+ from an acid.

C-terminal amino acid (Section 26.4): The amino acid with a free $-CO_2H$ group at the end of a protein chain.

Cahn–Ingold–Prelog sequence rules (Sections 6.5, 9.5): A series of rules for assigning relative priorities to substituent groups on a double-bond carbon atom or on a chirality center.

Cannizzaro reaction (Section 19.12): The disproportionation reaction of an aldehyde to yield an alcohol and a carboxylic acid on treatment with base.

Carbanion (Section 19.7): A carbon anion, or substance that contains a trivalent, negatively charged carbon atom ($R_3C{:}^-$). Carbanions are sp^3-hybridized and have eight electrons in the outer shell of the negatively charged carbon.

Carbene (Section 7.6): A neutral substance that contains a divalent carbon atom having only six electrons in its outer shell ($R_2C{:}$).

Carbinolamine (Section 19.8): A molecule that contains the $R_2C(OH)NH_2$ functional group. Carbinolamines are produced as intermediates during the nucleophilic addition of amines to carbonyl compounds.

Carbocation (Sections 5.5, 6.9): A carbon cation, or substance that contains a trivalent, positively charged carbon atom having six electrons in its outer shell (R_3C^+).

Carbohydrate (Section 25.1): A polyhydroxy aldehyde or ketone. Carbohydrates can be either simple sugars, such as glucose, or complex sugars, such as cellulose.

Carbonyl condensation reaction (Section 23.1): A reaction that joins two carbonyl compounds together by a combination of α-substitution and nucleophilic addition reactions.

Carbonyl group (Section 2.1): The C=O functional group.

Carboxyl group (Section 20.1): The $-CO_2H$ functional group.

Carboxylation (Section 20.5): The addition of CO_2 to a molecule.

Carboxylic acid (Chapter 20 introduction): A compound containing the $-CO_2H$ functional group.

Carboxylic acid derivative (Chapter 21 introduction): A compound in which an acyl group is bonded to an electronegative atom or substituent Y that can act as a leaving group in a substitution reaction, RCOY.

Catabolism (Section 29.1): The group of metabolic pathways that break down larger molecules into smaller ones.

Cation radical (Section 12.1): A reactive species formed by loss of an electron from a neutral molecule.

Chain-growth polymer (Section 31.1): A polymer whose bonds are produced by chain reactions. Polyethylene and other alkene polymers are examples.

Chain reaction (Section 5.3): A reaction that, once initiated, sustains itself in an endlessly repeating cycle of propagation steps. The radical chlorination of alkanes is an example of a chain reaction that is initiated by irradiation with light and then continues in a series of propagation steps.

Chair cyclohexane (Section 4.5): A three-dimensional conformation of cyclohexane that resembles the rough shape of a chair. The chair form of cyclohexane is the lowest-energy conformation of the molecule.

Chemical shift (Section 13.3): The position on the NMR chart where a nucleus absorbs. By convention, the chemical shift of tetramethylsilane (TMS) is set at zero, and all other absorptions usually occur downfield (to the left on the chart). Chemical shifts are expressed in delta units, δ, where 1 δ equals 1 ppm of the spectrometer operating frequency.

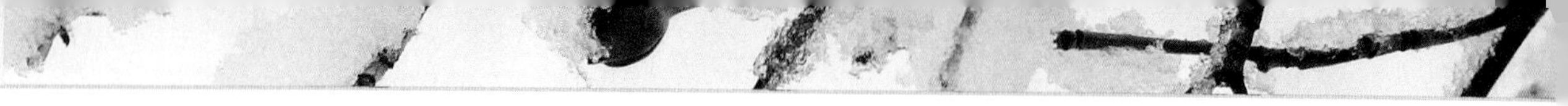

Chiral (Section 9.2): Having handedness. Chiral molecules are those that do not have a plane of symmetry and are therefore not superimposable on their mirror image. A chiral molecule thus exists in two forms, one right-handed and one left-handed. The most common cause of chirality in a molecule is the presence of a carbon atom that is bonded to four different substituents.

Chiral environment (Section 9.14): Chiral surroundings or conditions in which a molecule resides.

Chirality center (Section 9.2): An atom (usually carbon) that is bonded to four different groups.

Chlorohydrin (Section 7.3): A 1,2-disubstituted chloroalcohol; obtained by addition of HOCl to an alkene.

Chromatography (Chapter 12 *Focus On,* Section 26.7): A technique for separating a mixture of compounds into pure components. Different compounds adsorb to a stationary support phase and are then carried along it at different rates by a mobile phase.

Cis–trans isomers (Sections 4.2, 6.4): Stereoisomers that differ in their stereochemistry about a double bond or ring.

Citric acid cycle (Section 29.7): The metabolic pathway by which acetyl CoA is degraded to CO_2.

Claisen condensation reaction (Section 23.7): The carbonyl condensation reaction of an ester to give a β-keto ester product.

Claisen rearrangement reaction (Sections 18.4, 30.8): The pericyclic conversion of an allyl phenyl ether to an *o*-allylphenol by heating.

Coding strand (Section 28.4): The strand of double-helical DNA that contains the gene.

Codon (Section 28.5): A three-base sequence on a messenger RNA chain that encodes the genetic information necessary to cause a specific amino acid to be incorporated into a protein. Codons on mRNA are read by complementary anticodons on tRNA.

Coenzyme (Section 26.10): A small organic molecule that acts as a cofactor.

Cofactor (Section 26.10): A small nonprotein part of an enzyme that is necessary for biological activity.

Combinatorial chemistry (Chapter 16 *Focus On*): A procedure in which anywhere from a few dozen to several hundred thousand substances are prepared simultaneously.

Complex carbohydrate (Section 25.1): A carbohydrate that is made of two or more simple sugars linked together.

Concerted (Section 30.1): A reaction that takes place in a single step without intermediates. For example, the Diels–Alder cycloaddition reaction is a concerted process.

Condensed structure (Section 1.12): A shorthand way of writing structures in which carbon–hydrogen and carbon–carbon bonds are understood rather than shown explicitly. Propane, for example, has the condensed structure $CH_3CH_2CH_3$.

Configuration (Section 9.5): The three-dimensional arrangement of atoms bonded to a chirality center.

Conformation (Section 3.6): The three-dimensional shape of a molecule at any given instant, assuming that rotation around single bonds is frozen.

Conformational analysis (Section 4.8): A means of assessing the energy of a substituted cycloalkane by totaling the steric interactions present in the molecule.

Conformer (Section 3.6): A conformational isomer.

Conjugate acid (Section 2.7): The product that results from protonation of a Brønsted–Lowry base.

Conjugate addition (Section 19.13): Addition of a nucleophile to the β carbon atom of an α,β-unsaturated carbonyl compound.

Conjugate base (Section 2.7): The product that results from deprotonation of a Brønsted–Lowry acid.

Conjugation (Chapter 14 introduction): A series of overlapping *p* orbitals, usually in alternating single and multiple bonds. For example, 1,3-butadiene is a conjugated diene, 3-buten-2-one is a conjugated enone, and benzene is a cyclic conjugated triene.

Conrotatory (Section 30.2): A term used to indicate that *p* orbitals must rotate in the same direction during electrocyclic ring-opening or ring closure.

Constitutional isomers (Sections 3.2, 9.9): Isomers that have their atoms connected in a different order. For example, butane and 2-methylpropane are constitutional isomers.

Cope rearrangement (Section 30.8): The sigmatropic rearrangement of a 1,5-hexadiene.

Copolymer (Section 31.3): A polymer obtained when two or more different monomers are allowed to polymerize together.

Coupling constant, *J* (Section 13.11): The magnitude (expressed in hertz) of the interaction between nuclei whose spins are coupled.

Covalent bond (Section 1.5): A bond formed by sharing electrons between atoms.

Cracking (Chapter 3 *Focus On*): A process used in petroleum refining in which large alkanes are thermally cracked into smaller fragments.

Crown ether (Section 18.7): A large-ring polyether; used as a phase-transfer catalyst.

Crystallite (Section 31.5): A highly ordered crystal-like region within a long polymer chain.

Curtius rearrangement (Section 24.6): The conversion of an acid chloride into an amine by reaction with azide ion, followed by heating with water.

Cyanohydrin (Section 19.6): A compound with an –OH group and a –CN group bonded to the same carbon atom; formed by addition of HCN to an aldehyde or ketone.

Cycloaddition reaction (Sections 14.4, 30.6): A pericyclic reaction in which two reactants add together in a single step to yield a cyclic product. The Diels–Alder reaction between a diene and a dienophile to give a cyclohexene is an example.

Cycloalkane (Section 4.1): An alkane that contains a ring of carbons.

D Sugar (Section 25.3): A sugar whose hydroxyl group at the chirality center farthest from the carbonyl group points to the right when drawn in Fischer projection.

***d,l* form** (Section 9.8): The racemic modification of a compound.

Deactivating group (Section 16.4): An electron-withdrawing substituent that decreases the reactivity of an aromatic ring toward electrophilic aromatic substitution.

Debye, D (Section 2.2): A unit for measuring dipole moments; $1\ D = 3.336 \times 10^{-30}$ coulomb meter (C · m).

Decarboxylation (Section 22.7): The loss of carbon dioxide from a molecule. β-Keto acids decarboxylate readily on heating.

Degenerate orbitals (Section 15.2): Two or more orbitals that have the same energy level.

Degree of unsaturation (Section 6.2): The number of rings and/or multiple bonds in a molecule.

Dehydration (Sections 7.1, 11.10, 17.6): The loss of water from an alcohol. Alcohols can be dehydrated to yield alkenes.

Dehydrohalogenation (Sections 7.1, 11.8): The loss of HX from an alkyl halide. Alkyl halides undergo dehydrohalogenation to yield alkenes on treatment with strong base.

Delocalization (Section 10.5): A spreading out of electron density over a conjugated π electron system. For example, allylic cations and allylic anions are delocalized because their charges are spread out over the entire π electron system.

Delta scale (Section 13.3): An arbitrary scale used to calibrate NMR charts. One delta unit (δ) is equal to 1 part per million (ppm) of the spectrometer operating frequency.

Denaturation (Section 26.9): The physical changes that occur in a protein when secondary and tertiary structures are disrupted.

Deoxy sugar (Section 25.7): A sugar with one of its –OH groups replaced by an –H.

Deoxyribonucleic acid (DNA) (Section 28.1): The biopolymer consisting of deoxyribonucleotide units linked together through phosphate–sugar bonds. Found in the nucleus of cells, DNA contains an organism's genetic information.

DEPT-NMR (Section 13.6): An NMR method for distinguishing among signals due to CH_3, CH_2, CH, and quaternary carbons. That is, the number of hydrogens attached to each carbon can be determined.

Deshielding (Section 13.2): An effect observed in NMR that causes a nucleus to absorb downfield (to the left) of tetramethylsilane (TMS) standard. Deshielding is caused by a withdrawal of electron density from the nucleus.

Deuterium isotope effect (Section 11.8): A tool used in mechanistic investigations to establish whether a C–H bond is broken in the rate-limiting step of a reaction.

Dextrorotatory (Section 9.3): A word used to describe an optically active substance that rotates the plane of polarization of plane-polarized light in a right-handed (clockwise) direction.

Diastereomers (Section 9.6): Non–mirror-image stereoisomers; diastereomers have the same configuration at one or more chirality centers but differ at other chirality centers.

Diastereotopic (Section 13.8): Two hydrogens in a molecule whose replacement by some other group leads to different diastereomers.

1,3-Diaxial interaction (Section 4.8): The strain energy caused by a steric interaction between axial groups three carbon atoms apart in chair cyclohexane.

Diazonium salt (Section 24.8): A compound with the general structure $RN_2^+ X^-$.

Diazotization (Section 24.8): The conversion of a primary amine, RNH_2, into a diazonium ion, RN_2^+, by treatment with nitrous acid.

Dideoxy DNA sequencing (Section 28.6): A biochemical method for sequencing DNA strands.

Dieckmann cyclization reaction (Section 23.9): An intramolecular Claisen condensation reaction to give a cyclic β-keto ester.

Diels–Alder reaction (Sections 14.4, 30.6): The cycloaddition reaction of a diene with a dienophile to yield a cyclohexene.

Dienophile (Section 14.5): A compound containing a double bond that can take part in the Diels–Alder cycloaddition reaction. The most reactive dienophiles are those that have electron-withdrawing groups on the double bond.

Digestion (Section 29.1): The first stage of catabolism, in which food is broken down by hydrolysis of ester, glycoside (acetal), and peptide (amide) bonds to yield fatty acids, simple sugars, and amino acids.

Dipole moment, μ (Section 2.2): A measure of the net polarity of a molecule. A dipole moment arises when the centers of mass of positive and negative charges within a molecule do not coincide.

Dipole–dipole force (Section 2.13): A noncovalent electrostatic interaction between dipolar molecules.

Disaccharide (Section 25.8): A carbohydrate formed by linking two simple sugars through an acetal bond.

Dispersion force (Section 2.13): A noncovalent interaction between molecules that arises because of constantly changing electron distributions within the molecules.

Disrotatory (Section 30.2): A term used to indicate that *p* orbitals rotate in opposite directions during electrocyclic ring-opening or ring closing.

Disulfide (Section 18.8): A compound of the general structure RSSR′.

DNA (*See* Deoxyribonucleic acid; Section 28.1)

Double helix (Section 28.2): The structure of DNA in which two polynucleotide strands coil around each other.

Doublet (Section 13.11): A two-line NMR absorption caused by spin–spin splitting when the spin of the nucleus under observation couples with the spin of a neighboring magnetic nucleus.

Downfield (Section 13.3): Referring to the left-hand portion of the NMR chart.

***E* geometry** (Section 6.5): A term used to describe the stereochemistry of a carbon–carbon double bond. The two groups on each carbon are assigned priorities according to the Cahn–Ingold–Prelog sequence rules, and the two carbons are compared. If the high-priority groups on each carbon are on opposite sides of the double bond, the bond has *E* geometry.

E1 reaction (Section 11.10): A unimolecular elimination reaction in which the substrate spontaneously dissociates to give a carbocation intermediate, which loses a proton in a separate step.

E1cB reaction (Section 11.10): A unimolecular elimination reaction in which a proton is first removed to give a carbanion intermediate, which then expels the leaving group in a separate step.

E2 reaction (Section 11.8): A bimolecular elimination reaction in which both the hydrogen and the leaving group are lost in the same step.

Eclipsed conformation (Section 3.6): The geometric arrangement around a carbon–carbon single bond in which the bonds to substituents on one carbon are parallel to the bonds to substituents on the neighboring carbon as viewed in a Newman projection.

Eclipsing strain (Section 3.6): The strain energy in a molecule caused by electron repulsions between eclipsed bonds. Eclipsing strain is also called torsional strain.

Edman degradation (Section 26.6): A method for N-terminal sequencing of peptide chains by treatment with *N*-phenylisothiocyanate.

Eicosanoid (Section 27.4): A lipid derived biologically from 5,8,11,14-eicosatetraenoic acid, or arachidonic acid. Prostaglandins, thromboxanes and leukotrienes are examples.

Elastomer (Section 31.5): An amorphous polymer that has the ability to stretch out and spring back to its original shape.

Electrocyclic reaction (Section 30.3): A unimolecular pericyclic reaction in which a ring is formed or broken by a concerted reorganization of electrons through a cyclic transition state. For example, the cyclization of 1,3,5-hexatriene to yield 1,3-cyclohexadiene is an electrocyclic reaction.

Electromagnetic spectrum (Section 12.5): The range of electromagnetic energy, including infrared, ultraviolet, and visible radiation.

Electron configuration (Section 1.3): A list of the orbitals occupied by electrons in an atom.

Electron-dot structure (Section 1.4): A representation of a molecule showing valence electrons as dots.

Electron-transport chain (Section 29.1): The final stage of catabolism in which ATP is produced.

Electronegativity (Section 2.1): The ability of an atom to attract electrons in a covalent bond. Electronegativity increases across the periodic table from right to left and from bottom to top.

Electrophile (Section 5.4): An "electron-lover," or substance that accepts an electron pair from a nucleophile in a polar bond-forming reaction.

Electrophilic addition reaction (Section 6.7): The addition of an electrophile to a carbon–carbon double bond to yield a saturated product.

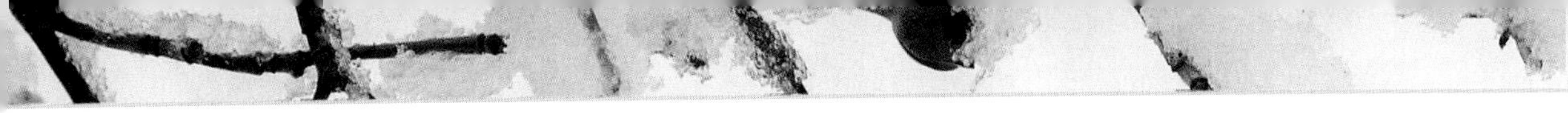

Electrophilic aromatic substitution (Chapter 16 introduction): A reaction in which an electrophile (E^+) reacts with an aromatic ring and substitutes for one of the ring hydrogens.

Electrophoresis (Section 26.2): A technique used for separating charged organic molecules, particularly proteins and amino acids. The mixture to be separated is placed on a buffered gel or paper, and an electric potential is applied across the ends of the apparatus. Negatively charged molecules migrate toward the positive electrode, and positively charged molecules migrate toward the negative electrode.

Electrostatic potential map (Section 2.1): A molecular representation that uses color to indicate the charge distribution in the molecule as derived from quantum-mechanical calculations.

Elimination reaction (Section 5.1): What occurs when a single reactant splits into two products.

Elution (Chapter 12 *Focus On*): The removal of a substance from a chromatography column.

Embden–Meyerhof pathway (Section 29.5): An alternative name for glycolysis.

Enamine (Section 19.8): A compound with the $R_2N{-}CR{=}CR_2$ functional group.

Enantiomers (Section 9.1): Stereoisomers of a chiral substance that have a mirror-image relationship. Enantiomers must have opposite configurations at all chirality centers.

Enantioselective synthesis (Chapter 19 *Focus On*): A reaction method that yields only a single enantiomer of a chiral product starting from an achiral substrate.

Enantiotopic (Section 13.8): Two hydrogens in a molecule whose replacement by some other group leads to different enantiomers.

3′ End (Section 28.1): The end of a nucleic acid chain with a free hydroxyl group at C3′.

5′ End (Section 28.1): The end of a nucleic acid chain with a free hydroxyl group at C5′.

Endergonic (Section 5.7): A reaction that has a positive free-energy change and is therefore nonspontaneous. In a reaction energy diagram, the product of an endergonic reaction has a higher energy level than the reactants.

Endo (Section 14.5): A term indicating the stereochemistry of a substituent in a bridged bicycloalkane. An endo substituent is syn to the larger of the two bridges.

Endothermic (Section 5.7): A reaction that absorbs heat and therefore has a positive enthalpy change.

Energy diagram (Section 5.9): A representation of the course of a reaction, in which free energy is plotted as a function of reaction progress. Reactants, transition states, intermediates, and products are represented, and their appropriate energy levels are indicated.

Enol (Sections 8.4, 22.1): A vinylic alcohol that is in equilibrium with a carbonyl compound.

Enolate ion (Section 22.1): The anion of an enol.

Enthalpy change, ΔH (Section 5.7): The heat of reaction. The enthalpy change that occurs during a reaction is a measure of the difference in total bond energy between reactants and products.

Entropy change, ΔS (Section 5.7): The change in amount of molecular randomness. The entropy change that occurs during a reaction is a measure of the difference in randomness between reactants and products.

Enzyme (Section 26.10): A biological catalyst. Enzymes are large proteins that catalyze specific biochemical reactions.

Epoxide (Section 7.8): A three-membered-ring ether functional group.

Equatorial bond (Section 4.6): A bond to cyclohexane that lies along the rough equator of the ring.

ESI (Section 12.4): Electrospray ionization, a mild method for ionizing a molecule so that fragmentation is minimized during mass spectrometry.

Essential oil (Chapter 6 *Focus On*): The volatile oil obtained by steam distillation of a plant extract.

Ester (Chapter 21 introduction): A compound containing the $-CO_2R$ functional group.

Estrogen (Section 27.6): A female steroid sex hormone.

Ether (Chapter 18 introduction): A compound that has two organic substituents bonded to the same oxygen atom, ROR′.

Exergonic (Section 5.7): A reaction that has a negative free-energy change and is therefore spontaneous. On a reaction energy diagram, the product of an exergonic reaction has a lower energy level than that of the reactants.

Exo (Section 14.5): A term indicating the stereochemistry of a substituent in a bridged bicycloalkane. An exo substituent is anti to the larger of the two bridges.

Exon (Section 28.4): A section of DNA that contains genetic information.

Exothermic (Section 5.7): A reaction that releases heat and therefore has a negative enthalpy change.

Fat (Section 27.1): A solid triacylglycerol derived from an animal source.

Fatty acid (Section 27.1): A long, straight-chain carboxylic acid found in fats and oils.

Fiber (Section 31.5): A thin thread produced by extruding a molten polymer through small holes in a die.

Fibrous protein (Section 26.9): A protein that consists of polypeptide chains arranged side by side in long threads. Such proteins are tough, insoluble in water, and used in nature for structural materials such as hair, hooves, and fingernails.

Fingerprint region (Section 12.7): The complex region of the infrared spectrum from 1500 to 400 cm^{-1}.

First-order reaction (Section 11.4): A reaction whose rate-limiting step is unimolecular and whose kinetics therefore depend on the concentration of only one reactant.

Fischer esterification reaction (Section 21.3): The acid-catalyzed nucleophilic acyl substitution reaction of a carboxylic acid with an alcohol to yield an ester.

Fischer projection (Section 25.2): A means of depicting the absolute configuration of a chiral molecule on a flat page. A Fischer projection uses a cross to represent the chirality center. The horizontal arms of the cross represent bonds coming out of the plane of the page, and the vertical arms of the cross represent bonds going back into the plane of the page.

Fmoc derivative (Section 26.7): A fluorenylmethyloxycarbonyl amide-protected amino acid.

Formal charge (Section 2.3): The difference in the number of electrons owned by an atom in a molecule and by the same atom in its elemental state.

Formyl group (Section 19.1): A –CHO group.

Frequency (Section 12.5): The number of electromagnetic wave cycles that travel past a fixed point in a given unit of time. Frequencies are expressed in units of cycles per second, or hertz.

Friedel–Crafts reaction (Section 16.3): An electrophilic aromatic substitution reaction to alkylate or acylate an aromatic ring.

Frontier orbitals (Section 30.1): The highest occupied (HOMO) and lowest unoccupied (LUMO) molecular orbitals.

FT-NMR (Section 13.4): Fourier-transform NMR; a rapid technique for recording NMR spectra in which all magnetic nuclei absorb at the same time.

Functional group (Section 3.1): An atom or group of atoms that is part of a larger molecule and that has a characteristic chemical reactivity.

Furanose (Section 25.5): The five-membered-ring form of a simple sugar.

Gabriel amine synthesis (Section 24.6): A method for preparing an amine by S_N2 reaction of an alkyl halide with potassium phthalimide, followed by hydrolysis.

Gauche conformation (Section 3.7): The conformation of butane in which the two methyl groups lie 60° apart as viewed in a Newman projection. This conformation has 3.8 kJ/mol steric strain.

Geminal (Section 19.5): Referring to two groups attached to the same carbon atom. For example, 1,1-dibromopropane is a geminal dibromide.

Gibbs free-energy change, ΔG (Section 5.7): The free-energy change that occurs during a reaction, given by the equation $\Delta G = \Delta H - T\Delta S$. A reaction with a negative free-energy change is spontaneous, and a reaction with a positive free-energy change is nonspontaneous.

Gilman reagent (Section 10.8): A diorganocopper reagent, R_2CuLi.

Glass transition temperature, T_g (Section 31.5): The temperature at which a hard, amorphous polymer becomes soft and flexible.

Globular protein (Section 26.9): A protein that is coiled into a compact, nearly spherical shape. These proteins, which are generally water-soluble and mobile within the cell, are the structural class to which enzymes belong.

Gluconeogenesis (Section 29.8): The anabolic pathway by which organisms make glucose from simple precursors.

Glycal assembly method (Section 25.11): A method for linking monosaccharides together to synthesis polysaccharides.

Glycerophospholipid (Section 27.3): A lipid that contains a glycerol backbone linked to two fatty acids and a phosphoric acid.

Glycoconjugate (Section 25.6): A biological molecule in which a carbohydrate is linked through a glycoside bond to a lipid or protein.

Glycol (Section 7.8): A diol, such as ethylene glycol, $HOCH_2CH_2OH$.

Glycolipid (Section 25.6): A biological molecule in which a carbohydrate is linked through a glycoside bond to a lipid.

Glycoprotein (Section 25.6): A biological molecule in which a carbohydrate is linked through a glycoside bond to a protein.

Glycolysis (Section 29.5): A series of ten enzyme-catalyzed reactions that break down glucose into 2 equivalents of pyruvate, $CH_3COCO_2^-$.

Glycoside (Section 25.6): A cyclic acetal formed by reaction of a sugar with another alcohol.

Graft copolymer (Section 31.3): A copolymer in which homopolymer branches of one monomer unit are "grafted" onto a homopolymer chain of another monomer unit.

Green chemistry (Chapter 11 *Focus On*): The design and implementation of chemical products and processes that reduce waste and minimize or eliminate the generation of hazardous substances.

Grignard reagent (Section 10.7): An organomagnesium halide, RMgX.

Ground state (Section 1.3): The most stable, lowest-energy electron configuration of a molecule or atom.

Haloform reaction (Section 22.6): The reaction of a methyl ketone with halogen and base to yield a haloform (CHX_3) and a carboxylic acid.

Halohydrin (Section 7.3): A 1,2-disubstituted haloalcohol, such as that obtained on addition of HOBr to an alkene.

Halonium ion (Section 7.2): A species containing a positively charged, divalent halogen. Three-membered-ring bromonium ions are implicated as intermediates in the electrophilic addition of Br_2 to alkenes.

Hammond postulate (Section 6.10): A postulate stating that we can get a picture of what a given transition state looks like by looking at the structure of the nearest stable species. Exergonic reactions have transition states that resemble reactant; endergonic reactions have transition states that resemble product.

Heat of hydrogenation (Section 6.6): The amount of heat released when a carbon–carbon double bond is hydrogenated.

Heat of reaction (Section 5.7): An alternative name for the enthalpy change in a reaction, ΔH.

Hell–Volhard–Zelinskii (HVZ) reaction (Section 22.4): The reaction of a carboxylic acid with Br_2 and phosphorus to give an α-bromo carboxylic acid.

Hemiacetal (Section 19.10): A functional group consisting of one –OR and one –OH group bonded to the same carbon.

Henderson–Hasselbalch equation (Sections 20.3, 24.5, 26.2): An equation for determining the extent of deprotonation of a weak acid at various pH values.

Heterocycle (Sections 15.5, 24.9): A cyclic molecule whose ring contains more than one kind of atom. For example, pyridine is a heterocycle that contains five carbon atoms and one nitrogen atom in its ring.

Heterolytic bond breakage (Section 5.2): The kind of bond-breaking that occurs in polar reactions when one fragment leaves with both of the bonding electrons: $A{:}B \rightarrow A^+ + B{:}^-$.

Hofmann elimination (Section 24.7): The elimination reaction of an amine to yield an alkene by reaction with iodomethane, followed by heating with Ag_2O.

Hofmann rearrangement (Section 24.6): The conversion of an amide into an amine by reaction with Br_2 and base.

HOMO (Sections 14.7, 30.2): An acronym for highest occupied molecular orbital. The symmetries of the HOMO and LUMO are important in pericyclic reactions.

Homolytic bond breakage (Section 5.2): The kind of bond-breaking that occurs in radical reactions when each fragment leaves with one bonding electron: $A{:}B \rightarrow A\cdot + B\cdot$.

Homopolymer (Section 31.3): A polymer made up of identical repeating units.

Homotopic (Section 13.8): Hydrogens that give the identical structure on replacement by X and thus show identical NMR absorptions.

Hormone (Section 27.6): A chemical messenger that is secreted by an endocrine gland and carried through the bloodstream to a target tissue.

Hückel's rule (Section 15.3): A rule stating that monocyclic conjugated molecules having $4n + 2$ π electrons (n = an integer) are aromatic.

Hund's rule (Section 1.3): If two or more empty orbitals of equal energy are available, one electron occupies each, with their spins parallel, until all are half-full.

Hybrid orbital (Section 1.6): An orbital derived from a combination of atomic orbitals. Hybrid orbitals, such as the sp^3, sp^2, and sp hybrids of carbon, are strongly directed and form stronger bonds than atomic orbitals do.

Hydration (Section 7.4): Addition of water to a molecule, such as occurs when alkenes are treated with aqueous sulfuric acid to give alcohols.

Hydride shift (Section 6.11): The shift of a hydrogen atom and its electron pair to a nearby cationic center.

Hydroboration (Section 7.5): Addition of borane (BH_3) or an alkylborane to an alkene. The resultant trialkylborane products are useful synthetic intermediates that can be oxidized to yield alcohols.

Hydrocarbon (Section 3.2): A compound that contains only carbon and hydrogen.

Hydrogen bond (Section 2.13): A weak attraction between a hydrogen atom bonded to an electronegative atom and an electron lone pair on another electronegative atom.

Hydrogenation (Section 7.7): Addition of hydrogen to a double or triple bond to yield a saturated product.

Hydrogenolysis (Section 26.7): Cleavage of a bond by reaction with hydrogen. Benzylic ethers and esters, for instance, are cleaved by hydrogenolysis.

Hydrophilic (Section 2.13): Water-loving; attracted to water.

Hydrophobic (Section 2.13): Water-fearing; repelled by water.

Hydroquinone (Section 17.10): A 1,4-dihydroxybenzene.

Hydroxylation (Section 7.8): Addition of two –OH groups to a double bond.

Hyperconjugation (Sections 6.6, 6.9): An interaction that results from overlap of a vacant *p* orbital on one atom with a neighboring C–H σ bond. Hyperconjugation is important in stabilizing carbocations and in stabilizing substituted alkenes.

Imide (Section 24.6): A compound with the –CONHCO– functional group.

Imine (Section 19.8): A compound with the $R_2C{=}NR$ functional group.

Inductive effect (Sections 2.1, 6.9, 16.4): The electron-attracting or electron-withdrawing effect transmitted through σ bonds. Electronegative elements have an electron-withdrawing inductive effect.

Infrared (IR) spectroscopy (Section 12.6): A kind of optical spectroscopy that uses infrared energy. IR spectroscopy is particularly useful in organic chemistry for determining the kinds of functional groups present in molecules.

Initiator (Section 5.3): A substance with an easily broken bond that is used to initiate a radical chain reaction. For example, radical chlorination of alkanes is initiated when light energy breaks the weak Cl–Cl bond to form Cl· radicals.

Integration (Section 13.10): A technique for measuring the area under an NMR peak to determine the relative number of each kind of proton in a molecule. Integrated peak areas are superimposed over the spectrum as a "stair-step" line, with the height of each step proportional to the area underneath the peak.

Intermediate (Section 5.10): A species that is formed during the course of a multistep reaction but is not the final product. Intermediates are more stable than transition states but may or may not be stable enough to isolate.

Intramolecular, intermolecular (Section 23.6): A reaction that occurs within the same molecule is intramolecular; a reaction that occurs between two molecules is intermolecular.

Intron (Section 28.4): A section of DNA that does not contain genetic information.

Ion pair (Section 11.5): A loose complex between two ions in solution. Ion pairs are implicated as intermediates in S_N1 reactions to account for the partial retention of stereochemistry that is often observed.

Isoelectric point, p*I* (Section 26.2): The pH at which the number of positive charges and the number of negative charges on a protein or an amino acid are equal.

Isomers (Sections 3.2, 9.9): Compounds that have the same molecular formula but different structures.

Isoprene rule (Chapter 6 *Focus On*): An observation to the effect that terpenoids appear to be made up of isoprene (2-methyl-1,3-butadiene) units connected head-to-tail.

Isotactic (Section 31.2): A chain-growth polymer in which the substituents are regularly oriented on the same side of the backbone.

Isotopes (Section 1.1): Atoms of the same element that have different mass numbers.

IUPAC system of nomenclature (Section 3.4): Rules for naming compounds, devised by the International Union of Pure and Applied Chemistry.

Kekulé structure (Section 1.4): A method of representing molecules in which a line between atoms indicates a bond.

Keto–enol tautomerism (Sections 8.4, 22.1): The rapid equilibration between a carbonyl form and vinylic alcohol form of a molecule.

Ketone (Chapter 19 introduction): A compound with two organic substituents bonded to a carbonyl group, $R_2C{=}O$.

Ketose (Section 25.1): A carbohydrate with a ketone functional group.

Kiliani–Fischer synthesis (Section 25.6): A method for lengthening the chain of an aldose sugar.

Kinetic control (Section 14.3): A reaction that follows the lowest activation energy pathway is said to be kinetically controlled. The product is the most rapidly formed but is not necessarily the most stable.

Kinetics (Section 11.2): Referring to reaction rates. Kinetic measurements are useful for helping to determine reaction mechanisms.

Koenigs–Knorr reaction (Section 25.6): A method for the synthesis of glycosides by reaction of an alcohol with a pyranosyl bromide.

Krebs cycle (Section 29.7): An alternative name for the citric acid cycle, by which acetyl CoA is degraded to CO_2.

L Sugar (Section 25.3): A sugar whose hydroxyl group at the chirality center farthest from the carbonyl group points to the left when drawn in Fischer projection.

Lactam (Section 21.7): A cyclic amide.

Lactone (Section 21.6): A cyclic ester.

eaving group (Section 11.2): The group that is replaced in substitution reaction.

evorotatory (Section 9.3): An optically active substance hat rotates the plane of polarization of plane-polarized ght in a left-handed (counterclockwise) direction.

ewis acid (Section 2.11): A substance with a vacant lownergy orbital that can accept an electron pair from a base. ll electrophiles are Lewis acids.

ewis base (Section 2.11): A substance that donates an lectron lone pair to an acid. All nucleophiles are Lewis ases.

ewis structure (Section 1.5): A representation of a moleule showing valence electrons as dots.

indlar catalyst (Section 8.5): A hydrogenation catalyst sed to convert alkynes to cis alkenes.

ine-bond structure (Section 1.5): A representation of a nolecule showing covalent bonds as lines between atoms.

$\rightarrow$4 Link (Section 25.8): An acetal link between the 1 $-$OH group of one sugar and the C4 $-$OH group of nother sugar.

Lipid (Section 27.1): A naturally occurring substance isoated from cells and tissues by extraction with a nonpolar olvent. Lipids belong to many different structural classes, ncluding fats, terpenes, prostaglandins, and steroids.

Lipid bilayer (Section 27.3): The ordered lipid structure that forms a cell membrane.

Lipoprotein (Chapter 27 *Focus On*): A complex molecule with both lipid and protein parts that transports lipids through the body.

Lone-pair electrons (Section 1.4): Nonbonding valenceshell electron pairs. Lone-pair electrons are used by nucleophiles in their reactions with electrophiles.

LUMO (Sections 14.4, 30.2): An acronym for lowest unoccupied molecular orbital. The symmetries of the LUMO and the HOMO are important in determining the stereochemistry of pericyclic reactions.

Magnetic resonance imaging, MRI (Chapter 13 *Focus On*): A medical diagnostic technique based on nuclear magnetic resonance.

MALDI (Section 12.4): Matrix-assisted laser desorption ionization; a mild method for ionizing a molecule so that fragmentation is minimized during mass spectrometry.

Malonic ester synthesis (Section 22.7): The synthesis of a carboxylic acid by alkylation of an alkyl halide, followed by hydrolysis and decarboxylation.

Markovnikov's rule (Section 6.8): A guide for determining the regiochemistry (orientation) of electrophilic addition reactions. In the addition of HX to an alkene, the hydrogen atom bonds to the alkene carbon that has fewer alkyl substituents.

Mass number, *A* (Section 1.1): The total of protons plus neutrons in an atom.

Mass spectrometry (Section 12.1): A technique for measuring the mass, and therefore the molecular weight (MW), of ions.

McLafferty rearrangement (Section 12.3): A mass-spectral fragmentation pathway for carbonyl compounds.

Mechanism (Section 5.2): A complete description of how a reaction occurs. A mechanism must account for all starting materials and all products and must describe the details of each individual step in the overall reaction process.

Meisenheimer complex (Section 16.7): An intermediate formed by addition of a nucleophile to a halo-substituted aromatic ring.

Melt transition temperature, T_m (Section 31.5): The temperature at which crystalline regions of a polymer melt to give an amorphous material.

Mercapto group (Section 18.8): An alternative name for the thiol group, $-$SH.

Meso compound (Section 9.7): A compound that contains chirality centers but is nevertheless achiral by virtue of a symmetry plane.

Messenger RNA (Section 28.4): A kind of RNA formed by transcription of DNA and used to carry genetic messages from DNA to ribosomes.

Meta- (Section 15.1): A naming prefix used for 1,3-disubstituted benzenes.

Metabolism (Section 29.1): A collective name for the many reactions that go on in the cells of living organisms.

Methylene group (Section 6.3): A $-CH_2-$ or $=CH_2$ group.

Micelle (Section 27.2): A spherical cluster of soaplike molecules that aggregate in aqueous solution. The ionic heads of the molecules lie on the outside, where they are solvated by water, and the organic tails bunch together on the inside of the micelle.

Michael reaction (Section 23.10): The conjugate addition reaction of an enolate ion to an unsaturated carbonyl compound.

Molar absorptivity (Section 14.7): A quantitative measure of the amount of UV light absorbed by a sample.

Molecular ion (Section 12.1): The cation produced in the mass spectrometer by loss of an electron from the parent molecule. The mass of the molecular ion corresponds to the molecular weight of the sample.

Molecular mechanics (Chapter 4 *Focus On*): A computer-based method for calculating the minimum-energy conformation of a molecule.

Molecular orbital (MO) theory (Section 1.11): A description of covalent bond formation as resulting from a mathematical combination of atomic orbitals (wave functions) to form molecular orbitals.

Molecule (Section 1.5): A neutral collection of atoms held together by covalent bonds.

Molozonide (Section 7.9): The initial addition product of ozone with an alkene.

Monomer (Section 7.10, Chapter 31 introduction): The simple starting unit from which a polymer is made.

Monosaccharide (Section 25.1): A simple sugar.

Monoterpenoid (Chapter 6 *Focus On,* Section 27.5): A ten-carbon lipid.

Multiplet (Section 13.11): A pattern of peaks in an NMR spectrum that arises by spin–spin splitting of a single absorption because of coupling between neighboring magnetic nuclei.

Mutarotation (Section 25.5): The change in optical rotation observed when a pure anomer of a sugar is dissolved in water. Mutarotation is caused by the reversible opening and closing of the acetal linkage, which yields an equilibrium mixture of anomers.

***n* + 1 rule** (Section 13.11): A hydrogen with n other hydrogens on neighboring carbons shows $n + 1$ peaks in its ^{1}H NMR spectrum.

N-terminal amino acid (Section 26.4): The amino acid with a free $-NH_2$ group at the end of a protein chain.

Newman projection (Section 3.6): A means of indicating stereochemical relationships between substituent groups on neighboring carbons. The carbon–carbon bond is viewed end-on, and the carbons are indicated by a circle. Bonds radiating from the center of the circle are attached to the front carbon, and bonds radiating from the edge of the circle are attached to the rear carbon.

Nitrile (Section 20.1): A compound containing the C≡N functional group.

Nitrogen rule (Section 24.10): A compound with an odd number of nitrogen atoms has an odd-numbered molecular weight.

Node (Section 1.2): A surface of zero electron density within an orbital. For example, a *p* orbital has a nodal plane passing through the center of the nucleus, perpendicular to the axis of the orbital.

Nonbonding electrons (Section 1.4): Valence electrons that are not used in forming covalent bonds.

Noncovalent interaction (Section 2.13): An interaction between molecules, commonly called intermolecular forces or van der Waals forces. Hydrogen bonds, dipole–dipole forces, and dispersion forces are examples.

Normal alkane (Section 3.2): A straight-chain alkane, as opposed to a branched alkane. Normal alkanes are denoted by the suffix *n*, as in n-C_4H_{10} (*n*-butane).

NSAID (Chapter 15 *Focus On*): A nonsteroidal anti-inflammatory drug, such as aspirin or ibuprofen.

Nuclear magnetic resonance, NMR (Chapter 13 introduction): A spectroscopic technique that provides information about the carbon–hydrogen framework of a molecule. NMR works by detecting the energy absorptions accompanying the transitions between nuclear spin states that occur when a molecule is placed in a strong magnetic field and irradiated with radiofrequency waves.

Nucleophile (Section 5.4): A "nucleus-lover," or species that donates an electron pair to an electrophile in a polar bond-forming reaction. Nucleophiles are also Lewis bases.

Nucleophilic acyl substitution reaction (Section 21.2): A reaction in which a nucleophile attacks a carbonyl compound and substitutes for a leaving group bonded to the carbonyl carbon.

Nucleophilic addition reaction (Section 19.4): A reaction in which a nucleophile adds to the electrophilic carbonyl group of a ketone or aldehyde to give an alcohol.

Nucleophilic aromatic substitution reaction (Section 16.7): The substitution reaction of an aryl halide by a nucleophile.

Nucleophilic substitution reaction (Section 11.1): A reaction in which one nucleophile replaces another attached to a saturated carbon atom.

Nucleophilicity (Section 11.3): The ability of a substance to act as a nucleophile in an S_N2 reaction.

Nucleoside (Section 28.1): A nucleic acid constituent, consisting of a sugar residue bonded to a heterocyclic purine or pyrimidine base.

Nucleotide (Section 28.1): A nucleic acid constituent, consisting of a sugar residue bonded both to a heterocyclic purine or pyrimidine base and to a phosphoric acid.

Nucleotides are the monomer units from which DNA and RNA are constructed.

Nylon (Section 21.9): A synthetic polyamide step-growth polymer.

Olefin (Chapter 6 introduction): An alternative name for an alkene.

Optical isomers (Section 9.4): An alternative name for enantiomers. Optical isomers are isomers that have a mirror-image relationship.

Optically active (Section 9.3): A substance that rotates the plane of polarization of plane-polarized light.

Orbital (Section 1.2): A wave function, which describes the volume of space around a nucleus in which an electron is most likely to be found.

Organic chemistry (Chapter 1 introduction): The study of carbon compounds.

Ortho- (Section 15.1): A naming prefix used for 1,2-disubstituted benzenes.

Oxidation (Sections 7.8, 10.9): A reaction that causes a decrease in electron ownership by carbon, either by bond formation between carbon and a more electronegative atom (usually oxygen, nitrogen, or a halogen) or by bond-breaking between carbon and a less electronegative atom (usually hydrogen).

Oxime (Section 19.8): A compound with the $R_2C{=}NOH$ functional group.

Oxirane (Section 7.8): An alternative name for an epoxide.

Oxymercuration (Section 7.4): A method for double-bond hydration using aqueous mercuric acetate as the reagent.

Ozonide (Section 7.9): The product formed by addition of ozone to a carbon–carbon double bond. Ozonides are usually treated with a reducing agent, such as zinc in acetic acid, to produce carbonyl compounds.

Para- (Section 15.1): A naming prefix used for 1,4-disubstituted benzenes.

Paraffin (Section 3.5): A common name for alkanes.

Parent peak (Section 12.1): The peak in a mass spectrum corresponding to the molecular ion. The mass of the parent peak therefore represents the molecular weight of the compound.

Pauli exclusion principle (Section 1.3): No more than two electrons can occupy the same orbital, and those two must have spins of opposite sign.

Peptide (Section 26.4): A short amino acid polymer in which the individual amino acid residues are linked by amide bonds.

Peptide bond (Section 26.4): An amide bond in a peptide chain.

Pericyclic reaction (Chapter 30 introduction): A reaction that occurs by a concerted reorganization of bonding electrons in a cyclic transition state.

Periplanar (Section 11.8): A conformation in which bonds to neighboring atoms have a parallel arrangement. In an eclipsed conformation, the neighboring bonds are syn periplanar; in a staggered conformation, the bonds are anti periplanar.

Peroxide (Section 18.1): A molecule containing an oxygen–oxygen bond functional group, ROOR′ or ROOH.

Peroxyacid (Section 7.8): A compound with the $-CO_3H$ functional group. Peroxyacids react with alkenes to give epoxides.

Phenol (Chapter 17 introduction): A compound with an $-OH$ group directly bonded to an aromatic ring, ArOH.

Phenyl (Section 15.1): The name for the $-C_6H_5$ unit when the benzene ring is considered as a substituent. A phenyl group is abbreviated as $-Ph$.

Phospholipid (Section 27.3): A lipid that contains a phosphate residue. For example, glycerophospholipids contain a glycerol backbone linked to two fatty acids and a phosphoric acid.

Phosphoric acid anhydride (Section 29.1): A substance that contains PO_2PO link, analogous to the CO_2CO link in carboxylic acid anhydrides.

Photochemical reaction (Section 30.3): A reaction carried out by irradiating the reactants with light.

Pi (π) bond (Section 1.8): The covalent bond formed by sideways overlap of atomic orbitals. For example, carbon–carbon double bonds contain a π bond formed by sideways overlap of two *p* orbitals.

PITC (Section 26.6): Phenylisothiocyanate; used in the Edman degradation.

Plane of symmetry (Section 9.2): A plane that bisects a molecule such that one half of the molecule is the mirror image of the other half. Molecules containing a plane of symmetry are achiral.

Plane-polarized light (Section 9.3): Ordinary light that has its electromagnetic waves oscillating in a single plane rather than in random planes. The plane of polarization is rotated when the light is passed through a solution of a chiral substance.

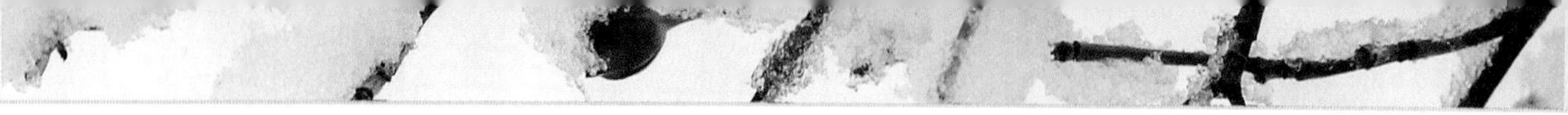

Plasticizer (Section 31.5): A small organic molecule added to polymers to act as a lubricant between polymer chains.

Polar aprotic solvent (Section 11.3): A polar solvent that can't function as a hydrogen ion donor. Polar aprotic solvents such as dimethyl sulfoxide (DMSO) and dimethylformamide (DMF) are particularly useful in S_N2 reactions because of their ability to solvate cations.

Polar covalent bond (Section 2.1): A covalent bond in which the electron distribution between atoms is unsymmetrical.

Polar reaction (Section 5.2): A reaction in which bonds are made when a nucleophile donates two electrons to an electrophile and in which bonds are broken when one fragment leaves with both electrons from the bond.

Polarity (Section 2.1): The unsymmetrical distribution of electrons in a molecule that results when one atom attracts electrons more strongly than another.

Polarizability (Section 5.4): The measure of the change in a molecule's electron distribution in response to changing electric interactions with solvents or ionic reagents.

Polycarbonate (Section 31.4): A polyester in which the carbonyl groups are linked to two −OR groups, [O=C(OR)$_2$].

Polycyclic (Section 4.9): A compound that contains more than one ring.

Polycyclic aromatic compound (Section 15.7): A compound with two or more benzene-like aromatic rings fused together.

Polymer (Sections 7.10, 21.9, Chapter 31 introduction): A large molecule made up of repeating smaller units. For example, polyethylene is a synthetic polymer made from repeating ethylene units, and DNA is a biopolymer made of repeating deoxyribonucleotide units.

Polymerase chain reaction, PCR (Section 28.8): A method for amplifying small amounts of DNA to produce larger amounts.

Polysaccharide (Section 25.1): A carbohydrate that is made of many simple sugars linked together by acetal bonds.

Polyunsaturated fatty acid (Section 27.1): A fatty acid that contains more than one double bond.

Polyurethane (Section 31.4): A step-growth polymer prepared by reaction between a diol and a diisocyanate.

Primary, secondary, tertiary, quaternary (Section 3.3): Terms used to describe the substitution pattern at a specific site. A primary site has one organic substituent attached to it, a secondary site has two organic substituents, a tertiary site has three, and a quaternary site has four.

	Carbon	Carbocation	Hydrogen	Alcohol	Amine
Primary	RCH_3	RCH_2^+	$RC\mathit{H}_3$	RCH_2OH	RNH_2
Secondary	R_2CH_2	R_2CH^+	$R_2C\mathit{H}_2$	R_2CHOH	R_2NH
Tertiary	R_3CH	R_3C^+	$R_3C\mathit{H}$	R_3COH	R_3N
Quaternary	$R_4\mathit{C}$				

Primary structure (Section 26.9): The amino acid sequence in a protein.

pro-R (Section 9.13): One of two identical atoms in a compound, whose replacement leads to an *R* chirality center.

pro-S (Section 9.13): One of two identical atoms in a compound whose replacement leads to an *S* chirality center.

Prochiral (Section 9.13): A molecule that can be converted from achiral to chiral in a single chemical step.

Prochirality center (Section 9.13): An atom in a compound that can be converted into a chirality center by changing one of its attached substituents.

Propagation step (Section 5.3): The step or series of steps in a radical chain reaction that carry on the chain. The propagation steps must yield both product and a reactive intermediate.

Prostaglandin (Section 27.4): A lipid derived from arachidonic acid. Prostaglandins are present in nearly all body tissues and fluids, where they serve many important hormonal functions.

Protecting group (Sections 17.8, 19.10, 26.7): A group that is introduced to protect a sensitive functional group toward reaction elsewhere in the molecule. After serving its protective function, the group is removed.

Protein (Section 26.4): A large peptide containing 50 or more amino acid residues. Proteins serve both as structural materials and as enzymes that control an organism's chemistry.

Protic solvent (Section 11.3): A solvent such as water or alcohol that can act as a proton donor.

Pyramidal inversion (Section 24.2): The rapid stereochemical inversion of a trivalent nitrogen compound.

Pyranose (Section 25.5): The six-membered-ring form of a simple sugar.

Quartet (Section 13.11): A set of four peaks in an NMR spectrum, caused by spin–spin splitting of a signal by three adjacent nuclear spins.

Quaternary (*See* Primary)

Quaternary ammonium salt (Section 24.1): An ionic compound containing a positively charged nitrogen atom with four attached groups, $R_4N^+ \ X^-$.

Quaternary structure (Section 26.9): The highest level of protein structure, involving a specific aggregation of individual proteins into a larger cluster.

Quinone (Section 17.10): A 2,5-cyclohexadiene-1,4-dione.

R group (Section 3.3): A generalized abbreviation for an organic partial structure.

***R,S* convention** (Section 9.5): A method for defining the absolute configuration at chirality centers using the Cahn–Ingold–Prelog sequence rules.

Racemic mixture (Section 9.8): A mixture consisting of equal parts (+) and (−) enantiomers of a chiral substance.

Radical (Section 5.2): A species that has an odd number of electrons, such as the chlorine radical, Cl·.

Radical reaction (Section 5.2): A reaction in which bonds are made by donation of one electron from each of two reactants and in which bonds are broken when each fragment leaves with one electron.

Rate constant (Section 11.2): The constant *k* in a rate equation.

Rate equation (Section 11.2): An equation that expresses the dependence of a reaction's rate on the concentration of reactants.

Rate-limiting step (Section 11.4): The slowest step in a multistep reaction sequence. The rate-limiting step acts as a kind of bottleneck in multistep reactions.

***Re* face** (Section 9.13): One of two faces of a planar, sp^2-hybridized atom.

Rearrangement reaction (Section 5.1): What occurs when a single reactant undergoes a reorganization of bonds and atoms to yield an isomeric product.

Reducing sugar (Section 25.6): A sugar that reduces silver ion in the Tollens test or cupric ion in the Fehling or Benedict tests.

Reduction (Sections 7.7, 10.9): A reaction that causes an increase of electron ownership by carbon, either by bond-breaking between carbon and a more electronegative atom or by bond formation between carbon and a less electronegative atom.

Reductive amination (Sections 24.6, 26.3): A method for preparing an amine by reaction of an aldehyde or ketone with ammonia and a reducing agent.

Refining (Chapter 3 *Focus On*): The process by which petroleum is converted into gasoline and other useful products.

Regiochemistry (Section 6.8): A term describing the orientation of a reaction that occurs on an unsymmetrical substrate.

Regiospecific (Section 6.8): A term describing a reaction that occurs with a specific regiochemistry to give a single product rather than a mixture of products.

Replication (Section 28.3): The process by which double-stranded DNA uncoils and is replicated to produce two new copies.

Replication fork (Section 28.3): The point of unraveling in a DNA chain where replication occurs.

Residue (Section 26.4): An amino acid in a protein chain.

Resolution (Section 9.8): The process by which a racemic mixture is separated into its two pure enantiomers.

Resonance effect (Section 16.4): The donation or withdrawal of electrons through orbital overlap with neighboring π bonds. For example, an oxygen or nitrogen substituent donates electrons to an aromatic ring by overlap of the O or N orbital with the aromatic ring *p* orbitals.

Resonance form (Section 2.4): An individual Lewis structure of a resonance hybrid.

Resonance hybrid (Section 2.4): A molecule, such as benzene, that can't be represented adequately by a single Kekulé structure but must instead be considered as an average of two or more resonance structures. The resonance structures themselves differ only in the positions of their electrons, not their nuclei.

Restriction endonuclease (Section 28.6): An enzyme that is able to cleave a DNA molecule at points in the chain where a specific base sequence occurs.

Retrosynthetic (Sections 8.9, 16.11): A strategy for planning organic syntheses by working backward from the final product to the starting material.

Ribonucleic acid (RNA) (Section 28.1): The biopolymer found in cells that serves to transcribe the genetic information found in DNA and uses that information to direct the synthesis of proteins.

Ribosomal RNA (Section 28.4): A kind of RNA used in the physical makeup of ribosomes.

Ring current (Section 15.8): The circulation of π electrons induced in aromatic rings by an external magnetic field. This effect accounts for the downfield shift of aromatic ring protons in the ^1H NMR spectrum.

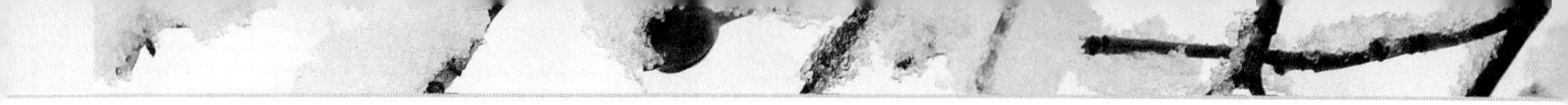

Ring-flip (Section 4.6): A molecular motion that converts one chair conformation of cyclohexane into another chair conformation. The effect of a ring-flip is to convert an axial substituent into an equatorial substituent.

RNA (*See* Ribonucleic acid; Section 28.1)

Robinson annulation reaction (Section 23.12): A synthesis of cyclohexenones by sequential Michael reaction and intramolecular aldol reaction.

s-**cis conformation** (Section 14.5): The conformation of a conjugated diene that is cis-like around the single bond.

Saccharide (Section 25.1): A sugar.

Salt bridge (Section 26.9): The ionic attraction between two oppositely charged groups in a protein chain.

Sandmeyer reaction (Section 24.8): The nucleophilic substitution reaction of an arenediazonium salt with a cuprous halide to yield an aryl halide.

Sanger dideoxy method (Section 2.6): The most commonly used method of DNA sequencing.

Saponification (Section 21.6): An old term for the base-induced hydrolysis of an ester to yield a carboxylic acid salt.

Saturated (Section 3.2): A molecule that has only single bonds and thus can't undergo addition reactions. Alkanes are saturated, but alkenes are unsaturated.

Sawhorse structure (Section 3.6): A manner of representing stereochemistry that uses a stick drawing and gives a perspective view of the conformation around a single bond.

Schiff base (Sections 19.8, 29.5): An alternative name for an imine, $R_2C{=}NR'$, used primarily in biochemistry.

Second-order reaction (Section 11.2): A reaction whose rate-limiting step is bimolecular and whose kinetics are therefore dependent on the concentration of two reactants.

Secondary (*See* Primary)

Secondary structure (Section 26.9): The level of protein substructure that involves organization of chain sections into ordered arrangements such as β-pleated sheets or α helices.

Semiconservative replication (Section 28.3): The process by which DNA molecules are made containing one strand of old DNA and one strand of new DNA.

Sequence rules (Sections 6.5, 9.5): A series of rules for assigning relative priorities to substituent groups on a double-bond carbon atom or on a chirality center.

Sesquiterpenoid (Section 27.5): A 15-carbon lipid.

Shell (electron) (Section 1.2): A group of an atom's electrons with the same principal quantum number.

Shielding (Section 13.2): An effect observed in NMR that causes a nucleus to absorb toward the right (upfield) side of the chart. Shielding is caused by donation of electron density to the nucleus.

***Si* face** (Section 9.13): One of two faces of a planar, sp^2-hybridized atom.

Sialic acid (Section 25.7): A group of more than 300 carbohydrates based on acetylneuramic acid.

Side chain (Section 26.1): The substituent attached to the α carbon of an amino acid.

Sigma (σ) bond (Section 1.6): A covalent bond formed by head-on overlap of atomic orbitals.

Sigmatropic reaction (Section 30.8): A pericyclic reaction that involves the migration of a group from one end of a π electron system to the other.

Simmons–Smith reaction (Section 7.6): The reaction of an alkene with CH_2I_2 and Zn–Cu to yield a cyclopropane.

Simple sugar (Section 25.1): A carbohydrate that cannot be broken down into smaller sugars by hydrolysis.

Skeletal structure (Section 1.12): A shorthand way of writing structures in which carbon atoms are assumed to be at each intersection of two lines (bonds) and at the end of each line.

S_N1 reaction (Section 11.4): A unimolecular nucleophilic substitution reaction.

S_N2 reaction (Section 11.2): A bimolecular nucleophilic substitution reaction.

Solid-phase synthesis (Section 26.8): A technique of synthesis whereby the starting material is covalently bound to a solid polymer bead and reactions are carried out on the bound substrate. After the desired transformations have been effected, the product is cleaved from the polymer.

Solvation (Sections 11.3): The clustering of solvent molecules around a solute particle to stabilize it.

***sp* Orbital** (Section 1.9): A hybrid orbital derived from the combination of an *s* and a *p* atomic orbital. The two *sp* orbitals that result from hybridization are oriented at an angle of 180° to each other.

sp^2 Orbital (Section 1.8): A hybrid orbital derived by combination of an *s* atomic orbital with two *p* atomic orbitals. The three sp^2 hybrid orbitals that result lie in a plane at angles of 120° to each other.

sp^3 Orbital (Section 1.6): A hybrid orbital derived by combination of an *s* atomic orbital with three *p* atomic orbitals. The four sp^3 hybrid orbitals that result are directed toward the corners of a regular tetrahedron at angles of 109° to each other.

Specific rotation, $[\alpha]_D$ (Section 9.3): The optical rotation of a chiral compound under standard conditions.

Sphingomyelin (Section 27.3): A phospholipid that has sphingosine as its backbone.

Spin–spin splitting (Section 13.11): The splitting of an NMR signal into a multiplet because of an interaction between nearby magnetic nuclei whose spins are coupled. The magnitude of spin–spin splitting is given by the coupling constant, *J*.

Staggered conformation (Section 3.4): The three-dimensional arrangement of atoms around a carbon–carbon single bond in which the bonds on one carbon bisect the bond angles on the second carbon as viewed end-on.

Step-growth polymer (Sections 21.9, 31.4): A polymer in which each bond is formed independently of the others. Polyesters and polyamides (nylons) are examples.

Stereochemistry (Chapters 3, 4, 9): The branch of chemistry concerned with the three-dimensional arrangement of atoms in molecules.

Stereoisomers (Section 4.2): Isomers that have their atoms connected in the same order but have different three-dimensional arrangements. The term *stereoisomer* includes both enantiomers and diastereomers.

Stereospecific (Section 7.6): A term indicating that only a single stereoisomer is produced in a given reaction rather than a mixture.

Steric strain (Sections 3.7): The strain imposed on a molecule when two groups are too close together and try to occupy the same space. Steric strain is responsible both for the greater stability of trans versus cis alkenes and for the greater stability of equatorially substituted versus axially substituted cyclohexanes.

Steroid (Section 27.6): A lipid whose structure is based on a tetracyclic carbon skeleton with three 6-membered and one 5-membered ring. Steroids occur in both plants and animals and have a variety of important hormonal functions.

Stork reaction (Section 23.11): A carbonyl condensation between an enamine and an α,β-unsaturated acceptor in a Michael-like reaction to yield a 1,5-dicarbonyl product.

Straight-chain alkane (Section 3.2): An alkane whose carbon atoms are connected without branching.

Substitution reaction (Section 5.1): What occurs when two reactants exchange parts to give two new products. S_N1 and S_N2 reactions are examples.

Sulfide (Section 18.8): A compound that has two organic substituents bonded to the same sulfur atom, RSR′.

Sulfone (Section 18.8): A compound of the general structure RSO_2R'.

Sulfonium ion (Section 18.8): A species containing a positively charged, trivalent sulfur atom, R_3S^+.

Sulfoxide (Section 18.8): A compound of the general structure RSOR′.

Suprafacial (Section 30.6): A word used to describe the geometry of pericyclic reactions. Suprafacial reactions take place on the same side of the two ends of a π electron system.

Symmetry-allowed, symmetry-disallowed (Section 30.2): A symmetry-allowed reaction is a pericyclic process that has a favorable orbital symmetry for reaction through a concerted pathway. A symmetry-disallowed reaction is one that does not have favorable orbital symmetry for reaction through a concerted pathway.

Symmetry plane (Section 9.2): A plane that bisects a molecule such that one half of the molecule is the mirror image of the other half. Molecules containing a plane of symmetry are achiral.

Syn periplanar (Section 11.8): Describing a stereochemical relationship in which two bonds on adjacent carbons lie in the same plane and are eclipsed.

Syn stereochemistry (Section 7.5): The opposite of anti. A syn addition reaction is one in which the two ends of the double bond react from the same side. A syn elimination is one in which the two groups leave from the same side of the molecule.

Syndiotactic (Section 31.2): A chain-growth polymer in which the substituents regularly alternate on opposite sides of the backbone.

Tautomers (Sections 8.4, 22.1): Isomers that are rapidly interconverted.

Template strand (Section 28.4): The strand of double-helical DNA that does not contain the gene.

Terpenoid (Chapter 6 *Focus On,* Section 27.5): A lipid that is formally derived by head-to-tail polymerization of isoprene units.

Tertiary (*See* Primary)

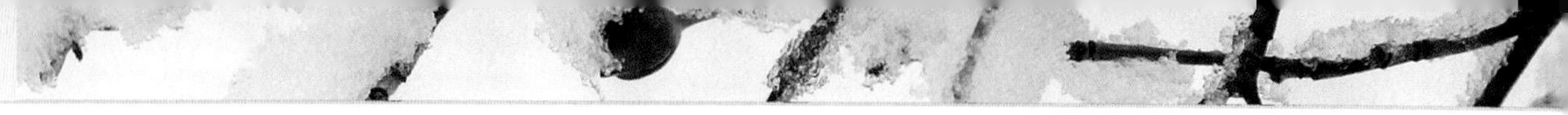

Tertiary structure (Section 26.9): The level of protein structure that involves the manner in which the entire protein chain is folded into a specific three-dimensional arrangement.

Thermodynamic control (Section 14.3): An equilibrium reaction that yields the lowest-energy, most stable product is said to be thermodynamically controlled.

Thermoplastic (Section 31.5): A polymer that has a high T_g and is therefore hard at room temperature, but becomes soft and viscous when heated.

Thermosetting resin (Section 31.5): A polymer that becomes highly cross-linked and solidifies into a hard, insoluble mass when heated.

Thioester (Section 21.8): A compound with the RCOSR′ functional group.

Thiol (Section 18.8): A compound containing the –SH functional group.

Thiolate ion (Section 18.8): The anion of a thiol, RS^-.

TMS (Section 13.3): Tetramethylsilane; used as an NMR calibration standard.

TOF (Section 12.4): Time-of flight mass spectrometry; a sensitive method of mass detection accurate to about 3 ppm.

Tollens' reagent (Section 19.3): A solution of Ag_2O in aqueous ammonia; used to oxidize aldehydes to carboxylic acids.

Torsional strain (Section 3.6): The strain in a molecule caused by electron repulsion between eclipsed bonds. Torsional strain is also called eclipsing strain.

Tosylate (Section 11.1): A *p*-toluenesulfonate ester; useful as a leaving group in nucleophilic substitution reactions.

Transamination (Section 29.9): The exchange of an amino group and a keto group between reactants.

Transcription (Section 28.4): The process by which the genetic information encoded in DNA is read and used to synthesize RNA in the nucleus of the cell. A small portion of double-stranded DNA uncoils, and complementary ribonucleotides line up in the correct sequence for RNA synthesis.

Transfer RNA (Section 28.4): A kind of RNA that transports amino acids to the ribosomes, where they are joined together to make proteins.

Transition state (Section 5.9): An activated complex between reactants, representing the highest energy point on a reaction curve. Transition states are unstable complexes that can't be isolated.

Translation (Section 28.5): The process by which the genetic information transcribed from DNA onto mRNA is read by tRNA and used to direct protein synthesis.

Tree diagram (Section 13.12): A diagram used in NMR to sort out the complicated splitting patterns that can arise from multiple couplings.

Triacylglycerol (Section 27.1): A lipid, such as those found in animal fat and vegetable oil, that is, a triester of glycerol with long-chain fatty acids.

Tricarboxylic acid cycle (Section 29.7): An alternative name for the citric acid cycle by which acetyl CoA is degraded to CO_2.

Triplet (Section 13.11): A symmetrical three-line splitting pattern observed in the ^1H NMR spectrum when a proton has two equivalent neighbor protons.

Turnover number (Section 26.10): The number of substrate molecules acted on by an enzyme per unit time.

Twist-boat conformation (Section 4.5): A conformation of cyclohexane that is somewhat more stable than a pure boat conformation.

Ultraviolet (UV) spectroscopy (Section 14.7): An optical spectroscopy employing ultraviolet irradiation. UV spectroscopy provides structural information about the extent of π electron conjugation in organic molecules.

Unimolecular reaction (Section 11.4): A reaction that occurs by spontaneous transformation of the starting material without the intervention of other reactants. For example, the dissociation of a tertiary alkyl halide in the S_N1 reaction is a unimolecular process.

Unsaturated (Section 6.2): A molecule that has one or more multiple bonds.

Upfield (Section 13.3): The right-hand portion of the NMR chart.

Urethane (Section 31.4): A functional group in which a carbonyl group is bonded to both an –OR group and an $-NR_2$ group.

Uronic acid (Section 25.6): The monocarboxylic acid resulting from enzymatic oxidation of the $-CH_2OH$ group of an aldose.

Valence bond theory (Section 1.5): A bonding theory that describes a covalent bond as resulting from the overlap of two atomic orbitals.

Valence shell (Section 1.4): The outermost electron shell of an atom.

Van der Waals forces (Section 2.13): Intermolecular forces that are responsible for holding molecules together in the liquid and solid states.

Vicinal (Section 8.2): A term used to refer to a 1,2-disubstitution pattern. For example, 1,2-dibromoethane is a vicinal dibromide.

Vinyl group (Section 6.3): An $H_2C{=}CH-$ substituent.

Vinyl monomer (Sections 7.10, 31.1): A substituted alkene monomer used to make chain-growth polymers.

Vinylic (Section 8.3): A term that refers to a substituent at a double-bond carbon atom. For example, chloroethylene is a vinylic chloride, and enols are vinylic alcohols.

Vitamin (Section 26.10): A small organic molecule that must be obtained in the diet and is required in trace amounts for proper growth and function.

Vulcanization (Section 14.6): A technique for cross-linking and hardening a diene polymer by heating with a few percent by weight of sulfur.

Walden inversion (Section 11.1): The inversion of configuration at a chirality center that accompanies an S_N2 reaction.

Wave equation (Section 1.2): A mathematical expression that defines the behavior of an electron in an atom.

Wave function (Section 1.2): A solution to the wave equation for defining the behavior of an electron in an atom. The square of the wave function defines the shape of an orbital.

Wavelength, λ (Section 12.5): The length of a wave from peak to peak. The wavelength of electromagnetic radiation is inversely proportional to frequency and inversely proportional to energy.

Wavenumber, $\tilde{\nu}$ (Section 12.6): The reciprocal of the wavelength in centimeters.

Wax (Section 27.1): A mixture of esters of long-chain carboxylic acids with long-chain alcohols.

Williamson ether synthesis (Section 18.2): A method for synthesizing ethers by S_N2 reaction of an alkyl halide with an alkoxide ion.

Wittig reaction (Section 19.11): The reaction of a phosphorus ylide with a ketone or aldehyde to yield an alkene.

Wohl degradation (Section 25.6): A method for shortening the chain of an aldose sugar.

Wolff–Kishner reaction (Section 19.9): The conversion of an aldehyde or ketone into an alkane by reaction with hydrazine and base.

Wood alcohol (Chapter 17 introduction): An old name for methanol.

Ylide (Section 19.11): A neutral dipolar molecule with adjacent positive and negative charges. The phosphoranes used in Wittig reactions are ylides.

***Z* geometry** (Section 6.5): A term used to describe the stereochemistry of a carbon–carbon double bond. The two groups on each carbon are assigned priorities according to the Cahn–Ingold–Prelog sequence rules, and the two carbons are compared. If the high-priority groups on each carbon are on the same side of the double bond, the bond has *Z* geometry.

Zaitsev's rule (Section 11.7): A rule stating that E2 elimination reactions normally yield the more highly substituted alkene as major product.

Ziegler–Natta catalyst (Section 31.2): A catalyst of an alkylaluminum and a titanium compound used for preparing alkene polymers.

Zwitterion (Section 26.1): A neutral dipolar molecule in which the positive and negative charges are not adjacent. For example, amino acids exist as zwitterions, $H_3\overset{+}{N}-CHR-CO_2^-$.

APPENDIX

D Answers to In-Text Problems

The following answers are meant only as a quick check while you study. Full answers for all problems are provided in the accompanying *Study Guide and Solutions Manual.*

CHAPTER 1

1.1 (a) $1s^2\,2s^2\,2p^4$ (b) $1s^2\,2s^2\,2p^6\,3s^2\,3p^2$ (c) $1s^2\,2s^2\,2p^6\,3s^2\,3p^4$

1.2 (a) 2 (b) 2 (c) 6

1.3

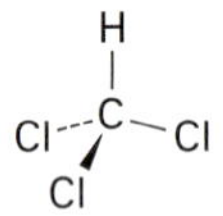

1.4

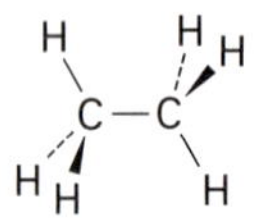

1.5 (a) $GeCl_4$ (b) AlH_3 (c) CH_2Cl_2 (d) SiF_4 (e) CH_3NH_2

1.6 (a)

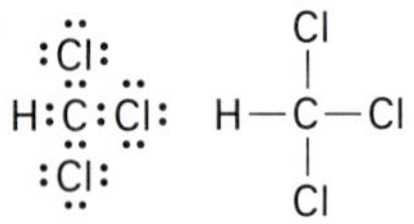

(b)

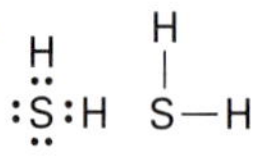

(c)

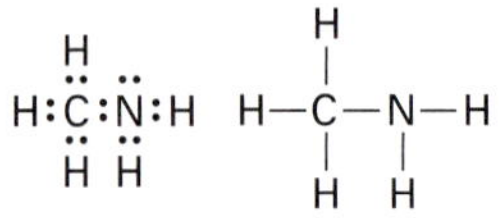

(d)

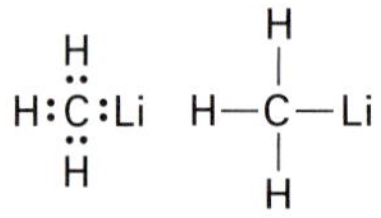

1.7 C_2H_7 has too many hydrogens for a compound with two carbons.

1.8

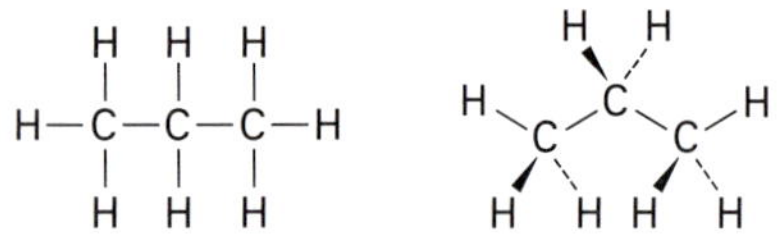

All bond angles are near 109°.

1.9

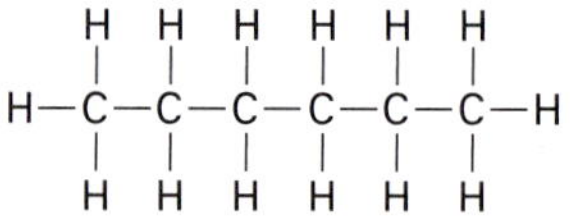

1.10 The CH_3 carbon is sp^3; the double-bond carbons are sp^2; the C=C–C and C=C–H bond angles are approximately 120°; other bond angles are near 109°.

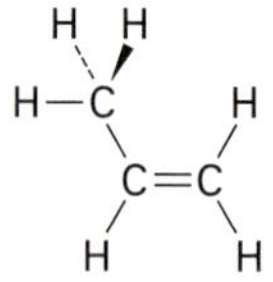

1.11 All carbons are sp^2, and all bond angles are near 120°.

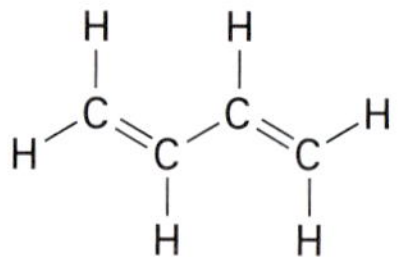

1.12 All carbons except CH_3 are sp^2.

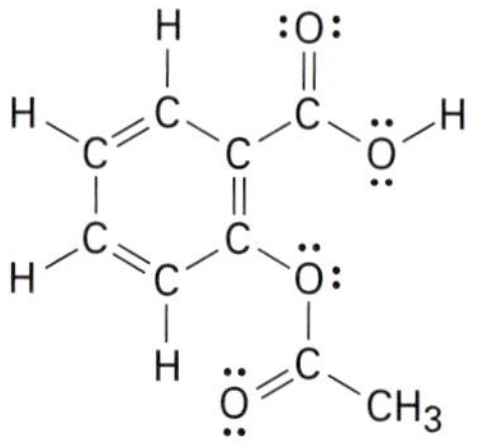

1.13 The CH_3 carbon is sp^3; the triple-bond carbons are *sp*; the C≡C–C and H–C≡C bond angles are approximately 180°.

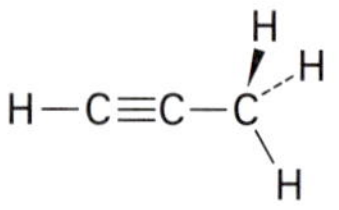

1.14 (a) O has 2 lone pairs and is sp^3-hybridized.
(b) N has 1 lone pair and is sp^3-hybridized.
(c) P has 1 lone pair and is sp^3-hybridized.
(d) S has 2 lone pairs and is sp^3-hybridized.

1.15 **(a)**

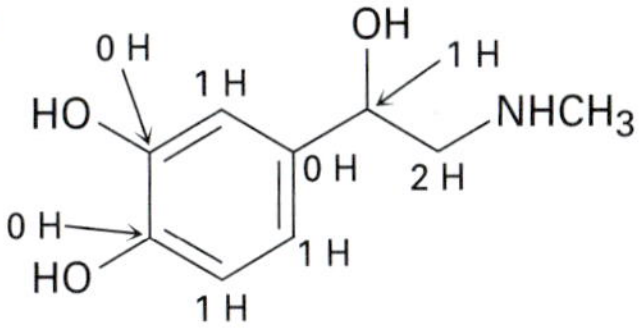

Adrenaline—$C_9H_{13}NO_3$

(b)

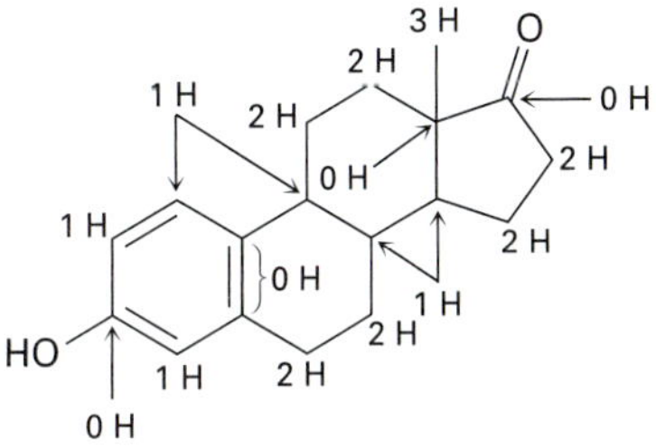

Estrone—$C_{18}H_{22}O_2$

1.16 There are numerous possibilities, such as:

(a) C_5H_{12} $CH_3CH_2CH_2CH_2CH_3$

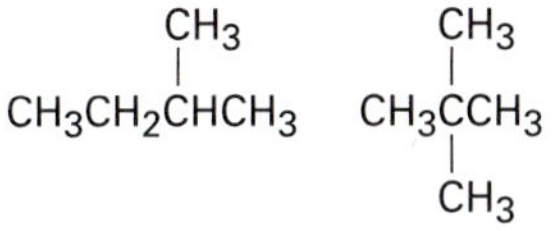

(b) C_2H_7N $CH_3CH_2NH_2$ CH_3NHCH_3

(c) C_3H_6O

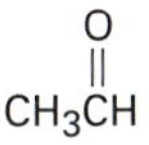

$H_2C{=}CHCH_2OH$ $H_2C{=}CHOCH_3$

(d) C_4H_9Cl $CH_3CH_2CH_2CH_2Cl$

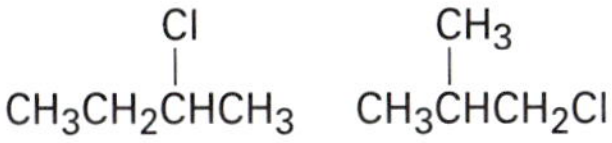

1.17

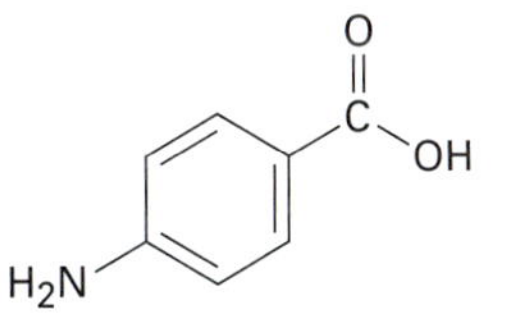

CHAPTER 2

2.1 (a) H (b) Br (c) Cl (d) C

2.2

(a) $\overset{\delta+}{H_3C}{-}\overset{\delta-}{Cl}$ **(b)** $\overset{\delta+}{H_3C}{-}\overset{\delta-}{NH_2}$ **(c)** $\overset{\delta-}{H_2N}{-}\overset{\delta+}{H}$

(d) $H_3C{-}SH$

Carbon and sulfur have identical electronegativities.

(e) $\overset{\delta-}{H_3C}{-}\overset{\delta+}{MgBr}$ **(f)** $\overset{\delta+}{H_3C}{-}\overset{\delta-}{F}$

2.3 $H_3C{-}OH < H_3C{-}MgBr < H_3C{-}Li = H_3C{-}F < H_3C{-}K$

2.4 The chlorine is electron-rich, and the carbon is electron-poor.

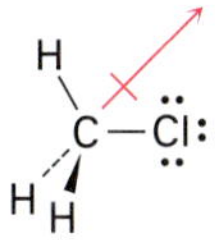

2.5 The two C–O dipoles cancel because of the symmetry of the molecule:

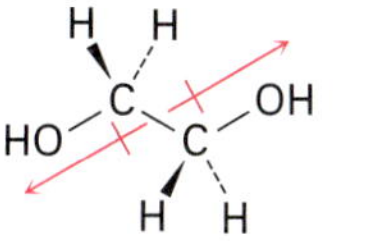

2.6 **(a)**

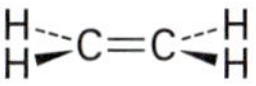

No dipole moment

(b)

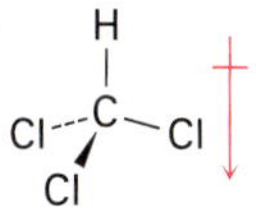

(c)

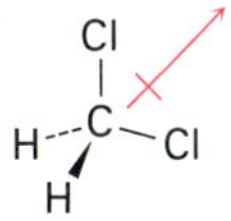

(d)

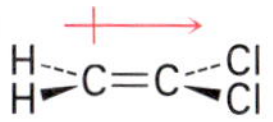

2.7 For nitrogen: FC = 5 − 8/2 − 0 = +1
For singly bonded oxygen: FC = 6 − 2/2 − 6 = −1

2.8 (a) For carbon: FC = 4 − 8/2 − 0 = 0
For the middle nitrogen: FC = 5 − 8/2 − 0 = +1
For the end nitrogen: FC = 5 − 4/2 − 4 = −1
(b) For nitrogen: FC = 5 − 8/2 − 0 = +1
For oxygen: FC = 6 − 2/2 − 6 = −1
(c) For nitrogen: FC = 5 − 8/2 − 0 = +1
For the end carbon: FC = 4 − 6/2 − 2 = −1

2.9

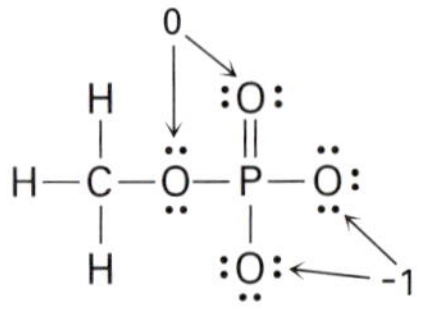

2.10

(a)

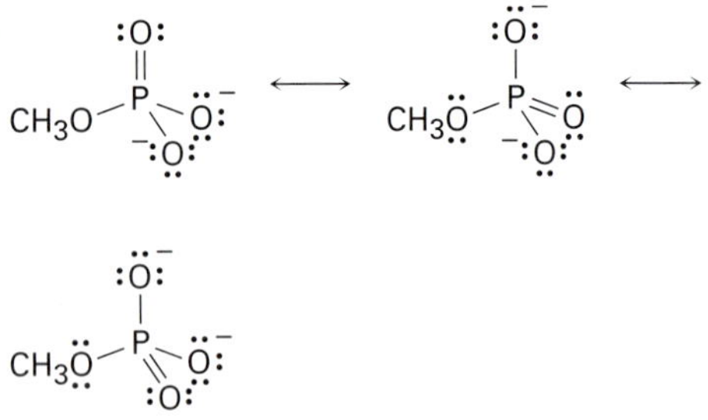

(b)

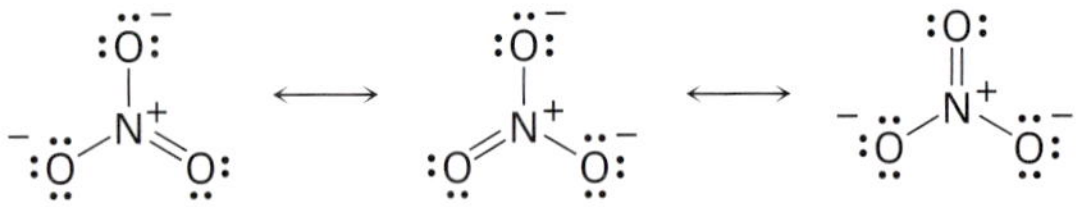

(c) $H_2C{=}CH{-}CH_2^+ \longleftrightarrow H_2\overset{+}{C}{-}CH{=}CH_2$

(d)

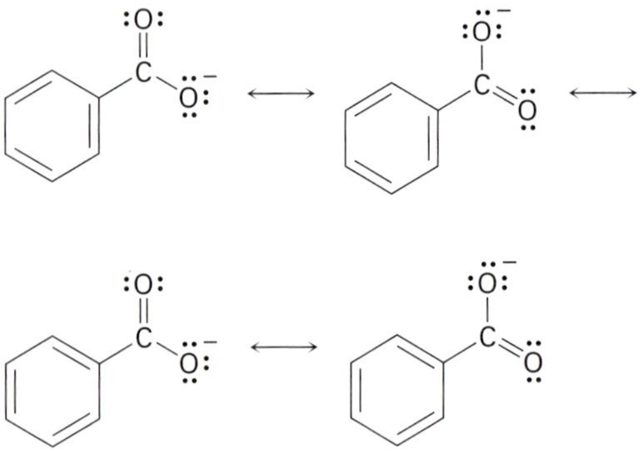

2.11

HNO_3	+	NH_3	$\longrightarrow$	NH_4^+	NO_3^-
Acid		Base		Conjugate acid	Conjugate base

2.12 Phenylalanine is stronger.

2.13 Water is a stronger acid.

2.14 Neither reaction will take place.

2.15 Reaction will take place.

2.16 $K_a = 4.9 \times 10^{-10}$

2.17

(a)

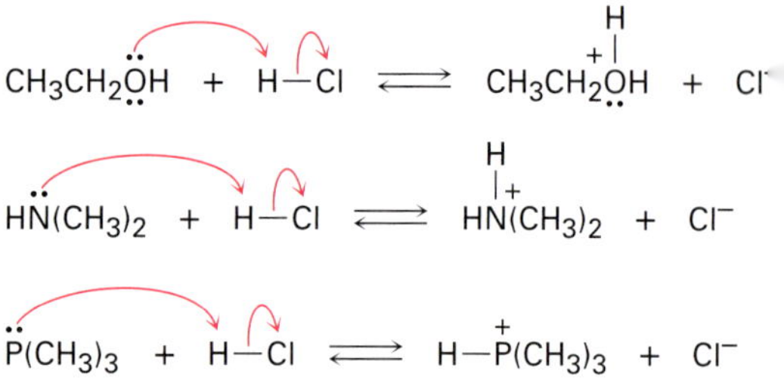

(b)

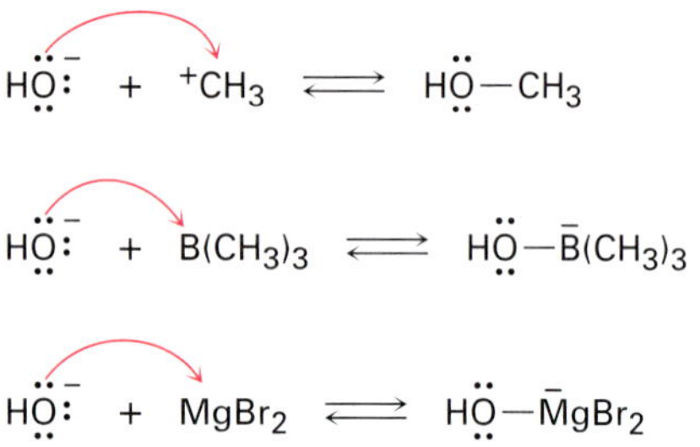

2.18

(a)

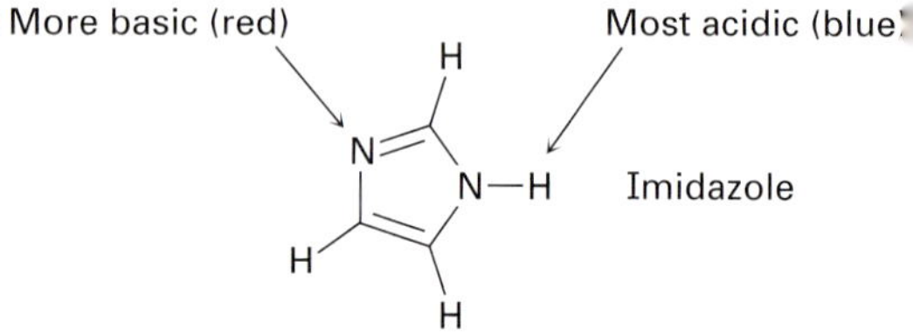

(b)

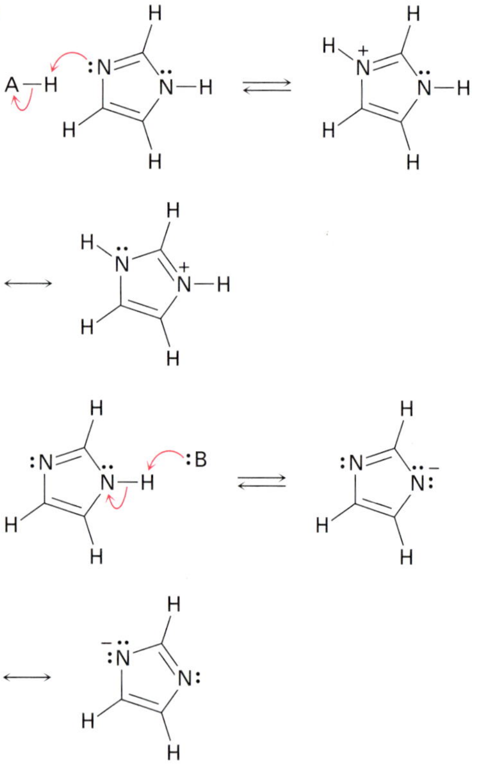

2.19 Vitamin C is water-soluble (hydrophilic); vitamin A is fat-soluble (hydrophilic).

CHAPTER 3

3.1 (a) Sulfide, carboxylic acid, amine
(b) Aromatic ring, carboxylic acid
(c) Ether, alcohol, aromatic ring, amide, C=C bond

3.2 (a) CH_3OH (b) 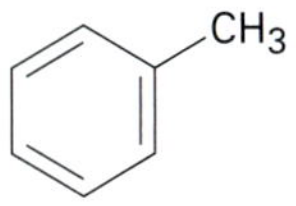 (c) 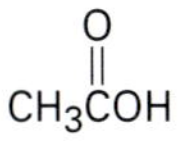

(d) CH_3NH_2 (e) 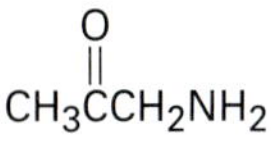 (f) 

3.3

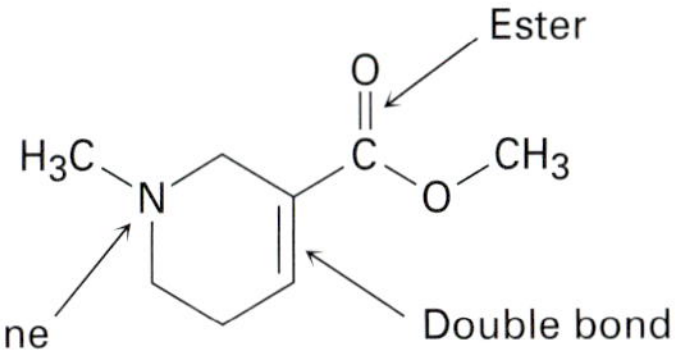

$C_8H_{13}NO_2$

3.4

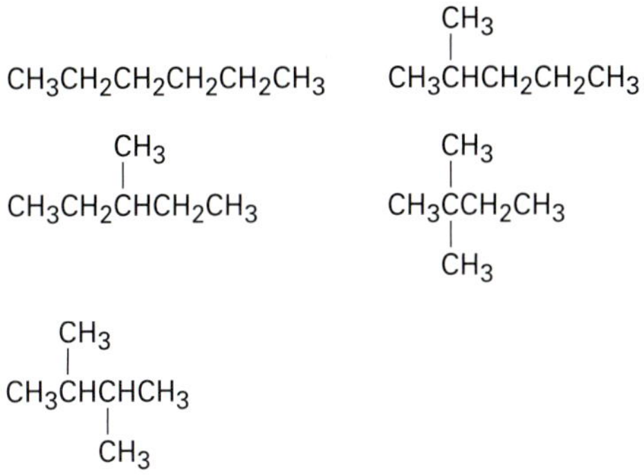

3.5 Part (a) has nine possible answers.

(a)

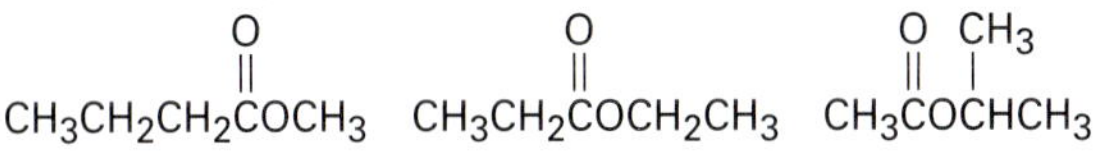

(b) 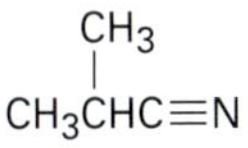 $CH_3CH_2CH_2C\equiv N$

(c) $CH_3CH_2SSCH_2CH_3$ $CH_3SSCH_2CH_2CH_3$ 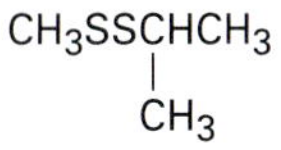

3.6 (a) Two (b) Four

3.7

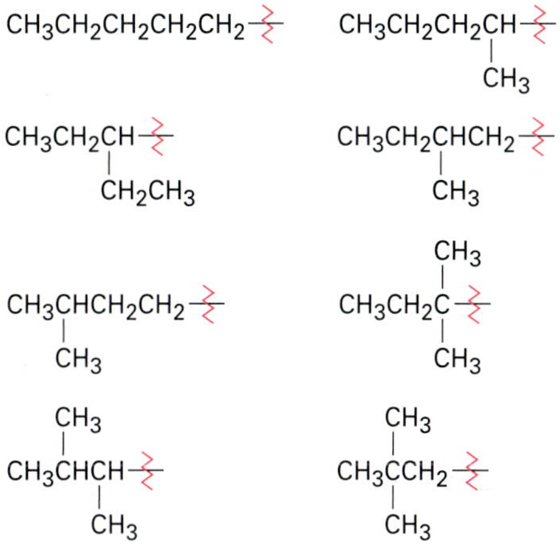

3.8 (a) 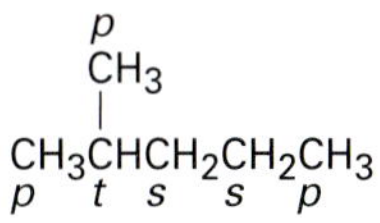 (b) 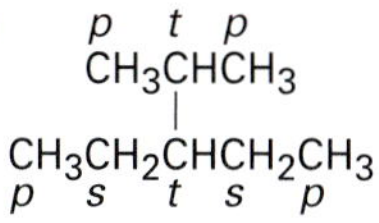

(c) 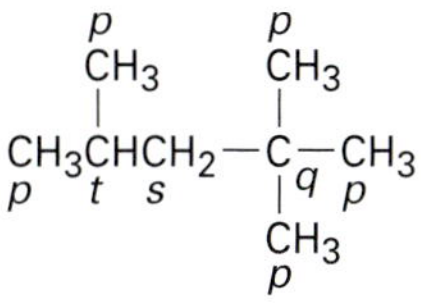

3.9 Primary carbons have primary hydrogens, secondary carbons have secondary hydrogens, and tertiary carbons have tertiary hydrogens.

3.10 (a) 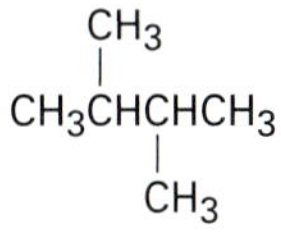 (b) 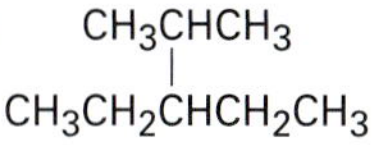

(c) 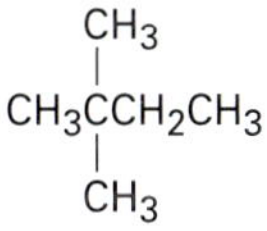

3.11 (a) Pentane, 2-methylbutane, 2,2-dimethylpropane
(b) 3,4-Dimethylhexane
(c) 2,4-Dimethylpentane
(d) 2,2,5-Trimethylheptane

3.12 (a)

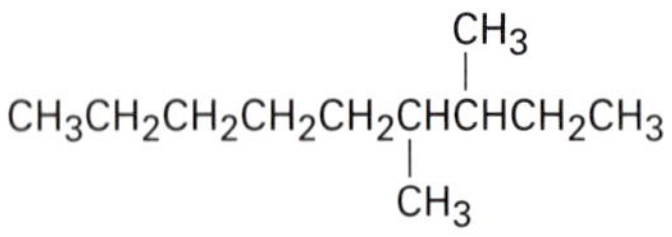

(b)

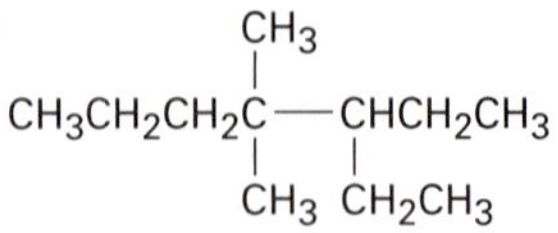

(c)

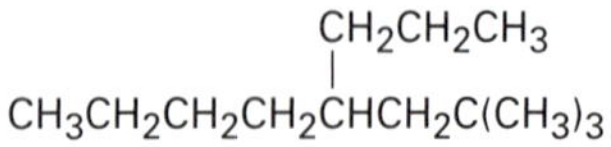

(d)

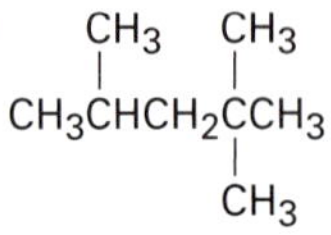

3.13 Pentyl, 1-methylbutyl, 1-ethylpropyl, 3-methylbutyl, 2-methylbutyl, 1,1-dimethylpropyl, 1,2-dimethylpropyl, 2,2-dimethylpropyl

3.14

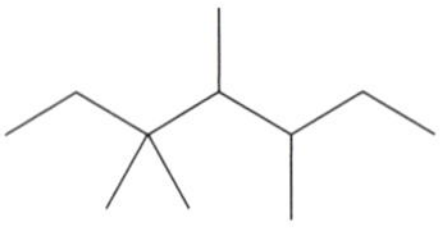

3,3,4,5-Tetramethylheptane

3.15

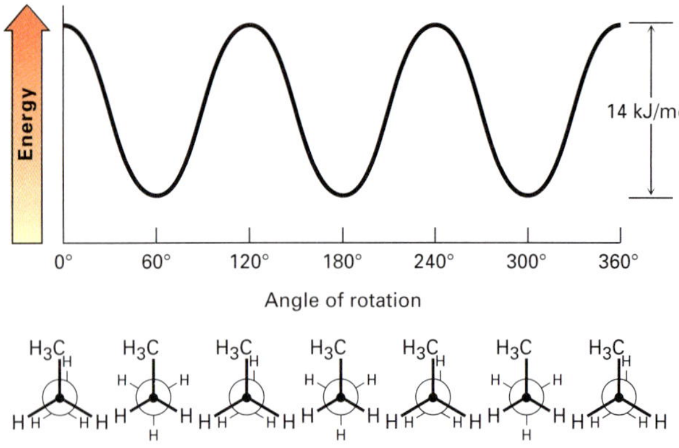

3.16

(a)

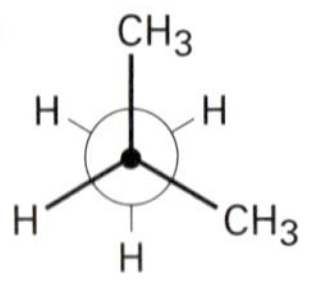

(b)

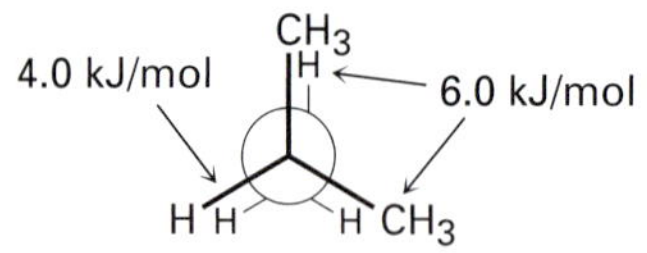

(c), (d)

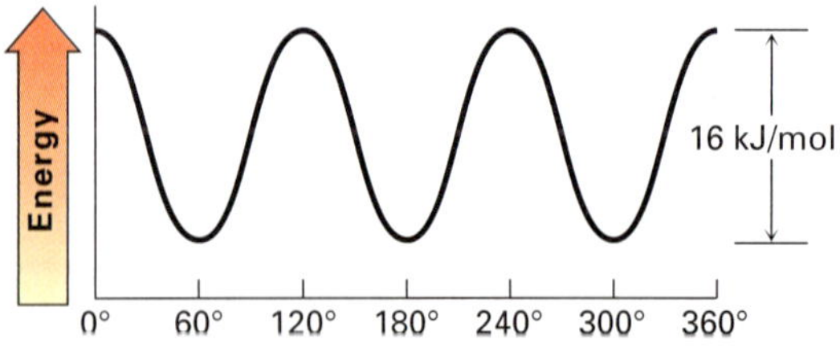

3.17

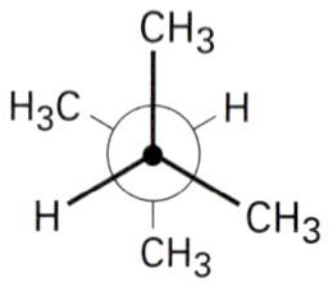

3.18

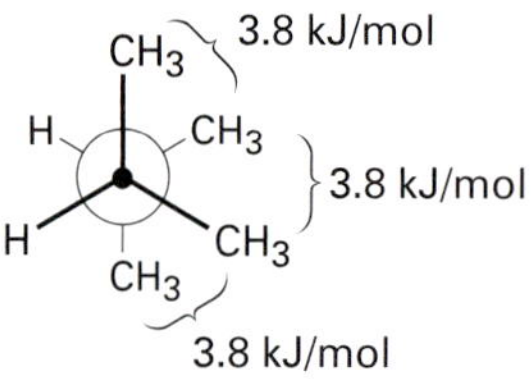

Total: 11.4 kJ/mol

CHAPTER 4

4.1 (a) 1,4-Dimethylcyclohexane
(b) 1-Methyl-3-propylcyclopentane
(c) 3-Cyclobutylpentane
(d) 1-Bromo-4-ethylcyclodecane
(e) 1-Isopropyl-2-methylcyclohexane
(f) 4-Bromo-1-*tert*-butyl-2-methylcycloheptane

4.2 (a)

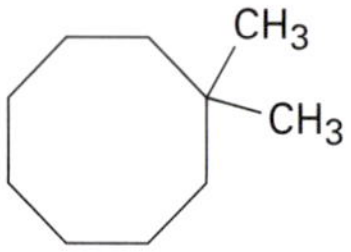

(b)

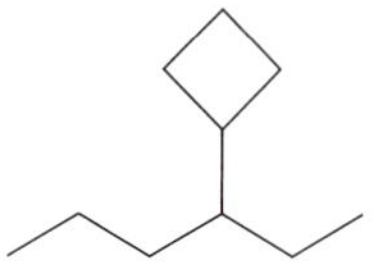

(c)

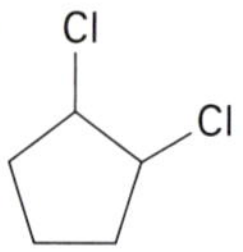

(d)

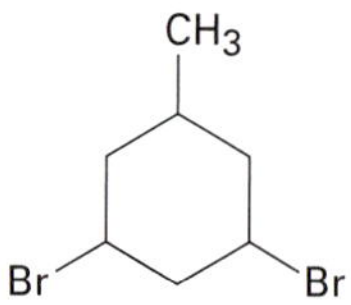

4.3 3-Ethyl-1,1-dimethylcyclopentane

4.4 (a) *trans*-1-Chloro-4-methylcyclohexane
(b) *cis*-1-Ethyl-3-methylcycloheptane

4.5 (a)

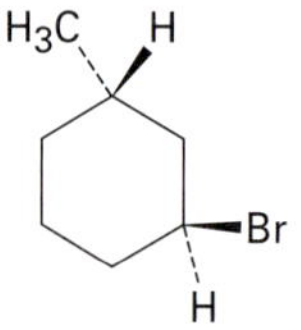

(b)

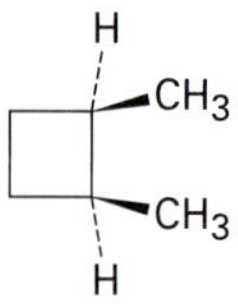

(c)

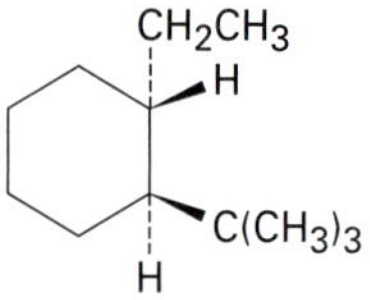

4.6 The two hydroxyl groups are cis. The two side chains are trans.

4.7 (a) *cis*-1,2-Dimethylcyclopentane
(b) *cis*-1-Bromo-3-methylcyclobutane

4.8 Six interactions; 21% of strain

4.9 The cis isomer is less stable because the methyl groups eclipse each other.

4.10 Ten eclipsing interactions; 40 kJ/mol; 35% is relieved.

4.11 Conformation **(a)** is more stable because the methyl groups are farther apart.

4.12

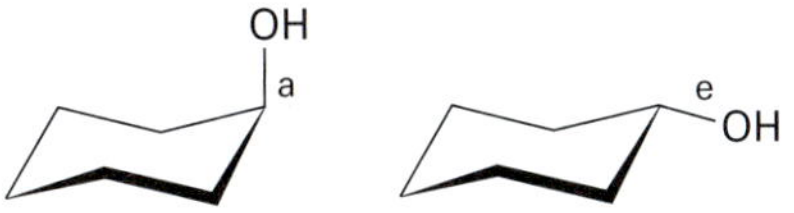

4.13

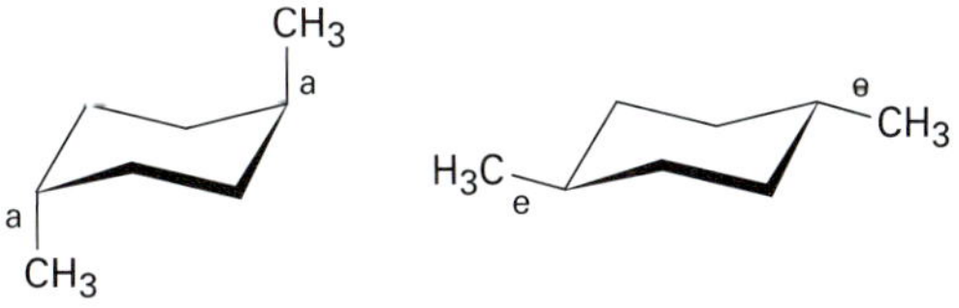

4.14 Before ring-flip, red and blue are equatorial and green is axial. After ring-flip, red and blue are axial and green is equatorial.

4.15 4.2 kJ/mol

4.16 Cyano group points straight up.

4.17 Equatorial = 70%; axial = 30%

4.18 (a) 2.0 kJ/mol (b) 11.4 kJ/mol (c) 2.0 kJ/mol (d) 8.0 kJ/mol

4.19

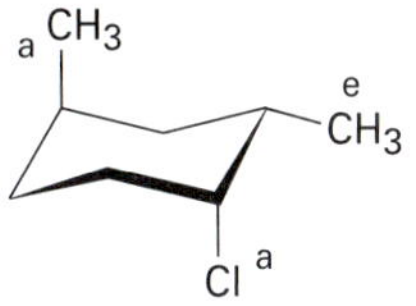

1-Chloro-2,4-dimethylcyclohexane (less stable chair form)

4.20 *trans*-Decalin is more stable because it has no 1,3-diaxial interactions.

CHAPTER 5

5.1 (a) Substitution (b) Elimination (c) Addition

5.2 1-Chloro-2-methylpentane
2-Chloro-2-methylpentane
3-Chloro-2-methylpentane
2-Chloro-4-methylpentane
1-Chloro-4-methylpentane

5.3 A radical addition reaction

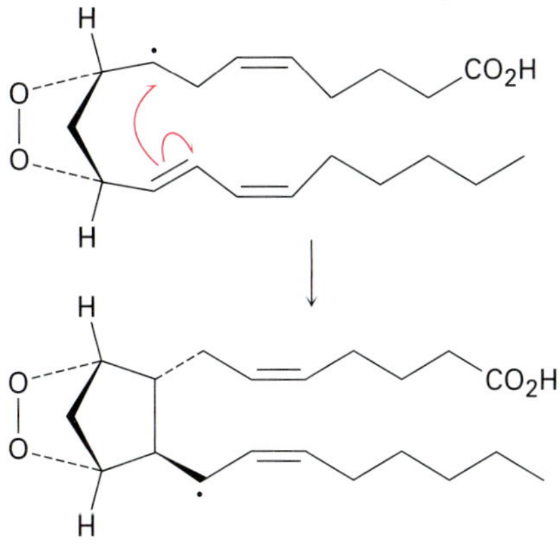

5.4 (a) Carbon is electrophilic.
(b) Sulfur is nucleophilic.
(c) Nitrogens are nucleophilic.
(d) Oxygen is nucleophilic; carbon is electrophilic.

5.5

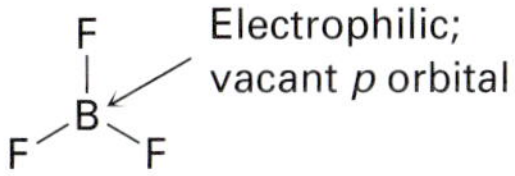

5.6 Bromocyclohexane; chlorocyclohexane

5.7

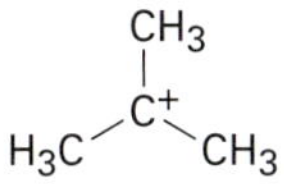

5.8

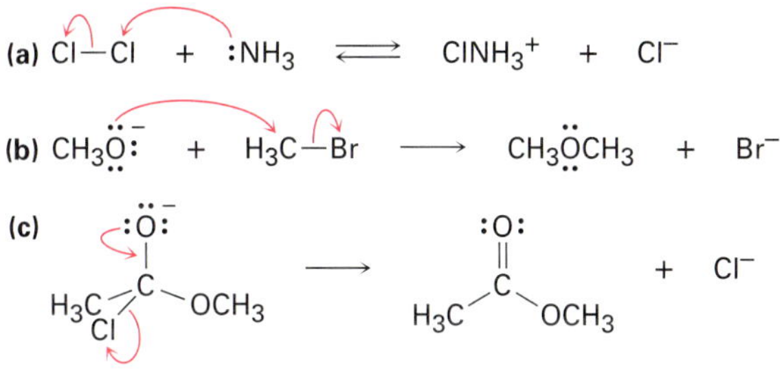

5.9

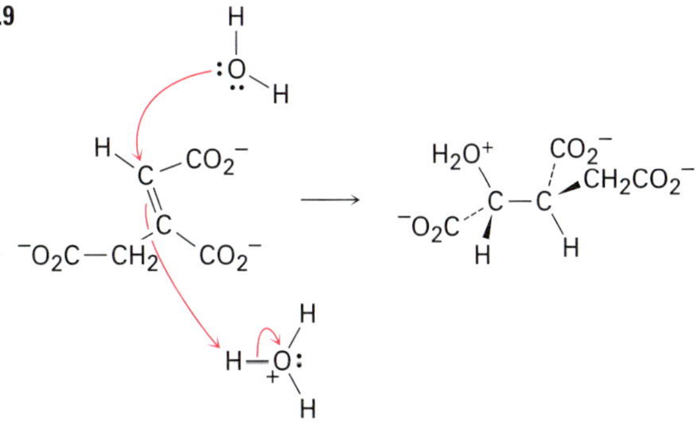

5.10 Negative $\Delta G°$ is more favored.

5.11 Larger K_{eq} is more exergonic.

5.12 Lower $\Delta G^{\ddagger}$ is faster.

5.13

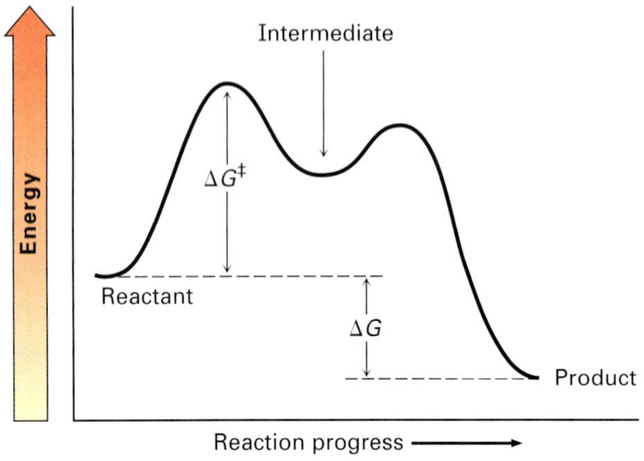

CHAPTER 6

6.1 (a) 1 (b) 2 (c) 2

6.2 (a) 5 (b) 5 (c) 3
(d) 1 (e) 6 (f) 5

6.3 $C_{16}H_{13}ClN_2O$

6.4 (a) 3,4,4-Trimethyl-1-pentene
(b) 3-Methyl-3-hexene
(c) 4,7-Dimethyl-2,5-octadiene
(d) 6-Ethyl-7-methyl-4-nonene

6.5 (a)

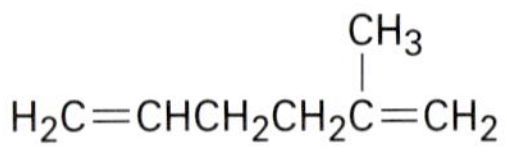

(b)

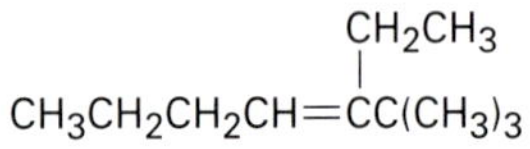

(c)

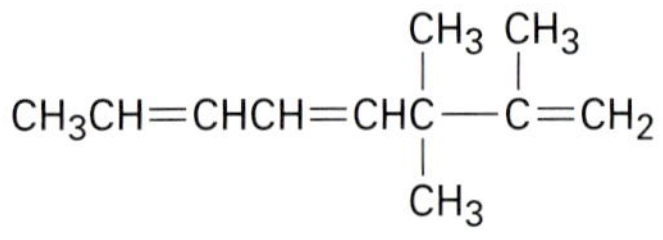

(d)

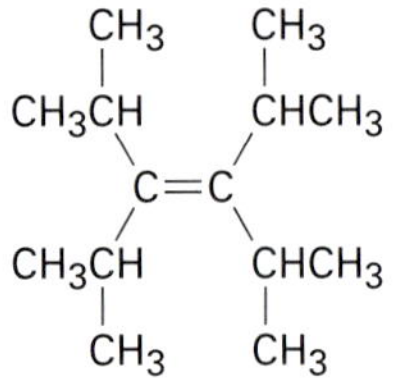

6.6 (a) 1,2-Dimethylcyclohexene
(b) 4,4-Dimethylcycloheptene
(c) 3-Isopropylcyclopentene

6.7 Compounds (c), (e), and (f) have cis–trans isomers.

6.8 (a) *cis*-4,5-Dimethyl-2-hexene
(b) *trans*-6-Methyl-3-heptene

6.9 (a) $-Br$ (b) $-Br$ (c) $-CH_2CH_3$
(d) $-OH$ (e) $-CH_2OH$ (f) $-CH{=}O$

6.10 (a) $-Cl$, $-OH$, $-CH_3$, $-H$
(b) $-CH_2OH$, $-CH{=}CH_2$, $-CH_2CH_3$, $-CH_3$
(c) $-CO_2H$, $-CH_2OH$, $-C{\equiv}N$, $-CH_2NH_2$
(d) $-CH_2OCH_3$, $-C{\equiv}N$, $-C{\equiv}CH$, $-CH_2CH_3$

6.11 (a) *Z* (b) *E* (c) *Z* (d) *E*

6.12

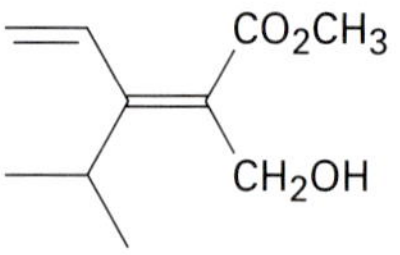

Z

6.13 (a) 2-Methylpropene more stable than 1-butene
(b) *trans*-2-Hexene more stable than *cis*-2-hexene
(c) 1-Methylcyclohexene more stable than 3-methylcyclohexene

6.14 (a) Chlorocyclohexane
(b) 2-Bromo-2-methylpentane
(c) 4-Methyl-2-pentanol
(d) 1-Bromo-1-methylcyclohexane

6.15 (a) Cyclopentene
(b) 1-Ethylcyclohexene or ethylidenecyclohexane
(c) 3-Hexene
(d) Vinylcyclohexane (cyclohexylethylene)

6.16 (a)

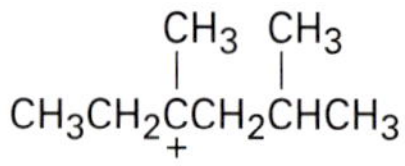

(b)

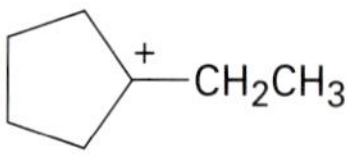

6.17 In the conformation shown, only the methyl-group C–H that is parallel to the carbocation *p* orbital can show hyperconjugation.

6.18 The second step is exergonic; the transition state resembles the carbocation.

6.19

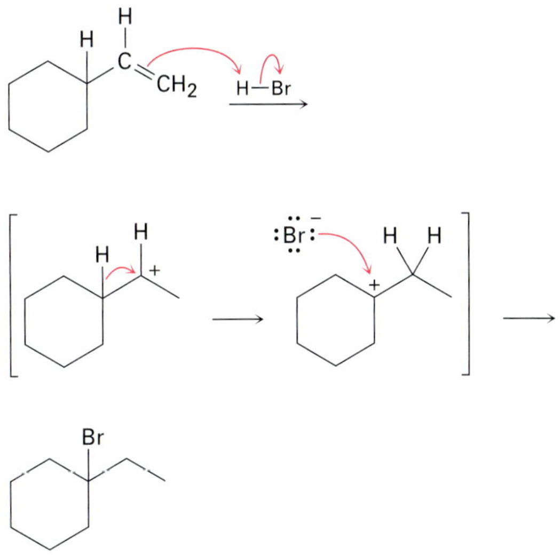

CHAPTER 7

7.1 2-Methyl-2-butene and 2-methyl-1-butene

7.2 Five

7.3 *trans*-1,2-Dichloro-1,2-dimethylcyclohexane

7.4

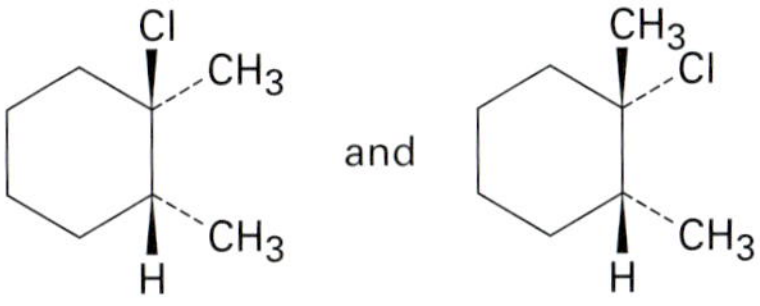

7.5 *trans*-2-Bromocyclopentanol

7.6 Markovnikov

7.7 (a) 2-Pentanol (b) 2-Methyl-2-pentanol

7.8 (a) Oxymercuration of 2-methyl-1-hexene or 2-methyl-2-hexene
(b) Oxymercuration of cyclohexylethylene or hydroboration of ethylidenecyclohexane

7.9 (a)

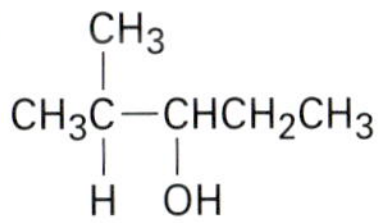

(b)

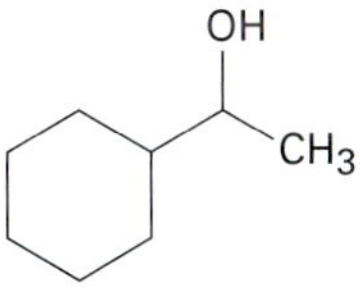

7.10 (a) 3-Methyl-1-butene
(b) 2-Methyl-2-butene
(c) Methylenecyclohexane

7.11

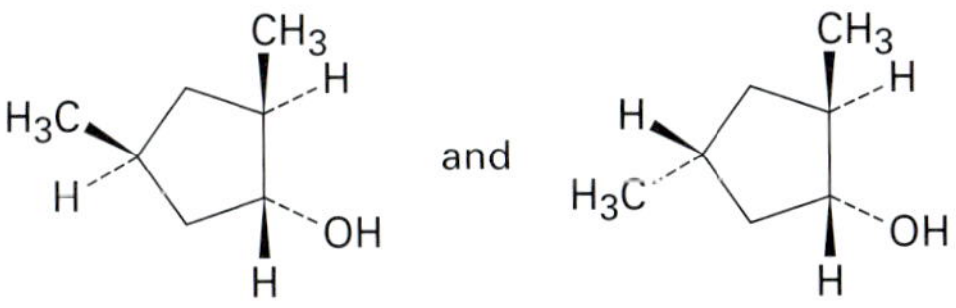

7.12 (a)

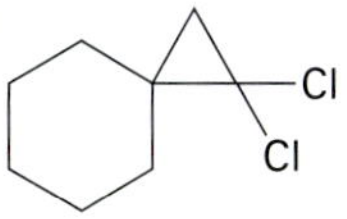

(b)

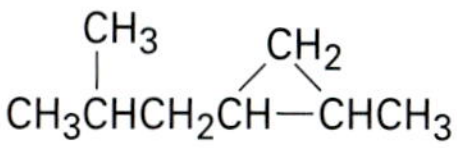

7.13 (a) 2-Methylpentane
(b) 1,1-Dimethylcyclopentane

7.14

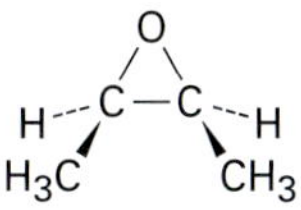

cis-2,3-Epoxybutane

7.15 (a) 1-Methylcyclohexene
(b) 2-Methyl-2-pentene
(c) 1,3-Butadiene

7.16 (a) $CH_3COCH_2CH_2CH_2CH_2CO_2H$
(b) $CH_3COCH_2CH_2CH_2CH_2CHO$

7.17 (a) 2-Methylpropene (b) 3-Hexene

7.18 (a) $H_2C{=}CHOCH_3$ (b) $ClCH{=}CHCl$

7.19

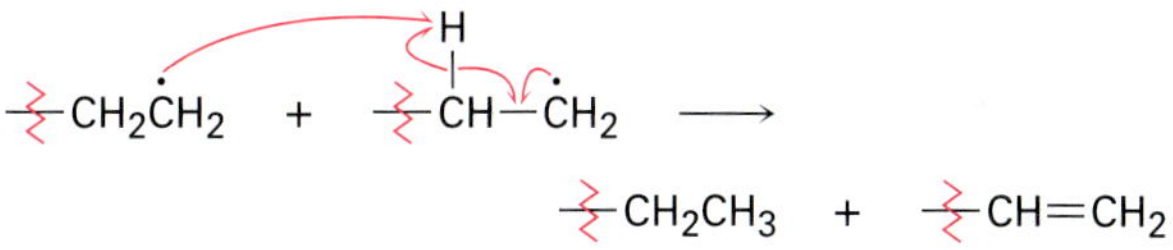

CHAPTER 8

8.1 (a) 2,5-Dimethyl-3-hexyne
(b) 3,3-Dimethyl-1-butyne
(c) 3,3-Dimethyl-4-octyne
(d) 2,5,5-Trimethyl-3-heptyne
(e) 6-Isopropylcyclodecyne
(f) 2,4-Octadiene-6-yne

8.2 1-Hexyne, 2-hexyne, 3-hexyne, 3-methyl-1-pentyne, 4-methyl-1-pentyne, 4-methyl-2-pentyne, 3,3-dimethyl-1-butyne

8.3 (a) 1,1,2,2-Tetrachloropentane
(b) 1-Bromo-1-cyclopentylethylene
(c) 2-Bromo-2-heptene and 3-bromo-2-heptene

8.4 (a) 4-Octanone
(b) 2-Methyl-4-octanone and 7-methyl-4-octanone

8.5 (a) 1-Pentyne (b) 2-Pentyne

8.6 (a) $C_6H_5C{\equiv}CH$ (b) 2,5-Dimethyl-3-hexyne

8.7 (a) Mercuric sulfate–catalyzed hydration of phenylacetylene
(b) Hydroboration/oxidation of cyclopentylacetylene

8.8 (a) Reduce 2-octyne with Li/NH_3
(b) Reduce 3-heptyne with H_2/Lindlar catalyst
(c) Reduce 3-methyl-1-pentyne

8.9 No: **(a)**, **(c)**, **(d)**; yes: **(b)**

8.10 **(a)** 1-Pentyne + CH_3I, or propyne + $CH_3CH_2CH_2I$
(b) 3-Methyl-1-butyne + CH_3CH_2I
(c) Cyclohexylacetylene + CH_3I

8.11 $CH_3C{\equiv}CH \xrightarrow[\text{2. } CH_3I]{\text{1. } NaNH_2} CH_3C{\equiv}CCH_3$

$\xrightarrow[\text{Lindlar cat.}]{H_2}$ *cis*-$CH_3CH{=}CHCH_3$

8.12 **(a)** $KMnO_4$, H_3O^+
(b) H_2/Lindlar
(c) 1. H_2/Lindlar; 2. HBr
(d) 1. H_2/Lindlar; 2. BH_3; 3. NaOH, H_2O_2
(e) 1. H_2/Lindlar; 2. Cl_2
(f) O_3

8.13 **(a)** 1. $HC{\equiv}CH$ + $NaNH_2$; 2. $CH_3(CH_2)_6CH_2Br$; 3. 2 H_2/Pd
(b) 1. $HC{\equiv}CH$ + $NaNH_2$; 2. $(CH_3)_3CCH_2CH_2I$; 3. 2 H_2/Pd
(c) 1. $HC{\equiv}CH$ + $NaNH_2$; 2. $CH_3CH_2CH_2CH_2I$; 3. BH_3; 4. H_2O_2
(d) 1. $HC{\equiv}CH$ + $NaNH_2$; 2. $CH_3CH_2CH_2CH_2CH_2I$; 3. $HgSO_4$, H_3O^+

CHAPTER 9

9.1 Chiral: screw, beanstalk, shoe

9.2 **(a)**

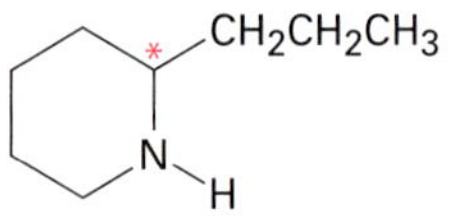

(b)

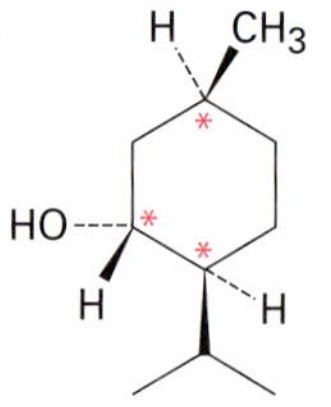

(c)

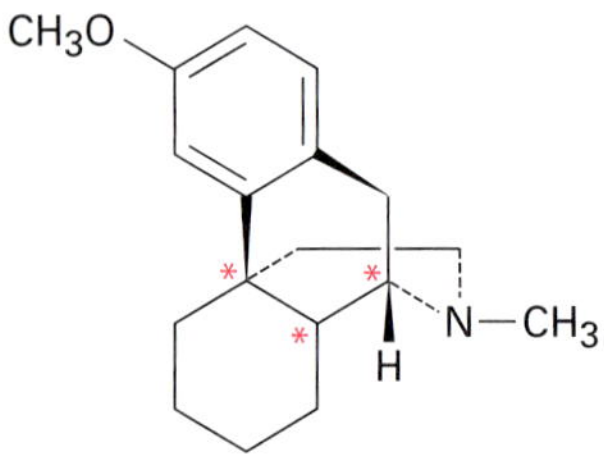

9.3

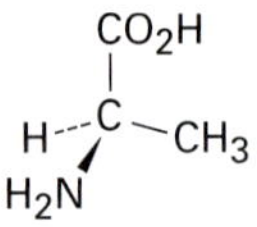

and

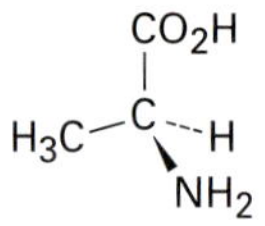

9.4 **(a)**

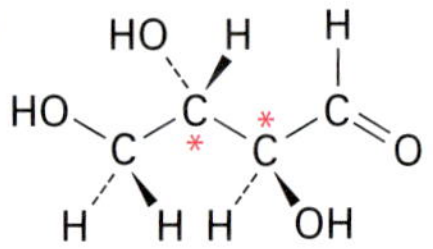

(b)

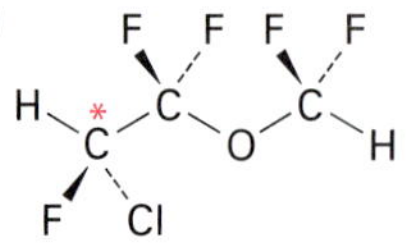

9.5 Levorotatory

9.6 +16.1°

9.7 **(a)** $-OH$, $-CH_2CH_2OH$, $-CH_2CH_3$, $-H$
(b) $-OH$, $-CO_2CH_3$, $-CO_2H$, $-CH_2OH$
(c) $-NH_2$, $-CN$, $-CH_2NHCH_3$, $-CH_2NH_2$
(d) $-SSCH_3$, $-SH$, $-CH_2SCH_3$, $-CH_3$

9.8 **(a)** *S* **(b)** *R* **(c)** *S*

9.9 **(a)** *S* **(b)** *S* **(c)** *R*

9.10

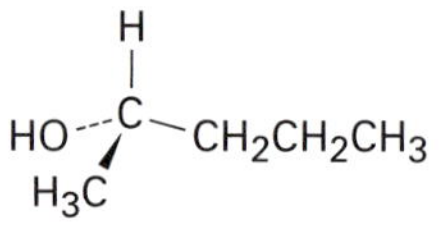

9.11 *S*

9.12 **(a)** *R,R* **(b)** *S,R* **(c)** *R,S* **(d)** *S,S*

Compounds **(a)** and **(d)** are enantiomers and are diastereomeric with **(b)** and **(c)**.

9.13 *R,R*

9.14 *S,S*

9.15 **(a)**, **(d)**

9.16 Compounds **(a)** and **(c)** have meso forms.

9.17

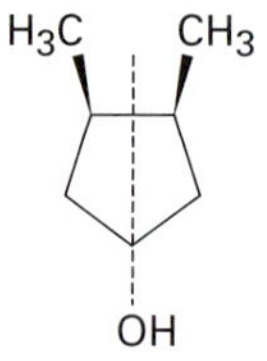

Meso

9.18 The product retains its *S* stereochemistry.

9.19 Two diastereomeric salts: (*R*)-lactic acid plus (*S*)-1-phenylethylamine and (*S*)-lactic acid plus (*S*)-1-phenylethylamine

9.20 **(a)** Constitutional isomers **(b)** Diastereomers

9.21 An optically inactive, non-50:50 mixture of two racemic pairs: (2*R*,4*R*) + (2*S*,4*S*) and (2*R*,4*S*) + (2*S*,4*R*)

9.22 Non-50:50 mixture of two racemic pairs: (1*S*,3*R*) + (1*R*,3*S*) and (1*S*,3*S*) + (1*R*,3*R*)

9.23 **(a)**

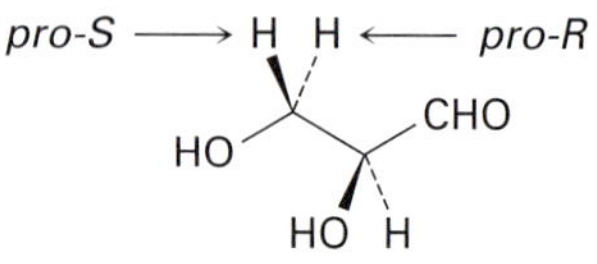

(b)

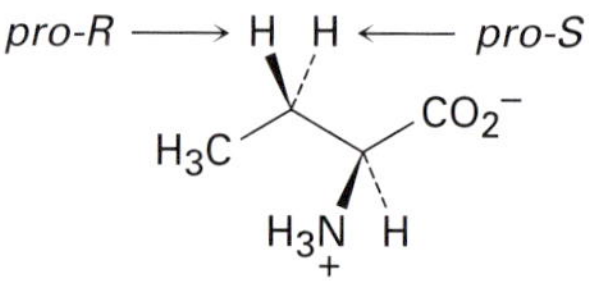

9.24 (a)

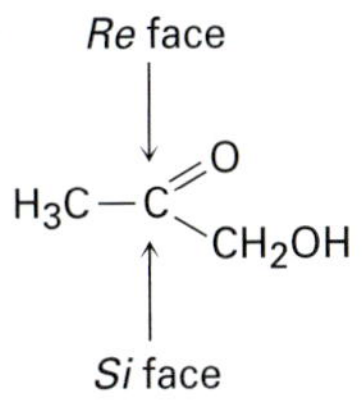

(b)

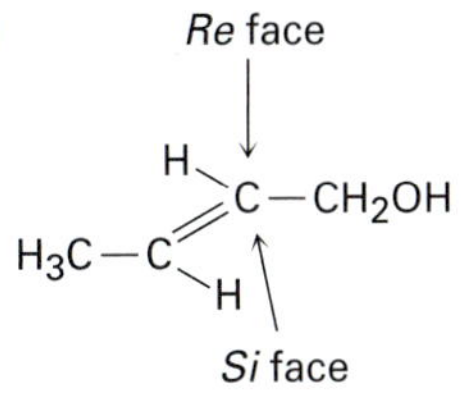

9.25 (*S*)-Lactate

9.26 The –OH adds to the *Re* face of C2, and –H adds to the *Re* face of C3. The overall addition has anti stereochemistry.

CHAPTER 10

10.1 (a) 1-Iodobutane
(b) 1-Chloro-3-methylbutane
(c) 1,5-Dibromo-2,2-dimethylpentane
(d) 1,3-Dichloro-3-methylbutane
(e) 1-Chloro-3-ethyl-4-iodopentane
(f) 2-Bromo-5-chlorohexane

10.2 (a) $CH_3CH_2CH_2C(CH_3)_2CH(Cl)CH_3$
(b) $CH_3CH_2CH_2C(Cl)_2CH(CH_3)_2$
(c) $CH_3CH_2C(Br)(CH_2CH_3)_2$

(d)

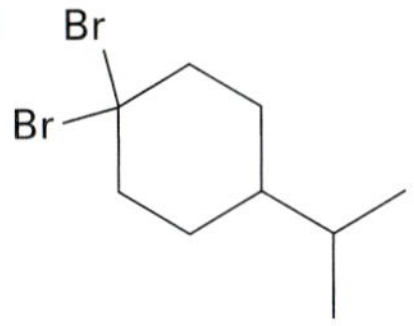

(e)

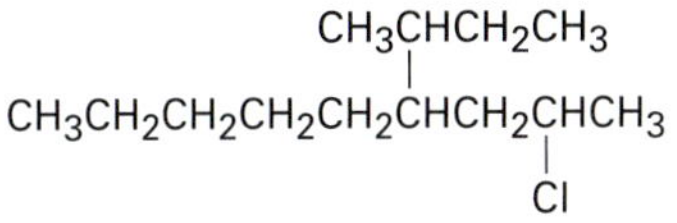

(f)

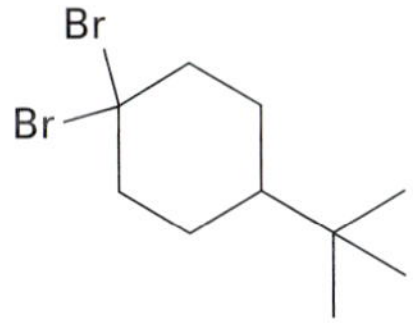

10.3 Chiral: 1-chloro-2-methylpentane, 3-chloro-2-methylpentane, 2-chloro-4-methylpentane

Achiral: 2-chloro-2-methylpentane, 1-chloro-4-methylpentane

10.4 1-Chloro-2-methylbutane (29%), 1-chloro-3-methylbutane (14%), 2-chloro-2-methylbutane (24%), 2-chloro-3-methylbutane (33%)

10.5

10.6 The intermediate allylic radical reacts at the more accessible site and gives the more highly substituted double bond.

10.7 (a) 3-Bromo-5-methylcycloheptene and 3-bromo-6-methylcycloheptene
(b) Four products

10.8 (a) 2-Methyl-2-propanol + HCl
(b) 4-Methyl-2-pentanol + PBr_3
(c) 5-Methyl-1-pentanol + PBr_3
(d) 2,4-Dimethyl-2-hexanol + HCl

10.9 Both reactions occur.

10.10 React Grignard reagent with D_2O.

10.11 (a) 1. NBS; 2. $(CH_3)_2CuLi$
(b) 1. Li; 2. CuI; 3. $CH_3CH_2CH_2CH_2Br$
(c) 1. BH_3; 2. H_2O_2, NaOH; 3. PBr_3; 4. Li, then CuI; 5. $CH_3(CH_2)_4Br$

10.12

(a)

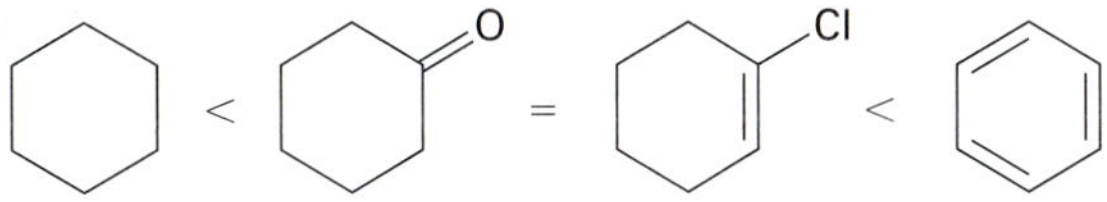

(b) $CH_3CH_2NH_2$ < $H_2NCH_2CH_2NH_2$ < $CH_3C{\equiv}N$

10.13 (a) Reduction (b) Neither

CHAPTER 11

11.1 (*R*)-1-Methylpentyl acetate, $CH_3CO_2CH(CH_3)CH_2CH_2CH_2CH_3$

11.2 (*S*)-2-Butanol

11.3 (*S*)-2-Bromo-4-methylpentane ⟶

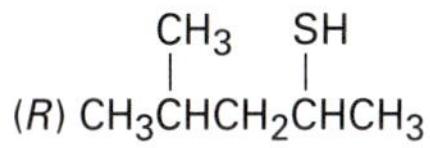

11.4 (a) 1-Iodobutane (b) 1-Butanol
(c) 1-Hexyne (d) Butylammonium bromide

11.5 (a) $(CH_3)_2N^-$ (b) $(CH_3)_3N$ (c) H_2S

11.6 $CH_3OTos > CH_3Br > (CH_3)_2CHCl > (CH_3)_3CCl$

11.7 Similar to protic solvents

11.8 Racemic 1-ethyl-1-methylhexyl acetate

11.9 90.1% racemization, 9.9% inversion

11.10

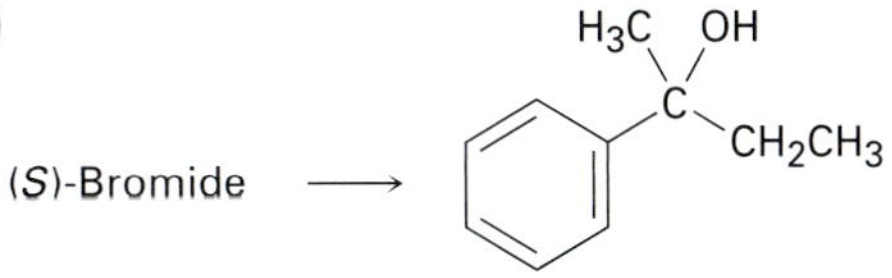

Racemic

11.11 $H_2C{=}CHCH(Br)CH_3 > CH_3CH(Br)CH_3 > CH_3CH_2Br > H_2C{=}CHBr$

11.12 The same allylic carbocation intermediate is formed.

11.13 **(a)** S_N1 **(b)** S_N2

11.14

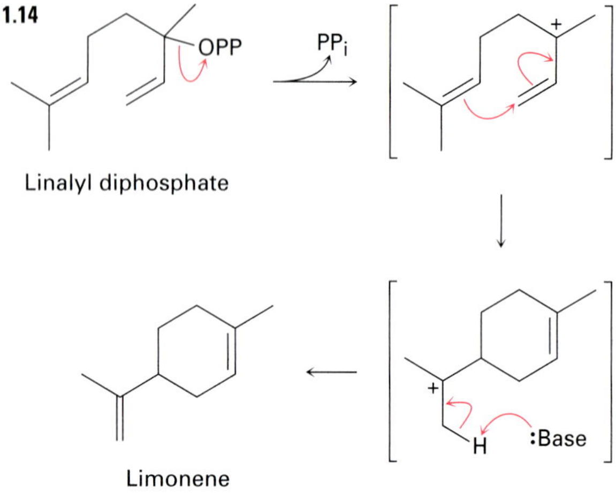

11.15 **(a)** Major: 2-methyl-2-pentene; minor: 4-methyl-2-pentene
(b) Major: 2,3,5-trimethyl-2-hexene; minor: 2,3,5-trimethyl-3-hexene and 2-isopropyl-4-methyl-1-pentene
(c) Major: ethylidenecyclohexane; minor: cyclohexylethylene

11.16 **(a)** 1-Bromo-3,6-dimethylheptane
(b) 4-Bromo-1,2-dimethylcyclopentane

11.17 (*Z*)-1-Bromo-1,2-diphenylethylene

11.18 (*Z*)-3-Methyl-2-pentene

11.19 Cis isomer reacts faster because the bromine is axial.

11.20 **(a)** S_N2 **(b)** E2 **(c)** S_N1 **(d)** E1cB

CHAPTER 12

12.1 $C_{19}H_{28}O_2$

12.2 **(a)** 2-Methyl-2-pentene **(b)** 2-Hexene

12.3 **(a)** 43, 71 **(b)** 82 **(c)** 58 **(d)** 86

12.4 102 (M^+), 84 (dehydration), 87 (alpha cleavage), 59 (alpha cleavage)

12.5 X-ray energy is higher; $\lambda = 9.0 \times 10^{-6}$ m is higher in energy.

12.6 **(a)** 2.4×10^6 kJ/mol **(b)** 4.0×10^4 kJ/mol
(c) 2.4×10^3 kJ/mol **(d)** 2.8×10^2 kJ/mol
(e) 6.0 kJ/mol **(f)** 4.0×10^{-2} kJ/mol

12.7 **(a)** Ketone or aldehyde **(b)** Nitro compound
(c) Carboxylic acid

12.8 **(a)** CH_3CH_2OH has an $-OH$ absorption.
(b) 1-Hexene has a double-bond absorption.
(c) $CH_3CH_2CO_2H$ has a very broad $-OH$ absorption.

12.9 1450–1600 cm^{-1}: aromatic ring; 2100 cm^{-1}: $C{\equiv}C$; 3300 cm^{-1}: $C{\equiv}C{-}H$

12.10 **(a)** 1715 cm^{-1} **(b)** 1730, 2100, 3300 cm^{-1}
(c) 1720, 2500–3100, 3400–3650 cm^{-1}

12.11 1690, 1650, 2230 cm^{-1}

CHAPTER 13

13.1 7.5×10^{-5} kJ/mol for ^{19}F; 8.0×10^{-5} kJ/mol for ^1H

13.2 1.2×10^{-4} kJ/mol

13.3 The vinylic C$-$H protons are nonequivalent.

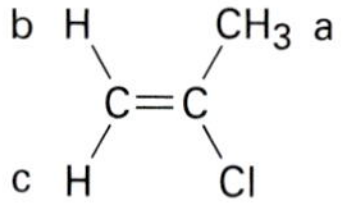

13.4 **(a)** 7.27 δ **(b)** 3.05 δ
(c) 3.46 δ **(d)** 5.30 δ

13.5 **(a)** 420 Hz **(b)** 2.1 δ **(c)** 1050 Hz

13.6 **(a)** 4 **(b)** 7 **(c)** 4 **(d)** 5 **(e)** 5 **(f)** 7

13.7 **(a)** 1,3-Dimethylcyclopentene
(b) 2-Methylpentane
(c) 1-Chloro-2-methylpropane

13.8 $-CH_3$, 9.3 δ; $-CH_2-$, 27.6 δ; C=O, 174.6 δ; $-OCH_3$, 51.4 δ

13.9

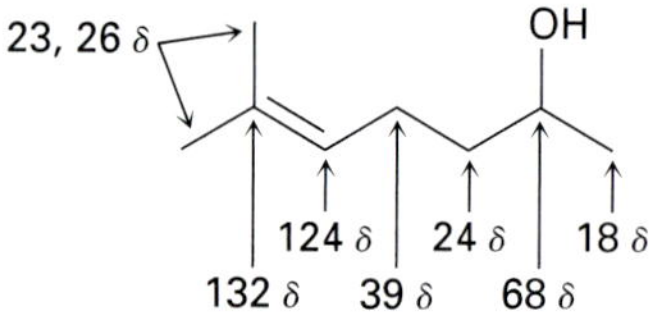

13.10

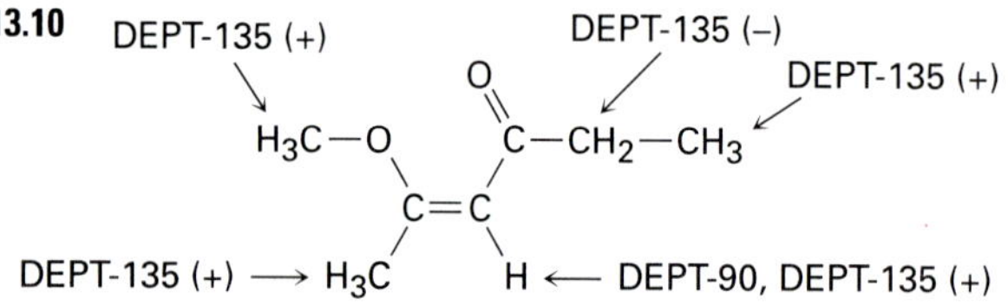

13.11

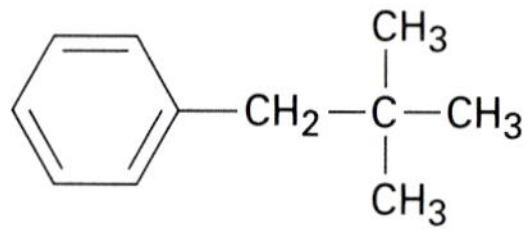

13.12 A DEPT-90 spectrum would show two absorptions for the non-Markovnikov product (RCH=CHBr) but no absorptions for the Markovnikov product ($RBrC{=}CH_2$).

13.13 (a) Enantiotopic (b) Diastereotopic (c) Diastereotopic (d) Diastereotopic (e) Diastereotopic (f) Homotopic

13.14 (a) 2 (b) 4 (c) 3 (d) 4 (e) 5 (f) 3

13.15 4

13.16 (a) 1.43 δ (b) 2.17 δ (c) 7.37 δ (d) 5.30 δ (e) 9.70 δ (f) 2.12 δ

13.17 Seven kinds of protons

13.18 Two peaks; 3:2 ratio

13.19 (a) $-CHBr_2$, quartet; $-CH_3$, doublet
(b) CH_3O-, singlet; $-OCH_2-$, triplet; $-CH_2Br$, triplet
(c) $ClCH_2-$, triplet; $-CH_2-$, quintet
(d) CH_3-, triplet; $-CH_2-$, quartet; $-CH-$, septet; $(CH_3)_2$, doublet
(e) CH_3-, triplet; $-CH_2-$, quartet; $-CH-$, septet; $(CH_3)_2$, doublet
(f) $=CH$, triplet, $-CH_2-$, doublet, aromatic C–H, two multiplets

13.20 (a) CH_3OCH_3 (b) $CH_3CH(Cl)CH_3$
(c) $ClCH_2CH_2OCH_2CH_2Cl$
(d) $CH_3CH_2CO_2CH_3$ or $CH_3CO_2CH_2CH_3$

13.21 $CH_3CH_2OCH_2CH_3$

13.22 J_{1-2} = 16 Hz; J_{2-3} = 8 Hz

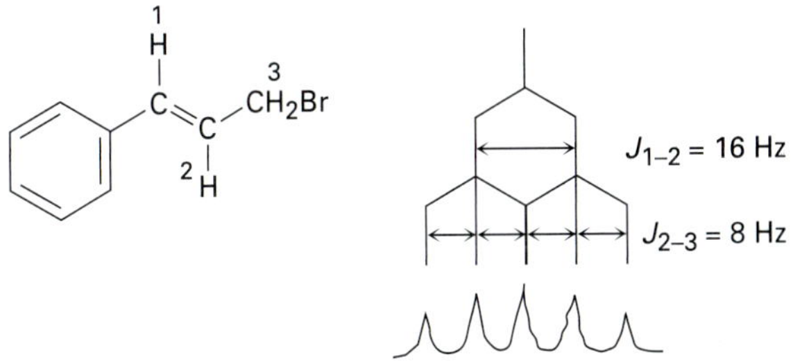

13.23 1-Chloro-1-methylcyclohexane has a singlet methyl absorption.

CHAPTER 14

14.1 Expected ΔH°_{hydrog} for allene is −252 kJ/mol. Allene is less stable than a nonconjugated diene, which is less stable than a conjugated diene.

14.2 1-Chloro-2-pentene, 3-chloro-1-pentene, 4-chloro-2-pentene

14.3 4-Chloro-2-pentene predominates in both.

14.4 1,2-Addition: 6-bromo-1,6-dimethylcyclohexene

1,4-Addition: 6-bromo-1,6-dimethylcyclohexene, 3-bromo-1,2-dimethylcyclohexene

14.5 Interconversion occurs by S_N1 dissociation to a common intermediate cation.

14.6 The double bond is more highly substituted.

14.7

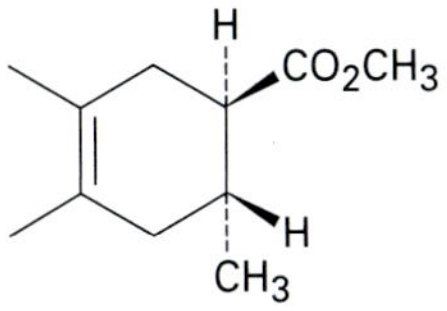

14.8 Good dienophiles: **(a)**, **(d)**

14.9 Compound **(a)** is *s*-cis. Compound **(c)** can rotate to *s*-cis.

14.10

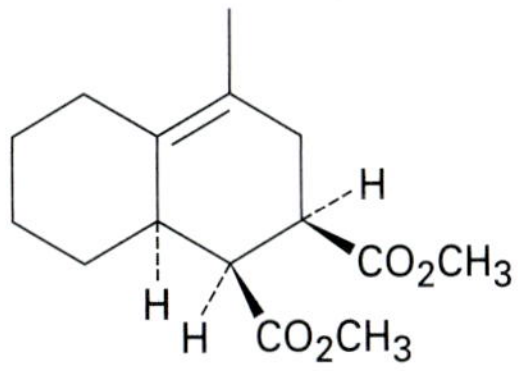

14.11

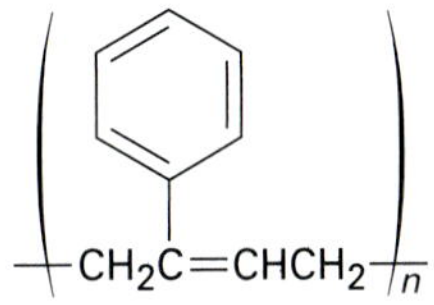

14.12

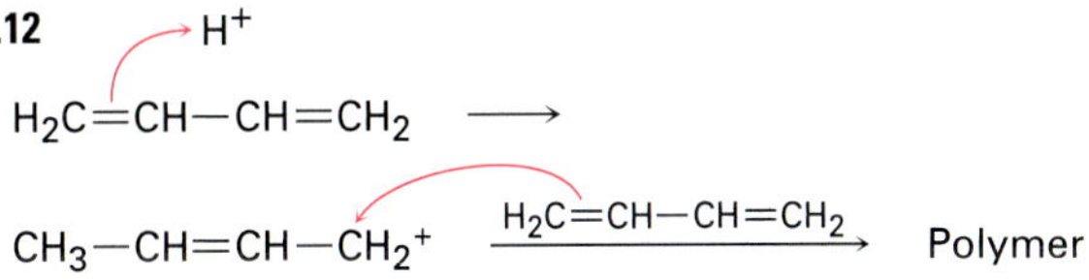

14.13 300–600 kJ/mol; UV energy is greater than IR or NMR energy.

14.14 1.46×10^{-5} M

14.15 All except **(a)** have UV absorptions.

CHAPTER 15

15.1 (a) Meta (b) Para (c) Ortho

15.2 (a) *m*-Bromochlorobenzene
(b) (3-Methylbutyl)benzene
(c) *p*-Bromoaniline
(d) 2,5-Dichlorotoluene
(e) 1-Ethyl-2,4-dinitrobenzene
(f) 1,2,3,5-Tetramethylbenzene

15.3 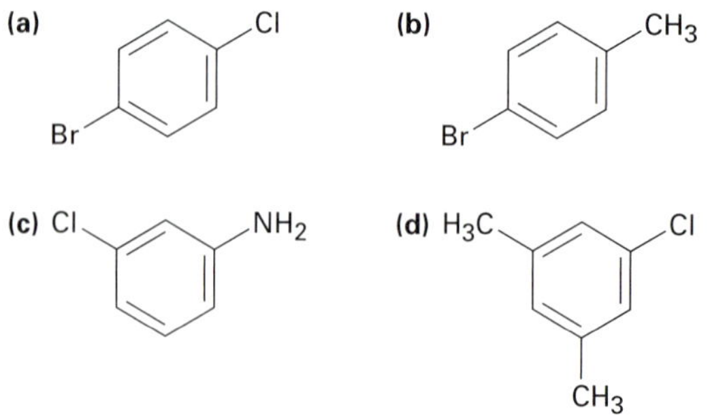

15.4 Pyridine has an aromatic sextet of electrons.

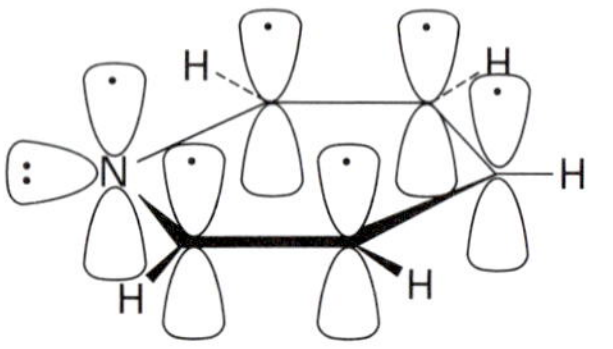

Pyridine

15.5 Cyclodecapentaene is not flat because of steric interactions.

15.6 All C–C bonds are equivalent; one resonance line in both ^{1}H and ^{13}C NMR spectra.

15.7 The cyclooctatetraenyl dianion is aromatic (ten π electrons) and flat.

15.8 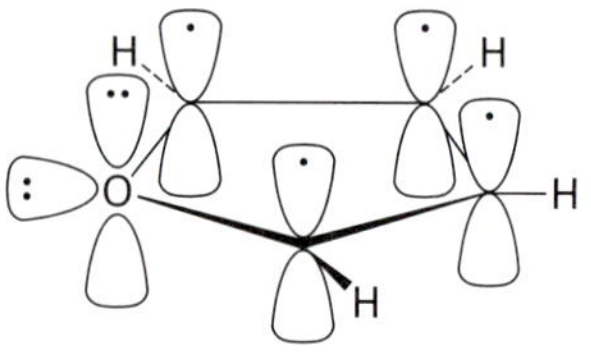

Furan

15.9 The thiazolium ring has six π electrons.

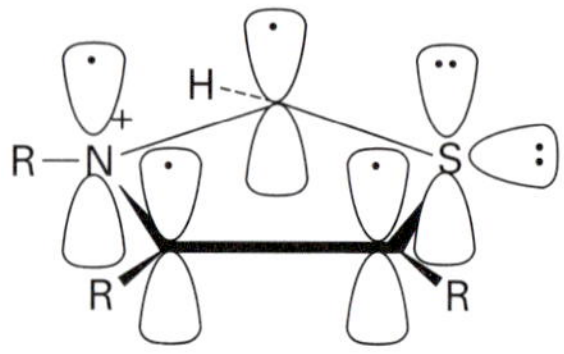

15.10 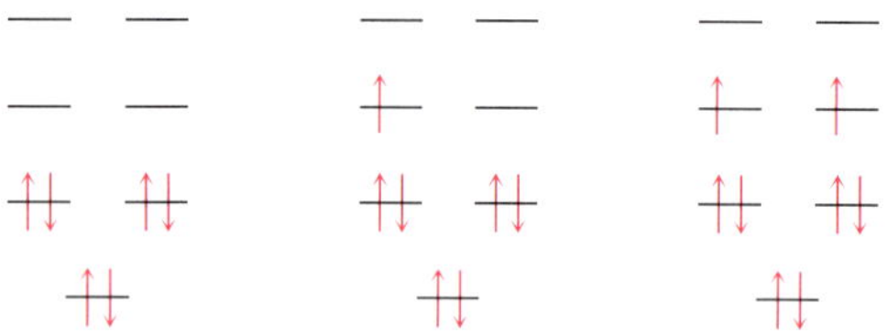

Cation **Radical** **Anion**

15.11 

15.12 The three nitrogens in double bonds each contribute one; the remaining nitrogen contributes two.

CHAPTER 16

16.1 *o*-, *m*-, and *p*-Bromotoluene

16.2 *o*-Xylene: 2; *p*-xylene: 1; *m*-xylene: 3

16.3 D^+ does electrophilic substitutions on the ring.

16.4 No rearrangement: **(a)**, **(b)**, **(e)**

16.5 *tert*-Butylbenzene

16.6 **(a)** $(CH_3)_2CHCOCl$ **(b)** PhCOCl

16.7 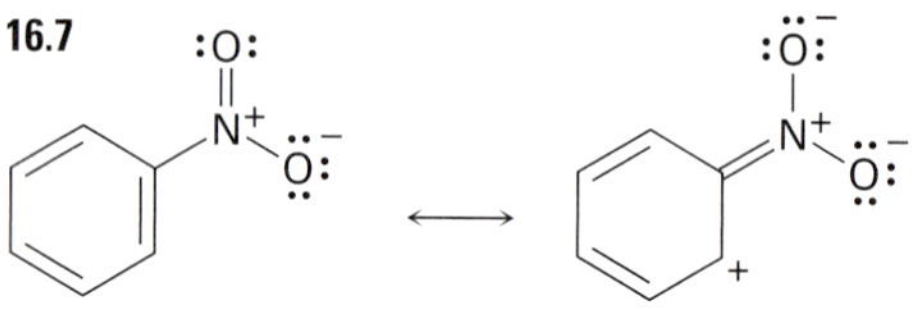

and others

16.8 

and others

16.9 **(a)** *o*- and *p*-Bromonitrobenzene
(b) *m*-Bromonitrobenzene
(c) *o*- and *p*-Chlorophenol
(d) *o*- and *p*-Bromoaniline

16.10 **(a)** Phenol > Toluene > Benzene > Nitrobenzene
(b) Phenol > Benzene > Chlorobenzene > Benzoic acid
(c) Aniline > Benzene > Bromobenzene > Benzaldehyde

16.11 Alkylbenzenes are more reactive than benzene itself, but acylbenzenes are less reactive.

16.12 Toluene is more reactive; the trifluoromethyl group is electron-withdrawing.

16.13 The nitrogen electrons are donated to the nearby carbonyl group and are less available to the ring.

16.14 The meta intermediate is most favored.

16.15 **(a)** Ortho and para to $-OCH_3$
(b) Ortho and para to $-NH_2$
(c) Ortho and para to $-Cl$

16.16 **(a)** Reaction occurs ortho and para to the $-CH_3$ group.
(b) Reaction occurs ortho and para to the $-OCH_3$ group.

16.17 The phenol is deprotonated by KOH to give an anion that carries out a nucleophilic acyl substitution reaction on the fluoronitrobenzene.

16.18 Only one benzyne intermediate can form from *p*-bromotoluene; two different benzyne intermediates can form from *m*-bromotoluene.

16.19 (a) *m*-Nitrobenzoic acid
(b) *p-tert*-Butylbenzoic acid

16.20 A benzyl radical is more stable than a primary alkyl radical by 52 kJ/mol and is similar in stability to an allyl radical.

16.21 1. CH_3CH_2Cl, $AlCl_3$; 2. NBS; 3. KOH, ethanol

16.22 1. PhCOCl, $AlCl_3$; 2. H_2/Pd

16.23 (a) 1. HNO_3, H_2SO_4; 2. Cl_2, $FeCl_3$
(b) 1. CH_3COCl, $AlCl_3$; 2. Cl_2, $FeCl_3$; 3. H_2/Pd
(c) 1. CH_3CH_2COCl, $AlCl_3$; 2. Cl_2, $FeCl_3$; 3. H_2/Pd; 4. HNO_3, H_2SO_4
(d) 1. CH_3Cl, $AlCl_3$; 2. Br_2, $FeBr_3$; 3. SO_3, H_2SO_4

16.24 (a) Friedel–Crafts acylation does not occur on a deactivated ring.
(b) Rearrangement occurs during Friedel–Crafts alkylation with primary halides; chlorination occurs ortho to the alkyl group.

CHAPTER 17

17.1 (a) 5-Methyl-2,4-hexanediol
(b) 2-Methyl-4-phenyl-2-butanol
(c) 4,4-Dimethylcyclohexanol
(d) *trans*-2-Bromocyclopentanol
(e) 4-Bromo-3-methylphenol
(f) 2-Cyclopenten-1-ol

17.2 (a) 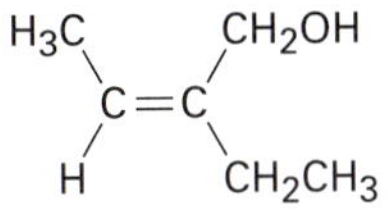
(b) 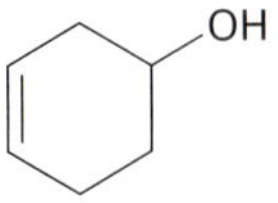
(c) 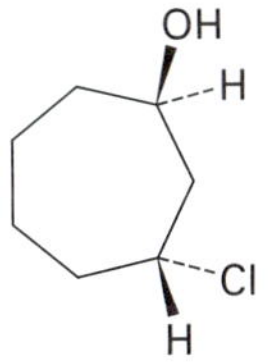
(d) $CH_3CH(OH)CH_2CH_2CH_2OH$
(e) 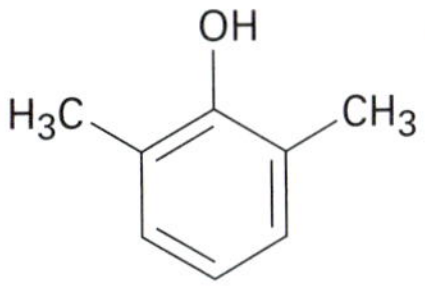
(f) 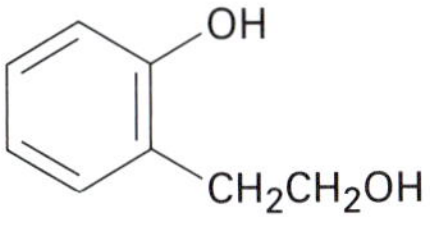

17.3 Hydrogen-bonding is more difficult in hindered alcohols.

17.4 (a) $HC{\equiv}CH < (CH_3)_2CHOH < CH_3OH < (CF_3)_2CHOH$
(b) *p*-Methylphenol < Phenol < *p*-(Trifluoromethyl)phenol
(c) Benzyl alcohol < Phenol < *p*-Hydroxybenzoic acid

17.5 The electron-withdrawing nitro group stabilizes an alkoxide ion, but the electron-donating methoxyl group destabilizes the anion.

17.6 (a) 2-Methyl-3-pentanol
(b) 2-Methyl-4-phenyl-2-butanol
(c) *meso*-5,6-Decanediol

17.7 (a) $NaBH_4$ (b) $LiAlH_4$ (c) $LiAlH_4$

17.8 (a) Benzaldehyde or benzoic acid (or ester)
(b) Acetophenone
(c) Cyclohexanone
(d) 2-Methylpropanal or 2-methylpropanoic acid (or ester)

17.9 (a) 1-Methylcyclopentanol
(b) 1,1-Diphenylethanol
(c) 3-Methyl-3-hexanol

17.10 (a) Acetone + CH_3MgBr, or ethyl acetate + 2 CH_3MgBr
(b) Cyclohexanone + CH_3MgBr
(c) 3-Pentanone + CH_3MgBr, or 2-butanone + CH_3CH_2MgBr, or ethyl acetate + 2 CH_3CH_2MgBr
(d) 2-Butanone + PhMgBr, or ethyl phenyl ketone + CH_3MgBr, or acetophenone + CH_3CH_2MgBr
(e) Formaldehyde + PhMgBr
(f) Formaldehyde + $(CH_3)_2CHCH_2MgBr$

17.11 Cyclohexanone + CH_3CH_2MgBr

17.12 1. *p*-TosCl, pyridine; 2. NaCN

17.13 (a) 2-Methyl-2-pentene
(b) 3-Methylcyclohexene
(c) 1-Methylcyclohexene
(d) 2,3-Dimethyl-2-pentene
(e) 2-Methyl-2-pentene

17.14 (a) 1-Phenylethanol (b) 2-Methyl-1-propanol
(c) Cyclopentanol

17.15 (a) Hexanoic acid, hexanal (b) 2-Hexanone
(c) Hexanoic acid, no reaction

17.16 S_N2 reaction of F^- on silicon with displacement of alkoxide ion.

17.17 Protonation of 2-methylpropene gives the *tert*-butyl cation, which carries out an electrophilic aromatic substitution reaction.

17.18 Disappearance of –OH absorption; appearance of C=O

17.19 (a) Singlet (b) Doublet (c) Triplet (d) Doublet (e) Doublet (f) Singlet

CHAPTER 18

18.1 (a) Diisopropyl ether
(b) Cyclopentyl propyl ether
(c) *p*-Bromoanisole or 4-bromo-1-methoxybenzene
(d) 1-Methoxycyclohexene
(e) Ethyl isobutyl ether
(f) Allyl vinyl ether

18.2 A mixture of diethyl ether, dipropyl ether, and ethyl propyl ether is formed in a 1:1:2 ratio.

18.3 (a) $CH_3CH_2CH_2O^- + CH_3Br$
(b) $PhO^- + CH_3Br$
(c) $(CH_3)_2CHO^- + PhCH_2Br$
(d) $(CH_3)_3CCH_2O^- + CH_3CH_2Br$

18.4

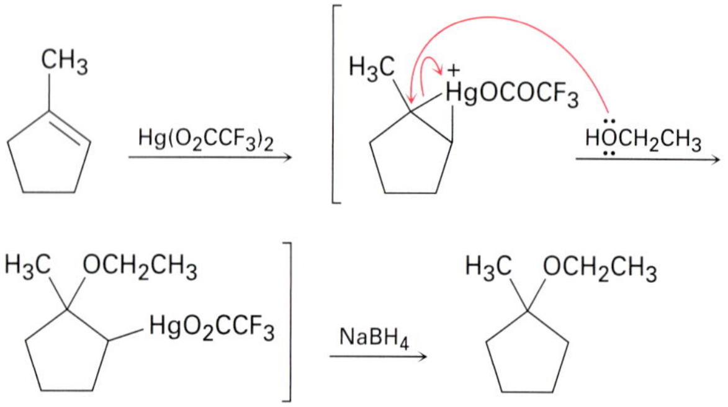

18.5 (a) Either method (b) Williamson
(c) Alkoxymercuration (d) Williamson

18.6 (a) Bromoethane > 2-Bromopropane > Bromobenzene
(b) Bromoethane > Chloroethane > 1-Iodopropene

18.7 (a)

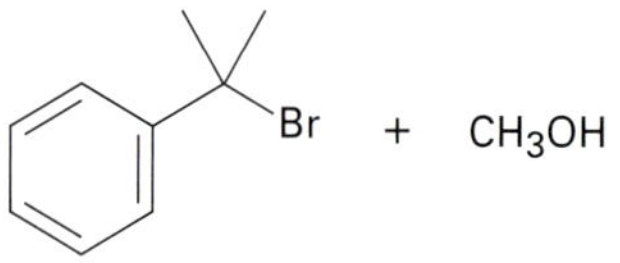

(b) $CH_3CH_2CH(CH_3)OH + CH_3CH_2CH_2Br$

18.8 Protonation of the oxygen atom, followed by E1 reaction

18.9 Br^- and I^- are better nucleophiles than Cl^-.

18.10 *o*-(1-Methylallyl)phenol

18.11 Epoxidation of *cis*-2-butene yields *cis*-2,3-epoxybutane, while epoxidation of *trans*-2-butene yields *trans*-2,3-epoxybutane.

18.12 (a)

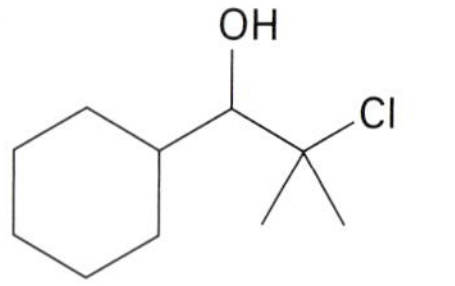

(b)

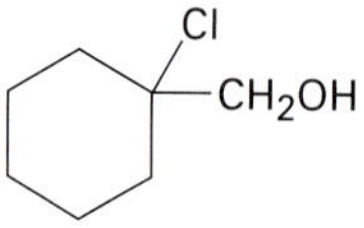

18.13 (a) 1-Methylcyclohexene + OsO_4; then $NaHSO_3$
(b) 1-Methylcyclohexene + *m*-chloroperoxybenzoic acid, then H_3O^+

18.14 (a) $CH_3CH_2C(CH_3)(OH^*)–CH_2OH$ (HO on C, * on the OH of CH_2)

(b) $CH_3CH_2C(CH_3)(O^*H)–CH_2OH$ (* on the HO of C)

(c)

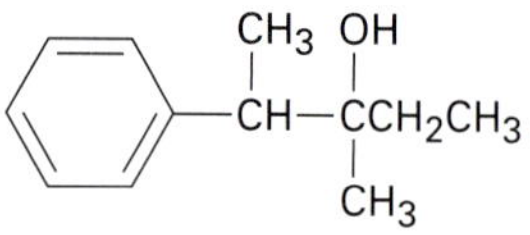

18.16 (a) 2-Butanethiol
(b) 2,2,6-Trimethyl-4-heptanethiol
(c) 2-Cyclopentene-1-thiol
(d) Ethyl isopropyl sulfide
(e) *o*-Di(methylthio)benzene
(f) 3-(Ethylthio)cyclohexanone

18.17 (a) 1. $LiAlH_4$; 2. PBr_3; 3. $(H_2N)_2C{=}S$; 4. H_2O, NaOH
(b) 1. HBr; 2. $(H_2N)_2C{=}S$; 3. H_2O, NaOH

18.18 1,2-Epoxybutane

PREVIEW OF CARBONYL CHEMISTRY

1. Acetyl chloride is more electrophilic than acetone.

2.

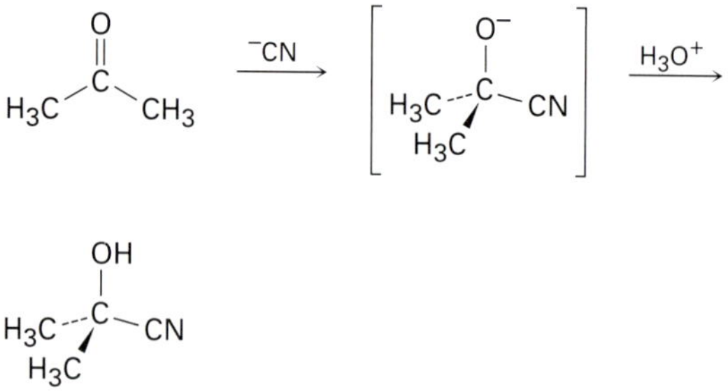

3. (a) Nucleophilic acyl substitution
(b) Nucleophilic addition
(c) Carbonyl condensation

CHAPTER 19

19.1 (a) 2-Methyl-3-pentanone
(b) 3-Phenylpropanal
(c) 2,6-Octanedione
(d) *trans*-2-Methylcyclohexanecarbaldehyde
(e) Pentanedial
(f) *cis*-2,5-Dimethylcyclohexanone

19.2

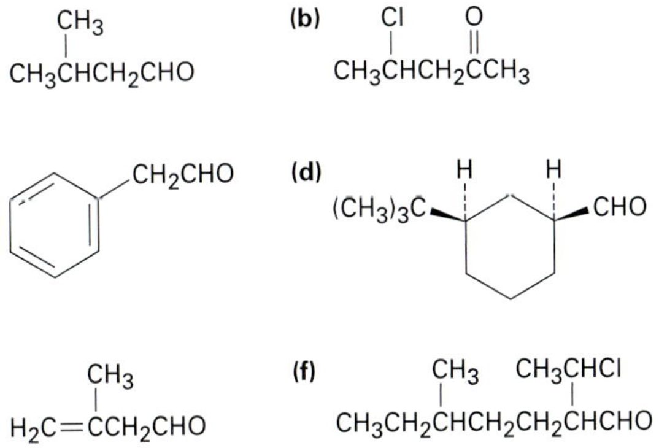

19.3 (a) PCC (b) 1. O_3; 2. Zn (c) DIBAH

19.4 (a) $Hg(OAc)_2$, H_3O^+
(b) 1. CH_3COCl, $AlCl_3$; 2. Br_2, $FeBr_3$
(c) 1. Mg; 2. CH_3CHO; 3. H_3O^+; 4. PCC
(d) 1. BH_3; 2. H_2O_2, NaOH; 3. PCC

19.5

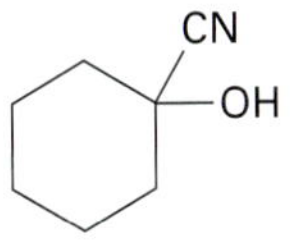

19.6 The electron-withdrawing nitro group in *p*-nitrobenzaldehyde polarizes the carbonyl group.

19.7 $CCl_3CH(OH)_2$

19.8 Labeled water adds reversibly to the carbonyl group.

19.9 The equilibrium is unfavorable for sterically hindered ketones.

19.10

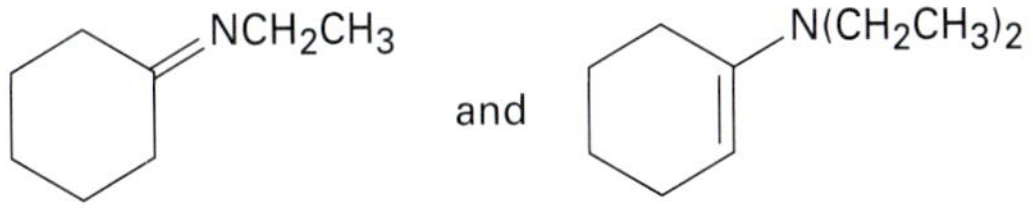

19.11 The steps are the exact reverse of the forward reaction.

19.12

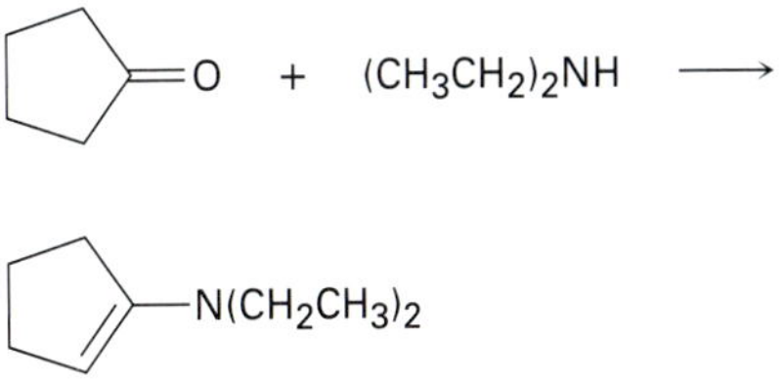

19.13 (a) H_2/Pd (b) N_2H_4, KOH
(c) 1. H_2/Pd; 2. N_2H_4, KOH

19.14 The mechanism is identical to that between a ketone and 2 equivalents of a monoalcohol (text Figure 19.12).

19.15

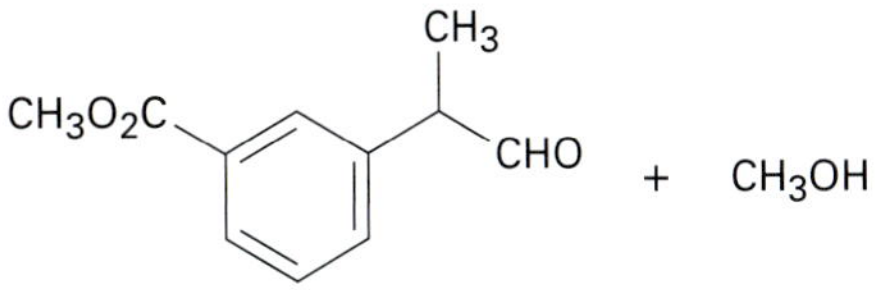

19.16 (a) Cyclohexanone + $(Ph)_3P{=}CHCH_3$
(b) 2-Cyclohexenone + $(Ph)_3P{=}CH_2$
(c) Acetone + $(Ph)_3P{=}CHCH_2CH_2CH_3$
(d) Acetone + $(Ph)_3P{=}CHPh$
(e) $PhCOCH_3$ + $(Ph)_3P{=}CHPh$
(f) 2-Cyclohexenone + $(Ph)_3P{=}CH_2$

19.17

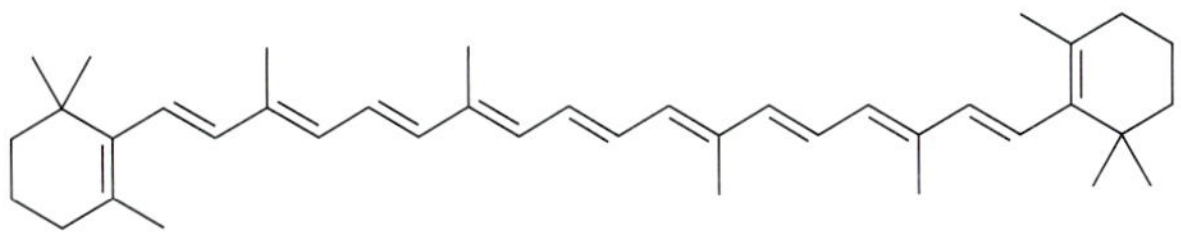

β-Carotene

19.18 Intramolecular Cannizzaro reaction

19.19 Addition of the *pro-R* hydrogen of NADH takes place on the *Re* face of pyruvate.

19.20 The −OH group adds to the *Re* face at C2, and −H adds to the *Re* face at C3, to yield (2*R*,3*S*)-isocitrate.

19.21

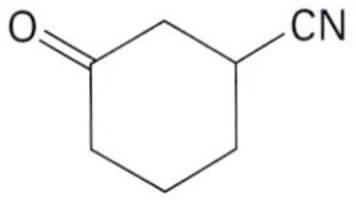

19.22 (a) 3-Buten-2-one + $(CH_3CH_2CH_2)_2CuLi$
(b) 3-Methyl-2-cyclohexenone + $(CH_3)_2CuLi$
(c) 4-*tert*-Butyl-2-cyclohexenone + $(CH_3CH_2)_2CuLi$
(d) Unsaturated ketone + $(H_2C{=}CH)_2CuLi$

19.23 Look for appearance of either an alcohol or a saturated ketone in the product.

19.24 (a) 1715 cm^{-1} (b) 1685 cm^{-1}
(c) 1750 cm^{-1} (d) 1705 cm^{-1}
(e) 1715 cm^{-1} (f) 1705 cm^{-1}

19.25 (a) Different peaks due to McLafferty rearrangement
(b) Different peaks due to α cleavage and McLafferty rearrangement
(c) Different peaks due to McLafferty rearrangement

19.26 IR: 1750 cm^{-1}; MS: 140, 84

CHAPTER 20

20.1 (a) 3-Methylbutanoic acid
(b) 4-Bromopentanoic acid
(c) 2-Ethylpentanoic acid
(d) *cis*-4-Hexenoic acid
(e) 2,4-Dimethylpentanenitrile
(f) *cis*-1,3-Cyclopentanedicarboxylic acid

20.2

(a) 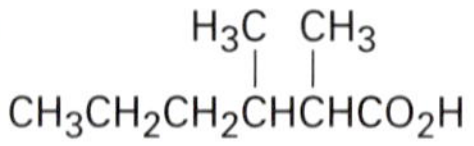

(b) 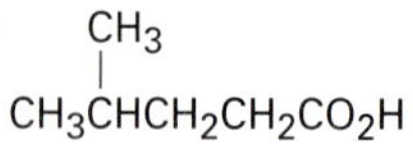

(c) 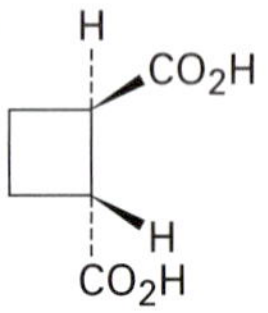

(d) 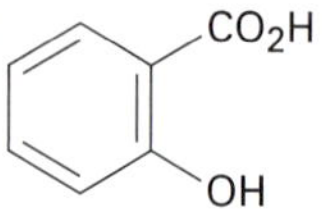

(e) 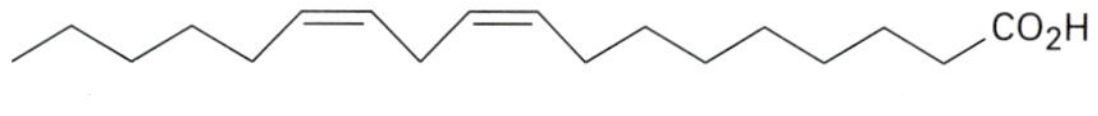

(f) $CH_3CH_2CH{=}CHCN$

20.3 Dissolve the mixture in ether, extract with aqueous NaOH, separate and acidify the aqueous layer, and extract with ether.

20.4 43%

20.5 (a) 82% dissociation (b) 73% dissociation

20.6 Lactic acid is stronger because of the inductive effect of the –OH group.

20.7 The dianion is destabilized by repulsion between charges.

20.8 More reactive

20.9 (a) *p*-Methylbenzoic acid < Benzoic acid < *p*-Chlorobenzoic acid
(b) Acetic acid < Benzoic acid < *p*-Nitrobenzoic acid

20.10 (a) 1. Mg; 2. CO_2; 3. H_3O^+
(b) 1. Mg; 2. CO_2; 3. H_3O^+ or 1. NaCN; 2. H_3O^+

20.11 1. NaCN; 2. H_3O^+; 3. $LiAlH_4$

20.12 1. PBr_3; 2. NaCN; 3. H_3O^+; 4. $LiAlH_4$

20.13 (a) Propanenitrile + CH_3CH_2MgBr, then H_3O^+
(b) *p*-Nitrobenzonitrile + CH_3MgBr, then H_3O^+

20.14 1. NaCN; 2. CH_3CH_2MgBr, then H_3O^+

20.15 A carboxylic acid has a very broad –OH absorption at 2500–3300 cm^{-1}.

20.16 4-Hydroxycyclohexanone: **H–C–O** absorption near 4 δ in ^1H spectrum and **C=O** absorption near 210 δ in ^{13}C spectrum. Cyclopentanecarboxylic acid: $-CO_2H$ absorption near 12 δ in ^1H spectrum and $-CO_2H$ absorption near 170 δ in ^{13}C spectrum.

CHAPTER 21

21.1 (a) 4-Methylpentanoyl chloride
(b) Cyclohexylacetamide
(c) Isopropyl 2-methylpropanoate
(d) Benzoic anhydride
(e) Isopropyl cyclopentanecarboxylate
(f) Cyclopentyl 2-methylpropanoate
(g) *N*-Methyl-4-pentenamide
(h) (*R*)-2-Hydroxypropanoyl phosphate
(i) Ethyl 2,3-Dimethyl-2-butenethioate

21.2

(a) $C_6H_5CO_2C_6H_5$ (b) $CH_3CH_2CH_2CON(CH_3)CH_2CH_3$

(c) $(CH_3)_2CHCH_2CH(CH_3)COCl$ (d) 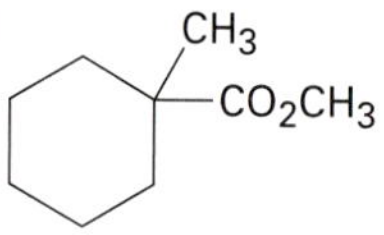

(e) 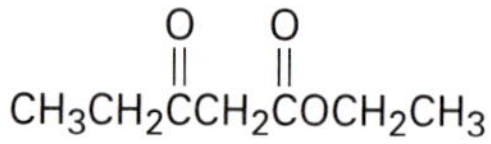

(f) 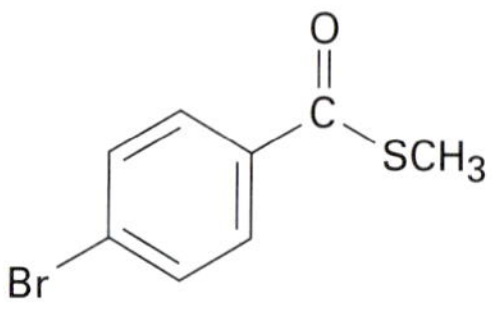

(g) 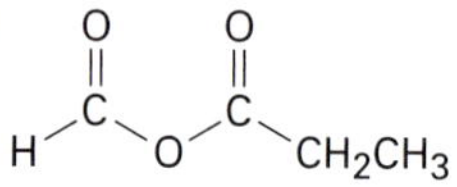

(h) 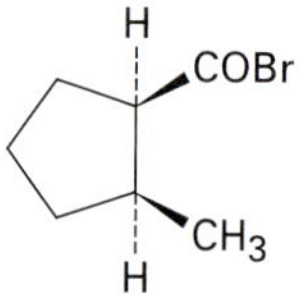

21.3 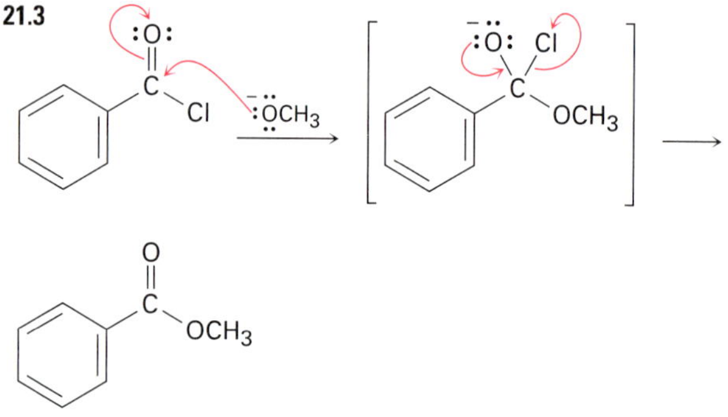

21.4 (a) Acetyl chloride > Methyl acetate > Acetamide
(b) Hexafluoroisopropyl acetate > 2,2,2-Trichloroethyl acetate > Methyl acetate

21.5 (a) $CH_3CO_2^- \; Na^+$ (b) CH_3CONH_2
(c) $CH_3CO_2CH_3 + CH_3CO_2^- \; Na^+$
(d) $CH_3CONHCH_3$

21.6

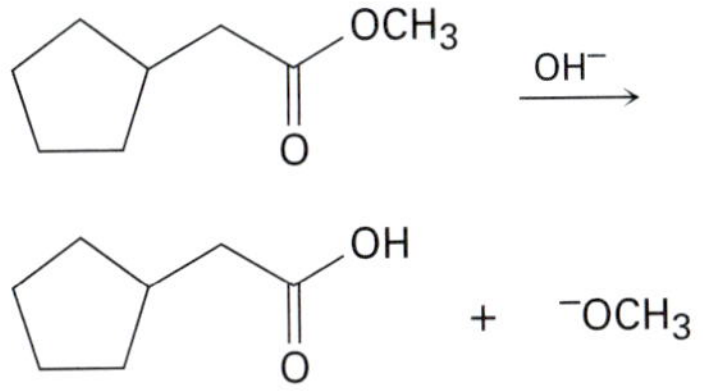

21.7 (a) Acetic acid + 1-butanol
(b) Butanoic acid + methanol
(c) Cyclopentanecarboxylic acid + isopropyl alcohol

21.8

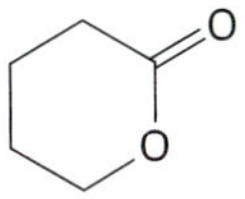

21.9 (a) Propanoyl chloride + methanol
(b) Acetyl chloride + ethanol
(c) Benzoyl chloride + ethanol

21.10 Benzoyl chloride + cyclohexanol

21.11 This is a typical nucleophilic acyl substitution reaction, with morpholine as the nucleophile and chloride as the leaving group.

21.12 (a) Propanoyl chloride + methylamine
(b) Benzoyl chloride + diethylamine
(c) Propanoyl chloride + ammonia

21.13 (a) Benzoyl chloride + $[(CH_3)_2CH]_2CuLi$, or 2-methylpropanoyl chloride + Ph_2CuLi
(b) 2-Propenoyl chloride + $(CH_3CH_2CH_2)_2CuLi$, or butanoyl chloride + $(H_2C{=}CH)_2CuLi$

21.14 This is a typical nucleophilic acyl substitution reaction, with *p*-hydroxyaniline as the nucleophile and acetate ion as the leaving group.

21.15 Monomethyl ester of benzene-1,2-dicarboxylic acid

21.16 Reaction of a carboxylic acid with an alkoxide ion gives the carboxylate ion.

21.17 $HOCH_2CH_2CH_2CHO$

21.18 (a) $CH_3CH_2CH_2CH(CH_3)CH_2OH$
(b) PhOH + $PhCH_2OH$

21.19 (a) Ethyl benzoate + 2 CH_3MgBr
(b) Ethyl acetate + 2 PhMgBr
(c) Ethyl pentanoate + 2 CH_3CH_2MgBr

21.20 (a) H_2O, NaOH
(b) Benzoic acid + BH_3
(c) $LiAlH_4$

21.21 1. Mg; 2. CO_2, then H_3O^+; 3. $SOCl_2$; 4. $(CH_3)_2NH$; 5. $LiAlH_4$

21.22

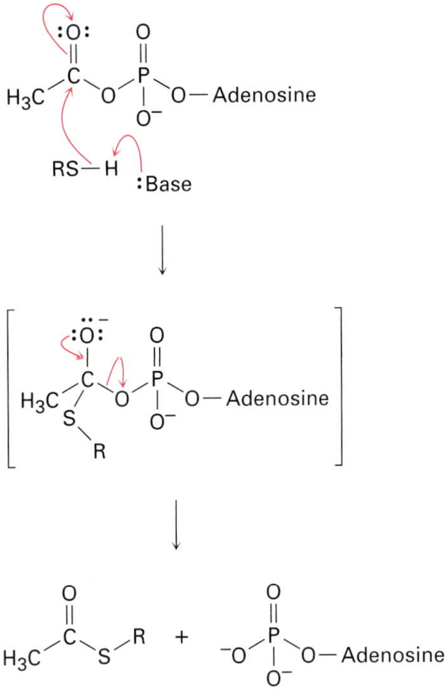

Acetyl CoA

21.23 (a)

$$-\!\!(\!-OCH_2CH_2CH_2OCH_2CH_2CH_2-\!)\!\!-_n$$

(b)

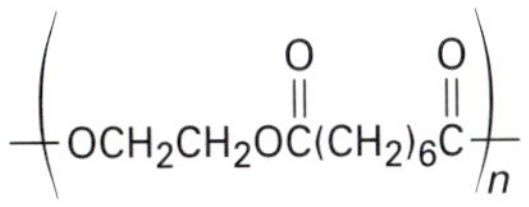

(c)

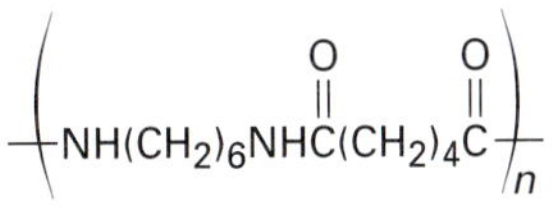

21.24

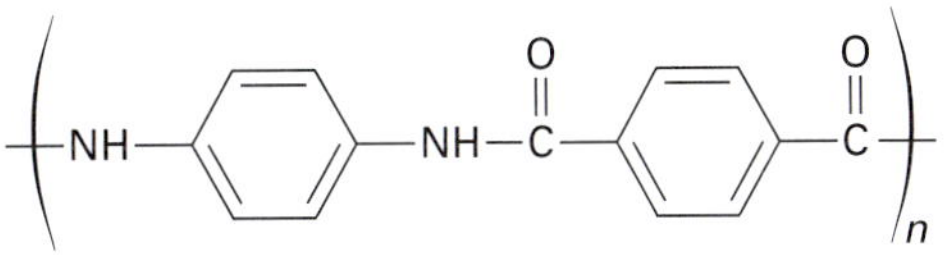

21.25 The product has a large amount of cross-linking.

21.26 (a) Ester (b) Acid chloride
(c) Carboxylic acid
(d) Aliphatic ketone or cyclohexanone

21.27 (a) $CH_3CH_2CH_2CO_2CH_2CH_3$ and other possibilities
(b) $CH_3CON(CH_3)_2$
(c) $CH_3CH{=}CHCOCl$ or $H_2C{=}C(CH_3)COCl$

CHAPTER 22

22.1 (a)
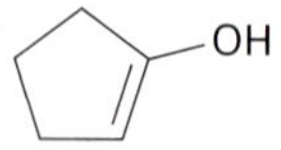

(b)
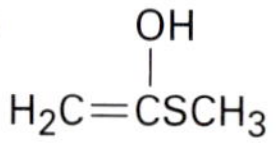

(c)
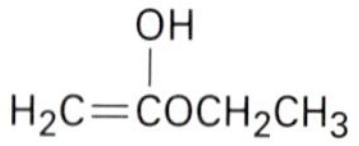

(d) $CH_3CH{=}CHOH$

(e)
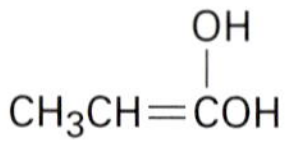

(f)
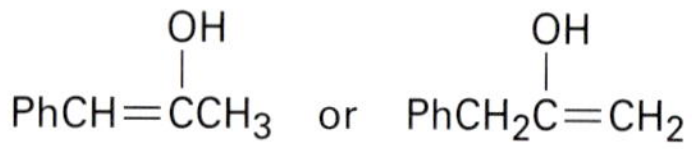

22.2 (a) 4 (b) 3 (c) 3 (d) 2 (e) 4 (f) 5

22.3
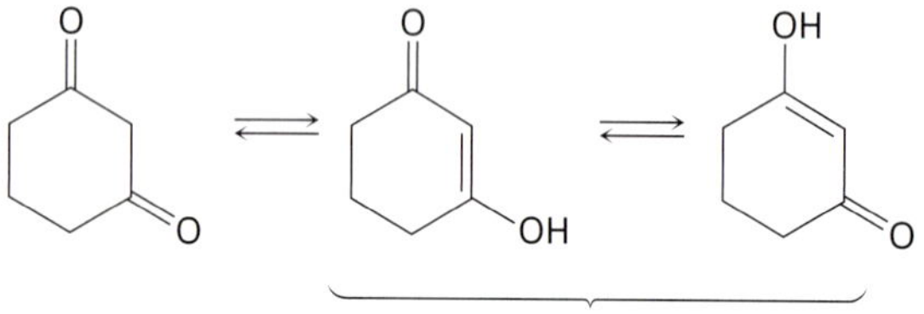

Equivalent;
more stable

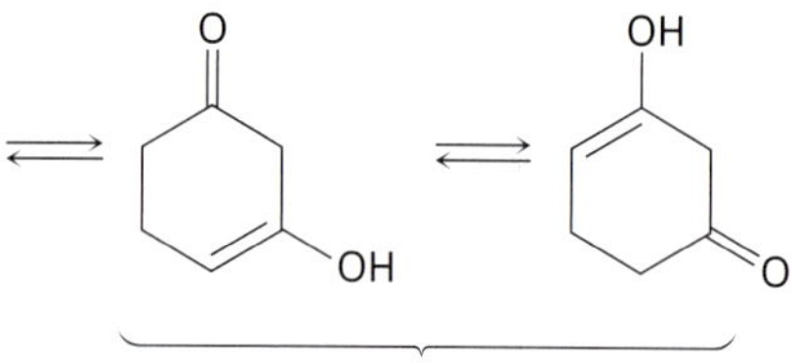

Equivalent;
less stable

22.4 Acid-catalyzed formation of an enol is followed by deuteronation of the enol double bond and dedeuteronation of oxygen.

22.5 1. Br_2; 2. Pyridine, heat

22.6 The intermediate α-bromo acid bromide undergoes a nucleophilic acyl substitution reaction with methanol to give an α-bromo ester.

22.7 (a) $CH_3C\underline{H}_2CHO$ (b) $(CH_3)_3CCOC\underline{H}_3$
(c) $CH_3CO_2\underline{H}$ (d) $PhCON\underline{H}_2$
(e) $CH_3CH_2C\underline{H}_2CN$ (f) $C\underline{H}_3CON(CH_3)_2$

22.8 $^-{:}CH_2C{\equiv}N{:} \longleftrightarrow H_2C{=}C{=}\ddot{N}{:}^-$

22.9 Acid is regenerated, but base is used stoichiometrically.

22.10 (a) 1. $Na^+\ ^-OEt$; 2. $PhCH_2Br$; 3. H_3O^+
(b) 1. $Na^+\ ^-OEt$; 2. $CH_3CH_2CH_2Br$; 3. $Na^+\ ^-OEt$; 4. CH_3Br; 5. H_3O^+
(c) 1. $Na^+\ ^-OEt$; 2. $(CH_3)_2CHCH_2Br$; 3. H_3O^+

22.11 Malonic ester has only two acidic hydrogens to be replaced.

22.12 1. $Na^+\ ^-OEt$; 2. $(CH_3)_2CHCH_2Br$; 3. $Na^+\ ^-OEt$; 4. CH_3Br; 5. H_3O^+

22.13 (a) $(CH_3)_2CHCH_2Br$ (b) $PhCH_2CH_2Br$

22.14 None can be prepared.

22.15 1. 2 $Na^+\ ^-OEt$; 2. $BrCH_2CH_2CH_2CH_2Br$; 3. H_3O^+

22.16 (a) Alkylate phenylacetone with CH_3I
(b) Alkylate pentanenitrile with CH_3CH_2I
(c) Alkylate cyclohexanone with $H_2C{=}CHCH_2Br$
(d) Alkylate cyclohexanone with excess CH_3I
(e) Alkylate $C_6H_5COCH_2CH_3$ with CH_3I
(f) Alkylate methyl 3-methylbutanoate with CH_3CH_2I

CHAPTER 23

23.1 (a)
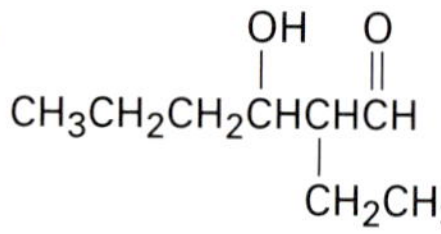

(b)
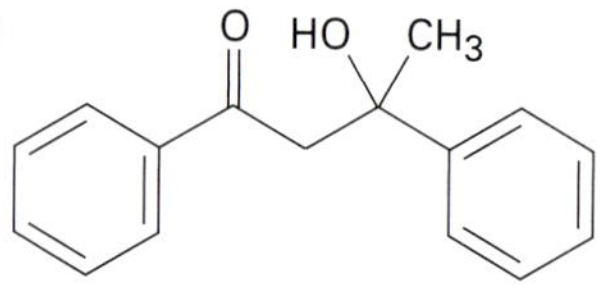

(c)
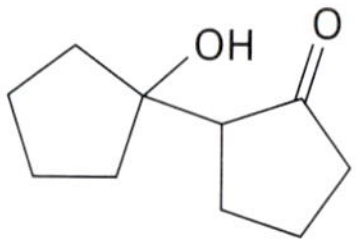

23.2 The reverse reaction is the exact opposite of the forward reaction.

23.3

(a)
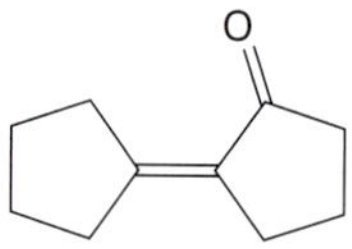

(b)
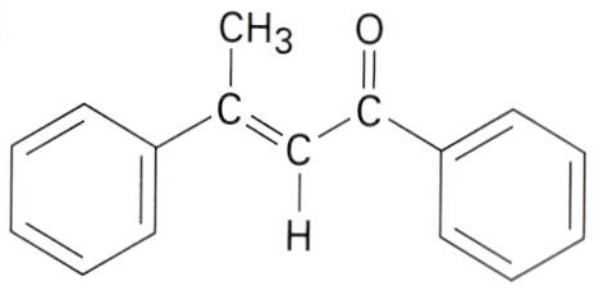

(c)
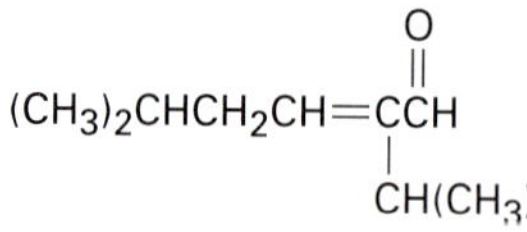

23.4

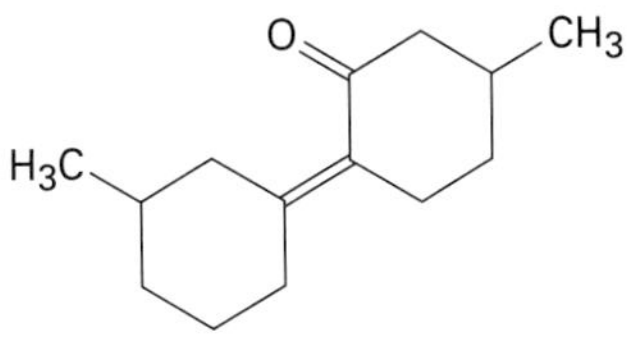

and

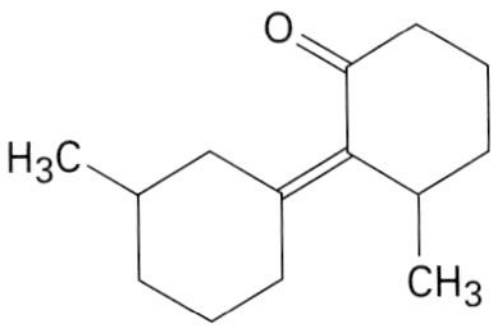

23.5 (a) Not an aldol product (b) 3-Pentanone

23.6 1. NaOH; 2. $LiAlH_4$; 3. H_2/Pd

23.7

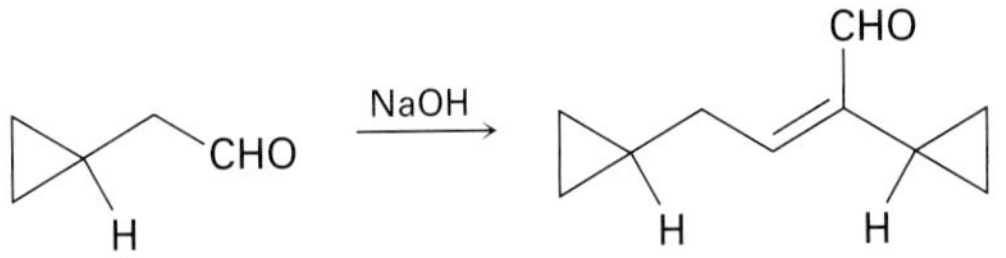

23.8 (a) $C_6H_5CHO + CH_3COCH_3$
(b), (c) Not easily prepared

23.9 The CH_2 position between the two carbonyl groups is so acidic that it is completely deprotonated to give a stable enolate ion.

23.10

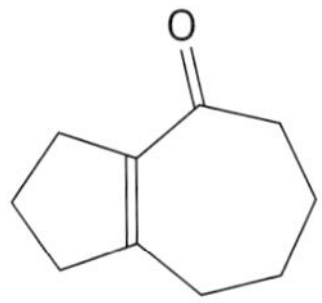

23.11 (a)

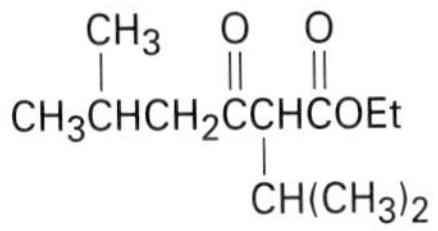

(b)

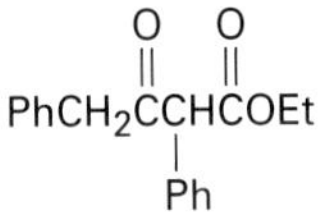

(c)

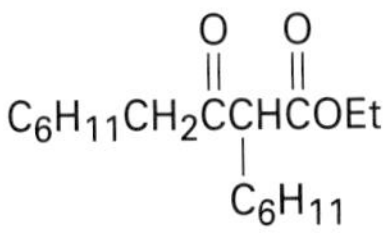

23.12 The cleavage reaction is the exact reverse of the forward reaction.

23.13

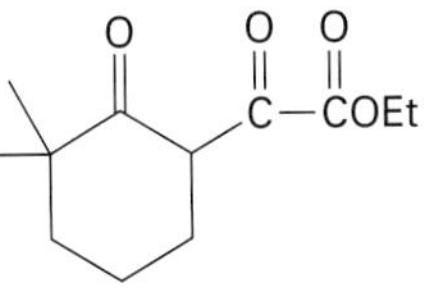

23.14

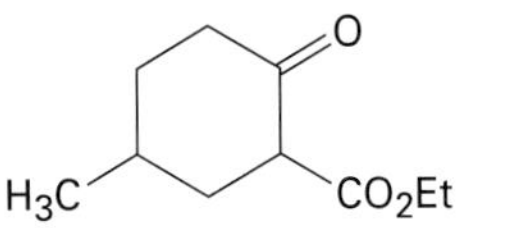

23.15

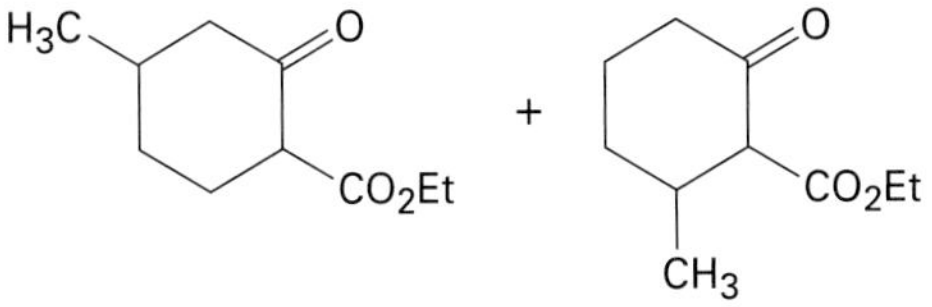

23.16 (a)

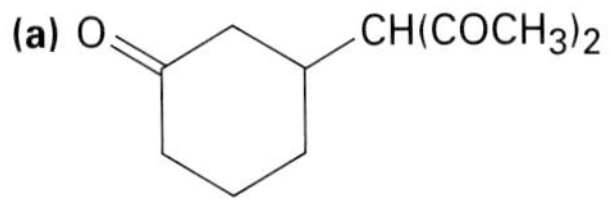

(b) $(CH_3CO)_2CHCH_2CH_2CN$

(c)

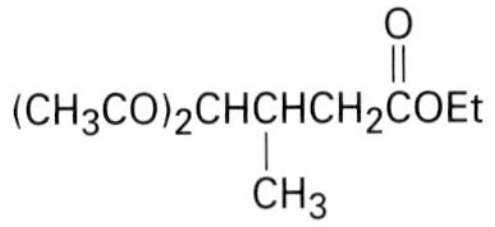

23.17

(a)

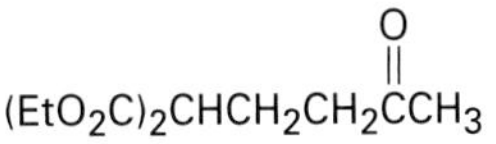

(b)

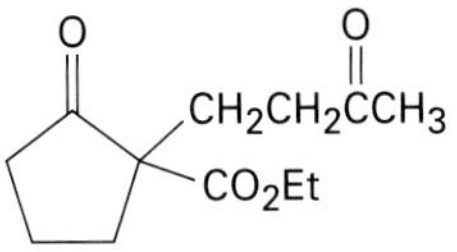

23.18 $CH_3CH_2COCH{=}CH_2 + CH_3CH_2NO_2$

23.19

(a)

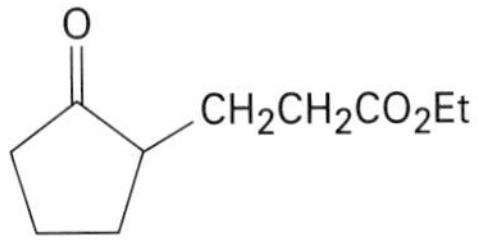

(b)

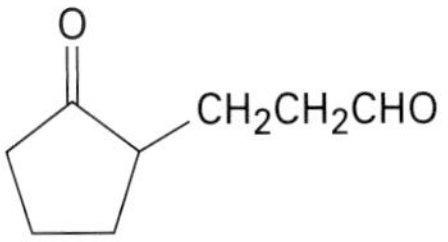

(c)

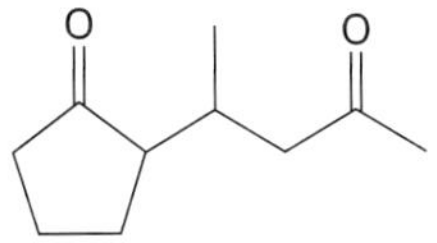

23.20 (a) Cyclopentanone enamine + propenenitrile
(b) Cyclohexanone enamine + methyl propenoate

23.21

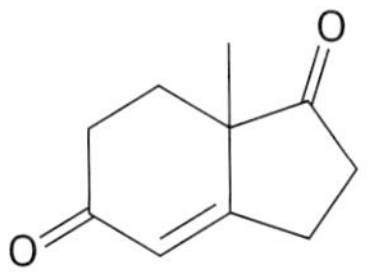

23.22 2,5,5-Trimethyl-1,3-cyclohexanedione + 1-penten-3-one

CHAPTER 24

24.1 (a) *N*-Methylethylamine
(b) Tricyclohexylamine
(c) *N*-Methyl-*N*-propylcyclohexylamine
(d) *N*-Methylpyrrolidine
(e) Diisopropylamine
(f) 1,3-Butanediamine

24.2

(a) $[(CH_3)_2CH]_3N$

(b) $(H_2C{=}CHCH_2)_2NH$

(c) 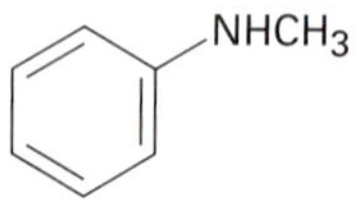

(d) 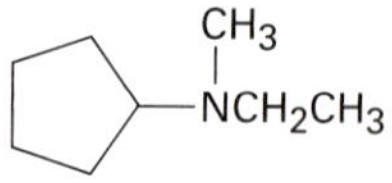

(e) 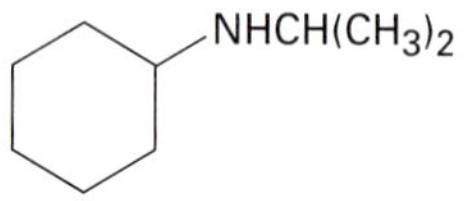

(f) 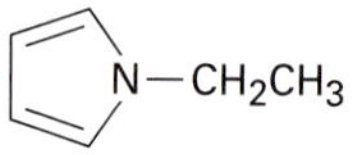

24.3

(a) 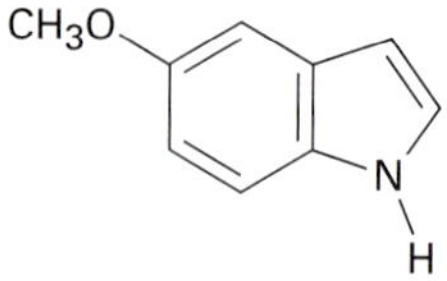

(b) 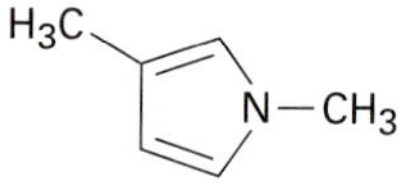

(c) 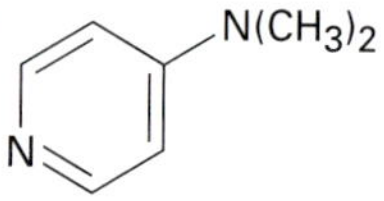

(d) 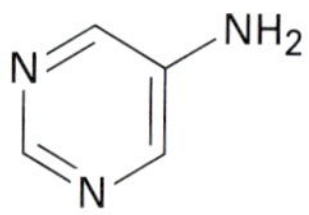

24.4 (a) $CH_3CH_2NH_2$ (b) NaOH
(c) CH_3NHCH_3

24.5 Propylamine is stronger; benzylamine $pK_b = 4.67$; propylamine $pK_b = 3.29$

24.6 (a) *p*-Nitroaniline < *p*-Aminobenzaldehyde < *p*-Bromoaniline
(b) *p*-Aminoacetophenone < *p*-Chloroaniline < *p*-Methylaniline
(c) *p*-(Trifluoromethyl)aniline < *p*-(Fluoromethyl)aniline < *p*-Methylaniline

24.7 Pyrimidine is essentially 100% neutral (unprotonated).

24.8 (a) Propanenitrile or propanamide
(b) *N*-Propylpropanamide
(c) Benzonitrile or benzamide
(d) *N*-Phenylacetamide

24.9 The reaction takes place by two nucleophilic acyl substitution reactions.

24.10 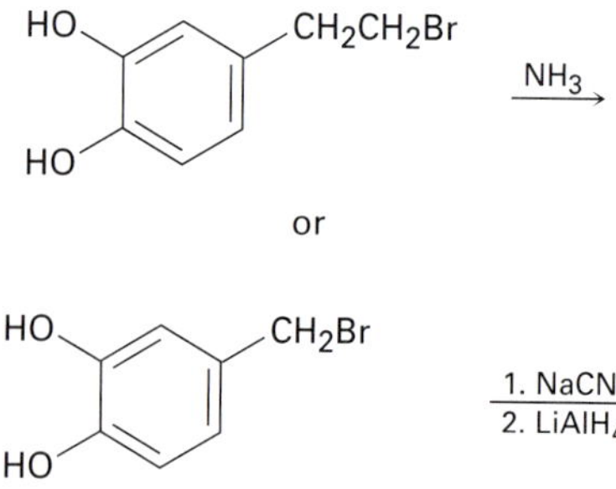

24.11 (a) Ethylamine + acetone, or isopropylamine + acetaldehyde
(b) Aniline + acetaldehyde
(c) Cyclopentylamine + formaldehyde, or methylamine + cyclopentanone

24.12 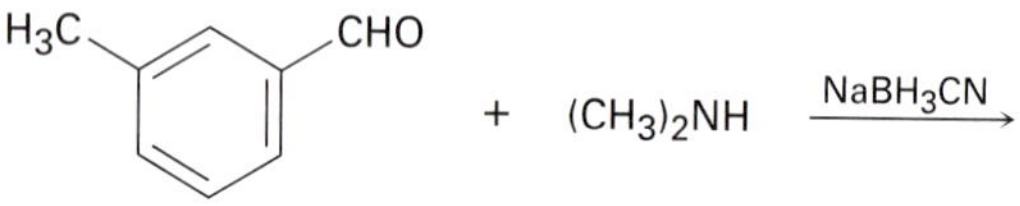

24.13 (a) 4,4-Dimethylpentanamide or 4,4-dimethylpentanoyl azide
(b) *p*-Methylbenzamide or *p*-methylbenzoyl azide

24.14 (a) 3-Octene and 4-octene
(b) Cyclohexene
(c) 3-Heptene
(d) Ethylene and cyclohexene

24.15 $H_2C{=}CHCH_2CH_2CH_2N(CH_3)_2$

24.16 1. HNO_3, H_2SO_4; 2. H_2/PtO_2; 3. $(CH_3CO)_2O$; 4. $HOSO_2Cl$; 5. aminothiazole; 6. H_2O, NaOH

24.17 (a) 1. HNO_3, H_2SO_4; 2. H_2/PtO_2; 3. 2 CH_3Br
(b) 1. HNO_3, H_2SO_4; 2. H_2/PtO_2; 3. $(CH_3CO)_2O$; 4. Cl_2; 5. H_2O, NaOH
(c) 1. HNO_3, H_2SO_4; 2. Cl_2, $FeCl_3$; 3. $SnCl_2$
(d) 1. HNO_3, H_2SO_4; 2. H_2/PtO_2; 3. $(CH_3CO)_2O$; 4. 2 CH_3Cl, $AlCl_3$; 5. H_2O, NaOH

24.18 (a) 1. CH_3Cl, $AlCl_3$; 2. HNO_3, H_2SO_4; 3. $SnCl_2$; 4. $NaNO_2$, H_2SO_4; 5. CuBr; 6. $KMnO_4$, H_2O
(b) 1. HNO_3, H_2SO_4; 2. Br_2, $FeBr_3$; 3. $SnCl_2$, H_3O^+; 4. $NaNO_2$, H_2SO_4; 5. CuCN; 6. H_3O^+
(c) 1. HNO_3, H_2SO_4; 2. Cl_2, $FeCl_3$; 3. $SnCl_2$; 4. $NaNO_2$, H_2SO_4; 5. CuBr
(d) 1. CH_3Cl, $AlCl_3$; 2. HNO_3, H_2SO_4; 3. $SnCl_2$; 4. $NaNO_2$, H_2SO_4; 5. CuCN; 6. H_3O^+
(e) 1. HNO_3, H_2SO_4; 2. H_2/PtO_2; 3. $(CH_3CO)_2O$; 4. 2 Br_2; 5. H_2O, NaOH; 6. $NaNO_2$, H_2SO_4; 7. CuBr

24.19 1. HNO_3, H_2SO_4; 2. $SnCl_2$; 3a. 2 equiv. CH_3I; 3b. $NaNO_2$, H_2SO_4; 4. product of 3a + product of 3b

24.20

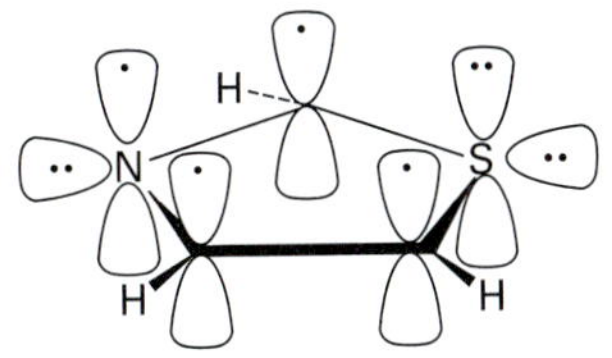

24.21 4.1% protonated

24.22

Attack at C2:

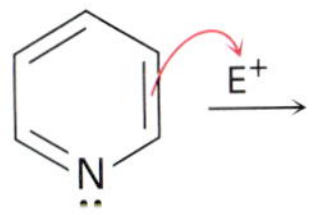

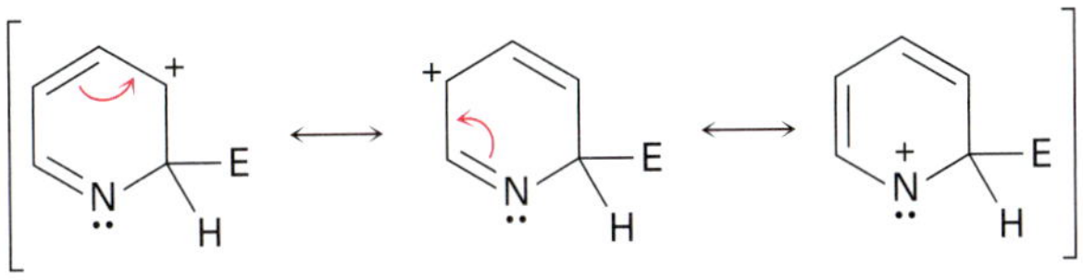

Unfavorable

Attack at C3:

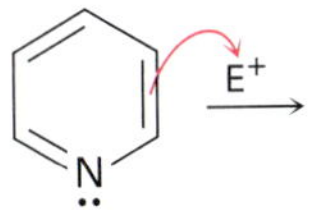

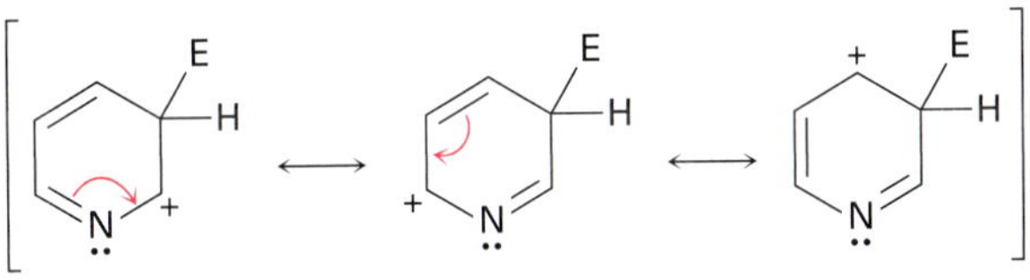

Attack at C4:

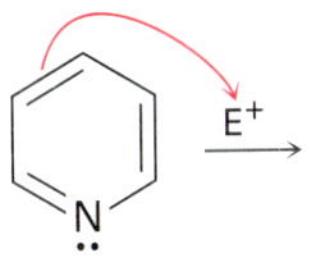

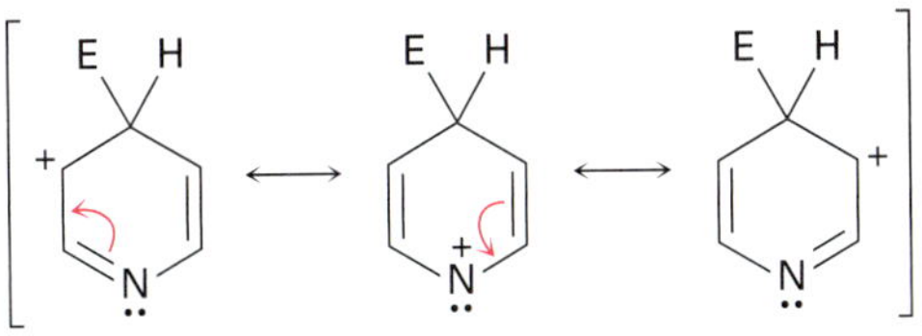

Unfavorable

24.23 The side-chain nitrogen is more basic than the ring nitrogen.

24.24 Reaction at C2 is disfavored because the aromaticity of the benzene ring is lost.

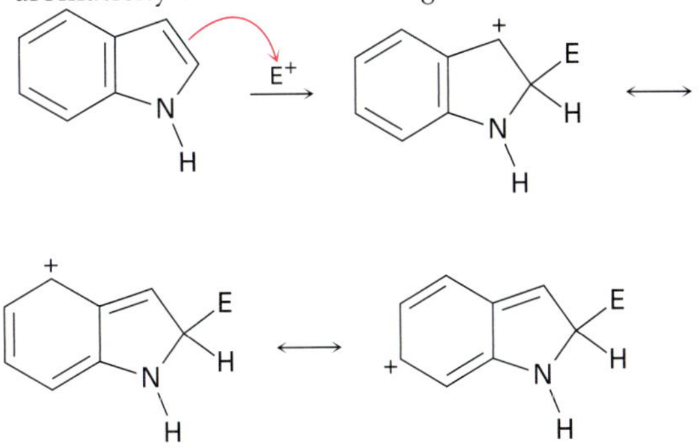

24.25 $(CH_3)_3CCOCH_3 \longrightarrow (CH_3)_3CCH(NH_2)CH_3$

CHAPTER 25

25.1 **(a)** Aldotetrose
(b) Ketopentose
(c) Ketohexose
(d) Aldopentose

25.2 **(a)** *S* **(b)** *R* **(c)** *S*

25.3 A, B, and C are the same.

25.4

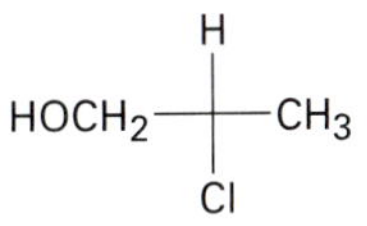

R

25.5

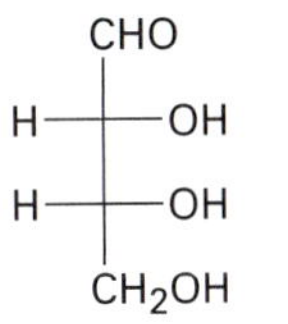

R
R

25.6 **(a)** L-Erythrose; 2*S*,3*S*
(b) D-Xylose; 2*R*,3*S*,4*R*
(c) D-Xylulose; 3*S*,4*R*

25.7

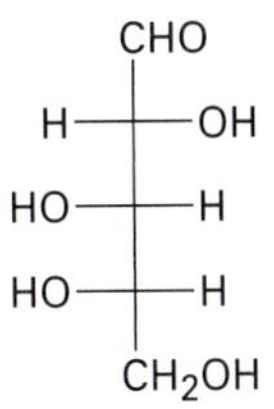

L-(+)-Arabinose

25.8

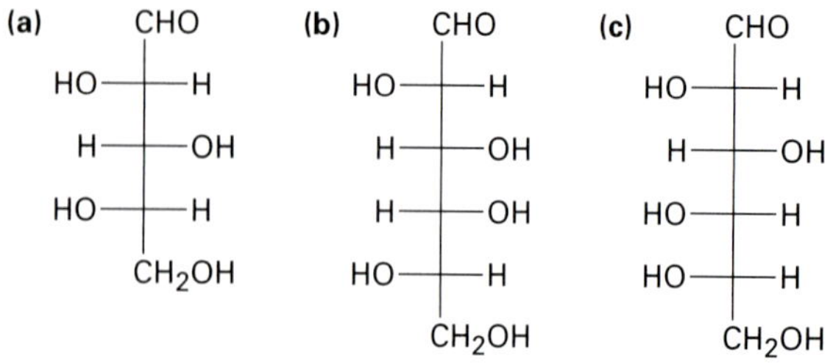

25.9 16 D and 16 L aldoheptoses

25.10

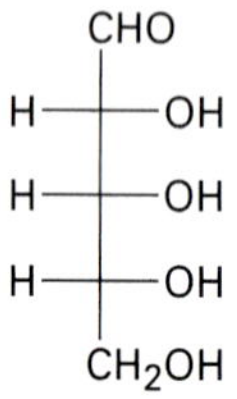

D-Ribose

25.11

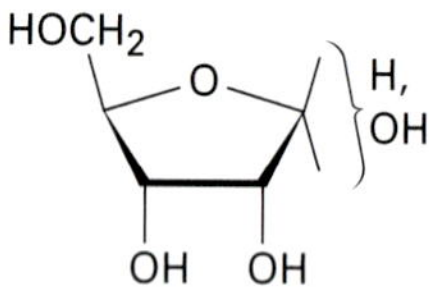

25.12

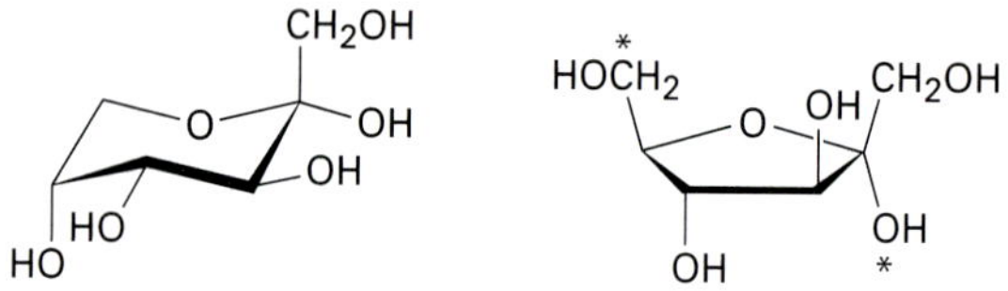

25.13

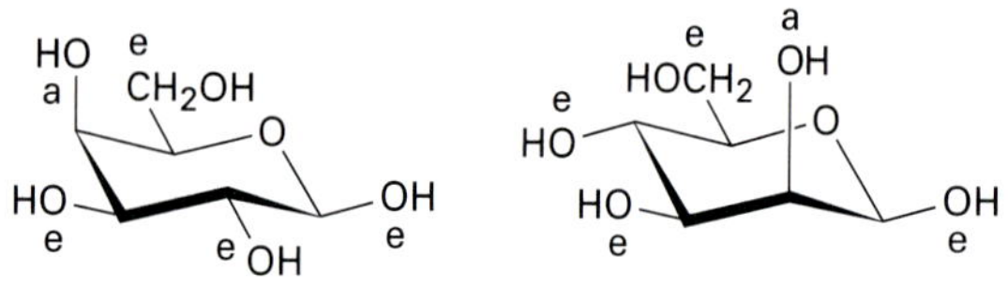

β-D-Galactopyranose β-D-Mannopyranose

25.14

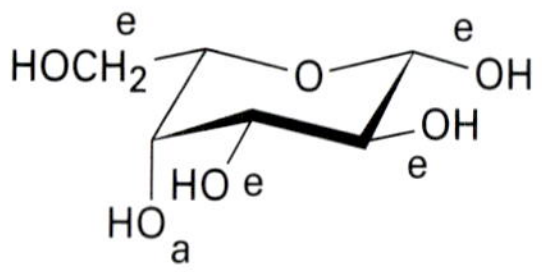

25.15 α-D-Allopyranose

25.16

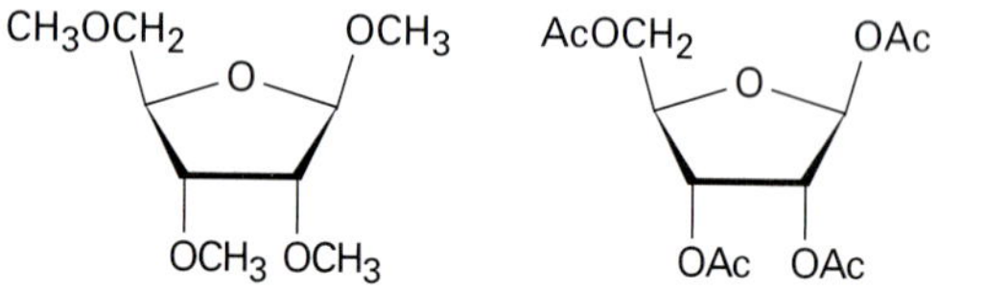

25.17 D-Galactitol has a plane of symmetry and is a meso compound, whereas D-glucitol is chiral.

25.18 The −CHO end of L-gulose corresponds to the $-CH_2OH$ end of D-glucose after reduction.

25.19 D-Allaric acid has a symmetry plane and is a meso compound, but D-glucaric acid is chiral.

25.20 D-Allose and D-galactose yield meso aldaric acids; the other six D-hexoses yield optically active aldaric acids.

25.21 D-Allose + D-altrose

25.22 L-Xylose

25.23 D-Xylose and D-lyxose

25.24

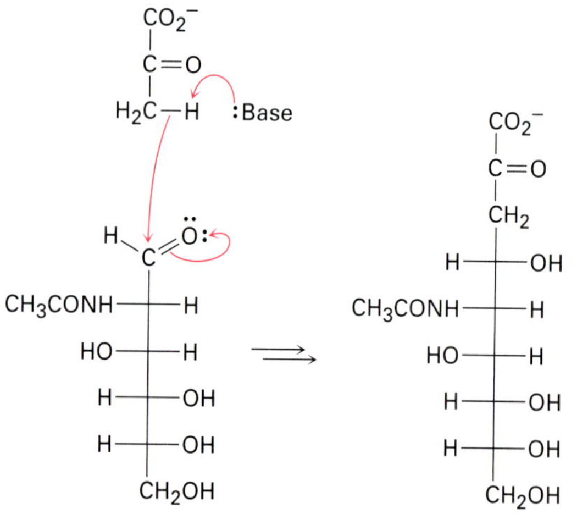

25.25 **(a)** The hemiacetal ring is reduced.
(b) The hemiacetal ring is oxidized.
(c) All hydroxyl groups are acetylated.

CHAPTER 26

26.1 Aromatic: Phe, Tyr, Trp, His; sulfur-containing: Cys, Met; alcohols: Ser, Thr; hydrocarbon side chains: Ala, Ile, Leu, Val, Phe

26.2 The sulfur atom in the $-CH_2SH$ group of cysteine makes the side chain higher in priority than the $-CO_2H$ group.

26.3

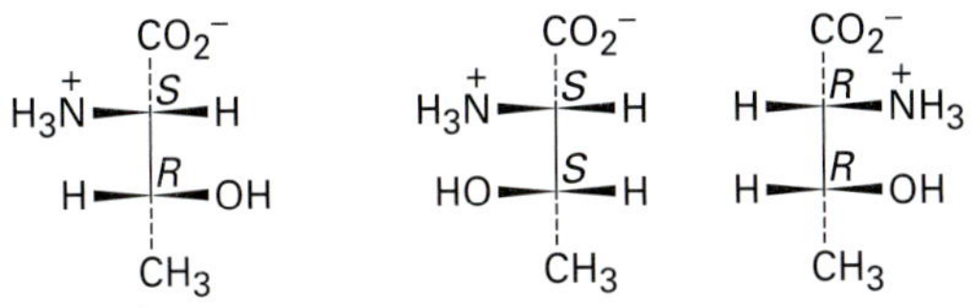

L-Threonine — Diastereomers of L-threonine

26.4 Net positive at pH = 5.3; net negative at pH = 7.3

26.5 (a) Start with 3-phenylpropanoic acid:
1. Br_2, PBr_3; 2. NH_3
(b) Start with 3-methylbutanoic acid:
1. Br_2, PBr_3; 2. NH_3

26.6 (a) $(CH_3)_2CHCH_2Br$

(b)

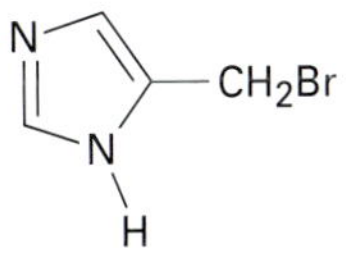

(c)

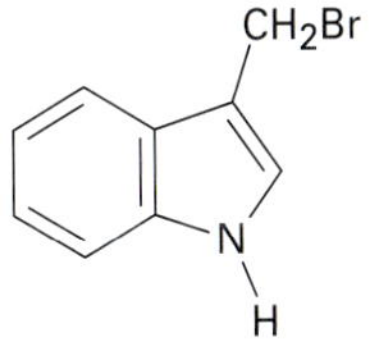

(d) $CH_3SCH_2CH_2Br$

26.7

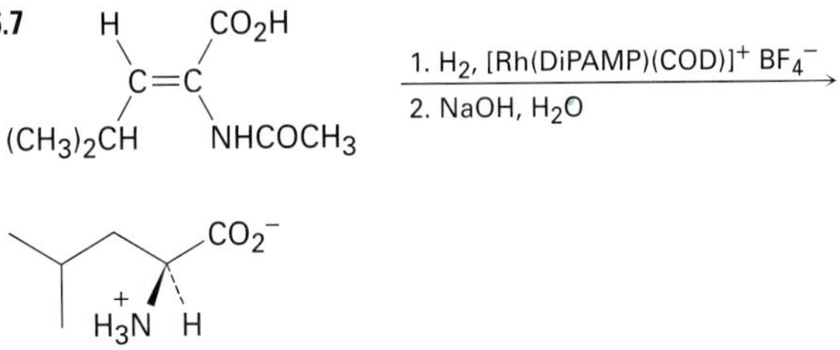

26.8 Val-Tyr-Gly (VYG), Tyr-Gly-Val (YGV), Gly-Val-Tyr (GVY), Val-Gly-Tyr (VGY), Tyr-Val-Gly (YVG), Gly-Tyr-Val (GYV)

26.9

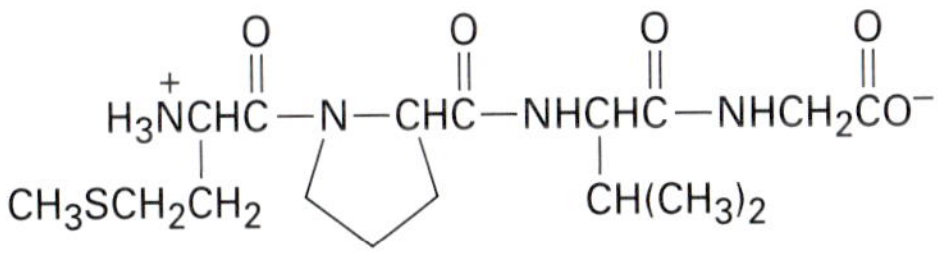

26.10

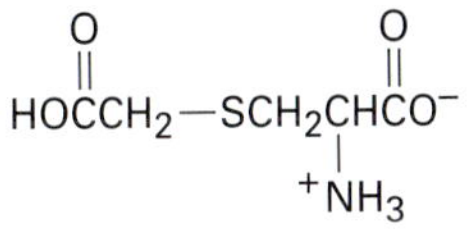

26.11

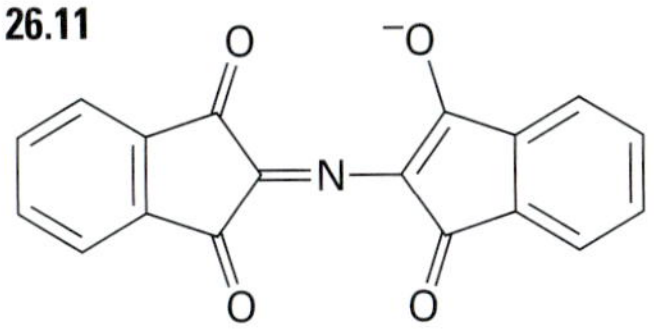

+ $(CH_3)_2CHCHO$ + CO_2

26.12 Trypsin: Asp-Arg + Val-Tyr-Ile-His-Pro-Phe
Chymotrypsin: Asp-Arg-Val-Tyr + Ile-His-Pro-Phe

26.13 Methionine

26.14

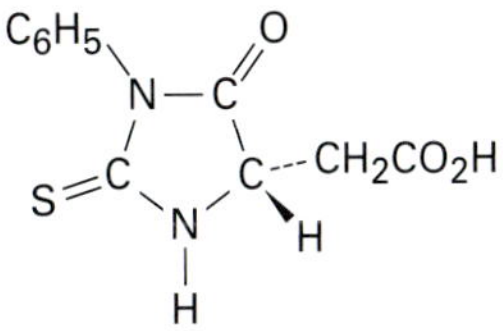

26.15 (a) Arg-Pro-Leu-Gly-Ile-Val
(b) Val-Met-Trp-Asp-Val-Leu (VMWNVL)

26.16 This is a typical nucleophilic acyl substitution reaction, with the amine of the amino acid as the nucleophile and *tert*-butyl carbonate as the leaving group. The *tert*-butyl carbonate then loses CO_2 and gives *tert*-butoxide, which is protonated.

26.17 (1) Protect the amino group of leucine.
(2) Protect the carboxylic acid group of alanine.
(3) Couple the protected amino acids with DCC.
(4) Remove the leucine protecting group.
(5) Remove the alanine protecting group.

26.18 (a) Lyase (b) Hydrolase (c) Oxidoreductase

CHAPTER 27

27.1 $CH_3(CH_2)_{18}CO_2CH_2(CH_2)_{30}CH_3$

27.2 Glyceryl tripalmitate is higher melting.

27.3 $[CH_3(CH_2)_7CH{=}CH(CH_2)_7CO_2^-]_2\ Mg^{2+}$

27.4 Glyceryl dioleate monopalmitate → glycerol + 2 sodium oleate + sodium palmitate

27.5

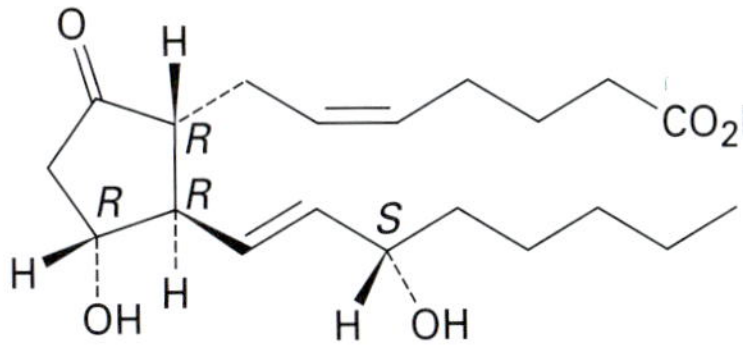

27.6 The *pro-S* hydrogen is cis to the $-CH_3$ group; the *pro-R* hydrogen is trans.

27.7

(a)

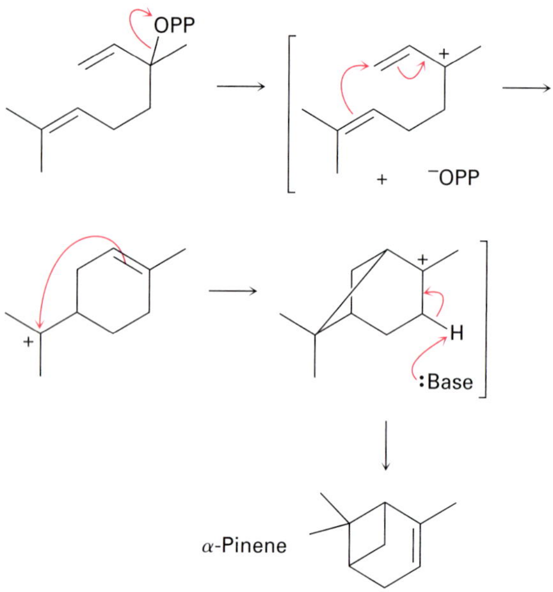

(b)

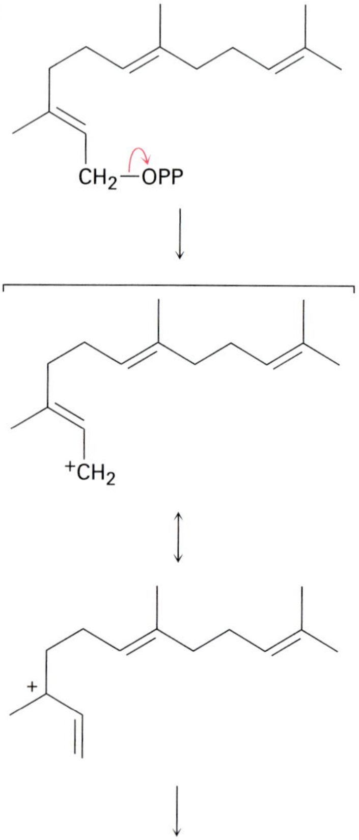

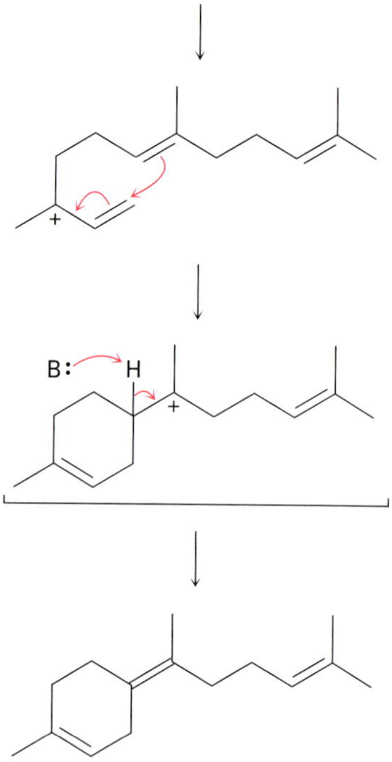

γ-Bisabolene

27.8

(a)

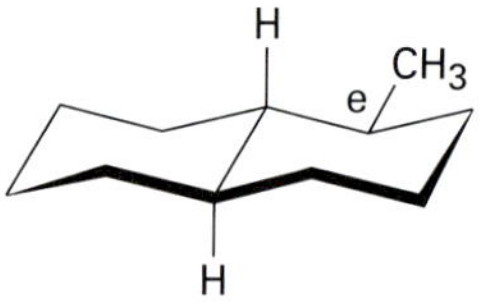

(b)

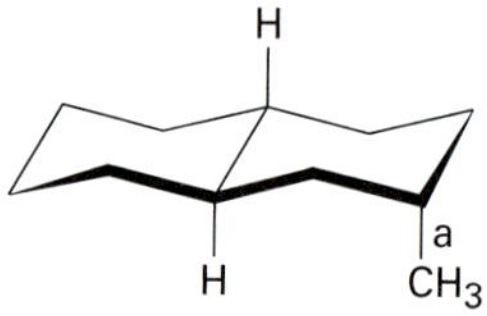

27.9

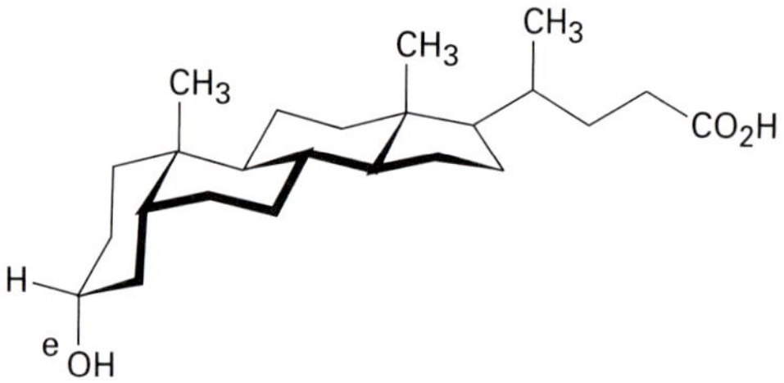

27.10 Three methyl groups are removed, the side-chain double bond is reduced, and the double bond in the B ring is migrated.

CHAPTER 28

28.3 (5′) ACGGATTAGCC (3′)

28.4

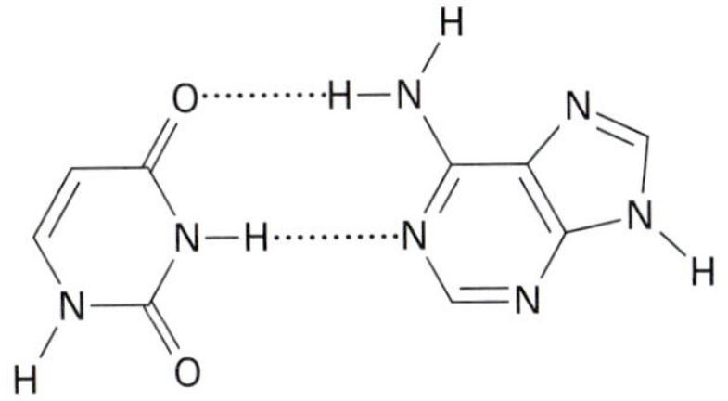

28.5 (3′) CUAAUGGCAU (5′)

28.6 (5′) ACTCTGCGAA (3′)

28.7 (a) GCU, GCC, GCA, GCG
(b) UUU, UUC
(c) UUA, UUG, CUU, CUC, CUA, CUG
(d) UAU, UAC

28.8 (a) AGC, GGC, UGC, CGC
(b) AAA, GAA
(c) UAA, CAA, GAA, GAG, UAG, CAG
(d) AUA, GUA

28.9 Leu-Met-Ala-Trp-Pro-Stop

28.10 (5′) TTA-GGG-CCA-AGC-CAT-AAG (3′)

28.11 The cleavage is an S_N1 reaction that occurs by protonation of the oxygen atom followed by loss of the stable triarylmethyl carbocation.

28.12

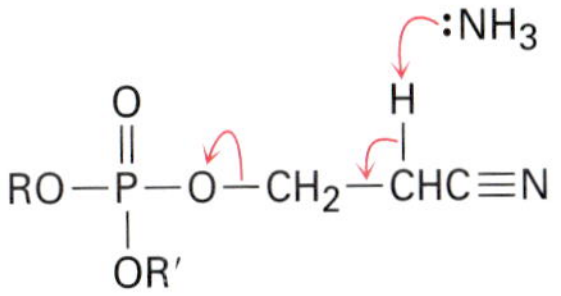

E2 reaction

CHAPTER 29

29.1 $HOCH_2CH(OH)CH_2OH + ATP \longrightarrow$
$HOCH_2CH(OH)CH_2OPO_3^{2-} + ADP$

29.2 Caprylyl CoA $\longrightarrow$ Hexanoyl CoA $\longrightarrow$
Butyryl CoA $\longrightarrow$ 2 Acetyl CoA

29.3 (a) 8 acetyl CoA; 7 passages
(b) 10 acetyl CoA; 9 passages

29.4 The dehydration is an E1cB reaction.

29.5 At C2, C4, C6, C8, and so forth

29.6 The *Si* face

29.7 Steps 7 and 10

29.8 Steps 1, 3: Phosphate transfers; steps 2, 5, 8: isomerizations; step 4: retro-aldol reaction; step 5: oxidation and nucleophilic acyl substitution; steps 7, 10: phosphate transfers; step 9: E2 dehydration

29.9 C1 and C6 of glucose become $-CH_3$ groups; C3 and C4 become CO_2.

29.10 Citrate and isocitrate

29.11 E2 elimination of water, followed by conjugate addition

29.12 *pro-R*; anti geometry

29.13 The reaction occurs by two sequential nucleophilic acyl substitutions, the first by a cysteine residue in the enzyme, with phosphate as leaving group, and the second by hydride donation from NADH, with the cysteine residue as leaving group.

29.14 Initial imine formation between PMP and α-ketoglutarate is followed by double-bond rearrangement to an isomeric imine and hydrolysis.

29.15 $(CH_3)_2CHCH_2COCO_2^-$

29.16 Asparagine

CHAPTER 30

30.1 Ethylene: ψ_1 is the HOMO and ψ_2^* is the LUMO in the ground state; ψ_2^* is the HOMO and there is no LUMO in the excited state. 1,3-Butadiene: ψ_2 is the HOMO and ψ_3^* is the LUMO in the ground state; ψ_3^* is the HOMO and ψ_4^* is the LUMO in the excited state.

30.2 Disrotatory: *cis*-5,6-dimethyl-1,3-cyclohexadiene; conrotatory: *trans*-5,6-dimethyl-1,3-cyclohexadiene. Disrotatory closure occurs.

30.3 The more stable of two allowed products is formed.

30.4 *trans*-5,6-Dimethyl-1,3-cyclohexadiene; *cis*-5,6-dimethyl-1,3-cyclohexadiene

30.5 *cis*-3,6-Dimethylcyclohexene; *trans*-3,6-dimethylcyclohexene

30.6 A [6 + 4] suprafacial cycloaddition

30.7 An antarafacial [1,7] sigmatropic rearrangement

30.8 A series of [1,5] hydrogen shifts occur.

30.9 Claisen rearrangement is followed by a Cope rearrangement.

30.10 (a) Conrotatory (b) Disrotatory
(c) Suprafacial (d) Antarafacial
(e) Suprafacial

CHAPTER 31

31.1 $H_2C{=}CHCO_2CH_3 < H_2C{=}CHCl < H_2C{=}CHCH_3 < H_2C{=}CH{-}C_6H_5$

31.2 $H_2C{=}CHCH_3 < H_2C{=}CHC_6H_5 < H_2C{=}CHC{\equiv}N$

31.3 The intermediate is a resonance-stabilized benzylic carbanion, Ph—C̈HR.

31.4 The polymer has no chirality centers.

31.5 No, the polymers are racemic.

31.6

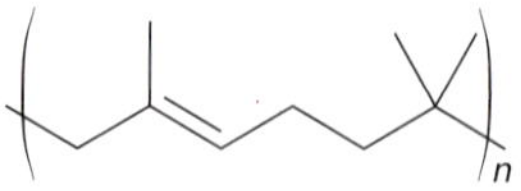

31.7

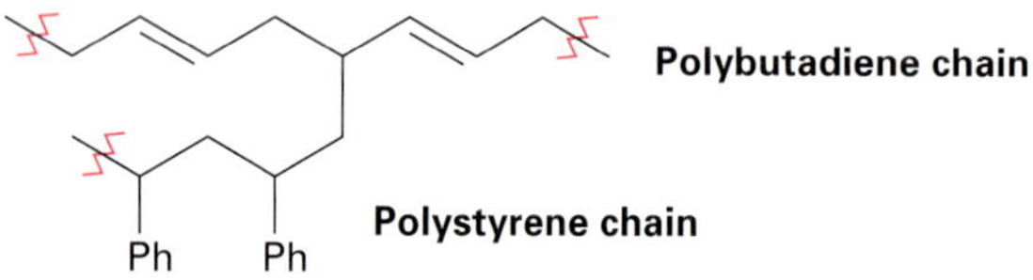

31.8

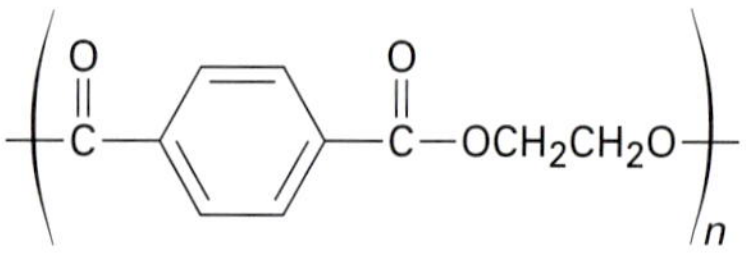

31.9

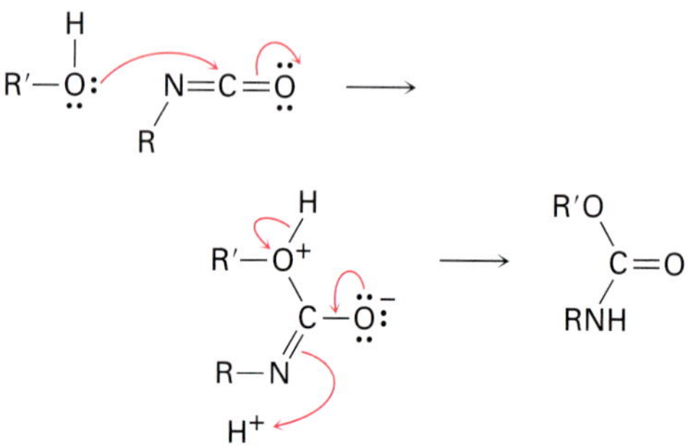

31.10

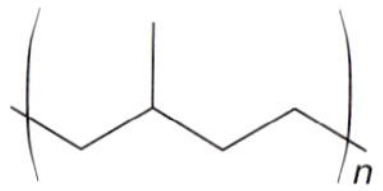

Atactic

31.11

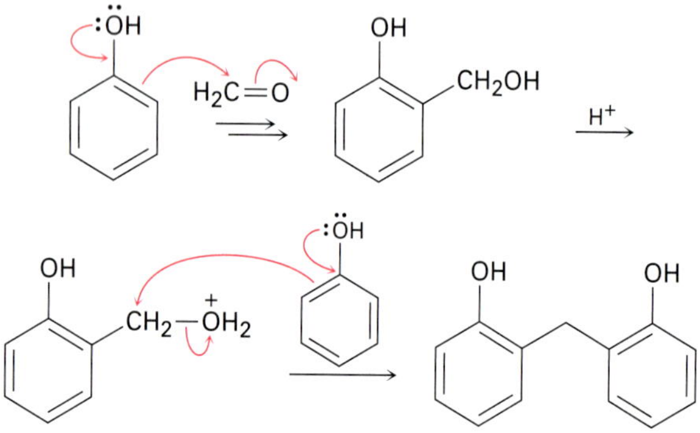

Index

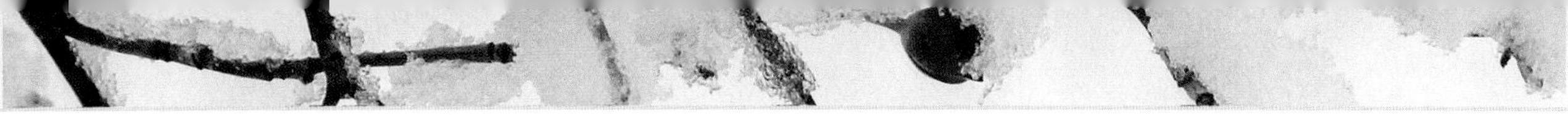

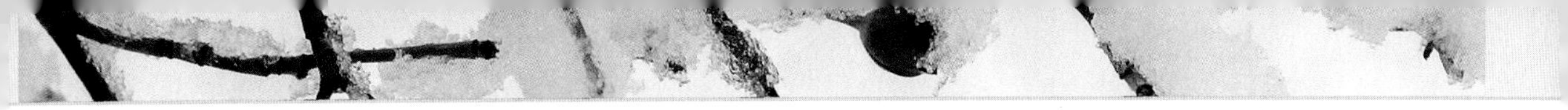

Structures of Some Common Functional Groups

Name	Structure*	Name ending	Example
Alkene (double bond)	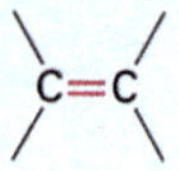	*-ene*	$H_2C{=}CH_2$ Ethene
Alkyne (triple bond)	$—C{\equiv}C—$	*-yne*	$HC{\equiv}CH$ Ethyne
Arene (aromatic ring)	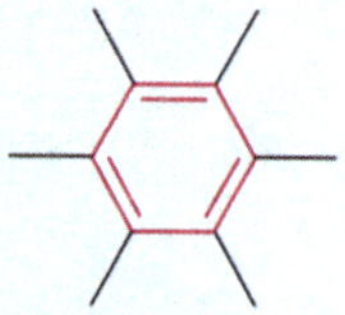	None	 Benzene
Halide	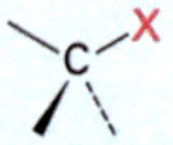 (X = F, Cl, Br, I)	None	CH_3Cl Chloromethane
Alcohol	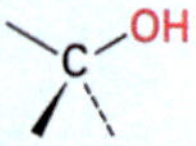	*-ol*	CH_3OH Methanol
Ether	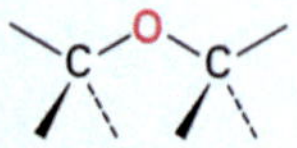	*ether*	CH_3OCH_3 Dimethyl ether
Monophosphate	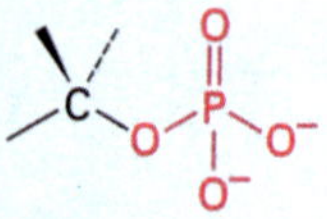	*phosphate*	$CH_3OPO_3^{2-}$ Methyl phosphate
Amine	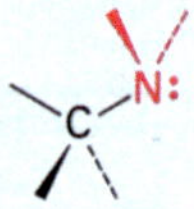	*-amine*	CH_3NH_2 Methylamine
Imine (Schiff base)	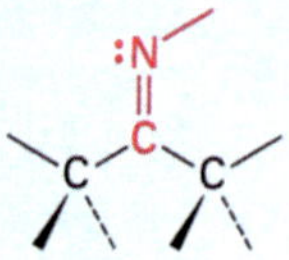	None	NH ‖ CH_3CCH_3 Acetone imine
Nitrile	$—C{\equiv}N$	*-nitrile*	$CH_3C{\equiv}N$ Ethanenitrile
Nitro	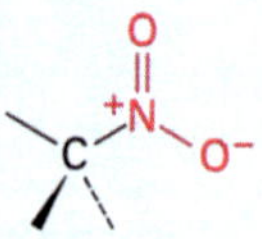	None	CH_3NO_2 Nitromethane
Thiol	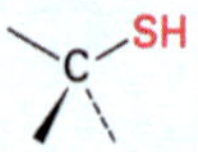	*-thiol*	CH_3SH Methanethiol

*The bonds whose connections aren't specified are assumed to be attached to carbon or hydrogen atoms in the rest of the molecule.

Name	Structure*	Name ending	Example
Sulfide	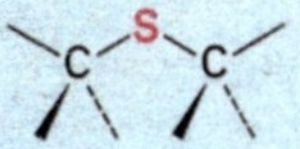	*sulfide*	CH_3SCH_3 Dimethyl sulfide
Disulfide	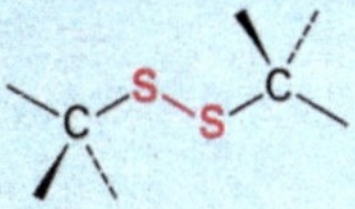	*disulfide*	CH_3SSCH_3 Dimethyl disulfide
Carbonyl	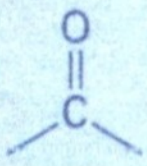		
Aldehyde	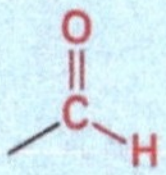	*-al*	$CH_3\overset{O}{\overset{\|}{C}}H$ Ethanal
Ketone	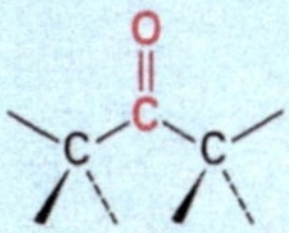	*-one*	$CH_3\overset{O}{\overset{\|}{C}}CH_3$ Propanone
Carboxylic acid	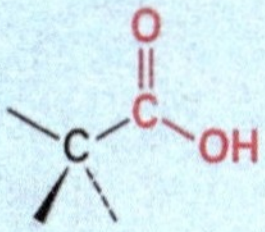	*-oic acid*	$CH_3\overset{O}{\overset{\|}{C}}OH$ Ethanoic acid
Ester	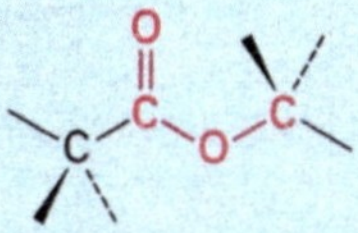	*-oate*	$CH_3\overset{O}{\overset{\|}{C}}OCH_3$ Methyl ethanoate
Amide	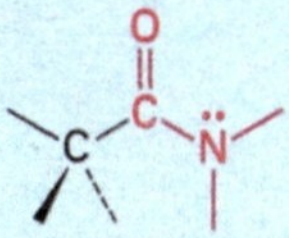	*-amide*	$CH_3\overset{O}{\overset{\|}{C}}NH_2$ Ethanamide
Carboxylic acid anhydride	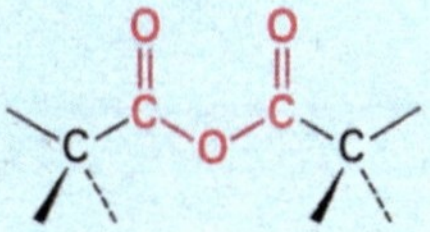	*-oic anhydride*	$CH_3\overset{O}{\overset{\|}{C}}O\overset{O}{\overset{\|}{C}}CH_3$ Ethanoic anhydride
Carboxylic acid chloride	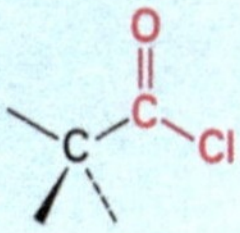	*-oyl chloride*	$CH_3\overset{O}{\overset{\|}{C}}Cl$ Ethanoyl chloride

*The bonds whose connections aren't specified are assumed to be attached to carbon or hydrogen atoms in the rest of the molecule.